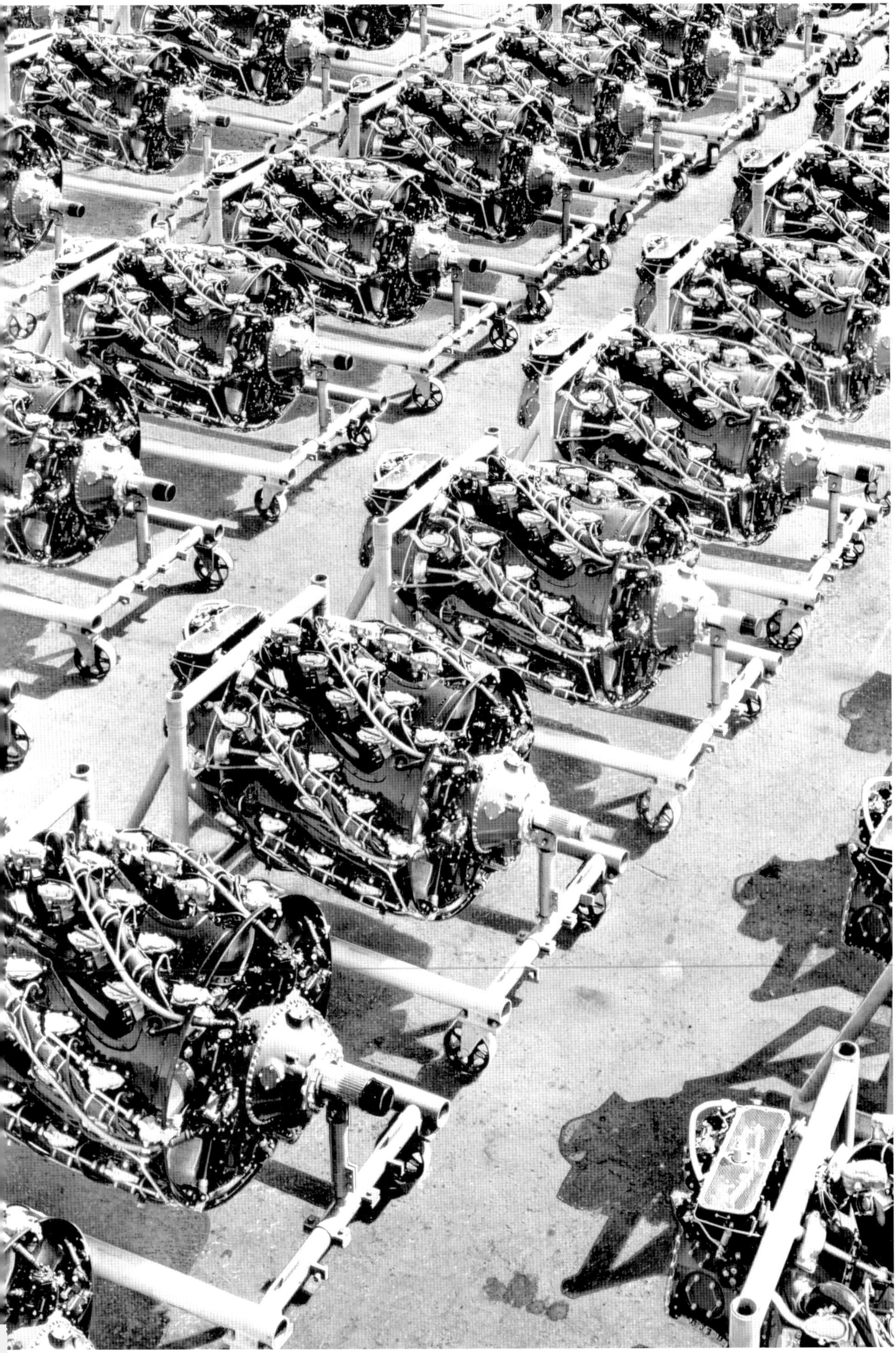

R-4360

Pratt & Whitney's Major Miracle

Graham White

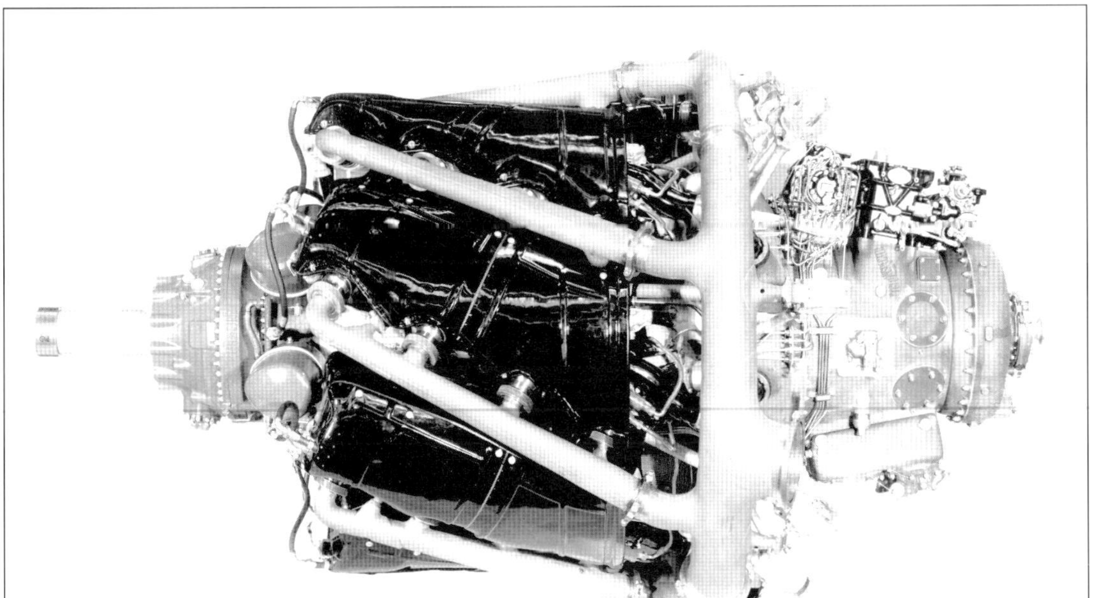

specialty press
PUBLISHERS AND WHOLESALERS

ISBN-13 978-1-58007-097-3
ISBN-10 1-58007-097-3

Item #SP097

39966 Grand Avenue
North Branch, MN 55056 USA
(651) 277-1400 or (800) 895-4585
www.specialtypress.com

Printed in China

Distributed in the UK and Europe by:
Midland Publishing
4 Watling Drive
Hinckley LE10 3EY, England
Tel: 01455 233 747 Fax: 01455 233 737
www.midlandcountiessuperstore.com

Cover: Almost time to run. Final preparations are made in the test cell. (United Technologies Corp./Pratt & Whitney)

Title page: R-4360-53. With this version, Pratt & Whitney had reached the definitive production version; fuel injection, hooded baffles, low-tension ignition, and rated at 3,800 hp. R-4360-53s powered B-36s.

Front Flap: A B-36 engine appears to have been tested and is ready for shipment. The hooded baffles still need to be installed. (United Technologies Corp./Pratt & Whitney)

Back Flap: Even though it was a proof-of-concept aircraft, the F4U WM was a nicely done conversion. The ring around the engine near the man's right hand represented the joining line between the original F4U-1 cowl and the extension for the R-4360. (Courtesy of Pratt & Whitney)

Library of Congress Cataloging-in-Publication Data

White, Graham, 1945- R-4360 : Pratt & Whitney's major miracle / by Graham White.
p. cm.
ISBN 1-58007-097-3
1. Airplanes--Motors--Design and construction--History. 2. Pratt & Whitney Aircraft Group--History. 3. Internal combustion engines, Spark ignition. I. Title.

TL703.P7W46 2006
629.134'35'0973--dc22

2005022961

Contents

Acknowledgments

It seems the more complex an engine, any book written on it follows suit. Therefore, attempting to acknowledge everyone who assisted in this project inevitably results in omissions. I offer my sincere apologies to those who have been omitted from this acknowledgments list.

The idea for this book started out with a huge amount of data on the R-4360 collected from the National Archives & Records Administration in College Park, Maryland, by Kimble McCutcheon. I sat on this goldmine of material, most of which had never seen the light of day since it was archived over half a century ago, for about two years. As other projects I was working on wound down, I figured it was time to put all that information to good use and write a book on this fascinating and misunderstood engine. Without Kim's relentless pursuit of long forgotten reports, memos, and other relevant information, my task would have been infinitely more difficult.

This book would have been out of the question without the cooperation of Pratt & Whitney. Specifically, Dick Wellman, since retired from Pratt & Whitney, greased the skids to get me access to their world-class archives. Two other Pratt & Whitney retirees, Jesse Hendershot and Jack Connors, took a week out of their schedules to guide me through the extensive archives. To these three gentlemen I owe a sincere debt of gratitude.

Aviation attorneys Richard Dean, Robert O'Hara, and Curtiss Isler jumped through the necessary hoops to obtain permission for use of the numerous illustrations copied from the Pratt & Whitney archives. Without this invaluable assistance, this book would not have been possible.

When writing a book, I always try to include as many anecdotes as I can. Glen Dye, a fellow member of the North American Speed Society, put me in touch with Luke Roy, a former B-36 mechanic. Luke's wonderful stories can be enjoyed in Chapter 10. It gives astonishing insight into how the unbelievably complex B-36 was kept on active duty. At the 2003 EAA Sun 'n Fun convention held in Lakeland, Florida, I ran into Barry Meacham. Like Luke Roy, he has some fascinating tales on the C-124 and KC-97. I am very grateful to these gentlemen for sharing their experiences in keeping R-4360s alive and running.

Air racing has to be one of the most exciting sports to watch. The *All Coast Super Corsair* was the first post-1949 racer to be powered by an R-4360. Steve Hinton, the driving force behind this remarkable aircraft, supplied a wealth of information. Another post-1949 racer is *Dreadnought*, campaigned by Dennis and Brian Sanders. Even though I bugged them for information just prior to the 2003 National Championship Air Races, Dennis and Brian answered a whole laundry list of questions I had. Their mother, Ruth, conveyed the information to me via several e-mails.

The world's expert on fuel systems for large piston engines is undoubtedly Pete Law, retired Lockheed Skunk Works engineer. Pete was always willing to share his wealth of information along

with his huge collection of fuel system manuals. All the illustrations in Chapter 6, Carburetors and Fuel Injection, came from Pete's collection.

Kevin Cameron is a modern-day Harry Ricardo, an individual who has more knowledge of the inner workings of the internal combustion engine than anyone else alive. Whenever I had a query, Kevin was right there with an answer.

Al Marcucci could be considered the equivalent of Pete Law, except Al's specialty is ignition systems. Al is probably one of the last people in the world with the expertise to overhaul an R-4360 ignition system. Thanks to Al's help and the use of his manuals, I was able to explain the complex system required to fire 56 spark plugs at exactly the correct time.

Bruce Vandermark has one of the world's largest private collections of aircraft manuals. He offered me the use of anything in his collection. This turned out to be a real boon as far as getting specific information on various R-4360-powered aircraft.

Warren Bodie has a huge collection of aircraft photographs, possibly one of the world's largest. On a couple of occasions I found myself with no suitable image of a particular aircraft, but Warren came through every time.

Hill Goodspeed contributed a large amount of material on Navy aircraft from the Museum of Naval Aviation at Pensacola.

To all of the above and those I may have left out, a collective thank you.

Introduction

As aviation technology progressed at a blindingly fast pace during the first half of the 20th century, new problems needing immediate resolution surfaced. From the Wright brothers' initial tentative steps towards powered flight, demands were soon placed upon aircraft to fly higher, fly faster, carry heavier loads, take off and land on shorter runways, fly greater distances, and consume less fuel with perfect dependability while doing all these things. Of course, all the aforementioned requirements placed greater demands upon the powerplant(s). For many years, it was an accomplishment if an aircraft engine would demonstrate even the most rudimentary requirements for reliability. Pilots were regarded as daredevils with lots of guts as the reliability of the engine was, oftentimes, unknown. With the onset of World War I, a whole new perspective was placed upon aircraft engines.

A worldwide conflict was at stake depending on who possessed the best aircraft, be it fighters, bombers, or reconnaissance aircraft. Naturally, this accelerated the pace of development, especially for engines. As aircraft became heavier, new types of construction appeared. At one time it would have seemed absurd to build an aircraft out of metal. After all, aren't aircraft supposed to be light? Yes, but when the existing technology—that of wooden frames covered in Irish linen—could not stand up to the loads being imposed upon them, the airframe designers were left little choice but to use metal. However, simply replacing structural members formerly manufactured from wood with metal was not the most efficient use of this material.

Consequently, the all-metal monocoque method of construction for high performance aircraft became almost universal by the late 1930s. Flaps, retractable landing gear, more efficient brakes, etc., increased the demands on the powerplant in order to drive the necessary hydraulic pumps and massive electric generators. And in the meantime, airframe manufacturers were demanding even more power from bigger and bigger engines. Not only more power, but higher specific power with lower specific weight—higher horsepower per cubic inch. Placing further demands on the engine were the requirements for lower specific fuel consumption, meaning less fuel burned for each horsepower delivered per hour. As if this wasn't enough to give the engine designers and developers ulcers, the holy grail of one-horsepower-per-pound of engine weight was always a top priority. These contradicting requirements had to be met, indeed exceeded, if the engine manufacturer was to remain in business. Propellers also took on a new degree of urgency for efficiency and operating flexibility. Improved efficiency was obtained through precision-made aluminum blades, and later steel. As engine powers increased, altitudes increased as did the demand for propellers that would be just as efficient at takeoff and cruise. Aircraft manufacturers demanded propellers that could change pitch according to operating conditions. These requirements led to variable pitch and later constant speed propellers. Without these propeller developments, the R-4360 would have been all but worthless.

Pratt & Whitney's first engine—the R-1340—was, by late 1920s standards, a relatively large engine, displacing 1,344 cubic inches. And yet in 10 years time, this displacement would appear to be minuscule, certainly insufficient for front-line military aircraft and even most commercial aircraft. Nevertheless, a new company embarking upon the manufacture of aircraft engines needs to start somewhere, and the R-1340, or Wasp as it was known in the commercial marketplace, was a good starting point. Although the R-1340 did not revolutionize engine design, it incorporated enough state-of-the-art features to make it a highly successful engine. Starting from the formative years of the R-1340, by the end of the piston engine era Pratt & Whitney had placed into mass production the largest and most powerful engine ever built in quantity. Obviously, this was not an easy road proceeding from the small and simple R-1340 to the mighty and complex R-4360. But like other engineering endeavors, dogged determination and first-class design work got Pratt & Whitney through difficult times when problems appeared insurmountable.

By the time the worst of the R-4360's problems had been resolved, the gas turbine was taking over aircraft propulsion duties. And yet the R-4360 proved invaluable during this transition period. It has often been argued that the R-4360 represented an anomaly, being so complex and temperamental that it hardly justified the immense engineering talent and expense pumped into its development. Maybe the argument has some validity. However, there was nothing else in the marketplace in the late 1940s that could achieve what the R-4360 did—a 3,000- to 4,000-hp powerplant. It has also been argued that it's a good thing gas turbines came along when they did otherwise the size and complexity of the R-4360 would have paled in comparison with what would have replaced it. As gas turbine development transcended piston engine development, piston engines were put on the back burner. From a historian's point of view, it's always fascinating to play "what-if" games. What if Pratt & Whitney had perfected the VDT theme (variable discharge turbine—further described in Chapter 4), then what would have followed this very advanced design? How would the requirement for a 10,000-hp engine have been met? Where would fuel and supercharger development have gone? These and many other questions go unanswered. All we can do is marvel at what was accomplished and be grateful for the fact we can ogle at the most wonderful hot rods ever conceived.

Last and by no means least, it is probably too much to hope that errors have not crept into this book despite the heroic efforts of the author. If you find any such errors, please let the publisher know.

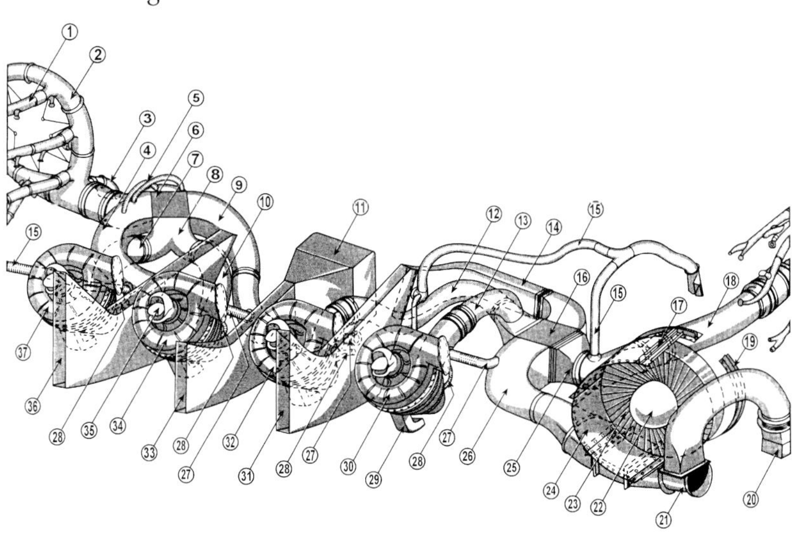

CHAPTER ONE

Whoppers!

If big is good, then bigger is better, right? Wrong! The Pratt & Whitney R-4360 was the only aircraft piston engine displacing over 4,000 cubic inches that entered into series production. However, the landscape is littered with attempts at producing a large-displacement, high-horsepower aircraft engine. This includes air-cooled radials, liquid-cooled inlines, and even liquid-cooled radials. History indicates that doubling the displacement of an engine causes the development problems to go up exponentially. These problems are associated with cooling, mechanical design, lubrication, and keeping weight under control.

Several methods of increasing displacement are available to the engine designer; increase the number of cylinders, increase the bore dimension, increase the stroke dimension, or combinations of all the above. So ideally, to get the most displacement, one simply gives it the most cylinders with the largest bore and stroke dimensions. This overly simplistic view of things has been pursued on numerous occasions. One of the earliest attempts at a large engine design was that of Bugatti. At the end of World War I, Bugatti developed the 16-cylinder King Bugatti *(Ref. 1-1).* The King Bugatti was later doubled into a 32-cylinder engine *(Ref. 2-2).* Better known for their automobile designs rather than aircraft engine design, Bugatti nevertheless developed a number of aircraft engines and at least one very advanced racing aircraft. But the success they enjoyed with their automobile designs never transferred to aviation. Other relatively large and/or complex World War I engines include the 2,290-ci Galloway Atlantic and the 3,914.3-ci Sunbeam Sikh *(Ref. 1-1).* However, it's doubtful if this latter engine flew prior to the cessation of World War I. Between World War I and World War II, engine development progressed at a rapid pace. Materials, fuels, manufacturing processes, supercharger design, engine displacement, and engine design matured into reliable and more powerful powerplants for the increasingly sophisticated aircraft they were installed in.

Doubling-Up

This was an easy, or so it was thought, method of increasing displacement. The idea was to simply double-up an existing successful engine. This concept was to continue throughout the piston era; however, few, if any, were successful. In retrospect, it would appear from historical precedent that the only way to design a new and larger engine was to start from scratch rather than working off an existing design. One of the few exceptions was the Pratt & Whitney R-1830, which was later developed into the R-2000. The 170-ci-increase was accomplished via the simple expedient of increasing the bore of the R-1830 from 5.5 inches to 5.75 inches *(Ref. 1-3).*

One of the first exponents of this method was Gnome, who simply doubled up the existing 9-cylinder, 80-hp rotary engine into a two-row

The idiosyncratic rotary engine powered many World War I aircraft thanks in no small part to its good power-to-weight ratio. Even though two-row rotary engines were developed, they did not see significant use. This photo shows an 18-cylinder Gnome "Delta-Delta" rated at 200 hp. It displaced 1,986.46 cubic inches–a huge size for the time. *(Photo courtesy of Joe Gertler)*

18-cylinder engine *(Ref. 1-4)*. Although its smaller 9-cylinder sibling saw huge production, the 18-cylinder engine was not as successful and relatively few were manufactured.

By the time war clouds were developing over Europe for the oncoming onslaught of World War II, 1,500 cubic inches was simply not going to cut it for frontline aircraft engines. Even so, some were built despite this deficiency in displacement. One example was the problematic Rolls-Royce Peregrine. Its displacement of 1,296 cubic inches was stretched to the limit, putting out a meager 800 hp *(Ref. 1-5)*. Even so, Rolls-Royce did not totally give up on the Peregrine. An engine featuring the Peregrine's 5.0-inch bore and 5.5-inch stroke was developed as a 24-cylinder "X"

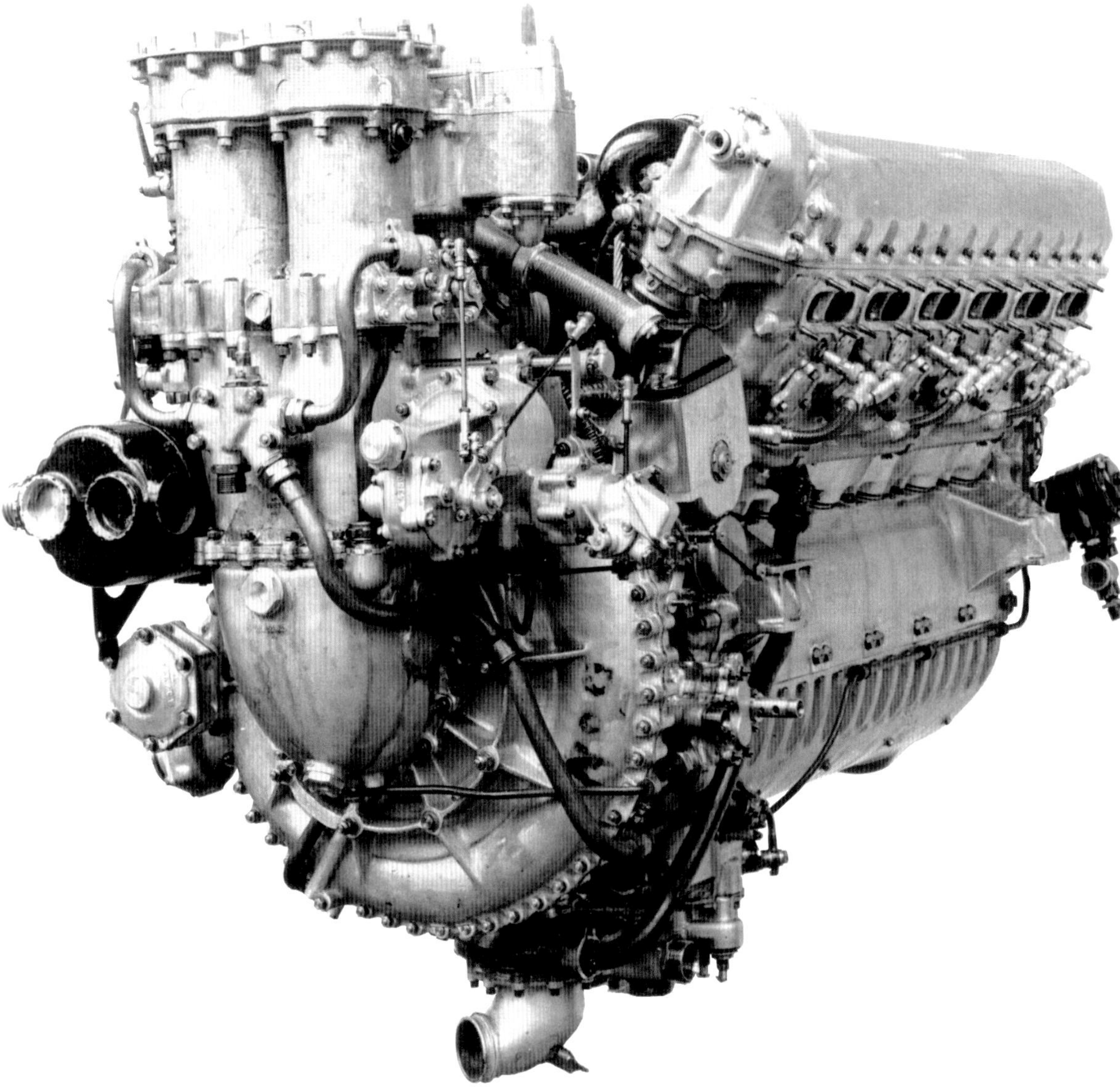

Rolls-Royce soon found out that trying to stress a small engine to its limit was not the way to increased horsepower. Based upon their successful Kestrel, the Rolls-Royce Peregrine with a displacement of 1,296 cubic inches was insufficient for reliable horsepower. *(Photo courtesy of the Rolls-Royce Heritage Trust)*

engine, essentially two V-12s grafted onto a common crankcase. Known as the Vulture, this engine represented one of Rolls-Royce's few failures. Vultures were used primarily to power Avro Manchester twin-engined bombers. Again, this engine followed the precedent of other doubled-up engines and turned out to be a miserable disappointment. Catastrophic failure of the connecting rods, main bearing failures, and other severe maladies forced Rolls-Royce to abandon Vulture development and its manufacture.

Even though the Vulture was a failure, the charismatic CEO (works manager) of Rolls-Royce at the time, Ernest, later Lord, Hives had the foresight to realize that attempting to develop the Vulture into a successful engine would take more perseverance and time than could be afforded during World War II. This difficult and brave decision was to pay immense dividends by freeing up additional manufacturing capacity plus engineering talent for Merlin improvements and Griffon development.

Nevertheless, Rolls-Royce's doubling-up efforts did not end with the Vulture. A design was completed in 1942 for an "H" configured Merlin with a proposed output of 3,500 to 4,000 hp *(Ref.*

1-6). This concept never proceeded beyond the design stage. An even more exciting concept was that of coupling a pair of Griffons in tandem. It's doubtful that this concept ever got as far as the design stage. Even so, it illustrates yet again the measures aircraft engine manufacturers were taking in order to obtain the necessary power for future aircraft with their greater demands for more speed, more altitude, and more range.

Germany's Double Jeopardy

Daimler-Benz

Germany's World War II strategic bombing campaign was seriously compromised due to abysmally unreliable engines. Their only attempt at a long-range strategic bomber was the Heinkel HE 177. Following a similar path taken by the British with the Rolls-Royce Vulture-powered Avro Manchester, the HE 177 was powered by various versions of the Daimler-Benz DB 600 series engine doubled-up to form one powerplant *(Ref. 1-7).* This ambitious effort was doomed from the start. Coupled DB 605s were known as DB 610s. This engine was rated at 2,950 hp at 2,800 rpm with a weight of 3,388 pounds. These specifics are actually fairly respectable until reliability is factored into the equation. It was the same story with the similar DB 613, made up from two DB 603s and rated at 3,800 hp at 2,700 rpm. The history of the DB 610s and 613s powering the HE 177 can only be described as awful. Engine fires and catastrophic engine failures plagued that program throughout its life. The

Jumping from the frying pan into the fire, Rolls-Royce doubled-up the Peregrine to produce the X-24 Vulture. This very problematic engine was soon withdrawn from service. *(Photo courtesy of the Rolls-Royce Heritage Trust)*

Germany's double jeopardy turned out to be a disaster for their strategic bombing campaign. Daimler-Benz doubled-up their DB 603 and DB 605 to produce the DB 613 and the DB 610. Other DB engines were doubled-up as well. They were all dismal failures. *(Photo courtesy of Professor Stefan Zima)*

mind just boggles at why the Germans did not see the light sooner and realize these engines were too seriously flawed for further development. Definitive information on what caused the fires and engine failures has not come to light.

One possible scenario is the consequence of flying the HE 177s at high altitudes. At high altitudes, lubricating oil starts to take on very different characteristics compared to sea level. Oil becomes aerated due to the much lower atmospheric pressure, and, exacerbating the situation, it congeals in oil coolers due to the frigid temperatures. This sets the scene for interrupted oil supply leading to connecting-rod failure and other bearing maladies. Apparently, the HE 177's oil tank resided next to the engine. If so, in the event a connecting rod failed and punched a hole in the side of the crankcase, hot oil would start spewing out at the rate of about 50 gallons per minute right onto the exhaust system. Once an oil fire is established, it's about the worst thing next to a magnesium fire, which incidentally is what the 610s and 613s were primarily constructed of. If Daimler-Benz had bitten the bullet when faced with the similar challenge facing Rolls-Royce with the Vulture, they would have abandoned it and redesigned the HE 177 as a four-engine bomber powered by BD 603s. In this scenario, Germany's strategic bombing capability would have had a very different outcome. But again, this is playing would haves, could haves, and should haves.

BMW

BMW developed the highly successful 801 series of engines during World War II. Coincidently, or perhaps not coincidently, this high-performance

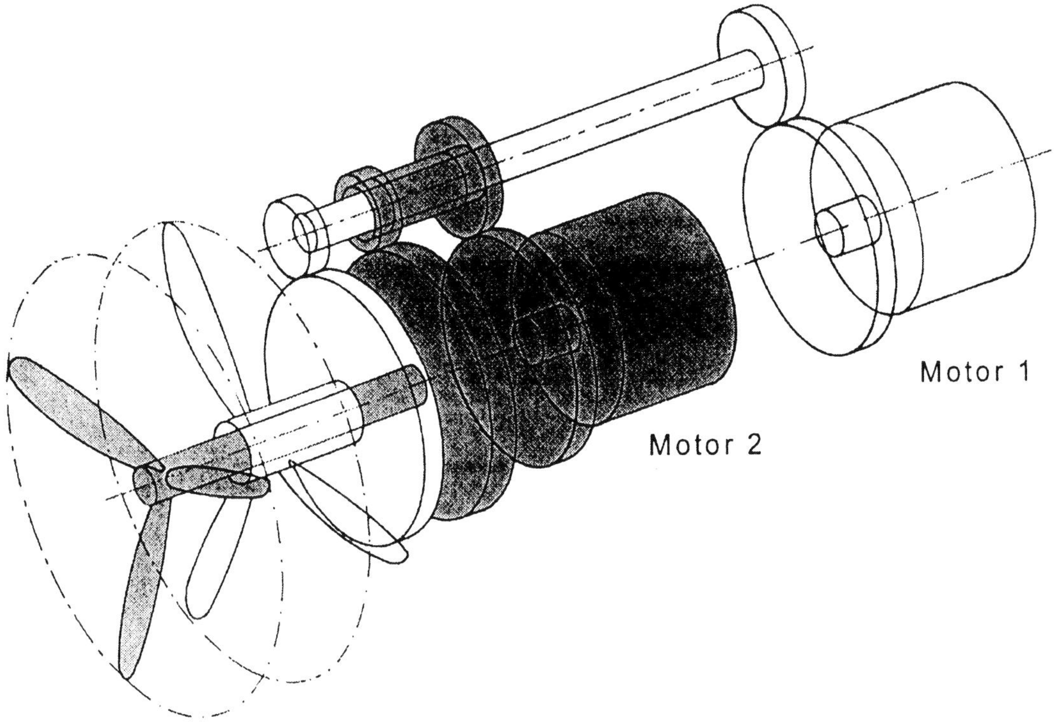

Based upon the success of their 14-cylinder 801 radial, BMW developed the final iteration on this theme, the mighty 803. It was only built in prototype form, so it's hard to say if it would have been a success had it been placed in production. Produced from two 14-cylinder modules, each module drove half of the contra-rotating propeller. *(Photo courtesy of Professor Stefan Zima)*

This photograph illustrates the complexity making up the BMW 803 contra-rotating propeller drive and its enormous length. This 28-cylinder behemoth never went into production. *(Photo courtesy of Professor Stefan Zima)*

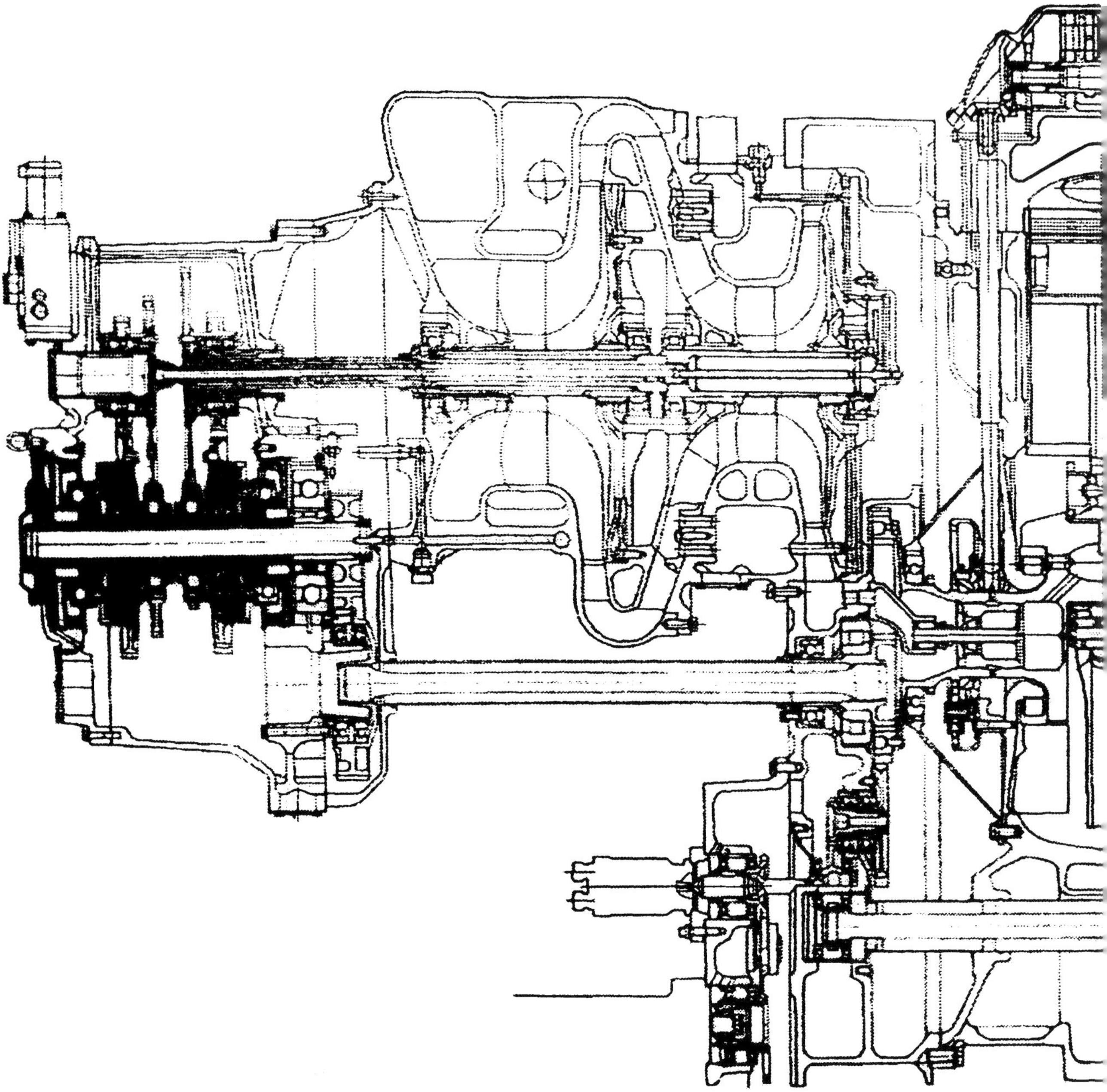

14-cylinder, two-row, air-cooled radial is similar to the Wright R-2600. BMW developed several larger engines using the same 6.12-inch bore and stroke: the 18-cylinder 802 and the 28-cylinder 803. Not unexpectedly, the 18-cylinder 802 was an evolution of the 801. It used 18 air-cooled cylinders in two-row format. Not so with the 28-cylinder 803. Although the 803 shared the same bore and stroke dimensions of the 801 and 802 at 6.15 inches, that's where the similarity ended. This highly complex engine was essentially a pair of liquid-cooled 14-cylinder units *(Ref. 1-7)*. A two-stage, four-speed supercharger provided boost requirements. The front 14-cylinder "module" drove the front component of a contra-rotating propeller assembly and the rear 14-cylinder module drove the rear propeller. Large-displacement, high-horsepower engines can suffer from long, "whippy" crankshafts. The usual solution is to simply beef them up, but at the cost of increased weight. The 803 solved this problem in an innovative way; the rear 14-cylinder

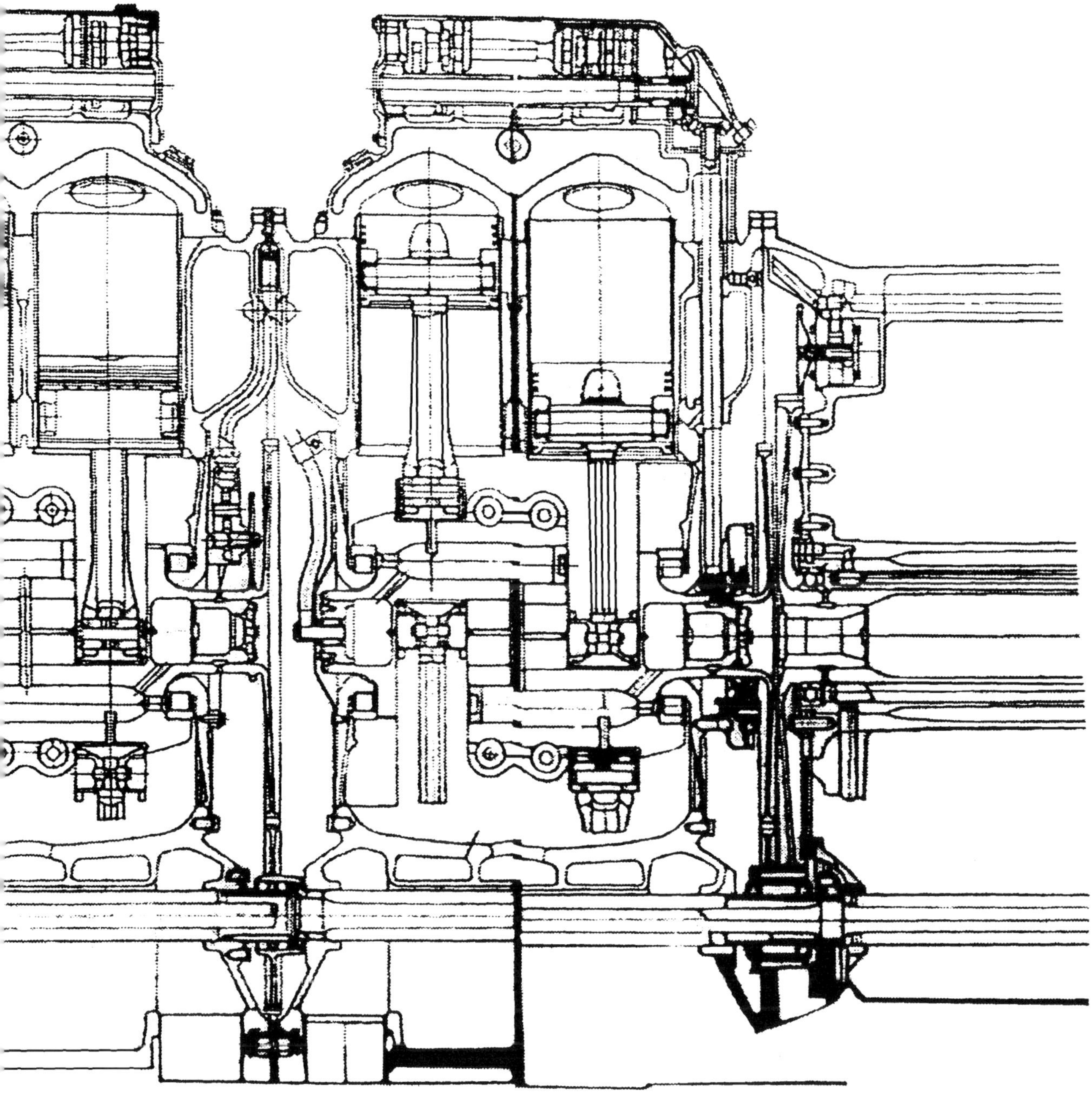

The 5,120-cubic-inch BMW 803. Note the two-stage, four-speed supercharger on the left of this cutaway drawing. Illustrated in this longitudinal cross section are the two separate crankshafts. Each one drives half the contra-rotating propeller. *(Photo courtesy of Professor Stefan Zima)*

module drove seven quill shafts, running between the cylinder skirts that extended to the reduction gearing. In this way the front 14-cylinder module was isolated from the rear module. Consequently, both modules behaved as though they were 14-cylinder units, and the cost of a heavy crankshaft was avoided. Interestingly, this concept was also incorporated into the 42-cylinder Wright Tornado. It may not be coincidence that a German engineer also designed the Tornado. Even though the 803 was capable of producing 4,000 hp, its weight of 9,080 pounds was unacceptably high. Neither the 802 nor 803 saw series production.

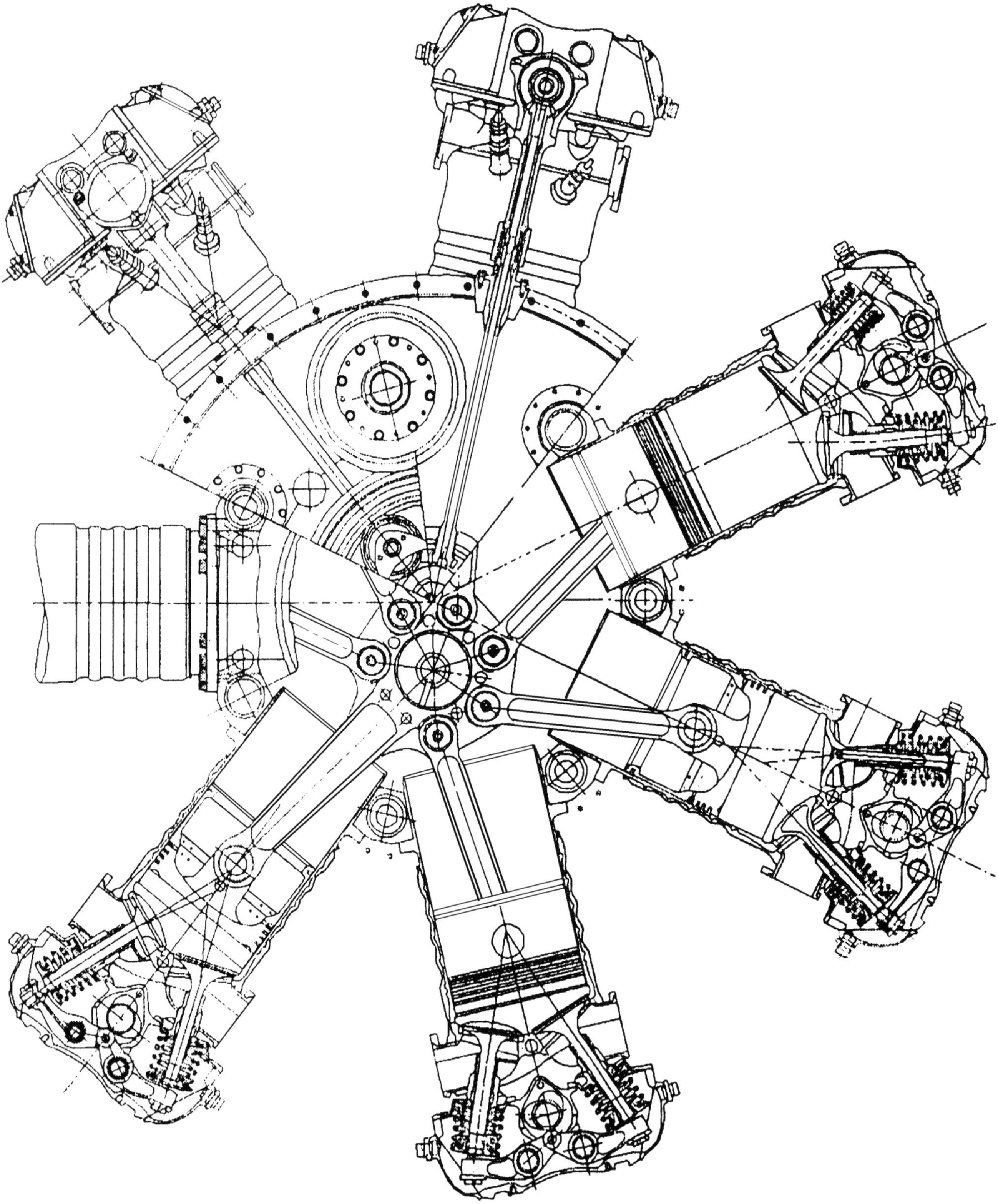

This lateral cross-section drawing of the BMW 803 shows typical radial engine design features with the exception of the overhead cams and liquid cooling. *(Photo courtesy of Professor Stefan Zima)*

One of the few doubled-up engines to show promise was the Allison V-3420 created from a pair of V-1710s. It was built in very limited numbers. Yet ironically this was probably the best shot at a doubled-up engine succeeding. We'll never know how much more capable the B-29 would have been when powered by this engine, but it's a tantalizing thought. *(Photo courtesy of the Rolls-Royce Heritage Trust)*

U.S. Endeavors

Allison

Allison developed the highly successful V-1710 series of engines that saw extensive service in World War II. When the worst of the development problems had been resolved with the V-1710, Allison doubled it up to create the V-3420 *(Ref. 1-8)*. Alas, this engine was probably one of the only coupled or doubled-up engines that actually had promise, yet it never found a permanent home. It was installed in a number of experimental aircraft but never entered mass production. This is really too bad, because it would have been an ideal engine to replace the problematic Wright R-3350s powering Boeing B-29s. One B-29 was actually converted, powered by V-3420s. Designated B-39, this aircraft offered far superior performance to the Wright-powered B-29 and no doubt exhibited far better reliability than the R-3350. The decision was made not to put the B-39 into production, in all likelihood because it would have disrupted production for the desperately needed B-29. In retrospect, the delay would probably have been worth it due to the horrendous problems with under-developed R-3350s at this time. But perhaps the V-3420's best shot at powering an aircraft was the Fisher XP-75. This cobbled-together abortion of an aircraft was soon cancelled after a few had been built and tested. However, its cancellation was no reflection on the performance of the V-3420.

Allison also developed another whopper, the X-4520. It's thought that only one was built and it never flew. Ordered by the Army Air Corps and built in 1924, this air-cooled behemoth was soon abandoned *(Ref. 1-9)*.

continued on page 22

Allison's one-off X-4520 air-cooled engine. Designed for air ships, it never saw production and never flew. *(National Archives & Records Administration)*

Even though the V-3420 showed promise, the same sentiment could not be expressed about the V-3420-powered Fisher XP-75 depicted in this photograph. This large and heavy fighter was no better, and in many respects was worse, than its contemporaries. *(Photo courtesy of the Rolls-Royce Heritage Trust)*

21
WHOPPERS!

Wright developed a 22-cylinder version of the R-3350 to create the R-4090. Little is known about this engine, so it's hard to say how successful it was or would have been. It's probable that Wright was stretched too thin working out the serious problems with early R-3350s and developing the 42-cylinder Tornado. *(New Jersey Aviation Hall of Fame Museum)*

continued from page 19

Wright

Wright was not going to be left out in the Whopper sweepstakes. Their entry consisted of a doctored-up R-3350. This two-row, air-cooled radial used 22 R-3350 sized cylinders arranged as 11 per row. This gave a total displacement of 4,090 cubic inches *(Ref. 1-10)*. The effort was soon dropped, but not before at least one prototype engine had been built and test run. One can only theorize that it was dropped because of ongoing development problems with the R-3350. If development had continued and had followed the same path as the R-3350, the R-4090 was a potential 4,000-hp engine. In a developed form such as this, it could have given the R-4360 serious competition. But this is pure conjecture, point being that development was stopped and it never went anywhere.

Lycoming

The all-time champ as the largest aircraft engine ever built resides with the Lycoming XR-7755. This gigantic 36-cylinder engineering tour de force included just about every conceivable bell

Lycoming's XR-7755 demonstrated just how big an aircraft piston-engine could get. This 36-cylinder behemoth never flew. Even so, it exhibited some very advanced features such as variable valve timing, two-speed reduction gearing, and contra-rotating propeller drive. *(National Archives & Records Administration)*

and whistle available in the late 1940s. Its four rows with 9-cylinders in each row, radial fashion, were liquid-cooled. Variable valve timing, two- speed propeller reduction gearing, and contra-rotating propeller shafts were just the tip of the iceberg of the engineering technology embedded within this true monster of an engine. Had gas turbines not come along, it's hard to say if Lycoming could have coaxed the requisite amount of reliability along with a decent power-to-weight ratio out of this true wonder of an engine. Prior to cancellation, over 5,000 hp had been persuaded out of the XR-7755 with the promise of more to come *(Ref. 1-11 and Ref. 1-12)*.

Studebaker's Colossus

Even Lycoming's XR-7755 was dwarfed by the Studebaker H-9350 *(Ref. 1-10)*. Although Studebaker is usually thought of as an automobile manufacturer, as with many auto companies they were gainfully employed during World War II by cranking out aircraft engines. It's a little known fact that the "Wagon Company" manufactured the majority of R-1820s powering Boeing B-17s.

Perhaps it was this experience that persuaded Studebaker to give developing its own aircraft engine a shot. Like many companies who thought that designing an aircraft engine would be a walk in the park, they had a rude awakening. Mercifully (for them and perhaps everyone else), Studebaker's efforts never proceeded beyond the design stage and the building of single and two-cylinder test rigs. Layout of Studebaker's H-9350 followed a design path pioneered by many others—a liquid-cooled H-24. It had two flat 12s pancaked on top of each other. Its bore of 8.0 inches and stroke of 7.75 inches yielded the 9,350 cubic inches. It was rated at a modest (for the displacement) 5,000 hp at 2,000 rpm. Apart from its dimensions, nothing distinguished this engine from many others. In fact, it was a rather mundane and conventional layout. After the single

The graceful lines of the Bristol Brabazon belie the fact that it was a huge white elephant that was doomed from the get-go. Underpowered even with eight Bristol Centauruses, the Brabazon was essentially a powered glider. The follow-on Brabazon II would have been powered by Bristol Proteus gas turbines, but this aircraft was never completed. *(Photo courtesy of the Bristol Branch of the Rolls-Royce Heritage Trust)*

Nice view of the Bristol Brabazon wing construction showing how the paired Centaurus 20s were mounted. Notice how the engines are mounted from the front rather than the rear. This is a good example of lightweight structural engineering. The circular hole in the middle accommodates the nacelle, which in turn mounts the huge reduction gearbox. However, the nacelle turned out to be the Brabazon's last nail in its coffin when fatigue cracks were discovered. *(Photo courtesy of the Bristol Branch of the Rolls-Royce Heritage Trust)*

cylinder development engines had been built, and possibly some two-cylinder engines as well, this monster disappeared into the sunset of ill-conceived aeronautical gaffes.

British Undertakings

Bristol

As World War II arrived to its inevitable conclusion, the British set their sights on postwar developments for commercial aviation. Being a leader at this time in aviation, the British did not want to be left at the starting gate when it came to selling up-to-date commercial aircraft to worldwide customers. To plan for this, a commission was put in place headed up by Lord Brabazon. A number of aircraft were planned—from small commuter aircraft to giants of the sky. The largest of these aircraft was named after the head of the commission and the company that built it: The Bristol Brabazon. British taxpayers footed the bill for this huge white elephant.

Spanning a gargantuan 240 feet, it would seem that the designers gave little thought as to what would power this aircraft. Furthermore, who could afford to pay for a ticket to ride in it? The

Above left: Although it turned out to be a failure, innovative engineering went into the Brabazon's design. This overhead view shows how each pair of Bristol Centaurus 20s was mounted. *(Photo courtesy of the Bristol Branch of the Rolls-Royce Heritage Trust) Above right:* In order to test out the Brabazon's unique powertrain, a wing section was made for this purpose. Convair did the same thing when developing the engine installation for the B-36. It's quite possible that Convair and Bristol engineers shared experiences developing their respective giants. *(Photo courtesy of the Bristol Branch of the Rolls-Royce Heritage Trust)*

only option for powerplants at the time of its design would be piston engines. But what British piston engine was available to haul the Brabazon's tremendous weight into the air? The largest and most powerful engine available to the British with any degree of reliability was the sleeve valve Bristol Centaurus. However, it would take a minimum of eight Centauruses to get the Brabazon into the air. Bristol came up with an innovative solution to this dilemma by coupling two Centauruses together, but not in the way we would be accustomed to *(Ref. 1-13 and Ref. 1-14).* Instead, the two Centauruses were buried within the thick wing of the Brabazon at an angle of 18 degrees to the longitudinal axis or 36 degrees included between engines. Each un-geared (direct drive) Centaurus 20 fed power, via a drive shaft, to a huge gearbox that transmitted power to a pair of contra-rotating

Opposite, Top: Note the unusual mounting structure for the direct-drive Bristol Centaurus 20. Two of these engines fed power into a common gearbox that drove contra-rotating propellers for the ill-fated Bristol Brabazon. It should be noted that this aircraft was bought and paid for by the poverty stricken British taxpayers during the austere postwar years. *(Photo courtesy of the Bristol Branch of the Rolls-Royce Heritage Trust) Opposite, Bottom:* Gearbox for the pair of Centaurus 20s. The Bristol Brabazon employed four of these gearboxes. Two Centaurus 20s were coupled to each gearbox. One of the problems experienced by this huge white elephant was fatigue cracking of the support structure for these massive and heavy gearboxes. Mounted on the end of the nacelle, fatigue cracks were soon generated. After just a few flights, this terrible waste of British taxpayer money was grounded and soon after scrapped. *(Photo courtesy of the Bristol Branch of the Rolls-Royce Heritage Trust)*

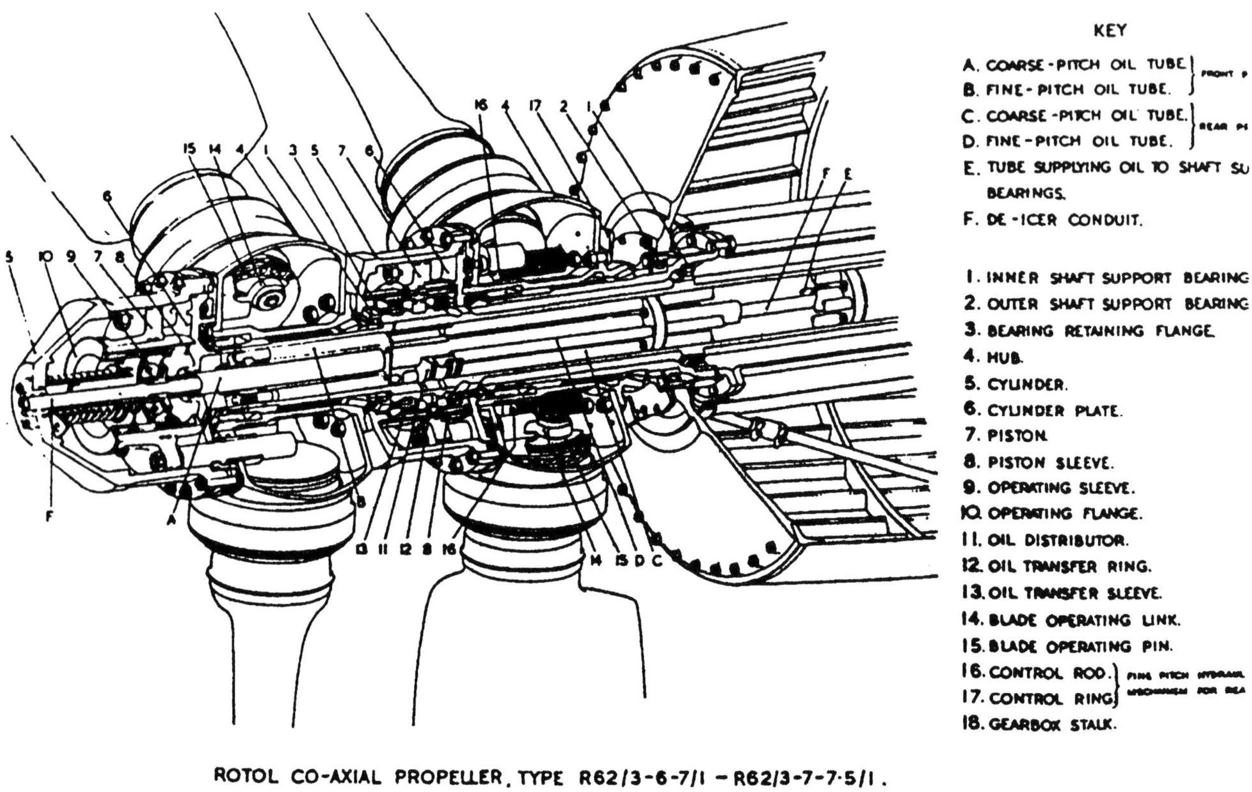

Sectional view of the Rotol contra-rotating propellers used for the Brabazon. The British were the only ones who developed a successful contra-rotating propeller that went into mass production. *(Line drawing courtesy of the Bristol Branch of the Rolls-Royce Heritage Trust)*

propellers. Each of the two engines forming a pair was independent of the other. In this way, it was possible to shut one engine down in flight and keep the other one running. The innovative gearbox design allowed for this.

The combined power of the paired Centauruses amounted to 5,100 hp, but at a weight of 8,390 pounds. Even with an excess of 20,000 hp, the Brabazon was essentially a powered glider. Its days were short lived, and after a brief flight test schedule it was grounded due to fatigue cracks in the support structure for the reduction gearboxes. Another Bristol whopper was the Orion, a further 18-cylinder sleeve valve engine. Roy Feddon, Bristol's enigmatic chief designer, drew up the design of this huge 18-cylinder air-cooled sleeve valve radial shortly before the Bristol Board of Directors fired him. This engine would have displaced 4,142 cubic inches. It was never built. After the piston era, Bristol went on to manufacture some of the worlds' most successful gas turbines.

Japanese Experiments

With the ever-tightening noose encircling the Japanese Empire during World War II, they came up with a bomber concept to attack mainland America from bases in Japan. Of course, this demanded a considerable range. This requisite range, in turn, demanded a large fuel capacity. Getting all this fuel airborne, along with the aircraft and its bomb load, demanded an extraordinarily powerful engine. The Japanese solution was to use a pair of 18-cylinder Nakajima Homare engines attached in a longitudinal fashion. Cooling this 5,000-hp engine must have been a nightmare. Even so, the Japanese came up with

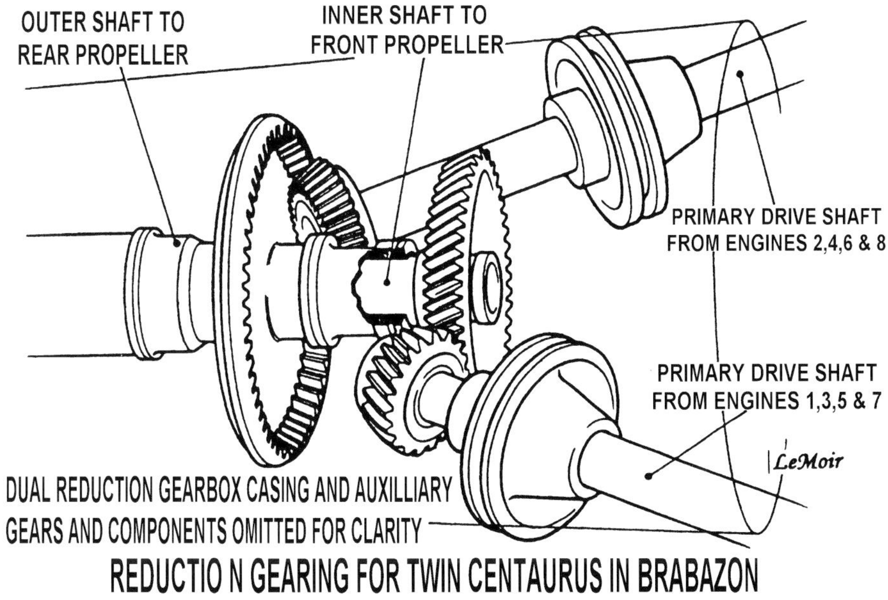

This line drawing shows the ingenuity displayed in the Brabazon's reduction gearbox. An engine could be shut down if necessary without affecting the other engine. *(Line drawing courtesy of the Bristol Branch of the Rolls-Royce Heritage Trust)*

an innovative solution. The cowl was divided into a pair of compartments. One compartment cooled the front 18 cylinders and the other cooled the rear 18 cylinders. The drawback to this solution was the increased diameter of the cowl. It would appear that three variations on cooling were considered. All variations used a fan—a front-mounted fan, a rear-mounted fan, and a combination of a front and a rear fan. It's not known if any of these engines were built.

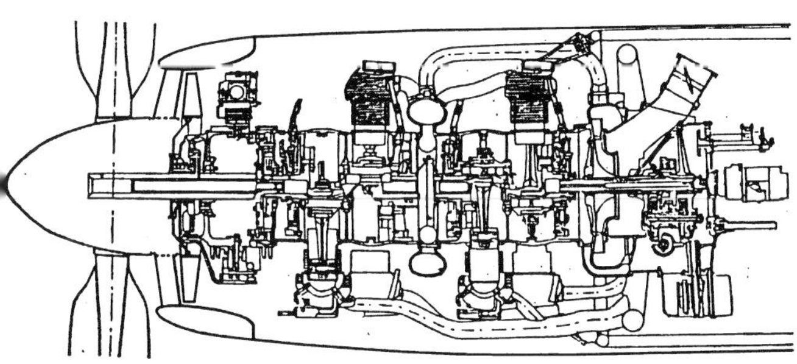

Even with fan-assisted cooling it's hard to imagine that this huge Japanese engine did not overheat. Studying available drawings, it appears to be a pair of 18-cylinder engines coupled together. Designed for the "America Bomber," it's uncertain if any of these engines were built. *(Japanese World War II engineering concept drawing)*

French Frolics

Arsenal

Not to be outdone, especially by the British or Americans, the French came up with a pair of cobbled-together whoppers. The Arsenal 24H took advantage of German World War II technology, not surprising considering the French were forced to manufacture German aircraft during the occupation. This postwar development was based upon four Junkers Jumo 213 cylinder banks, giving a total displacement of 4,265 cubic inches. Rated at a respectable 4,000 hp at a remarkably fast 3,000 rpm, this engine powered a few experimental aircraft and then disappeared from the scene in the late 1940s *(Ref. 1-12)*. Perhaps its porky weight of 4,190 pounds did it in. By the time essentials such as a cooling system were added, the weight almost certainly climbed to an unacceptable 5,000 pounds...or greater.

Hispano-Suiza

Yet another French double jeopardy effort was the Hispano-Suiza 24Z. Again, using off-the-shelf components, four Hispano 12Z cylinder banks were employed. Conceptually similar to the Arsenal 24H, it displaced a similar 4,394 cubic inches. Rated at 3,600 hp at 2,800 rpm, its weight

The French Arsenal 24H, which used four Junkers Jumo cylinder banks configured as an H-24. Developed postwar, it's likely that the French had a supply of leftover Jumo engines plus tooling from the German occupation.

was a svelte 3,197 pounds, or 883 pounds lighter than the Arsenal 24H *(Ref. 1-12).*

The foregoing examples offer a taste of the treacherous ground being trodden when displacement exceeds 4,000 cubic inches and/or power approaches or exceeds 4,000 hp. The fact that some were built meeting these parameters does not indicate whether these engines would have been a commercial or military success. In most cases, I fear not. Nothing has been mentioned of Russian attempts. I'm sure the Russians built large engines; however, the information is not forthcoming. Perhaps some dusty and long ignored archive in Moscow or Monino holds a treasure trove of Russian experiments, but until that information sees the light of day again, not much can be said about their attempts.

Looking back with 60 years perspective it's easy to pass judgment on these efforts. However, we need to look at what was available to the aircraft designer in the 1940s. The gas turbine had not yet proven its worth and their tremendous appetite for fuel all but doomed them for anything except military aircraft. It was not until the late 1950s that jet powered aircraft entered the commercial arena. Of course, this statement does not include the ill-fated deHavilland Comet. With their inherently superior fuel consumption compared to pure jets, turbo props were the first gas turbine powered aircraft to infiltrate the commercial arena. Aircraft such as the Vickers Viscount powered by Rolls-Royce Darts and Lockheed Electras powered by Allison T56s pioneered the trail for jet powered commercial airliners.

With the foregoing in mind, the brave efforts to manufacture monster piston engines, that for the most part failed, are in retrospect perfectly understandable. It's also an additional accolade to Pratt & Whitney engineers who managed to pull off the seemingly impossible—that of developing and putting into production the world's largest and most powerful aircraft piston engine when the path they trod was strewn with so many failures. And making things even more difficult was the rate at which talented engineers were siphoned off the R-4360 project in order to concentrate on gas turbine designs. This left the remaining engineers with an additional burden, which was carried with fortitude and skill demonstrated by the over 16,000 R-4360s manufactured.

References

1-1 *Jane's All the World's Aircraft 1919*

1-2 *Torque Meter Volume 2, Number 4*

1-3 *Index of Twin Wasp & R-1830 Designated Engines. Revised 5/15/55*

1-4 Foxworth, Thomas G., *The Speed Seekers.* Doubleday & Company, Inc., New York.

1-5 Rubbra, A. A., *Rolls-Royce Piston Aero Engines.* Rolls-Royce Heritage Trust Historical Series No. 16.

1-6 Jones, Lyndon., *Sectioned Drawings of Piston Aero Engines.* Rolls-Royce Heritage Trust.

1-7 Jane, Fred, *Jane's All the World's Aircraft 1945/46*, published by Sampson Low Marston & Company Limited, 1946.

1-8 Whitney, Daniel., *Vees for Victory.* Schiffer, 1998.

1-9 Sonnenberg, Paul, and William A. Schoneberger, *Allison, Power of Excellence 1915-1990.* Coastline Publishers, Malibu, Calif., 1990.

1-10 *Model Designations of USAF Aircraft Engines*, 9th ed., by authority of Commanding General, Air Material Command, U.S. Air Force, 1 Jan. 1949.

1-11 *Introducing Lycoming XR-7755 the World's Largest Reciprocating Aircraft Engine*, Lycoming Div., The Aviation Corp., Williamsport, Pa., 31 Oct. 1946.

1-12 Wilkinson, Paul H., *Aircraft Engines of the World, 1948*, Paul H. Wilkinson, New York 1948.

1-13 *Aeroplane*, September 23, 1949.

1-14 *Sleeve Notes*, Bristol branch, Rolls-Royce Heritage Trust.

Another French whopper, the Hispano-Suiza 24Z was based upon the successful 12Y design. Four sets of 12Y cylinder banks were used on a common crankcase. Possibly due to German influence, it features direct fuel injection.

CHAPTER TWO

R-4360 Development

Initial Development

The prior chapter gives an inkling of how difficult it was/is to develop aircraft engines displacing over 4,000 cubic inches developing 4,000 hp or greater. Of course, engines were developed that could produce in excess of 4,000 hp, but for how long? There is little point in developing a hot-rod that lasts 200 hours between overhauls. Commercial aviation, in particular, demands reliability with reasonable time between overhauls. Bottom line economics reign supreme in the unforgiving world of commercial aviation—then and now. Even Pratt & Whitney experienced difficulty meeting these demands with the R-4360. After overcoming innumerable obstacles, they managed to meet these demands and furthermore were the only ones to achieve this lofty and admirable goal. The R-4360 excelled as a transition to gas turbine power, but it was never smooth sailing. It took the combined efforts of Pratt & Whitney's best engineers to arrive at concepts that would work. And as the gas turbine became prominent, more engineering talent was siphoned off for this new form of motive power leaving the R-4360 program bereft of engineering talent at critical times during its development. The following is a history of what it took to develop a four-row, 28-cylinder, air-cooled radial known as the R-4360.

MOCK UP #1 3 ROWS, 9 CYLINDERS PER ROW
R-2800 CYLINDERS

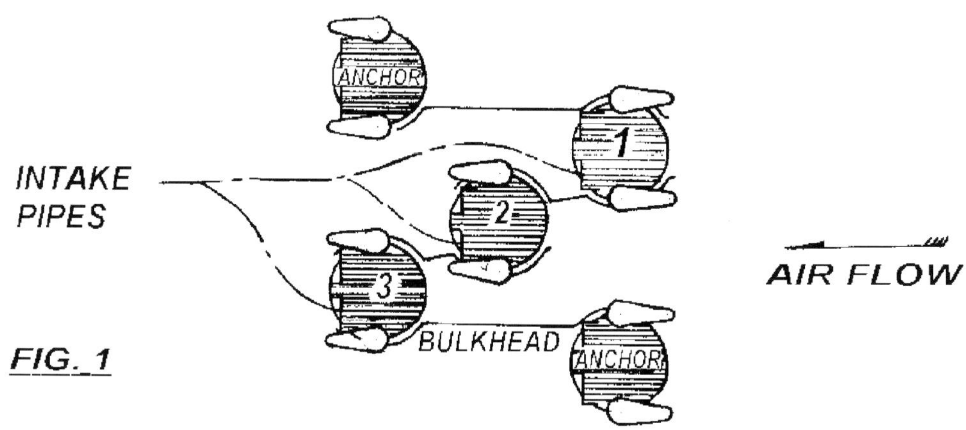

FIG. 1

MOCK UP #2 4 ROWS - SPIRAL LEFT HANDED
7 CYLINDERS PER ROW
R-2800 REAR CYLINDERS

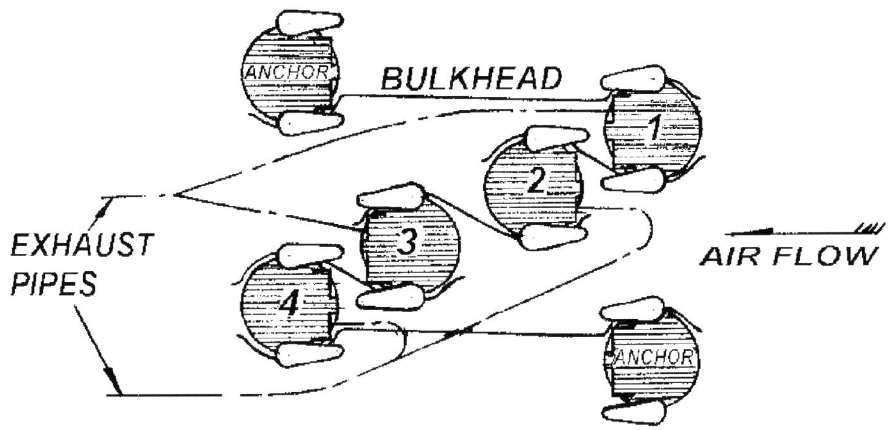

FIG. 2

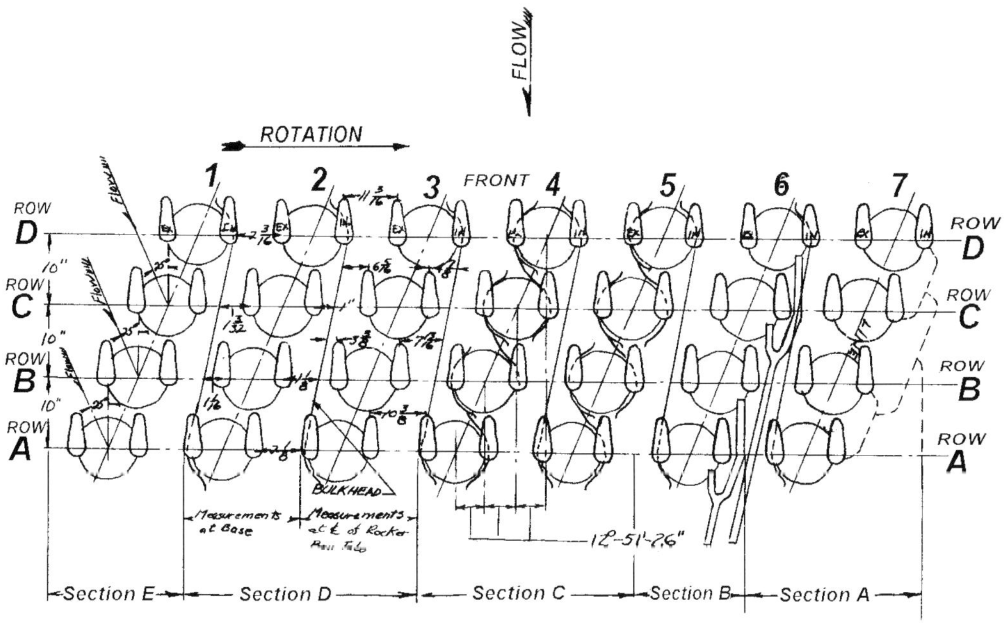

Detail Development Study

The foregoing is a brief chronological history of R-4360 development. The following is a detailed examination of what it took to arrive at the various models of the R-4360 engine. The question now arises: How to achieve 4,360 cubic inches, or thereabouts? This was the first of many challenges facing Pratt & Whitney engineers. After their disastrous foray into liquid-cooled sleeve valve engines, led by George Mead, Pratt & Whitney's brilliant engine designer, Luke Hobbs, chose to go the path they were familiar with—an air-cooled radial. But this then begged the question of how many cylinders, what should the bore and stroke be, and in what arrangement? Pratt & Whitney looked into six rows of six for 36 cylinders, three rows of nine for 27 cylinders, six rows of five cylinders for 30 cylinders, five rows of seven cylinders for 35 cylinders, plus other variations and permutations. The eight accompanying illustrations give

MOCK UP #3, 2 ROWS, 9 CYLINDERS PER ROW
R-2800 CYLINDERS

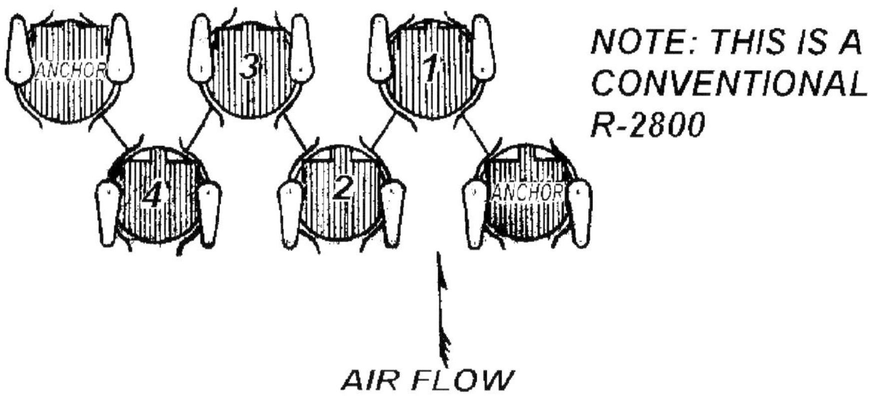

FIG. 3

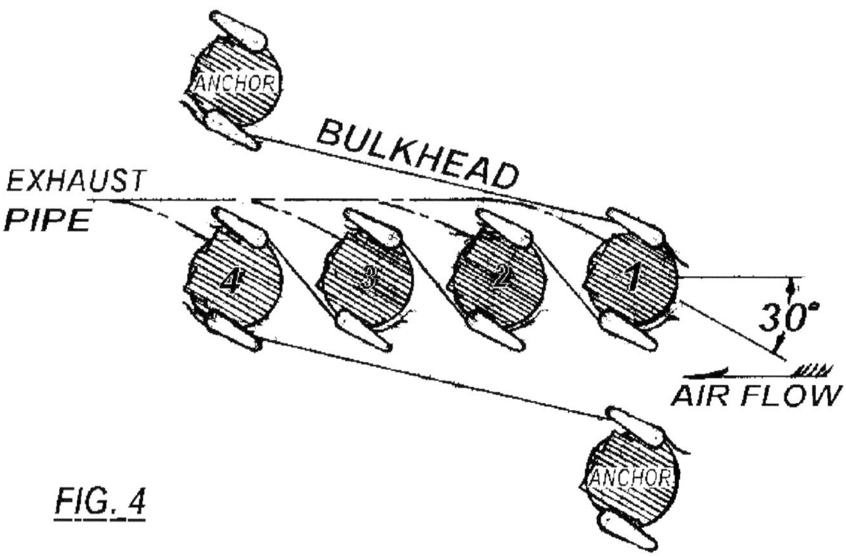

FIG. 4

MOCK UP #4 4 ROWS - IN LINE, 7 CYLINDERS PER ROW
R-2180 COMPOSITE CYLINDERS

an idea of the variations tested and/or investigated *(Ref. 2-1).*

The one combination that appeared doable and offered the most promise was four rows of seven cylinders for a total of 28 cylinders. Pratt & Whitney wisely chose to use the same cylinder dimensions used on the successful R-2800—5.75-inch bore and 6.0-inch stroke. The cylinder bore was probably as large as one would want to go with a high-performance air-cooled engine. True, companies such as Wright Aeronautical used cylinders with bores exceeding six inches, but they paid the price in the form of dependability and susceptibility to detonation. Twenty-eight R-2800 sized cylinders resulted in a displacement of 4,363 cubic inches. But how should these cylinders be arranged, in line or in a spiral? And if in a spiral, which way should the spiral twist—to the left, or to the right? And what should the offset be between rows? A number of considerations dic-

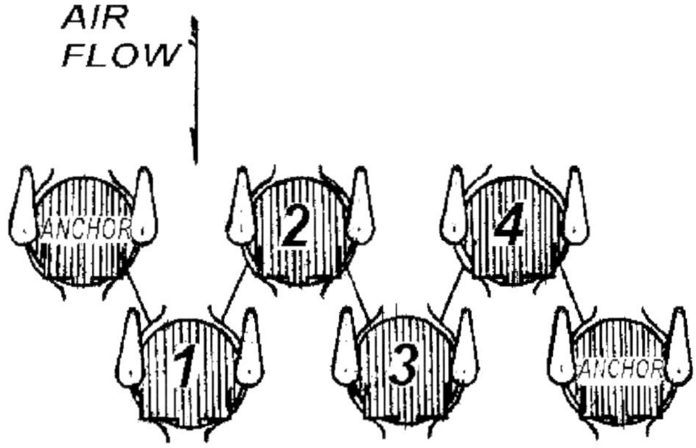

MOCK UP #5 2 ROWS, 7 CYLINDERS PER ROW
R-2180 CYLINDERS

NOTE: THIS SIMULATES AN R-2180; HOWEVER
CYLINDER SPACING IS ON A DUMMY R-2800 CRANKCASE

FIG. 5

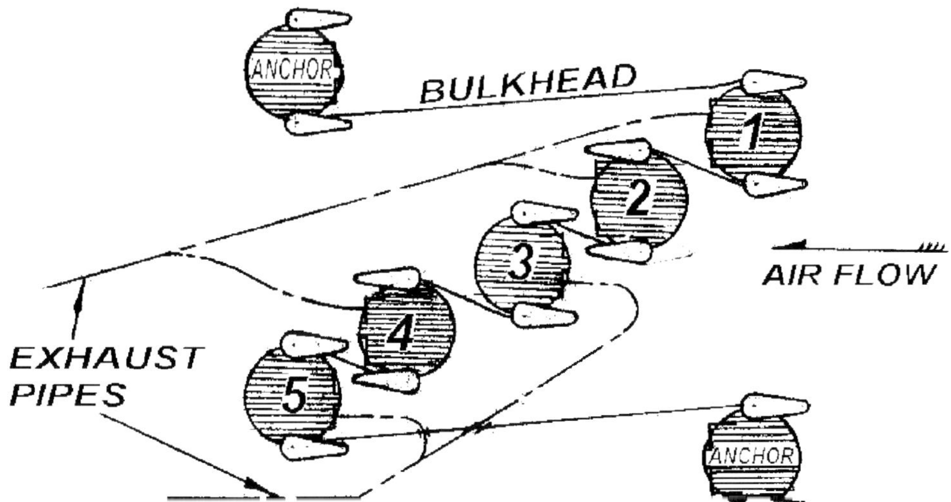

MOCK UP #6 5 ROWS - SPIRAL LEFT HAND, 7 CYLINDERS
PER ROW, R-2800 CYLINDERS

FIG. 6

tated which combination would prevail, firing and cooling alone being the primary ones. Intensive studies were carried out, initially using existing R-2800 cylinders. However, this cylinder design was not intended to accommodate four rows and 28 cylinders. Nevertheless, the first proof of concept engine featured 28 R-2800 "B" series front cylinders arranged in a right-hand spiral. Apart from the cylinders, other off-the-shelf components were used for the initial proof of concept engine including a modified H-3130 nose case and reduction gear, supercharger and Bendix PT-13 carburetor from a "B" series R-2800, and R-2180 connecting rods *(Ref. 2-2)*.

Serious design began on November 11, 1940, with the first proof of concept engine previously described running on April 28, 1941. Unusual for a radial, the fore and aft valve positions were evolving, even though the first attempts placed the exhaust port on top of the cylinder and the

MOCK UP #7 6 ROWS - IN LINE, 5 CYLINDERS PER ROW, R-2180 COMPOSITE CYLINDERS

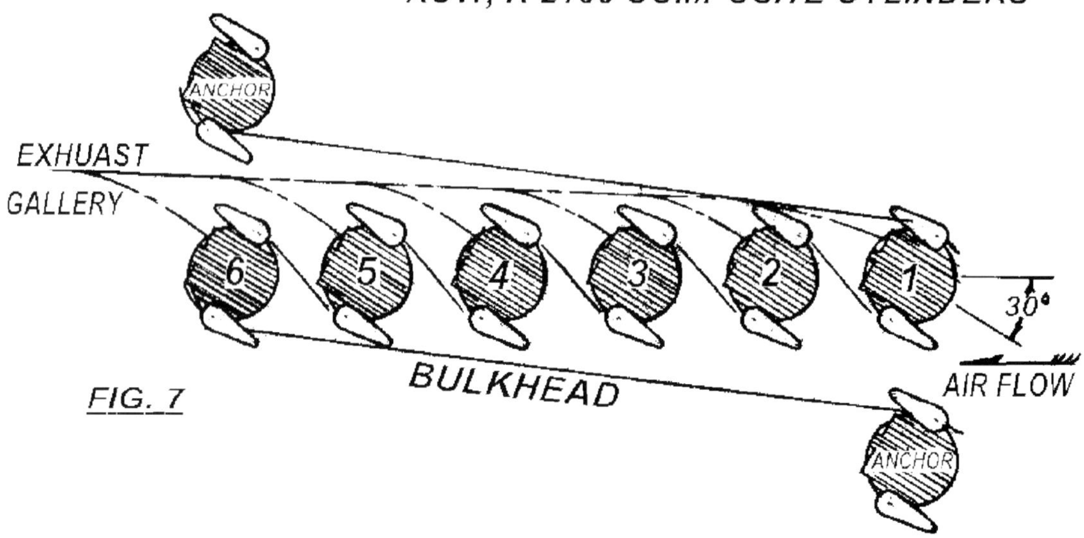

FIG. 7

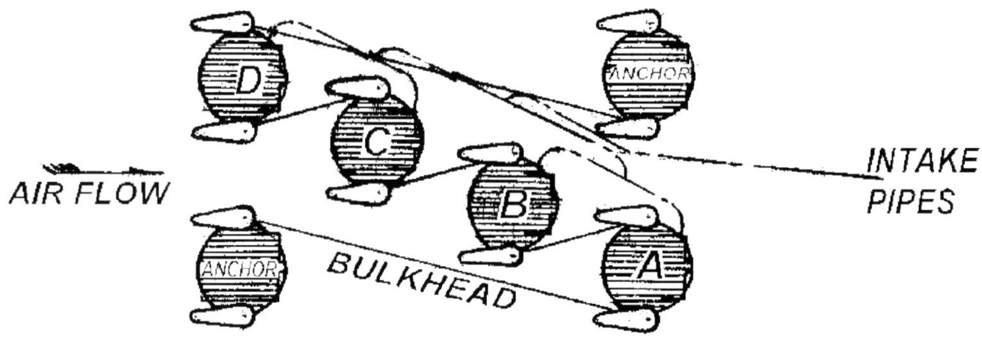

FIG. 8

MOCK UP #8 4 ROW - SPIRAL RIGHT HAND, 7 CYLNDERS PER ROW R-2800 FRONT CYLNDERS

intake off to the side. In retrospect, this may have been a better arrangement than the one finally chosen *(Ref. 2-3)*. Even though the United States was not yet enmeshed in World War II, the writing was on the wall—it was only a matter of time. Luke Hobbs was committed to having the R-4360 ready for combat. As events unraveled, this did not occur but Hobbs only missed this ambitious goal by a short period of time. As if further proof was necessary to reinforce the decision to put the liquid-cooled projects on the back burner, Pratt & Whitney took a series of photographs to illustrate the difference in frontal area. The R-4360 was clearly superior in every aspect: reduced frontal area for a significantly greater displacement.

1941 saw a flurry of activity with the building of three complete engines, two single cylinder engines, and a single-row 7-cylinder engine. In addition, parts for another six engines had been machined. Five hundred hours of testing the full size engines plus 176 hours of single-row and 1,412 hours of single cylinder testing had been completed *(Ref. 2-4)*.

1941 also saw the transition to forged cylinder heads, a feature that all production R-4360s would have. It was at this time in 1941 that Pratt & Whitney were getting into the production of forged heads for their "C" series R-2800. The advantages of this manufacturing technique were far superior to casting. Forging allowed closer and deeper cooling fins to be machined in rather than relying on a casting foundry to perform this critical operation.

First flight of a new engine is always fraught with apprehension; however, all was well with the R-4360's maiden flight on April 25, 1942. Powering a modified Vultee Vengeance dive-bomber and designated V-85, the first flight went smoothly. It wasn't long, however, before problems bubbled to the surface. One of these

continued on page 41

MOCK UP #9 4 ROW - SPIRAL RIGHT HAND 7 CYLINDERS PER ROW R-2800 FRONT CYLINDERS

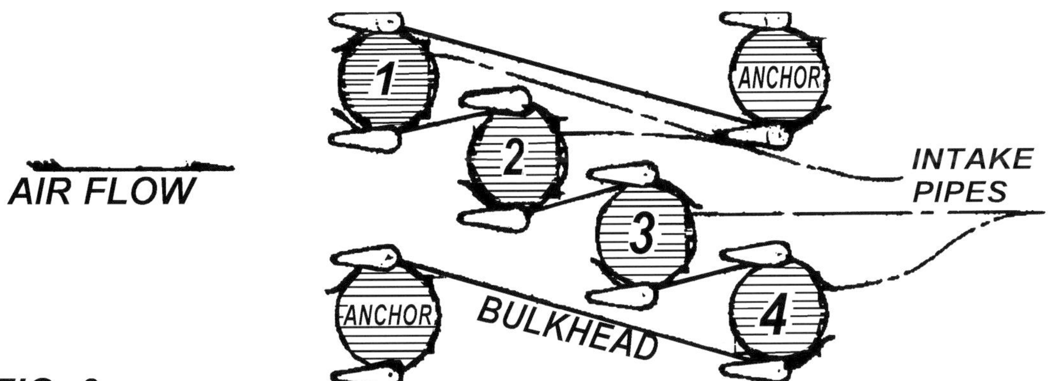

FIG. 9

MOCK UP #10 6 ROW - IN LINE, 6 CYLINDERS PER ROW R-2180 COMPOSITE CYLINDERS

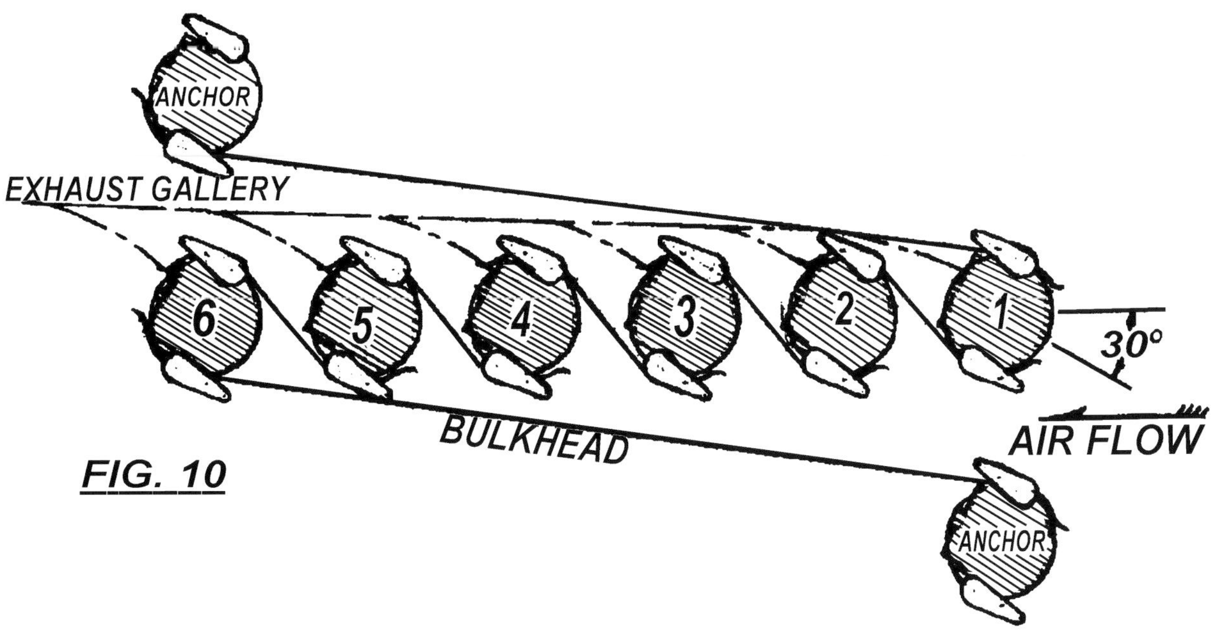

FIG. 10

Part of the evolution to arrive at the fore and aft pushrod concept is shown here. It still exhibits its B series R-2800 heritage; nevertheless, it clearly represents part of the thought process that led to the definitive R-4360 cylinder design. *(United Technologies Corp./Pratt & Whitney)*

The first of many. Made up from available off-the-shelf components, this proof of concept engine paved the way to further R-4360 development. This three-quarter right front view shows the B series R-2800 front cylinders and the H-3130 nose case. *(United Technologies Corp./Pratt & Whitney)*

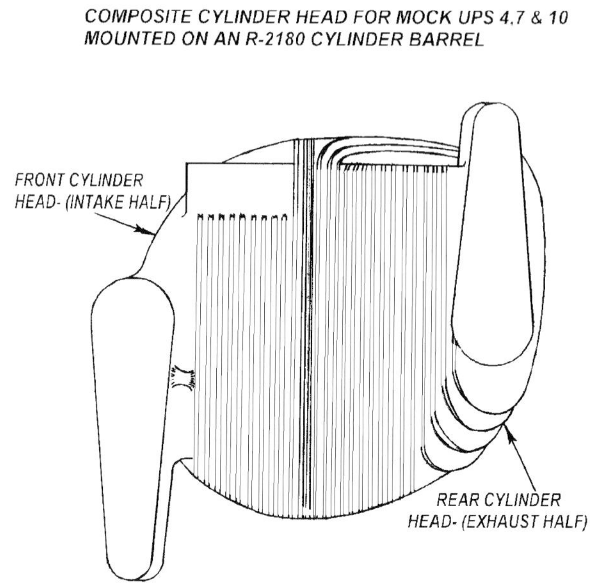

FIG. 11

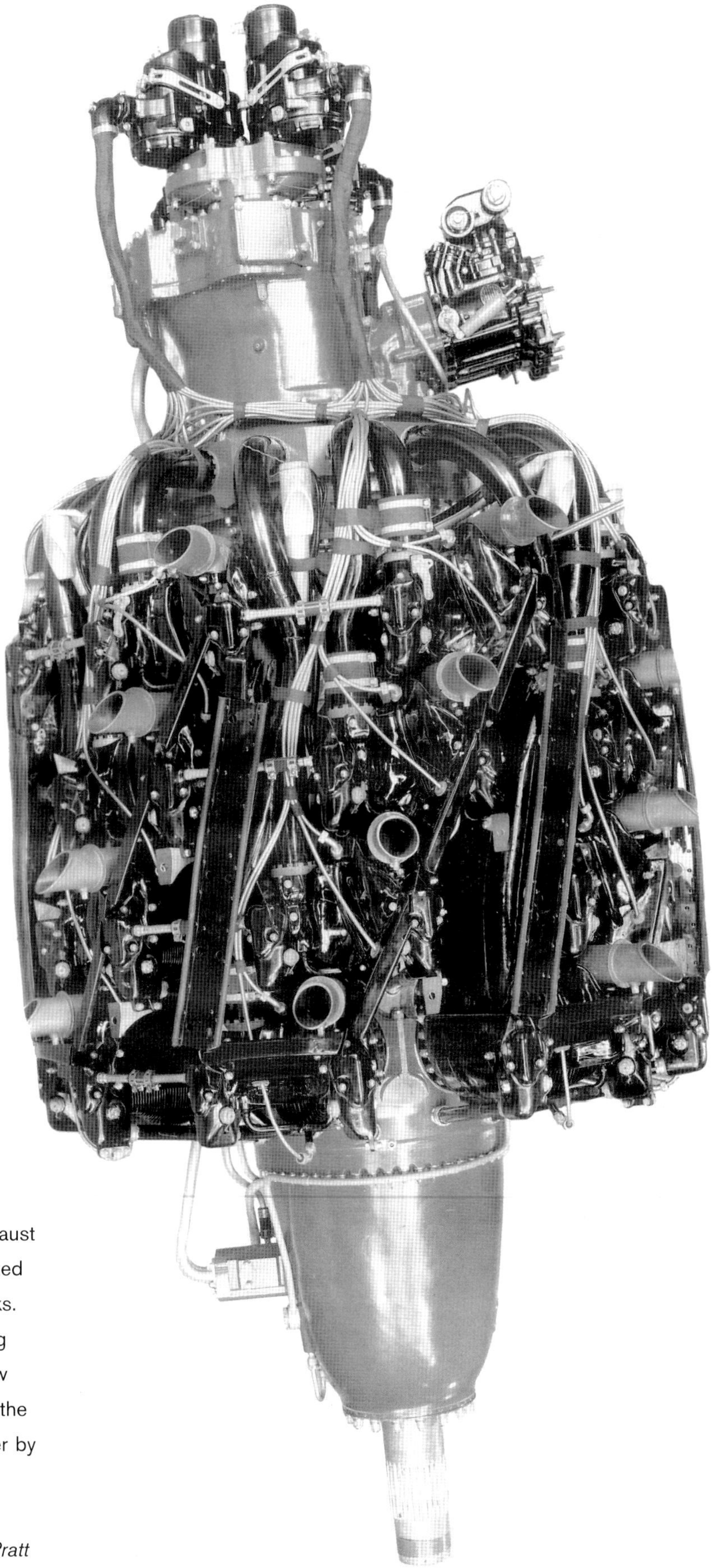

This right side view shows the simple exhaust system, which consisted of 28 very short stacks. These were for testing purposes only to allow the tester to observe the health of each cylinder by "reading" the exhaust flame. *(United Technologies Corp./Pratt & Whitney)*

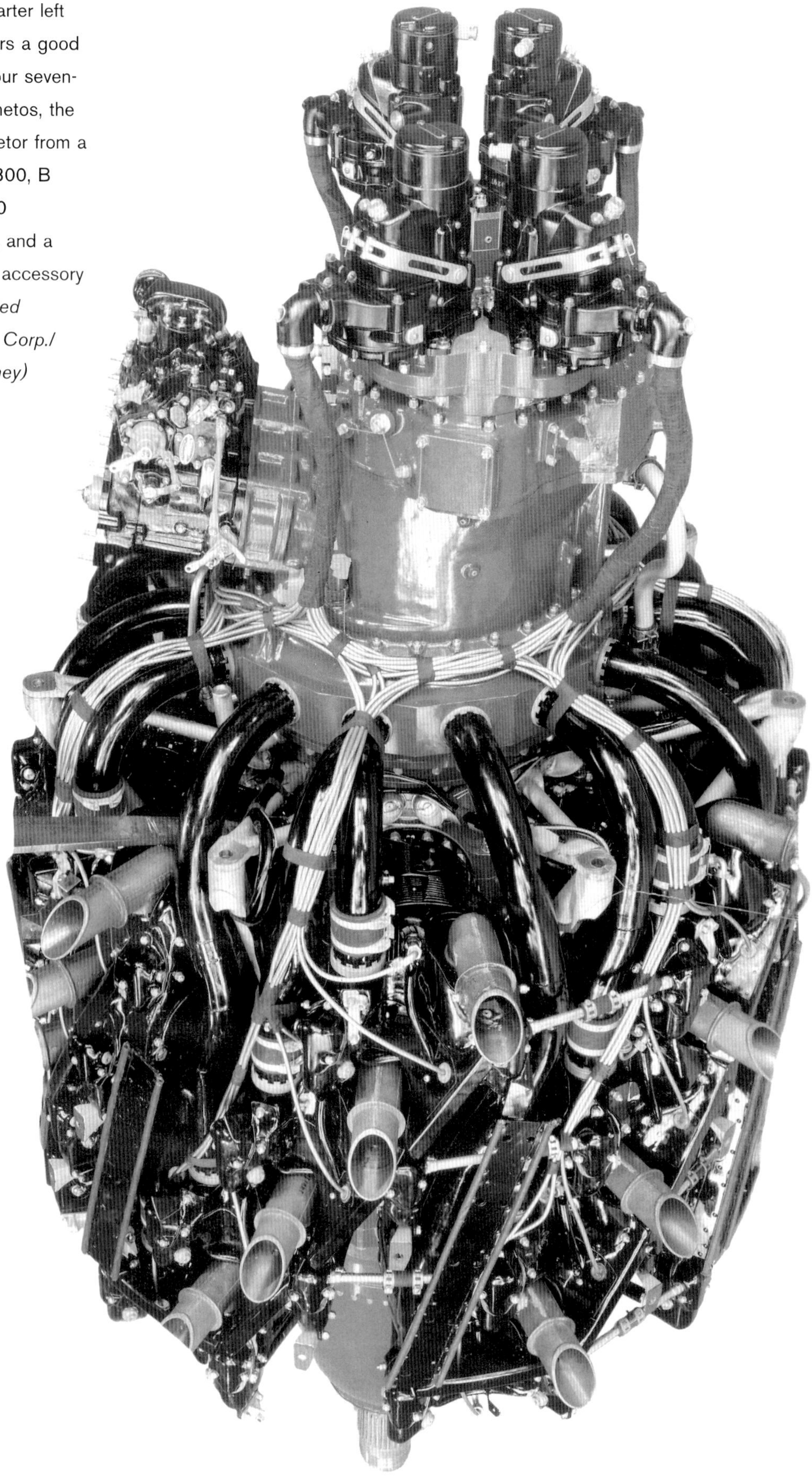

This three-quarter left rear shot offers a good view of the four seven-cylinder magnetos, the PT-13 carburetor from a B series R-2800, B series R-2800 supercharger, and a purpose-built accessory section. *(United Technologies Corp./ Pratt & Whitney)*

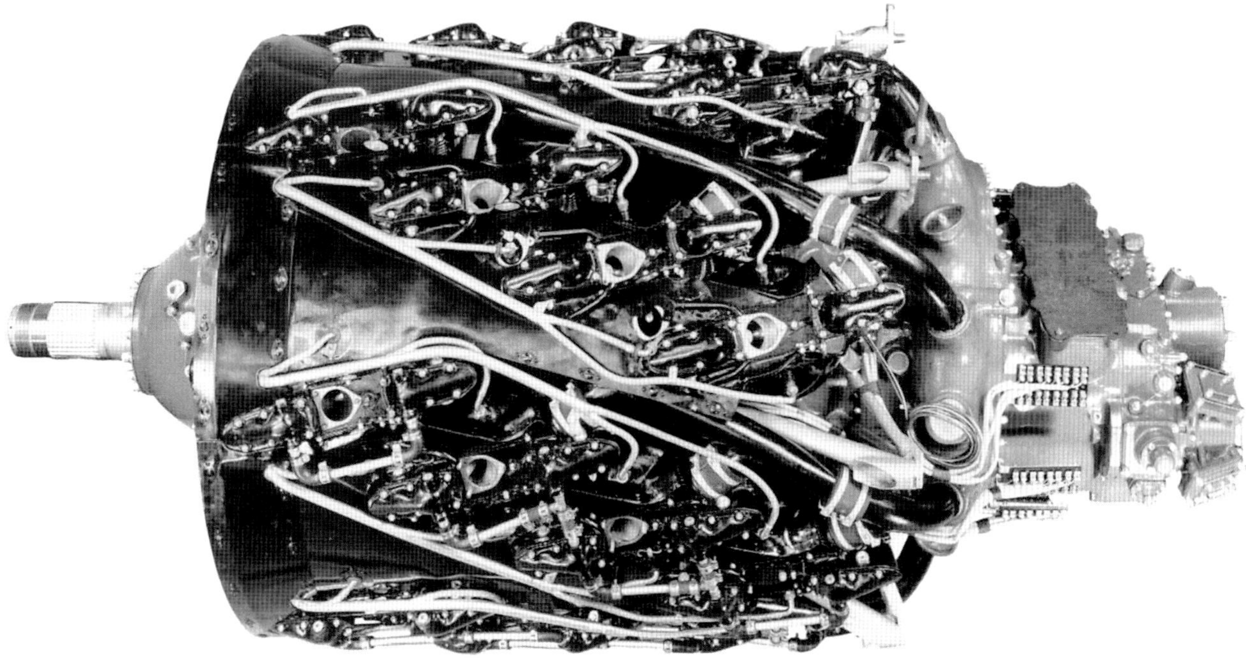

This three-quarter top left view shows the early top-exhaust and side intake arrangement. This was prior to the final design of a top intake and side exhaust. *(United Technologies Corp./Pratt & Whitney)*

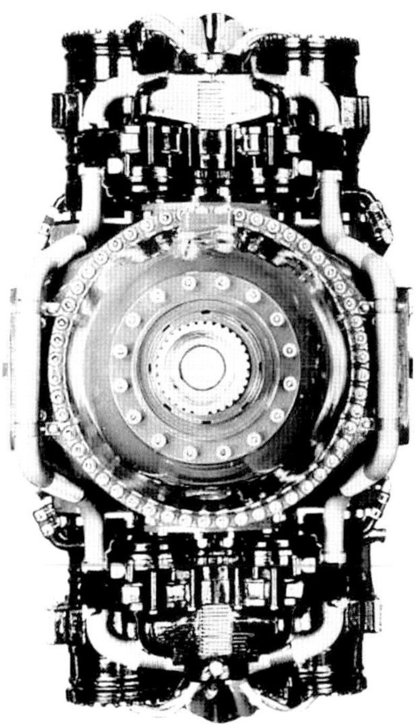

An interesting comparison of the R-4360 and the doomed H-3130. Despite having over 1,000 more cubic inches and capability of developing considerably more power than the H-3130, the R-4360 looks quite svelte next to its porkier sibling. Little wonder that Pratt & Whitney threw in the towel with this ill-conceived liquid-cooled sleeve valve effort. *(United Technologies Corp./Pratt & Whitney)*

continued from page 37

problems resulted in the write-off of the V-85 due to a ruptured fuel feed nozzle diaphragm. This aircraft met its ignominious end in a tobacco field north of Hartford, Connecticut, fortunately with no injury to the pilot. Pratt & Whitney test pilot Howard H. Sargent Jr. performed all initial flights. *(Ref. 2-3A).*

1942 also saw 10 semi-production engines delivered as experimental models for test aircraft. As one could imagine, aircraft designers were chomping at the bit to get their hands on a

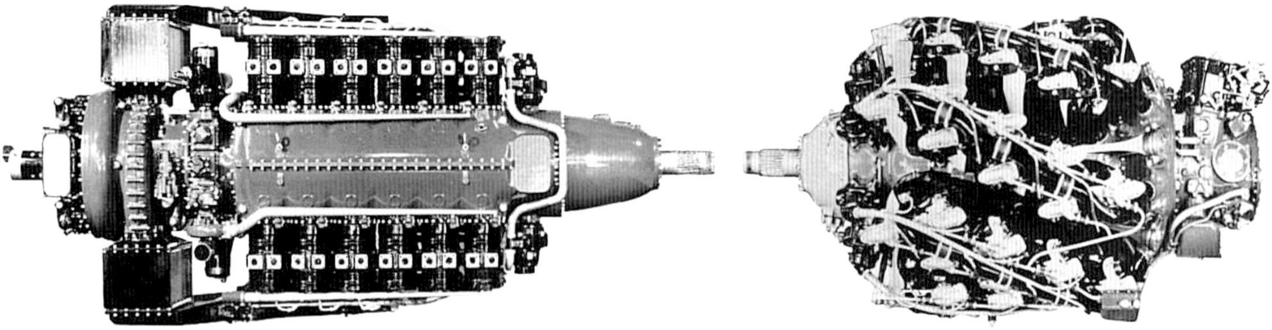

This side view comparison offers more proof of why Pratt & Whitney threw in the towel with its liquid-cooled ventures. Like they say, stick to your knitting and leave the liquid-cooled ventures to companies whose core competency lay in this arena. *(United Technologies Corp./Pratt & Whitney)*

3,000-hp engine. Flexibility was a key requirement that Pratt & Whitney had to keep in mind when designing this engine. Pusher, tractor, and pusher with remote reduction gears and dual rotation propeller drive configurations all had to be accommodated in the R-4360 design. Pusher designs were buried installations, meaning they were buried within the wing structure. This type of installation did not allow a free flow of cooling air as compared to a conventionally mounted tractor installation. Consequently, this demanded a cooling fan driven off the rear accessory case.

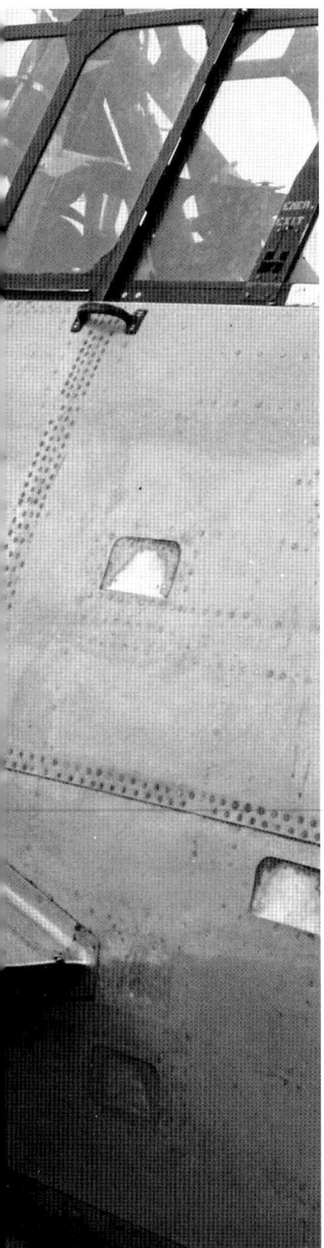

Modified from a Vultee Vengeance dive-bomber, the V-85 served as a good test platform for the XR-4360. This shot shows the cowl flaps wide open and the ram-air induction scoop. The round object behind the gear leg is one of the two oil coolers. *(United Technologies Corp./Pratt & Whitney)*

Pratt & Whitney resolved this issue by simply mounting accessories such as starter, generator, hydraulic pump, etc., on the periphery of the rear case, thus giving unrestricted access for a cooling fan and its drive. These requirements also meant that the engines should be capable of being cooled regardless of which direction cooling-air came in from—front or rear. Most, if not all, air-cooled radials have their cylinder and cooling-fin design optimized for a certain direction for cooling-air inflow—the R-4360 flew in the face of this prior convention. A similar degree of flexibility was demonstrated by the ability to simply remove the rear cover plate and bolt on an additional supercharger stage. In this way, a standard design could be used for single-stage, single-speed, variable-speed, or two-stage.

Development Timeline

The following is a brief timeline of R-4360 development from 1940 to 1949.

1940 *(Ref. 2-4)*

Design studies, for what finally resulted in the R-4360, were initiated during the summer of 1940. Early investigations centered on the optimum arrangement of a radial multi-bank engine

Several aircraft powered by the R-4360 used the so-called buried installation, or the engine was so tightly cowled that fan-assisted cooling was required. This shot shows an early B-36 unit. The cooling fan is driven off the rear case. Some installations, such as the Republic X-12 Rainbow, drove the fan from the nose case. *(United Technologies Corp./Pratt & Whitney)*

The ram-air induction scoop above the cowl is shown to good advantage in this shot. It's also interesting to note the metalized rudder. Clearly, this aircraft was modified to handle the 3,000-plus horsepower that powered it. Too bad it only lasted a short time before being written off in an accident due to engine failure caused by a failed fuel pump. Pratt & Whitney test pilot Howard H. Sargent Jr. flew the aircraft. *(United Technologies Corp./Pratt & Whitney)*

to provide around 3,000 hp. As a result of these studies, design of a 28-cylinder, four-row radial air-cooled engine of 4,360 cubic inch displacement started on November 11, 1940.

1941 *(Ref. 2-4)*

Built up from off-the-shelf parts including the use of 28 "B" series R-2800 front cylinders, the first engine ran on April 28, 1941. The initial ratings on this type of engine were 2,800 hp at 2,600 rpm. Treading carefully, Pratt & Whitney was simply aiming for 100 hp per cylinder—the same rating as the current "B" series R-2800s then in production. This was known as the "A" model of the R-4360. During 1941 a total of 500 hours was accumulated on R-4360 multi-cylinder engine testing and a total of $1,500,000 was expended. It should be kept in mind that, $1.5 million in 1941 was a huge sum of money, especially when it is taken into account that the United States was still embroiled in the Great Depression, even though the worst of it was over.

1942 *(Ref. 2-4)*

The first military model designation of the R-4360 was assigned on March 1, 1942. This was known as the XR-4360-2 and was also the first engine purchased by the services. The Navy

A single-ring gear drove all the accessories mounted around the periphery of the rear accessory case. This freed up the rear case to drive larger components such as cooling fans and auxiliary superchargers. It also had the added advantage of shortening the engine. *(United Technologies Corp./Pratt & Whitney)*

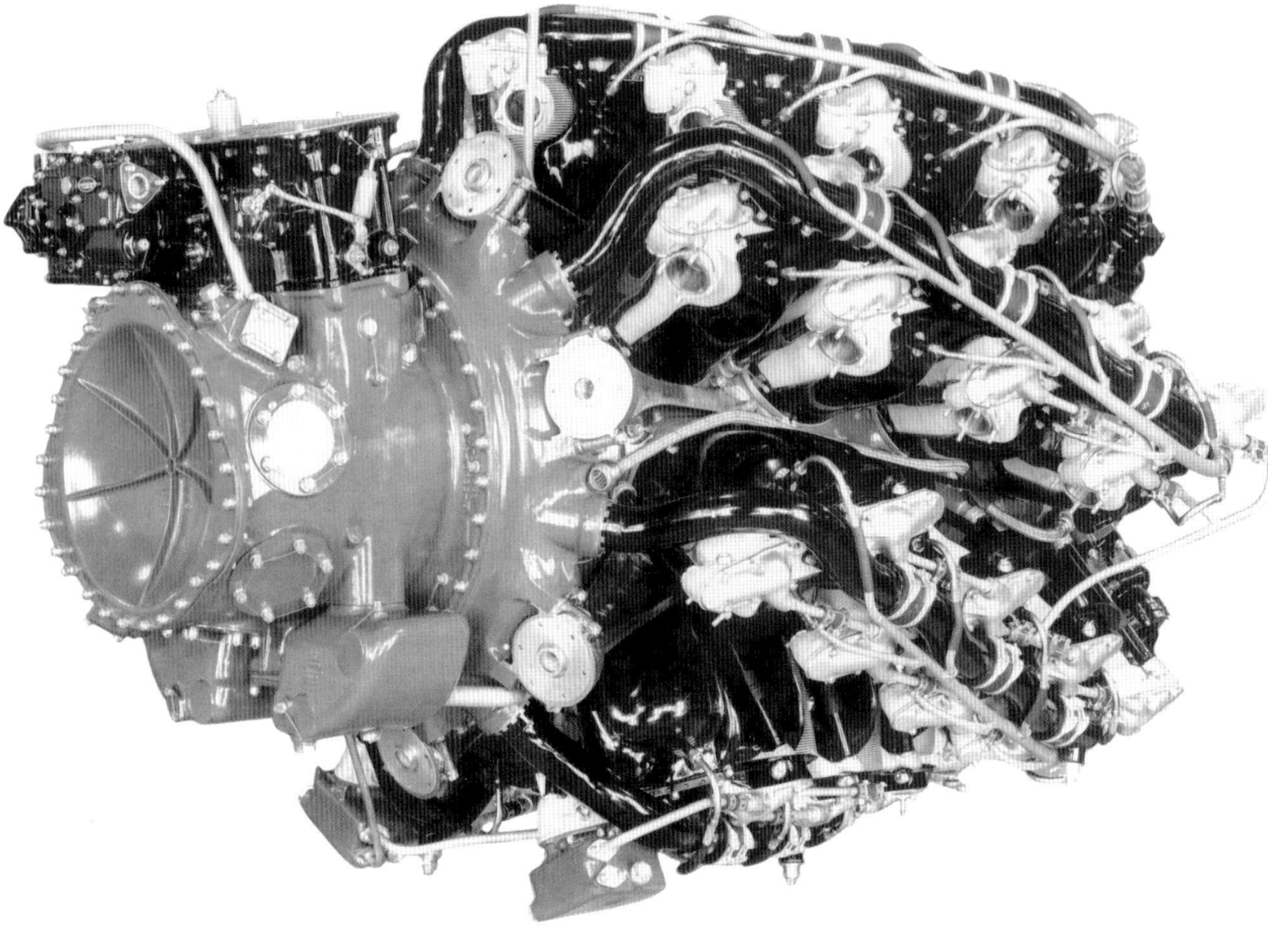

On March 1, 1942, the XR-4360-2 was the first variant of this magnificent engine delivered to a customer–in this case the U.S. Navy. Even though this is still an early prototype engine, from an appearance perspective it differs little from the last ones delivered. *(United Technologies Corp./Pratt & Whitney)*

procured it for the running of a type test. As Navy engine models carry even number designations, "-2" was used, and the "-1" designation was not assigned to any model. Because the Navy was the first service to request the R-4360, it made sense that the first model should be referred to as a -2.

April 1942

In April of 1942, a program of semi-production engine models for use in prototype aircraft was set up, and on April 30, 1942, model designations were set up to cover these various configurations. These engine models and their major features were as follows:

R-4360-4
Navy designation. Single-stage, variable-speed tractor engine for use in P&W flight-test plane. This was considered the basic model and other models were variations of it.

R-4360-3
Army designation. Single-stage, single-speed dual-rotation tractor or pusher engine for use in Curtiss XP-71 and Republic XP-72.

R-4360-5
Army designation. Single-stage, single-speed supercharger, plus two speed, single reduction gear for pusher installation in Consolidated XB-36.

R-4360-6
Navy designation having two-stage variable speed supercharger. None built.

Showing off its huge cooling fan, the R-4360-5 was another early delivery, this time to the Army Air Force for use in the XB-36. Being a buried and pusher installation, augmented cooling was required. The fan achieved this goal. *(United Technologies Corp./Pratt & Whitney)*

R-4360-7
Army designation. Single-stage, single-speed supercharger and remotely mounted two speed, dual-rotation reduction gear for pusher outboard installation in Northrop XB-35.

R-4360-8
Navy designation. Dual-rotation, single-speed reduction gear for use in Douglas XTB2D and Curtiss XBTC.

R-4360-9
Army designation. .500:1 reduction gearing in place of .381:1 basic gear for use in Vultee XA-41.

R-4360-10
Navy designation having two-stage variable speed supercharger and dual-rotation two speed reduction gear for use in Boeing XF8B-1.

R-4360-11
Army designation similar to -7 except for length of extension shafting to accommodate inboard nacelle of Northrop XB-35.

R-4360-13
Army designation having remotely mounted two-stage variable-speed supercharger and .425:1 single-rotation reduction gear for Republic XP-72 in place of –3 engine originally used in study of XP-72.

By the time this group of model designations was assigned, testing had shown that the 3,000 hp for

In what would turn out to be a lost cause, a total waste of resources, and worst of all cause loss of life, the first XB-35 engine—an R-4360-11—was delivered in 1942. Note the lack of a nose case and reduction gearing. Pratt & Whitney performed exhaustive testing with a complete XB-35 drivetrain. Not a lick of a problem surfaced during this testing, and yet, when installed in the XB-35 serious problems arose. This offers further testimony to the flawed installation in this problem-plagued aircraft. *(United Technologies Corp./Pratt & Whitney)*

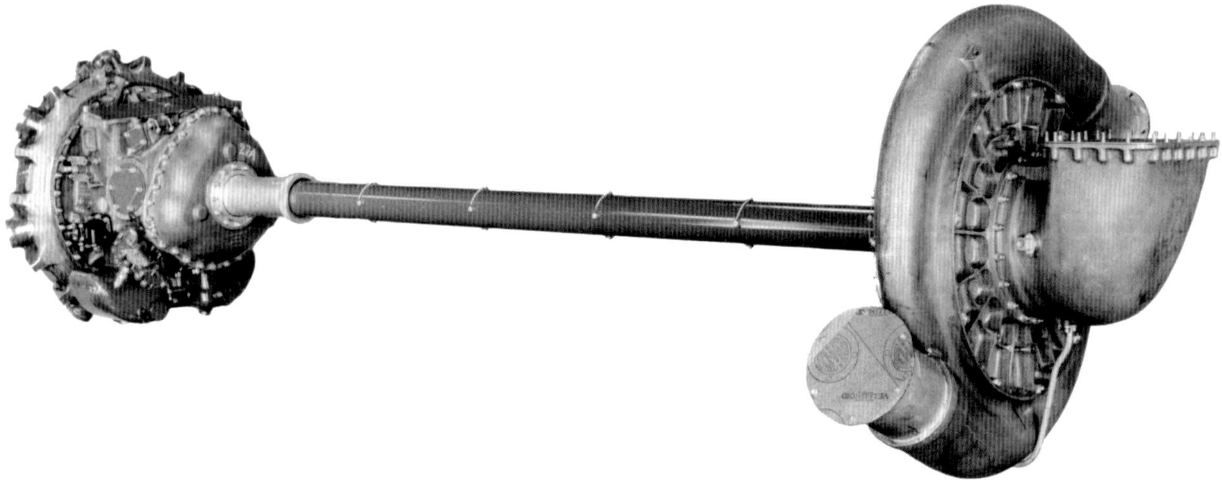

Considering how early it was in the development cycle, it's surprising to see that Pratt & Whitney was developing a two-stage engine in 1942. This is the auxiliary blower drive to be used in the Republic XP-72. By this time Pratt & Whitney had achieved considerable experience building two-stage superchargers primarily for the R-2800. *(United Technologies Corp./Pratt & Whitney)*

In 1943, Douglas was hard at work designing the C-74, which was powered by the R-4360-15. Too bad this very versatile transport aircraft was not developed in time for World War II service. *(United Technologies Corp./Pratt & Whitney)*

which the engine had been originally designed for could be realized. Therefore, this rating was used for all models subsequent to theXR-4360-2.

The first flight with an R-4360-powered aircraft was made on May 25, 1942.

During the next year additional entire models were established for new aircraft projects or to provide production model designations for engines to be used in R-4360-powered aircraft scheduled for production.

July 1942

R-4360-12

Navy designation having a dual-rotation, two-speed reduction gear for use in Curtiss XBTC-2 replacing R-4360-8 previously planned for this plane. It's doubtful if the two-speed feature was ever implemented. Pratt & Whitney records indicate that engines fitted with this feature were locked in one speed.

By the conclusion of 1942, a total of 3,500 hours of test stand operation had been accumulated and a total of $5,700,000 had been spent. Of course, by now the United States had been mobilized for World War II and consequently development dollars went up considerably. Adding to the R-4360's development effort was the fact that the ill-fated liquid-cooled H-3130 program was now on the back burner. Pratt & Whitney bit the bullet and ate most of the development costs of this project in order to start with a clean slate for the R-4360.

R-4360-25s were the production version of the R-4360-5. Note the six-inch extension incorporated into the nose case–a feature of all B-36 engines. This extension accommodated the propeller spinner after body. *(United Technologies Corp./Pratt & Whitney)*

1943 *(Ref. 2-4)*

February 1943

R-4360-15

Army designation having .425:1 single rotation reduction gear for use in Douglas C-74, predecessor of the C-124 Globemaster.

When the first production models were established, it was decided to identify these as "B" series engines. Therefore, the "B" series identified the production bill as distinguished from the semi-production engines.

February 1943

R-4360-17

Army designation for production version of R-4360-7 used in Northrop YB-35 (outboard nacelles, i.e., #1 and #4 positions).

February 3, 1943

R-4360-19

Army designation for production version of R-4360-13 to be used in Republic XP-72. However, this version of the XP-72 was never built.

February 1943

R-4360-21

Army designation for production version of R-4360-11 used in Northrop YB-35 (inboard nacelles, i.e., #2 and #3 positions).

March 1943

R-4360-23

Army designation of production version of R-4360-3 engine planned to power the Curtiss XP-71. No R-4360-23s or Curtiss XP-71s were built.

This R-4360-27 was the production version of the R-4360-15. -27s were used to power Douglas C-74 transports. *(United Technologies Corp./Pratt & Whitney)*

March 1943

R-4360-25

Army designation of production version of -5 engine used in Consolidated B-36 and C-99. This engine, however, did not feature two-speed reduction gearing.

R-4360-27

Army designation of production version of -15 engine used in Douglas C-74.

The first engine shipped to the military services was an engine for type test shipped to A.E.L., (Army Engineering Lab) Philadelphia, on June 7, 1943.

June 1943

R-4360-29

Army designation having two-stage variable speed supercharger for use in developing an R-4360-powered prototype of a Boeing B-29 later designated XB-44. No R-4360-29s were manufactured. The sole XB-44 was powered by R-4360-33s.

August 1943

R-4360-14

Navy designation having single-speed, dual-rotation reduction gear with #60 and #70 spline propeller shafts in place of #60 and #80 spline propeller shafts replacing R-4360-8 in Curtiss XBTC.

Semi-production R-4360 engines were first shipped in September 1943. These were an R-4360-5 to Consolidated for B-36 test rig operation and an R-4360-13 to Republic for the XP-72. The R-4360-5 had the reduction gear locked

The -33 was another two-stage variation used to power the XB-44, which was a B-29 converted to R-4360 power. In retrospect, it may have been better to have powered production B-50s with this engine instead of the problem-prone turbosupercharged R-4360-35. *(United Technologies Corp./Pratt & Whitney)*

into one speed and could not be used as a two-speed reduction gear. A section of B-36 wing was built to ground-test the R-4360-5 in a similar manner in which Bristol ground tested the paired Centaurus 20s for use in the Bristol Brabazon.

An R-4360-7 was shipped in October 1943 to Martin for use on the B-35 program. Even though this was a Northrop project, after cancellation of B-26 contracts Martin engineers were assigned to B-35 development. In the same month an R-4360-8 was shipped to Curtiss for use on the XBTC-2 program.

In November 1943, an R-4360-9 engine was shipped to Vultee for use in the XA-41. In December 1943 an additional R-4360-9 engine was shipped to Wright Field, an R-4360-15 was shipped to Douglas, and an R-4360-8 was shipped to Curtiss-Wright Corporation, Columbus, Ohio.

There is no record of the Navy designation R-4360-16 ever having been assigned.

December 1943

R-4360-18

Navy designation having single-stage, single-speed supercharger for use in Lockheed C-89 Constitution.

December 1943

R-4360-31

Army designation having single-stage, single-speed supercharger and single-speed dual-rotation reduction gear for use in Hughes XF-11 and Republic XF-12. R-4360-31s were converted to single rotation R-4360-37s by the Army, possibly due to Howard Hughes' near fatal crash of the XF-11 fitted with dual-rotation R-4360-31s.

At the conclusion of 1943, a total of eight semi-production R-4360 engines had been shipped (six Army and two Navy). 7,200 hours of test stand operation had been documented, and Pratt & Whitney expended a total of $10,000,000 on the R-4360.

1944 (Ref. 2-4)

During January and February of 1944, a total of eight semi-production engines were shipped, including four R-4360-4s, one R-4360-7, one R-4360-8, one R-4360-11, and one R-4360-13.

March 1944

R-4360-33

Army designation having two-stage, variable-speed supercharger for use in Boeing XB-44.

During the remainder of 1944, a total of 19 more semi-production engines were shipped. These included six R-4360-5s, one R-4360-13, four R-4360-10s, four R-4360-4s, one R-4360-9, and two R-4360-8s.

By the close of 1944, a total of 30 semi-production engines were shipped (17 Army and 13

By 1945, Boeing was well advanced with the design of the B-50. This interim bomber was powered by R-4360-35s. Note the use of hooded baffles for the first time. These baffles were employed to keep the ignition harnesses and coils cool. Although the base engine was satisfactory, the same could not be said for its very problematic GE turbosupercharger. *(United Technologies Corp./Pratt & Whitney)*

Navy.) Test stand operation had increased to a total of 12,000 hours and Pratt & Whitney engineering expenses on the R-4360 now totaled $15,300,000.

1945 *(Ref. 2-4)*

In January 1945 the first production engine was shipped—an R-4360-4. All prior engine shipments to this date were of a pre-production status.

The R-4360 completed its official 150-hour type test on February 6, 1945, at the A.E.L.

During the summer of 1945, it became necessary to provide additional takeoff power for the B-50. To accomplish this the B3 series of engines was developed. This engine differed principally from the "B" series in that compression ratio was reduced from 7.0:1 to 6.7:1 to allow greater manifold pressure, cylinder baffles were changed to protect the rocker boxes, and ignition system from exhaust heat (hooded baffles) and power section and reduction gearing parts were strengthened.

July 1945

R-4360-35

Army designation having single-stage, single-speed supercharger and 3,500-hp rating for use in Boeing B-50. This application was augmented with a GE CH-7 turbosupercharger.

November 1945

R-4360-37

Army designation having single-stage supercharger and single-speed reduction gear for use in Republic F-12 and Hughes XF-11 in place of R-4360-31. However, as it turned out, most -31s were converted to -37s.

Late in 1945 it was realized that to have an engine that could be rated in excess of 3,500 hp and display some degree of reliability, it would be necessary to redesign the "B" series engine to provide additional strength throughout. To accomplish this, the series "C" engine was laid out. It differed from the "B" series in all major parts; the crankshaft was changed to permit strengthening of the crankcases, the crankcases were made heavier and stronger, the mounting system was changed to permit the use of stronger attachments to the main cases, and the cylinder was changed to provide for better cooling. For the first time direct fuel injection was introduced. This required that the rear case be changed to accommodate fuel injection and its associated drive requirements. Last but not least, the reduction gear was strengthened to permit the transmission of higher powers. All of the changes constituted the "C" series engines. However, not all "C" series engines were fuel injected—many retained the Bendix PR-100 or Chandler Evans CECO 100CPB-9 carburetor.

During 1945, engine shipments totaled 30 semi-production engines and 81 production engines, and a total of 141 R-4360 engines were shipped (62 Army and 79 Navy). Development test time reached 17,000 hours and engine expenditures reached a total of $19,700,000.

1946 *(Ref. 2-4)*

January 1946

R-4360-39

Army designation of "C" series single-stage, single-speed engine having a 3,800 bhp rating to use for development of "C" series R-4360.

March 1946

R-4360-20

Navy designation of basic model engine having 3,250 bhp rating (subsequently raised to 3,500 hp) for use in Douglas C-124A (-20A wet), Fairchild R4Q-1 (-20A wet), Fairchild C-119B (-20A wet), Fairchild C-119C (-20A wet), Fairchild C-120, Fairchild XC-120 (-20A, dry, no ADI) Martin P4M-1 (-20A, dry), and Martin XP4M-1 (-20A, dry).

In July of 1946 the 150-hour model tests of the R-4360-35 engine (for B-50) and the R-4360-25 engine for the B-36 were completed. It was therefore decided that B-36 engines should also have those features incorporated that would permit operation at 3,500 hp, such as the "C" series standards.

Douglas C-74s were re-engined with R-4360-49s offering 3,500 hp as opposed to the 3,000 of the -27s. An additional 2,000 hp never hurt any aircraft, especially in a transport category. However, the nacelle design of the C-74 and follow-on C-124 can only be described as awful. Andrew Wilgoos was incensed when he saw the C-74's nacelle design. Nevertheless, this aircraft soldiered on for many years. *(United Technologies Corp./Pratt & Whitney)*

October 1946

R-4366-41A

Army designation of engine for B-36 having 3,500-hp rating used in place of R-4360-25.

During 1946 a total of 383 engines were shipped bringing the total to 524, of which 323 were shipped to the Air Force and 221 were shipped to the Navy. Development test time on the test stand had been increased to 24,000 hours, and engine development expenditures had reached $25,700,000.

1947 (Ref. 2-4)

During "C" series development, studies of various means of compounding the engine had shown that by removing the supercharger drive requirements from the engine and using the exhaust energy in a turbosupercharger to do all the engine supercharging, an increase in takeoff rating of 500 hp could be realized. Relieving the engine of its supercharger drive requirements also resulted in a reduction in cruising fuel consumption of about eight percent.

Development work on the "C" series engine was therefore carried out to permit its use with such an arrangement. This was the genesis of the VDT program (variable discharge turbine, further described in Chapter 4).

April 1947

R-4360-43

Army designation of "C" series VDT engine with external supercharging, i.e., turbosupercharging only, having a 4,300-hp rating for use in Boeing YB-50C and B-54.

April 1947

R-4360-45

Army designation for single-rotation reduction gear engines for use in Northrop B-35 in place of R-4360-17 (outboard nacelle).

April 1947

R-4360-47

Army designation for single rotation reduction gear engine for Northrop B-35 in place of R-4360-21 (inboard nacelle).

May 1947

R-4360-22

Navy designation of 3,500 bhp single-speed, single-stage engine for use in Lockheed C-89 Constitution in place of R-4360-18.

June 1947

R-4360-51

Army designation of externally supercharged tractor "C" series engine (VDT) having remotely mounted .3125:1 reduction gear and cooling fan, with 4,300-hp rating for use in Consolidated B-36C. As we now know, this was the start of the ill-fated VDT B-36C program.

During 1947, 790 engines were shipped bringing the total to 1,314, of which 1,026 were shipped to the Air Force and 289 were shipped to the Navy. Test stand operation had reached 31,600 hours and total engineering expenditures now totaled $33,200,000.

1948 *(Ref. 2-4)*

March 1948

R-4360-49

Army designation for 3,500-hp engine for use in Douglas C-74 in place of R-4360-27.

During 1948, 1,005 engines were shipped bringing total R-4360 shipments to 2,319, of which 2,009 were Air Force and 310 were Navy engines. Test stand operation had reached 40,000 hours and engineering expenditures were at $40,600,000.

1949 *(Ref. 2-4)*

June 1949

R-4360-53

Army designation for single-stage, single-speed "C" series pusher engine with 3,800-hp rating and fan-assisted cooling for use in Consolidated B-36 in place of R-4360-41.

June 1949

R-4360-55

Army designation for externally supercharged pusher "C" series engine with 4,300-hp rating having externally mounted cooling fan for use in later version of Consolidated B-36 in place of either R-4360-41 or R-4360-53.

June 1949

R-4360-57

Army designation for advanced development of R-4360-55, having higher compression ratio (7.5:1) to obtain 20 percent reduction in fuel consumption compared to engines that were currently installed in B-36s. Although some were built, this high-compression engine was never implemented.

Through June 1949, 719 engines were shipped for the year, bringing the grand total of R-4360 shipments to 3,038. Of this total 2,844 went to the Air Force and 394 went to the Navy. Engineering expenditures as of that date were at $42,900,000. By this date over

Opposite: In 1949 the fuel-injected R-4360-53 was introduced for the B-36. Each engine developed an additional 300 hp, now rated at 3,800 hp. Fuel injection also eased some maintenance difficulties and allowed for better fuel mixture management. *(United Technologies Corp./Pratt & Whitney)*

R-4360 DEVELOPMENT

What might have been. This is an R-4360-57 VDT engine with high-compression pistons. The VDT project was cancelled after huge sums of money had been expended. Even so, it came tantalizingly close to entering production to power B-36Cs and possibly Republic XF-12 Rainbows. *(United Technologies Corp./Pratt & Whitney)*

40,000 hours of testing had been completed. It's always interesting to do the arithmetic and figure out how much fuel was consumed. It's a safe bet to say that 100 gallons per hour, on average, was consumed by the R-4360. This would amount to well over 4,000,000 gallons!

After 1949, Pratt & Whitney's emphasis was predominately aimed at the now more established gas turbine field. In 1950 the Andrew Wilgoos Turbine Laboratory was opened. This occasion, although not realized at the time, marked the watershed for R-4360 development. Like other major aircraft engine manufacturers, with the exception of Curtiss-Wright, Pratt & Whitney placed all their faith in this new and obviously superior technology. Superior, from a power-to-weight ratio, although one could argue the charismatic value of the gas turbine.

What's In a Name?

All Pratt & Whitney piston engines used the term "Wasp" as part of its designation for commercial applications. In the case of the R-4360, the name "Wasp Major" was bestowed upon it *(Ref. 2-5)*. In a similar fashion to Pratt & Whitney's other piston engines, this moniker never really stuck; people referred to it as "the R-4360" or simply "the 4360." Although it has been explained many times it may be worth reiterating where the alphanumeric designation R-4360 came from. In the early 1920s the Army Air Corps wanted to develop a consistent and easily understood convention for naming aircraft engines. This requirement was accomplished by calling out the cylinder configuration representing the first letter. The basic configurations include **R**adial, **O**pposed, **V**ee...etc. A hyphen followed the letter representing the cylinder configuration. After the hyphen, a number representing the displacement in cubic inches rounded out to the nearest ten gave immediate information on the cubic displacement of the engine. Another hyphen after the displacement followed by a so-called dash number, indicated the degree of development, or power rating of that particular model of engine. Even dash numbers were engines developed for the Navy and odd dash number engines were developed for the Army and/or Air Force. Occasionally one would see an engine, for instance, with an odd dash number installed in a Navy aircraft and vice versa; however, these instances are quite rare. Inter-service rivalry oftentimes prevented the use of engines funded by one branch of the military being used by another branch. Oh, how things have changed since!

With 28 cylinders, things get confusing when simply numbered one through 28. Prior radial engine practice had been to just number the cylinders; however, this would not have been the most practical naming convention with the R-4360's 28 cylinders. For instance, if cylinder number 18 is called out, one would need to go through some mental gyrations in order to figure out which row and what position that cylinder resided. Pratt & Whitney overcame this issue by using an alphanumeric convention. Each row was given the alpha designation. Starting at the rear with row "A" and moving forward to the front with row "D." Each cylinder within a row was given the numeric designation one through seven. Looking from the rear of the engine, the uppermost cylinder is number one and going clockwise they are numbered up to seven (Ref. 2-6).

R-4360 Design Features

The following six illustrations display the locations of various components called out in the text. These illustrations are of an R-4360-59, which is a later development version. However, unlike the R-2800, which went through major redesigns, the R-4360 only went through incremental improvements during its production life span, therefore these illustrations serve their purpose well for describing the pertinent parts *(Ref. 2-7)*.

After the basic concept of 28 cylinders in four rows had been established, Pratt & Whitney's next goal was putting all the parts together, making it work, putting out decent horsepower, and cooling effectively. It should be remembered that no one had, before or since, successfully put a radial engine in production with more than two rows. It had been tried, but all development efforts faded into the sunset and deserved obscurity. With 28 cylinders, even firing dictates that a cylinder fires every 25.714285 degrees (720 divided by 28). This number would have a profound influence on the R-4360's design. With seven cylinders per row, there is little choice for angular spacing between the cylinders on a particular row; it is 51.428571 degrees (360 divided by 7). Now the question of what offset angle to give between cylinder rows arises. This is arrived at by dividing 25.714285 degrees (half of 51.428571 degrees) by two, which gives 12.857142 degrees. This 12.857142-degree offset is what gives the R-4360 its characteristic spiral

continued on page 66

(Wasp Major Models B6 and CB2
Overhaul Manual Issued June 1953)

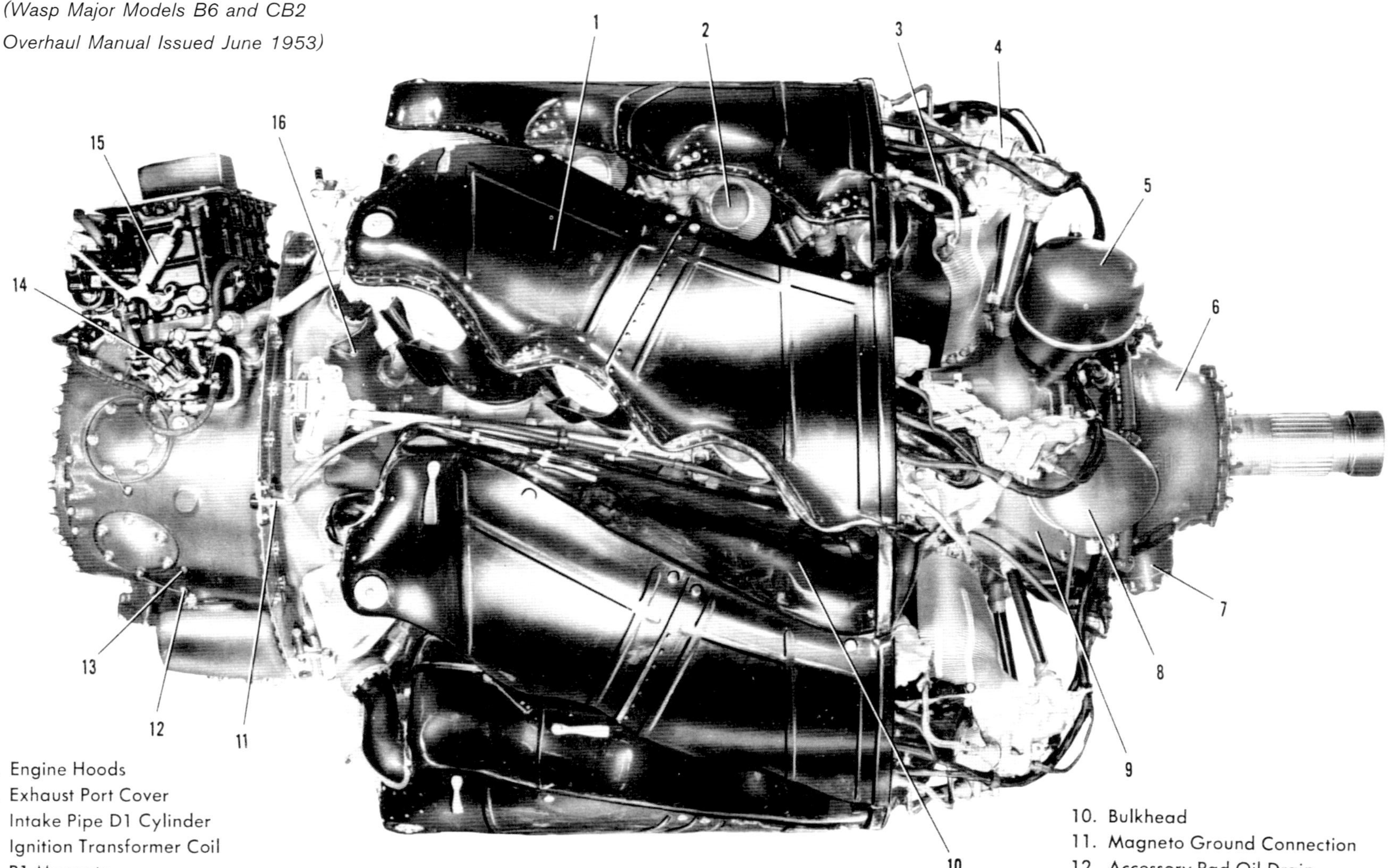

1. Engine Hoods
2. Exhaust Port Cover
3. Intake Pipe D1 Cylinder
4. Ignition Transformer Coil
5. R1 Magneto
6. Prop Shaft Case
7. Front Oil Pump
8. R2 Magneto
9. Magneto Drive Case
10. Bulkhead
11. Magneto Ground Connection
12. Accessory Pad Oil Drain
13. Low Pressure Oil Gage Connection
14. Water Regulator
15. Throttle Anti-Creep Mechanism
16. Manifold Temperature Connection

Right Side View of Engine

(*Wasp Major Models B6 and CB2 Overhaul Manual Issued June 1953*)

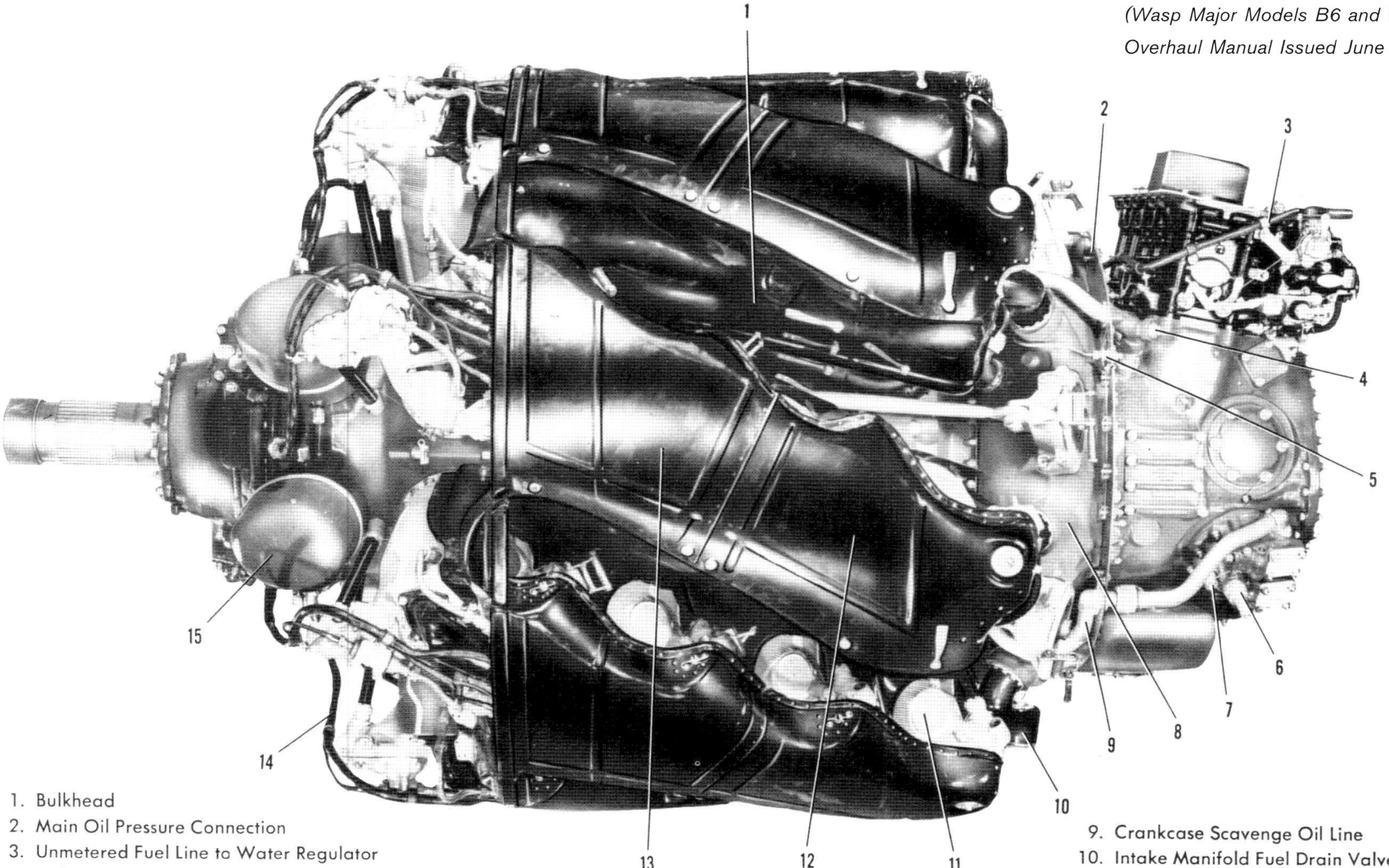

1. Bulkhead
2. Main Oil Pressure Connection
3. Unmetered Fuel Line to Water Regulator
4. Collector Case Breather
5. Manifold Pressure Connection
6. Main Oil Pressure Relief Valve
7. Main Oil Pressure By-Pass Valve
8. Collector Case
9. Crankcase Scavenge Oil Line
10. Intake Manifold Fuel Drain Valve
11. A5 Exhaust Port Cover
12. Trailing Hoods
13. Leading Hoods
14. Ignition Branch Manifold
15. L2 Magneto

Left Side View of Engine

(Wasp Major Models B6 and CB2 Overhaul Manual Issued June 1953)

7. Propeller Shaft 60A Splines
8. Propeller Shaft Thrust Nut
9. Ignition Branch Manifold
10. Front Oil Pump
11. Magneto Oil Drain Manifold
12. Tube-Flooded Rockerbox Oil Drain
13. Ignition Ground Wire Conduit
14. Ignition Manifold Ring
15. D3 Cylinder Deflector
16. R2 Magneto Cover
17. R1 Magneto
18. D1 Spark Plug
19. D1 Intake Pipe

1. Magneto Spark Advance Solenoid Pad
2. Propeller Shaft Oil Transfer Tube Plug
3. Rockershaft Cap
4. Governor Pad
5. L1 Magneto Ventilator
6. Propeller Shaft Thrust Cover

Right Front View of Engine

(Wasp Major Models B6 and CB2 Overhaul Manual Issued June 1953)

1. Ignition Ring Manifold
2. Front Lifting Link
3. Torquemeter Transmitter Pad
4. Ignition Transformer Coil
5. Ignition Branch Manifold
6. Magneto Ventilator
7. L1 Magneto Cover
8. Cowl Bracket Attachment
9. Magneto Cover Clamp
10. Ventilator
11. Torquemeter Pressure Relief Valve
12. Bulkhead
13. Thermocouple Attachment
14. Intake Manifold Fuel Drain Plug
15. Magneto Pad Oil Drain
16. Front Oil Pump
17. Thrust Bearing Cover
18. Thrust Nut
19. Thread Protector Cap
20. Propeller Shaft
21. Thrust Bearing Cover Slushing Oil Plug
22. Oil Transfer Tube Plugs
23. Governor Pad

Left Front View of Engine

(Wasp Major Models B6 and CB2 Overhaul Manual Issued June 1953)

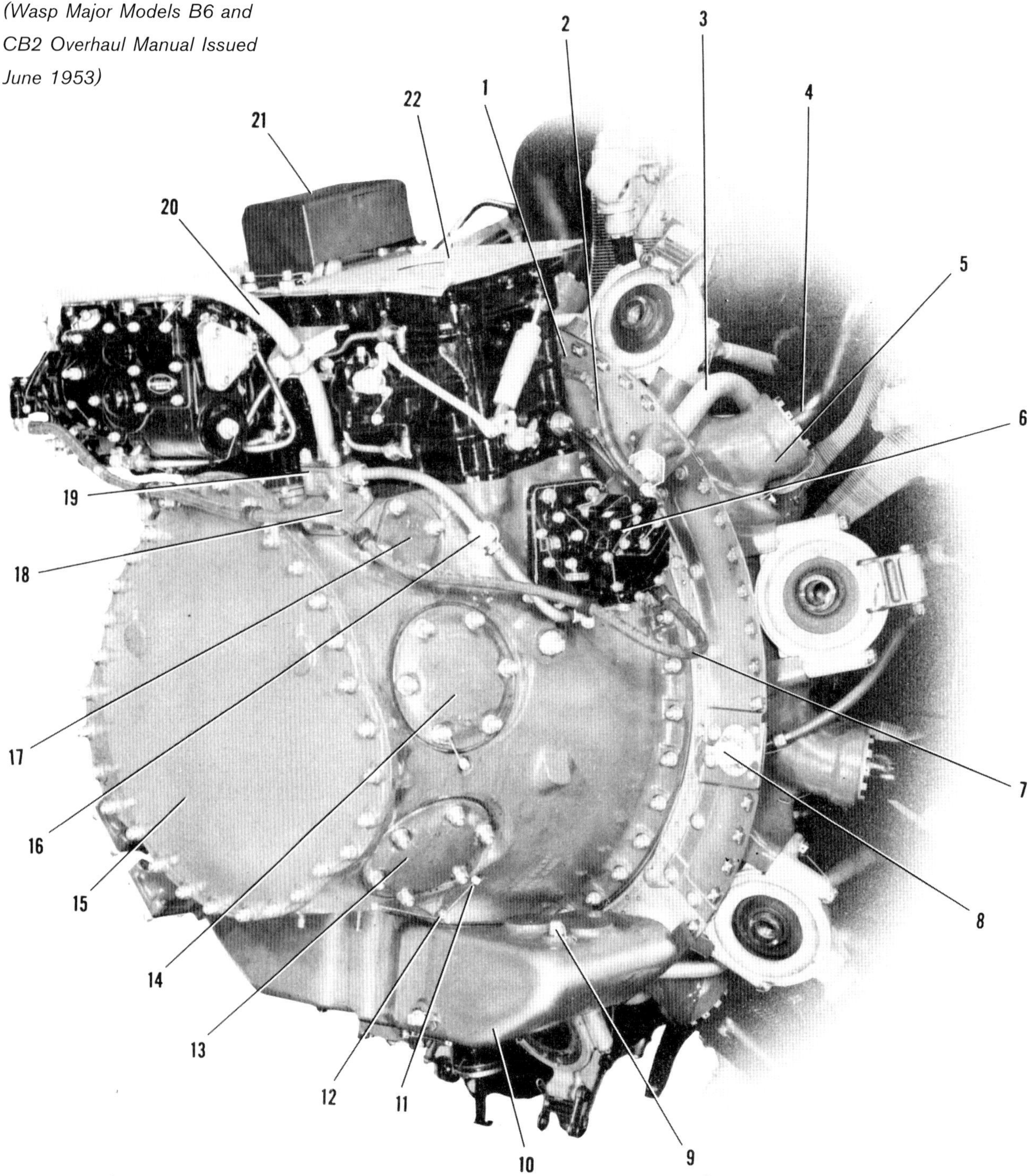

1. Fire Seal
2. Unmetered Fuel Line to Water Regulator
3. Collector Case Right Side Breather
4. Intake Manifold Spanner Nut
5. Collector Case
6. Water Regulator
7. Water Line to Carburetor Derichment Valve
8. Magneto Ground Connection
9. Accessory Oil Return
10. Sump
11. Low Pressure Oil Gage Connection
12. Accessory Pad Oil Drain
13. Generator Drive Pad
14. Starter Drive Pad
15. Accessories Drive Case Cover
16. Metered Water Line
17. Right Side Fuel Pump Drive Pad
18. Right Side Tachometer Adapter Pad
19. Fuel/Water Transfer Adapter to Fuel Feed Valve
20. Metered Fuel Line
21. Carburetor Shipping Cover
22. Carburetor Shipping Board

Right Rear View of Engine

Left Rear View of Engine

1. Mixture Control
2. Metered Fuel Line
3. Discharge Nozzle Pressure Gage Connection
4. Vapor Vent Plug
5. Primer Solenoid
6. Accelerating Pump
7. Fuel Gage Connection
8. Derichment Valve
9. Left Side Tachometer Adapter
10. Left Side Fuel Pump Pad
11. Accessory Drive Pad
12. Main Oil Pump Body
13. Oil Outlet
14. Oil Inlet
15. Pressure Oil Relief Valve
16. Oil Pressure By-Pass Valve Cap
17. Crankcase Scavenge Line
18. Fireseal
19. Automatic Engine Control Pad
20. Manifold Pressure Connection
21. Collector Case Breather
22. Exhaust "H" Line
23. Unmetered Fuel Line to Water Regulator

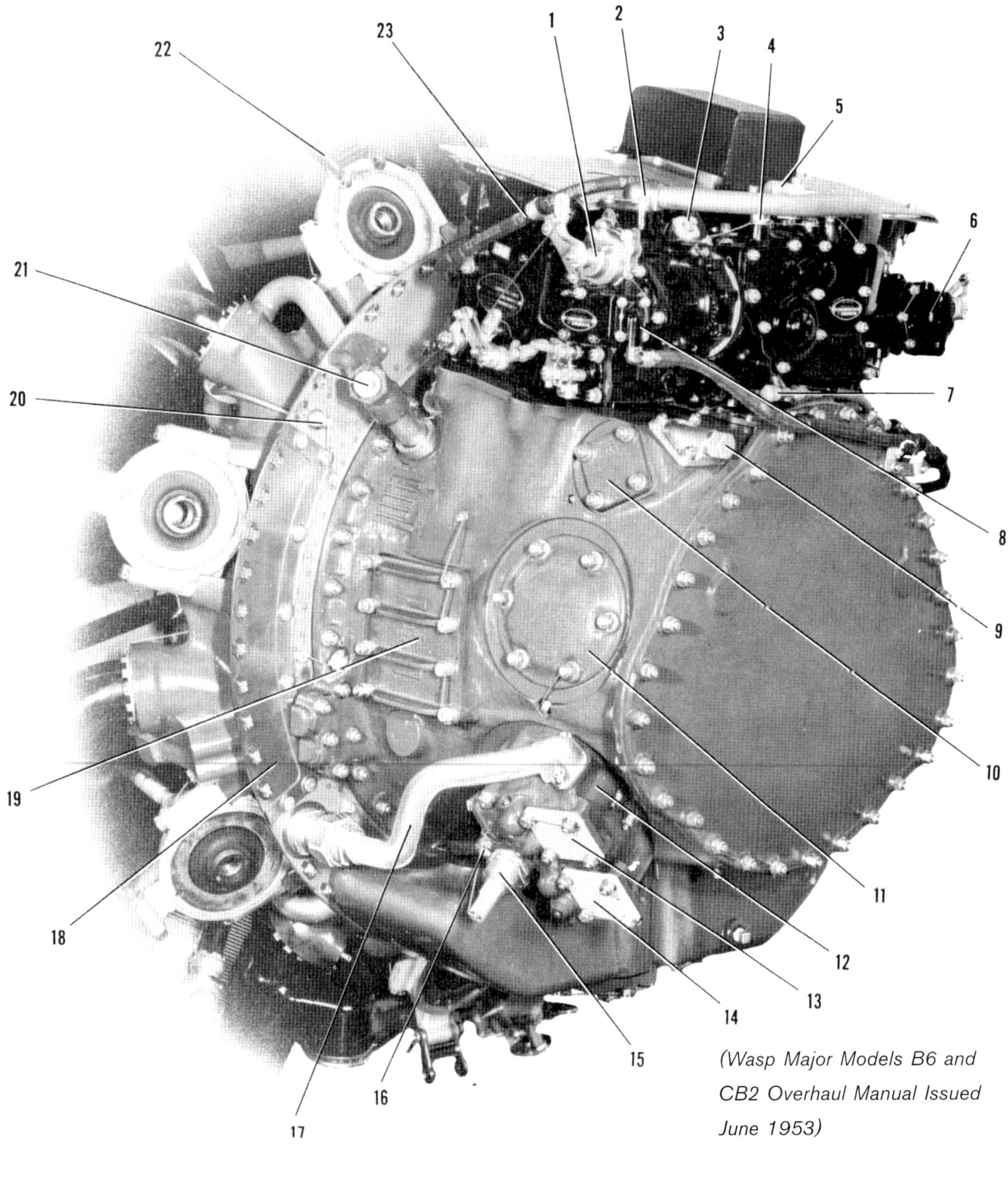

(Wasp Major Models B6 and CB2 Overhaul Manual Issued June 1953)

continued from page 59
shape; 12.857142 is also 1/28 of 360. The four-throw crankshaft has a similar offset as the cylinders plus 180 degrees or 192.857142 degrees. In other words, starting at row "A" with the master rod journal at zero degrees, the next master rod journal for row "B" is displaced at 192.857142 degrees, "twisting" in the same direction as the cylinder arrangement and so on.

This arrangement gave a number of advantages, including the requisite even firing and cylinders fired in order for each cylinder bank. In other words, when cylinder A1 (for instance) fires on "A" row, the next cylinder to fire in that cylinder bank is B1, then C1, and finally D1. Furthermore, each bank of cylinders was essentially a four-cylinder inline engine. Typical four-cylinder inline engines have the pistons running in pairs. For example, when the two end cylinders are at bottom dead center, the two middle cylinders are at top dead center. The R-4360 was similar in this regard, but instead of the two end cylinders and two middle cylinders running in pairs, the R-4360 ran A1 with C1 and B1 with D1. Therefore, when cylinders A1 and C1 are at top dead center, cylinders B1 and D1 are at bottom dead center. In this way, the R-4360 can be considered as seven four-cylinder engines *(Ref. 2-8, 2-9, and 2-10)*.

Not only did this concept give perfectly even firing, but it also had an additional advantage. In the event of a failed magneto (more on ignition later), one bank (i.e., four cylinders) would be lost, reducing vibration due to uneven firing. The foregoing gives the impression that even firing is essential. Though desirable, that's not always the case. A number of engines have been designed that did not feature even firing. One prime example is the World War I V-12 Liberty. With a 45 degree included angle between cylinder banks and a 120-degree crankshaft, firing was anything but even; however, the engine clearly survived. Another modern day example is Chrysler's V-10 automobile engines. For even firing, a V-10 should ideally have a 72 degree included cylinder bank angle; Chrysler's V-10 uses 90 degrees. The probable explanation for this is the fact that Chrysler's V-10 is simply a V-8 with two extra cylinders tacked on. Chrysler got by the issue by simply regarding this engine as two inline 5-cylinder engines. In this way, pairs of cylinders are fired at the same time on opposite banks, giving rise to a rather ugly exhaust note. Like they say, if it doesn't sound right, it isn't right. For smoothness and reduced engine stress, even firing is always advantageous.

Firing order is as follows reading across:

A1	B5	C2	D6
A3	B7	C4	D1
A5	B2	C6	D3
A7	B4	C1	D5
A2	B6	C3	D7
A4	B1	C5	D2
A6	B3	C7	D4

The illustration opposite shows the firing order diagrammatically.

Crankin' Around

Crankshafts tend to be one of the heavier components of internal combustion engines. They are also difficult components to get right from a balance and vibration perspective. To design a lightweight yet relatively vibration free crankshaft demands highly skilled engineering.

A key challenge to the designer is how to lay out the crankshaft and master connecting rods, i.e., one piece versus two piece versus built-up. For radials, two basic schools of thought existed for multi-row radials: One-piece master rod with built-up crankshaft, or two-piece master rod and one-piece crankshaft. Pratt & Whitney had experience with both methods prior to the R-4360. The two-row R-1535 and R-1830 used two-piece master rods and one-piece crankshafts. The follow-on R-2800, despite going through a number of total redesigns, always used a one-piece maser rod and built-up crankshaft. One-piece master

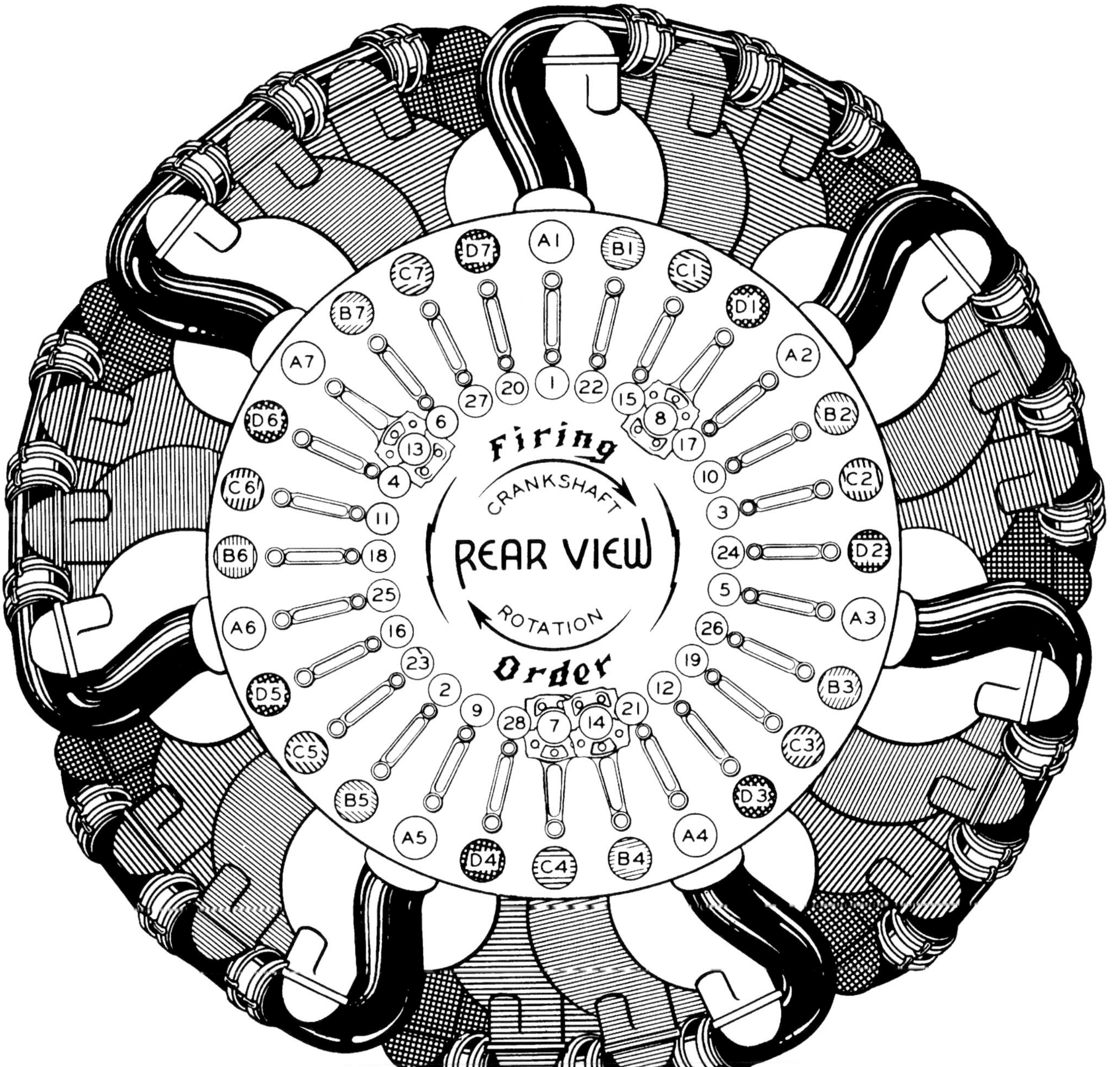

This diagram shows the rear of the engine looking forward toward the propeller. It's a handy reference that not only shows the position of the individual cylinders but also the location of the master rods. This latter feature is important, especially when disassembling the engine. It is critical that a master rod cylinder is not removed prior to any others in that row, or a holding fixture needs to be put in place if it's necessary to remove a master rod cylinder. The reverse is also true; the master rod cylinder is the first to go on during reassembly. *(Handbook Erection and Maintenance Instructions USAF Series B-36F Aircraft. March 27, 1951)*

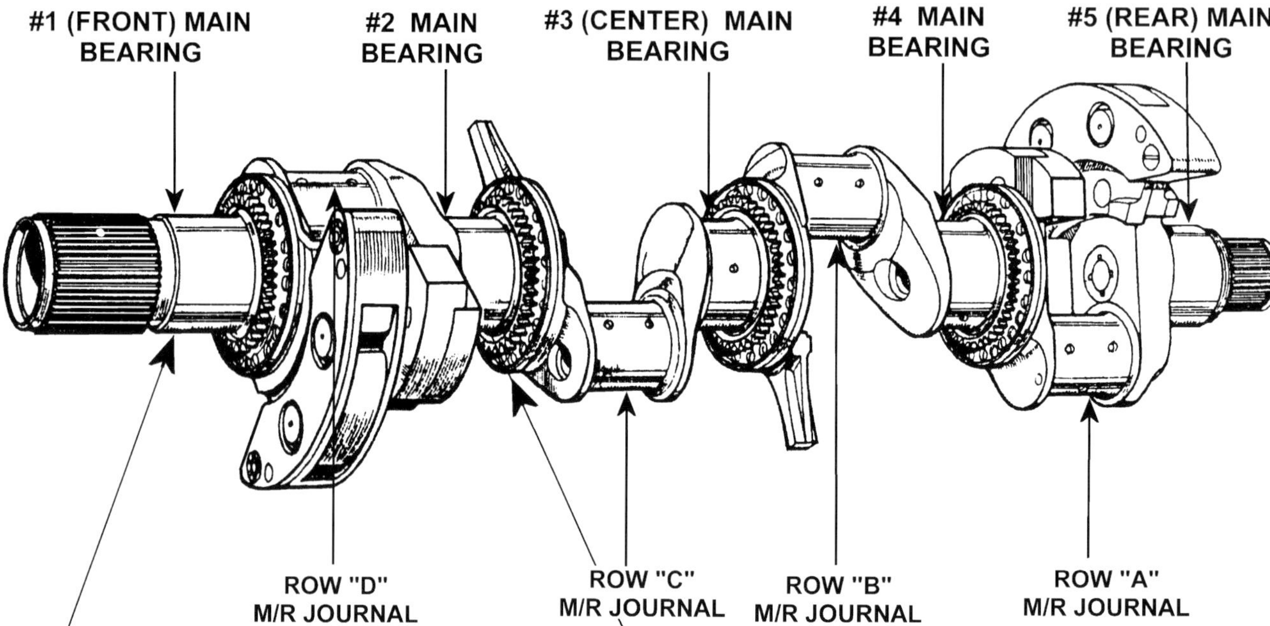

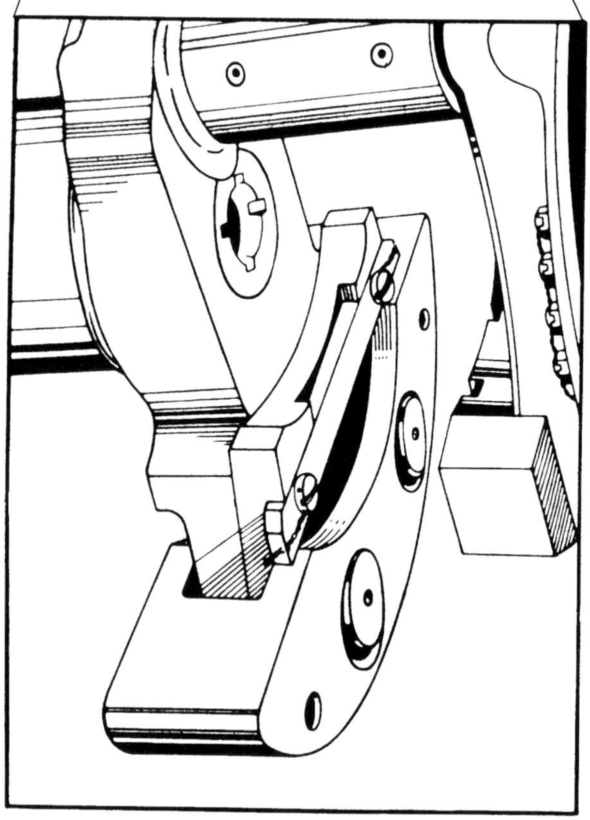

Engineering elegance at its finest–the R-4360 crankshaft. This view also shows to advantage the bifilar dynamic counterweight, the master rod journals, and the main bearing journals. Considerable engineering time was spent in developing this key component in order to achieve the lightest weight commensurate with good vibration characteristics. *(Drawn by author based on parts manual illustration)*

rods are the preferable method, but it's also fraught with major design and manufacturing challenges, as Pratt & Whitney found out when redesigning the "B" series R-2800 into the "C" series R-2800. The "B" engines used male/female joints to join the crankshaft sections, while the "C" engine used face splines. For the R-4360, Pratt & Whitney chose to take the more prudent route of a one-piece crankshaft with split master rods. While this introduced other design difficulties, it was the lesser of two evils. However, based on their R-1830 experience, it was not surprising to see a similar design concept followed for the all-important two-piece master rod. Therefore, a pair of link rods straddled the top half of the master rod, and the master-rod cap contained the remaining four link rods. Four through-bolts retained the cap to the upper half.

With the decision made to use a one-piece crank manufactured from a steel forging, development proceeded to make it run smoothly, reliably, and with a low weight—tough and opposing requirements. It used five main bearings and, of course, four master rod journals. The next question was how to make it light, resist torsional and linear vibration, and hold up to the power and abuse it would suffer during its working life. As

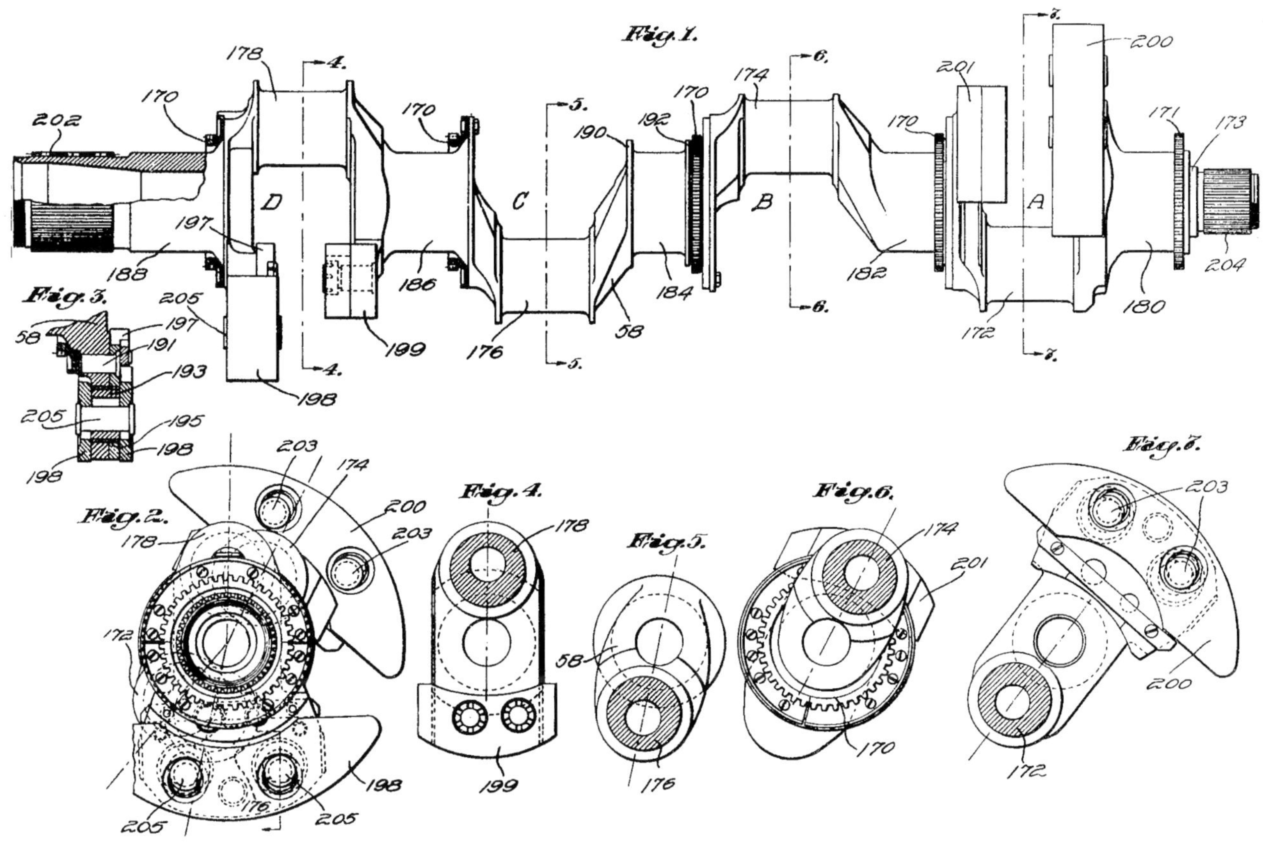

This patent drawing illustrates the key features of the R-4360 crank and how the two end dynamic weights are displaced from each other. *(United States Patent 2,426,876)*

previously mentioned, the four-throw crankshaft had the master rod journals spaced at 192.857142 degrees between throws. The R-4360 was top secret during its early development, and some innovative red herrings were implemented to make sure it stayed this way. One of these red herrings was to call out five throws on the crankshaft drawings; this is how the first few crankshaft forgings were manufactured by one of Pratt & Whitney's vendors. Of course, as soon as Pratt & Whitney received the crank, the extra master rod and main bearing journals were simply machined off. Along similar lines, during World War II when the R-4360 was still top secret and confidential, a crankshaft was accidentally displayed with other engine components for a display. Pratt & Whitney engineers soon explained it away by claiming R-2800 cranks were made in pairs and then separated!

Keeping Things Smooth

With the complex dynamics of a multi-row radial engine, attenuating torsional and dynamic vibration is a tough assignment. Fortunately, by the time the R-4360 was developed a wealth of experience had been gained from previous engine developments. Conflicting requirements must be met—those of counterweighting the crankshaft in order to achieve smooth running commensurate with attenuation of torsional and dynamic vibration. In addition to meeting these difficult requirements, weight had to be kept to a minimum.

With the R-4360's four-throw crankshaft, each throw and its associated weight acts as a balance weight for an adjacent throw. Its associated weight and intermediate counterweights opposite the crankpins are therefore unnecessary. The pairs of crank throws—A, B and C, D—each produce

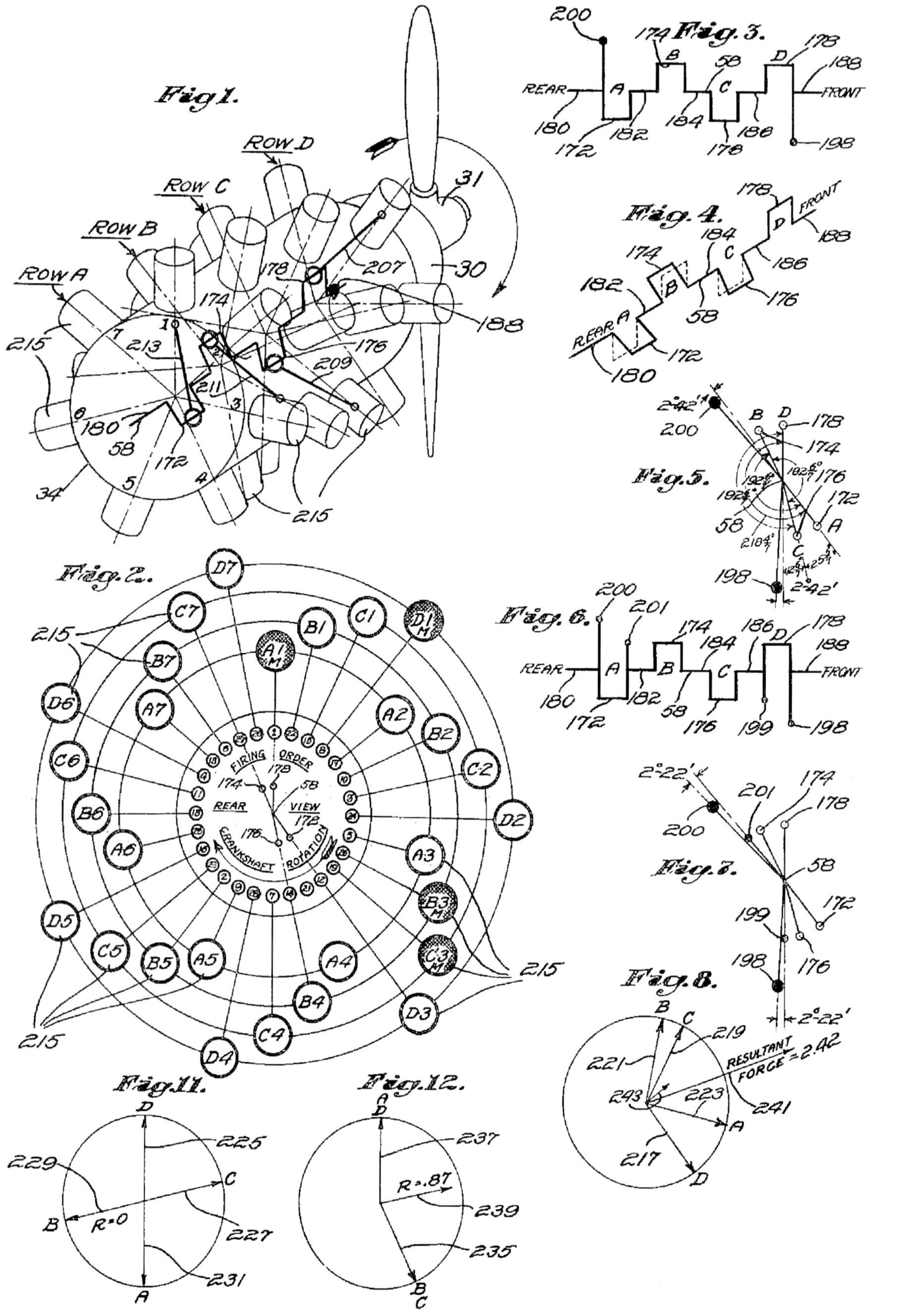

Shown in a diagrammatic manner is the relationship between the crank throws and the individual cylinders. *(United States Patent 2,426,876)*

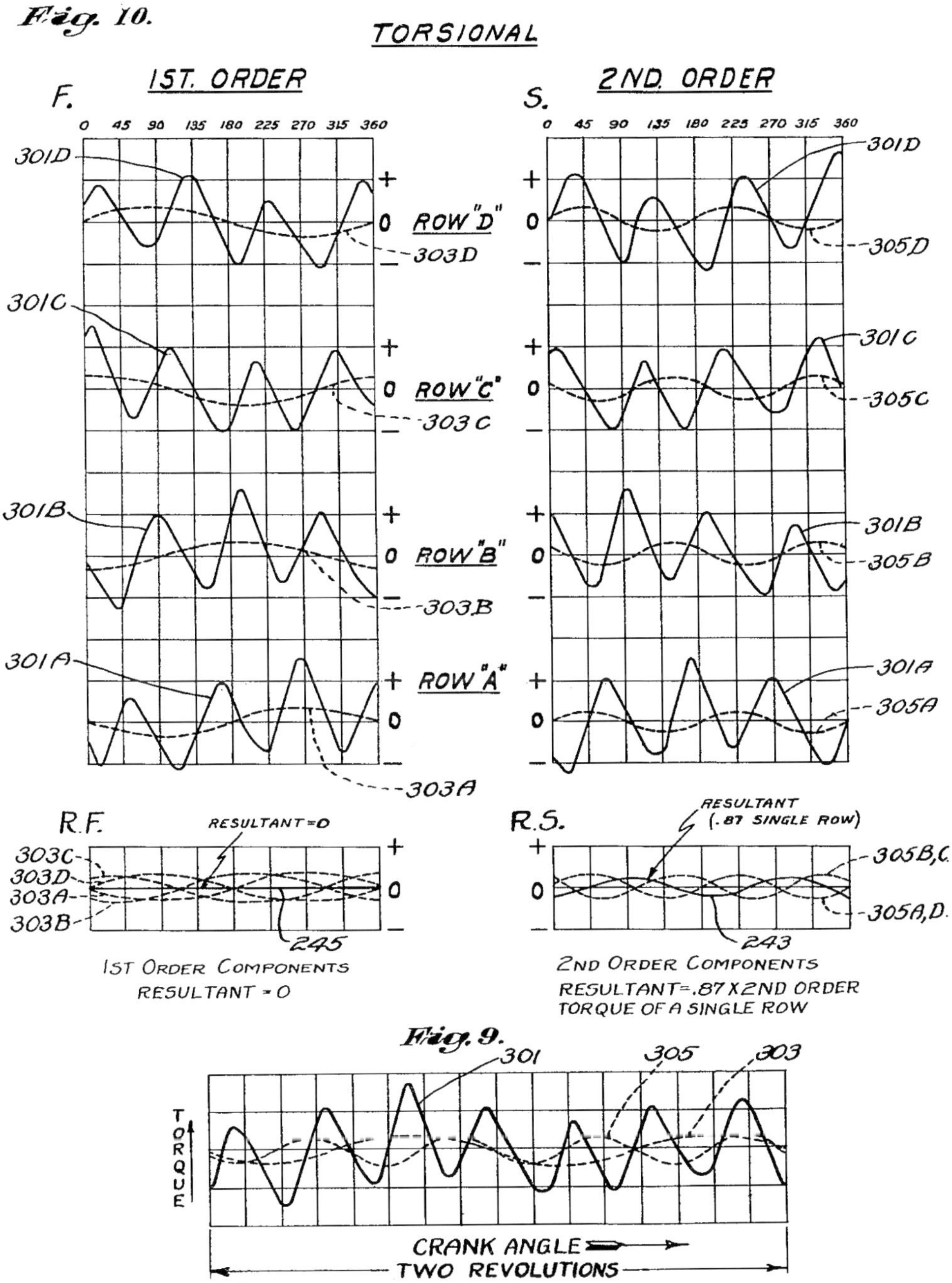

This illustrates the torsional amplitudes for first order and second order vibration. *(United States Patent 2,426,876)*

additive rocking couples, which tend to rotate the crankshaft in a counterclockwise direction. However, these couples, and any other unbalanced primary force of material size, may be balanced by a pair of dynamic counterweights located at extreme ends of the crank at the number one and number five main bearing crank webs. They are of the "bifilar" type. In other words, it's a floating dynamic counter weight supported on two floating bushings and rollers. They are tuned to attenuate the firing frequency of a row or 3½ order.

Crank throws A, B, C, and D do not lay in the same plane, but are angularly displaced to the same degree as the cylinder offset; in other words, 12.857142 degrees (360 dived by 28), plus 180 degrees or 192.857142 degrees. This angular displacement or twist of throws is compensated for with regards to balancing by displacing the end (dynamic) counterweights angularity with respect to the crank throws. This negates the necessity of intermediate or additional counterweights. The front dynamic counterweight is dis-

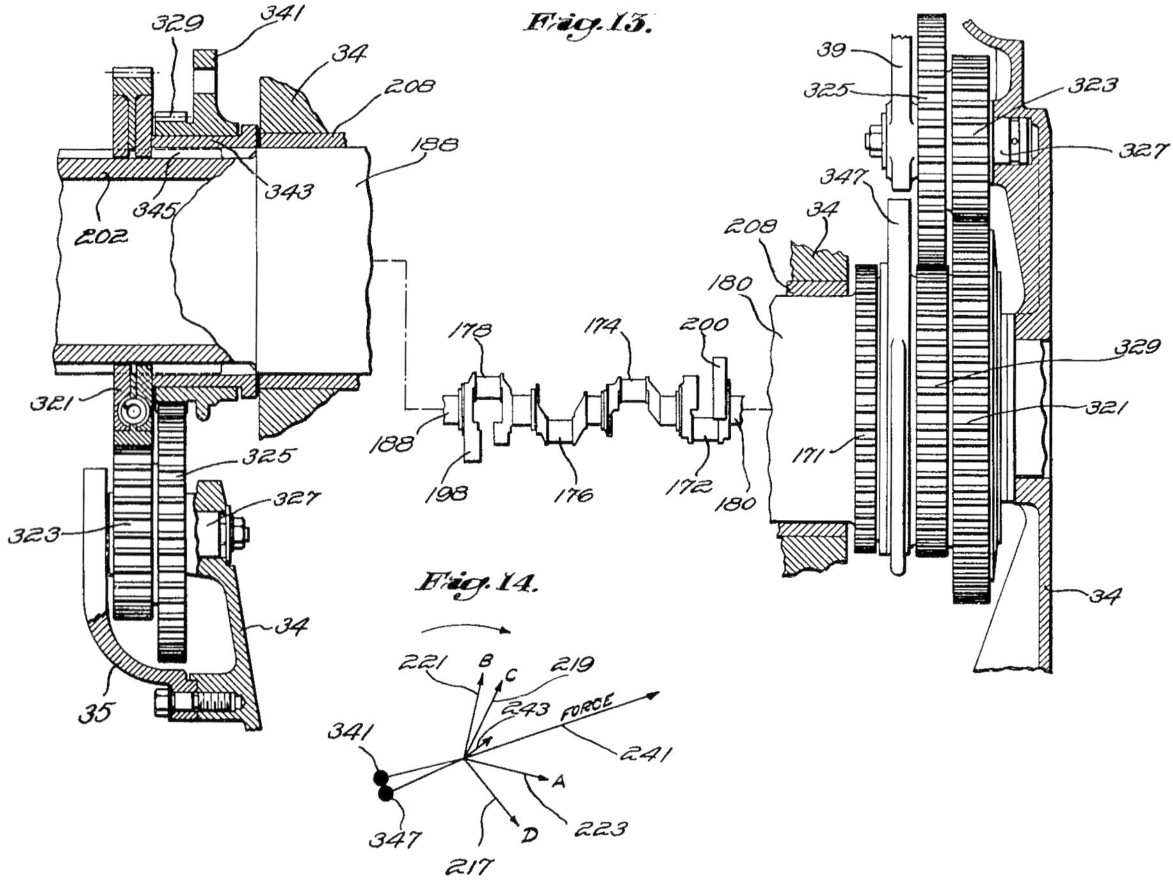

This patent drawing shows the torsional oscillations that the blower drive is subjected to. After seeing this diagram it's easy to understand why Pratt & Whitney put an inordinate amount of effort into cushioning the blower drive via hydraulics and springs. *(United States Patent 2,426,876)*

placed by two degrees, 42 minutes in a clockwise direction from the front crankpin (D). In a similar fashion, the rear dynamic counterweight is displaced a similar amount but counterclockwise from the plane of the rear (A) crank throw. To further assist in attenuating vibration, fixed, bolt-on counterweights are attached at the ends of the crank, inboard of the D and A crank throws. Again, these counter weights are displaced from the place of the crank throw in the same direction as their dynamic counterparts. However, this time the displacement is two degrees, 42 minutes *(Ref. 2-11, 2-12, and 2-13)*. As with the R-2800 crankshaft development, Pratt & Whitney called in the cavalry in the form of J. P. Den Hartog, the world's most respected vibration expert. Early crankshafts were breaking at the crank cheek. In a counterintuitive move, Den Hartog recommended removing material in order to make the crankshaft less rigid. Once again, Den Hartog worked his magic.

Master and Link Rods

By the time the R-4360 development came around, Pratt & Whitney had gained extensive experience in the design of two-piece master rods. So why reinvent the wheel? Split along the centerline of the master rod bearing, the upper half contained the master rod and a link rod on both sides. The bearing cap contained the four remaining link rods. Bronze bearings are pressed into both ends of each link rod, which ride on hardened steel pins. Four precision bolts attach the cap to its mater rod. Each pair of bolts straddles the link rods. Holes drilled through the link-rod pins closest to the bearing parting line allow the bolts to pass through.

Having to contain the immense loading of the master rod, its bearing requires careful design consideration. On later models of the R-4360, well over 1,000 hp is transmitted through this critical bearing. Of course, not all that 1,000 plus horsepower ends up at the propeller; several hundred horsepower is required to drive the supercharger and accessories, plus a significant amount of power is absorbed through frictional losses. Nevertheless, the R-4360's master rod bearing doesn't know this; as far as this bearing is concerned, upwards of 1,100 hp is going through it. Numerous operating requirements including corrosion resistance, erosion resistance, and compatibility with the hardened-steel-bearing journal need to be accommodated. Additionally, a good plain bearing should be sufficiently hard and strong to withstand the heavy pressures and hammering without significant distortion, which are some of the occupational hazards of a master rod bearing. As if these requirements were not enough, good bearing design dictates that it should have sufficient elasticity and malleability to conform to minor irregularities and misalignment of the journal plus oil wetting and anti-friction characteristics. One inevitable byproduct of running a high-performance engine is debris entering the oil system. This debris can take the form of carbon as a byproduct of combustion or wear particles from other engine comments. Bearings need to have "embedibility." In other words, they need to absorb small particles of debris without compromising bearing performance.

Based on Pratt & Whitney's pioneering work in the 1930s, a lead/indium silver bearing is employed not only for the master rod, but other key load bearing components as well, such as the crankshaft main bearings and reduction gear pinion bearings. Luke Hobbs, Pratt & Whitney's engineering manager, led an intensive program to resolve continuing master rod bearing failures in R-1535s. The end result was a steel-backed bearing with a silver overlay, twenty to thirty thousandths of an inch thick.

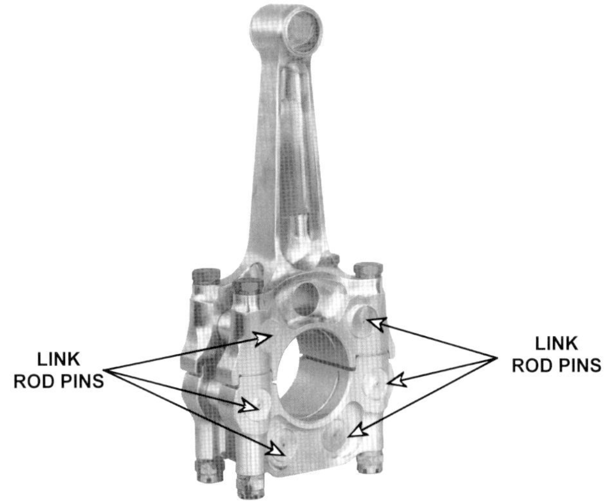

With its prior two-piece master rod experience, Pratt & Whitney chose to use a conventional design for this highly stressed component. R-1535, R-1830, and R-2000 influence is apparent, which also employed two-piece master rods. *(Illustration by author generated from Pratt & Whitney archival material)*

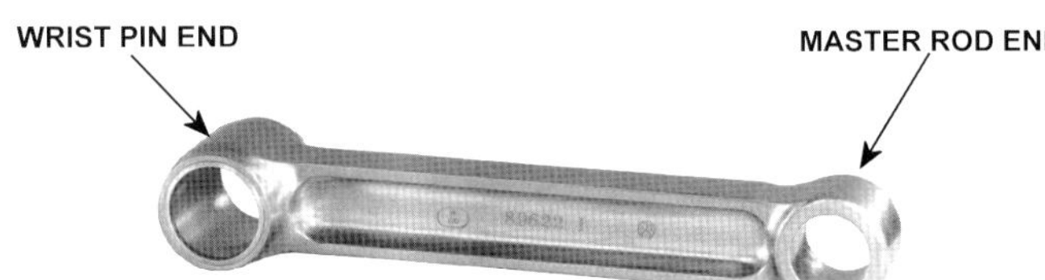

A conventional I-beam design was employed for the six (per row) link rods. *(United Technologies Corp./Pratt & Whitney)*

A thin flash of lead/indium is plated on top of the silver. The lead flash varies in thickness, but is typically one to one-and-one-half thousandths of an inch thick *(Ref. 2-14)*. Clean room conditions are required for the critical plating processes in order to ensure the future health of this heavily loaded component. Any sign of master rod bearing distress demands an engine be removed from an aircraft.

Crankcase Design

Pratt & Whitney found out from its very first engine, the R-1340 Wasp, an aluminum forging offered the best material and mechanical properties

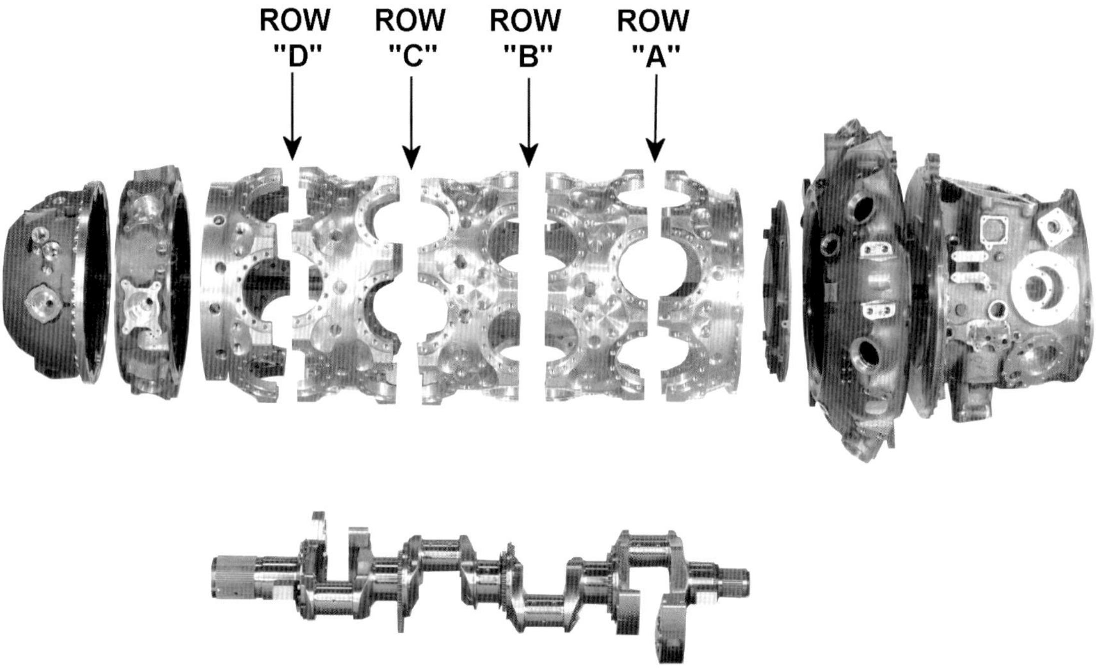

Five piece forged aluminum crankcase. One can only imagine the tolerances required to avoid a huge tolerance stack-up problem when manufacturing this assembly. A considerable amount of machining was required at all the tapped holes, cylinder pads, main bearing supports, and so on. *(Illustration by author generated from Pratt & Whitney archival material)*

for radial engine crankcase construction. This method of crankcase construction was developed by the Bristol Aeroplane Company, specifically by their enigmatic chief engineer Roy Feddon *(Ref. 2-15).* Forging offers considerably higher levels of fatigue resistance and improved material properties over cast aluminum. With the foregoing in mind, it was an easy decision to use forged aluminum as the crankcase material. Tying everything together so the engine was manufacturable and durable presented unprecedented challenges. Pratt & Whitney's solution was to use a five-piece construction with the five sections pancaked on top of each other. The centerline of the cylinders represented the parting line between sections. In this way, the middle of each crankcase section of the three center sections supports a crankshaft main bearing. With a one-piece crankshaft design, a hole was required in the three center sections of the crankcase that would accommodate the main bearing carrier. In this way, the crankcase sections could be slipped over the one-piece crankshaft. The three center main bearing carriers are manufactured from magnesium and located in a steel ring pressed into the forged aluminum crankcase section. Under operating conditions the magnesium carrier expands more than the steel ring, thus ensuring an interference fit and the requisite rigidity. Similar concepts have been used before. For instance, the famous Offenhauser Indy racecar engine used a similar concept to feed the crankshaft, with its main bearings attached, into the crankcase.

What Makes the Valves Go Up & Down?

As long as the intake and exhaust valves open and close at the correct times with the correct acceleration and lift, the engine does not care how these events occur. In the case of the R-4360, it used typical radial engine technology—that of cam rings. Each cam ring has an intake and exhaust cam track machined into it. Three lobes per cam track are used, which dictates 6:1 reduction gearing between the crank and the cam ring rotating in the opposite direction to the crank. On later "C" series engines this arrangement was revamped by utilizing four cam lobes per cam track and rotating in the same direction as the crank. The unique design of the R-4360's fore and aft cylinder design (an intake valve at the front and an exhaust valve at the rear of the cylinder) introduces a new set of problems. With conventional radials, a pair of pushrods resides either at the front or the rear of the cylinder. Therefore, one cam ring accommodates an entire row's intake and exhaust valve requirements via two cam tracks. Not so with the R-4360.

Pratt & Whitney overcame this problem by using five cam rings, all identical. The front cam ring operates the intake valves of the "D" row of

Fig. 9.

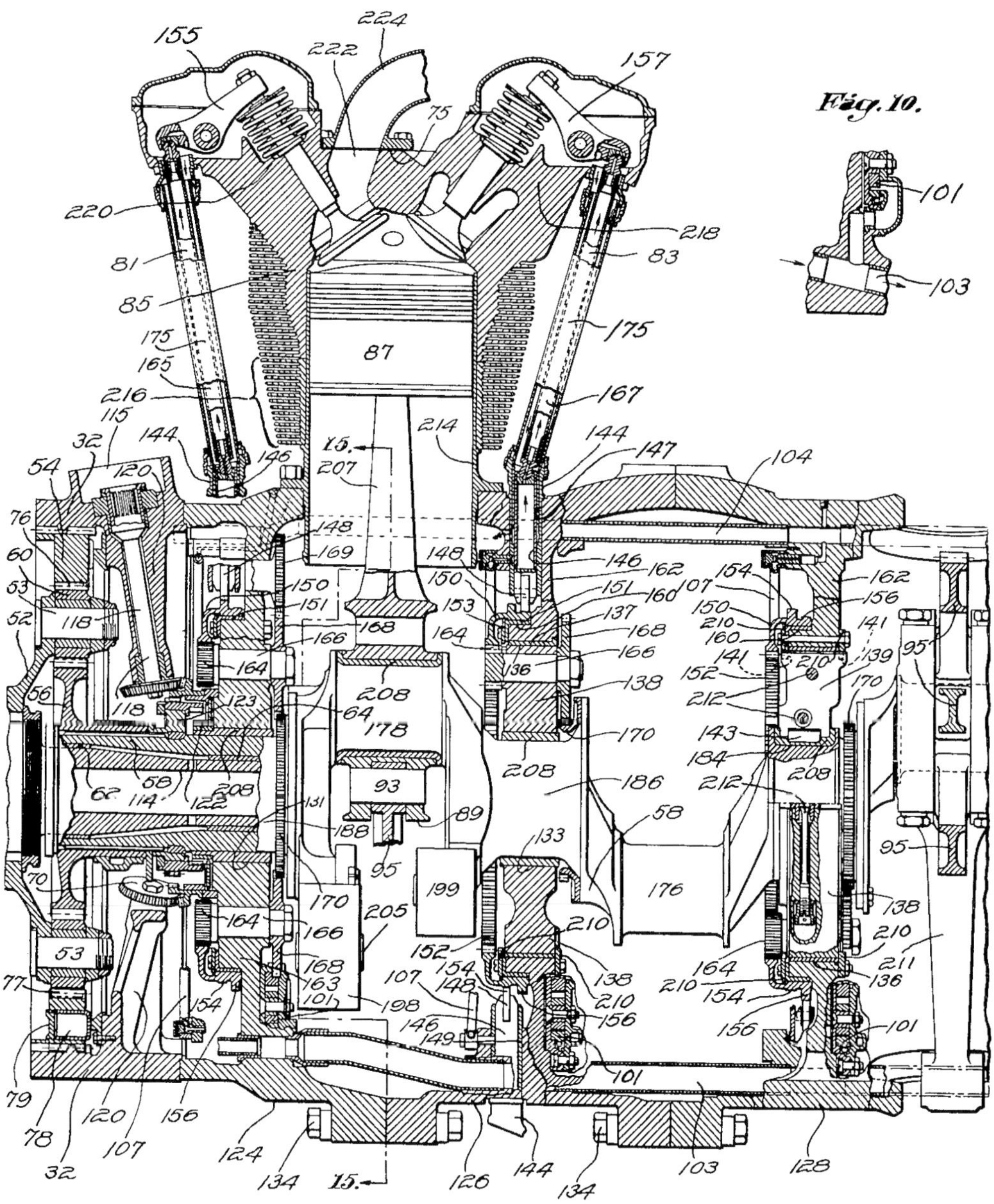

Fig. 10.

With its unique fore and aft valve locations, the R-4360 required the use of two cam rings per row–five in total. A split gear mounted on the crankshaft transmits power via a train of gears mounted in the main bearing cap and the crankcase section. *(United States Patent 2,426,879)*

Fig. 15.

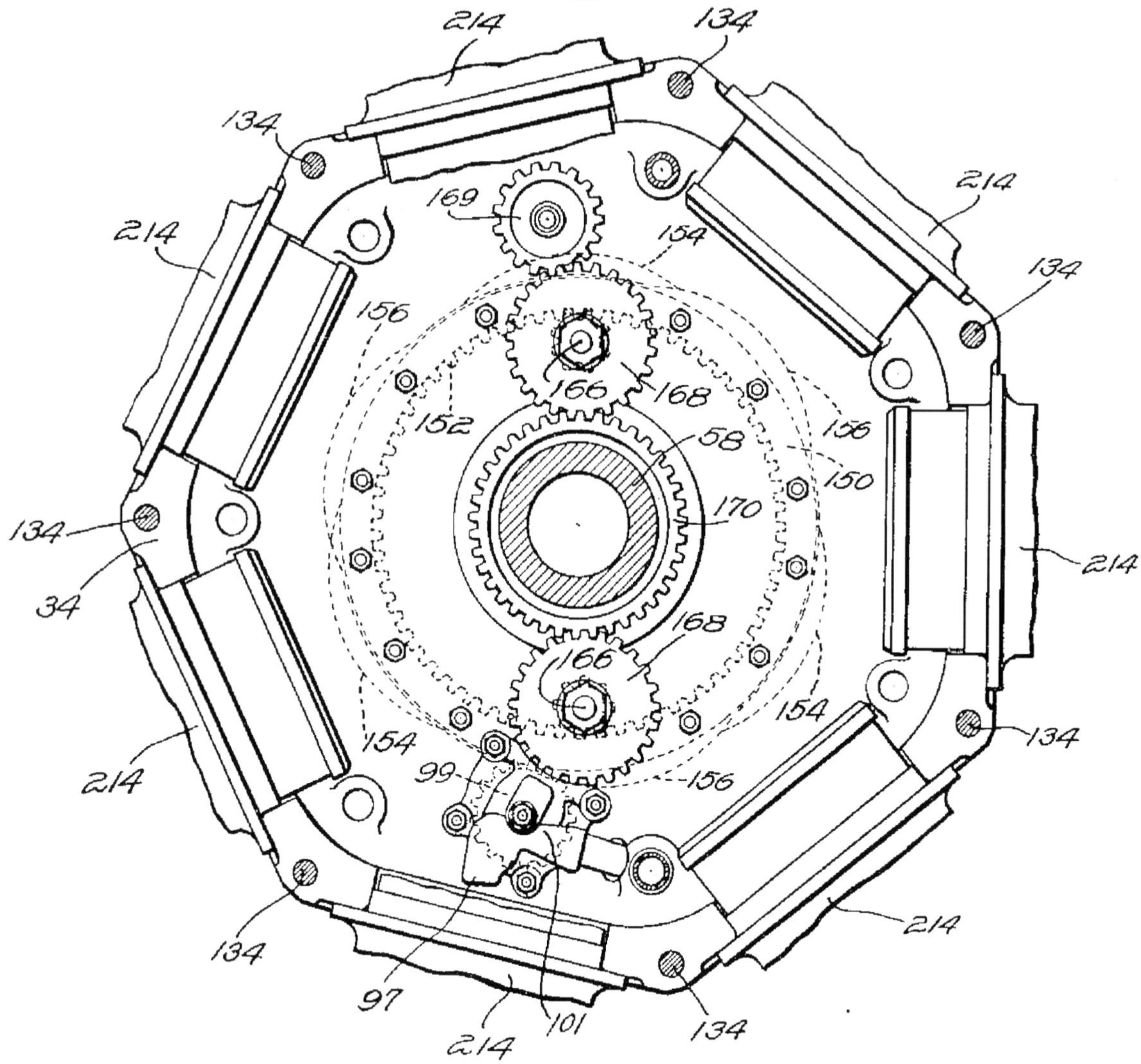

Fig. 16.

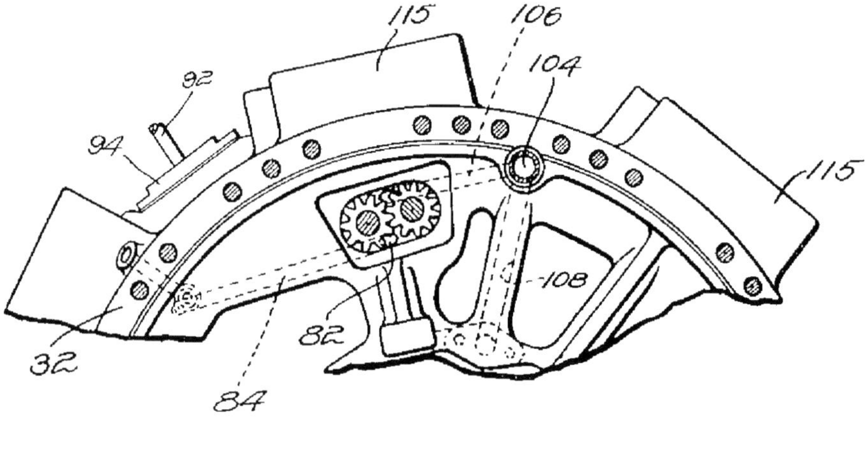

This patent drawing shows the gear train for driving the cams and gerotor-type pump that scavenges each crankcase section. Scavenging turned out to be a major development headache for the R-4360. *(United States Patent 2,426,879)*

cylinders. This means the exhaust cam track is redundant—it doesn't do anything. The next cam, which resides between the "D" row and "C" row, operates the exhaust valves of the "D" row and the intake valves of the "C" row. Between rows "C" and "B" another cam ring resides, operating in a similar fashion to the one that resides between rows "D" and "C." The fifth and last cam ring resides at the rear of the engine. Like the front cam ring, it has a redundant cam track. This time it's the intake, that is, the rearmost cam ring operates the exhaust valves of row "A." Now that the cam ring setup has been established, the question is now begged, how are they driven? With a single or two-row radial, a set of reduction gears driven off the crankshaft performs the cam ring drive duties. But a number of difficulties are introduced with the R-4360's five cam rings. The one-piece crankshaft negates the possibility of using one-piece reduction gears driven off the crankshaft.

Pratt & Whitney got around this problem by splitting the primary drive gear along its axis, at least for the three center ones. The two end cam-rings presented no problems as far as drive requirements went because one-piece gears are used. The three center split gears are mounted and clamped on the crankshaft. Each crankshaft main bearing cap is bored through to support a gear driven off the split gear. A shaft that goes through the bored hole in the main bearing cap has a gear mounted on both ends. The gear on one end is driven off the split gear, and the gear on the other end drives an internal gear in the cam ring. The two end cam rings are driven in a similar fashion except with no need to split the crankshaft gear. The three center main bearing carriers are the same with the exception of the center one which is shaped to fit within the crankshaft locating flanges. This also offers longitudinal thrust support for the crank. All five main bearing carriers are offset by 12.857142 degrees between rows. This allows commonality of parts and makes for easier valve timing when assembling the engine. Additionally, cam drive gears serve double duty by driving a gerotor-type oil pump, five in all, which scavenge each crankcase section.

So How On Earth Do We Cool This Thing?!

As related previously, the first concept engine simply consisted of bolting 28 "B" series front cylinders to a crankcase made up from modified R-2180 components. It quickly became apparent that production engines would need something significantly more sophisticated than this proof of concept engine. At the time of its early design in 1941, Pratt & Whitney were developing a forged cylinder head for the "C" series R-2800. With the capability of unlimited depth and any number of cooling fins, it became a no-brainer to use forged heads. Commensurate with the requirement to keep head temperatures at something below a value that would not cause undue stress to the engine, i.e., less than 500 degrees, cooling drag had to be kept to a minimum.

Contrary to popular belief, the drag of a radial engine is not so much related to its frontal area but rather the pressure drop across the cowl. In other words, a large pressure drop would indicate a large mass airflow throughout the cowl, which is not good for low drag. Pratt & Whitney performed exhaustive testing with dummy cylinder configurations in a purpose-built wind tunnel. One row of simulated cylinders along with one additional adjacent front and one additional adjacent rear row cylinder, referred to as "anchor" cylinders in Pratt & Whitney memorandum reports, made up the test configuration. It had been determined that a pressure drop of approximately 12 inches of water would be required. Although this may not sound like much, after performing the math, the mass airflow requirements get into gas turbine territory. Testing had indicated that a mass airflow of 6,250 pounds per hour, per cylinder was required to cool the R-4360 at takeoff power. This amounts to a staggering rate of 175,000 pounds of air per hour at takeoff power for the complete engine, or

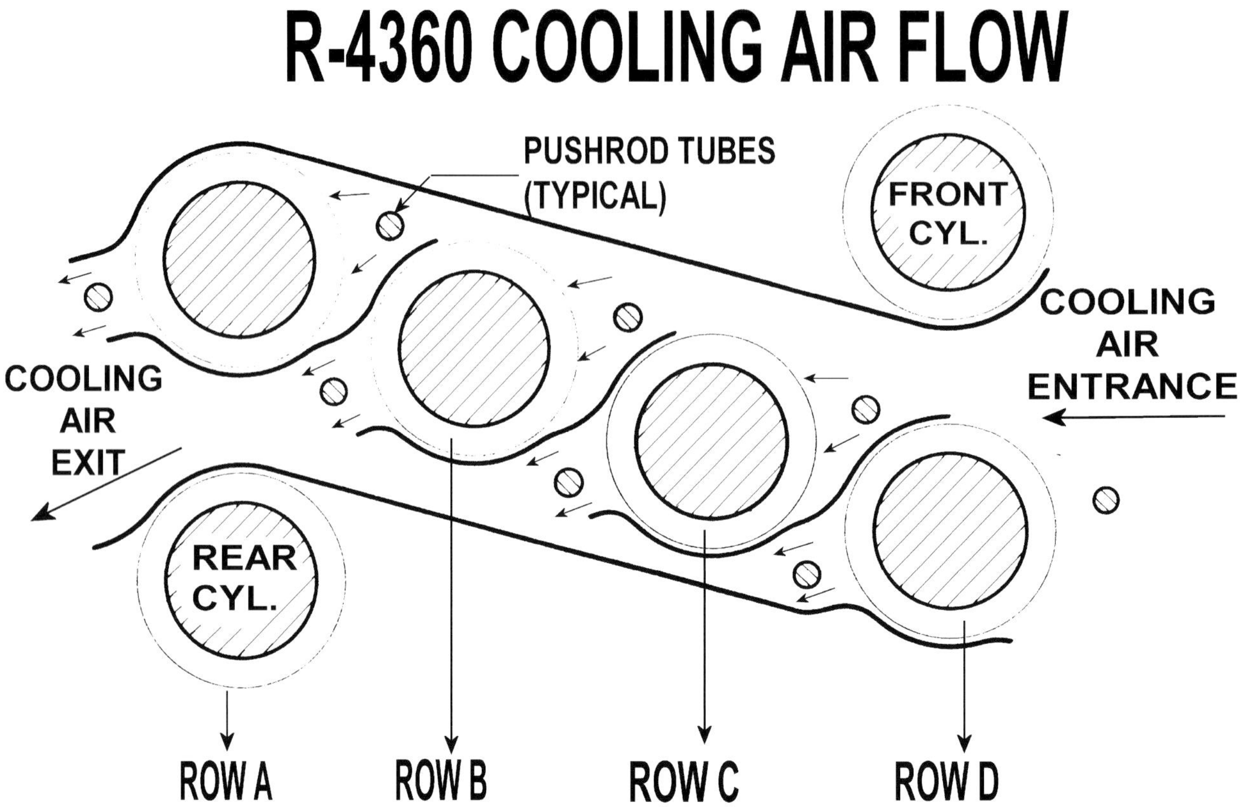

The secret to how the R-4360 keeps its cool. Each bank of four cylinders is made up of a plenum system as shown in this line drawing. As can be seen, cooling flow is quasi cross-flow. With its four rows, Pratt & Whitney engineers were forced to re-write the rulebook for cooling this complex engine. (Illustration by author generated from Pratt & Whitney archival material) *Opposite:* Cooling air that enters the cowl is a precious commodity not to be wasted. Therefore it is critical that the outside diameter of the engine is sealed against the inside diameter of the cowl. Any air that leaks past the seals results in wider cowl flap openings to compensate for the leakage. This then leads to a dramatic rise in drag due to parasitic drag of the flaps and, more significantly, the greater mass airflow through the cowl. *(Installation Handbook, Section III Supplement F Wasp Major B Series Engines. February 1953. Modified by author.)*

48.6 pounds of cooling air per second. Or, to put it in different terms, heat rejection requirements operating at takeoff power are a staggering 16,225 BTUs per minute, not including the heat rejection from the oil *(Ref. 2.3)*.

The foregoing offers good insight into just how difficult the cooling problems can be. The fact that they were resolved is a testament to the ingenuity that dominated R-4360 development. It quickly became apparent that all previous ideas and concepts of radial engine design had to be thrown out the window. Simply bolting more cylinders onto a crankcase did not cut it. And perhaps herein lies the reason the R-4360 succeeded where others had fallen by the wayside. Normal radial engine cooling, whether it's for a single-row or two-row, requires the cooling air to follow a straight-through path. Cooling air enters the front of a cylinder and exits through the rear, oftentimes aided by baffles wrapped around the cylinder in order to take full advantage of mass airflow. The R-4360 disregarded this concept. Instead, seven plenums are created between rows of cylinders. Air enters the plenum and is forced through the cylinders at an angle of approximately 30 degrees, in other words, a quasi cross-flow cooling path *(Ref. 2-6, 2-16, and 2-17)*. Cooling airflow is helped with extensive baffling of the cylinders so every ounce of mass airflow is put to good use. Furthermore, it was

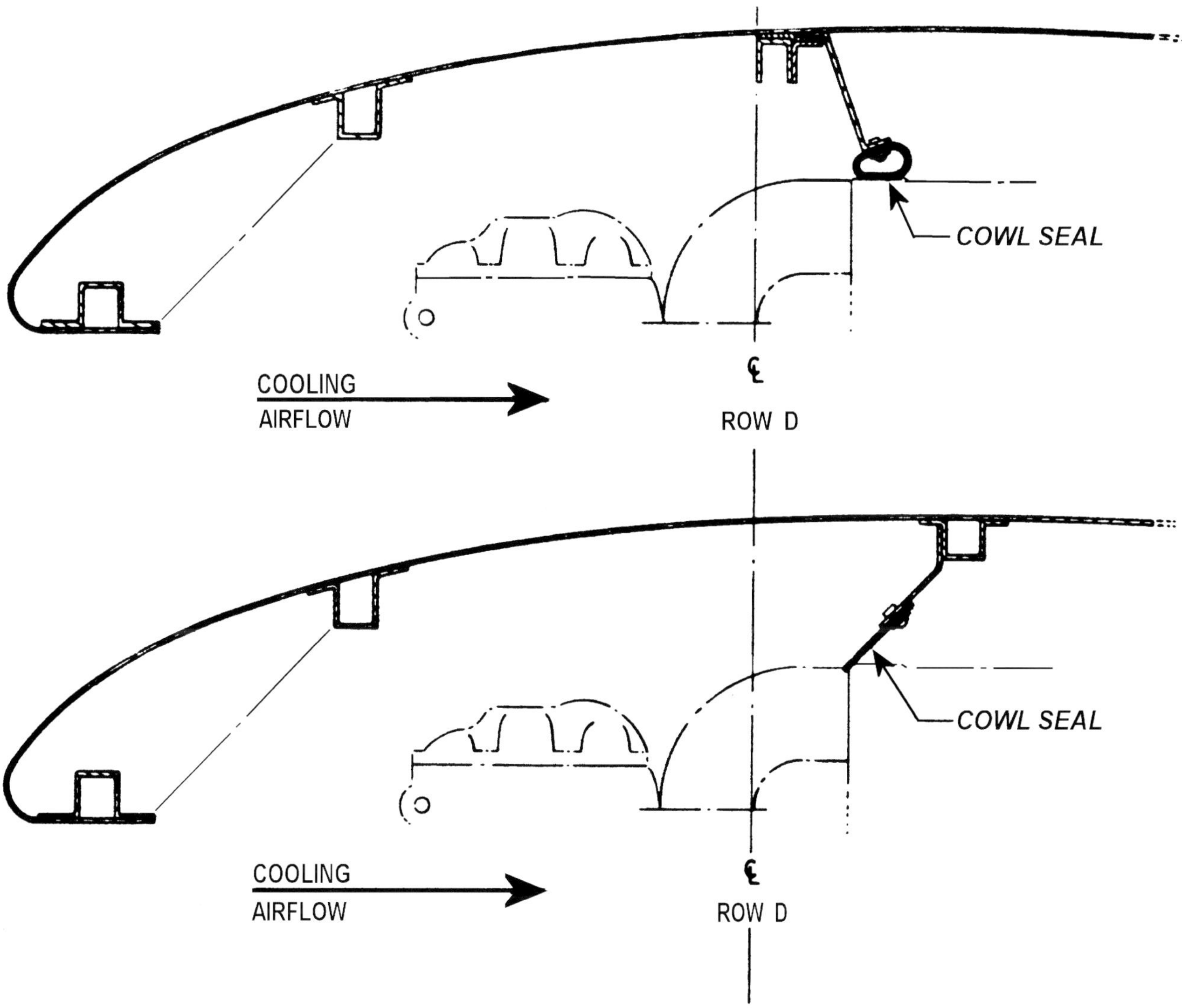

imperative that any R-4360 application sealed the cooling air to avoid leakage. Estimated cooling requirements are shown here graphically. Up to 58 pounds per second mass airflow at takeoff power puts it in the same league as a gas turbine!

Cylinder Design

After the successful design of the R-2800 combustion chamber, the R-4360 followed suit with a two-valve hemispherical design *(Ref. 2-18).* Intake and exhaust manifolds would normally take up the volume required for the seven cooling plenums. Pratt & Whitney again arrived at an innovative solution that would be copied extensively in the automotive racing world years later.

Now the problem arose of how to pack all the manifolds into a very compact space. The solution was to use a "downdraft" intake port. In other words, run the intake manifold over the top of the four cylinder heads that constituted one bank. In this way, the intake manifold was tucked out of the way and offered excellent volumetric efficiency, at least for the port. Interestingly, this concept was copied by the BRM F1 engine design of the early '60s. The only time BRM won the constructor's championship was with an R-4360-style intake design for their 1.5-liter V-8. Likewise, Ford Motor Company copied the intake design for their "289"-based DOHC V-8 Indy

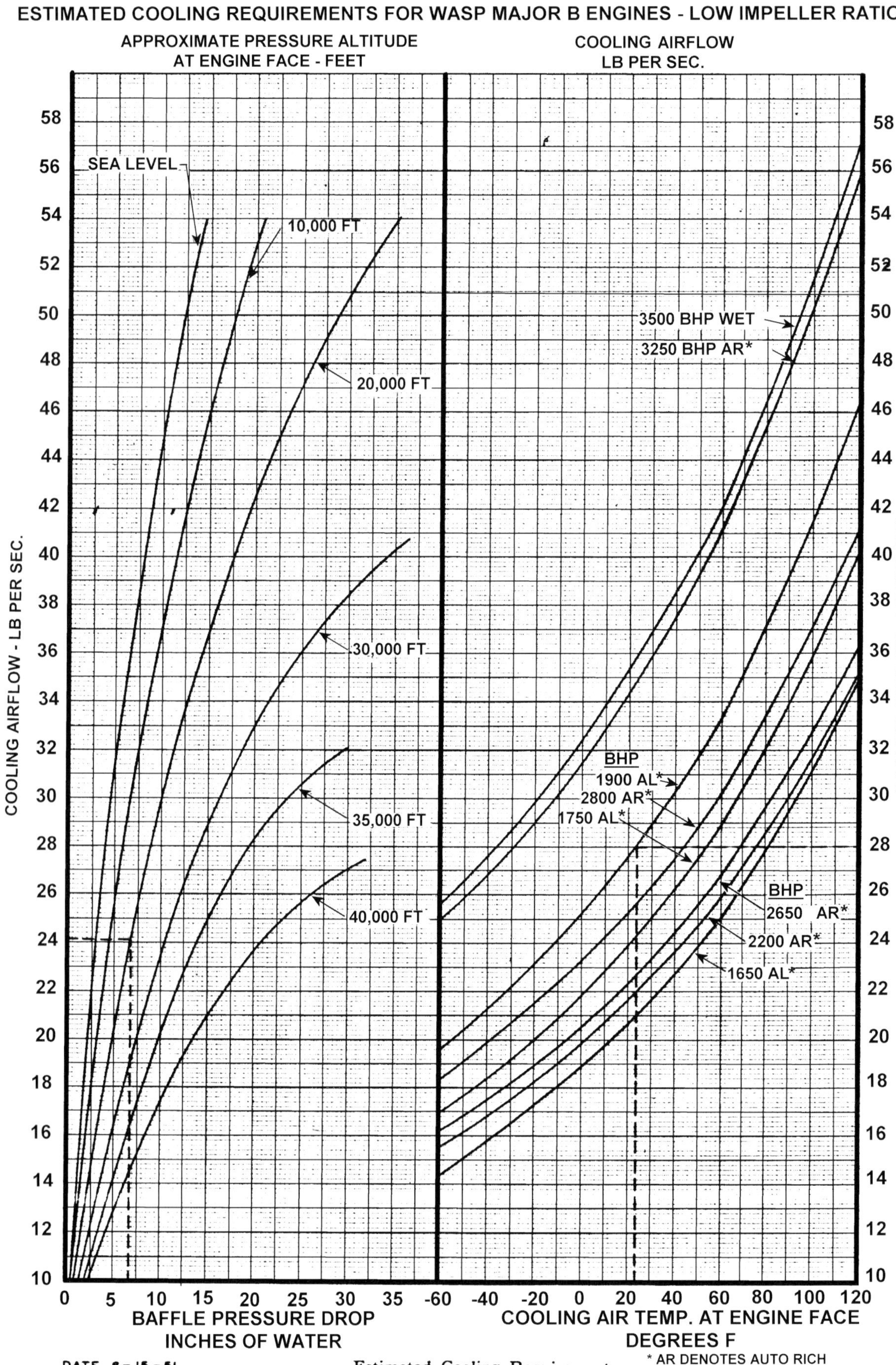

Estimated Cooling Requirements

Opposite: This graph gives an inkling of the huge mass airflow required to keep the temperature of the R-4360 under control. As an example, over 50 pounds per second is required to keep things cool at takeoff power on a 100-degree day. Of course, at the start of the take-off roll this kind of mass airflow is not possible. Instead, the engine relies on its 3,500 pound plus mass to dissipate heat until sufficient air speed has built up to generate the required airflow through the cooling fins. *(Installation Handbook, Section III Supplement F Wasp Major B Series Engines. February 1953. Modified by author.)*

Fig. 1.

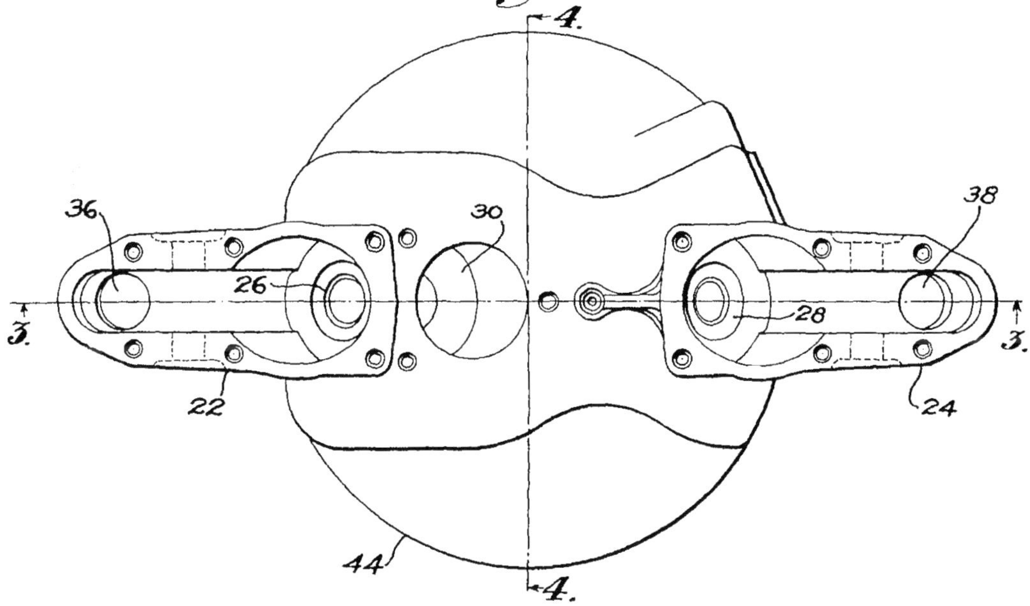

Fig. 3.

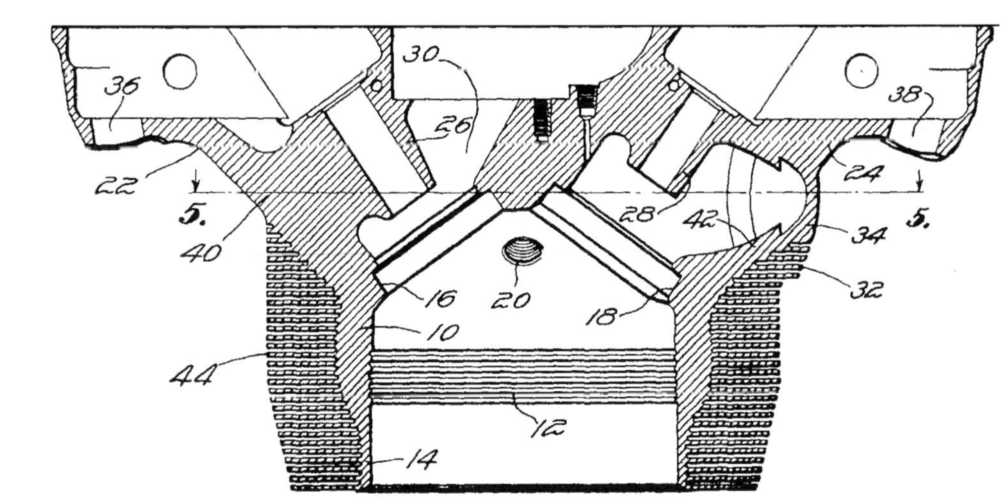

This longitudinal view of the cylinder illustrates to good advantage the hemispherical combustion chamber and the downdraft intake port. Use of a downdraft intake port was unique in radial engine design practice. *(United States Patent 2,401,210)*

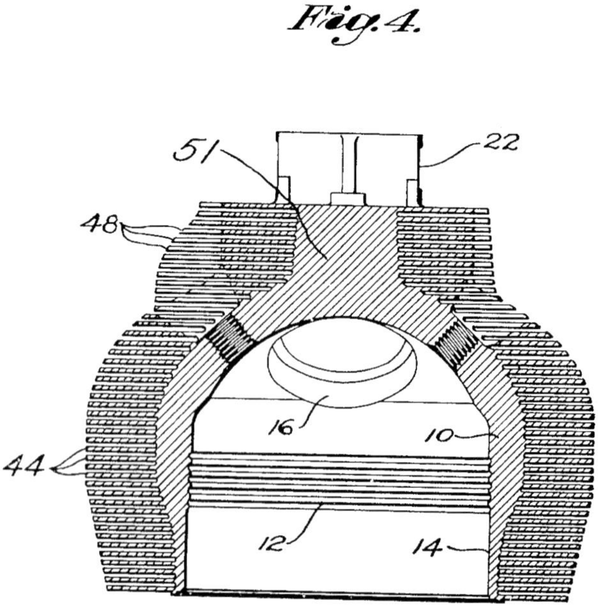

This lateral cross-section illustrates the position of the dual spark plugs. Also worth mentioning is the depth and number of cooling fins. At take-off power, a mass airflow of almost 60 pounds per second was required to keep temperatures under control. *(United States Patent 2,401,210)*

racecar engine. The fact that Ford built R-4360s under license in the 1950s may have influenced their design decisions on the Indy V-8 engine. Interestingly, one of the first prototype engines, the X-109, reversed the convention described and used a side intake and top exhaust. This arrangement was not used on any other R-4360. However, when Pratt & Whitney developed the R-2180, often described as half R-4360, the top exhaust and side intake was, once again, employed.

If Pratt & Whitney had simply used the same concept for pushrod location as that used on other radial engines (i.e., intake and exhaust pushrods located at the front or rear of the cylinder), the R-4360 would have experienced problems with its intake and exhaust port design not to mention some awkward cam ring configurations. The intake in particular, owing to the right-hand spiral, would have required doglegs designed into it. To overcome this, Pratt & Whitney rotated the cylinder about the engine's axis so the intake pushrod was at the front and the exhaust pushrod was at the rear of the cylinder.

As previously stated, R-4360s used forged cylinder heads. But due to the severe cooling requirements of this engine, materials other than aluminum were investigated for the cylinder head. Copper is a well-known conductor of heat; in fact it's about three times greater than aluminum, but accompanied by a corresponding weight penalty. A Pratt & Whitney Memorandum Report issued in 1941 summarized the pros and cons of aluminum versus copper *(Ref. 2-19)*:

For a fixed fin length, irrespective of weight and pumping power limitations, copper fins give the greatest heat transfer.

For a given fin weight, irrespective of length and pumping power limitations, aluminum fins give greater heat transfer than copper.

For a specified pumping power, irrespective of fin length and weight limitations, copper fins give greater heat transfer than aluminum.

Nothing about this analysis is earth shattering; however, it articulates in a very succinct manner the pros and cons of copper versus aluminum. The decision as to which material to use is not an easy one. Successful examples abound for the use of both materials. For instance, when Bristol developed its 14-cylinder Hercules sleeve valve engine into the 100 series rated at over 2,000 hp, it went from aluminum to copper for the cylinder head in order to control head temperatures. The same was true of the smaller deHavilland Gypsy series of light aircraft engines when their specific powers were increased—copper heads were their salvation.

Flexibility of the basic engine was a design goal from the initial stages. To this end, all cylinders were interchangeable, unlike any other multi-row radial, which require different cylinders for the front and rear rows. Even more remarkable was the fact that cylinders could be rotated 180 degrees to accommodate a pusher or tractor installation. However, different cam rings would be required depending on whether it's a pusher or tractor. Even so, this clearly helped with spare parts inventories for the military, which operated R-4360s in both configurations.

Construction of the R-4360 cylinder followed advanced air-cooled radial practice, i.e., the forged head is screwed and shrunk onto a steel barrel. Prior to this operation, a forged-aluminum cooling-muff is shrunk on to the barrel and all cooling fins machined in. The degree of shrink for the head to barrel and cooling muff to barrel dialed-in the requisite cylinder choke. At operating temperature, the choke almost goes away due to thermal expansion. Interestingly, all cooling fins on the barrel, as well as the cylinder head, are in the horizontal axis. Not only does this improve cooling, but it makes for an easier manufacturing job by not having to worry about machining fins in two (horizontal and vertical) axes.

Ears That Wave in the Breeze

Another serious problem plaguing R-4360 development was that of cylinders flexing due to valve opening forces *(Ref. 2-20, 2-21, and 2-22).* The exhaust valve rocker box was the primary culprit. In order to obtain good volumetric efficiency and exhaust scavenging, it's important to open the exhaust valve well before bottom dead center of the power stroke. But in doing so, considerable residual pressure is still present inside the cylinder—up to 100 pounds per square inch. Somewhat like trying to pull the plug in a full bathtub, the water pressure has to be overcome before the plug can be pulled. This residual exhaust pressure bears against the closed exhaust valve, just like the bathtub plug, and with a valve area of about 5.4 square inches this means a force of 540 pounds is required just to overcome exhaust gas pressure. In addition to exhaust gas pressure is the valve spring pressure of 900 pounds. And that's not all; additional force is required to accelerate the rocker arm assembly, which has a negative mechanical advantage. When it's all added up, Pratt & Whitney engineers figured out that the initial opening force for the exhaust valve was over 4,220 pounds.

If weight were not an issue it would simply be a case of making the cylinder out of lots of cast iron or adding considerable amounts of aluminum. But an aircraft engine does not enjoy the luxury of simply adding material to make up for a design shortfall. Under a test in which the entire valvetrain was under such stress, the entire system took on the characteristics of a rubber band. As the tappet roller rode up the cam profile, residual exhaust gas pressure resisted opening of the exhaust valve. Something had to give, so the exhaust rocker box "ear" would flex until sufficient force had been built up in the system to release the pent-up pressure and force the valve open. When it did open, it opened with a bang, bouncing the exhaust tappet off its cam track. Simply adding stiffer valvesprings was not a viable solution. In fact, it only exacerbated the situation. Carefully beefing up the valve rocker box areas where they overhung the cylinder helped, but it was always a marginal situation. Additional fixes included modifying the cam profile. As a starting point, the same profile used on the R-2800 was employed. Then a gentle ramp was designed to ease the exhaust valve off its seat. To overcome this and other problems, the Air Force undertook numerous tests with various valve clearances ranging from 0.045 inch to 0.060 inch.

With the foregoing in mind, it's easy to see why the R-4360 demanded strict adherence to valve clearance checks. If they weren't adhered to, serious problems such as valve bounce would ensue. For several years during the early 1940s, Pratt & Whitney investigated various forms of automatic valve adjustment. These investigations took the form of hydraulic tappets and hydraulic pushrods. Despite many hours of endurance testing, none of these concepts were adopted. Although it would have been nice to eliminate a time-consuming maintenance chore, none of the automatic valve clearance ideas worked out. Thompson Products made the automatic pushrod, but Pratt & Whitney determined that they were impractical for the following reasons:

continued on page 86

Fig. 2.

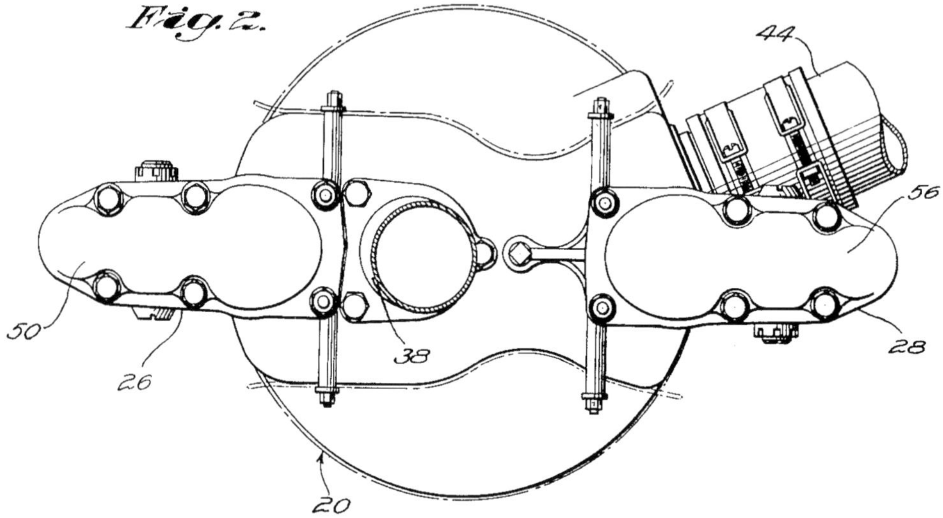

Fig. 3.

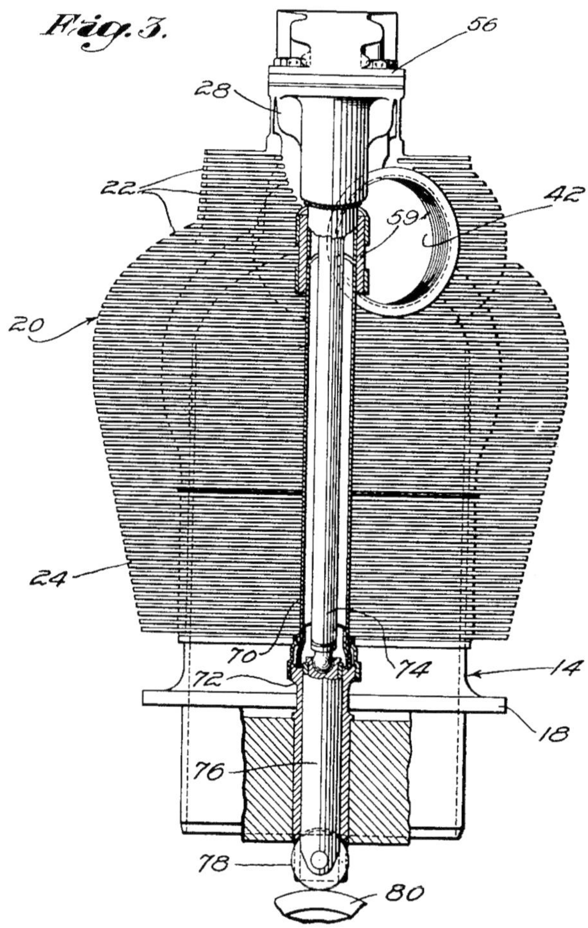

Above and opposite page: Construction of the cylinder assembly followed Pratt & Whitney's "C" series R-2800 practice–a forged head screwed and shrunk onto a steel barrel. The barrel has an aluminum muff with cooling fins shrunk on. The enormous cooling fin area is evident in this illustration. *(Above: United States Patent 2,401,210; Opposite: Drawn by author based on parts manual illustration)*

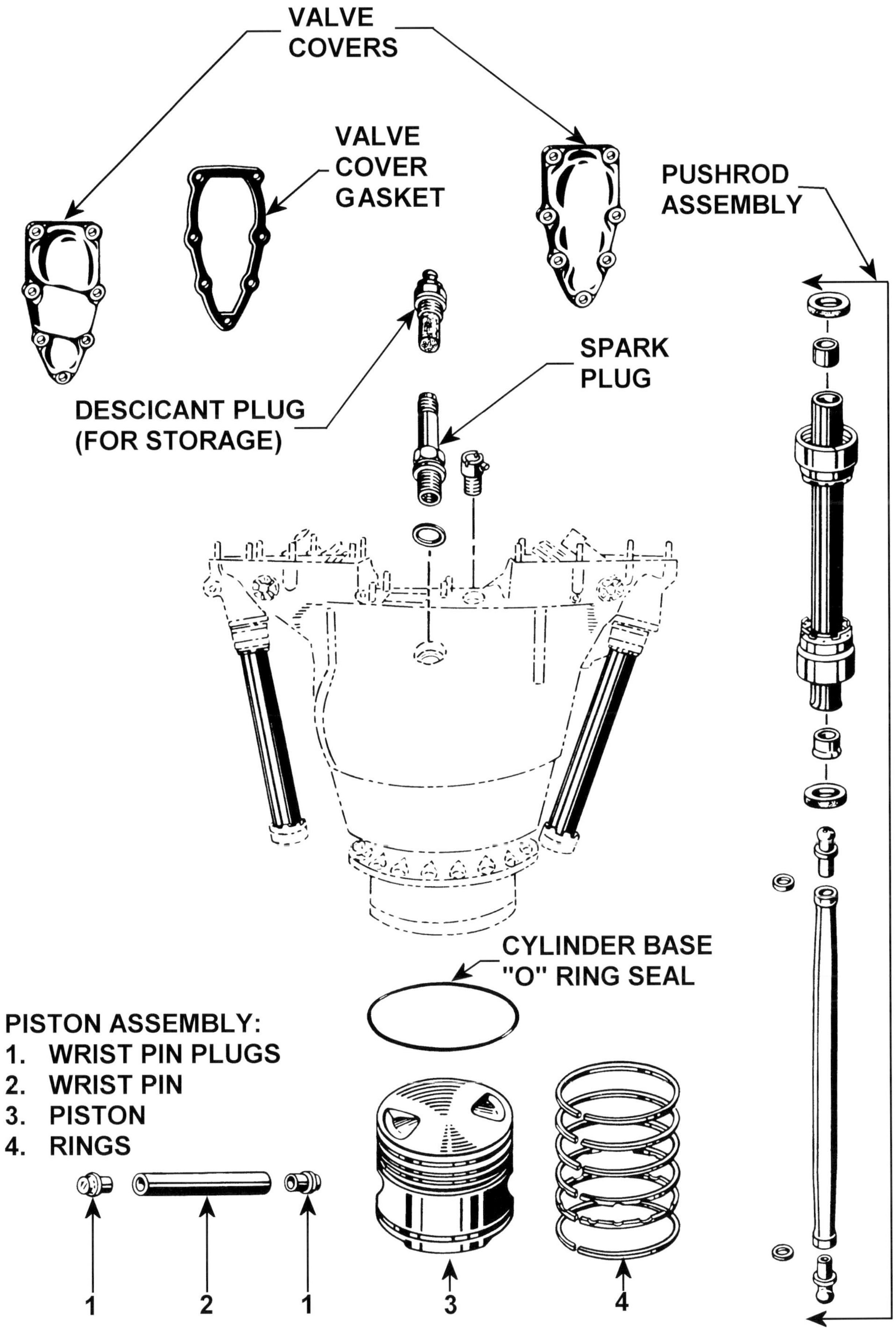
VALVE
COVERS
VALVE
COVER
GASKET
PUSHROD
ASSEMBLY
SPARK
PLUG
DESCICANT PLUG
(FOR STORAGE)
CYLINDER BASE
"O" RING SEAL
PISTON ASSEMBLY:
1. WRIST PIN PLUGS
2. WRIST PIN
3. PISTON
4. RINGS
1
2
1
3
4

continued from page 83

Too complicated and too frail. Even worse, under some conditions, if a valve stick opens temporarily, the automatic feature tries to compensate by extending to its full length. A serious failure would occur on the next cycle resulting in a bent pushrod. The Wilcox-Rich hydraulic tappets deserved closer inspection. Nevertheless, under the harsh operating environment of an R-4360, they experienced severe wear such as galling of the tappet pinholes. While these and other problems appeared to be fixable, other more serious consequences appeared that defied any easy solution. These included pitted cam profiles, particularly on the exhaust side and burned valve faces. Incredibly, five years, from 1940 to 1945, were spent testing these devices. In a final report issued on March 20, 1945, the concluding remarks stated *(Ref. 2-23, and 2-24):*

"From the record it appears that the hydraulic tappets in their present state of development have little to recommend them. The primary purpose of automatic valve gear is to obtain a combination good starting and idling characteristics and low valve gear acceleration at high engine speeds. These hydraulic tappets appear to have accomplished neither."

Intake Manifold Design

Once again, the R-4360 re-wrote the rules as they pertain to radial engine design. Typical single-row or two-row engines feature intake manifolds made up of thin-walled tubing emanating from the blower housing and terminating at the cylinder head's intake port. On some two-row radials,

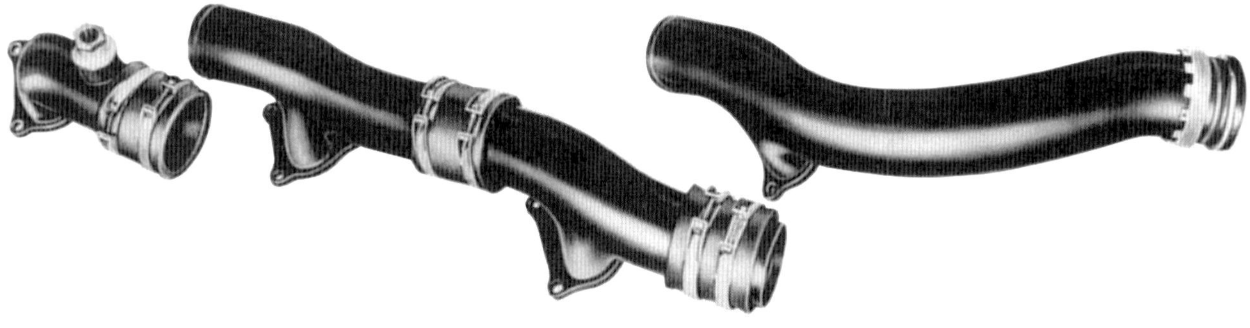

Early intake manifolds simply used rubber couplings sealed with hose clamps. This design was clearly deficient due to the relative motion of adjoining cylinders, which caused fretting and, ultimately, leaks. *(Service Instructions for Model R-4360-31 Aircraft Engine. May 1, 1945)*

Later intake manifolds used metal couplings sealed with chevron seals–a far better design than the previous one. It was also more tolerant of cylinder "weave" and any other relative motion that could lead to fretting. *(Handbook Overhaul Instructions Model R-4360-53 Aircraft Engines. May 15, 1952)*

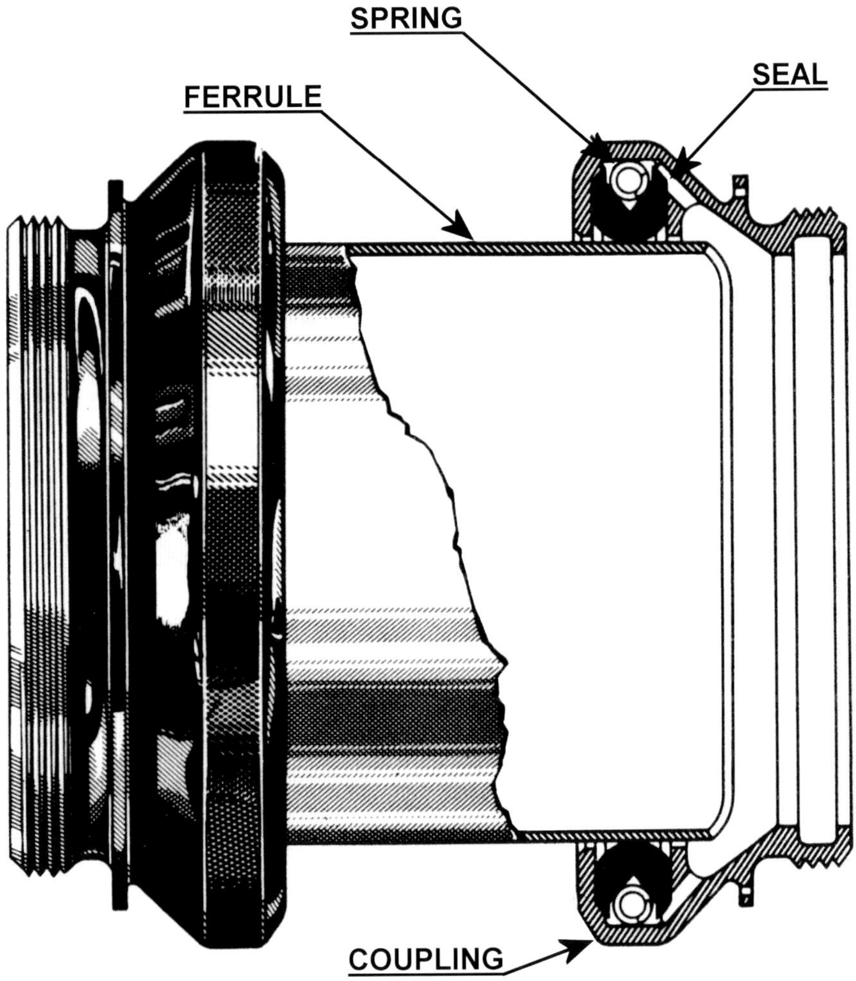

Detail view of the later intake manifold coupling. *(Handbook Overhaul Instructions Model R-4360-53 Aircraft Engines. May 15, 1952)*

the intake pipe is siamesed so one discharge from the blower splits off and feeds a front and rear cylinder. In the case of the R-4360, seven radial outlets from the blower housing connect to tubes, which then reach over the top of each bank of four cylinders *(Ref. 2-6, 2-7, and 2-25).* Each cylinder "tees off" at the intake manifold. Each manifold is made up of four pieces including three "tees." It terminates at the "D" (front) row of cylinders. In this way, a bank of four cylinders is fed by one of the seven intake manifold assemblies. Early manifolds simply joined the sections with three rubber hoses and hose clamps. This soon proved to be impractical, in part because of the amount of flexing that occurred between the cylinders. The definitive fix was the use of a coupling fitted with rubber chevron seals. This allowed the intake manifold sections to flex and at the same time seal the pressurized fuel/air mixture.

Lubrication

By their very nature, radial engines demand dry sump lubrication. That is, engine oil is contained in a separate container rather than being integral with the engine, as would be the case with an automobile. Early radials were relatively simple in that they simply had a single pressure pump that forced oil through the various galleries and a single scavenge pump that sucked oil out of the engine and returned it to the oil tank via an oil cooler. Of course, a simple system such as that described wouldn't cut it for an engine like the R-4360. Nevertheless, the basic principle remained the same—a pressure system and a scavenge system. At takeoff power, oil flow is 520 pounds per minute. (See pags 97–104 for color pictures of lubrication systems.)

To Scavenge or Not to Scavenge

As previously described, a dry sump lubrication system was employed. Two main lines connect the tank to the engine—oil-in and oil-out. As their names suggest, the oil-in line supplies oil to the suction side of the main engine pressure pump. After performing its lubrication duties, the oil is sucked out of the engine via scavenge pumps through the oil-out line. Scavenge oil (oil-out) is pumped through a cooler and then back to the oil tank. The preceding is an overly simplified description of oil flow, but it gives a basic idea of the dry sump concept. This description is also true for dry sump inline engines.

Getting lubricating oil to bearings, cylinders, and other components that come into intimate contact is, of course, essential to an engine's health. But getting the oil out after it has performed its lubricating duties is just as important. Pratt & Whitney learned this valuable lesson when developing the R-2800. Early "A" series R-2800s were rated at 1,850 hp. Follow-on "B" series R-2800s were rated at 2,000 hp. Pratt & Whitney picked up the additional 150 hp by simply improving the scavenging efficiency of the oil system. And this improvement was obtained through the ridiculously simple expedient of opening up drain holes in the crankcase sections. As an added bonus, the "B" series engine imparted less heat in the oil so a smaller oil cooler could be used. When oil is thrashed around inside an engine it not only creates a significant amount of drag, but it also causes oil temperatures to go through the roof. Without a doubt, scavenging is key to maintaining good power and keeping heat rejection to the oil to a minimum. But it's not always a question of simply bolting on bigger pumps, turning them faster, or offering better drainage paths for the oil: although these remedies obviously have their place. As a Pratt & Whitney executive succinctly put it in a 1940s report *(Ref. 2-26)*:

"In plain English it can be stated that the initial R-4360 wouldn't scavenge nohow."

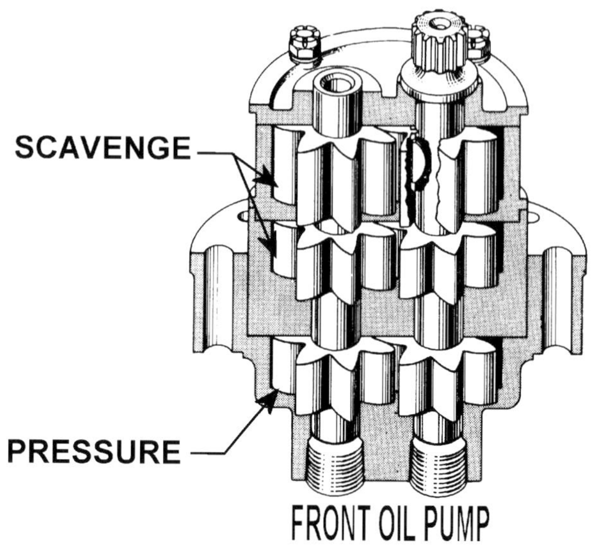

Mounted at the bottom of the nose case, this pump served double duty as a pressure and scavenge pump. It served to provide pressure lubrication to the reduction gearing, magneto drive, and other nose case requirements. Additionally, it provided pressurized oil to the propeller governor, which then boosted the pressure to over 400 psi. *(Overhaul Instructions Wasp Major B [R-4360] Engines)*

It took two and half years of intensive work to get the scavenging problem under control. This work included the obvious solutions of increasing scavenge pump capacity, installing more scavenge pumps, revising inlet conditions of scavenge pumps, etc. The problem was exacerbated when testing the initial proof-of-concept R-4360-powered F4U WM Corsair. Due to the dramatic power-to-weight ratio of this aircraft, it was capable of very steep climb angles. This capability imposed an additional burden on the scavenge system.

When oil is picked up by the scavenge system, it has been well and truly beaten into a froth. And herein lies part of the R-4360's scavenging woes—oil was hopelessly aerated when picked up by the scavenge pumps. Like many great ideas, the solution was deceptively simple yet worked like a charm. Just adding sheet-metal-perforated screens, especially in the rear section, removed a

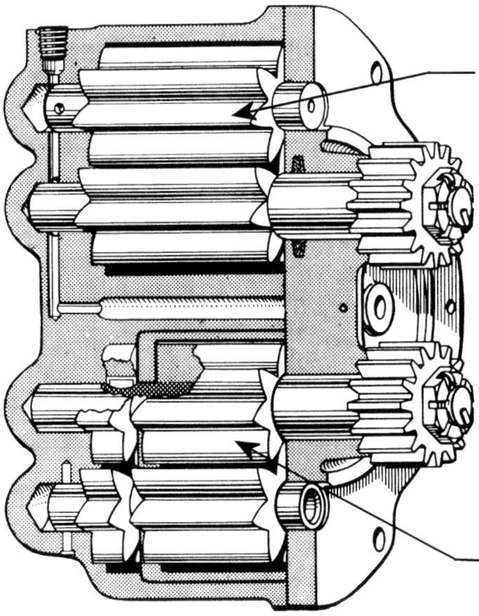

Like the nose case pump, the rear pump assembly served as a pressure and scavenge pump. It provided pressurized oil to the power section and rear section. It also scavenged the rear section. Each power section compartment had its own gerotor-type scavenge pump. *(Overhaul Instructions Wasp Major B [R-4360] Engines)*

significant portion of the entrapped air before the oil was allowed to enter the suction side of the scavenge pumps. This is a technique used to this day in high-performance engines regardless of whether they are dry sump or wet sump. The analogy used by Pratt & Whitney engineers when they arrived at this solution was that of the kitchen sink. If soapy water is beaten into a froth and then the plug is pulled, the water does not drain. Put a strainer in and the frothy water drains much easier.

In another departure from previous practice, the R-4360 used a gerotor-type scavenge pump in each crankcase section. These four pumps were driven off the cam drive gears. It's unclear when the gerotor-type pump was first developed, but the R-4360 must have been one of the first applications of this type of pump. Gerotor pumps are now used almost universally in automobiles *(Ref. 2-27)*.

Light My Fire

Two basic types of ignition systems were employed by the R-4360—high tension and low tension. The better of the two was demonstrably low tension; however, this superior system was still under development during the R-4360's embryonic years. A significant problem when operating an ignition system at high altitude was that of crossfire. This phenomenon was due to the low dielectric strength of the upper atmosphere. With the reduced insulation value of the upper atmosphere, it was possible for the 20,000-volt high tension intended for the spark plug to take the path of least resistance. This resulted in cross firing inside the magneto. Low-tension ignition solved these and other ignition related woes. Instead of generating 20,000 or greater voltage, the magneto generates 100 to 150 volts. High tension is generated at the cylinder head via a step-up transformer or coil—one for each of the 56 spark plugs. Early B-36s fitted with high-tension ignition suffered greatly with ignition problems because of the high altitude capability of this remarkable aircraft. More on the B-36 anon.

High Tension

All early R-4360s used high-tension ignition in the form of seven Bendix D-4RN-2 magnetos arranged radial fashion around the periphery of the magneto drive case. Each magneto fired eight spark plugs in the four cylinders making up a bank. Unlike many ignition systems, both magneto and distributor requirements are assembled in one unit. In the event a magneto failed, four cylinders would drop out leaving the remaining 24 cylinders to pick up the slack *(Ref. 2-28)*. With the high altitude capability of the R-4360, it was only inevitable that high-altitude flashover would occur. The usual remedy in this scenario is to pressurize the magnetos. And so it was, a pressurization pump ensured that a sea level atmosphere was maintained within the magneto regardless of altitude. Even this Band-Aid fix had its problems. It was critical that pressurized air

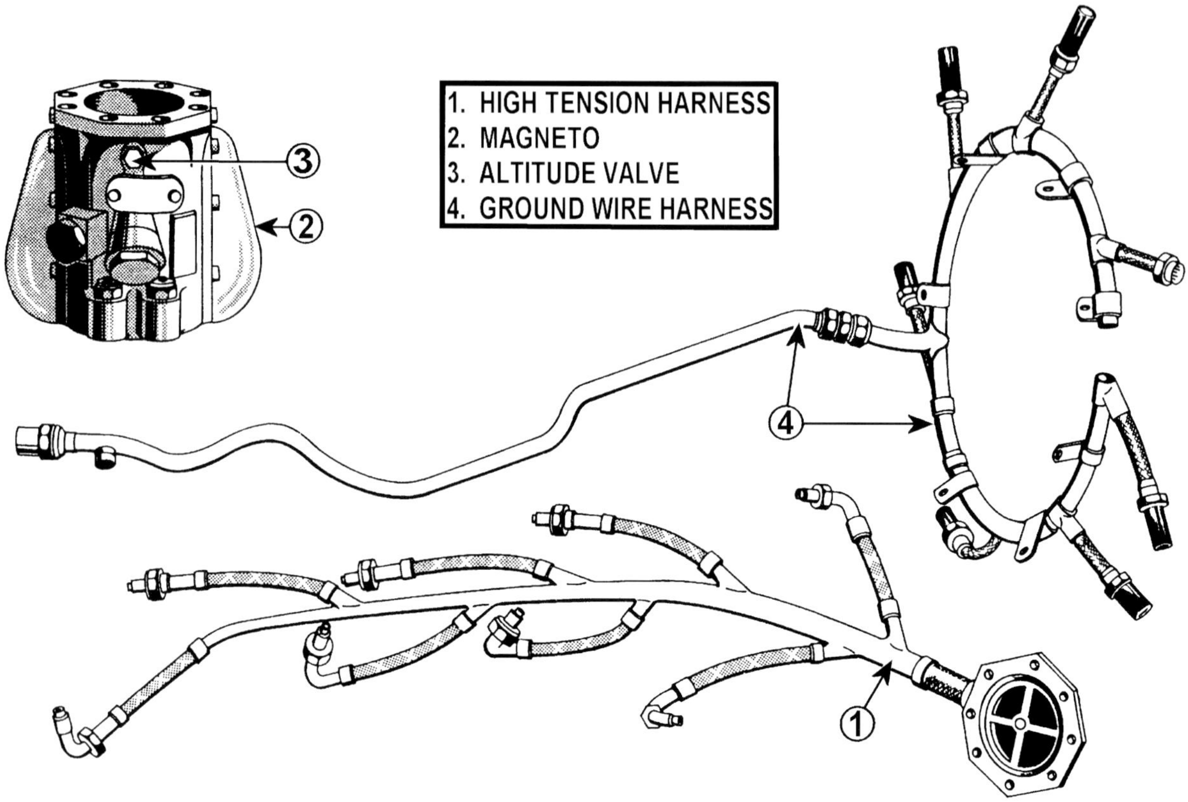

One of seven harnesses and magnetos that make up a high-tension ignition system. High-tension systems could suffer from flashover at elevated altitudes. Pressurizing the systems offered some relief from this problem, but this was never a definitive fix. Item 3 in the illustration is the so-called altitude valve, which bled off some of the pressurized air. Without this feature, ozone would build up inside the magneto, causing its own set of problems including corrosion. *(Service Instructions Bendix-Scintilla Ignition System Used On Pratt & Whitney R-4360 Wasp Major Aircraft Engines. Revised October 1948)*

was allowed to bleed off; otherwise dangerous amounts of nitric acid would build up in the distributor.

Low Tension

Once the clear advantage of low-tension ignition had been proved with some models of the "C" series R-2800, the R-4360 followed suit. Still arranged radial fashion around the front accessory case, the number of magnetos was reduced to four. Each magneto fired the seven cylinders making up a row *(Ref. 2-29).* Unlike other engines employing low-tension ignition, the R-4360 used a boost coil for each spark plug, 56 in all. Normal practice was to use a single boost coil per cylinder with two outputs, one for the front plug and one for the rear.

Front Section

Following a similar concept to the "C" series R-2800, the front section was made up of two major assemblies—the "propeller shaft case" and the "magneto drive case." However, this terminology changed on the R-2800. This engine referred to the propeller shaft section as the "front section" and the magneto drive section as the "front accessory section." No telling why these terms were changed when they performed the same function *(Ref. 2-7).*

Again, as with the R-2800, the propeller shaft section houses a massive rolling element bearing to take care of the immense radial and axial loads of the propeller shaft. This bearing has a split inner race to assist in engine assembly. A large nut retains the inner race and also transmits the

Photograph of one of the seven magnetos used in a high-tension ignition system. Mounted radially around the front accessory case, it resulted in a very complex ignition system that was susceptible to misfiring at high altitude. Bleed air tapped off the turbosupercharger was used to pressurize the inside of the magneto, which offered some relief to the high altitude misfiring maladies. *(United Technologies Corp./Pratt & Whitney)*

Photograph of one of the four magnetos that make up a low-tension ignition system. This was a much simpler, albeit still complex, system. Its high-altitude capability was a savior for aircraft such as the B-36. Unlike low-tension schemes used on other engines, the R-4360's system employed a boost coil for each spark plug. *(United Technologies Corp./Pratt & Whitney)*

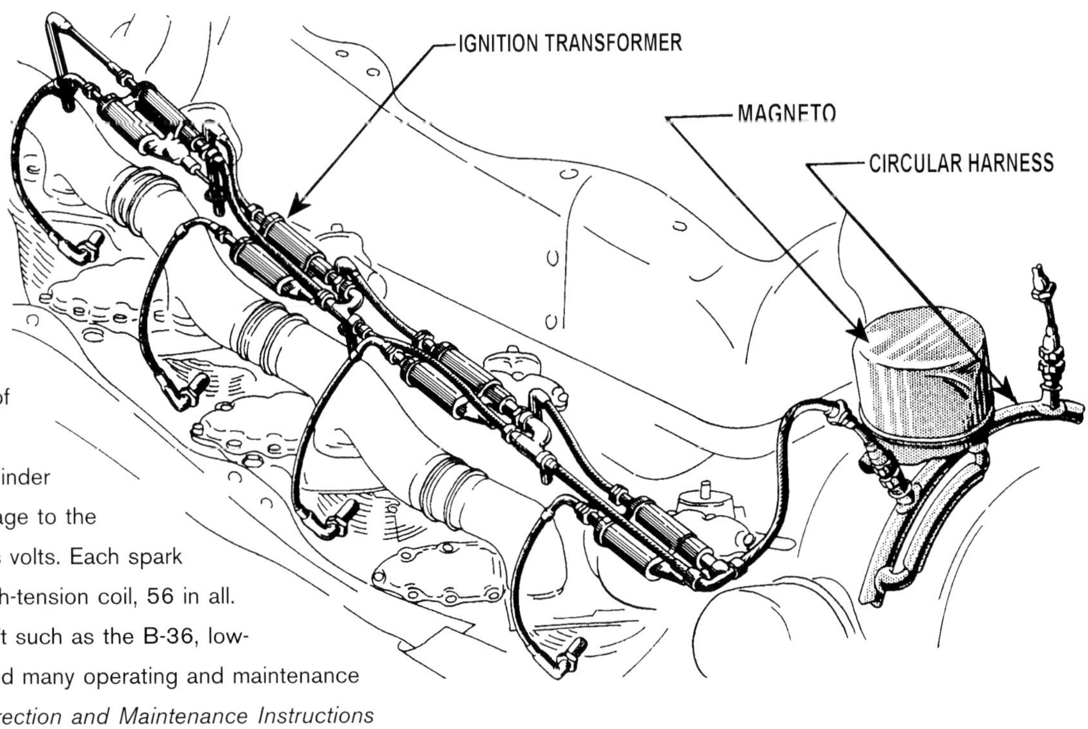

This line drawing shows one of the low-tension magnetos feeding power, via a fully shielded manifold, to the four cylinders that make up a bank. Note the pair of step-up transformers mounted on each cylinder head that boost voltage to the required 20,000-plus volts. Each spark plug had its own high-tension coil, 56 in all. For high-flying aircraft such as the B-36, low-tension ignition solved many operating and maintenance woes. *(Handbook Erection and Maintenance Instructions USAF Series B-36F Aircraft. March 27, 1951)*

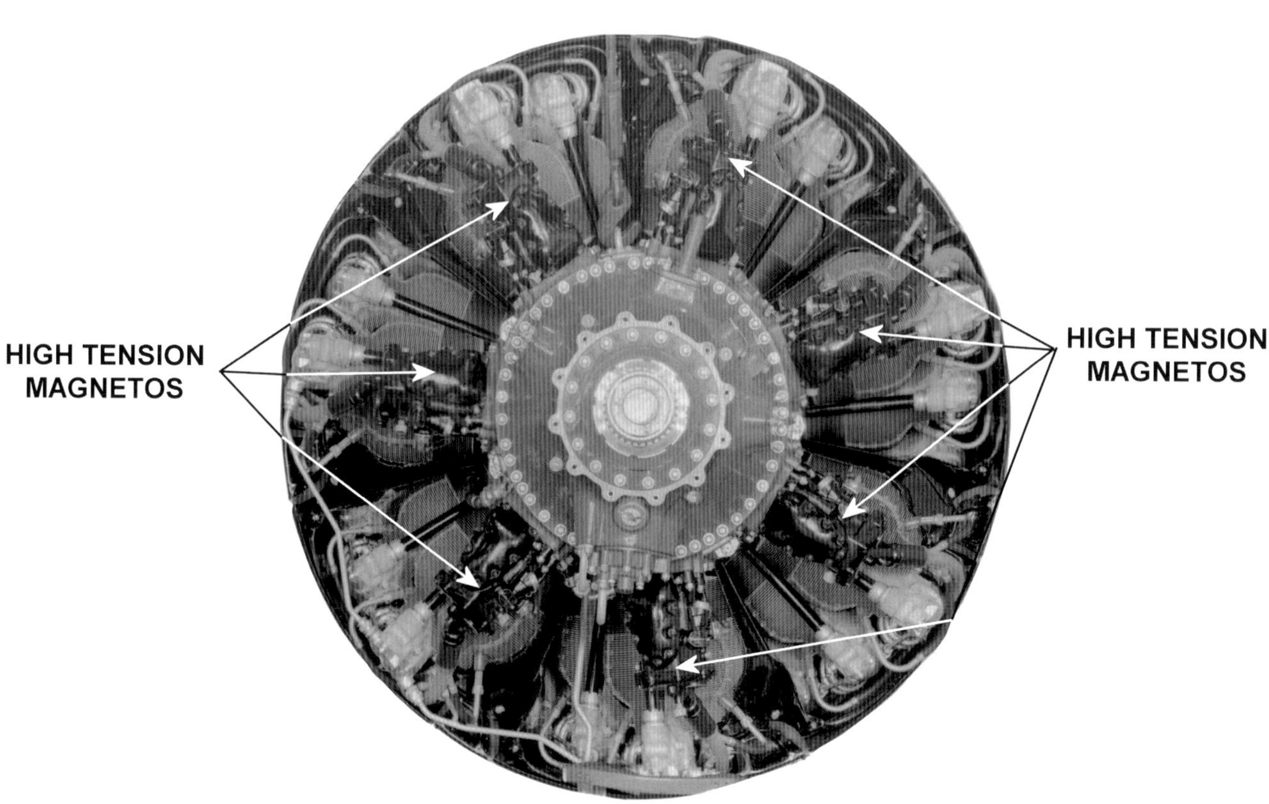

This is how the front of the engine appears when fitted with a high-tension ignition system. *(United Technologies Corp./Pratt & Whitney)*

This is how the front of the engine looks when fitted with a low-tension ignition system. *(United Technologies Corp./Pratt & Whitney)*

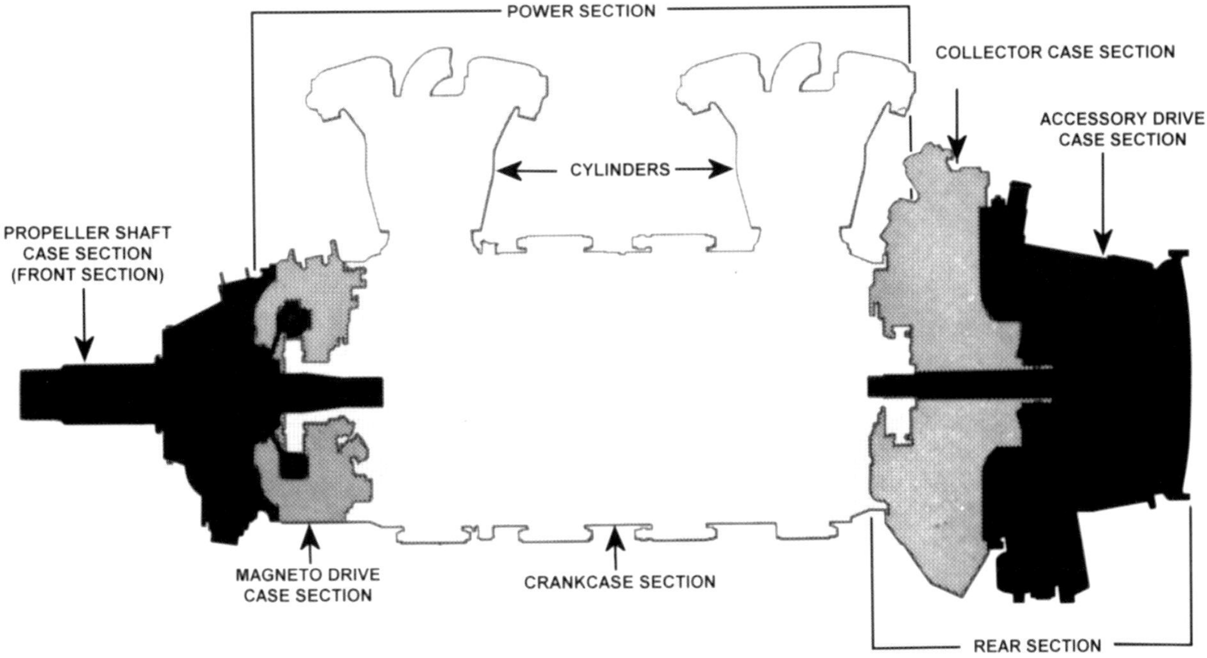

Naming conventions for the R-4360 differed somewhat from prior Pratt & Whitney radial engines. This illustration shows the main sections. *(Wasp Major Models B6 and CB2 Overhaul Manual Issued June 1953)*

engine's thrust through the engine, engine mount, and into the airframe. In helicopter terminology, this nut would be referred to as the "Jesus Nut." In other words, should this part fail, it would result in dire consequences. A plate bolted to the propeller shaft case retains the outer race. As with "C" series R-2800s, this bearing represents the only rolling element bearing in the entire engine, all others are plain.

Another "C" series R-2800 influenced design was the propeller shaft. Most, but not all, single-rotation shafts featured an SAE #60 spline. There were several exceptions to this convention: The early "X Wasp" engines and the R-4360-63/A installed in the Douglas Globemaster, both of which employed an SAE #70 spline shaft. As previously explained, the front of the propeller shaft is supported by a rolling element bearing, while the shaft's rear, or pilot end, is supported by two steel-backed bronze plain bearings. These bearings are inserted in the front of the crankshaft. Although piloting the rear of the propeller shaft in the crankshaft sounds convenient, as with any action, consequences result. These consequences include additional load placed upon the front main crankshaft bearing. Two compartments separated by a plug make up the interior of the propeller shaft. The front accepts oil from the oil transfer bearing and provides oil for hydraulically operated propellers. Lubricating oil for the reduction gearing comes from the rear compartment.

A multi-pinion epicyclical reduction gearing is also housed in the propeller shaft case. Another scaled-up R-2800 design, the 11 or 16 small pinions, depending on reduction ratio, ride in a pinion shaft support integral with the propeller shaft. A sun gear, driven off the crankshaft, engages the pinions on their inner diameter. Their outer diameter runs in a stationary gear. Following good gear design practice, the pinions were phased to engage sequentially. In other words, they do not engage simultaneously. It should be understood that a straight cut involute gear does not transmit power in a smooth, constant velocity manner. But this shortcoming can be mitigated via the aforementioned sequential engagement. In this

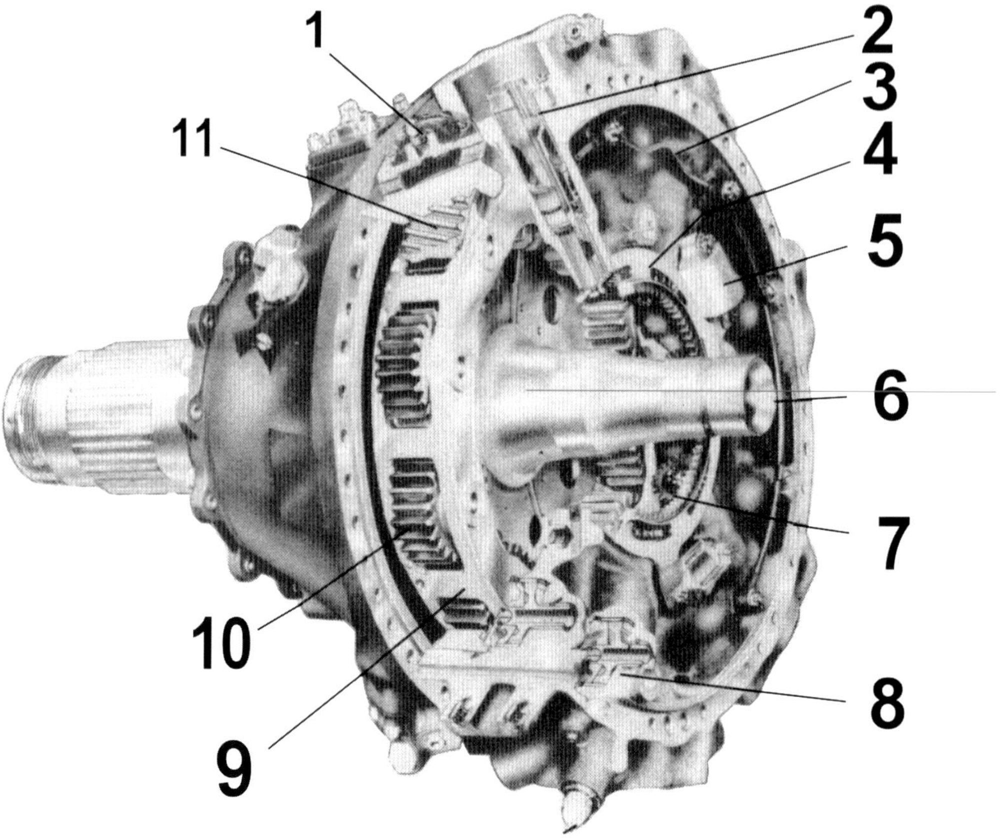

Front nose case assembly. 1. Torquemeter oil pressure transmitter; 2. Magneto driveshaft; 3. Spark advance oil feed tube; 4. Magneto drive fixed gear; 5. Spark advance cylinder; 6. Propeller shaft; 7. Magneto driveshaft gear; 8. Torquemeter master piston; 9. Propeller shaft reduction pinion support; 10. Propeller shaft reduction pinion; 11. Propeller shaft reduction drive fixed gear. *(Wasp Major Models B6 and CB2 Overhaul Manual Issued June 1953)*

way, potentially destructive vibration modes are avoided. Even in today's environment, it would require some serious computing power to figure this out. In the days of the R-4360's development, it was all done on a slide rule. A bevel gear, attached to the pinion shaft support, drives the propeller governor and the front section scavenge pump, which mounts at the bottom of the propeller shaft case. This scavenge pump is made up of three chambers. The top one scavenges oil from the propeller shaft case and magneto drive housing, the middle chamber scavenges the rocker box front sump, and the bottom chamber scavenges the front cam compartment.

The magneto drive housing, sandwiched between the propeller shaft section and the front of the crankcase, is packed full of components. These include the magneto drive requirements, seven magnetos, spark advance mechanism, and the torquemeter. Not all R-4360s were equipped with a torquemeter. The magneto drive housing could accommodate a torquemeter or simply have a bolted-on fixed gear that would not have the capability of registering a torque reading. Design of the torquemeter followed "C" series R-2800 practice except more pistons are used in order to handle the additional power of the R-4360—42 in all. The principle of the torquemeter is quite simple. The fixed gear in the epicyclic reduction gearing reacts to the power being transmitted. The outer diameter of the fixed gear has a helical spline machined on its periphery. Therefore, as power is transmitted, if the fixed gears were not restrained they would try to ride forward due to the resultant force generated. Pistons located around the fixed gear's front face offer an equal but opposing reaction. Engine oil, whose pressure is boosted by a torquemeter pump, offers the appropriate restraining force. Boosted oil is fed to all 42 pistons. Reading the pressure generated gives a cockpit reading of torquemeter pressure.

Again, "C" series R-2800 concepts were followed for the spark advance unit except, as one would expect, they were much more complex and

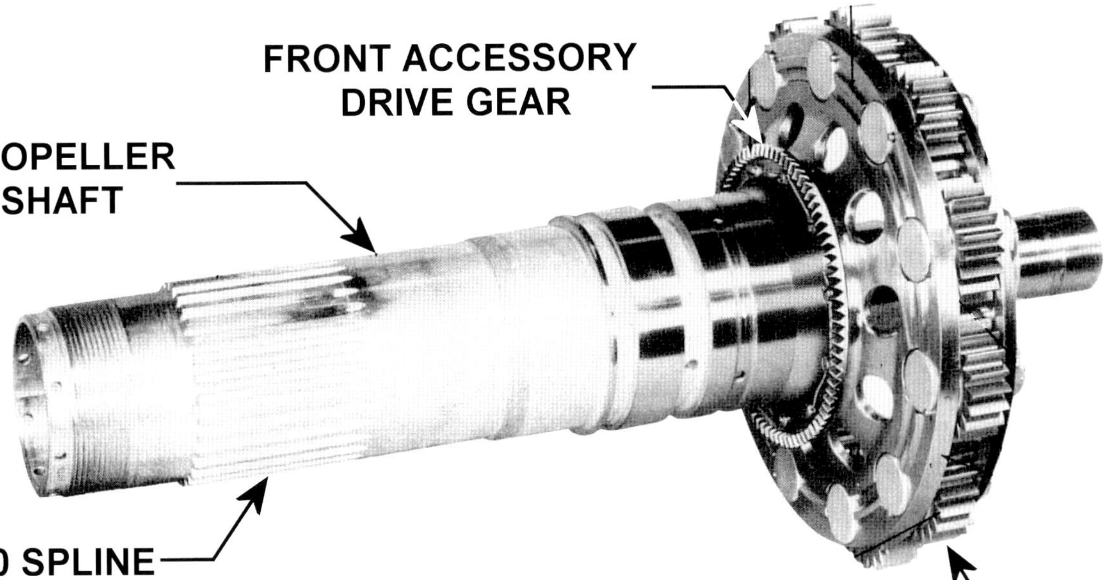

Propeller shaft and reduction gear pinion assembly. Exquisite workmanship is evident. Without this degree of finish and workmanship this heavily loaded component would fail. Light weight and reliability are always the opposing requirements aircraft engine designers strive for. This assembly personifies that requirement. *(Wasp Major Models B6 and CB2 Overhaul Manual Issued June 1953)*

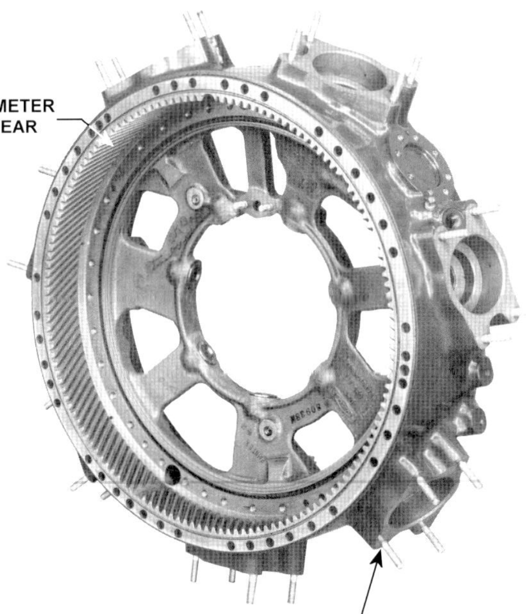

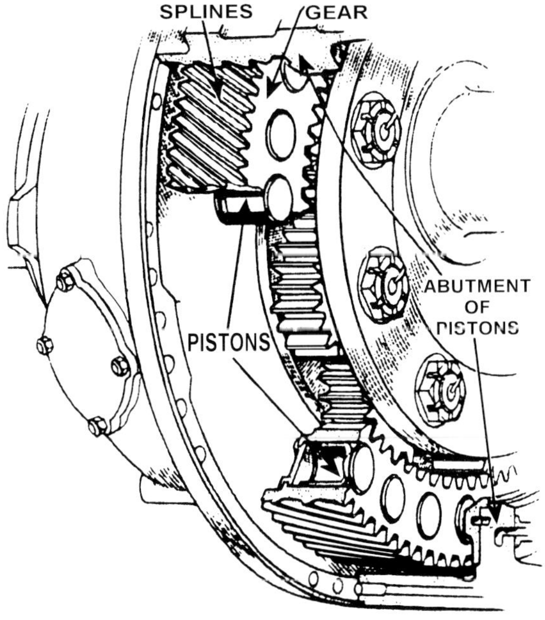

Above left: Magnesium casting that makes up the magneto drive case. This one is for a high-tension ignition system with seven magnetos. This complex yet lightweight component supported the critical ignition system. Also of note is the torquemeter fixed gear. All the torque reaction generated by the engine went through this gear. *(United Technologies Corp./Pratt & Whitney) Above right:* This partial cutaway line drawing of the torquemeter shows the key components and their relationship to each other. The combination of straight-cut gears and a fixed gear with helical teeth on its periphery allowed torque measurements to be made. *(Wasp Major Models B6 and CB2 Overhaul Manual Issued June 1953)*

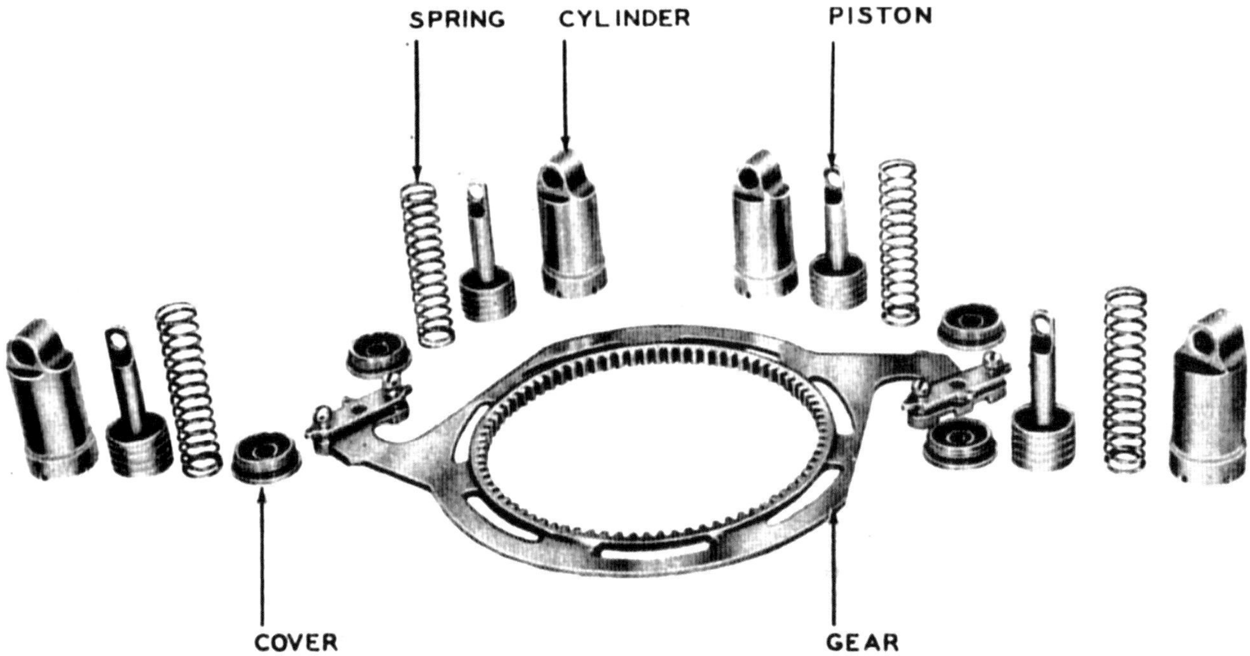

Four oil-operated servo pistons adjust ignition timing according to manifold pressure. Like many engines fitted with automatic ignition timing, this feature was disabled in later years and the engine ran on fixed timing. *(Wasp Major Models B6 and CB2 Overhaul Manual Issued June 1953)*

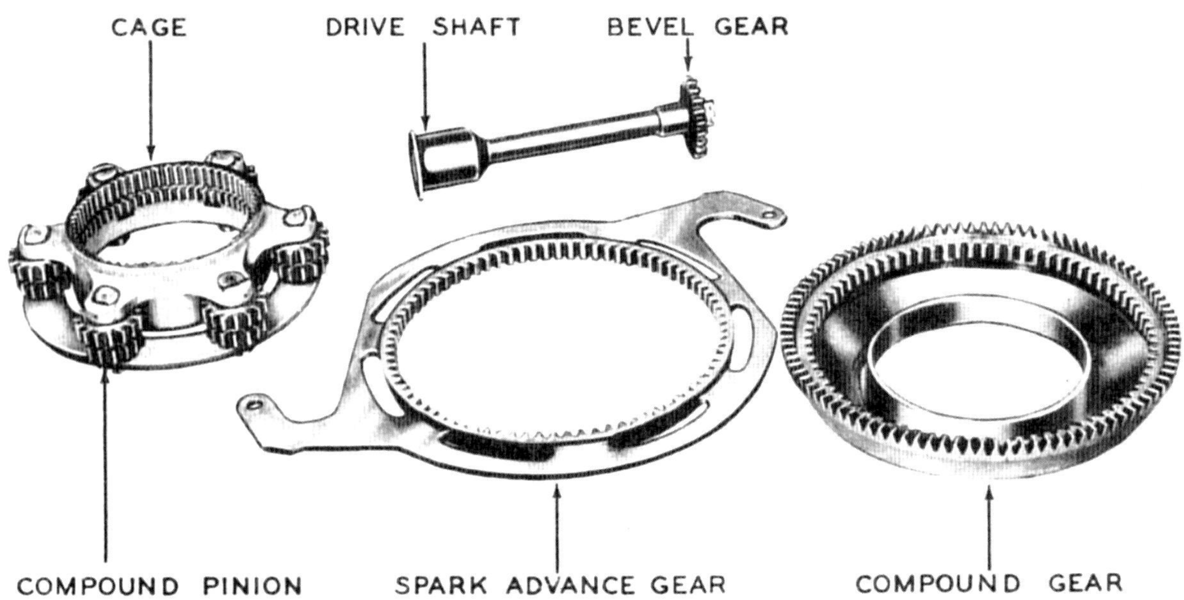

Magneto drivetrain disassembled. After all the design effort put into the automatic spark advance it seems a terrible waste of engineering talent to have it all disabled. *(Wasp Major Models B6 and CB2 Overhaul Manual Issued June 1953)*

sophisticated. An example of exquisite workmanship and engineering of the automatic spark advance unit consisted of: four spark advance cylinders, a compound epicyclic drive, stationary spark advance gear, and a compound gear. The spark advance gear, which performs the function of changing the relationship of the magnetos to the cranks and consequently the ignition timing, has two arms. Each of these arms is connected to a pair of pistons, which are allowed to pivot on both ends. A link made up from a straight beam is pivoted in the middle and attached to the spark advance gear. Each piston is spring loaded with

continued on page 113

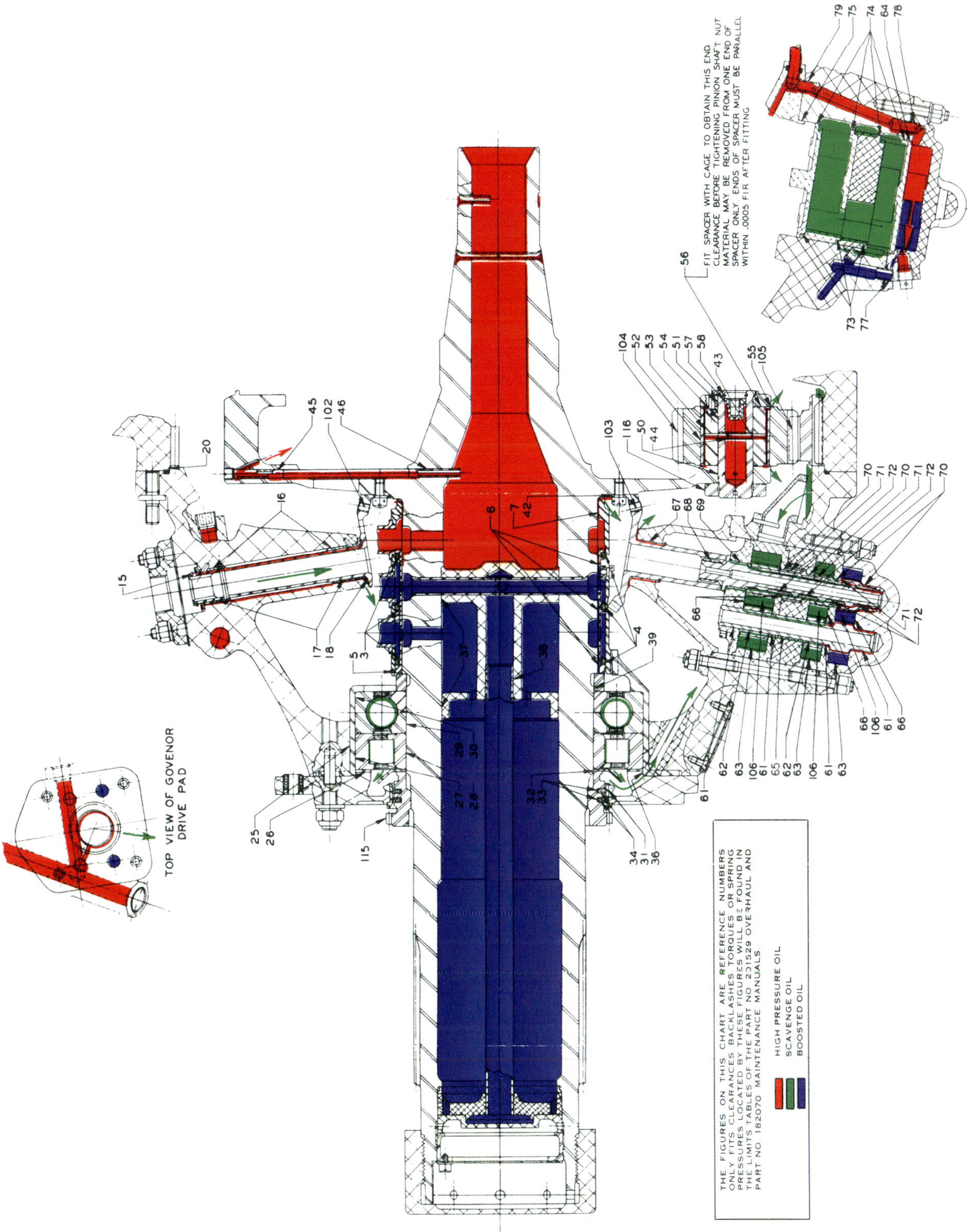

Longitudinal cross section through the nose case. Note how the propeller shaft is extended (to the right in this view) to be supported in the crankshaft. Scavenge and boosted pressure oil pumps are mounted at the bottom. Boosted oil, shown in blue, was used for hydraulically controlled propellers. Two massive rolling element bearings support the propeller shaft–one being a ball thrust bearing and the other being a roller bearing to accommodate axial loads. *(Overhaul Instructions Wasp Major B [R-4360] Engines)*

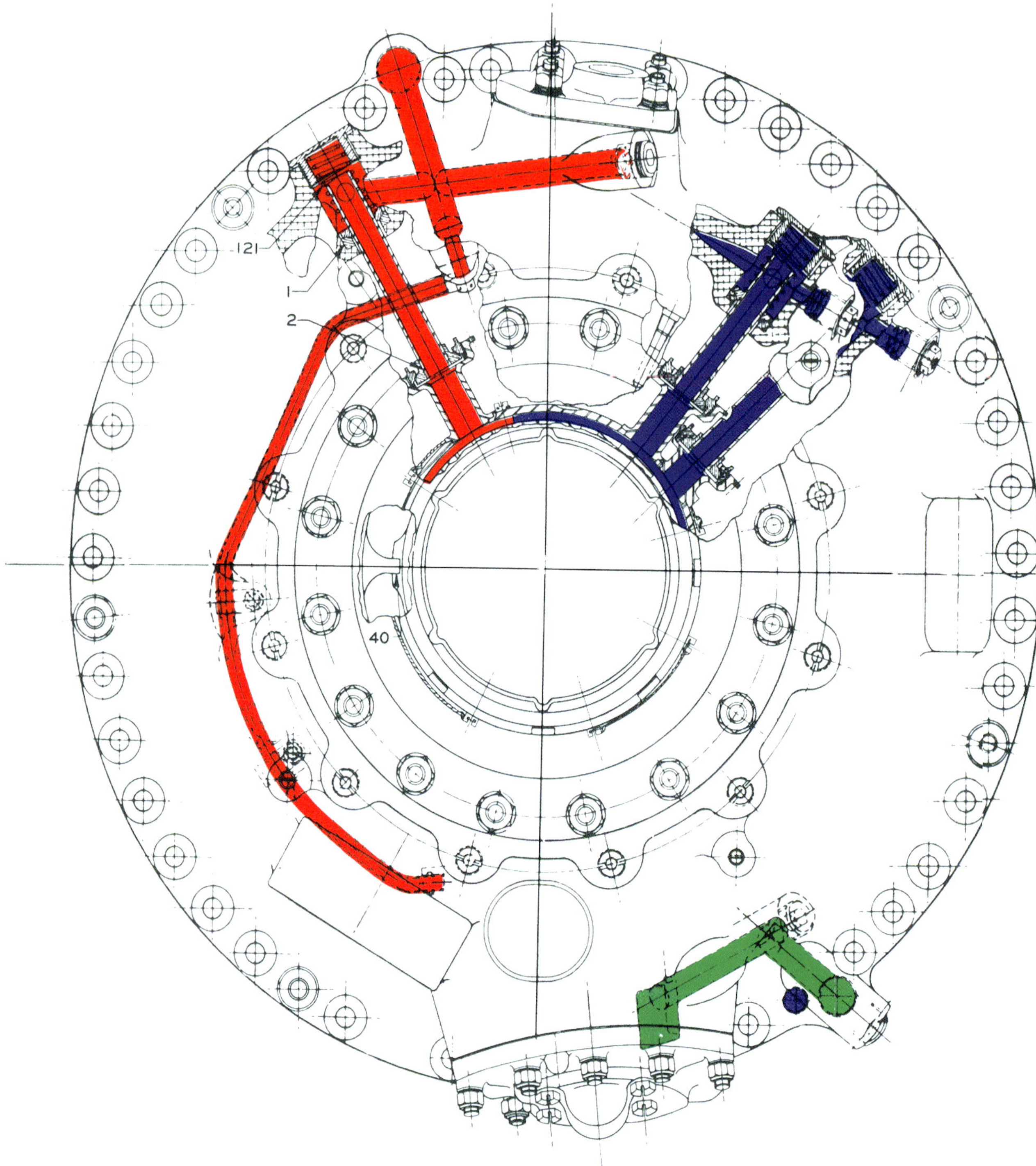

THE FIGURES ON THIS CHART ARE REFERENCE NUMBERS ONLY. FITS, CLEARANCES, BACKLASHES, TORQUES, OR SPRING PRESSURES LOCATED BY THESE FIGURES WILL BE FOUND IN THE LIMITS TABLES OF THE PART NO. 201529 OVERHAUL AND PART NO. 182070 MAINTENANCE MANUALS.

HIGH PRESSURE OIL
SCAVENGE OIL
BOOSTED OIL

Lubrication system: Lateral view of the nose case lubrication system. Scavenge (green), high-pressure (red), and boosted (blue) oil galleries are shown. *(Overhaul Instructions Wasp Major B [R-4360] Engines)*

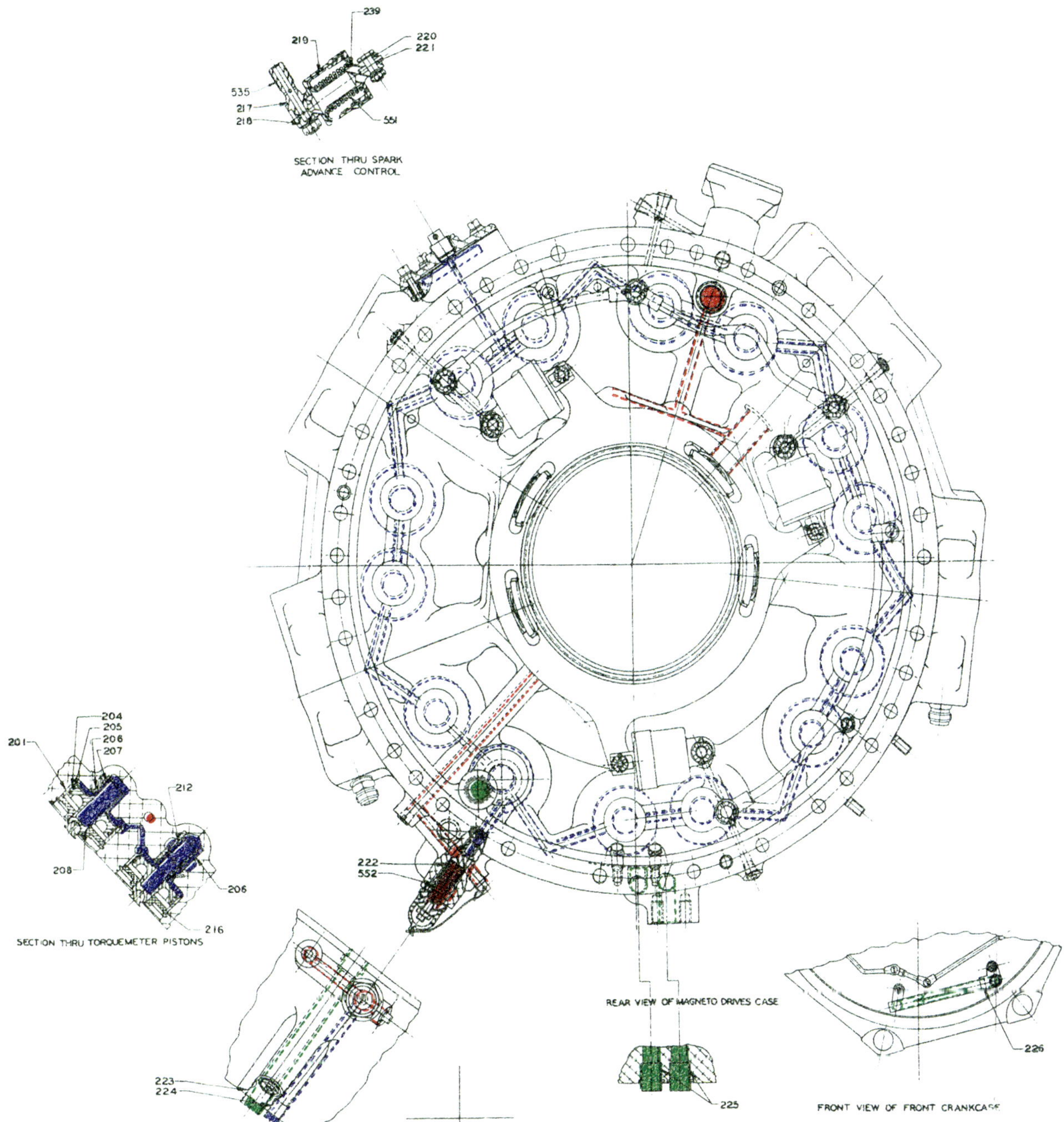

THE FIGURES ON THIS CHART ARE REFERENCE NUMBERS ONLY. FITS, CLEARANCES, BACKLASHES, TORQUES, OR SPRING PRESSURES LOCATED BY THESE FIGURES WILL BE FOUND IN THE LIMITS TABLES OF THE PART NO. 201529 OVERHAUL AND PART NO. 182070 MAINTENANCE MANUALS.

HIGH PRESSURE OIL
SCAVENGE OIL
BOOSTED OIL

Lubrication system: Lateral cross section of magneto drive case. This illustration shows a low-tension system with four magnetos. The mounting pads can be seen on the periphery. Oil supply to the 14 torquemeter pistons can also be seen. *(Overhaul Instructions Wasp Major B [R-4360] Engines)*

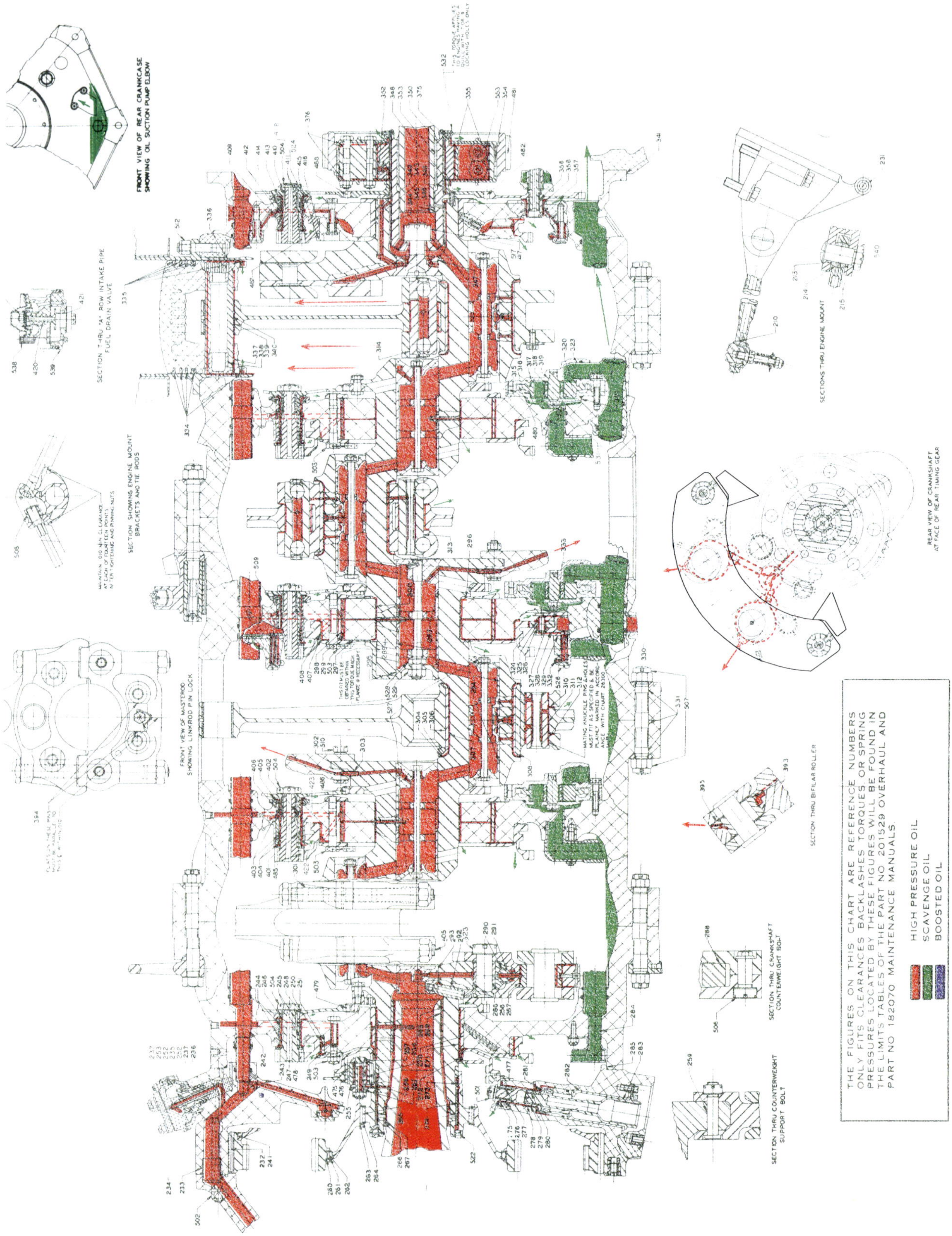

Lubrication system: Longitudinal cross section through the power section. The tail end of the propeller shaft can be seen to the left of this drawing supported in the front of the crankshaft. Each crankcase section is provided with its own gerotor scavenge pump. *(Overhaul Instructions Wasp Major B [R-4360] Engines)*

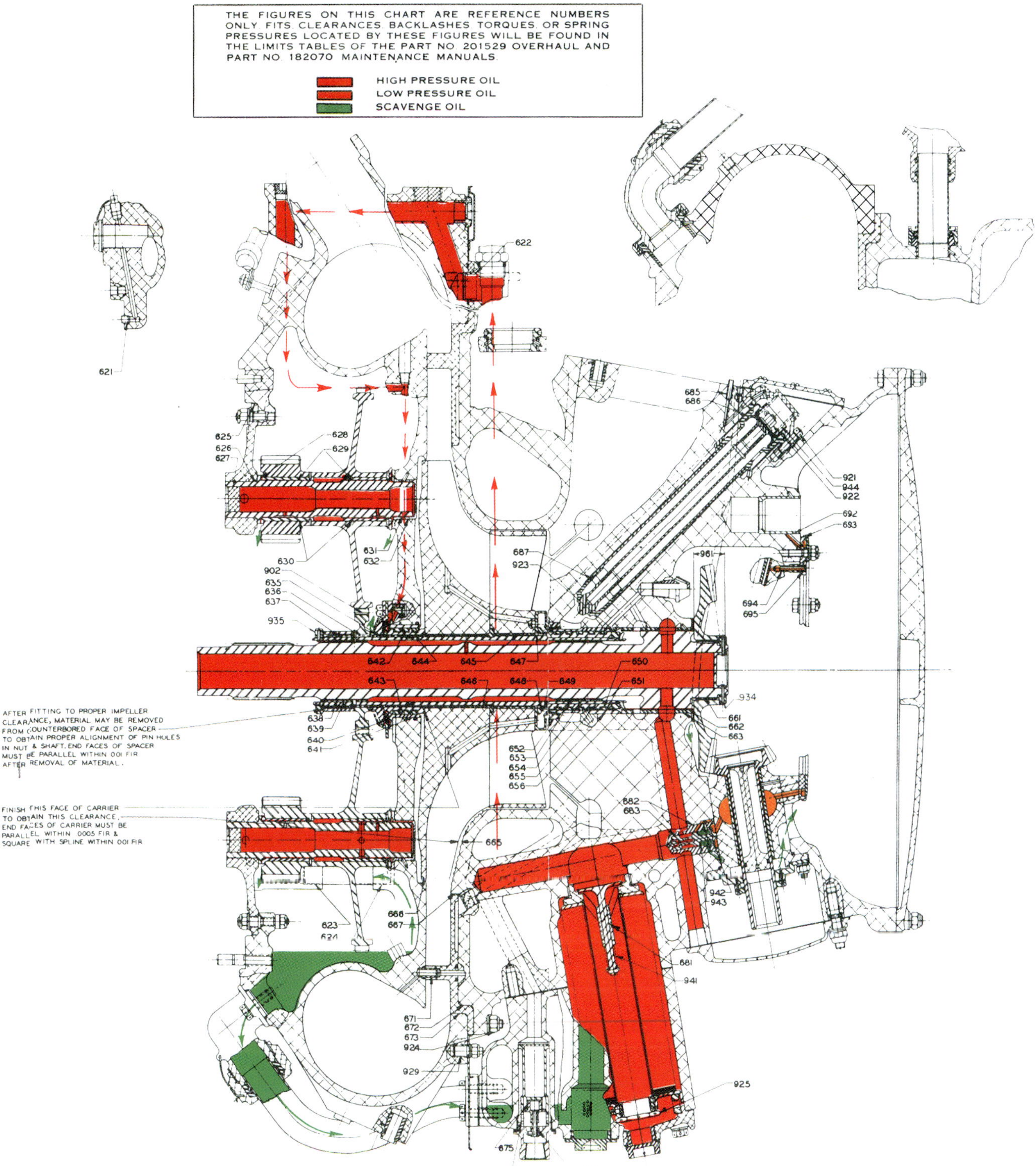

Lubrication system: Longitudinal cross section through rear section. The fuel feed valve shown to the right at approximately 45 degrees is shown to good advantage as is the supercharger impeller. *(Overhaul Instructions Wasp Major B [R-4360] Engines)*

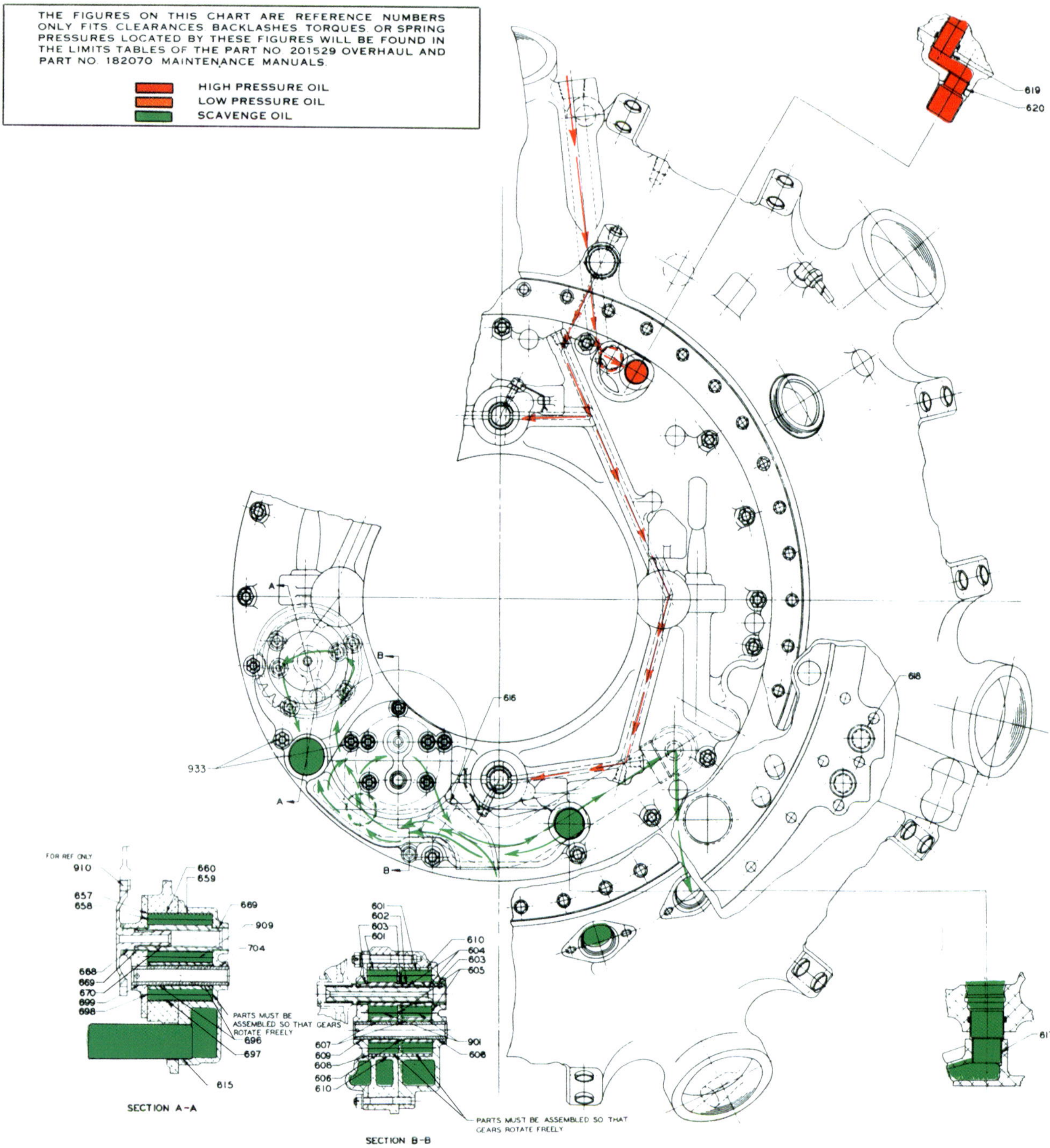

Lubrication system: Lateral cross section through supercharger housing. Three of the seven intake manifold discharges can be seen. *(Overhaul Instructions Wasp Major B [R-4360] Engines)*

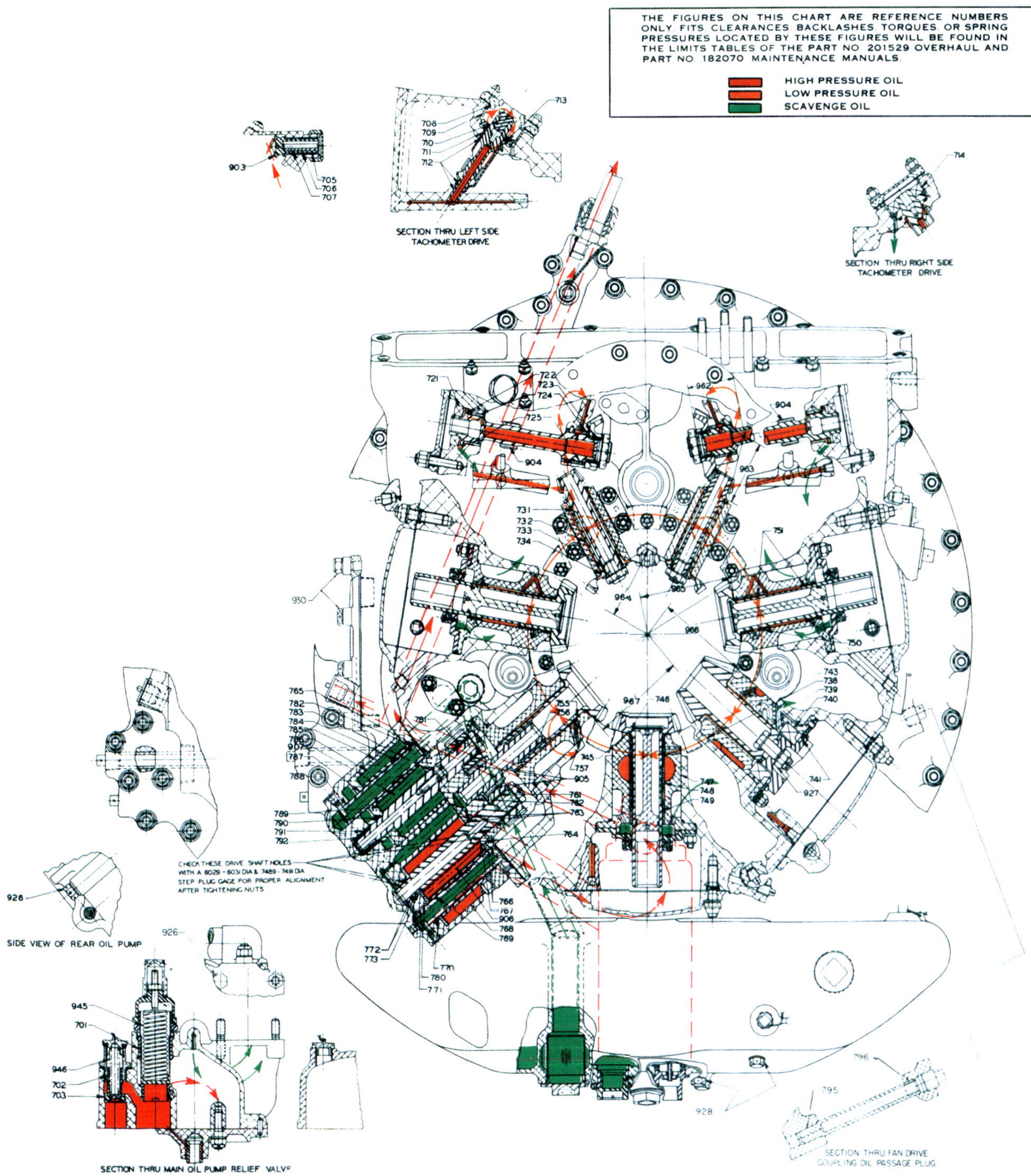

Lubrication system: Lateral cross section through rear section. This view shows pressure and scavenge pumps. Of particular note in the cross section is the radial arrangement of the accessory drives. This is in stark contrast to prior radial engine practice. *(Overhaul Instructions Wasp Major B [R-4360] Engines)*

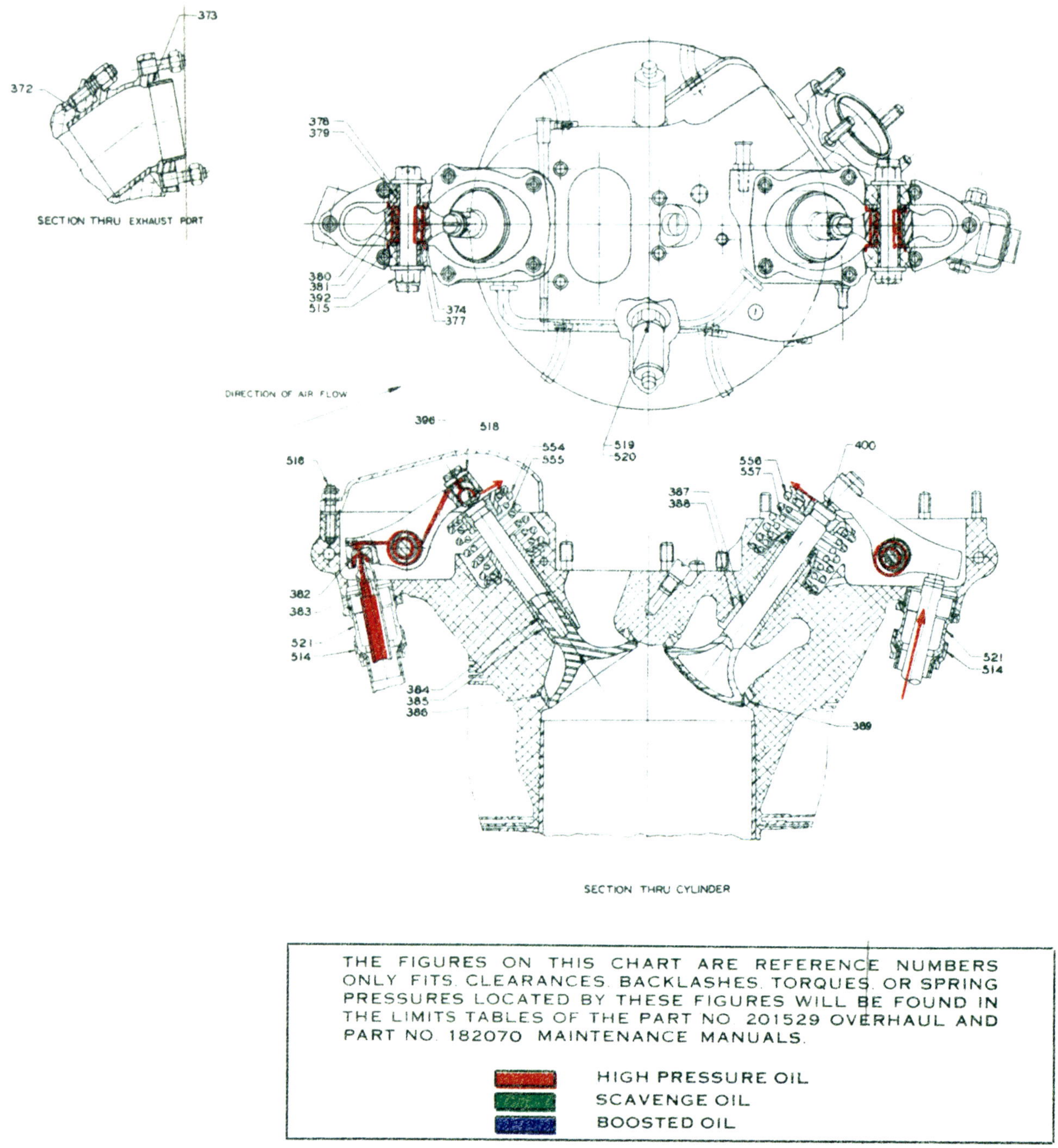

Lubrication system: Longitudinal cross section and top view of cylinder. Due to the high temperatures experienced in this part of the engine, severe "cooking" of the oil was a significant headache, particularly on the exhaust side shown to the right in these views. The overhung nature of the rocker boxes is also of note. This feature could cause flexing of the cylinder. *(Overhaul Instructions Wasp Major B [R-4360] Engines)*

Opposite: Diagram of the fuel and air circuits for the CECO carburetor. Refer to the text for a detailed explanation of how it all works. *(Handbook of Overhaul Instructions. Chandler Evans Carburetor Model 100-CPB9)*

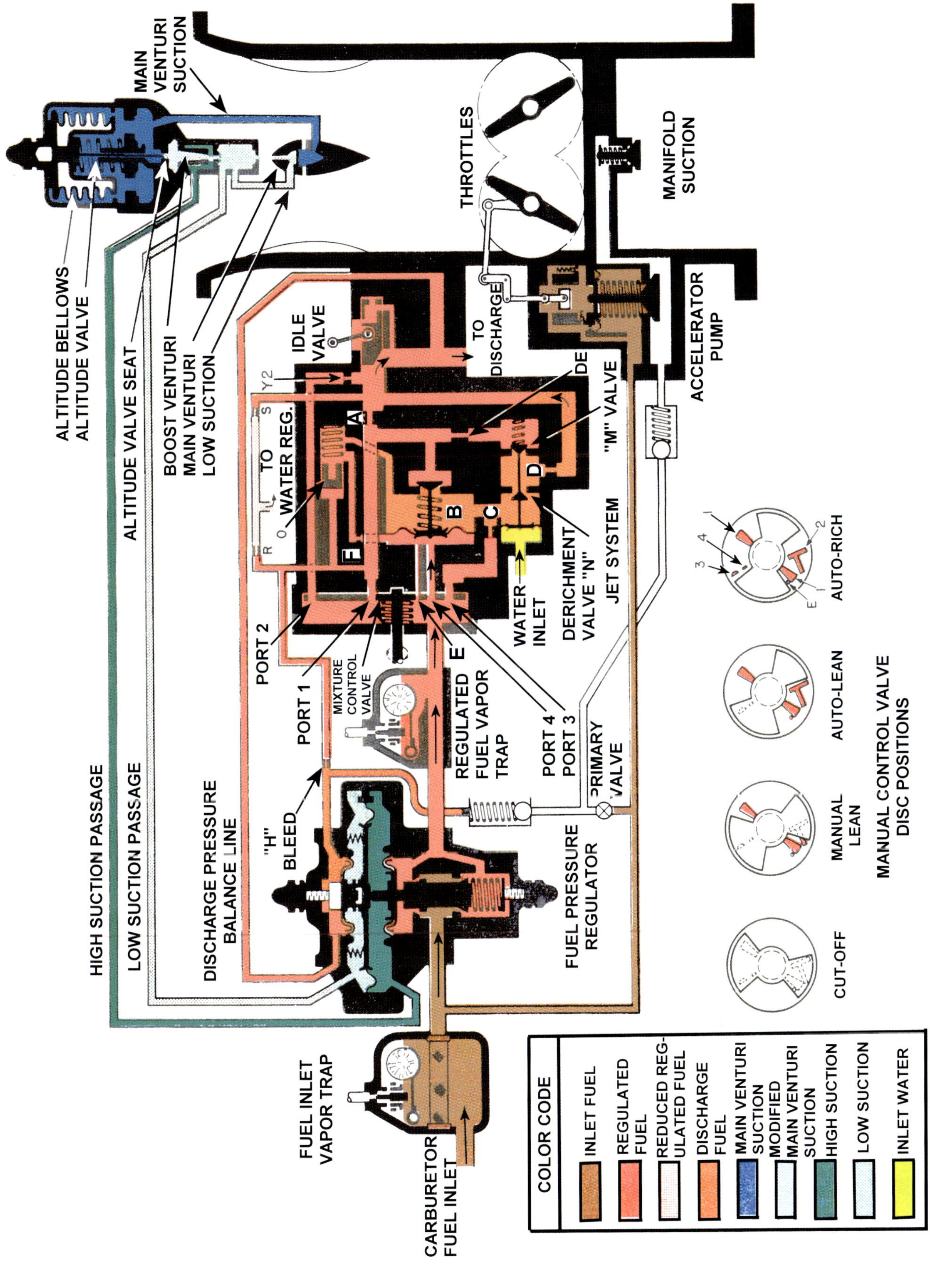

ALTITUDE BELLOWS
ALTITUDE VALVE
ALTITUDE VALVE SEAT
BOOST VENTURI
MAIN VENTURI
LOW SUCTION
MAIN VENTURI SUCTION
THROTTLES
MANIFOLD SUCTION
TO DISCHARGE
ACCELERATOR PUMP
IDLE VALVE
Y2
S
A
TO WATER REG.
R
O
F
B
C
D
DE
"M" VALVE
DERICHMENT VALVE "N"
JET SYSTEM
WATER INLET
HIGH SUCTION PASSAGE
LOW SUCTION PASSAGE
DISCHARGE PRESSURE BALANCE LINE
PORT 2
PORT 1
MIXTURE CONTROL VALVE
E
REGULATED FUEL VAPOR TRAP
PORT 4
PORT 3
PRIMARY VALVE
"H" BLEED
FUEL PRESSURE REGULATOR
FUEL INLET VAPOR TRAP
CARBURETOR FUEL INLET
AUTO-RICH
AUTO-LEAN
MANUAL LEAN
CUT-OFF
MANUAL CONTROL VALVE DISC POSITIONS
COLOR CODE
INLET FUEL
REGULATED FUEL
REDUCED REGULATED FUEL
DISCHARGE FUEL
MAIN VENTURI SUCTION
MODIFIED MAIN VENTURI SUCTION
HIGH SUCTION
LOW SUCTION
INLET WATER

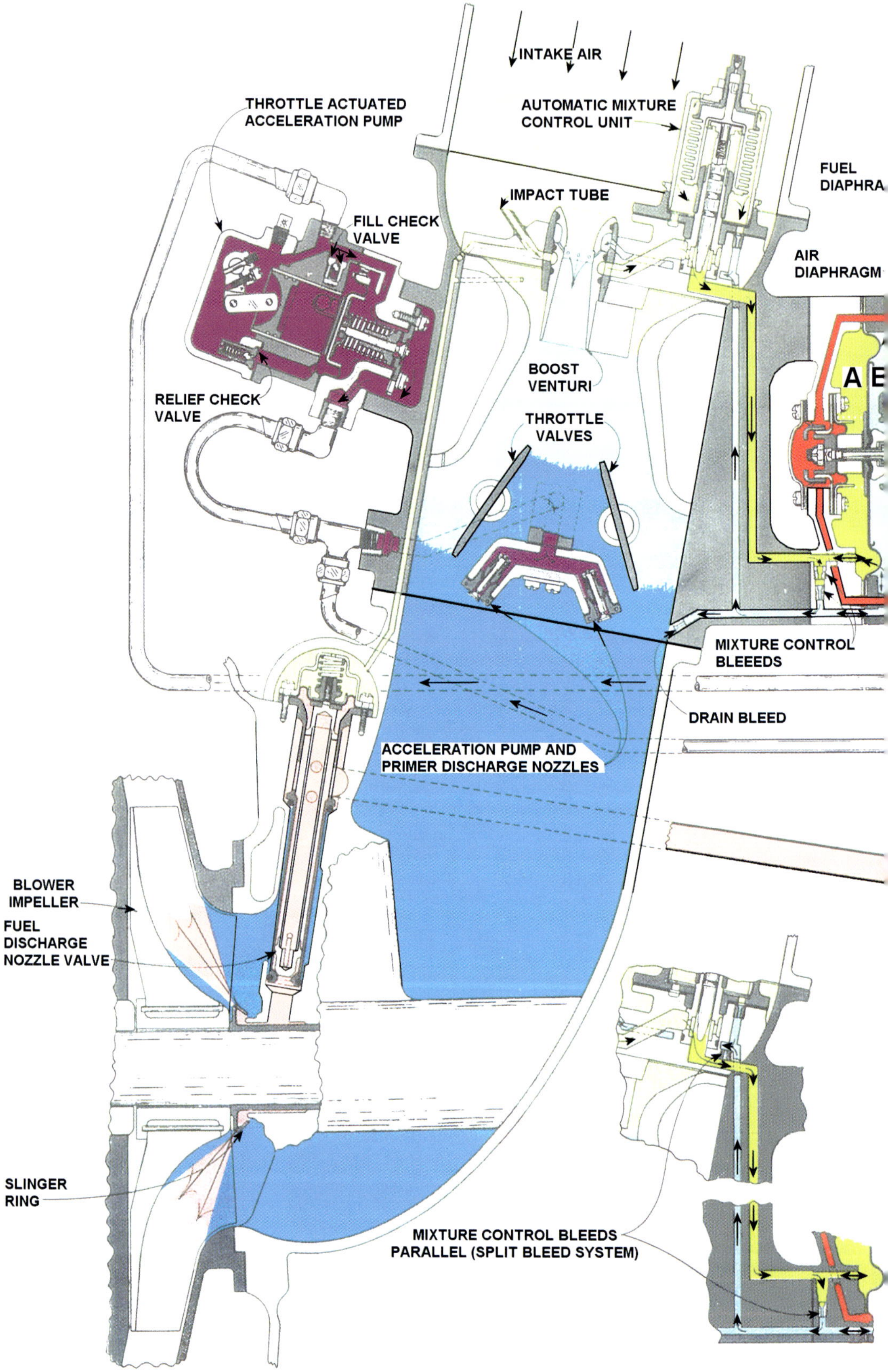
INTAKE AIR
THROTTLE ACTUATED
ACCELERATION PUMP
AUTOMATIC MIXTURE
CONTROL UNIT
FUEL
DIAPHRA
IMPACT TUBE
FILL CHECK
VALVE
AIR
DIAPHRAGM
BOOST
VENTURI
A E
RELIEF CHECK
VALVE
THROTTLE
VALVES
MIXTURE CONTROL
BLEEEDS
DRAIN BLEED
ACCELERATION PUMP AND
PRIMER DISCHARGE NOZZLES
BLOWER
IMPELLER
FUEL
DISCHARGE
NOZZLE VALVE
SLINGER
RING
MIXTURE CONTROL BLEEDS
PARALLEL (SPLIT BLEED SYSTEM)

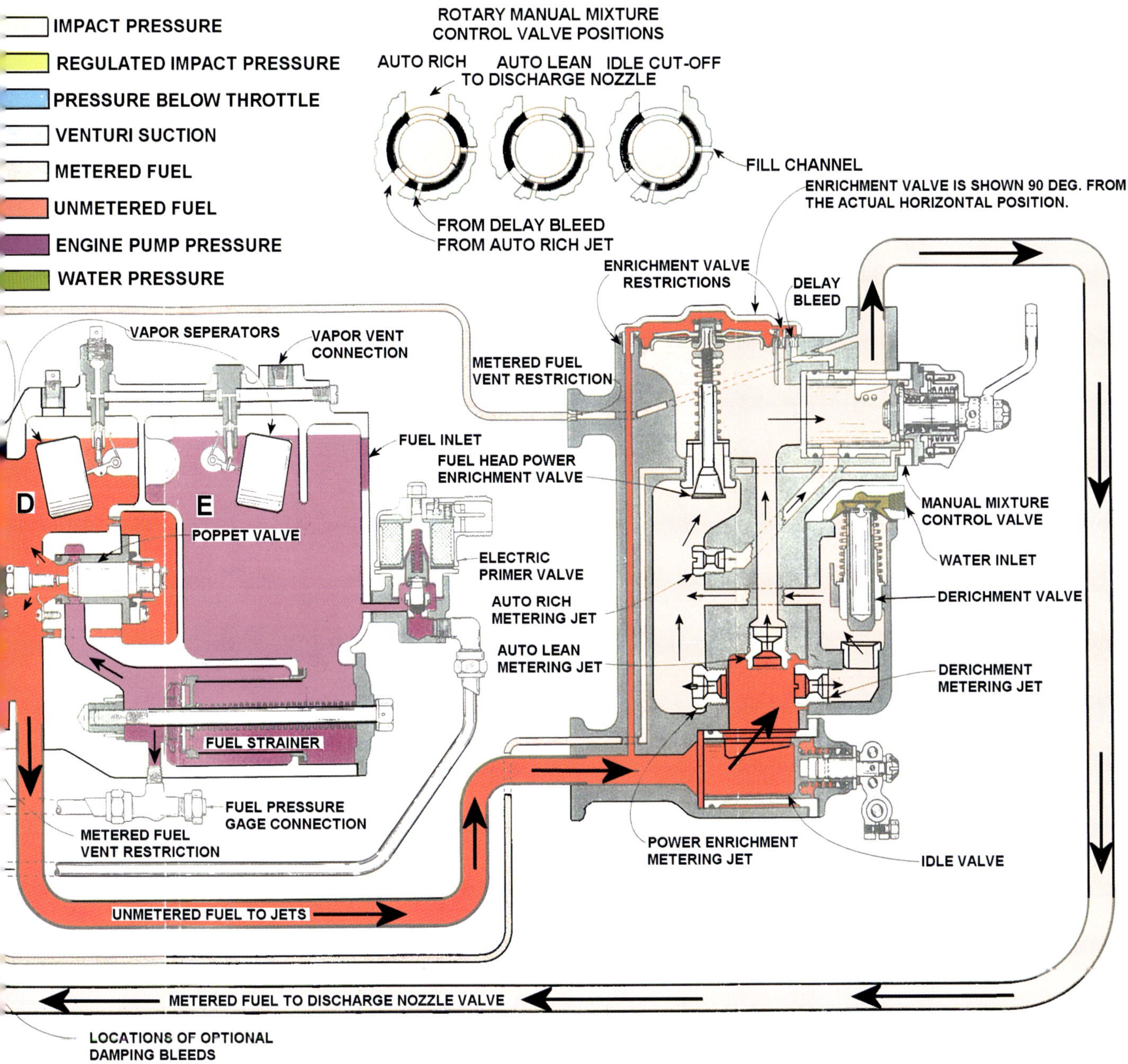

Diagram of the fuel and air circuits for the Bendix PR-100 carburetor. Refer to the text for a detailed explanation of how it all works. *(Handbook of Overhaul Instructions. Bendix Carburetor Model PR-100)*

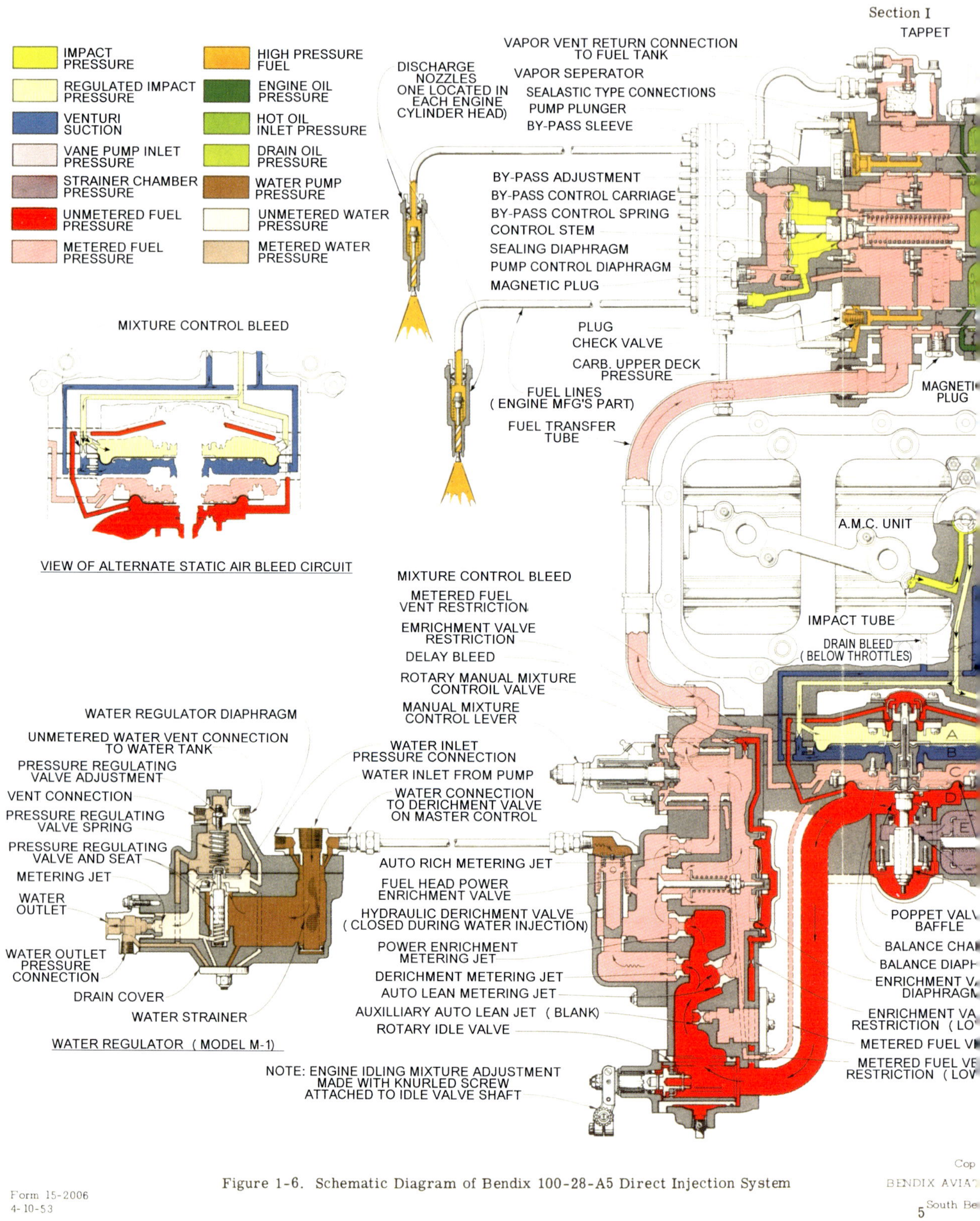

This color schematic shows how the entire 100-28 fuel injection system works. *(Preliminary Instructions on Stromberg 100-28-A3 Direct Fuel Injection System. Courtesy of Pete Law)*

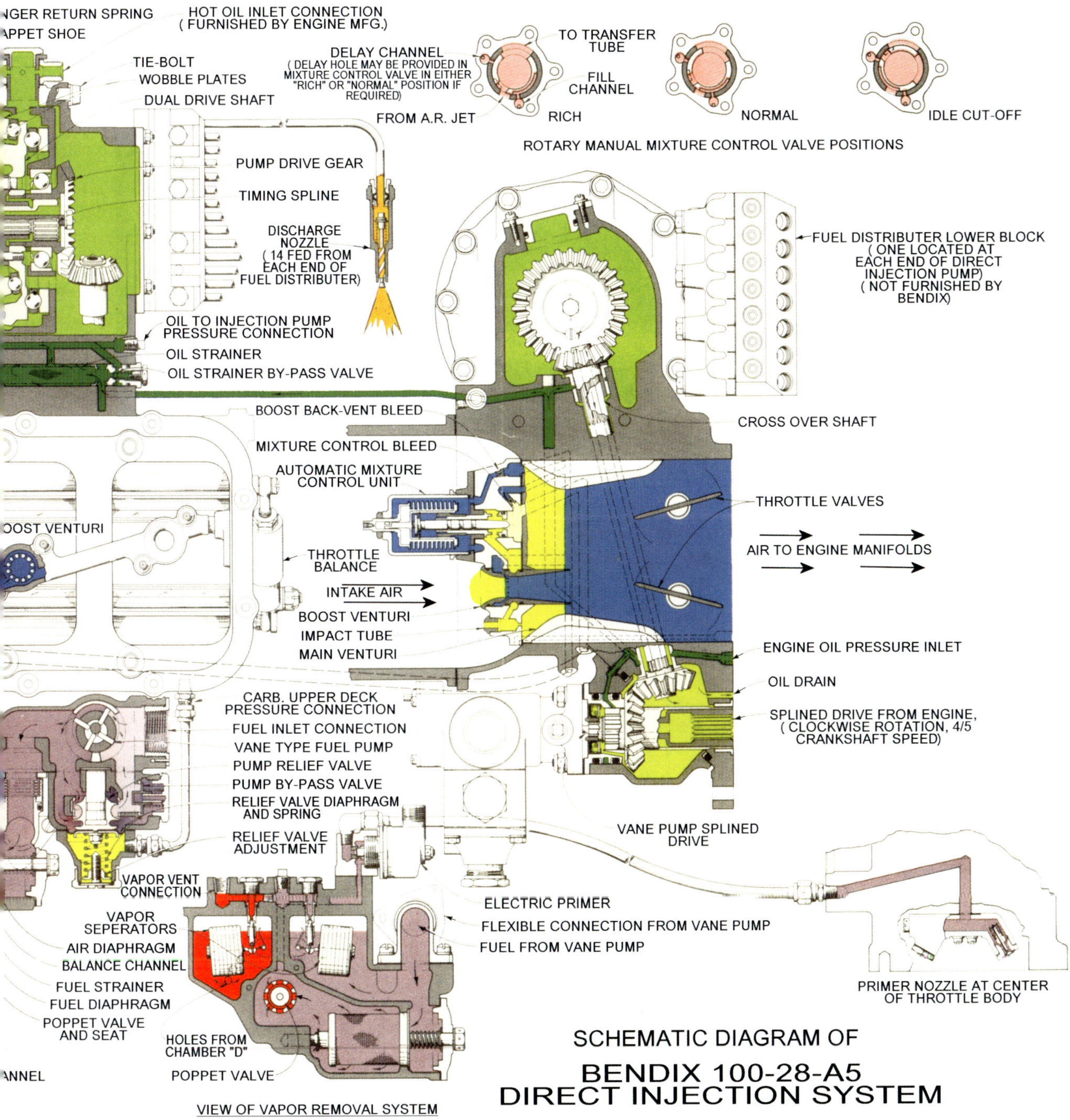

SCHEMATIC DIAGRAM OF
BENDIX 100-28-A5
DIRECT INJECTION SYSTEM

953
RPORATION
ndiana

Printed in U.S. A.

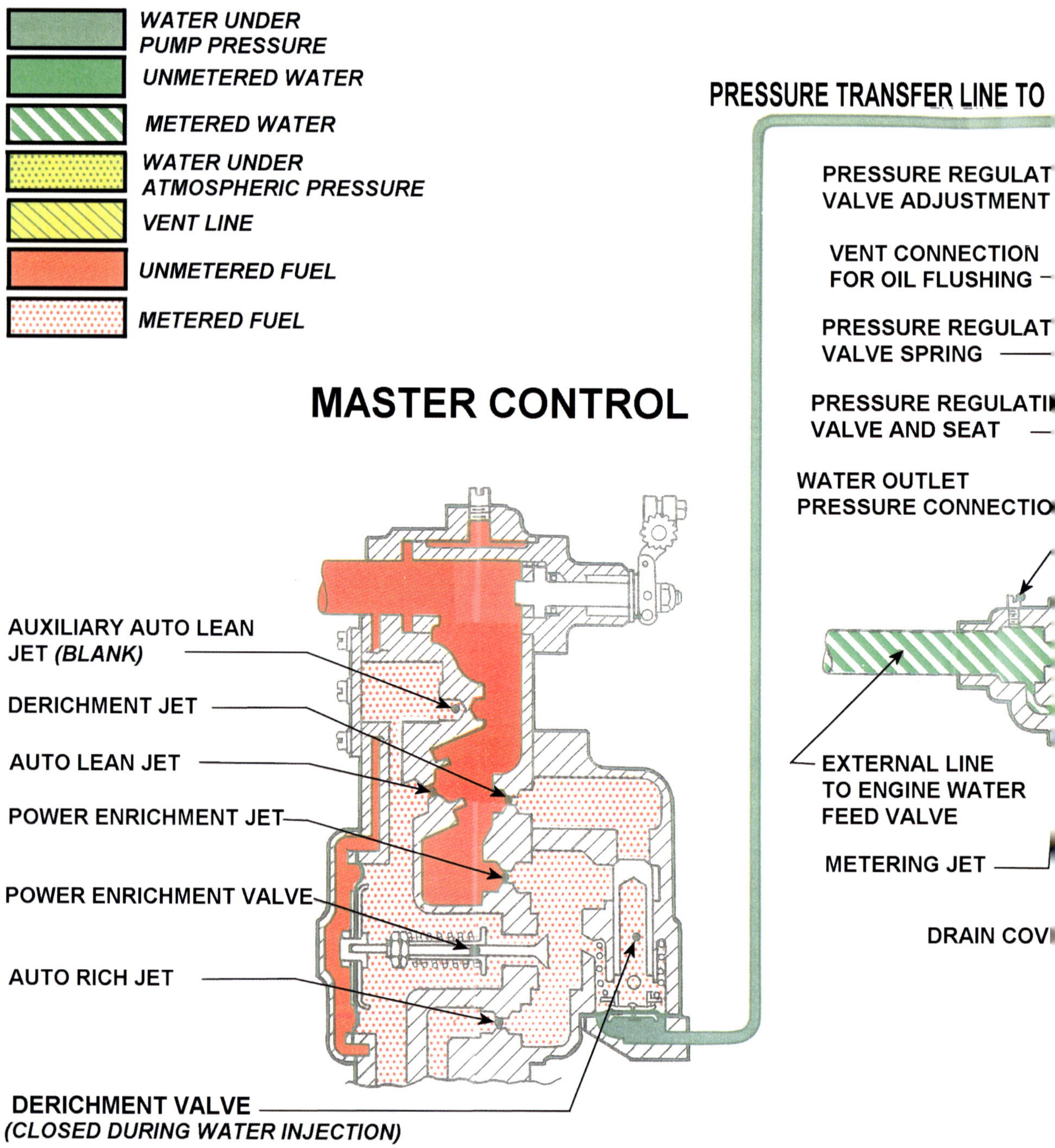

Schematic of the water regulator. Refer to the text for detailed description of how it operates. *(Water Injection Systems For Pratt & Whitney Aircraft Reciprocating Engines. Courtesy of Pete Law)*

ATER REGULATOR

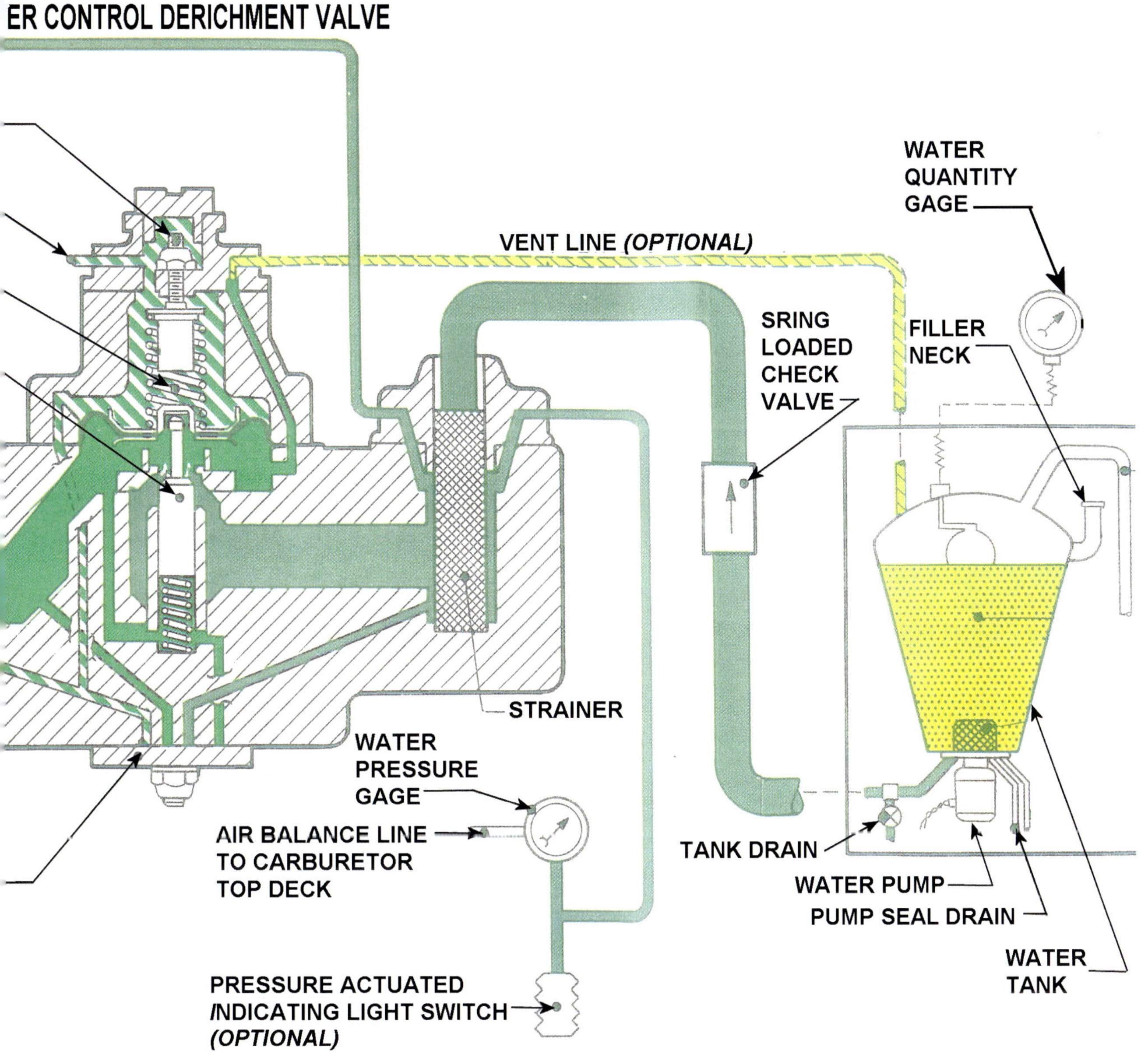

PROPOSED B-36 POWER PLANT WITH ADVANCED R-4360

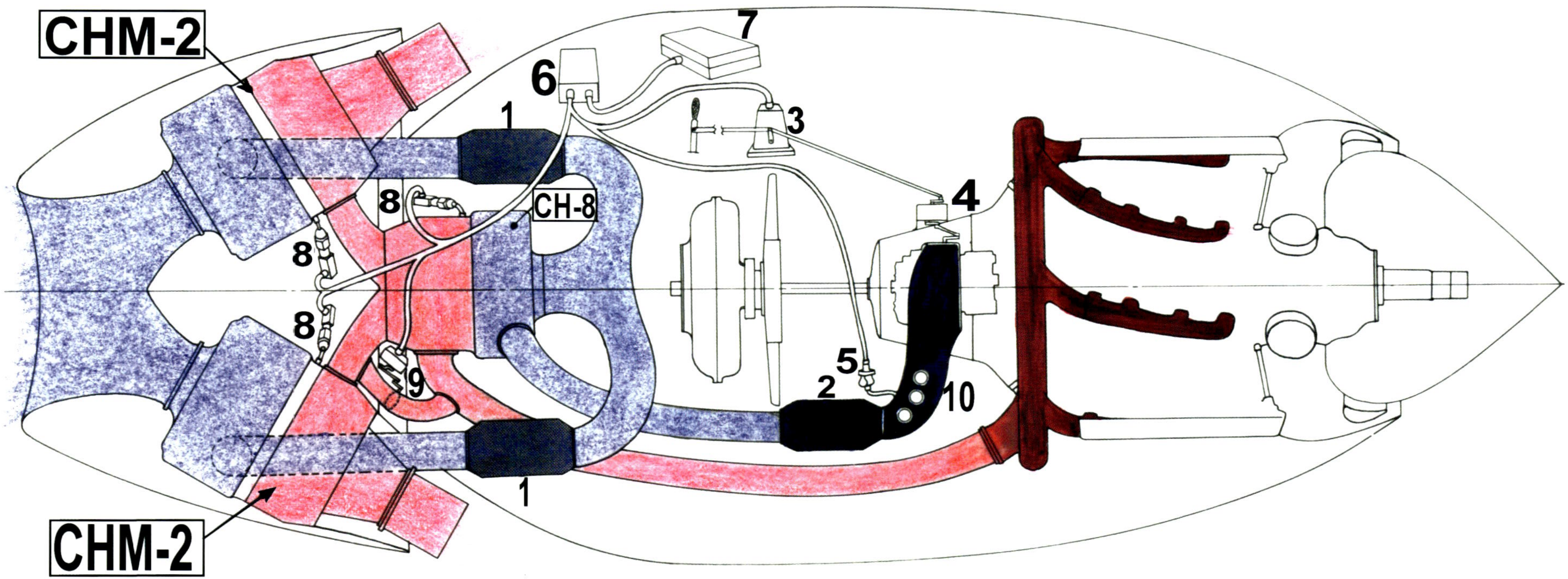

1. INTERCOOLER
2. AFTERCOOLER
3. BOOST SELECTOR
4. THROTTLE REGULATOR
5. DECK PRESSURE TRANSMITTER
6. NACELLE JUNCTION BOX
7. AMPLIFIER
8. TACHOMETER
9. TURBINE CONTROL ACTUATOR
10. OVERBOOST VALVES

Opposite: Even though the VDT-powered B-36C was stillborn, other proposals were made to enhance its performance. The scheme illustrated here would have employed a pair of two-stage GE CHM-2 turbosuperchargers running in parallel with a GE CH-8. It would have also featured a crossover system for exhaust energy. This represented just part of the innovative thinking that went into proposed B-36 powerplants. Even engineers today are astounded that such forward thinking and ideas were incorporated over 50 years ago. *(Courtesy of Pratt & Whitney)*

continued from page 96

the spring tension forcing the spark advance gear into minimum spark advance. This places the magneto timing in the correct position for starting, which is five degrees before top dead center. As oil pressure builds up after engine start, the pistons are allowed to travel out and advance the spark timing to 20 degrees before top dead center.

Moving the stationary gear under the influence of the four pistons changes the relationship of the magneto drive to the crankshaft, consequently changing the ignition timing. The magneto drive pinion cage, splined to the front of the crankshaft, supports six magneto drive pinions. The front and larger gear of the compound pinion engages with the spark advance gear and the smaller gear of the compound pinion drives the magneto intermediate drive gear. Driven off the crankshaft, the pinion cage rotates within the spark advance gear, which in turn drives the magneto intermediate drive gear supported on a journal formed on the front of the crankshaft. The bevel gear formed integral with the magneto inter-

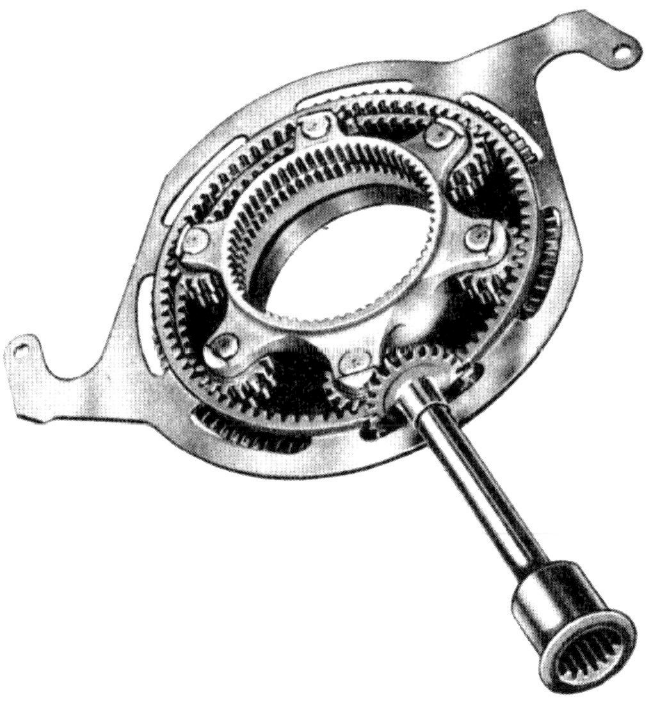

Above left: Magneto drive assembly shown assembled with one magneto driveshaft shown. A total of seven would be used in a high-tension ignition system. *(Wasp Major Models B6 and CB2 Overhaul Manual Issued June 1953) Above right:* This is how it looks when assembled in the magneto drive case. With all the components installed very little room was left for anything else; such was the complexity of the high-tension ignition system. *(Wasp Major Models B6 and CB2 Overhaul Manual Issued June 1953)*

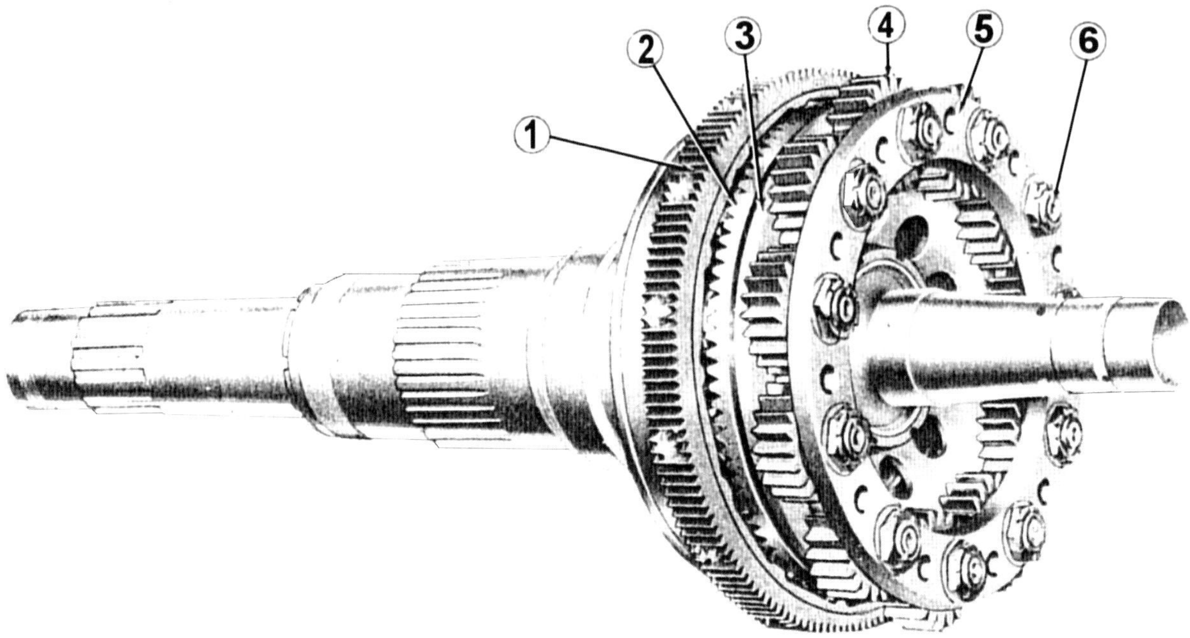

A number of prototype aircraft were fitted with dual-rotation propellers. Pratt & Whitney arrived at an innovative solution for the drive requirements. 1. Reverser pinion cage; 2. Reverse drive gear; 3. Reduction drive pinion shaft front support; 4. Reduction drive pinion; 5. Reduction drive pinion shaft rear support; 6. Reduction drive pinion shaft. *(Handbook Service Instructions Models R-4360-17, -21, - 45, and -47 Aircraft Engines August 1, 1948)*

mediate drive gear drives the magneto drive gears, and is attached to each magneto via a drive shaft.

A beautifully designed unit, the spark advance "reads" manifold pressure, which in turn sends a signal to the four pistons allowing a higher or lower oil pressure in the cylinders, thus changing ignition timing.

Dual/Contra-Rotating Nose Cases

A 3,000- to 4,000-hp engine sounds very attractive to the airframe manufacturers. That's why all the major aircraft companies designed at least one aircraft around the R-4360. However, that prodigious power comes with liabilities, mainly in the form of torque reaction. One solution to this problem was to use dual-rotation or contra-rotating propellers. This arrangement would eliminate any torque reaction from the engine. Like everything else in life, there is no such thing as a free lunch. With contra-rotating propellers, the costs are additional weight from the increased gearing and the weight of an additional propeller. Adding to the weight considerations are added complexity in the nose case and even more so in the propeller. A good example of the propeller issue was the ill-fated Hughes XF-11, which featured a Hamilton Standard "Super" Hydromatic propeller. Failure of one of the propellers almost cost Hughes his life. Apart from the XF-11, the R-4360's driving contra-rotating propellers powered a number of aircraft. These included the Northrop XB-35, Boeing XF8B-1, and Douglas XTB2D-1 Skypirate. None of these aircraft entered full-scale production.

Inner Workings

Looking at photographs of R-4360s with dual-rotation propeller shafts, it's hard to envision how it was done, as the nose case hardly looks any larger than a single rotation nose case. The secret was a beautifully designed and compact arrangement that fully utilized available space. Typically, the two propeller shafts would employ an SAE #60 spline (inner) and a larger SAE #80 spline (outer) shaft. The inner (SAE #60 spline) propeller shaft is driven in the normal way, through planetary reduction gearing. Drive to the outer (SAE #80 spline) shaft is derived from a large diameter "Reverse Drive Gear." This reverse drive gear is a bevel gear that drives 14 small fixed bevel pinion gears. This is where the reversing motion takes place. The 14 small bevel gears drive an outer propeller shaft drive gear *(Ref. 2-30, 2-31, 2-32, and 2-33)*. All U.S. built high-horsepower engines used SAE propeller splines. The numbering system can be confusing because the SAE spline number does not relate to how splines are actually incorporated. For instance, an SAE #60 spline shaft does not have 60 splines. Table 2-1 shows how shaft size relates to horsepower being transmitted. However, some "B" series R-2800s were capable of producing far more than 2,000 hp, yet used a 50-spline shaft. When the "C" series R-2800 was introduced, a 60-spline shaft was employed.

Table 2-1

SPLINE	HORSEPOWER CAPABILITY
#50	2,000 Horsepower
#60	3,500
#70	4,500
#80	OVER 5,000*

Although R-4360s were not capable of producing 5,000 hp, the #80 spline configuration was a necessity for the outer propeller shaft on dual-rotation applications.

It's possible that Pratt & Whitney engineers got the inspiration for the dual-rotation nose case from their earlier R-1830, which used a bevel planetary gear reduction design, albeit for single rotation. Even so, the design similarities are there. Henri Farmen was the original inventor of the bevel reduction gear, a design used extensively not only by Pratt & Whitney, but by the Bristol Aeroplane Company on their radial engine reduction gears as well.

Rear Section

The rear section can be considered everything behind the crankcase. This means the supercharger, accessory drive housing, intake ducting from the carburetor, and auxiliary drive housing.

Supercharger

With the exception of the VDT engines (more on that one later), all R-4360s used some form of gear-driven supercharger. Right from the initial design studies, flexibility was designed into the R-4360. Depending on the mission profile, R-4360s could have the following supercharger

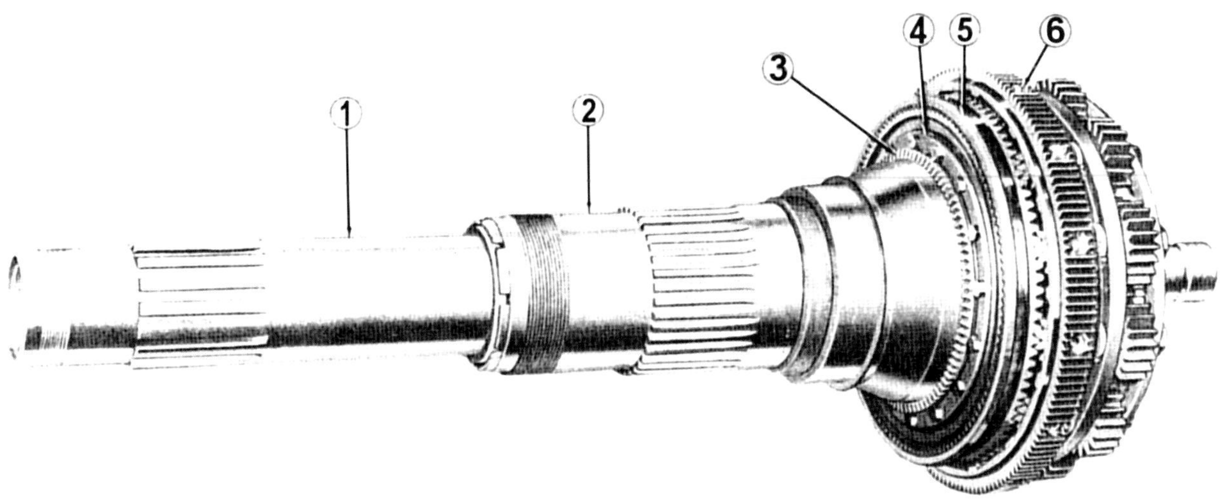

Another perspective on the dual-rotation gear drive. 1. Inner propeller shaft; 2. Outer propeller shaft; 3. Front accessory drive gear; 4. Outer propeller shaft coupling; 5. Outer propeller shaft drive gear; 6. Reverser pinion cage. *(Handbook Service Instructions Models R-4360-17, -21, - 45, and -47 Aircraft Engines August 1, 1948)*

Right: Supercharger casing made from a magnesium casting. This is where the intake manifolds join the supercharger. It also provides pads for the seven engine mounts. *(Technical Manual Illustrated Parts Breakdown USAF Models R-4360-20WA and -35B Aircraft Engines)* *Below:* Photograph of the supercharger spring drive gear. Purpose-built tooling was required to install the 12 powerful cushioning springs. *(Handbook Service Instructions Models R-4360-17, -21, - 45, and -47 Aircraft Engines August 1, 1948)*

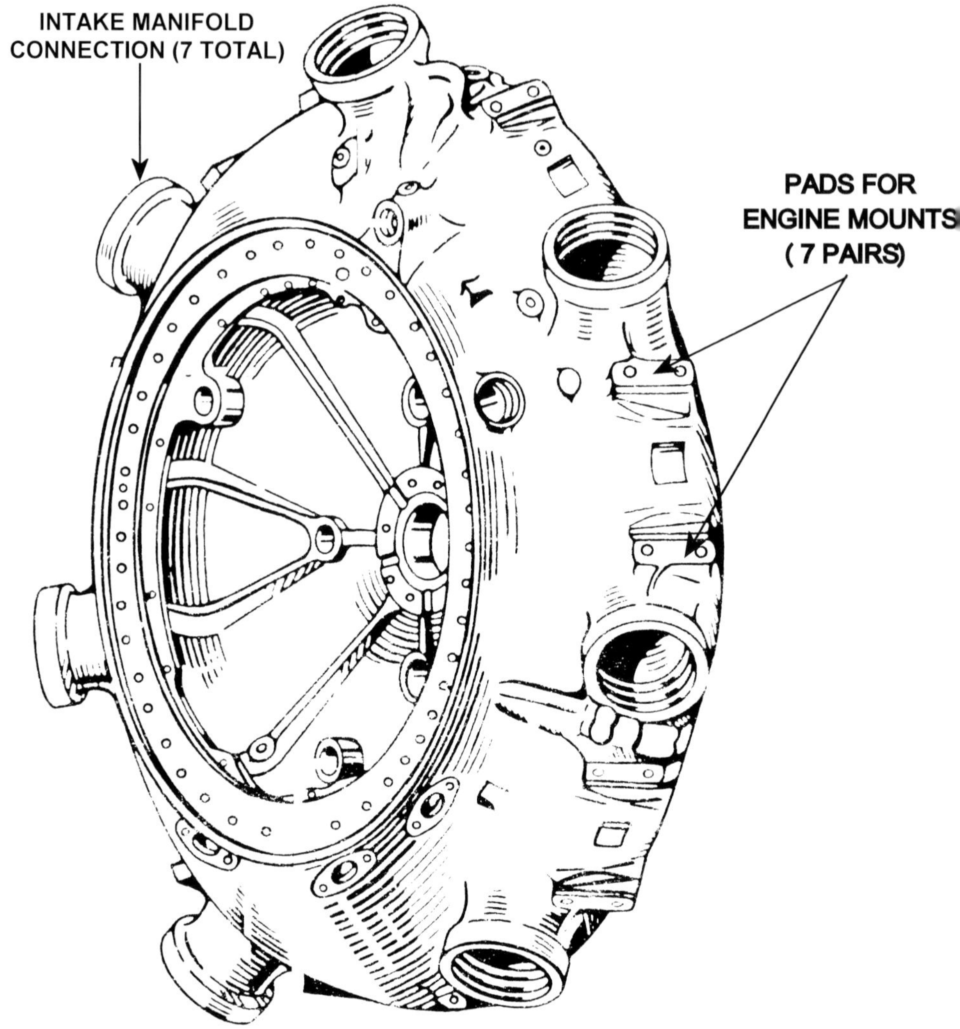

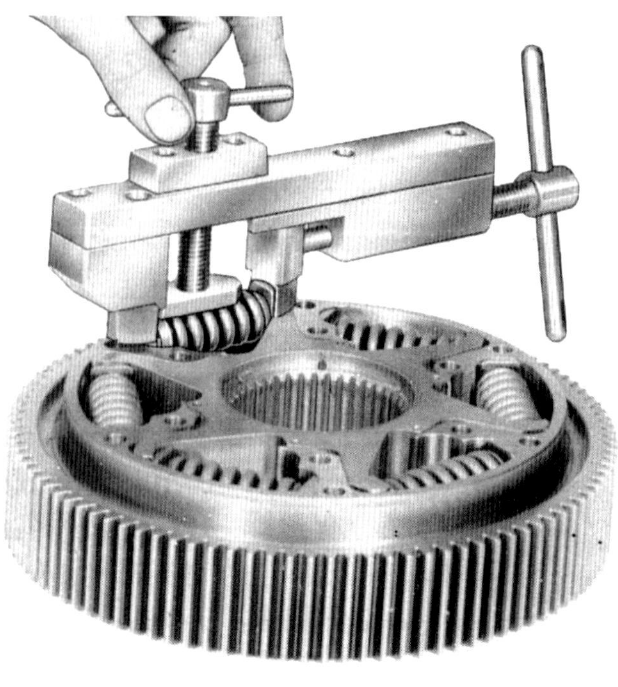

configurations: (i) relatively simple and basic single-stage, single-speed supercharger, (ii) single-stage, two speed, (iii) single-stage—variable speed, (iv) single-stage with an auxiliary intercooled gear driven supercharger, and (v) single-stage augmented by turbosupercharging. The foregoing does not include the VDT, which utilized turbosupercharging for all its boost requirements.

The supercharger casing is made from a complex and intricate magnesium casting bolted to the crankcase via a ring of wasted studs. Residing inside the supercharger casing are the step-up gears for the supercharger. As previously stated, this drivetrain could take on a number of flavors. With the simple single stage, single speed blower, a two-stage step-up is used to obtain the appropriate ratio. All the engines augmented with turbosupercharging used this setup. Drive for all blower variations emanates from the rear of the crankshaft through a device called a "spring" drive made up from two assemblies. Mounted on the rear of the crankshaft, its drive features six coil springs in compression mounted inside the assembly. A spider, driven off the crank, makes up the central portion of this assembly. Power is transmitted to the gear via the six compression springs and spider to the gear, which makes up the outer part of the assembly. The entire assembly is sealed on both sides via plates bolted to the gear. High-pressure oil introduced into the spring drive offers additional hydraulic damping.

As can be gathered from this brief description, driving high-speed superchargers is a major design challenge. Part of the challenge is protecting gears from torsional vibration originating from the crank. The single-stage, two-speed engines used a similar design, except cone clutches were incorporated. One clutch would be used for low blower and, due to the increased power requirements to drive it, two clutches are used for high blower. Single stage, variable speed blowers incorporated some quite innovative engineering. A pair of hydraulic couplings, a.k.a. fluid flywheels or couplings, offer a low ratio and another pair of hydraulic couplings offer a high ratio. Although fluid couplings offer tremendous flexibility with respect to input speed compared to output speed, there is a penalty to pay. When the input speed varies by more than 10 percent of the output speed, efficiency goes down the tubes. This inefficiency or slip is transferred into heat. This places a greater burden on the entire lubrication system, particularly with regards to cooling. However, if slippage rates are contained to less than 10 percent of input speed compared to output speed, efficiency is acceptable.

With the use of two fluid couplings there is a seamless transition between the slowest speed in low ratio and the highest speed in the high ratio. This is due to the operating characteristics of this type of drive. The couplings used for the R-4360 blower drive were typical of their genre. The drives were divided into two basic halves, one being the driver, or impeller, and the other

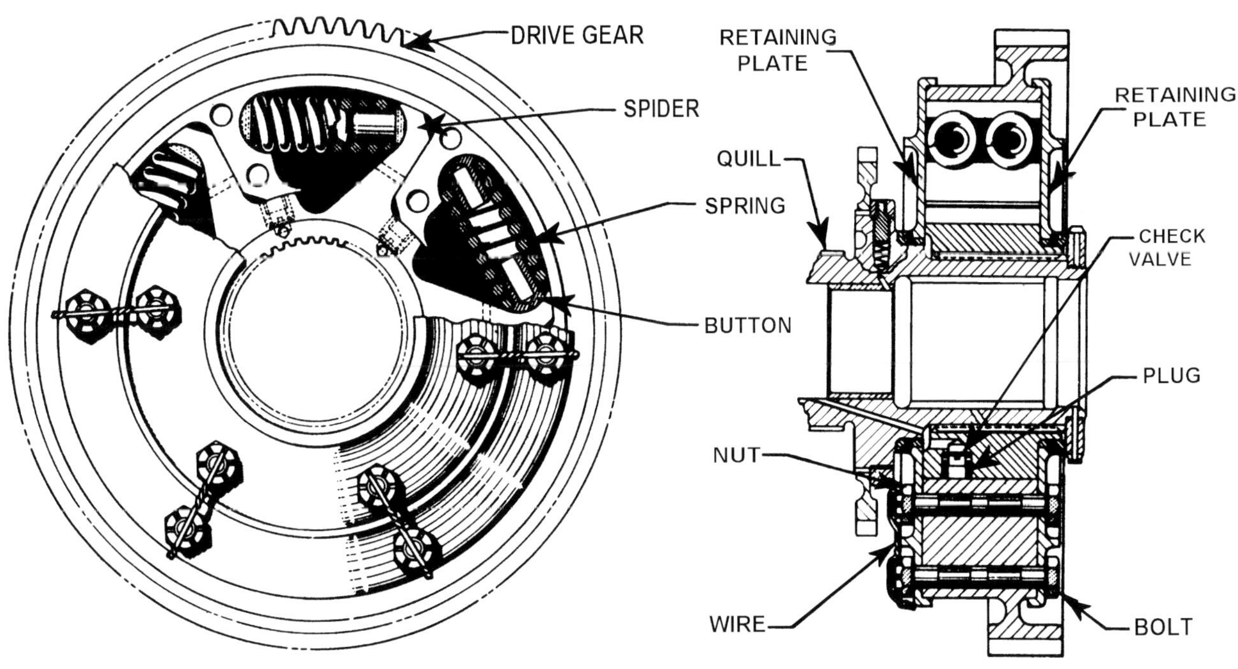

One of the key engineering challenges in any supercharged engine is designing a cushioned drive that can transmit hundreds of horsepower to the supercharger impeller. For the R-4360 a combination of springs and hydraulics were used. *(Handbook Service Instructions Models R-4360-17, -21, -45, and -47 Aircraft Engines August 1, 1948)*

being the driver or runner. Drive for the impeller comes from the spring drive.

A sleeve-type valve with a leaded silver bearing surface on its outer diameter and front face is installed on the shaft of each (two) low ratio coupling pinion between the coupling impeller and runner. A spacer, instead of a valve, occupies a similar position in each high-ratio coupling. Bronze friction ring segments, which contact the bore of the runner, are installed in the ring groove located in the outer diameter of each valve. Pins in the bottom of the groove prevent the bronze ring segments from turning with respect to the valve. Pins in the hub of the coupling impeller engage slots in the rear face of the valve, thereby limiting movement of the valve with respect to the coupling impeller.

While the speed of the coupling impeller is greater than the speed of the runner, the valve turns with the coupling impeller and the valve's oil holes are aligned with the oil holes in the pinion's shaft. Under the conditions just described, oil flows from the shafts (two per ratio) into the pair of couplings. The low ratio coupling impellers, hydraulically engaged by the low ratio coupling impellers, drive the supercharger impeller drive gear at a fixed ratio in a similar fashion to the low ratio of a two-speed engine.

When high-pressure oil is directed to the high-ratio couplings, the high-ratio runners, hydraulically engaged by the high-ratio coupling impellers, drive the supercharger impeller drive gear at variable ratios up to a fixed maximum ratio. Oil metered into the high-ratio coupling determines the drive ratio. Under these conditions and when the low-ratio runners, which are meshing with the supercharger impeller drive gear, rotate faster than the low-coupling impellers, the friction between each low-ratio coupling valve and runner causes the valve to turn on the shaft of the pinion in the direction of coupling rotation. When the valve turns, its oil holes move out of alignment with the oil holes in the pinion shaft, thereby cutting off the flow of high-pressure oil into the low-ratio coupling and thus preventing low-ratio coupling operation from interfering with high-ratio coupling operation. In other words, it ensures two blower speeds are not engaged at the same time.

High-pressure oil is cut off from the high-ratio couplings to such an extent that the speed of the low-ratio coupling runners becomes less than the speed of the low-ratio coupling impellers. This results in friction between each low-ratio coupling valve and coupling runner, causing the valve to turn on the shaft of the pinion in a direction opposite to that of coupling rotation. When the valves turn, their oil holes line up with the oil holes in the shafts of the pinions and oil fills the low-ratio coupling, thus providing drive to the supercharger drive gear.

Supercharger Impeller Assembly

Although typical of its ilk, the centrifugal supercharger employed by the R-4360 benefited from years of prior experience using this method of boosting manifold pressure. Although ostensibly a centrifugal supercharger, the impeller employed a significant axial flow component. A noteworthy change from other Pratt & Whitney impellers was the use of a shroud for the inducer *(Ref. 2-34).* Although early impellers were cast in one piece, Pratt & Whitney designed an impeller with a pressed-on shroud. This introduced a number of difficulties. Due to the high rotational speed of the impeller, up to 27,000 rpm for later engines, hoop stresses expanded the shroud and thus it lost its interference fit. An intensive development program followed where a number of alternatives were tried including various plastics, fiberglass, steel, and combinations of these and other materials. A shroud made from 4340 chrome moly steel shrunk onto the impeller was finally determined to be the best combination. Although only slightly, an interference fit still existed at an overload speed of 30,000 rpm.

The impeller assembly is made up of a shaft assembly, impeller, and drive gear. As described earlier, all four hydraulic coupling gears are in con-

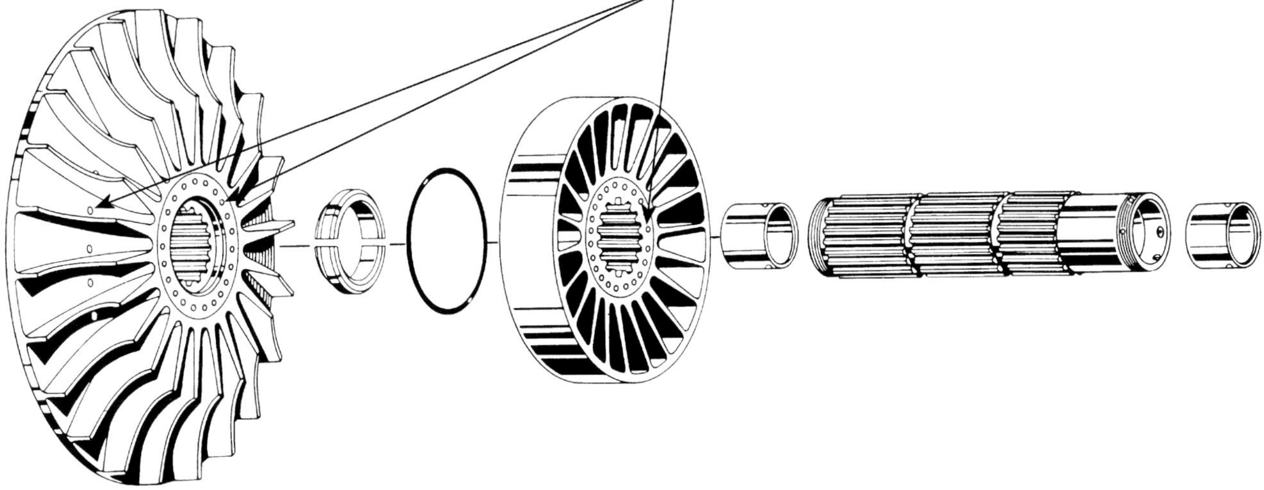

Supercharger shaft and impeller assembly. Due to its shape and the fact that they were forgings, it was necessary to manufacture the impeller in two parts. *(Technical Manual Illustrated Parts Breakdown USAF Models R-4360-20WA and -35B Aircraft Engines)*

stant mesh with the supercharger impeller gear. Of course, one pair of hydraulic coupling gears only provides drive to the supercharger impeller at any one time. Support for the supercharger impeller shaft comes from steel backed bronze bearings.

In 1942 Pratt & Whitney patented its "slinger ring" concept of fuel distribution *(Ref. 2-35)*. Wright Aeronautical copied this unique method of getting fuel accurately distributed for its radials such as the later R-1820s and R-3350s. Fuel enters the supercharger via the fuel feed valve, essentially a low-pressure pintle-type valve. Fuel is injected into an annular groove machined into the impeller shaft. To avoid fuel leaking into the blower section, sealing rings are employed. Therefore, fuel injected into the annular groove can only go one place—into the impeller throat. A ring, the so-called slinger ring, picks up fuel from the annular groove. With multiple small holes drilled into its periphery, fuel is centrifuged into the eye of the impeller where it gets slung out under the powerful influence of centrifugal force. Galleries drilled between each vane of the supercharger impeller further assist fuel distribution. Thus some fuel enters these galleries to be slung out through exit holes positioned at about half the diameter of the impeller.

Centrifugal impellers simply impart immense kinetic energy to the fuel/air mixture. The trick in good supercharger design is how to convert this kinetic energy into pressure energy, which is manifold pressure. This task is accomplished by the diffuser and to a lesser extent, the supercharger housing. Surrounding the impeller, the diffuser is made up from a number of vanes with an airfoil section. In this way, fuel/air flung out by the impeller is picked up by the diffuser's vanes and directed into the annulus of the blower rim. Seven circular discharges then feed fuel/air mixture into the seven intake manifolds.

Accessory Drive Case

As previously stated, the R-4360 was designed to be as flexible as possible. This means that with relatively minor modifications, it could power everything from a single-engined fighter to multi-engine strategic bombers. Part of this flexibility was the use of auxiliary superchargers and pusher installations with fan-assisted cooling. Typical radial engine design locates accessories such as starters and generators on the rear, mounted in the same longitudinal axis as the engine. Not only

Fuel atomization is key for good combustion. Pratt & Whitney patented their "fuel slinger ring" for this purpose. A ring with peripheral holes slung fuel out as the impeller rotated. Additional holes lined up with galleries drilled in the impeller. *(United States Patent 2,287,021)*

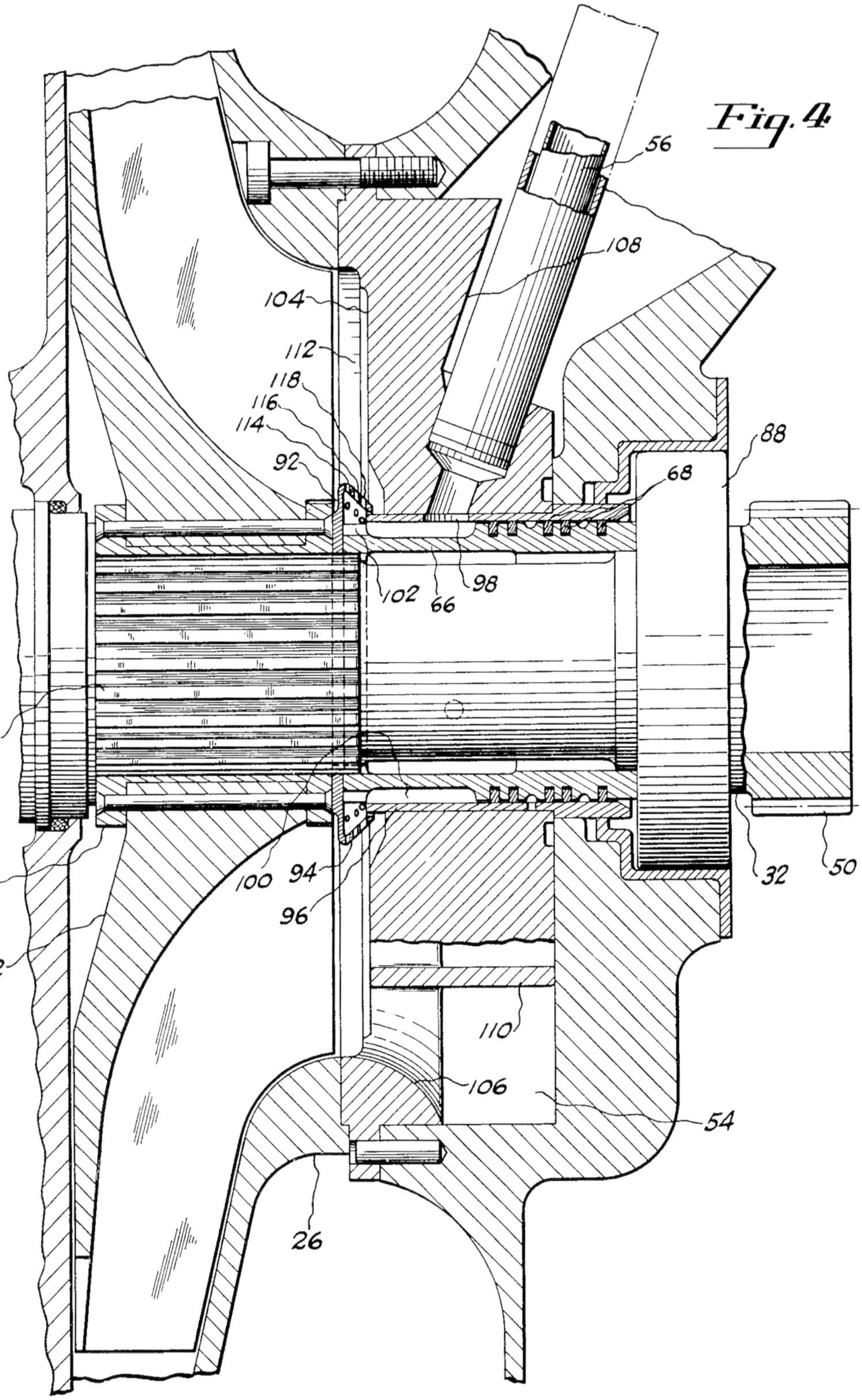

did this increase the length of the engine, it also negated the possibility of easily mounting a cooling fan or a bolt-on auxiliary supercharger on the rear face of the engine.

Again, ingenuity prevailed when Pratt & Whitney engineers thought outside the box. A large ring gear, integral with the accessory drive shaft driven off the rear of the crankshaft, was essentially a one-stop shop for all accessory drive requirements. Somewhat reminiscent of the ring gear in the rear axle of a truck, numerous gears engaged this central ring gear. Disposed radially

This cross section shows the fuel feed galleries drilled in the supercharger impeller. These galleries lined up with the rotating inlet guide vane galleries. *(Drawn by author based on overhaul manual illustration)*

around the periphery of the rear accessory case, pads were supplied for essentials such as the starter, fuel pump, hydraulic pump(s), tachometer, generator, etc. In some cases, the speed output would not have been correct if the drive gear engaging the accessory drive gear drove only the requisite accessory. To overcome this, intermediate jackshafts were interspersed so the correct output speed was obtained via appropriate intermediate gearing. Another key requirement of the accessory drive case was that of providing a mounting surface for the truly massive carburetor. Following prior Pratt & Whitney practice, the carburetor is mounted on a rectangular pad set at an angle. This pad opens up into the inlet throat of the supercharger. The fuel feed valve is installed behind the carburetor and intersects with the annular groove for the slinger ring, which is integral with the supercharger impeller.

Power Control Unit

From the dawn of aviation, the Eclipse Corporation supplied many aircraft accessories such as starters and generators. With this wealth of aviation experience, they delved into the design and manufacture of automatic controls. One aspect of many U.S.-built aircraft engines that created some problems was the lack of an automatic boost control. In other words, a pilot could, and often did, over boost an engine. Situations such as high

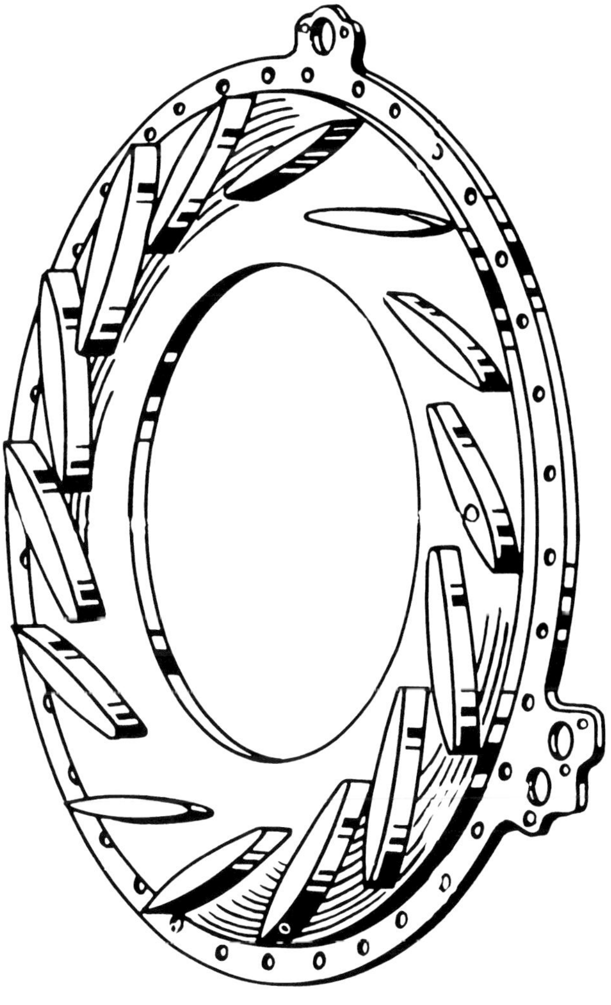

A supercharger impeller simply imparts kinetic energy to the fuel/air mixture. That kinetic energy needs to be converted into potential, or pressure, energy to create manifold pressure. The diffuser performs this critical function. *(Technical Manual Illustrated Parts Breakdown USAF Models R-4360-20WA and -35B Aircraft Engines)*

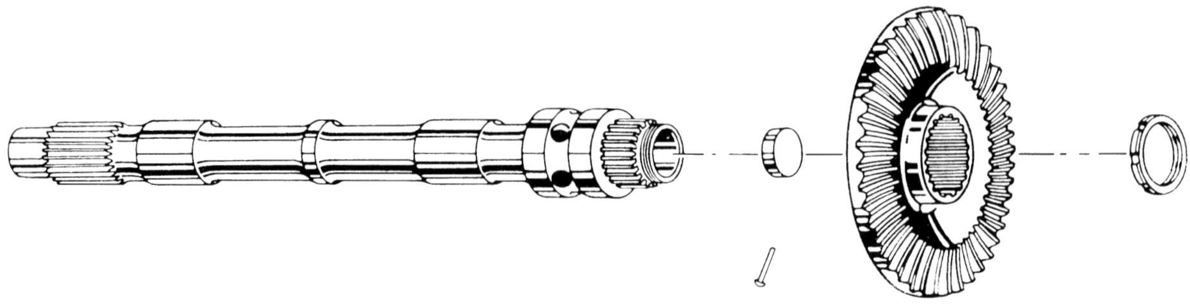

Accessories such as vacuum pumps, tachometer generators, etc., are driven in an innovative manner. A single ring gear drove all accessories around the perimeter of the rear case. This served at least two functions: (i) it freed up the rear of the engine for devices such as cooling fans or auxiliary superchargers, and (ii) it reduced the installed length of the engine. This is how aircraft such as the P-47 and Corsair, both designed around the R-2800, could easily accommodate the considerably larger R-4360. *(Technical Manual Illustrated Parts Breakdown USAF Models R-4360-20WA and -35B Aircraft Engines)*

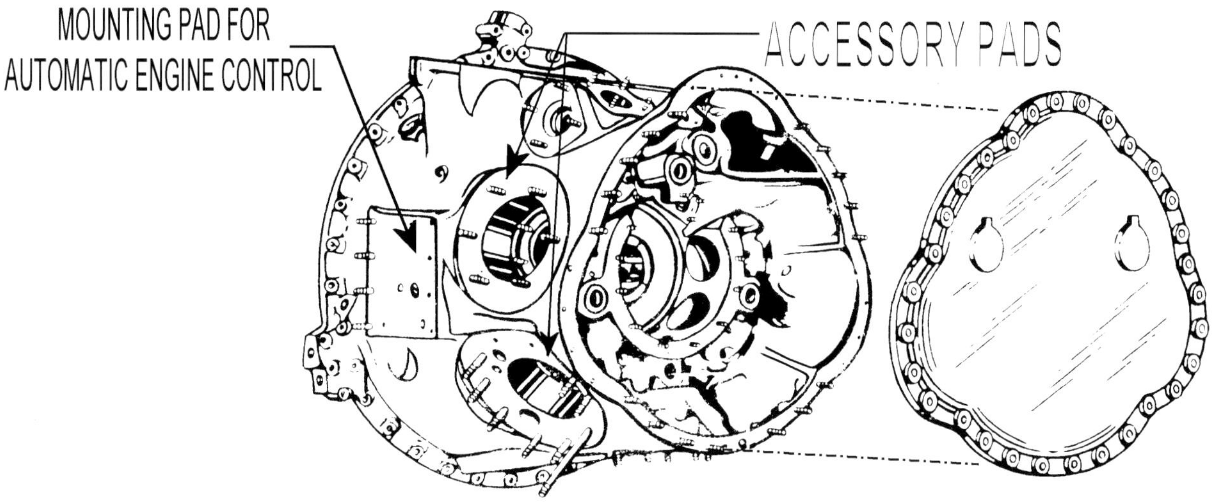

Manufactured from a magnesium casting, the rear accessory case and its cover are shown in this line drawing. *(Technical Manual Illustrated Parts Breakdown USAF Models R-4360-20WA and -35B Aircraft Engines)*

gross-weight takeoffs and combat situations often saw manifold pressures soar to astronomical values. If the engine did not suffer severe damage, it often needed intensive maintenance after an incursion into forbidden manifold pressure territory. However, airlines in particular did not embrace the concept of automation. Their argument was that pilots are paid to fly, and also that the automatic systems often proved to be another maintenance headache. The military did not always go along with this philosophy; therefore, Eclipse developed an automatic boost control for the R-4360. The Eclipse power control unit is attached to a pad on the left side of the accessory drive case. It performs the task of automatic boost control, in other words, even if the throttle was pushed all the way forward, particularly at low altitude, there would be no danger of over-boosting the engine. With this control unit, one requires just one cockpit control, which operates and coordinates the carburetor throttle and supercharger fluid drive coupling selector valves.

As an automatic power manifold pressure regulator, it maintains a constant manifold pressure regardless of change in altitude. This is accomplished via an aneroid bellows setup, which

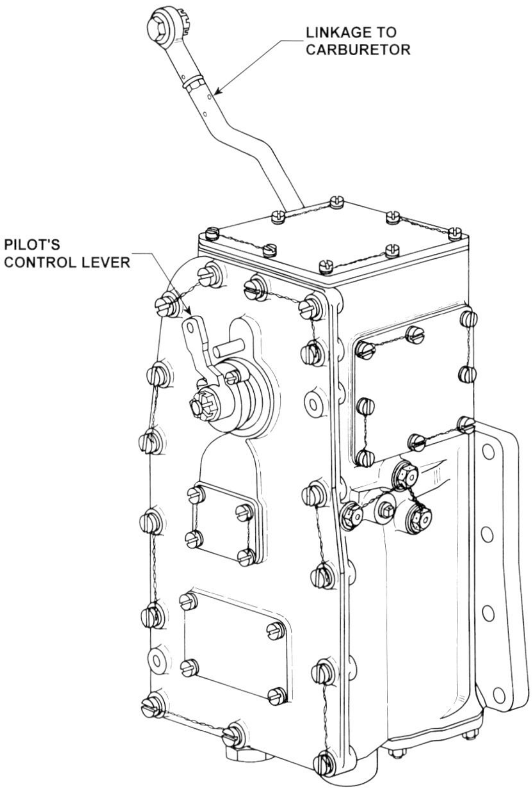

The United States was somewhat late in incorporating automatic boost controls. This is an Eclipse unit, which attaches to the side of the accessory case. When an automatic boost control was fitted, the pilot's throttle lever was not attached to the carburetor. Instead, it was linked to the boost control, which then determined what throttle setting to apply to the engine–a forerunner to FADEC (full authority digital engine control) now universally used in gas turbines and some high end piston engines. *(Wasp Major Models B6 and CB2 Overhaul Manual Issued June 1953)*

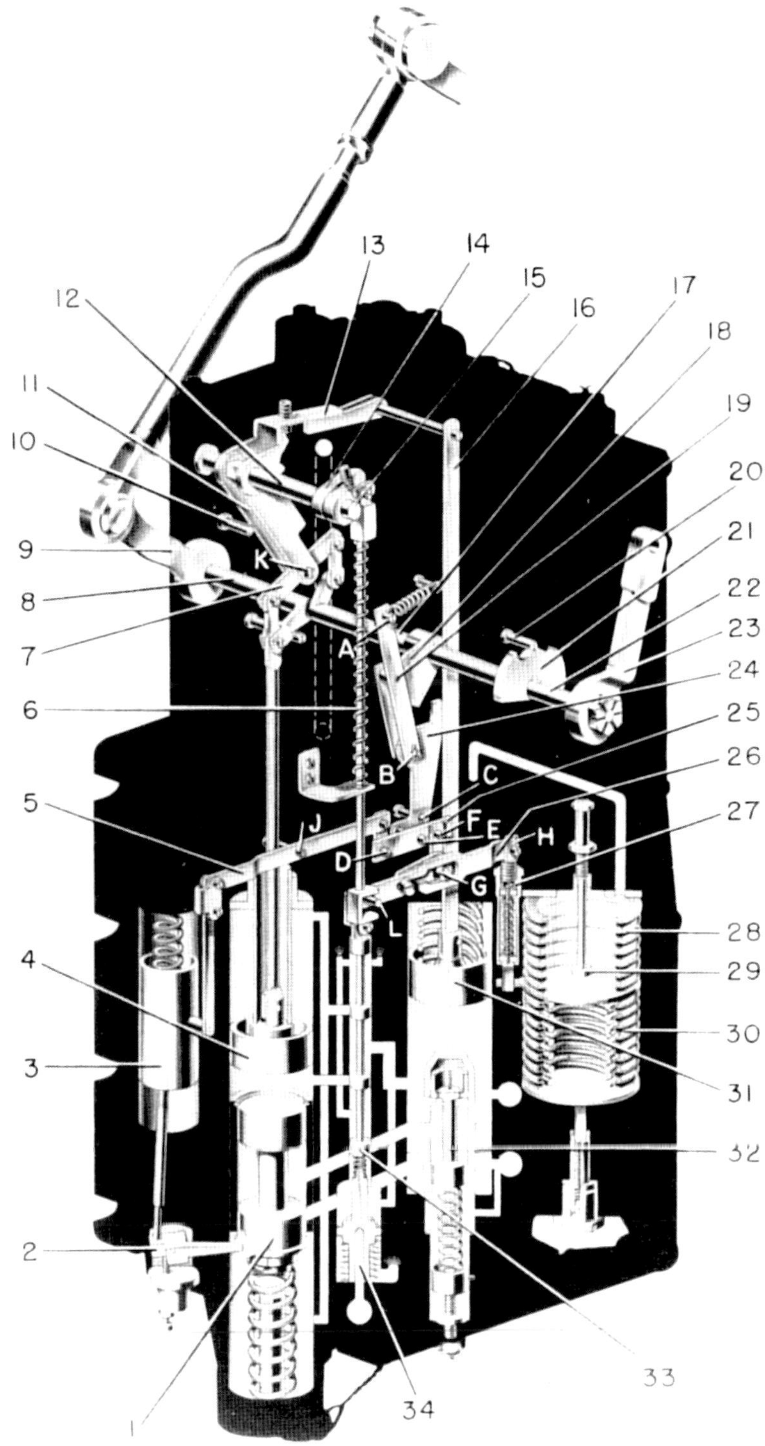

1. Flow regulating valve plunger; 2. Follow-up arm; 3. Follow-up piston; 4. Operating piston; 5. Follow-up lever; 6. Push rod; 7. Walking beam; 8. Throttle shaft; 9. Throttle lever; 10. Stop pin; 11. Throttle linkage arm; 12. Manual control shaft; 13. Manual control arm; 14. Manual control shaft driving arm; 15. Push rod end; 16. Manual control piston rod; 17. Rear pilot's shaft; 18. Selector cam; 19. Cam follower arm assembly; 20. Stop pin; 21. Pilot's control shaft stop; 22. Front pilot's shaft; 23. Pilot's control lever; 24. Cam follower plate; 25. Reset walking beam; 26. Bellows walking beam; 27. Collapsible link assembly; 28. Pressure bellows; 29. Bellows stop; 30. Evacuated bellows; 31. Manual control piston; 32. Pressure reducing valve; 33. Pilot valve; 34. Drain valve. *(Wasp Major Models B6 and CB2 Overhaul Manual Issued June 1953)*

regulates the throttle in conjunction with the supercharger impeller speed. Sufficient engine oil is metered into the low-ratio supercharger fluid drive couplings to maintain maximum low-ratio impeller speed during operations over the part throttle power range. With increased supercharger demand, such as increased manifold pressure requirements beyond full throttle, the pair of high-ratio fluid drive couplings replace the low-ratio drive fluid drive couplings in driving the supercharger impeller. Supercharging requirements in high ratio are controlled by metering the appropriate amount of pressure oil to the high-ratio fluid drive couplings. This arrangement permits, in high ratio, the ideal conditions for operating at full throttle and at the same time ensuring the supercharger impeller is driven at the appropriate speed without overboosting. Additionally, the control ensures the supercharger impeller is driven at the lowest speed that maintains the desired manifold pressure *(Ref. 2-36)*. This latter point is key when it is understood that the drive requirements of an R-4360 supercharger can run into the hundreds of horsepower.

Auxiliary Cooling Fan

Several R-4360 applications demanded an auxiliary cooling fan. These applications included the Convair B-36 and Northrop's B-35. Both aircraft were pusher installations with the engines buried in the wing. This type of installation demanded a forced airflow over the engine to augment cooling, particularly on the ground. To drive the fan, Pratt & Whitney used a design very similar to that used on the variable speed supercharger. The fan, mounted on the rear accessory case, has two ratios, high and low, with fluid couplings being employed. Due to the increased drive requirements of the high ratio, two fluid couplings were engaged to drive the fan in high ratio and one fluid coupling for low fan speed. All couplings were driven off the accessory shaft. Step-up gears engaged the couplings from the accessory drive shaft. Couplings were fed with engine oil via hollow shafts and conduits. The drivetrain employed the use of sun and planet gears to drive the fan, which ran co-axially with the crankshaft. In this way, cooling for the R-4360 could be optimized according to operating conditions. Of course, the extra cooling comes at a price—that price being power absorption to the tune of over 200 hp under some conditions. Depending on the installation, some of that 200 hp could be recovered through optimizing cooling airflow to augment thrust. A sophisticated fan brake was installed for when the engine was shut down. *(Ref. 2-36 and 2-37)*.

Aircraft such as the B-35 and B-36 employed fan-assisted cooling. It wasn't always needed, so the flight engineer could shut it down. A brake mechanism was incorporated to prevent the fan from windmilling, which would induce additional drag. 1. Fan Brake Adjusting Screw; (3) 2. Fan Brake Adjusting Screw, Washer, and Nut; (5) 3. Fan Brake Adjusting Screw Bracket; 4. Fan Brake Band; 5. Fan Brake Drum; 6. Fan Brake Drum Screws (10) and Lockwire; 7. Fan Holding Studs (12); 8. Power Take-Off Drive Shaft Thrust Cover Seal; 9. Power Take-Off Drive Shaft Thrust Cover Bolt Seal; 10. Power Take-Off Drive Shaft Thrust Cover Bolt (6); 11. Power Take-Off Drive Shaft Thrust Cover; 12. Power Take-Off Drive Shaft; 13. Power Take-Off Drive Shear Coupling Cover Snapring; 14. Power Take-Off Drive Shear Coupling; 15. Power Take-Off Drive Shear Coupling Cover; 16. Power Take-Off Drive Shear Coupling Cover Seal; 17. Power Take-Off Drive Shaft Seal Ring (3); 18. Power Take-Off Drive Shaft Thrust Bearing; 19. Fan Drive Shaft; 20. Fan Drive Shaft Thrust Cover; 21. Fan Drive Shaft Bearing Liner Seal; 22. Fan Drive Shaft Seal Ring (4); 23. Fan Drive Shaft Thrust Bearing Seal; 24. Fan Drive Shaft Thrust Bearing Nut; 25. Fan Drive Shaft Oil Baffle; 26. Power Take-Off Drive Shaft Gear; 27. Fan Drive Shaft Thrust Bearing Nut Tabwasher; 28. Fan Drive Shaft Thrust Bearing; 29. Fan Drive Shaft Thrust Bearing Oil Slinger; 30. Fan Drive Shaft Thrust Cover Washer (4) and Nut (7); 31. Pan Brake Case; 32. Fan Drive Shaft Thrust Cover Nut Seal (7). *(Handbook Overhaul Instructions Model R-4360-53 Aircraft Engines. May 15, 1952)*

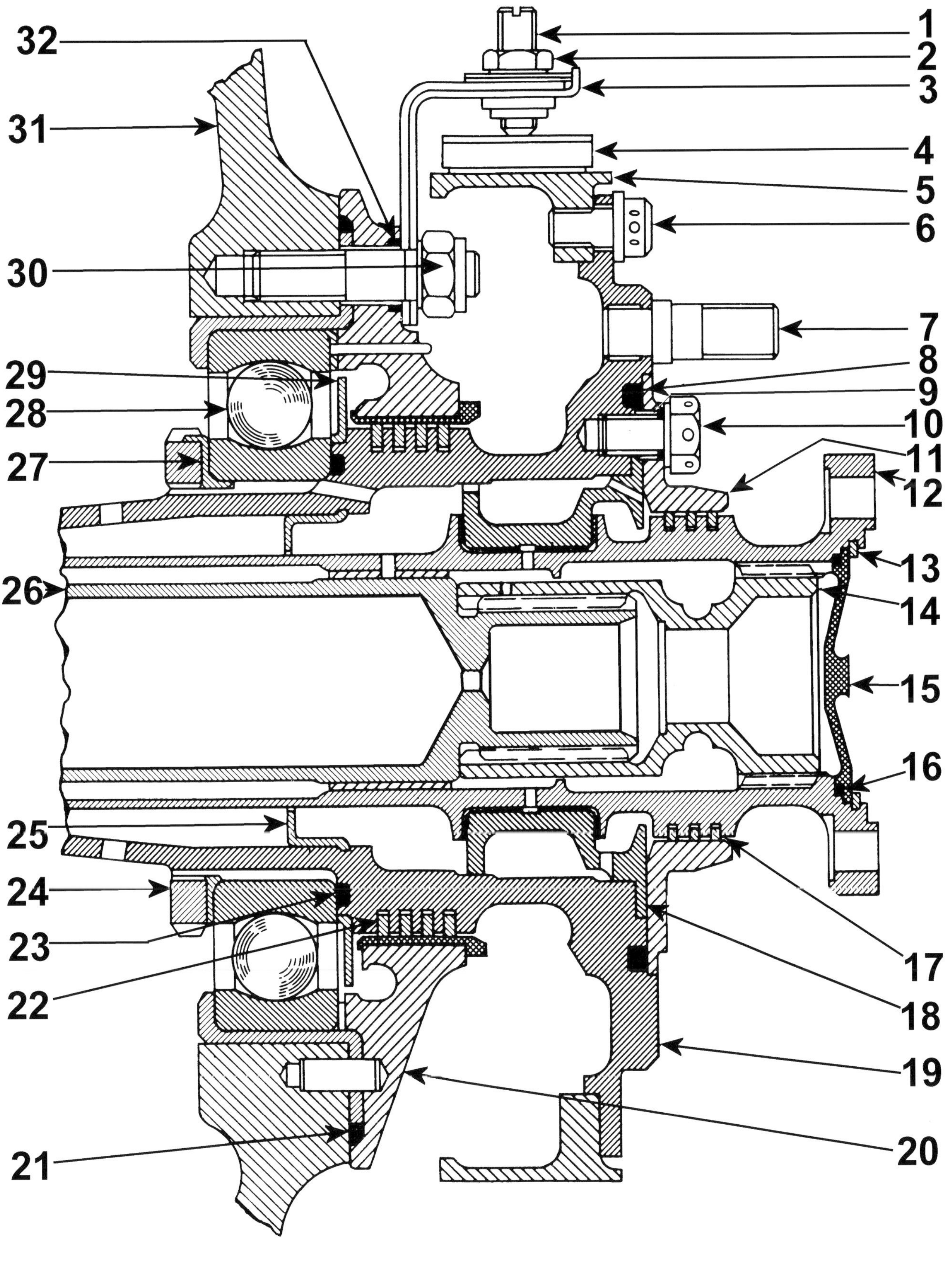
1
2
3
4
5
6
7
8
9
10
11
12
13
14
15
16
17
18
19
20
21
22
23
24
25
26
27
28
29
30
31
32

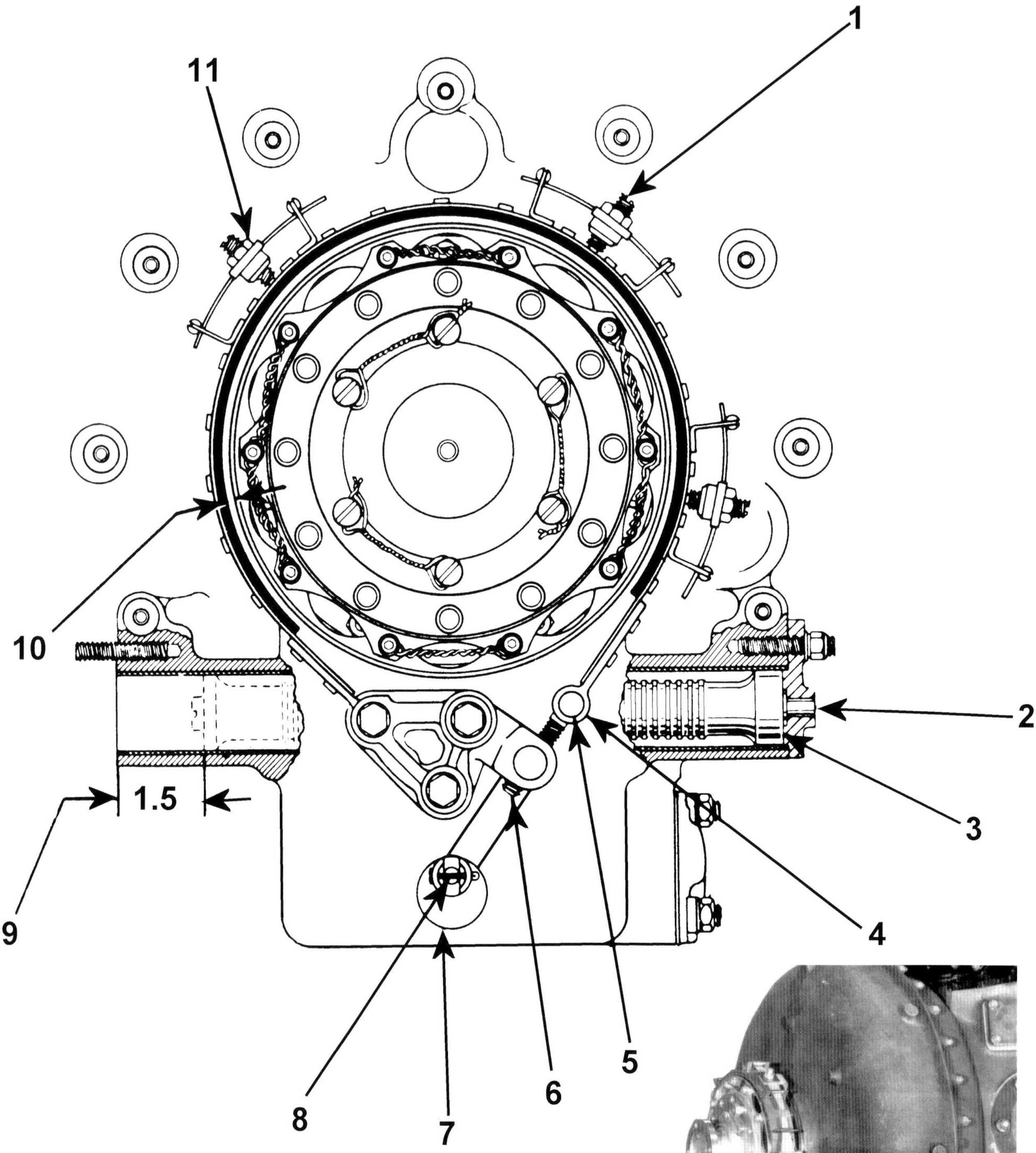

Fan brake. 1. Fan Brake Adjusting Screw; 2. Fan Brake Piston Cover; 3. Fan Brake Piston; 4. Fan Brake Band Clevis; 5. Fan Brake Band Adjusting Eyebolt Clevis Pin; 6. Fan Brake Band Adjusting Eyebolt; 7. Fan Brake Gear Shaft; 8. Fan Brake Band Link Pin Cotter Pin; 9. Fan Brake Piston Minimum Operating Clearance; 10. Fan Brake Lining Clearance; 11. Fan Brake Adjusting Screw Nut. *(Handbook Overhaul Instructions Model R-4360-53 Aircraft Engines. May 15, 1952)*

Close-up photo of the fan brake assembly and fluid coupling drive. *(United Technologies Corp./Pratt & Whitney)*

This close-up view shows the auxiliary supercharger. It's open to debate whether or not a gear-driven two-stage supercharger would have been preferable to a turbosupercharger for altitude performance. Based upon the reliability of the turbo'd applications, one would be drawn to the conclusion that, yes, it would. *(United Technologies Corp./Pratt & Whitney)*

Auxiliary Supercharger

If mission profiles demanded it, it was easy to install an auxiliary supercharger stage onto the rear accessory case due to the flexibility designed into the R-4360. Again, in a similar fashion to the main supercharge stage, drive was provided through fluid couplings. Several variations were possible. A fixed ratio driven by a two-stage step-up gear train in conjunction with a variable speed drive, or two sets of fluid couplings could be employed, each set giving a variable speed within a defined range. Yet another variation on this theme was a remotely mounted auxiliary supercharger driven off the accessory drive shaft. An example of this configuration was the Republic XP-72. This aircraft mounted its auxiliary stage supercharger behind the pilot. In all probability

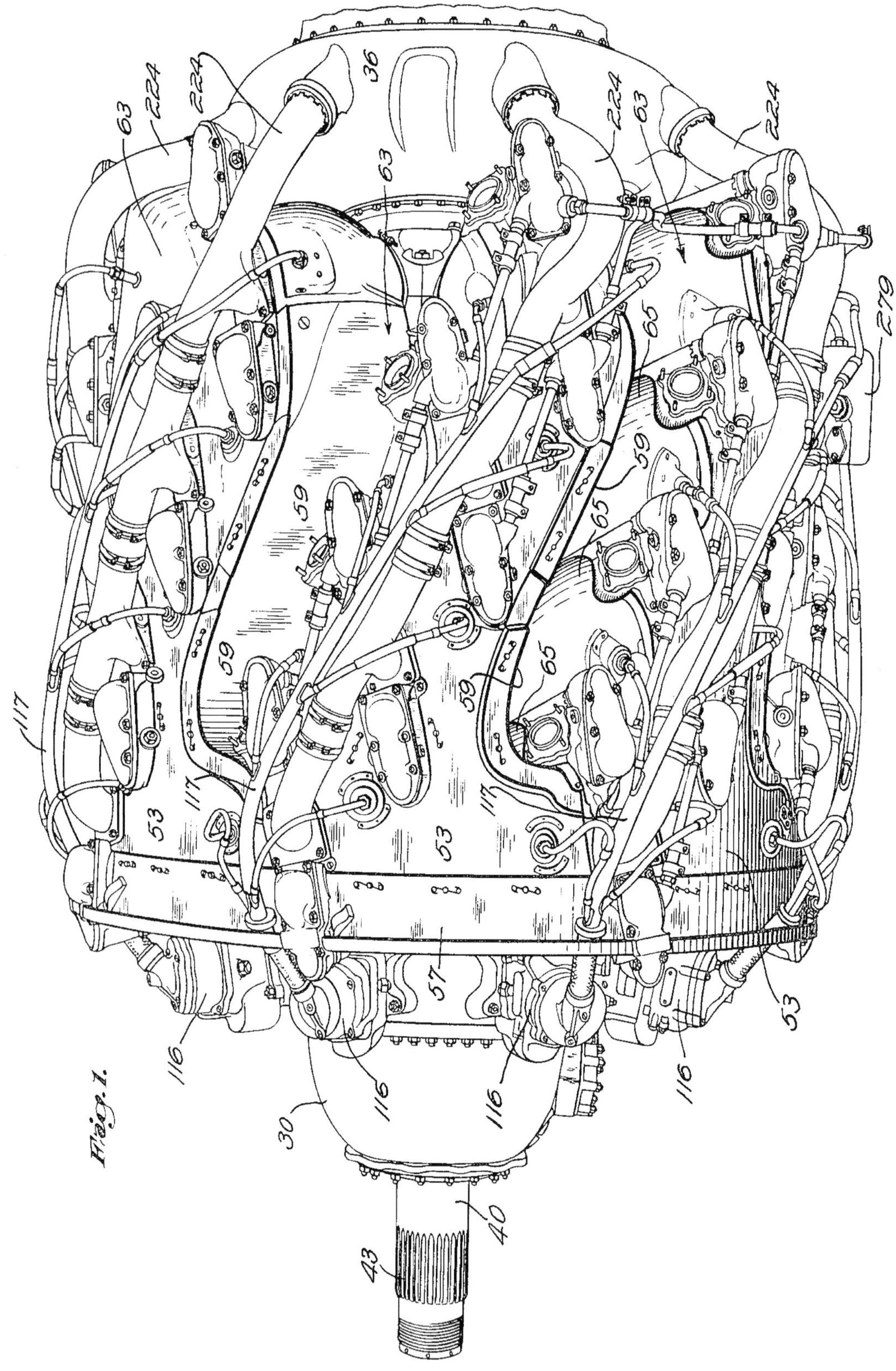

The patent illustrations start with the overall engine (shown), and then work through the entire engine from front to rear. This is a side view with part of the rear section omitted. *(U.S. Patent No. 2,426,879)*

Fig. 2.

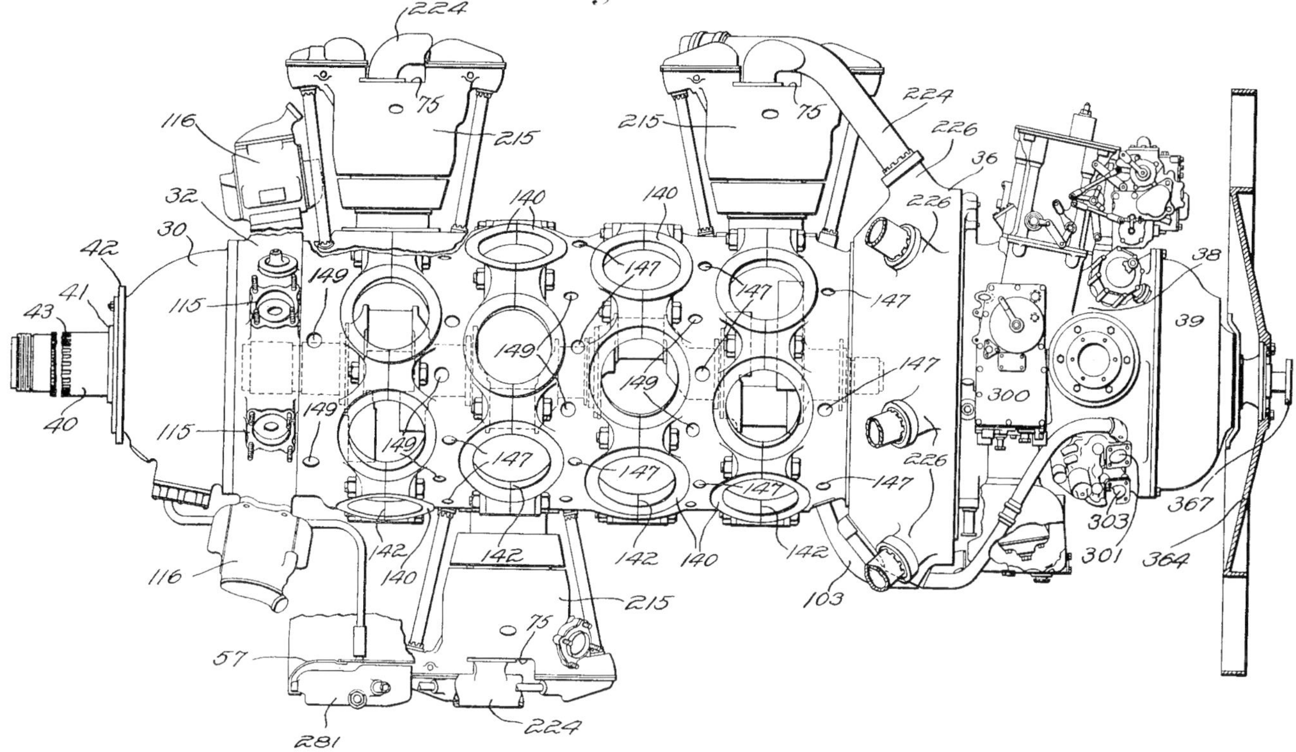

Another side view of the overall engine with cylinders magnetos, manifolds, and other parts removed to show crankcase design features. *(U.S. Patent No. 2,426,879)*

this was done to maintain the aircraft's center of gravity. The XP-72 was based upon the P-47, which has a large and heavy General Electric Type C turbosupercharger mounted in the rear fuselage. None of the gear-driven two-stage engines went into production; however, they did power a number of interesting experimental aircraft. Apart from the Republic XP-72 mentioned previously, the Boeing XB-44 and Boeing XF8B-1 are two other examples of two-stage supercharging. Applications that demanded multiple stages of supercharging relied on turbosupercharging as the preferred method of boosting, particularly at high altitudes. In retrospect, it could be argued that the two-stage engines may have proved to be more dependable. It appears from records that the most problematic R-4360s were those that employed turbosupercharging. "Open stack" engines seemed to have enjoyed a far more peaceful existence.

Patent Files

Protecting the design of a mechanism, particularly one as complex and innovative as the R-4360, demands protection from those who would attempt to steal ideas. Ideas such as these are referred to as "intellectual property," that is, the property of the original designer. To put it in legalese: "Works of the mind." The usual path taken for intellectual property protection of a mechanical device is via a patent. This is oftentimes a misunderstood process. Many times it's assumed that a corporation such as Pratt & Whitney owns the patents. In fact, patents can only be assigned to individuals, not companies. However, as a condition of employment, employees typically assign all rights to their patents to their employer. This was and still is standard practice throughout corporate America and the rest of the industrialized world.

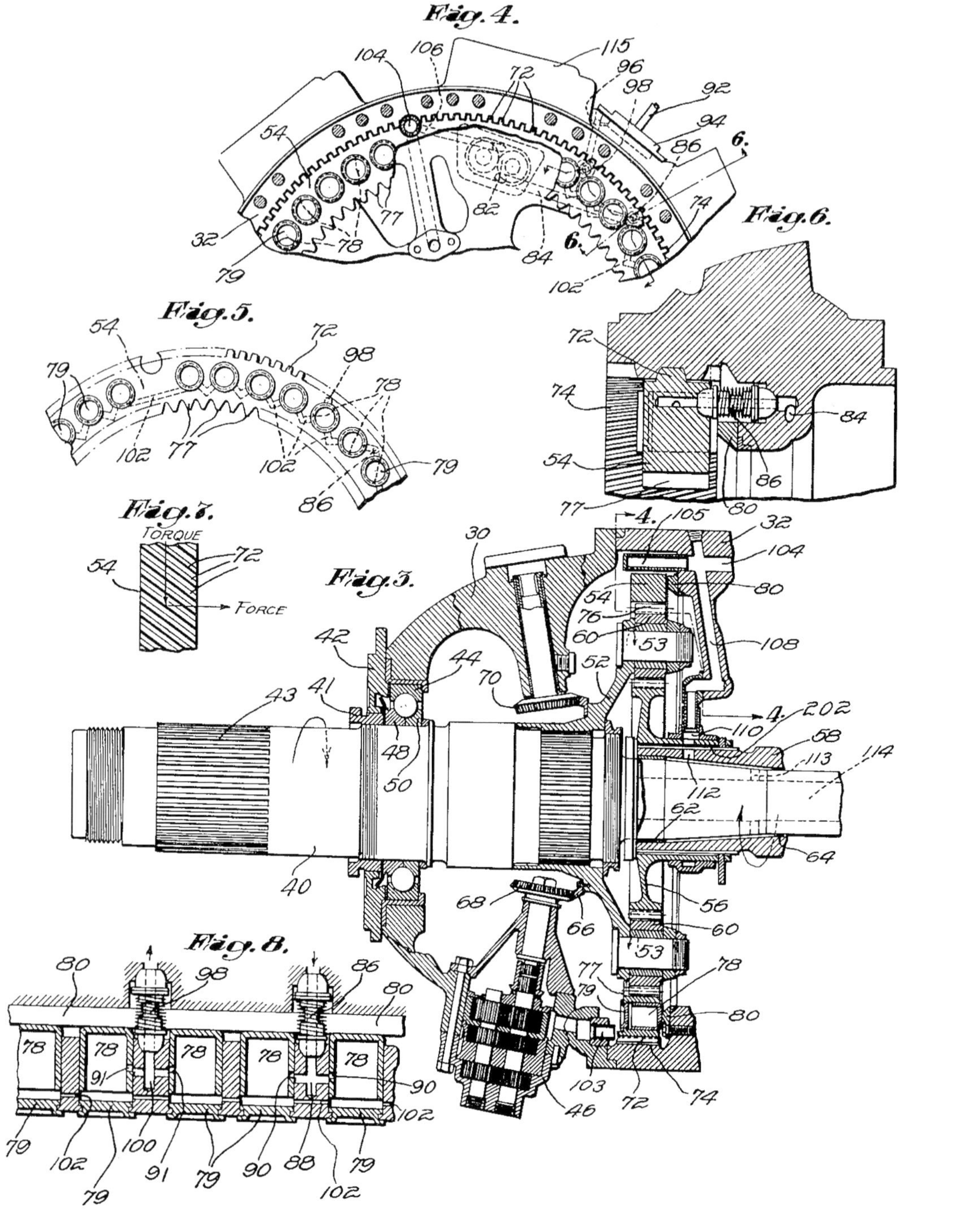

Nose case details showing torque meter, reduction gearing, pumps, and propeller shaft. *(U.S. Patent No. 2,426,879)*

Typical patent format is to submit a patent application via a filing. If the application is successful, a patent number is assigned. A typical patent has a title such as "Radial Aircraft Engine" with the inventor(s) listed. After the list of inventors, the patent then calls out whom the patent is being assigned to. In the case of Pratt & Whitney, specifically the R-4360 patents, it was United Technologies. The invention is then described and supported by whatever drawings and illustrations are necessary. Many patents filed for the R-4360 were due to the unique problems and their innovative solutions. Therefore, patents were filed and issued for various aspects such as cooling, cam design, balance, torquemeter, intake design, bearing design, fuel injection, etc.

One of the more encompassing patents for the R-4360 was U.S. Patent 2,426,879. It was filed on

Fig. 9. *Fig. 10.*

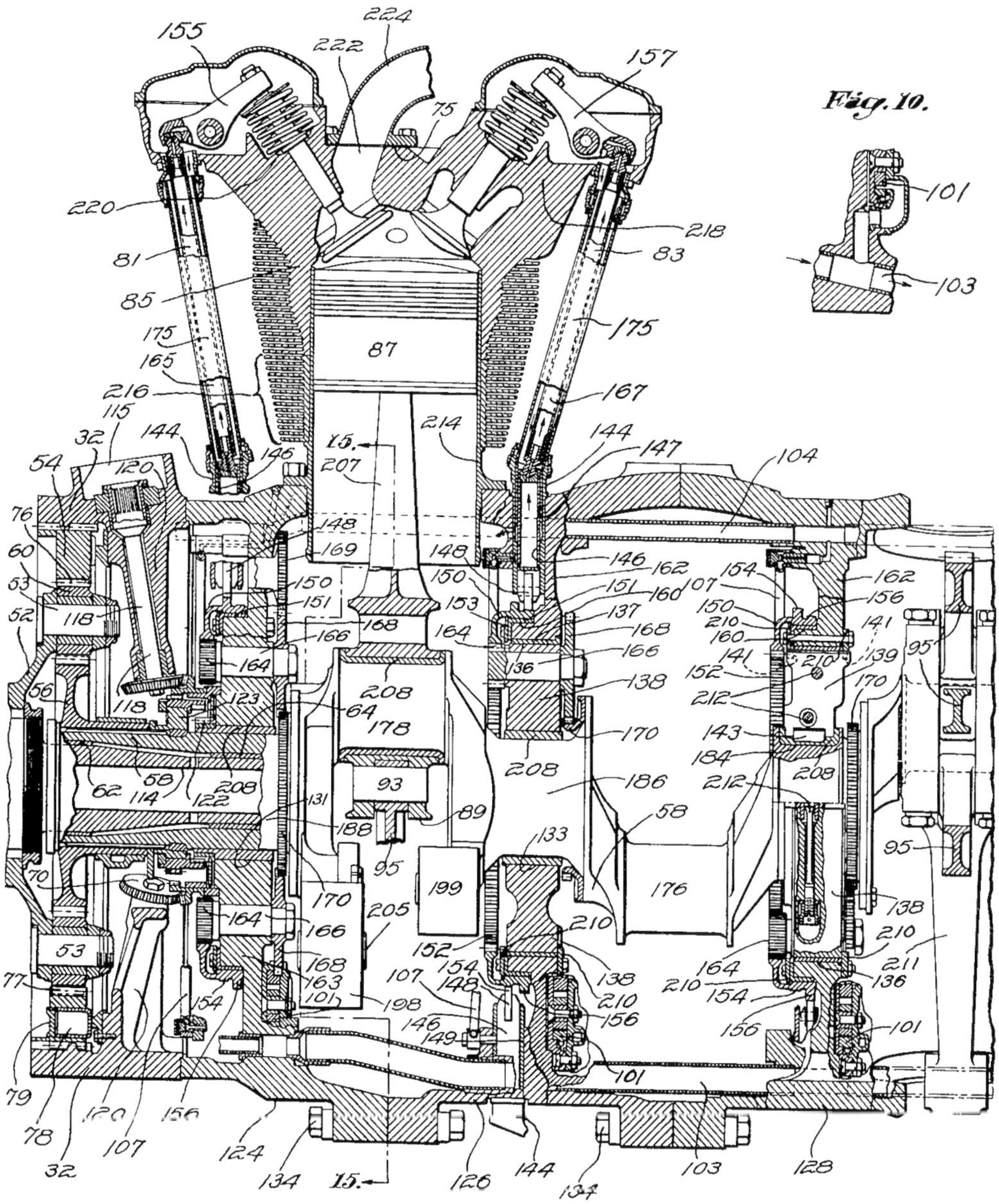

Typical crankcase section showing cylinder and intake port design. The cylinder shown in this view is in the "D" (front) row. The propeller tail shaft can be seen to the left. Also shown are the numbers 5, 4, and 3 main bearings. *(U.S. Patent No. 2,426,879)*

September 1, 1944, and issued on September 2, 1947. Patents make for pretty brutal reading due to their unique grammatical style. However, the patent drawings from 2,426,879 offer a good idea of what it took to design the R-4360 and protect its inherent intellectual property. Each aspect of the engine from the nose case to the rear section is covered. It's also interesting to note that the only two names on the patent are Leonard S. (Luke) Hobbs and Andrew Wilgoos. No one denies the fact that without these two brilliant individuals the R-4360 would never have happened. It is also fair to say that many other unsung heroes were instrumental in the design and success of the R-4360.

The letter from W. G. Gwinn signifying the end of R-4360 production says it all!

Fig. 11.

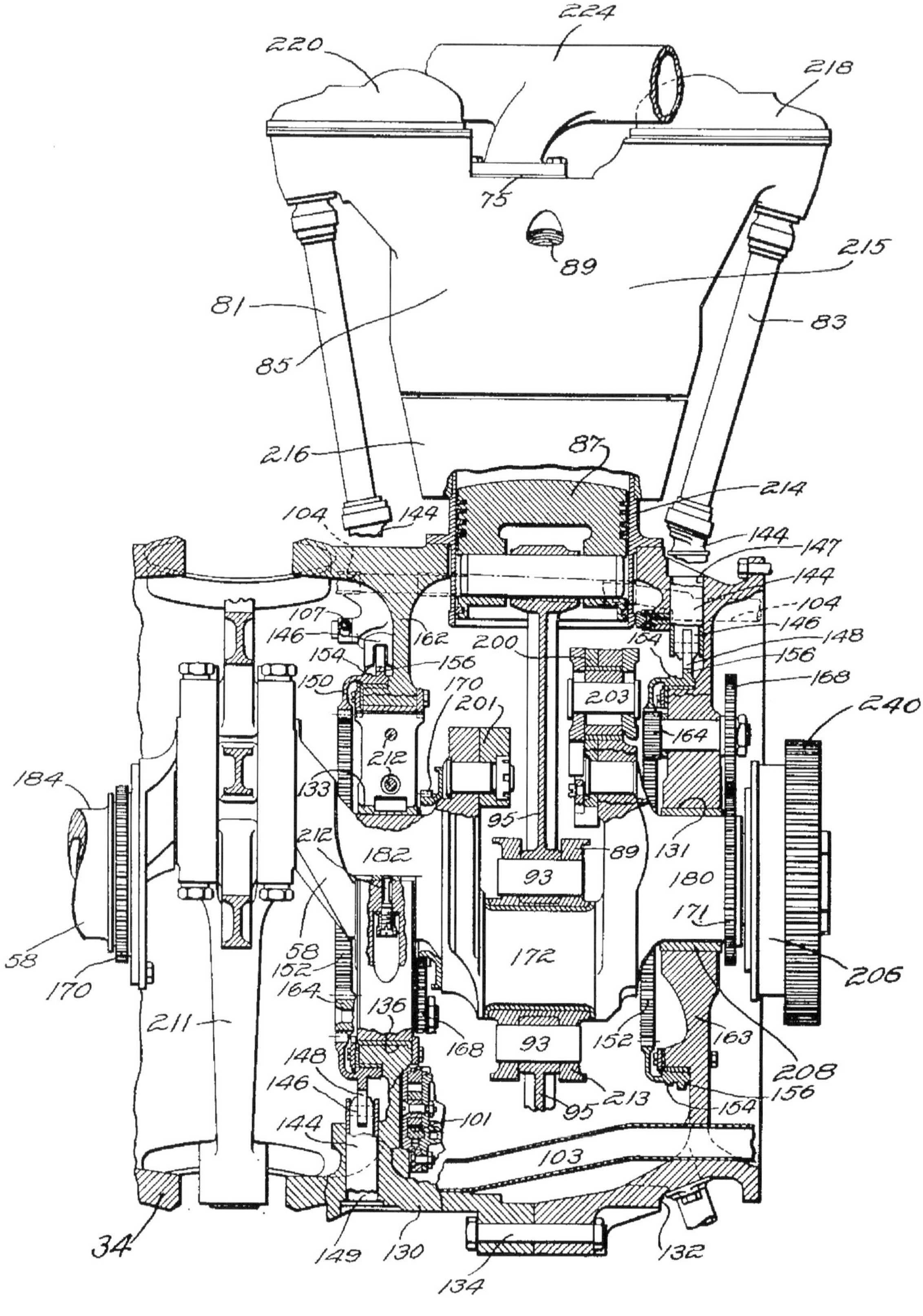

Rear crankcase section showing rear dynamic counterweight, piston, and master rod. This is the "A" row and also shows the numbers 1 and 2 main bearings. *(U.S. Patent No. 2,426,879)*

Fig. 12.

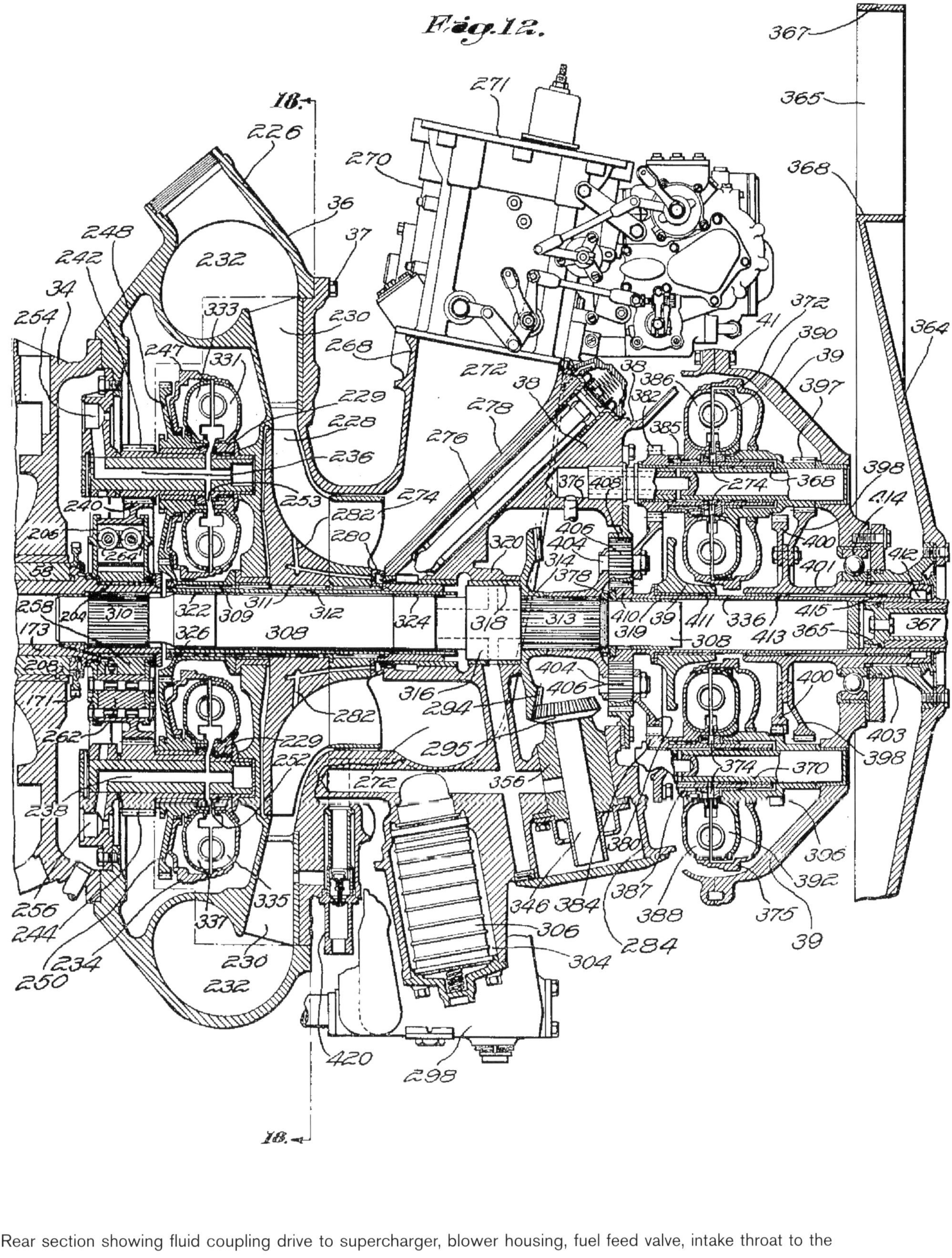

Rear section showing fluid coupling drive to supercharger, blower housing, fuel feed valve, intake throat to the supercharger, and fluid coupling drive to the auxiliary cooling fan. *(U.S. Patent No. 2,426,879)*

Fig. 13.

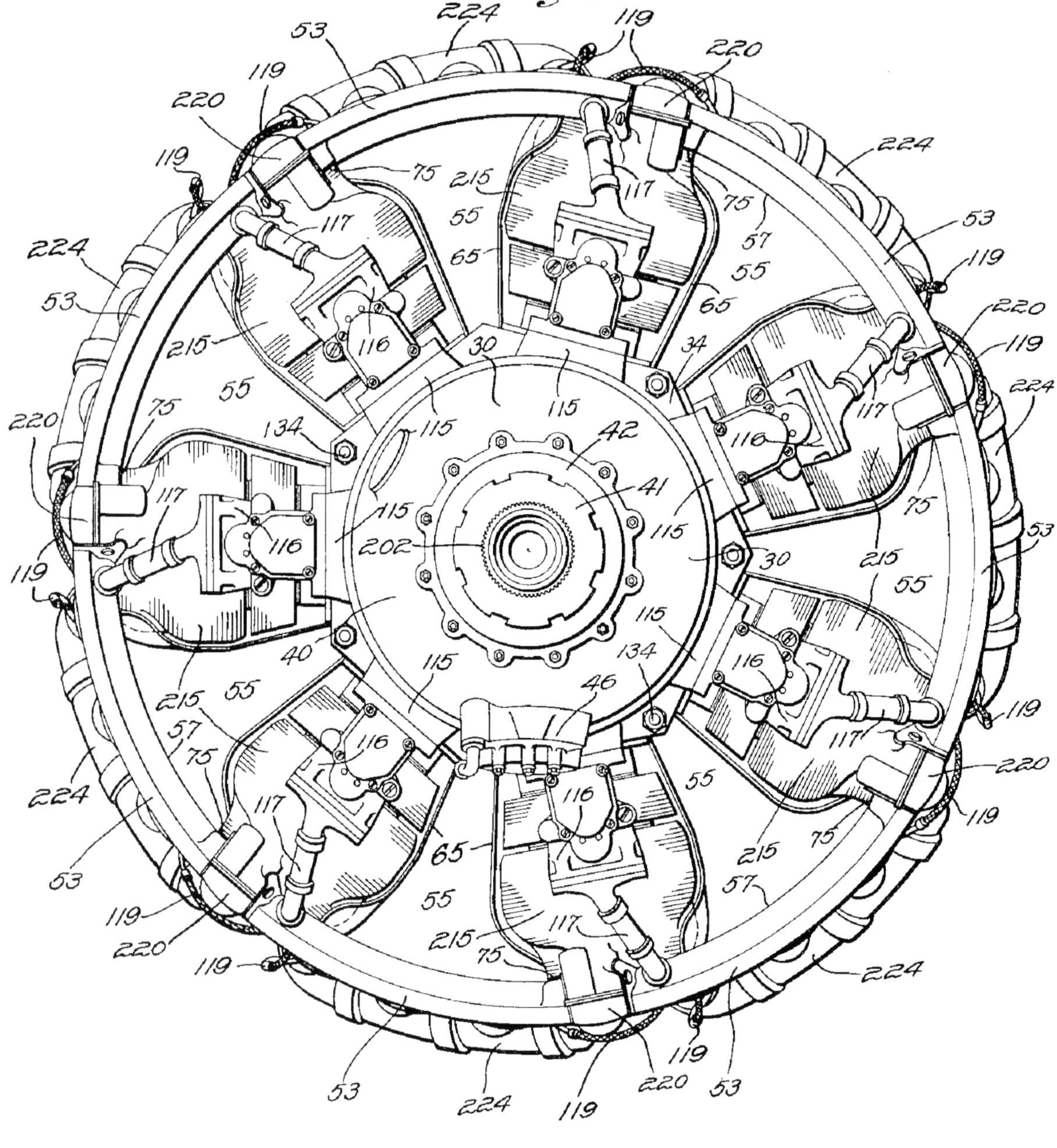

Front lateral view showing seven high-tension magnetos. For clarity the three rear rows ("A," "B," and "C") have been omitted. Intake manifolds are visible around the periphery. *(U.S. Patent No. 2,426,879)*

References

2-1 Hersey, Donald S., and Dumarrs. *Memorandum Report – Cooling Airflow Investigation of Ten Dummy Air Cooled Radial Engines of Various Cylinder Combinations.* Pratt & Whitney Aircraft, East Hartford, Connecticut. October 2, 1940, through January 13, 1941.

2-2 Armbruster, G. E., *History Of The Development Of R-4360 Engines.* Pratt & Whitney Aircraft, East Hartford, Connecticut.

2-3 Hersey, Donald S., *Memorandum Report – Dynamometer Cooling Investigation of Wasp Major Engine X-109.* Pratt & Whitney Aircraft, East Hartford, Connecticut. April 20, 1942.

Fig. 14.

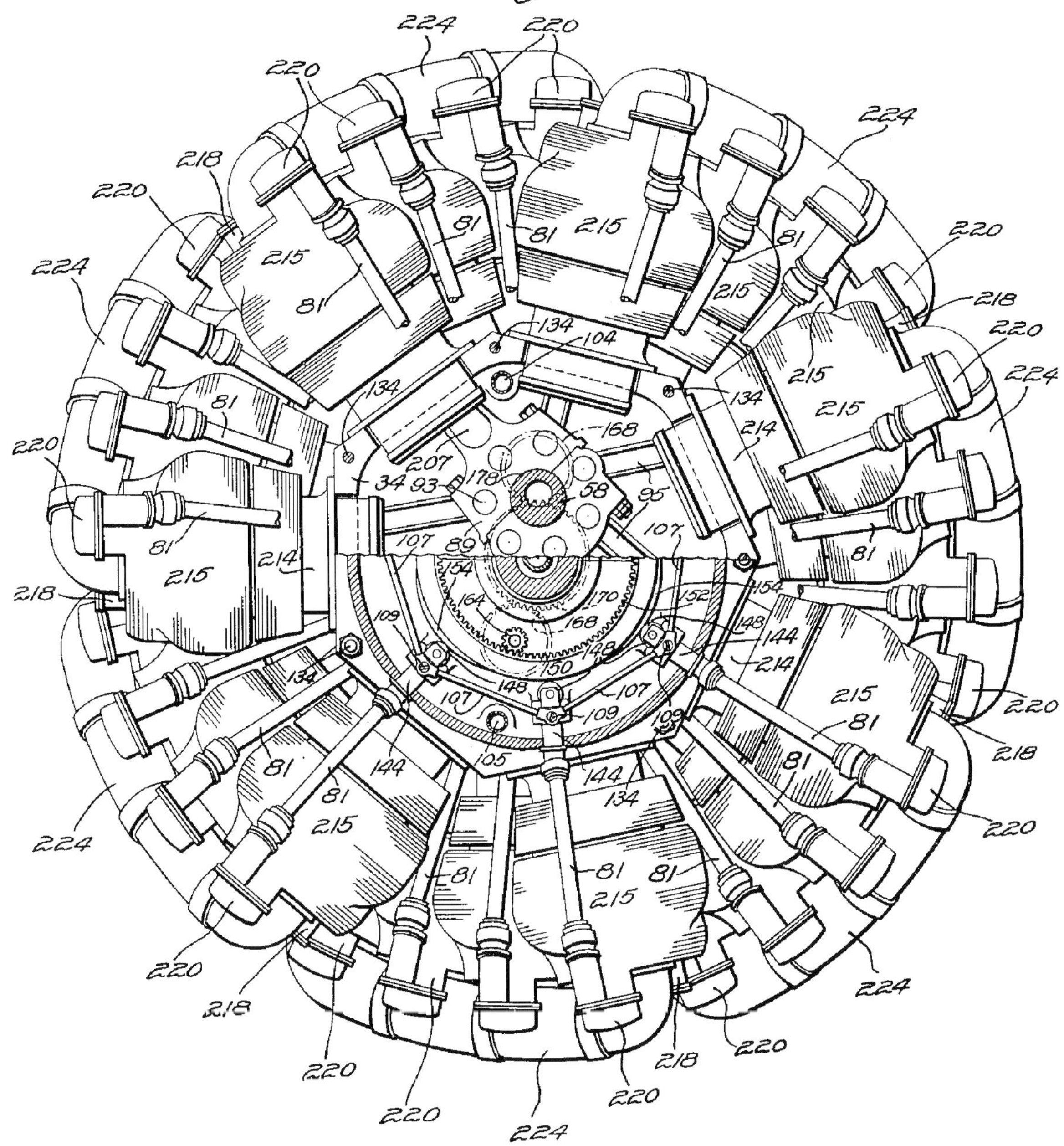

Lateral section taken through row "A" showing master rod/link rod relationship and cam ring. Also shown, at the bottom, are the front cam ring, cam ring drive, and oil feed to the push rods. *(U.S. Patent No. 2,426,879)*

2-3A Certificate issued by Pratt & Whitney acknowledging the first flight of the Wasp Major.

2-4 Armbruster, G. E., *Problems Encountered In The Development Of The R-4360 Engine Peculiar Only To That Engine*. Pratt & Whitney Aircraft, East Hartford, Connecticut.

2-5 *The Pratt & Whitney Aircraft Story*. Pratt & Whitney Aircraft, East Hartford, Connecticut. 1952.

2-6 *Wasp Major B Series Engines, Pratt & Whitney Aircraft Installation Handbook*. Pratt & Whitney Aircraft, East Hartford, Connecticut. February 1953.

2-7 *Wasp Major Models B6 and CB2 Overhaul Manual*. Pratt & Whitney Aircraft, East Hartford, Connecticut. Reissued June 1953. Revised December 1957.

Fig. 15.

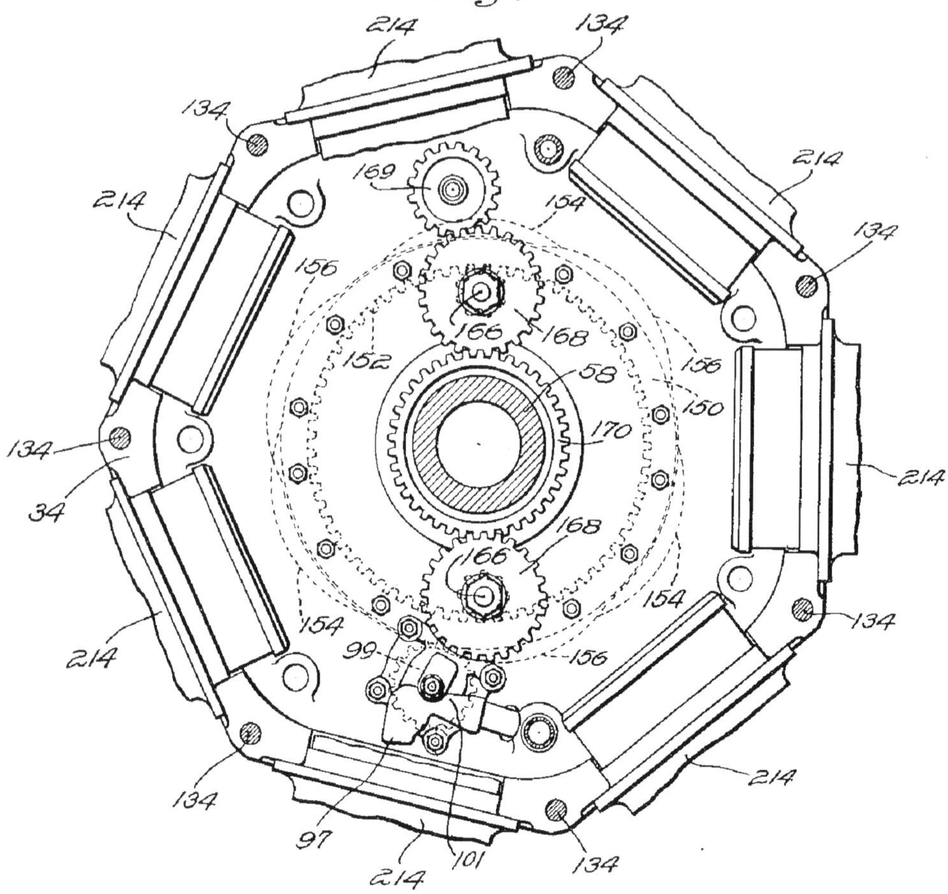

Fig. 16.

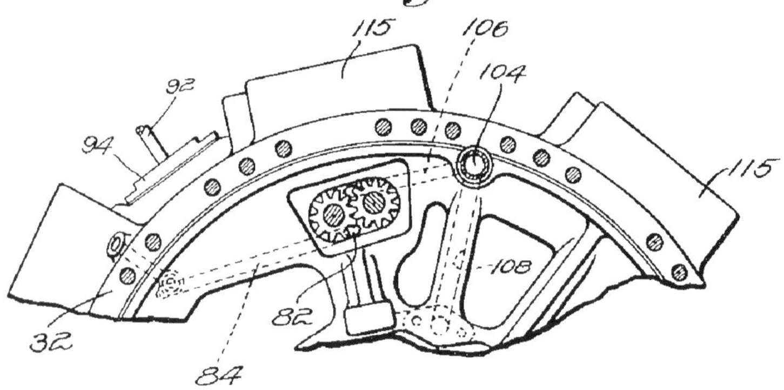

Lateral section showing cam drive, torquemeter gear drive, scavenge pump drive, and scavenge pump inlet connection. *(U.S. Patent No. 2,426,879)*

2-8 Patent Number 2,426,877 *Radial Aircraft Engine*. United States Patent Office. Patented Sept. 2, 1947.

2-9 Patent Number 2,426,879 *Radial Aircraft Engine*. United States Patent Office. Patented Sept. 2, 1947.

2-10 Patent Number 2,666,418 *Means For Balancing Piston Engines*. United States Patent Office. Patented Jan. 19, 1954.

2-11 Patent Number 2,426,874 *Radial Aircraft Engine*. United States Patent Office. Patented Sept. 2, 1947.

2-12 Patent Number 2,426,875 *Radial Aircraft Engine*. United States Patent Office. Patented Sept. 2, 1947.

2-13 Patent Number 2,426,876 *Radial Aircraft Engine*. United States Patent Office. Patented Sept. 2, 1947.

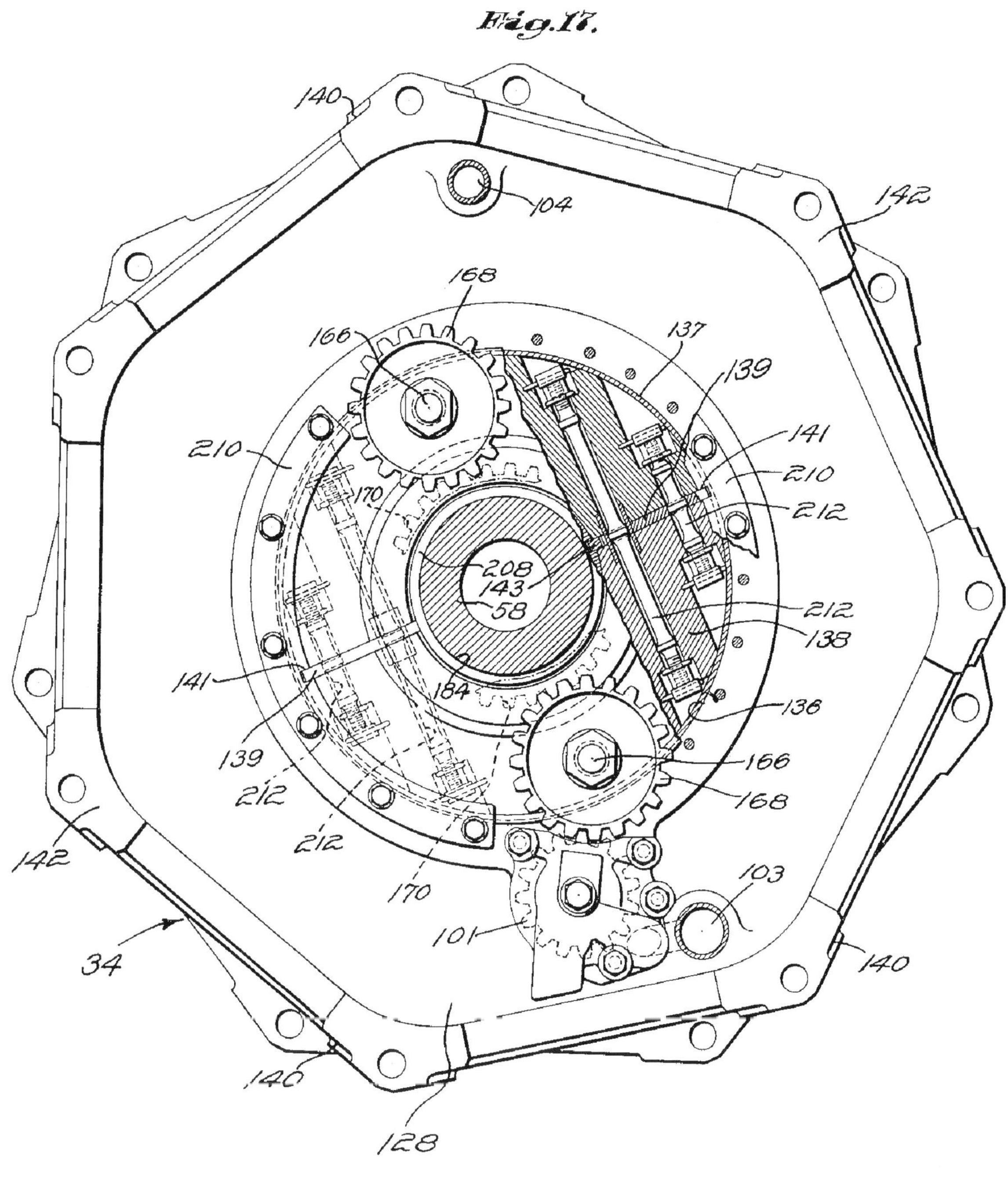

Lateral section, partially broken away, taken immediately behind the center crankcase partition looking forward. It shows the crankshaft supporting means plus cam and scavenge pump drives. *(U.S. Patent No. 2,426,879)*

2-14 Patent Number 2,266,320 *Antifriction Bearing*. United States Patent Office. Patented Dec. 16, 1941.

2-15 Gunston, W. T., *By Jupiter*. R.Ae.S., London 1978.

2-16 Patent Number 2,426,871 *Air Cooling For Radial Engines*. United States Patent Office. Patented Sept. 2, 1947.

2-17 Patent Number 2,426,872 *Air Cooling For Radial Engines*. United States Patent Office. Patented Sept. 2, 1947.

2-18 Patent Number 2,401,210 *Cylinder Head*. United States Patent Office. Patented May 28, 1946.

2-19 Hersey, Donald S., *Memorandum Report #403. Fin Cooling Analysis for Altitude*

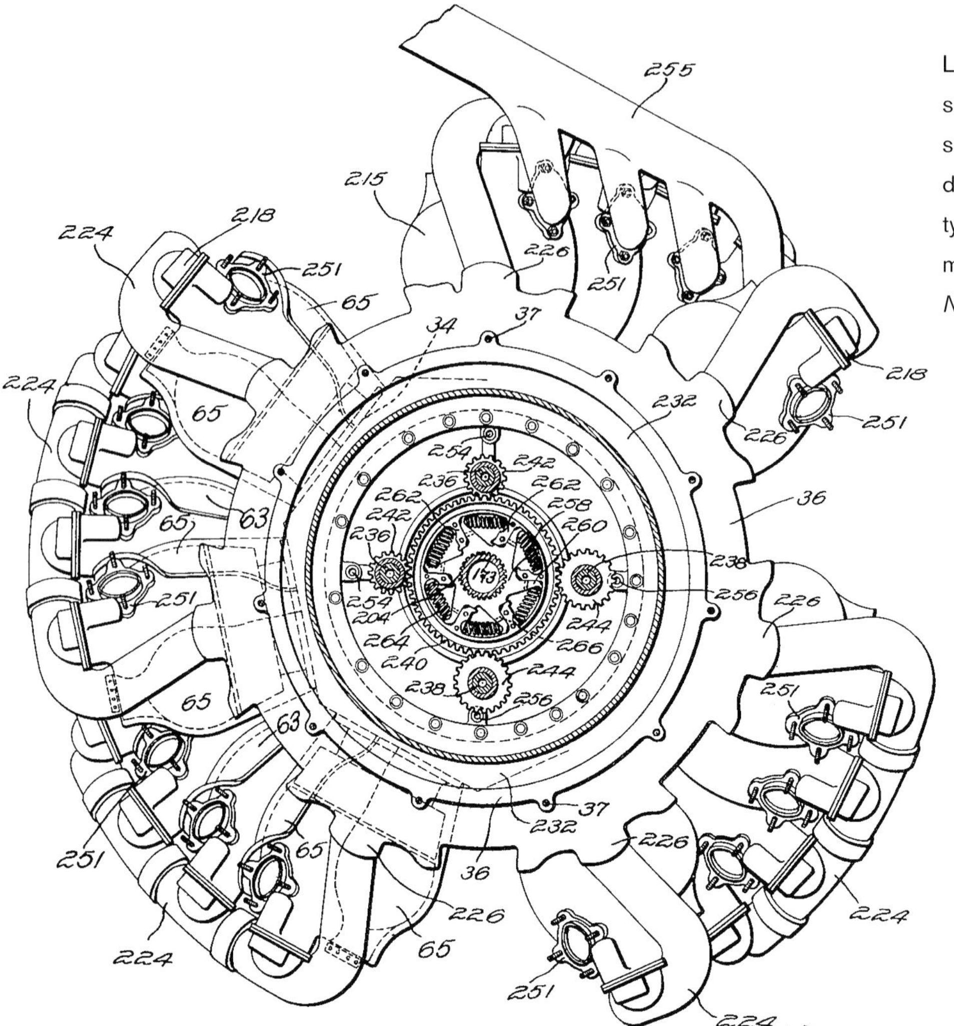

Lateral cross section showing the supercharger spring drive. It also illustrates typical intake and exhaust manifolds. *(U.S. Patent No. 2,426,879)*

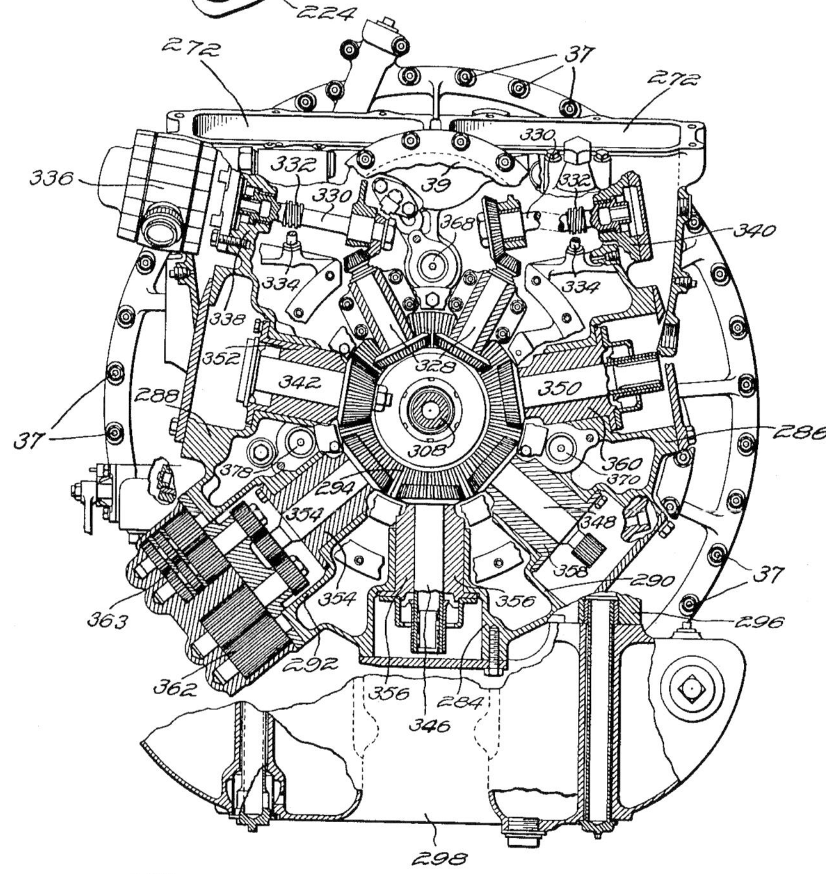

Lateral section through the rear accessory section looking forward showing the main ring drive gear and various jackshaft drives to accessories. *(U.S. Patent No. 2,426,879)*

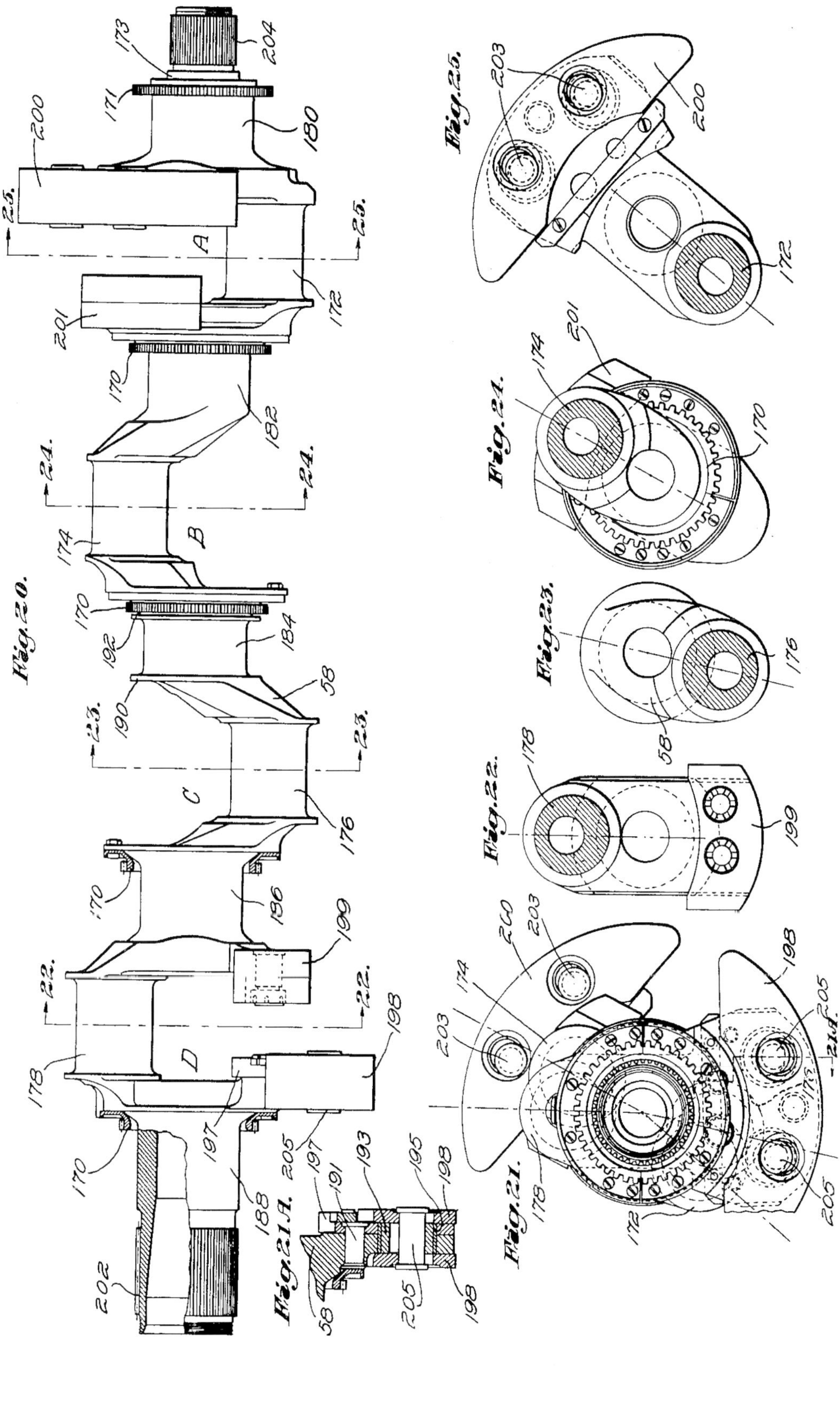

Side view of the crankshaft. *(U.S. Patent No. 2,426,879)*

Fig. 26.

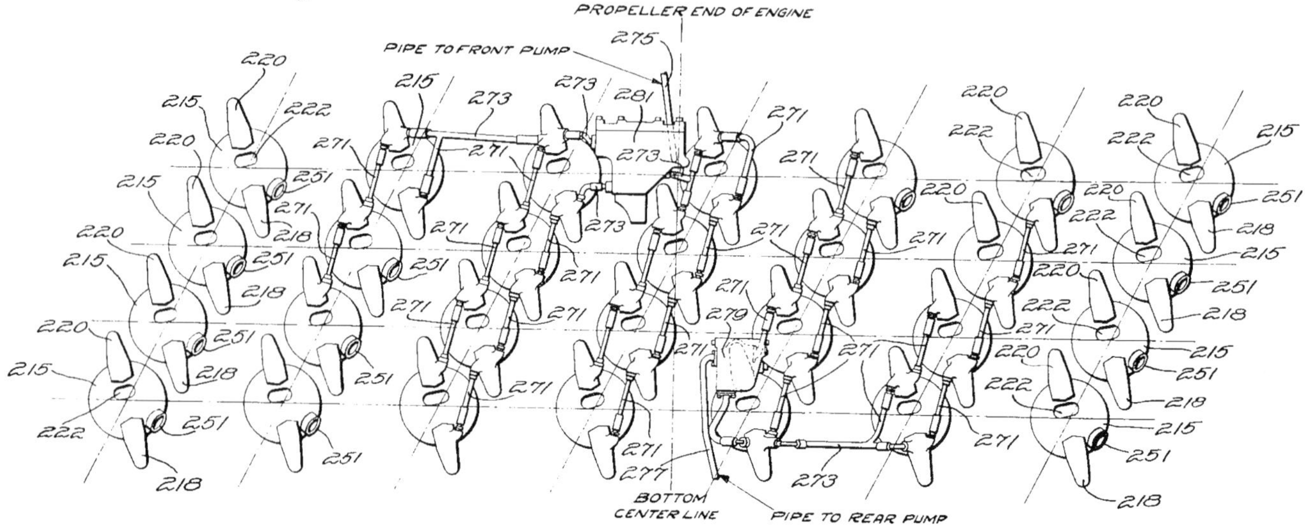

This shows a developed view of the rocker box oil drainage system. *(U.S. Patent No. 2,426,879)*

Operation of the Wasp Major Cylinder. Pratt & Whitney Aircraft, East Hartford, Connecticut. June 3, 1942.

2-20 Beckwith, C. Gordon and R. E. Gorton, *Memorandum Report #399 Valve Motion Characteristics of the X Wasp Engine No. X-109*. Pratt & Whitney Aircraft, East Hartford, Connecticut. March 21, 1942, through March 29, 1942.

2-21 Beckwith, C. Gordon and R. E. Gorton, *Memorandum Report #406 Valve Motion Characteristics of the X Wasp Engine No. X-110*. Pratt & Whitney Aircraft, East Hartford, Connecticut. May 27, 1942, through May 30, 1942.

2-22 Beckwith, C. Gordon and R. E. Gorton, *Memorandum Report #409 Valve Motion Characteristics of Wasp Major Single No. S-12*. Pratt & Whitney Aircraft, East Hartford, Connecticut. February 21, 1942, through March 19, 1942.

2-23 Toft, R. L., Inter-Office Correspondence. *Subject: Valve Actuating Devices used in R-4360 X-108*. Pratt & Whitney Aircraft, East Hartford, Connecticut. March 6, 1945.

2-24 Phinney, D. G., Inter-Office Correspondence. *Subject: R-4360 Hydraulic Tappet Testing*. Pratt & Whitney Aircraft, East Hartford, Connecticut. March 8, 1945.

2-25 Carlson, W. D., Experimental Test Department Short Memorandum Report. *Subject: X-Wasp Single Cylinder Volumetric Efficiency Investigation*. Pratt & Whitney Aircraft, East Hartford, Connecticut. April 2, 1942.

2-26 Breck, M. J., and A. B. Swenson. Experimental Test Department Short Memorandum Report. *Subject: Results of Power Section Scavenge Study on Wasp Major Engine X-116*. Pratt & Whitney Aircraft, East Hartford, Connecticut. December 1, 1942.

2-27 Esmeir, E., Experimental Test Department Short Memorandum Report. *Subject: A Comparison of Pressure Drops across a Wasp Major Scavenge Pump #55221 after Several Changes had been made in the Pump Pas-*

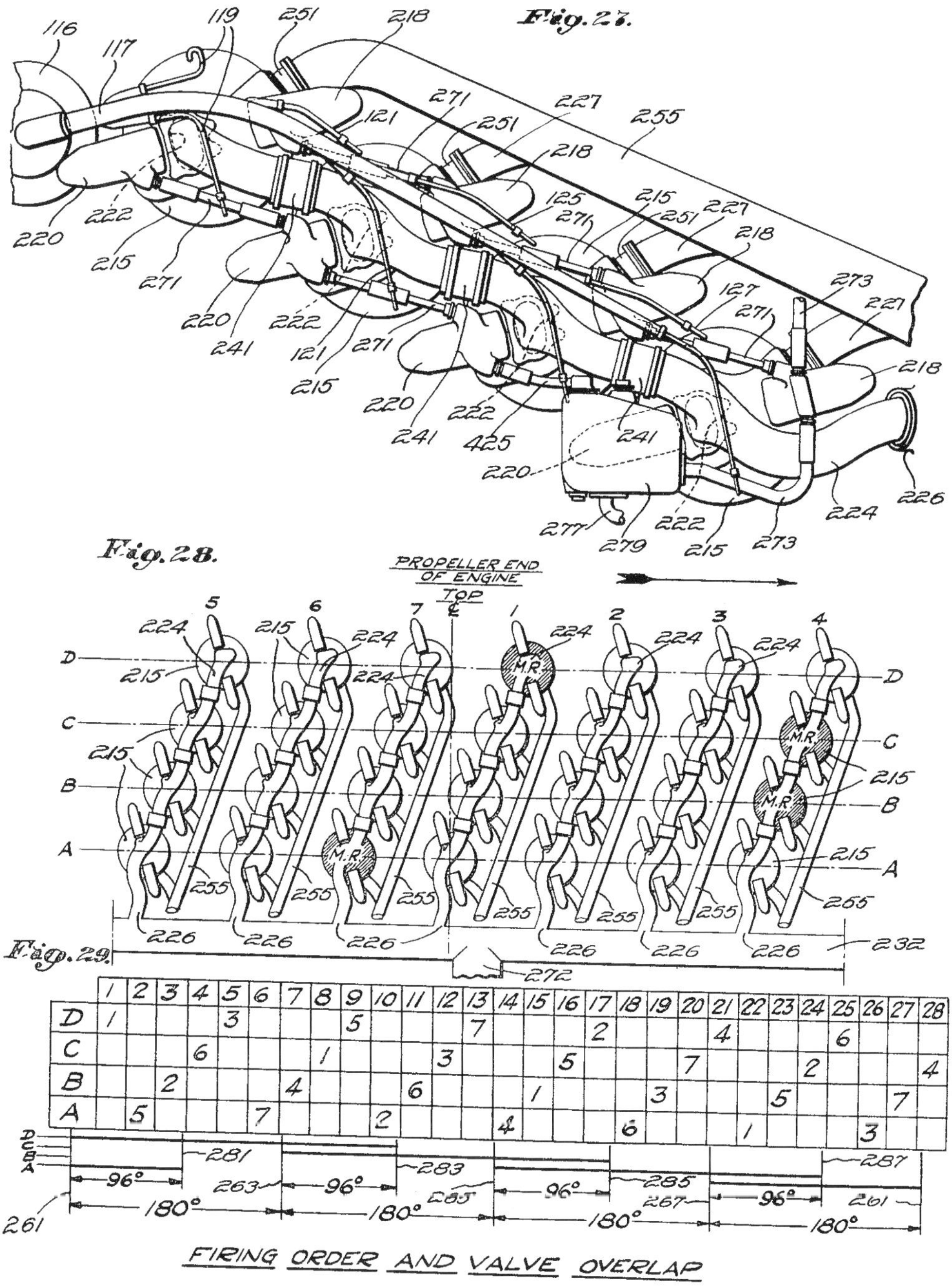

The top portion of this figure shows a developed view of the number five cylinder bank including the intake and exhaust manifolds, ignition, oil drainage pipes, and rear sump. The lower portion of this illustration is a developed view showing diagrammatically the arrangement of the cylinders, intake, and exhaust manifolds. *(U.S. Patent No. 2,426,879)*

sages. Pratt & Whitney Aircraft, East Hartford, Connecticut. October 19, 1942.

2-28 Service Instructions. *Bendix-Scintilla Ignition System Used on Pratt & Whitney R-4360 Wasp Major Aircraft Engines*. Scintilla Magneto Division, Sidney, New York, U.S.A. Revised October 1948.

2-29 Installation, Service and Maintenance Instructions. Bendix Low Tension – High Altitude Ignition System. Used on Pratt & Whitney Aircraft Engine Model R-4360-B6 or CB2. Scintilla Magneto Division, Sidney, New York, U.S.A. Printed December 1952.

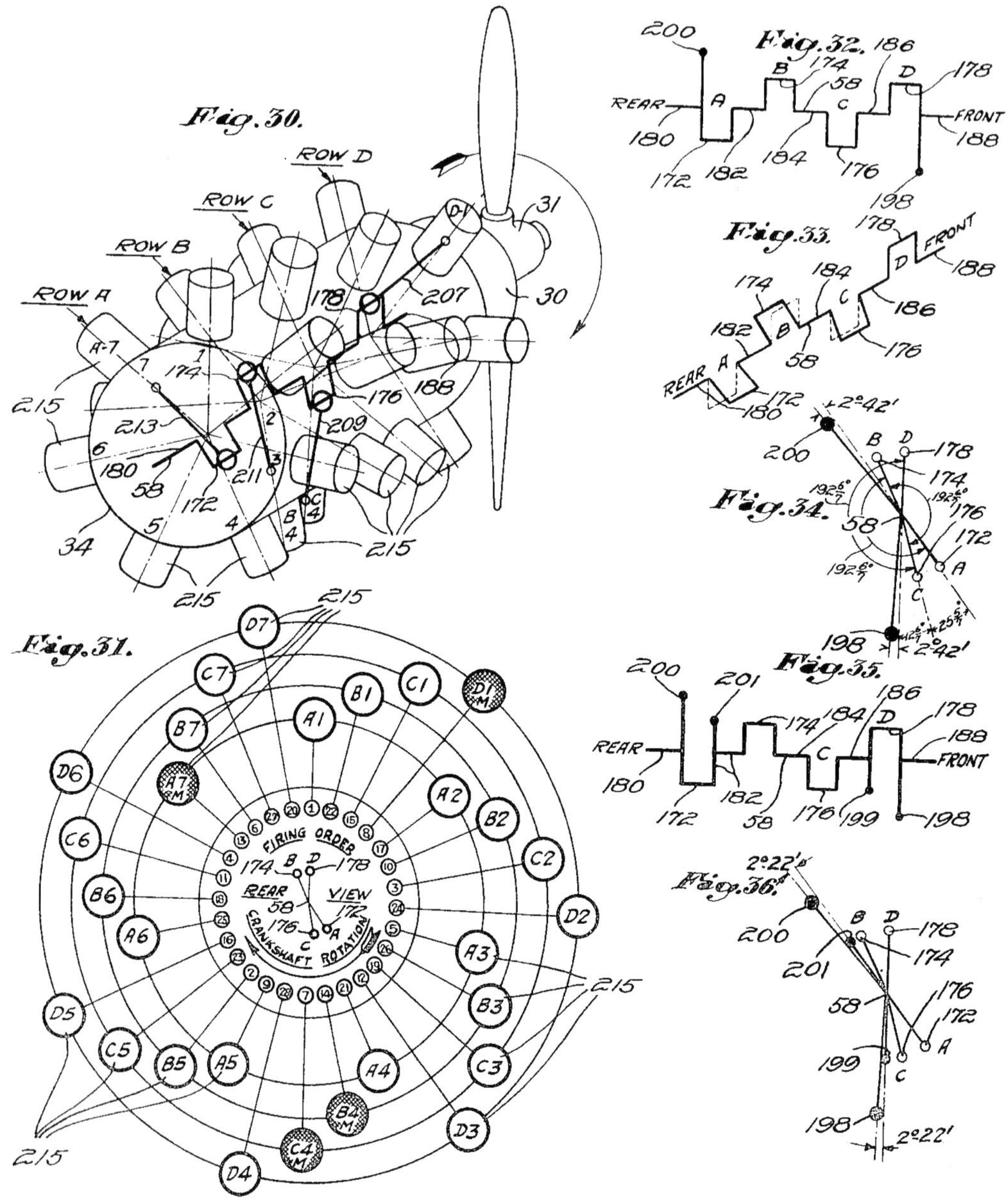

This is a diagram showing the relationship of the cylinder spacing to crank pinion spacing and the firing order. *(U.S. Patent No. 2,426,879)*

2-30 Williams, G. L., Report number CG-47-6. *Planetary Reduction Gears. Useful Relations for Compound Planetary*. Pratt & Whitney Aircraft, East Hartford, Connecticut. Revised March 23, 1948.

2-31 Williams, G. L., Report number CG-47-9. Dual *Rotation Reduction Gears (4360 Compounded)*. Pratt & Whitney Aircraft, East Hartford, Connecticut. Revised March 5, 1947.

2-32 AN 02-10HD-2. Service Instructions for Model R-4360-31 Aircraft Engine. May 1, 1945.

2-33 AN 02A-10HC-2. Handbook Service Instructions Models R-4360-17, -45, and -47 Aircraft Engines. August 1, 1948.

PWA F 51.

INTER-OFFICE CORRESPONDENCE
PERMANENT MEMORANDUM FOR FILE
PRATT & WHITNEY AIRCRAFT ENGINES

To Mr. E. M. Benham — *Date* June 23, 1955

From Mr. W. P. Gwinn

cc: Mr. L. C. Mallet
Mr. P. W. Fisher

Subject

As you probably know, next month sees the end of the 4360 production.

I think we ought to certainly have some pictures of the last engine for future use as well as possibly using them now, for after all this is the largest piston type engine ever really put into production and a considerable quantity have been built and it has powered some pretty important commercial and military airplanes.

I will leave this in your hands to follow through on.

W. P. Gwinn

WPG:rk

The end of the line! W. G. Gwinn was a high-level executive with Pratt & Whitney. In today's environment, executives don't even know what their corporations are building, never mind recognizing the end of production of a significant product. But then again, the R-4360 was built in different times with a different culture that had a sense of morality and sense of purpose. *(United Technologies Corp./Pratt & Whitney)*

2-34 Newkirk, Burt L., Report No. 410. Memorandum Report. *The Design of Inducers for the Wasp Major Engine.* Pratt & Whitney Aircraft, East Hartford, Connecticut. July 9, 1942.

2-35 Patent Number 2,287,021 *Fuel Supply System.* United States Patent Office. Patented June 23, 1942.

2-36 AN 02A-10HC-2. Handbook Service Instructions Models R-4360-17, -45, and -47 Aircraft Engines. August 1, 1948.

2-37 AN 02A-10HF-3. Overhaul Instructions Models R-4360-25 and -41 Aircraft Engines. Revised January 15, 1949.

CHAPTER THREE

Into Production & Variations

Designing and developing an aircraft engine as complex as the R-4360 is difficult at best, but once the worst of these problems have been overcome it's time to mass-produce it. Much of the development of the R-4360 was done in parallel with the R-2800. Therefore, it's not surprising to see that similar nomenclature was used to describe the various families of R-4360. Unlike the R-2800, which went through a total redesign when developed from the "B" series to the "C" series, the R-4360 was more evolutionary. Nevertheless, it still used the "B" series and "C" series nomenclature to designate significant design changes. Early experimental engines were designated as "X" Wasp engines. Those fell into the "A" series development phase. Engines from the R-4360-2 through R-4360-37 were "B" series engines. The R-4360-39 represented the first "C" series, although some "B" series were developed after the -39. For the most part, however, "B" engines with designations higher than -39 were simply adaptations or developments of earlier engines. Finally, as with the R-2800, the ultimate development was the "CB" series. In simple terms, "B" engines were rated at 3,000 hp, "C" engines were rated at 3,500 hp, and "CB" engines were rated at 3,800 hp.

Two-Stage Variations

Pratt & Whitney's first production version of a gear-driven two-stage supercharger was fitted to the R-1830-65, which powered F4F-4 Wildcats *(Ref. 1-3)*. This late-1930s engine design was ahead of its time in some respects, resulting in a steep learning curve regarding the idiosyncrasies of two-stage supercharging. One serious problem that afflicted early two-stage engines was surging at high altitudes, or pressure pulsations within the induction system. The follow-on R-2800 had a number of versions with two-stage supercharging, but despite the early lessons from the R-1830, surging could still occur. Two-stage R-2800s, which powered aircraft such as Vought Corsairs, Grumman Hellcats, and Northrop Black Widows, could fall victim to this malady *(Ref. 3-1)*.

With this knowledge base, it was only natural that the R-4360 should also be the beneficiary of gear-driven two-stage supercharging. Interestingly, even though several innovative two-stage supercharger designs were developed for the R-4360, none saw mass-production. Instead they were built for aircraft not awarded production contracts. It was felt that high-altitude operation could be accomplished via turbosupercharging. Single-stage variable speed was the usual choice for lower altitudes. Although the auxiliary blower housing looks the same for all two-stage engines, it could be mounted integral with the engine. This was used on an R-4360-10 used on the Boeing XF8B-1, or the R-4360-33 used to power the Boeing XB-44. Another variation on this theme was to mount the auxiliary supercharger behind the engine and drive it via an

Perhaps based on their successful deployment of two-stage gear-driven superchargers with the R-2800, Pratt & Whitney also developed them for use on the R-4360. All the gear-driven two-stage engines were for Army Air Force aircraft. This photograph shows an R-4360-33 developed for the Boeing XB-44. *(United Technologies Corp./Pratt & Whitney)*

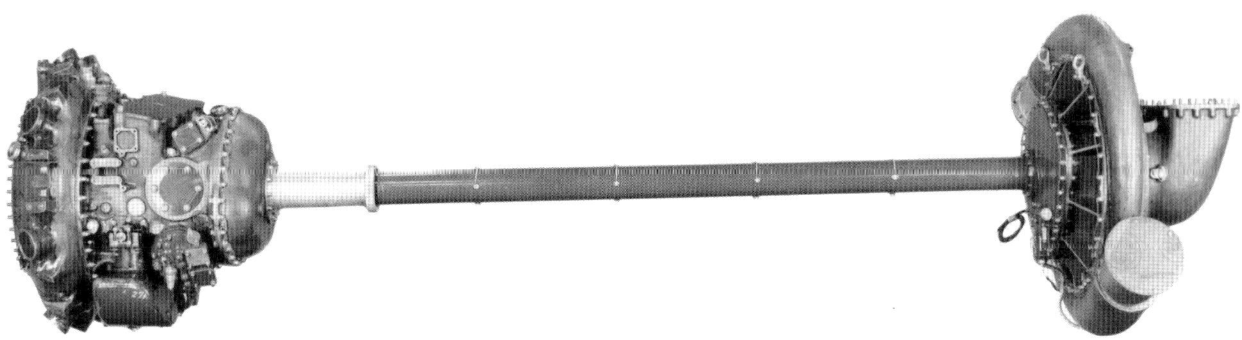

Republic Aviation enjoyed remarkable success with the P-47, so it was only natural that an R-4360 should be shoehorned into it. The resulting XP-72 showed great promise; however, after two prototypes were built, the program was dropped. The auxiliary supercharger was mounted behind the pilot and driven via an extension shaft. In all probability this was done to maintain the correct center of gravity. *(United Technologies Corp./Pratt & Whitney)*

extension shaft. This was the method of choice for the Republic XP-72.

It would appear to make sense to have the engine and its supercharger stages in one neat assembly; however, the auxiliary blower was moved back for other requirements and considerations such as maintaining the correct aircraft center of gravity. In all probability, this was the rationale for the use of an extension shaft to drive the auxiliary supercharger of the R-4360-13 for use in the XP-72. Despite the fact that the XP-72 was designed around the two-stage blower driven

A direct-drive variation on the R-4360 was built for the Northrop XB-35. Whether it was an inboard or outboard engine determined the length of the drive shaft, which terminated at a reduction gearbox mounted above the trailing edge of the wing. *(United Technologies Corp./Pratt & Whitney)*

by an extension shaft, it's doubtful it was fitted. Photos and what little information has survived indicate that the R-4360-13 powering the XP-72 was boosted with a General Electric turbosupercharger.

The foregoing describes two-stage engines with gear driven auxiliary superchargers. Another alternative is to use turbosuperchargers driven by exhaust gasses. This aspect of supercharging is covered in Chapter 4 (Turbocharging, VDT & Other Exhausting Ventures).

Direct Drive

Only one application called for a direct drive—the Northrop B-35. Having the engine buried in the wing demanded that some form of extension shaft was used to drive the reduction gears. It would have been possible to incorporate the propeller reduction gears with the engine, but then the torque multiplication resulting from the reduction gears would have required a far heavier drive shaft. An accessory nose section was still retained for the magnetos and scavenge pump drive requirements. However, instead of the usual nose case, a so-called crankshaft extension ball bearing support—basically a blanking plate—replaced the reduction gear assembly and nose case *(Ref. 2-32 and 2-33).* At one time Sikorsky had entertained the idea of using an R-4360 to power a helicopter. If this idea had ever matured into fruition, it is quite probable that the B-35 type power arrangement would have been used to transmit power to the rotor head and its reduction gearing.

Cooling (Baffling Variations)

Despite untold hours developing a complex baffling system, the R-4360 went through a number of variations after it entered production. Perhaps the most noticeable change was the use of the so-called "hooded baffle" *(Ref. 2-6).* Early engines had a tendency to fry the ignition harness. To overcome this deficiency, additional baffles were placed over the tops of the cylinder heads of the last three rows. For a tractor configuration, rows A, B, and C had hooded baffles. For a pusher installation such as the B-36 this would mean rows B, C, and D. This improvement forced additional cooling air over the cylinder heads, at the same time giving the ignition harnesses some relief from the blast furnace environment they operated in. The accompanying figures show a comparison of the two styles of cooling baffle. It is also worth noting the difference in the early style nose of the R-4360-15 compared to the later R-4360-63. The early nose cases exhibited a distinct bell mouth shape, whereas later nose cases such as shown in the R-4360-63 photograph used a straight taper from the accessory case to the propeller shaft. No satisfactory explanation has been forthcoming as to why this seemingly minor change was made.

Early R-4360s, this one a -15, did not have the so-called "hooded baffles." In-service problems revealed a propensity for engines without hooded baffles to cook the ignition system. *(United Technologies Corp./Pratt & Whitney)*

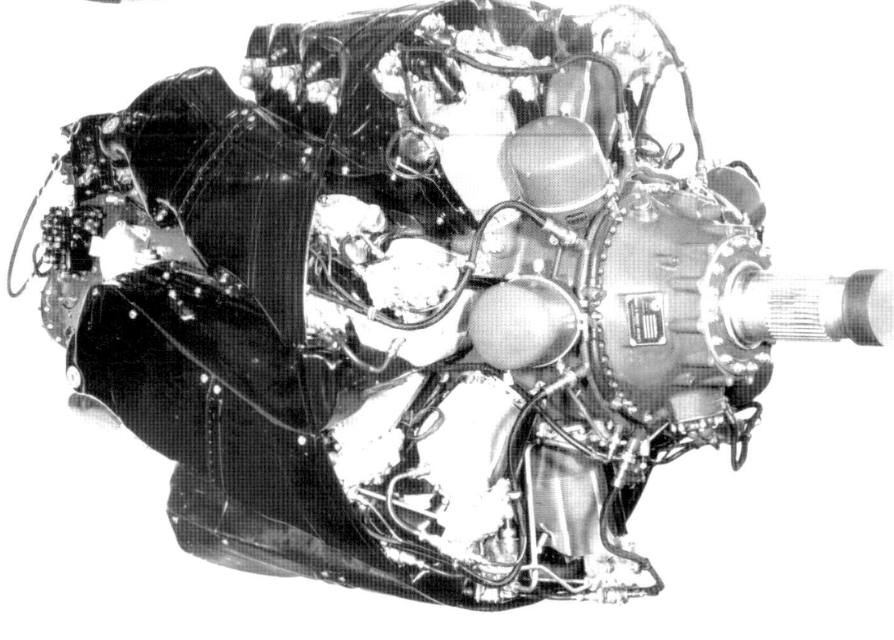

Later engines featured "hooded baffles" to keep ignition components cool–including the 56 boost coils used on low-tension ignition models. *(United Technologies Corp./Pratt & Whitney)*

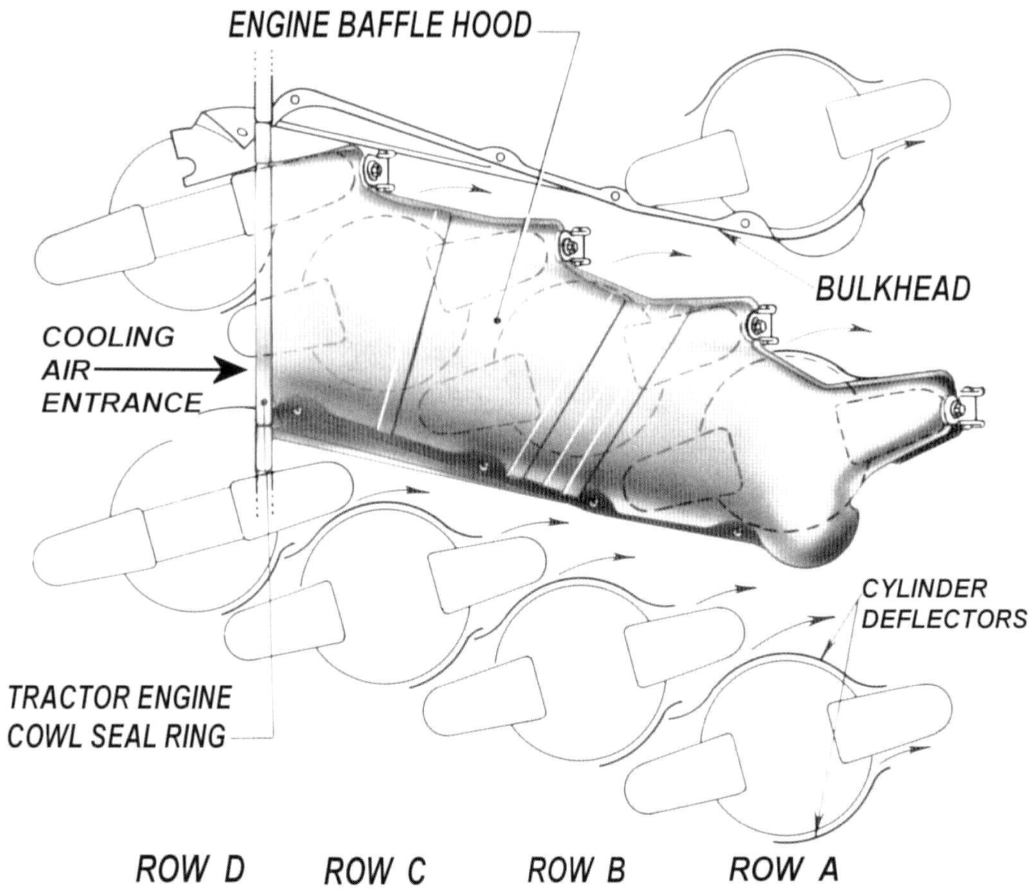

This illustration shows hooded baffles fitted to a bank of cylinders. *(Drawing by author based on an illustration in Installation Handbook Wasp Major B Series Engines)*

Hanging By A Thread

When you think about it, it's a miracle that Pratt & Whitney engineers managed to design a lightweight yet strong mount for the R-4360. But like everything else on this huge engine, it was never smooth sledding. Cantilevering 5,000 plus pounds (engine and propeller) from a lightweight wing spar posed some pretty daunting challenges to the engineers. Traditional radial engine designs incorporated six or more dynafocal[1] mounts cast into the blower housing. A lightweight tubular structure then picks up these engine mounts and transfers the loads into the airframe. Due to the

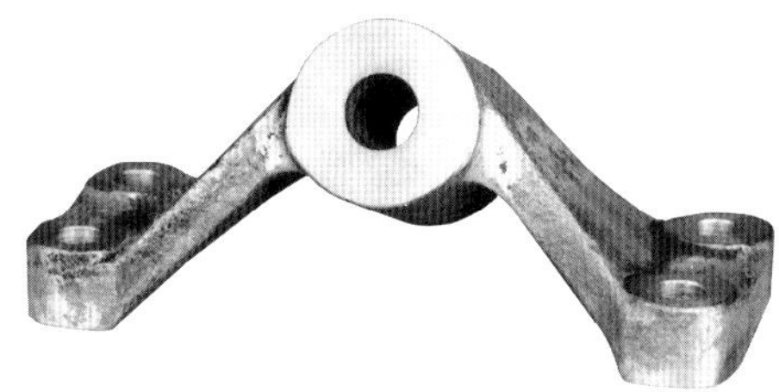

This steel forged bracket served as the mount for installing R-4360s in shipping containers. *(United Technologies Corp./Pratt & Whitney)*

R-4360's immense weight and length, this practice would have overloaded the blower housing or resulted in an unacceptably heavy blower

[1] The term "dynafocal" is a mounting system whereby if an imaginary line were drawn through each mount, these imaginary lines would converge at the center of gravity of the mass being supported.

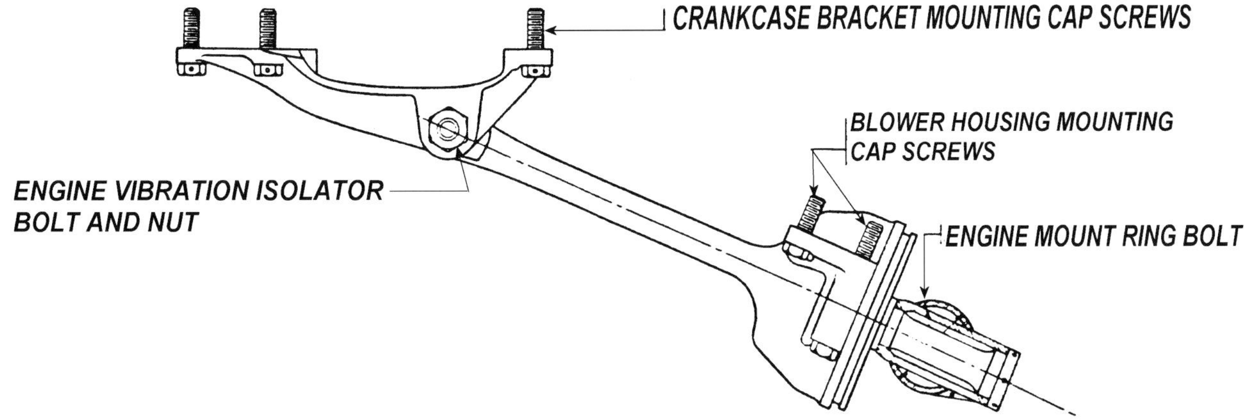

It was soon realized that cantilevering R-4360s way out, especially with a huge propeller hung on the front, would require something more substantial than the usual radial engine mount. A bracing bar terminating at a bracket bolted to the crankcase served the purpose of offering additional support. *(Installation Handbook Wasp Major B Series Engines)*

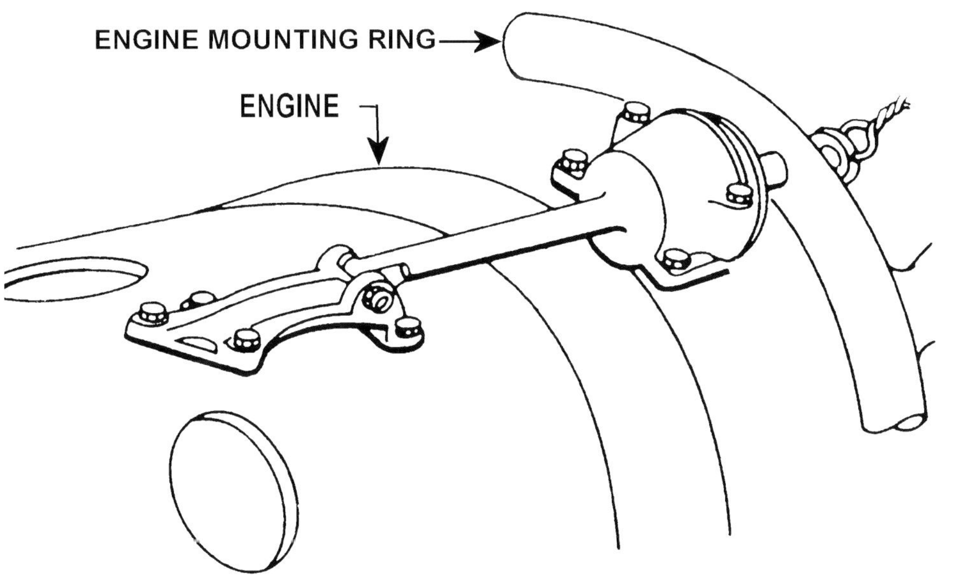

This line drawing illustrates the relationship between the engine mount, bracing bar, and tubular aircraft mount. *(Martin Mauler Erection & Maintenance Manual)*

housing. For the R-4360, seven mounts were employed, lined up between each cylinder on the A row. An additional strut designed as an extension to each of the seven mounts was incorporated as part of the mounting scheme. A simple steel bracket was used to attach the engine to shipping containers. An additional bracket at the front supported the propeller shaft within the container. The seven struts extended from the mount integrated into the blower housing. They then extended to a bracket bolted to the crankcase. The mounting design went through a number of significant changes. However, all variations used seven mounts—one between each bank of cylinders. In this way it was possible to extend the struts well into the power section between banks. On early engines the struts extended to a bracket bolted to the crankcase situated between cylinders on row A.

Progressing through an evolutionary process, the strut was extended to a bracket located between rows A and B. The last of the strut designs extended even further forward, terminating between rows B and C. Along the way, ball joints were introduced to offer some flexibility. It should be remembered that under power, the R-4360's crankcase would flex to a remarkable degree. Prototype engines utilized a forged-steel I-beam for

This photograph shows the early bracing bar setup with the crankcase bracket terminating at the "A" row of cylinders. *(United Technologies Corp./Pratt & Whitney)*

Progressing from the mount in the prior illustration, ball joints were incorporated at both ends of the bracing bar to offer a greater degree of flexibility to handle vibration and heat expansion. *(United Technologies Corp./Pratt & Whitney)*

the strut. All subsequent designs used a simple bar. As the strut was extended ever forward to the front of the engine, it became necessary to angle it in relation to the longitudinal axis of the engine. This was to avoid interference with cylinders in rows A and B as well as to conform to the helix angle of the cylinder banks. This may have been a fortuitous necessity. The angularity of the strut would also assist in resisting torque reaction.

One exception to these mounting systems was the one employed by the Northrop B-35. A buried installation allowed Northrop engineers to get away from having to cantilever the engine with its resulting enormous loads. Instead, the B-35 applications used the seven existing mounts at the rear but without the struts. An additional circular mount, situated at the front and co-axial with the output shaft, was used to support the forward end of the engine. Remember, the B-35 applications used direct-drive engines with no reduction gear integral with the engine. Multi-engine applications required an additional space frame in order to tie into the wing spar. Single-engine applications were mounted directly to the firewall.

Vibration Isolation

The foregoing describes the engine mounts. Incorporated within each mount are a vibration isolator and damper unit. The "MB" company of

This is the first variation on the braced mount. This unit is from a very early R-4360-2. *(United Technologies Corp./Pratt & Whitney)*

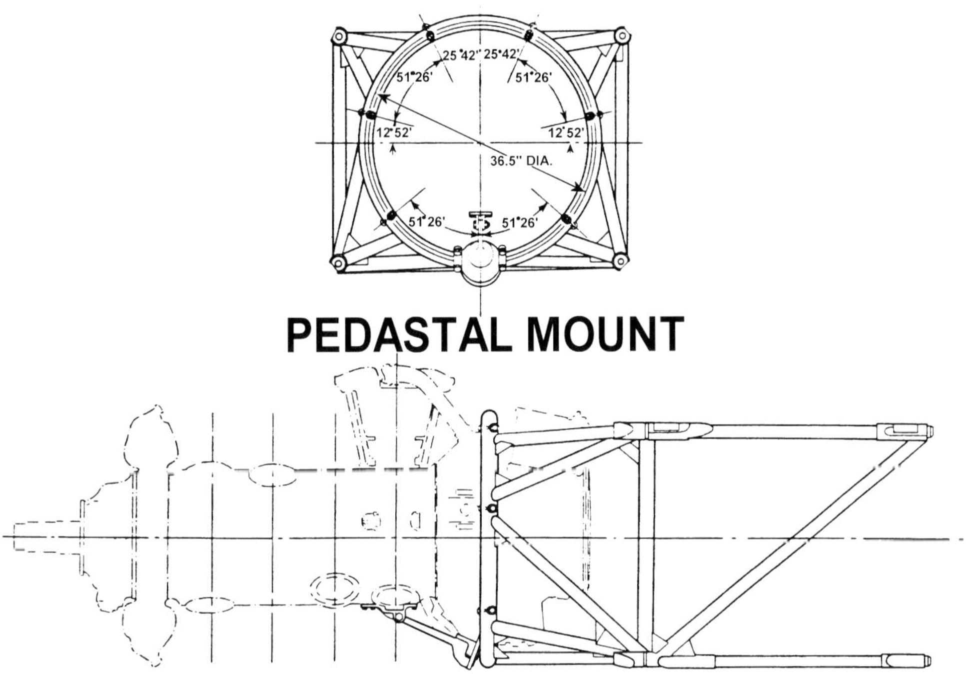

Once the engine is attached to the engine mounts, the entire assembly needs to be attached to the aircraft. This line drawing shows how a typical multi-engine application ties in the mount to the wing spar. The angle of 51 degrees, 26 minutes is a significant one–it represents the angular displacement between cylinders. *(Installation Handbook Wasp Major B Series Engines)*

New Haven, Connecticut, developed and manufactured R-4360 mounts and vibration isolators. MB vibration isolators, shown in Fig. 3-12A, were used to mount the engine in the aircraft *(Ref. 3-2)*. Their function was to isolate powerplant vibration from the aircraft. Each isolator of the type used on the R-4360 consisted of a laminated steel and rubber core assembly, a housing assembly, a crankcase bracket, and a brake and bearing assembly, in addition to bolts, nuts, washers, and other small parts. The complete unit was attached directly to the engine blower case and to the engine crankcase. The stem of the laminated core was attached to the engine mounting-ring in the aircraft with a mount-ring bolt. The vibration isolator assembly was furnished either with or without the mount-ring bolt, depending on the requirements of the particular installation.

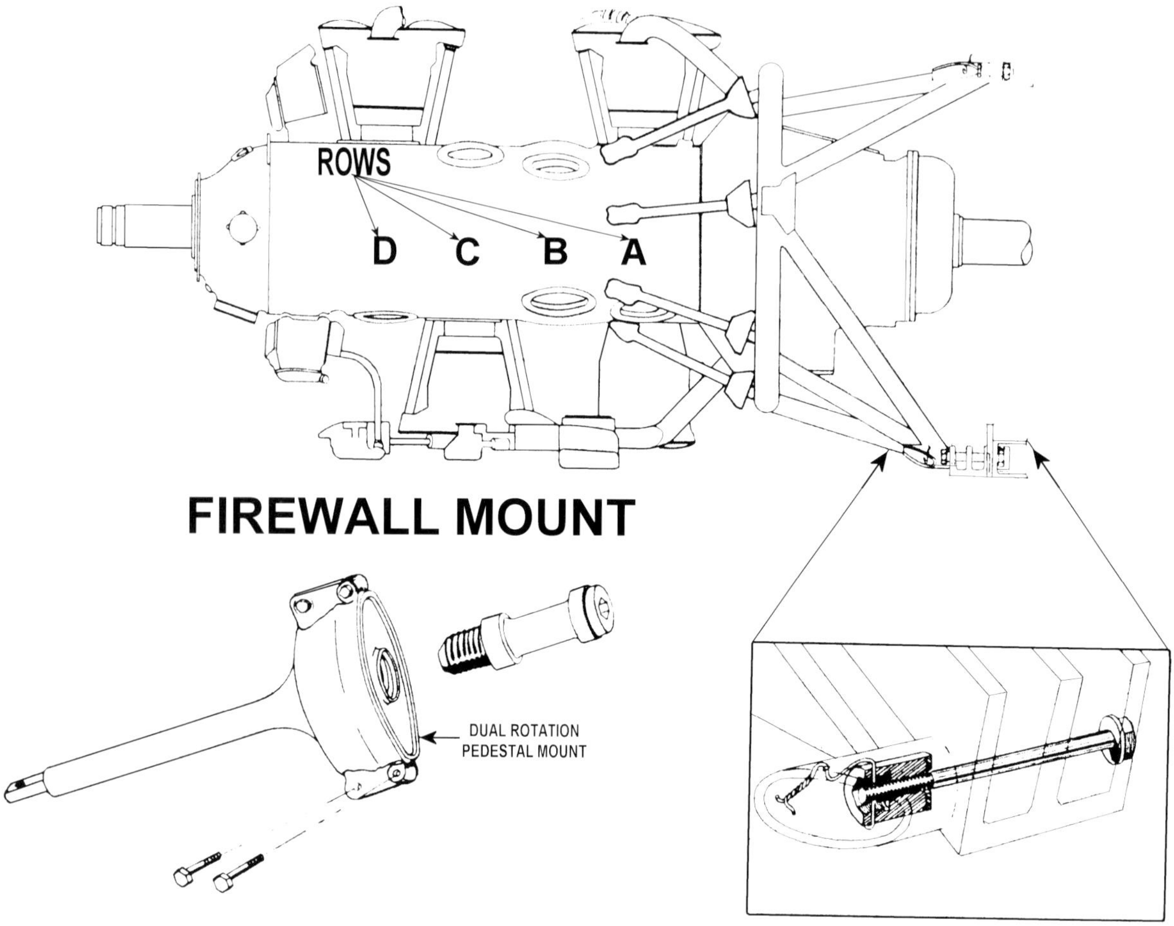

Single-engined aircraft typically bolt the tubular mount structure to the firewall, which then transfers the loads through the monocoque fuselage structure. *(Republic XP-72 Erection & Maintenance Manual)*

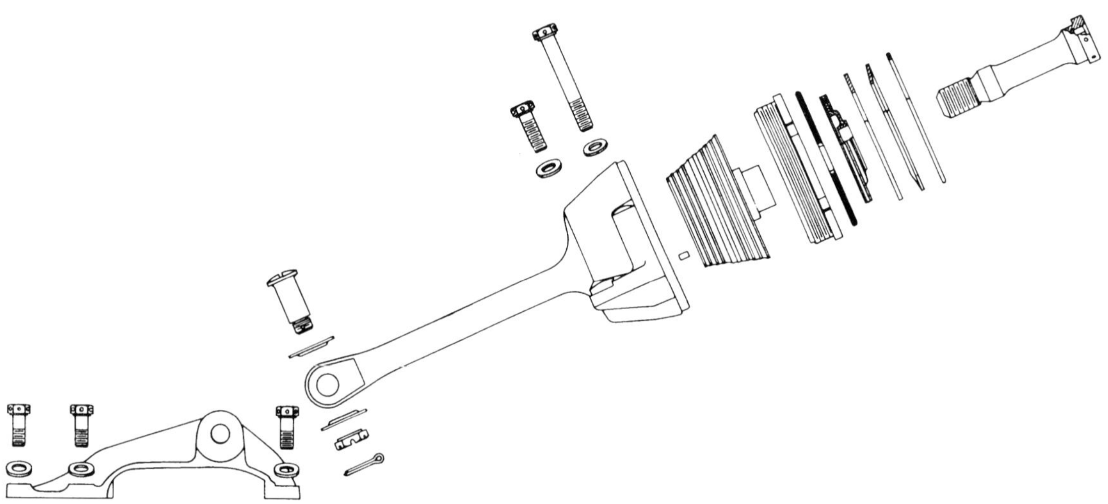

Separating the airframe from engine vibration is the engine mount isolator. Rubber disks attenuate vibration and a diaphragm spring-loaded clutch facing dampens the amplitude. *(MB Engine Vibration Isolators for Pratt & Whitney Wasp Major [R-4360] Aircraft Engines)*

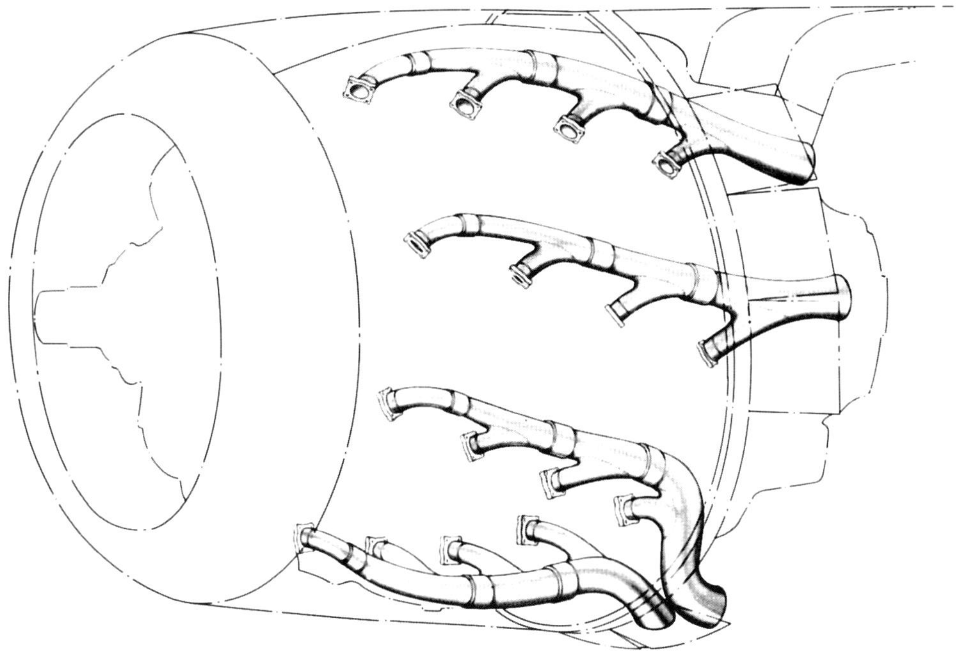

Some early applications of the R-4360 did not take advantage of exhaust scavenging and therefore simply manifolded each bank of four cylinders into one discharge pipe. *(Installation Handbook Wasp Major B Series Engines)*

Blowin' in the Wind

High-performance engines demand an efficient exhaust system, otherwise valuable horsepower is squandered. With 28 massive cylinders pumping huge amounts of spent combustion products overboard, it was especially important to ensure no restriction would interfere with this task. Furthermore, by taking advantage of exhaust scavenging, further improvements in volumetric efficiency could be gained. In turbosupercharged applications, an efficient exhaust system enhances the performance of the gas turbine.

Four basic types of exhaust system were employed by the R-4360: (i) manifolded, (ii) siamesed, (iii) collector for feeding into a turbosupercharger, and (iv) collector ring with no turbo. The term "open stack" is often used in aircraft exhaust systems. It simply means that no turbosupercharger or power recovery turbine is employed. In other words, once the exhaust valve opens, the products of combustion are dumped overboard via some type of exhaust system. Exhaust systems (i), (ii), and (iv) would be termed open stack. Some applications used a combination of the methods listed. For instance, the Douglas C-124 used a siamesed system and some exhaust was collected in a quasi collector ring setup before being dumped overboard. A well-designed open stack system could enhance engine cooling and aircraft performance. Cooling can be enhanced by the powerful pumping effect of the exhaust as it exits the tailpipe. By placing the tailpipe at the optimum position in relation to the cowl flaps, additional cooling air is forced through the cowl. Furthermore, the powerful jet effect adds thrust to that already produced by the propeller. Generally speaking, open stack systems would be used on aircraft whose mission profile did not require operation at altitudes higher than 15,000 feet. Conversely, aircraft operating in the upper atmosphere benefited enormously from turbosuperchargers. The higher the altitude for a turbo, the more efficient it becomes. Furthermore, bleed air from the turbo provided pressurization without resorting to engine driven cabin blowers *(Ref. 2-6)*.

I. Manifolded

A manifolded design, also known as four port headers, collects exhaust from a bank of four cylinders and dumps it overboard via seven pipes. No exhaust tuning or scavenging is possible with this system. Of all the open stack systems, this

After studying the timing diagram of the R-4360 it quickly becomes apparent that if D and B cylinders were siamesed and C and A cylinders were siamesed, as shown in this illustration, a significant exhaust scavenging effect could be gained. *(Installation Handbook Wasp Major B Series Engines)*

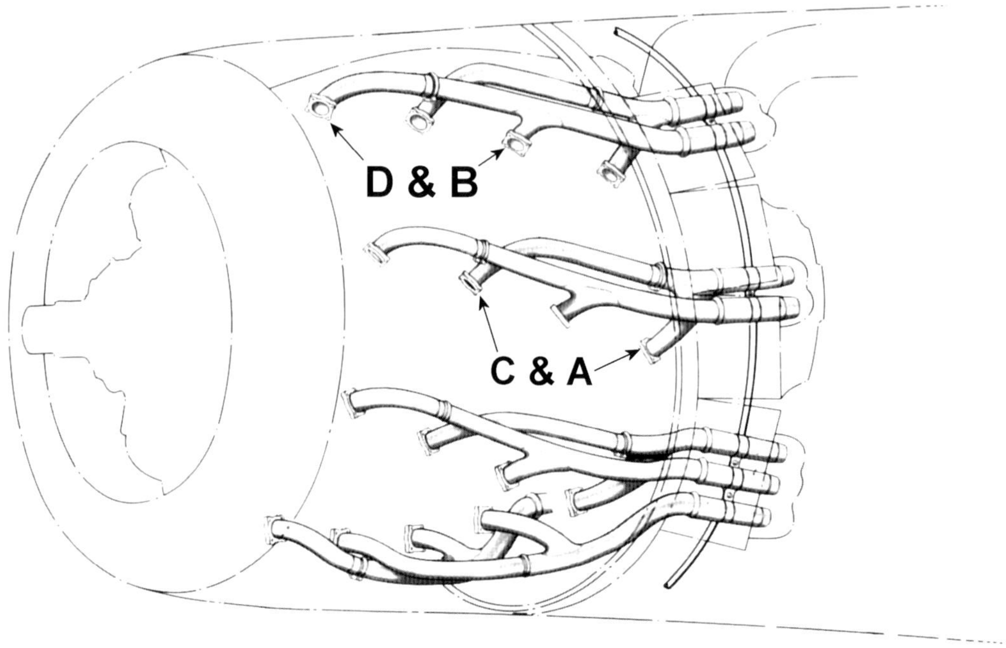

one offered the least performance gain. Although less desirable from a scavenging perspective, it did see some applications.

II. Siamesed

By utilizing the exhaust pulsations due to the opening and closing of the exhaust valve, the exhaust scavenging can render significant gains. This translates into overall improved volumetric efficiency—the premise behind the siamesed system. On any bank, cylinders from rows A and C would be siamesed. Likewise, cylinders from rows B and D would be siamesed. The trick is to ensure that only one exhaust valve is open at one time. This was accomplished with the siamesing sequence explained previously. In fact, for a period of 80 crank degrees, both exhaust valves were closed. This was not the case for the manifolded design where at least two exhaust valves would be open for 100 degrees of crank rotation.

III. Collector Ring in Conjunction With a Turbo

Rather than simply dumping combustion products overboard, the large amount of energy contained in the exhaust can be utilized to drive a turbosupercharger. When the exhaust valve opens, upwards of 100 psi residual pressure resides within the cylinder. This is the driving force for the gas turbine that gives motive power to the turbosupercharger. A turbo'd system demands some form of collector ring in order to gather the spent gasses from the seven banks of cylinders. The seven manifolds discharge into the common collector ring made from a large diameter pipe. A gas turbine, integral with the turbosupercharger compressor, is joined to the collector ring discharge via a ball joint. Some applications, such as the Convair B-36, Hughes XF-11, and Northrop B-35, utilized two turbosuperchargers per engine; however, the principle remained the same. Instead of having one discharge from the collector ring feed the gas turbine, two discharges are employed. Every turbo'd system needs a method to control exhaust gas flow through the turbine section. This is accomplished via a waste gate—a simple butterfly valve. Without a waste gate, unacceptably high manifold pressure is generated, particularly at high power settings at low altitudes. The waste gate system is integral with the exhaust system. It diverts exhaust gases around the turbine thus reducing the energy it imparts to the compressor.

IV. Collector Ring With No Turbo

Less common than the previously mentioned

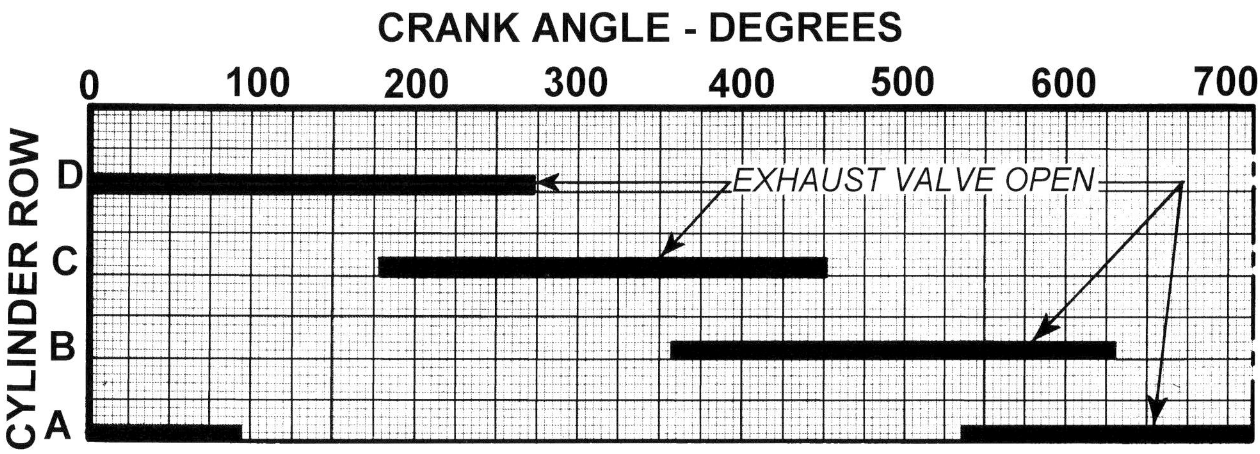

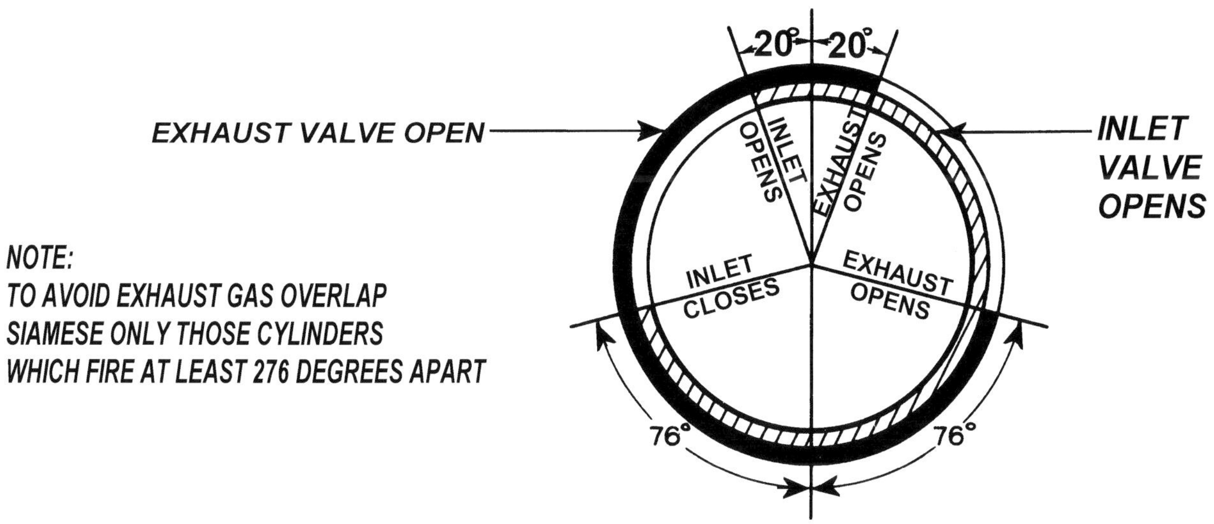

This chart graphically illustrates in very simple terms why D and B should be siamesed and A and C should be siamesed. With this arrangement only one exhaust valve is open at any one time therefore maximizing the scavenging effect. *(Installation Handbook Wasp Major B Series Engines)*

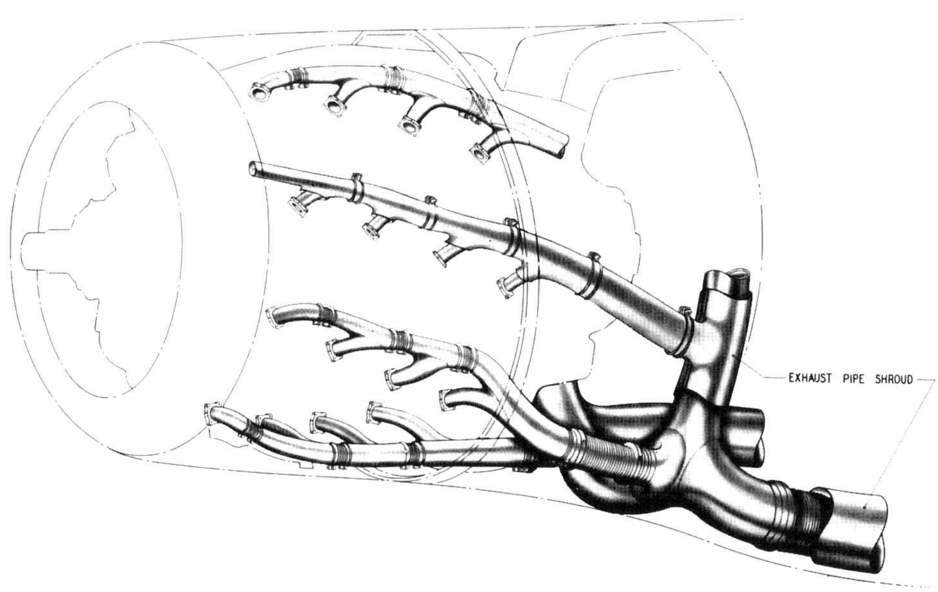

With a turbo'd installation there is little choice regarding exhaust systems. All seven manifolded banks discharge into a common collector ring, which then feeds the turbosupercharger(s). *(Installation Handbook Wasp Major B Series Engines)*

exhaust systems, some applications used a collector ring. These were similar to those used on a turbo'd system except that the exhaust gasses were dumped overboard rather than being utilized to drive a gas turbine for a turbosupercharger. Of course, this system lost any advantage that could be gained from siamesing. Yet this was the method chosen, for example, by Fairchild for the C-119.

Into Production

Ford Motor Company

Ford was chosen by Pratt & Whitney to produce "B" series R-2800s during World War II. Starting from nothing, not even a manufacturing facility, Ford managed to produce 57,637 of these complex engines in three years *(Ref. 3-3 and 3-4).* And for the most part, their quality was first class. After World War II, the Rouge plant, where the R-2800 had been manufactured, was converted to auto production. Ford's production feat left an indelible mark on Pratt & Whitney and the Air Material Command (AMC). When the Air Material Command mandated that defense contracts would be based on a "licensor/licensee" arrangement, it was no surprise that Ford would have a good shot at being awarded a production contract. In other words, the designer of a particular product was not necessarily awarded a production contract. If the development (in this case the R-4360 aircraft engine) was funded by the government, then the government would own the intellectual property rights to that product. This arrangement allowed the Air Material Command to award production contracts to whomever it wanted. With this understanding, it's easy to see that the contract for the R-4360 was between Ford Motor Company and the Air Material Command. Not, as one would think between Pratt & Whitney and the Air Material Command.

With the postwar boom in full swing, Ford was committed to its other products such as cars and trucks. So the big concern for Ford was:

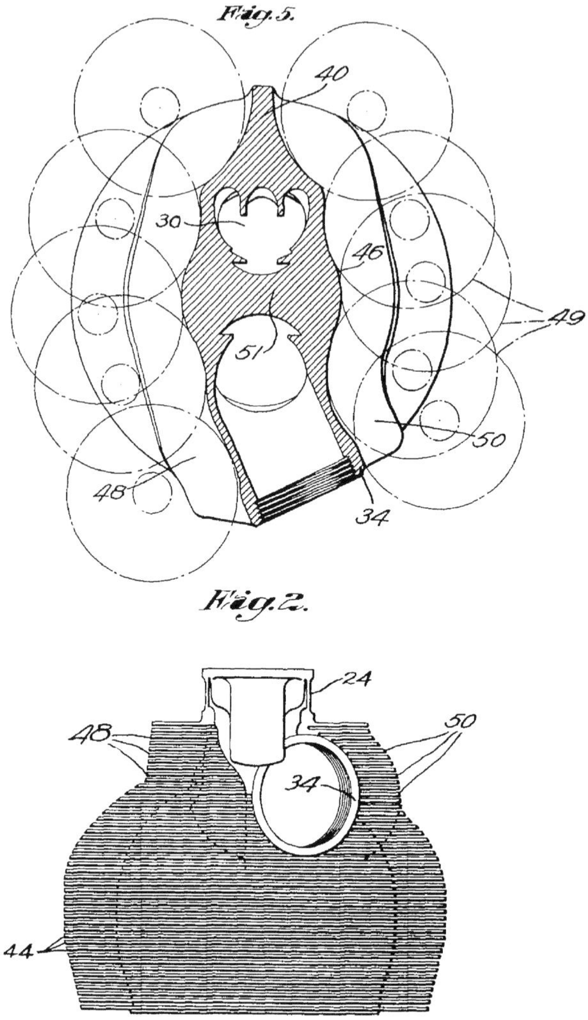

This patent drawing shows the tool path for machining in cylinder head cooling fins from a raw forging. Pratt & Whitney pioneered this cylinder design with "C" series R-2800s. *(United States Patent 2,401,210)*

where to manufacture the R-4360? During World War II, the government built a huge facility run by Dodge Chicago for the manufacture of Wright R-3350s to power Boeing B-29s. It's a little known fact that Dodge manufactured more B-29 engines than Wright. At the conclusion of the war, Dodge moved out of the facility located at 7401 South Cicero Avenue in Melrose Park, a suburb of Chicago. The facility lay empty for a couple of years until an enterprising Preston Tucker happened upon the scene. Part con artist and part genius, Tucker envisioned the highways of the U.S. being populated with his innovative

This Footburt automatic drill was a valuable asset once Ford Motor Company could get it on line. The Korean War took precedence over Ford's production of the R-4360; consequently they had to wait for equipment such as this. Note the raw cylinder head forgings coming off the Footburt after going through a number of drilling, tapping, and reaming operations. A considerable amount of work still remains on this complex manufacturing job, not the least of which is machining in the cooling fins. *(From the Collections of Henry Ford Museum & Greenfield Village)*

Pallet loads of cylinder head forgings ready to have their cooling fins machined in. This particular photograph was taken at Ford Motor Company's facility. Note the crankcase sections in the background also ready for machining operations. *(From the Collections of Henry Ford Museum & Greenfield Village)*

car. He talked the War Assets Board into leasing the old Dodge Chicago plant to him. After just 51 cars had been built, Tucker's empire collapsed *(Ref. 3-5)*. Once again, the old Dodge Chicago facility was left empty and unused. The fact that the facility had originally been built to manufacture aircraft engines made it an ideal location for Ford's R-4360 production plans. However, years of disuse and abuse by Tucker had left the building in sad shape; it was in disrepair and needed a major cleanup.

Tucker Tantrums

Although Tucker had vacated the old Melrose Park production facility after his futile bid to overthrow the major auto manufacturers, he was not going to lie down and let Ford march into his old facility. Notwithstanding the fact he never even owned these facilities (they were leased to him under the War Assets Act), Tucker, ever the con man, tried every trick in the book to hang on to his manufacturing site. To protect Ford from possible litigation, an unusual concession was made by the government in the contract. Through the government, the Air Material Command offered full indemnity and protection to Ford in the event Tucker would ever attempt to sue Ford for its use of Melrose Park. Knowing that Tucker was the consummate con artist, it's probably fair to say that the Ford attorneys requested this clause in the contract *(Ref. 3-6, 3-7)*.

The task facing Ford was a daunting one. Of the 2,000 unique parts that went into an R-4360, Ford would make 227, the rest being farmed out to sub contractors. The remaining dilemma was how to manufacture these 227 high-precision parts. It would take 10,872 machining operations and entail the use of a staggering 6,000 machine tools to accomplish this task. Just as Ford was trying to gear up for production, the Korean War broke out. This put a different emphasis on machine tool requisition. Rather than wait the 12 to 14 months estimated for the specialized tooling to be delivered, Ford took advantage of the huge inventory of tools left over from World War II. Although the World War II tools would not offer the mass production capability Ford was wishing for, at least it offered a viable stop gap. As an example, Ford had ordered a 40-spindle machine tool that drilled, tapped, and reamed. During the interim the solution was to simply use a battery of radial arm drills that could handle one hole at a time. As they say: Whatever it takes. By April 1951, 3,000 people were employed at the Chicago plant. This number would soon rise to 18,000. Not surprisingly, the manufacturing endeavor

Ford Motor Company personnel inspecting cylinder heads prior to having their cooling fins machined in. It should be noted that with any aircraft engine component, each manufacturing step is inspected. By the time a complex part, such as a cylinder head, is complete it has undergone numerous inspections. *(From the Collections of Henry Ford Museum & Greenfield Village)*

facing Ford must have seemed like an impossible mission. During late 1951, Ford had over 20,000 R-4360 parts finished yet could not assemble one complete engine *(Ref. 3-8, 3-9, 3-10 and 3-11)*.

In September 1950, Henry Ford II issued a memo announcing the establishment of the "Organization of Defense Production Activities." This new division, named Aircraft Engine Division, was formed to concentrate on lucrative defense contracts. For the R-4360 job, thousands of sub-contractors were taken on board, including Harley-Davidson, for the production of magnesium and aluminum castings and forgings. Speed Queen Corporation, a washing machine manufacturer, and Sterling Window Company located in Newcastle, Indiana, were engaged for the production of aluminum stampings for cooling baffles, etc. It can be seen that Ford was not in this alone even though they were the primary contractor. Cylinder head manufacturing, in particular, demanded significant manufacturing capability. Starting out with a raw billet of aluminum, numerous forging operations were necessary to arrive at the rough shape of the cylinder head. Machining the cylinder head required sophisticated equipment such as precision slitting saws to create the thin yet deep cooling fins. Pratt & Whitney patented its own method for this tricky machining operation.

Ford technicians test running a Ford-built R-4360. It can be seen through the glass in the test cell. Judging by the extended nose case, the engine appears to be one for a B-36. *(From the Collections of Henry Ford Museum & Greenfield Village)*

After the test run, engines were typically stripped down to check for anomalies. *(United Technologies Corp./Pratt & Whitney)*

The R-4360 production contract was signed on September 15, 1950. Ford had an excellent track record and had pulled off production miracles before, some of the more notable being the prodigious World War II production of Consolidated B-24 Liberators and R-2800s. Notwithstanding their stellar performance on prior projects, production of the R-4360 must have been daunting. A total of 12,269 process sheets, 2,211 summaries of operation sheets, and 10,507 special tool drawings would be required. Over 3,000 machine tools were obtained through the Air Force—and this was simply to get production rolling. As experience was gained, specialized

This is a photograph taken in Pratt & Whitney's East Hartford facility. The crankshaft is the heart of the engine and this is where final assembly begins. *(United Technologies Corp./Pratt & Whitney)*

tooling was brought on line to facilitate production. The R-4360 project turned out to be a tough nut to crack—even with the formidable manufacturing capability of Ford Motor Company. By the early 1950s Ford had a pretty good handle on this monumental manufacturing challenge and R-4360s were rolling off the assembly line. As per usual practice with a new aircraft engine, each engine was disassembled and checked for anomalies after the obligatory break-in and test run. Initially, Ford built R-4360-53s for Convair B-36s; however, by the time they built their last engine in 1954, the following engines had been manufactured: R-4360-53, R-4360-59, R-4360-61, R-4360-63, and R-4360-63A.

East Hartford Production

The only other facility to produce R-4360s was Pratt & Whitney's East Hartford facility. Ford Motor Company was thrust into the production of the R-4360 with no manufacturing facility and little understanding of an R-4360. On the other hand, Pratt & Whitney's progression to production was evolutionary. Starting out with experimental engines and then the limited production that commenced just prior to the ending of hostilities in World War II, Pratt & Whitney were slowly weaned into producing this enormous engine. It appears from records that Ford only built engines for military contracts and Pratt &

continued on page 170

Left: This photograph gives an excellent perspective of the assembly procedure. Starting with a vertically mounted crankshaft, the crankcase sections are assembled onto it until a complete power section is ready. *(United Technologies Corp./Pratt & Whitney)*

Below: After the power section is assembled, the nose case and magnetos are installed. *(United Technologies Corp./Pratt & Whitney)*

Right: When the power section, nose case, and ignition system are built up, it's time to drop this assembly onto the supercharger. *(United Technologies Corp./Pratt & Whitney)*

Below: Final assembly prior to test. Note that rocker covers are removed. Checking valve clearances will be one of the last requirements before test running. *(United Technologies Corp./Pratt & Whitney)*

Left: B-36 engines being inspected prior to crating for shipment to the customer–the U.S. Air Force. *(United Technologies Corp./Pratt & Whitney)*

Below: This photo shows an engine being prepped for the test cell. Note the air duct mounted on the carburetor. This duct would lead to a fresh air supply outside the test cell. *(United Technologies Corp./Pratt & Whitney)*

Right: Almost time to run. Final preparations are made in the test cell. *(United Technologies Corp./Pratt & Whitney)*

Below: Running under its own power for the first time in the test cell. *(United Technologies Corp./Pratt & Whitney)*

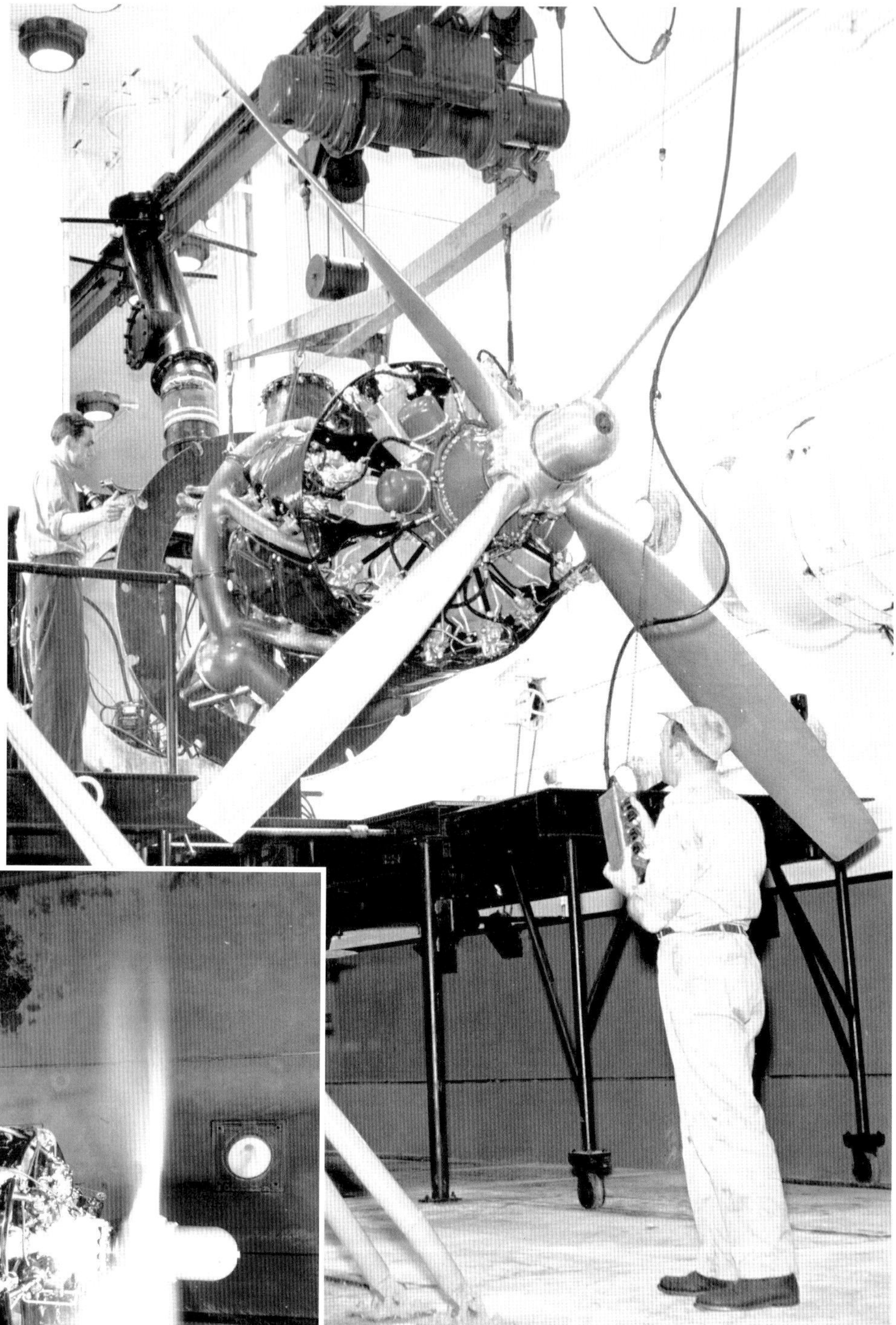

Opposite top: At the height of R-4360 production, engines were churned out on a production line. *(United Technologies Corp./Pratt & Whitney)* *Opposite bottom:* A B-36 engine appears to have been tested and is ready for shipment. The hooded baffles still need to be installed. *(United Technologies Corp./Pratt & Whitney)*

At least 32 new R-4360s are visible in this photograph. *(United Technologies Corp./Pratt & Whitney)*

Summary of R-4360/Wasp Major Production *(Ref. 3-12)*

(Table 3-1)

Engine Model (Navy)	Engine Model (USAF/Army)	Engine Model (Commercial)	Quantity (Navy)	Quantity (USAF/Army)	Quantity (Commercial)	Quantity (Total)
	TSB-1G			4	4	
-4	-4	-4	197	1	2	200
	-4A			16		16
-4W			130			130
		TSB-11G				
	-5			8		8
	-7			2		2
-8			5			5
-8A			2			2
	-9			2		2
		VSB-11G			25	25
		B13			46	46
	-9T			1		1
-10		-10	2		4	6
	-11			1		1
	-13			3		3
-14			2			2
	-15			1		1
	-17			28		28
-18			13			13
-20	-20		78	197		275
-20W	20W		37	546		583
-20WA	-20WA		155	1,948		2,053
	-21			28		28
-22W			24			24
	-25			221		221
	-27			84		84
	TSB-3G			391	391	
	-31	-31		27	1	28
	-33			6		6
	-35	-35		1,929	2	1,931
	-35A			382		382
	-35B			661		661
	-35C			598		598
	-41			947		947
	-41A			269		269
	-43			8		8
	-45			13		13
	-47			13		13
	-53			2,240		2,240
	-55			16		16
	-59			8		8
	-59B	-59B		4,260	2	4,262
	-61			6		6
	-63			6		6
		CB2			2	2
GRAND TOTAL:						**15,600**

Production Totals By Year

(Table 3-2)

Year	Navy	Army/USAF	Commercial	Total
1943	2	6	3	11
1944	11	11	5	27
1945	66	45	3	114
1946	142	238	7	387
1947	68	722	195	985
1948	21	984	215	1,220
1949	144	1,425	7	1,576
1950	36	1,565	12	1,613
1951	102	2,227	5	2,334
1952	53	2,750	26	2,829
1953	0	2,258	0	2,258
1954	0	1,608	1	1,609
1955	0	637	0	637
Sub Totals:	645	14,476	479	
GRAND TOTAL:				**15,600**

continued from page 162
Whitney produced both commercial and military engines. The following figures go through the various stages of manufacturing starting with a bare crankshaft to delivery to the customer.

R-2180

Arguably the finest piston engine developed by Pratt & Whitney was the R-2180. But like many things in life, it was a day late and a dollar short. Often described as half an R-4360, this unique 14-cylinder engine was in fact quite different. However, from an appearance perspective it certainly looked like an R-4360 chopped in half. The history of the R-2180 goes back to the 1930s when Pratt & Whitney developed the R-2180 Twin Hornet intended for the DC-4. However, the first DC-4 sporting triple-tails bore no resemblance to the later production models powered by Pratt & Whitney R-2000s. At the onset of World War II, the R-2180 Twin Hornet was shelved. After the cessation of hostilities, the R-2180 concept was resurrected. But this time around considerable experience had been gained on engine design, reliability, and serviceability.

Starting with a clean sheet of paper, the R-2180 Twin Wasp E was developed. R-4360 influence was written all over it. Nevertheless, it incorporated many new and unique features. Like the R-4360, a fore and aft valve arrangement was incorporated. Cylinder design also closely followed that of the R-4360—with one big difference. R-4360 cylinders had a downdraft intake

This B-36 engine is being packaged for shipment. *(United Technologies Corp./Pratt & Whitney)*

Often described as half an R-4360, the R-2180, although very similar in many respects, diverted from R-4360 design philosophy in key areas–the main difference being the opposite orientation of the ports. The intake was now off to the side and the exhaust exited on top of the head. *(United Technologies Corp./Pratt & Whitney)*

port arrangement with the intake port located on top of the cylinder. The exhaust port was off to one side of the cylinder. Although the R-2180 cylinder looked identical, the port arrangement was swapped. The exhaust port was now located on top of the cylinder where the intake port was for the R-4360 and the intake port was off to the side where the exhaust port was on the R-4360. Variable timing was incorporated in the R-2180 design.

The updraft carburetor used a fuel-metering device mounted on the side of the rear case. In

Front view of a cowled R-2180. Even though it could have been Pratt & Whitney's finest piston engine, relatively few were manufactured. It would have made an excellent powerplant for the DC-4/C-54; however, by the time the R-2180 was in production no one cared, so the DC-4 soldiered on with R-2000s. *(United Technologies Corp./Pratt & Whitney)*

other words, it was separate from the throttle body. The fuel-metering device also incorporated the water regulator and fuel pump. One-piece maser rods in conjunction with a built-up crank were another differentiator from the R-4360. Many other innovative design features distinguished the R-2180 from the R-4360 (or any other radial engine for that matter). It's often tempting to play what-if games. What if the R-4360 had incorporated the top exhaust, side intake system? Would it have cooled better? Probably so. What if the variable valve timing had been incorporated? How much additional power would have been available? Based on claims for the R-2180, about another 200 hp. If one-piece master rods and a built-up crank had been incorporated into production R-4360s, more power certainly could have been extracted.

So if the R-2180 was such a great engine, how come it never went anywhere? The answer is easy. Pratt & Whitney was now fully engaged in gas turbine development, plus there were plenty of

other engines available in the horsepower range offered by the R-2180. In fact, the R-2180 only saw two applications—the Piasecki H-16 helicopter and the Svenska Aeroplan Aktiebolaget Model 90A-2 and Model 90B-2. The Svenska Aeroplan Aktiebolaget Model was a twin-engined commercial aircraft somewhat akin to a Martin or Convair twin. Neither aircraft was built in any significant numbers. In fact it appears that only about 60 R-2180 Twin Wasp Es were ever built.

References

3-1 *Pilot's Manual For Grumman F6F Hellcat.* Aviation Publications, 1973.

3-2 *MB Engine Vibration Isolators for Pratt & Whitney Wasp Major (R-4360) Aircraft Engines.* New Haven 11, Conn. Copyright 1948.

3-3 Folder titled "*Misc. Data*" located in Henry Ford Museum archives, Greenfield Village.

3-4 Folder titled "*Misc. Data*" #596, Box 2 located in Henry Ford Museum archives, Greenfield Village.

3-5 http://www.tuckerclub.org/, http://www.freespeaker.org/Americanhistory/tucker-greg.html

3-6 Nevins, Allen and Frank Ernest Hill. *Ford, Decline and Rebirth 1933 – 1962.* Charles Scribner's Sons.

3-7 Ford Motor Company Aircraft Engine Division Press Conference. December 18, 1952, Sheraton Hotel, Chicago, Illinois.

3-8 Wirry, Anthony. *Tools For War Being Readied In Vast Plant.* Chicago Sunday Tribune. March 18, 1951.

3-9 *Moving Toward Production.* Ford Aircraft Engine News. March 15, 1951.

3-10 *Wasp Major: Plane Engine's Story Typifies Defense Output Problems.* Wall Street Journal, May 1951.

3-11 *Wasp Major: Huge Plane Engine's Tells Tale of Some Defense Output Delays. Ford Division Finds It's A 17 Months Battle From Contracts to Production.* Wall Street Journal, October 30 1951.

3-12 Pratt & Whitney Aircraft—East Hartford, Connecticut. Total R-4360 *Wasp Major Engine Shipments.*

CHAPTER FOUR

Turbocharging, VDT & Other Exhausting Ventures

Turbocharging & Turbo Compounding

Exhaust energy is a precious commodity with a high-performance piston engine. Squander it in the case of the R-4360 and several hundred horsepower is wasted. Now the question arises of how to utilize this valuable energy source. Options available to designers are: (i) simple jet stacks for the exhaust system, (ii) turbosupercharging including VDT, (iii) turbo compounding, and (iv) afterburning. All the foregoing methods were, at a minimum, investigated. Some were built in prototype form only and others entered production. Combinations were also implemented. For example, even the depleted exhaust energy from a turbosupercharger was utilized for jet thrust. Afterburning, in combination with turbosupercharging, was another avenue of investigation.

General Electric supplied all the turbosuperchargers that went into production versions of the R-4360. They came in various models depending on the degree of development and mission profile. However, at least two other companies developed turbosuperchargers for the R-4360: Turbo Engineering Corporation and Lockheed Aircraft. Below is a list of the types of GE turbosuperchargers fitted to various models of the R-4360.

BH-1	basic model initially fitted to R-4360 and all production B-36s
BH-3	turbosupercharger fitted to Lockheed Constitution
BH-4	modified BH-1 for use in Boeing Model 377 Stratocruiser
B-32	turbine wheel and housing for use in turbo compounding
7SBH-4B1	modified BH-1 for use in Boeing C-97
CHM-1	two-stage used for VDT applications
CHM-2	same as above but improved
CHM-3	same as above but improved
CH-5	P-47 Thunderbolt turbosupercharger modified to CH-7 standards
CH-6	high altitude B-50 proposal
CH-7	Boeing B-50 and C-97 unit
CH-7A1	modified CH-7 to correct problems
CH-7B1	modified CH7A1 to correct its problems
CH-7B1	Unmodified CH-7B1s that needed problems corrected "red dot"
CH-7B1	Modified CH-7B1s with problems corrected "green dot"
CH-8	used in conjunction with CHM-2 for VDT applications
CH-9	same as above
GT15-2	purpose built turbosupercharger for Lockheed Constitution.

The alphanumeric designations of GE turbosuperchargers are as follows:

"B" or "C" represents the series of GE turbos and indicates the horsepower it is capable of handling. A "B" is rated at 1,000 hp and a "C" is rated at 1,500 hp.

"H" indicates that this turbo is intended for high altitude operation, which in fact all R-4360 turbos were.

"M" means that the turbo has multiple compressor stages, i.e., two stages of compression.

The number following the alpha designation indicates the level of development.

The foregoing offers a taste of the many variations of turbosuperchargers installed with R-4360s. In a similar fashion to R-4360 development, General Electric suffered its own development problems with its R-4360 turbosuperchargers. So much so in fact that the Air Material Command became increasingly irritated with GE and its slow response to problems along with cost overruns. When their vast

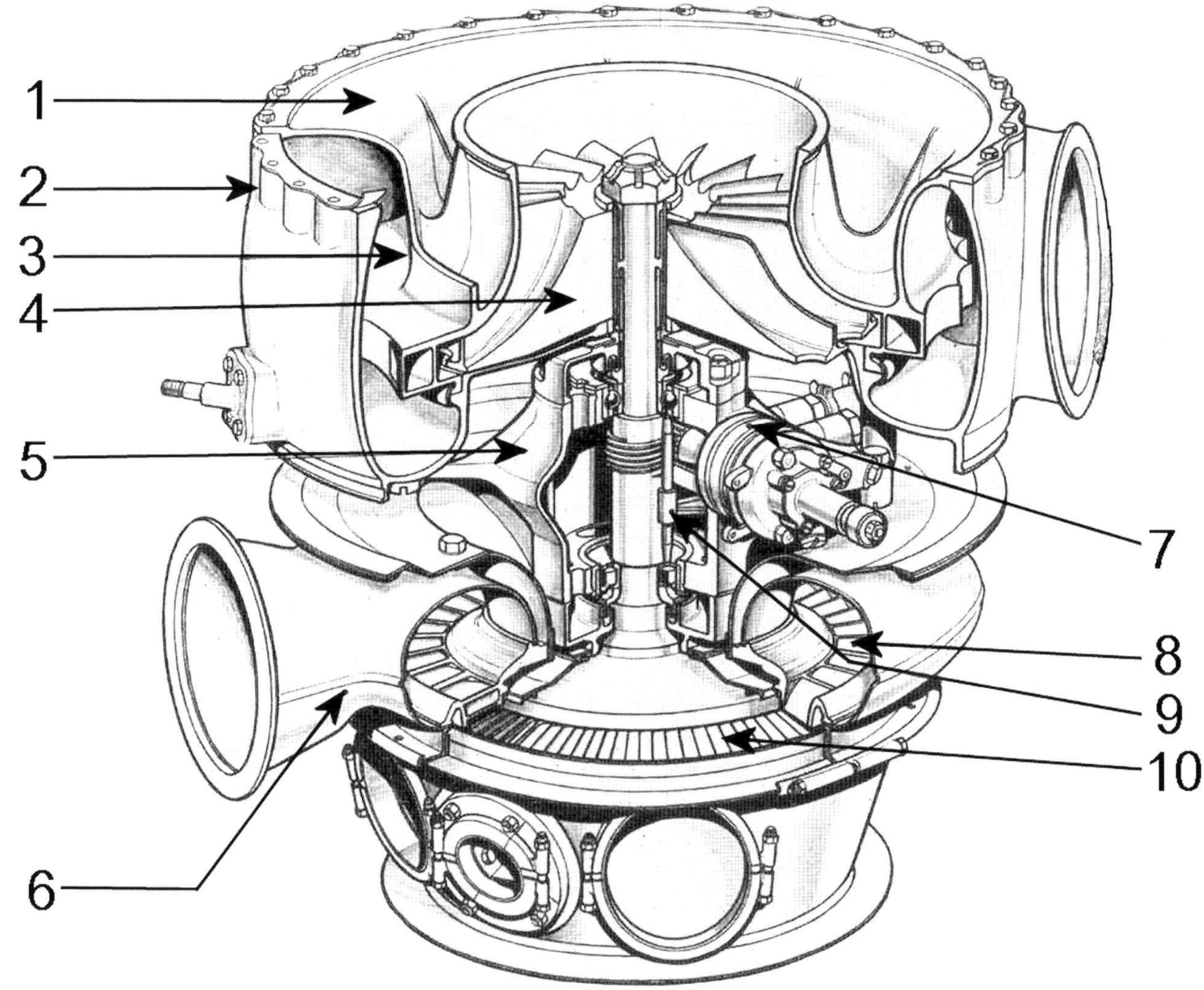

BH-4 cutaway. This turbosupercharger was used on commercial Boeing 377s. 1. Rear compressor housing; 2. Front compressor housing; 3. Diffuser; 4. Impeller; 5. Bearing and pump housing; 6. Nozzle box; 7. Oil pump assembly; 8. Turbine nozzle diaphragm; 9. Oil jet; 10. Turbine wheel.

experience with turbo superchargers dating back to 1917 is taken into consideration, it could be argued that it's surprising that with this wealth of knowledge under its belt GE should have fallen down on its responsibilities. As portend of the future, cost overruns and implied accusations of cost padding emerged. In their reports, the AMC became increasingly frustrated and visibly angry about GE's performance in respect to its turbo-supercharger responsibilities. The answers to the above are complex; numerous reasons surfaced as to why GE suffered such grief with this (presumably) proven technology. Whenever the Air Force felt that a program was suffering problems, whether it was financial, technical, or both, a "Case History" was written. Several Case Histories were written regarding GE's problems with their turbo-superchargers for various R-4360 applications. In other words, if you were the subject of an Air Force Case History study, it was not a good sign. The following details some of the issues facing GE.

BH-1 *(Ref. 4-1)*

The first indication that requirements for a turbosupercharger capable of handling the demands of a 2,500- to 3,000-hp engine came up in August 1943 at a supercharger symposium held in Lynn, Massachusetts. The BH-1 carried on with the

Nicely designed nacelle for the Boeing C/KC-7 showing the relationship of the BH-4 in the assembly. Interestingly, even though the B-50 looks identical, the turbo was mounted on the opposite side of the nacelle. *(Boeing C-97 T/O 1C-97A-10)*

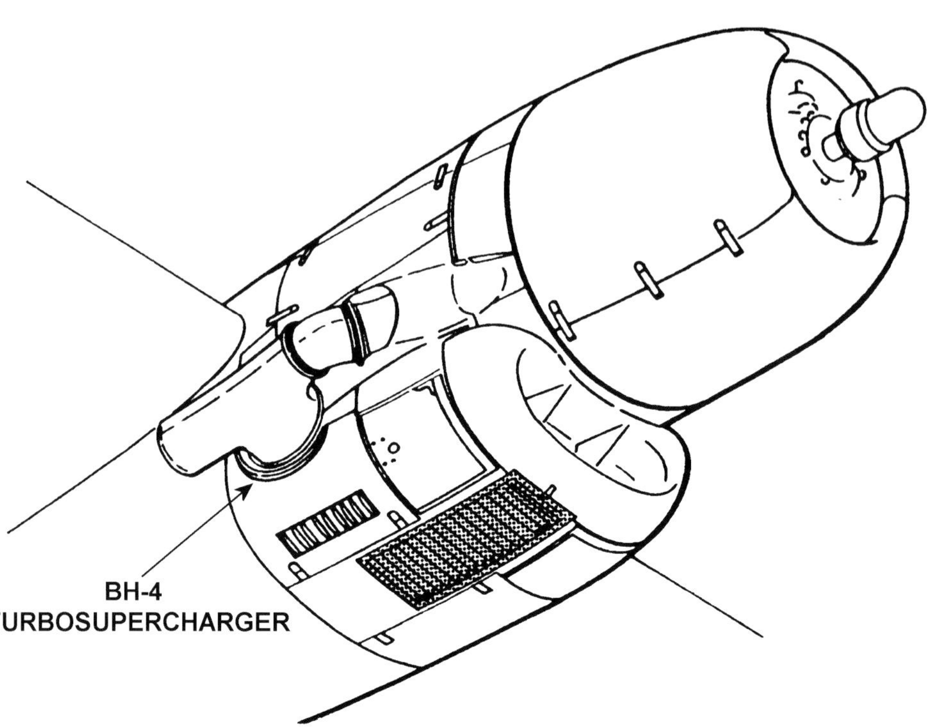

tradition of GE's prior experience with developing and manufacturing turbosuperchargers for the Army Air Force for such applications as the B-17, B-24, P-47, etc. It featured a single-stage turbine and single-stage centrifugal flow compressor. As it turned out, the BH-1 was one of the few, if not the only one, not to give GE fits with quality problems, service problems, and poor reliability. Being the first of many R-4360 turbosuperchargers, it was used in Convair B-36As, the Convair C-99, and all models of the Northrop B-35 flying wing.

The BH-1 design was approved in 1944 for use in the Northrop B-35A *(Ref. 4-2)*. A sealing plate at the back of the turbine provided sealing and prevented mixing of air with exhaust gasses. In this way flame suppression, a critical concern during World War II, could be achieved. Lubrication was provided via a dry sump system that included oil filtration. The initial contract was for the procurement of 2,505 BH-1s for use in B-35s. As this problematic program got bogged down in delays and it became apparent that the B-35 would make no contribution to World War II, this order was drastically cut back. However, an order for 916 was placed for use in Convair B-36s plus an additional 206 for use as spares. This initial order was subsequently increased to satisfy the ever-growing demand for B-36s. Numerous delays caused by the demands placed upon General Electric resulted in a seven-month period during 1950 when no BH-1s were shipped. Primary causes of these delays were the almost insurmountable problems GE was enduring with the CH-7 as related later. BH-1s weighed 200 pounds and were capable of delivering 160 pounds of air per minute. It could deliver 31.67 in. Hg. pressure at the carburetor deck up to 35,000 feet.

BH-3

Turbosupercharger specifically developed for the Lockheed Constitution *(Ref. 4-3)*.

B-32

Turbine wheel and housing for use in turbo compounding *(Ref. 4-1)*.

BH-4

BH-4s were a modified and improved version of the BH-1 intended as a cruising turbosupercharger for the R-4360. Optimized for long-range aircraft where good fuel economy was paramount, it was approved as standard equipment in March 1947. First recipient of the BH-4 was the Boeing Model 377 Stratocruiser. Designed to handle 3,000 hp at 6,000 feet, it could also furnish cruising air requirements at altitudes of up to 30,000. BH-4s were the first turbosuperchargers to be used in a commercial application. This turbo weighed 220 pounds and could deliver 224 pounds of air per minute. At 25,000 feet 27.8 in. Hg. it was still available at the carburetor deck *(Ref. 4-1 and 4-4).*

7SBH-4B1

Modified BH-1s for use in Boeing C-97s.

CHM-1

No information is available on this turbosupercharger, but it has to be assumed that being the predecessor to the CHM-2, it was of a similar layout. However, its designation tells us that it employed a two-stage compressor.

CHM-2

One of the more complex and interesting of the

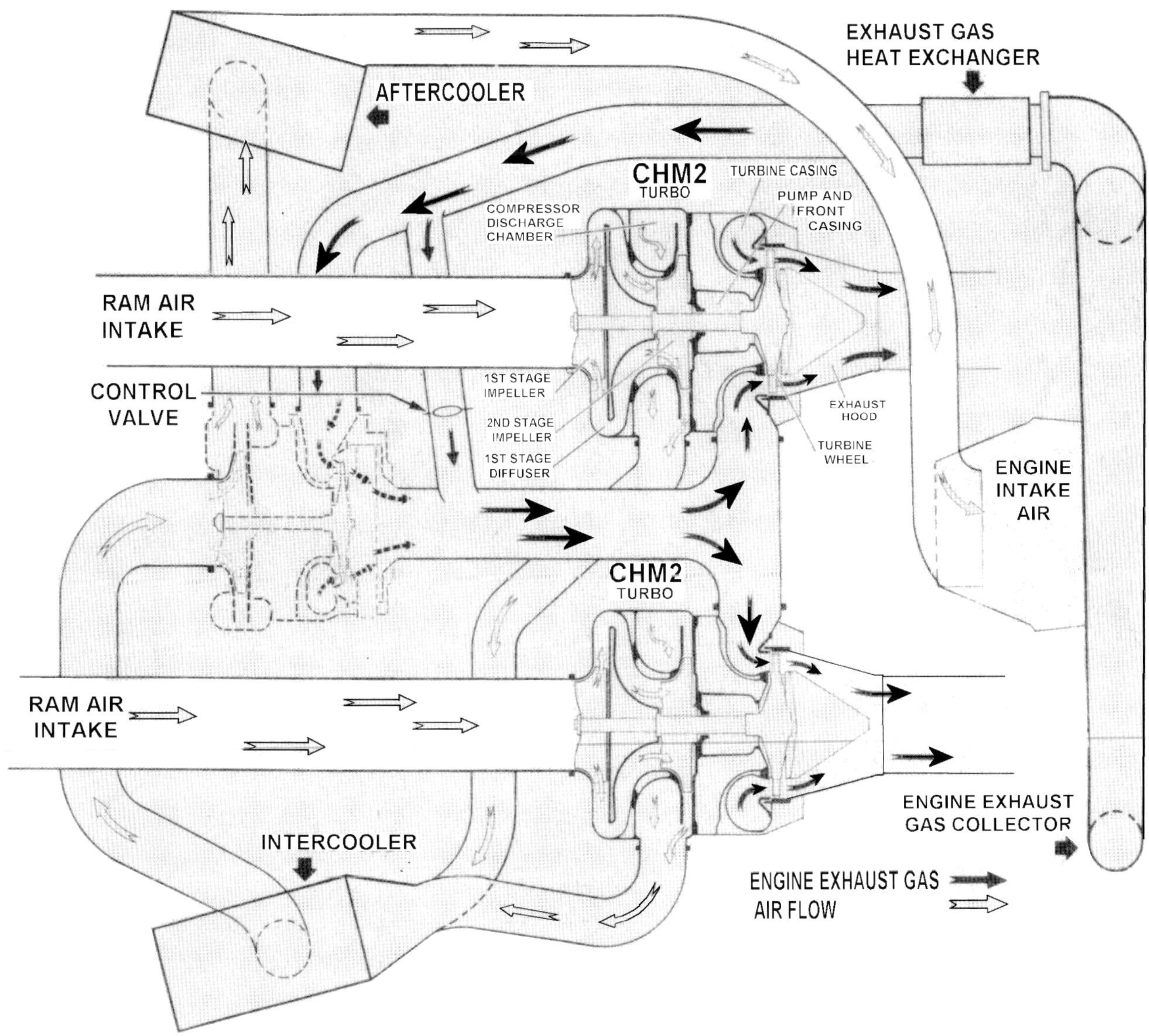

This diagram shows a pair of two-stage CHM-2s running in series with a single stage CH-8. *(GEI-32056 Preliminary Operation and Service Instructions for Turbosuperchargers Models CHM-2C, CHM2-2D, CHM-2E, CH-8C, CH8-D. April 15, 1950)*

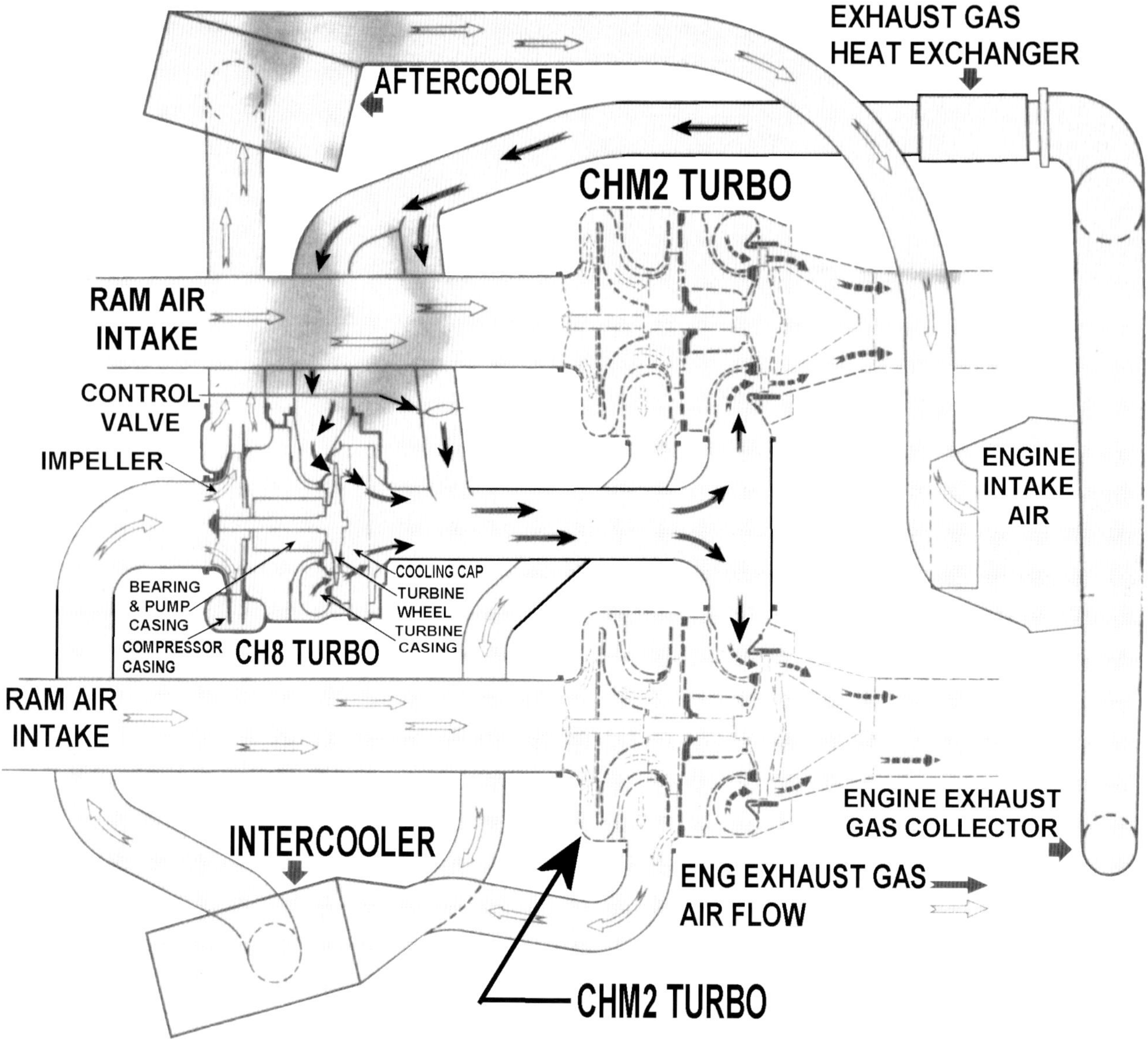

This diagram shows how the CH-8 runs in series with the pair of CHM-2s. *(GEI-32056 Preliminary Operation and Service Instructions for Turbosuperchargers Models CHM-2C, CHM2-2D, CHM-2E, CH-8C, CH8-D. April 15, 1950)*

R-4360 turbosuperchargers, the CHM-2 featured a single stage turbine driving a two-stage centrifugal compressor. A pair of CHM-2s was designed to run in parallel with a third turbosupercharger. A CH-8 ran in series with the other two. A cutaway of the CHM2 *(Ref. 4-5)* is shown above.

The air circuit is as follows: ram air from the aircraft ram scoop enters the suction side of the first stage compressors. The pair of CHM-2s then discharge into the single CH-8 via an intercooler, which then discharges through an aftercooler, and finally into the engine intake system.

The exhaust circuit is as follows: Exhaust gasses leaving the engine first enter the CH-8 turbine, although a small percentage is tapped off into the CHM-2 turbines. Exhaust gasses leaving the CH-8 then drive the CHM-2 turbines. The engine supercharging requirements at a particular altitude and throttle setting determines the amount of exhaust gas that bypasses the CH-8. Exhaust gas discharged from the engine collector ring drives the CH-8 turbine; however, some exhaust gas is bled off and bypassed around the turbine nozzle box via an adjustable valve. The

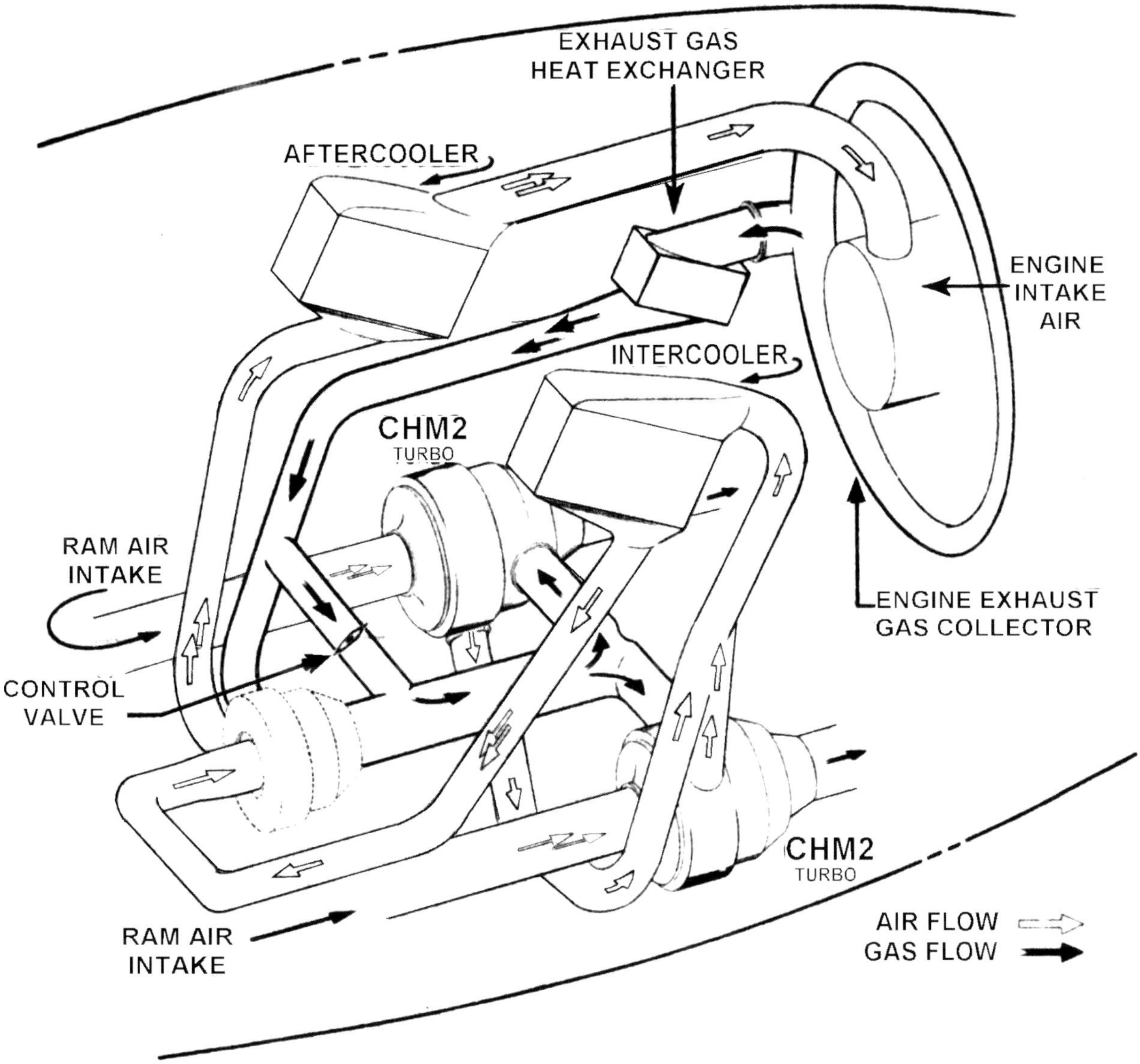

This illustration offers another perspective on the CHM-2s and CH-8. This was one of the many proposals for the B-36 that never went into production. *(GEI-32056 Preliminary Operation and Service Instructions for Turbosuperchargers Models CHM-2C, CHM2-2D, CHM-2E, CH-8C, CH8-D. April 15, 1950)*

amount of exhaust gas bypassed around the CH-8 determines the degree of supercharge for the entire system. It is analogous to a waste gate in a simpler system *(Ref. 4-6)*. The illustration on page 181 shows the nicely designed exhaust hood.

This set-up was clearly designed for high altitude. In fact, with this triple turbosupercharger arrangement, the R-4360-55, the VDT engine it was intended for, could maintain 3,000 hp to a staggering 50,000 feet—far higher than any propeller could be used with any degree of efficiency. With today's computerized FADEC (full authority digital engine control), it would have been a breeze to control this complex supercharging system. As it was, GE had serious problems with controllers that were satisfactory to meet these requirements. Residual exhaust pressure exiting the CHM-2 was converted into jet thrust thus maximizing to the fullest extent every available aspect of exhaust energy. The trade-off for this triple turbo arrangement is size, weight, and complexity. The only known photograph of a CHM is shown here.

continued on page 182

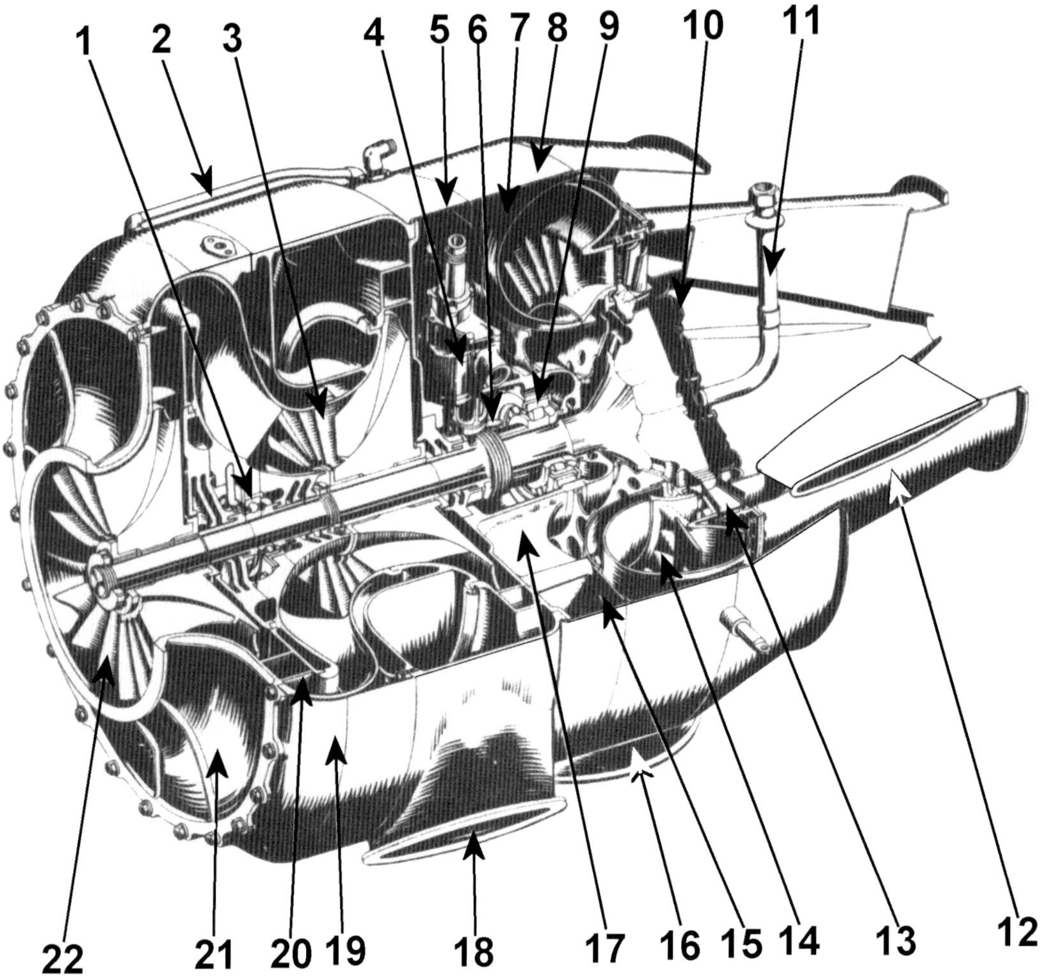

Above: Two-stage CHM-2 cutaway. 1. Rear Bearing; 2. Rear Bearing Scavenge Oil; 3. Second Stage Impeller; 4. Lubrication Pump; 5. Cooling-air Shroud; 6. Thrust Bearing Oil Jet; 7. Turbine Casing; 8. Turbine Casing Shroud; 9. Front Bearing; 10. Cooling Cap; 11. Cooling-air Tube; 12. Exhaust Hood; 13. Turbine Wheel Buckets; 14. Turbine Nozzle Diaphragm; 15. Baffle Ring; 16. Turbine Casing Inlet; 17. Pump & Front Casing; 18. Compressor Discharge; 19. Interstage Casing; 20. First Stage Diffuser; 21. Rear Casing; 22. First Stage Impeller.

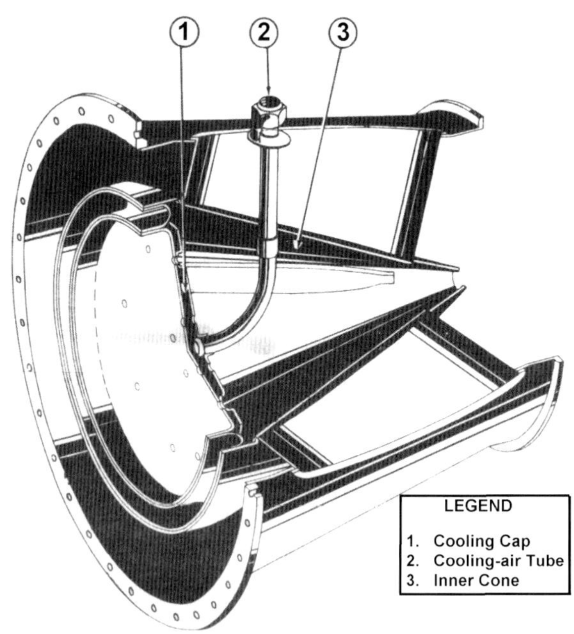

Exhaust Hood Cutaway, Type CHM2 Turbosupercharger

Right: It's easy to see how the transition from turbosuperchargers to gas turbines took place. This flight hood for a CHM-2 bears all the hallmarks of a jet exhaust nozzle.

Opposite: The all-important turbine casing and nozzle box for CHM-2. 1. Turbine casing shroud; 2. Outer nozzle diaphragm ring; 3. Turbine casing; 4. Turbine casing support; 5. Inner diaphragm ring; 6. Turbine wheel; 7. Turbine wheel buckets; 8. Nozzle diaphragm blades; 9. Turbine casing inlet. *(All diagrams GEI-32056 Preliminary Operation and Service Instructions for Turbosuperchargers Models CHM-2C, CHM2-2D, CHM-2E, CH-8C, CH8-D. April 15, 1950)*

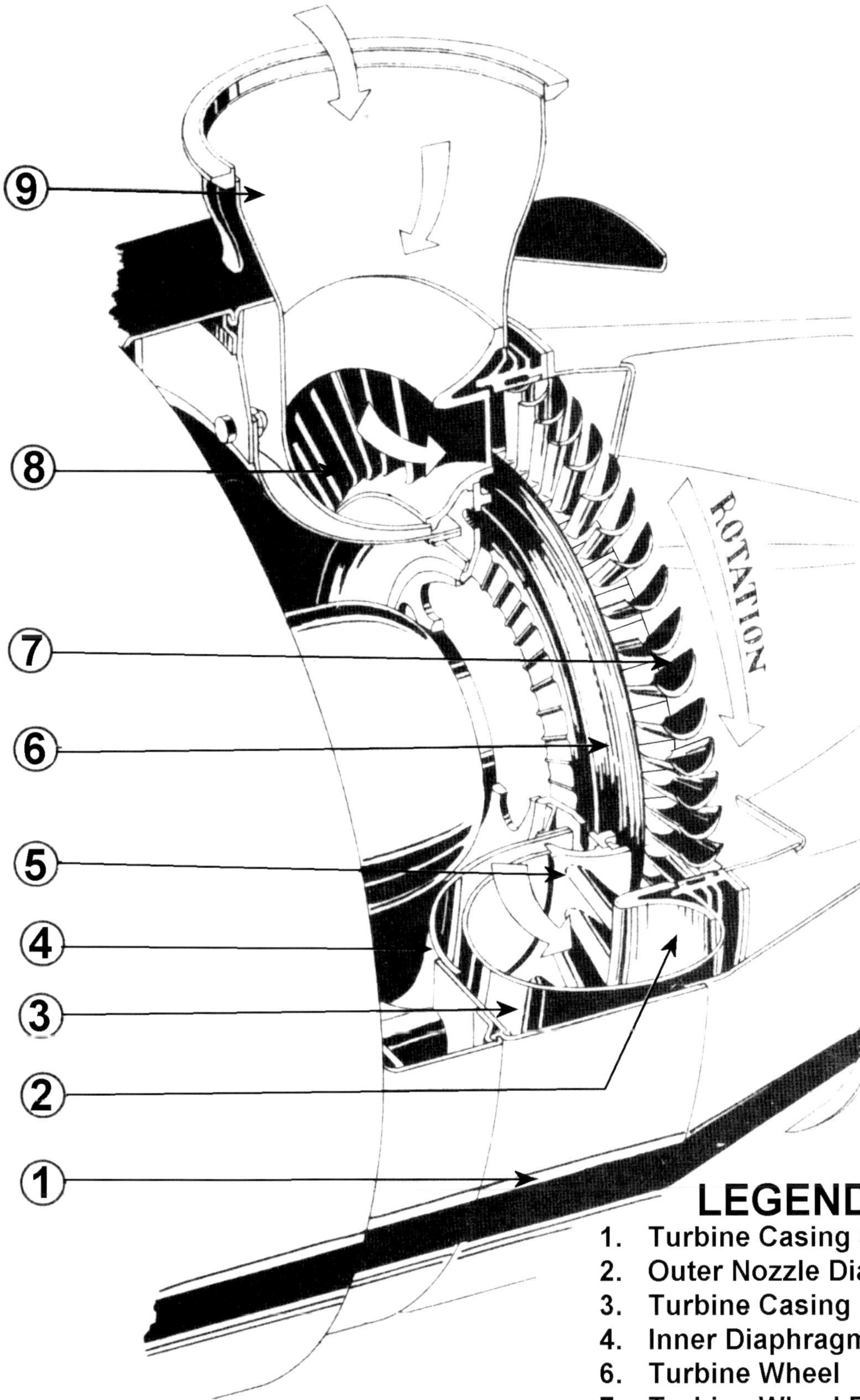

LEGEND

1. Turbine Casing Shroud
2. Outer Nozzle Diaphragm Ring
3. Turbine Casing
4. Inner Diaphragm Ring
6. Turbine Wheel
7. Turbine Wheel Buckets
8. Nozzle Diaphragm Blades
9. Turbine Casing Inlet

Turbine Casing and Nozzle Diaphragm Cutaway
Type CHM2 Turbosupercharger

Designed for the VDT program, the CHM-2 represented the state of the art for turbosuperchargers– before or since. *(United Technologies Corp./Pratt & Whitney)*

CHM-2 parameters

Max. continuous turbine speed	18,500 rpm
Max. compressor discharge temperature	450 degrees F
Max. continuous exhaust gas temperature at nozzle inlet	1,650 degrees F
Lube oil circulation	6 quarts per minute at 18,500 rpm
Max. diameter	29 inches
Max. length	42 11/16 inches
Weight	575 lbs.

CHM-3

One positive outcome of the ill-fated VDT program was the "C" series R-4360, a far more robust engine than its predecessor the "B" series.

At the request of Air Force General Wolfe in April 1948, Pratt & Whitney outlined a proposal covering the engineering work entailed in adapting the B-36 for R-4360 "C" series power. Then-current B-36s were capable of operating at 35,000 feet. By adapting the "C" series, it was estimated that the B-36 could attain 50,000 feet. This was done through better supercharging with the more robust "C" series engine in conjunction with a three-turbo setup using the CHM-3 and CH-8, in a similar way as the CH-2 and CH-8. None of the three-turbo arrangements entered production *(Ref. 4-5, 4-6)*.

CH-5

CH-5s were P-47 Thunderbolt turbosuperchargers used to modify CH-7 turbos *(Ref. 4-1)*.

CH-6

In 1950, as a quick fix to obtain more altitude from the B-50, Boeing submitted a report suggesting the use of a "C" series R-4360 with two turbos in series—a CH-6 and BH-1. However, the Powerplant Lab felt it would be considerably easier to develop a purpose-built turbo. General Electric was, at that time, designing a new family of turbos with multi-stage compressors driven by multi-stage turbines running on concentric shafts. This never happened *(Ref. 4-1)*.

CH-7

Of all the turbosuperchargers fitted to the R-4360, it's probably fair to say that the CH-7 and its subsequent variations were the most problematic. Like all other R-4360 turbos, with the exception of the CHM series, it featured a single-stage axial-flow turbine and a single-stage centrifugal compressor. Somewhat similar to the CH-5 except it could handle a higher air flow and 3,000 hp at an impressive 35,000 feet—at least in theory (as we shall see). For low-power cruising, rather than use a simple waste gate, the CH-7 incorporated a so-called "cruising valve" that made some of the nozzle inactive. Approval for the CH-7 was granted in June 1946 *(Ref. 4-1)*.

Like the BH-4, the CH-7 was specifically designed for the military in conjunction with the R-4360. However, unlike the BH-4, only the military employed the CH-7. They used it to get the most out of the "C" series R-4360, except for compounded and VDT versions. Attempting to build the CH-7 on the cheap, the government furnished GE with surplus CH-5 parts to be utilized in CH-7 production. CH-7s manufactured from CH-5 parts were designated CH-7A1s. This turned out to be a major mistake due to the mismatch of a turbo intended for the R-2800 used with the R-4360. Not surprisingly, it was realized that the turbine would not be efficient for an R-4360 application at high power settings and high manifold pressures *(Ref. 4-1)*.

After the first CH-7A1s were installed in Boeing B-50s, it quickly became apparent that fundamental problems existed—problems that were to plague the CH-7 for its production life. Right off the bat it was found that the turbine bucket area was inadequate. Worse was to come. Pressure buildup between the turbine wheel and nozzle diaphragm caused serious overheating, which led to a weakened turbine disk. Consequently, the first of many restrictions was placed on the CH-7A1 in B-50A applications. Although designed to run at 20,000 rpm, it was restricted to 18,000 rpm, which resulted in the critical altitude of the engine being reduced from 35,000 to 31,000 feet. Realizing the seriousness of the situation, the Powerplant Laboratory advised against further procurement of CH-7A1s. Despite opposition, this finally happened in 1948. In the

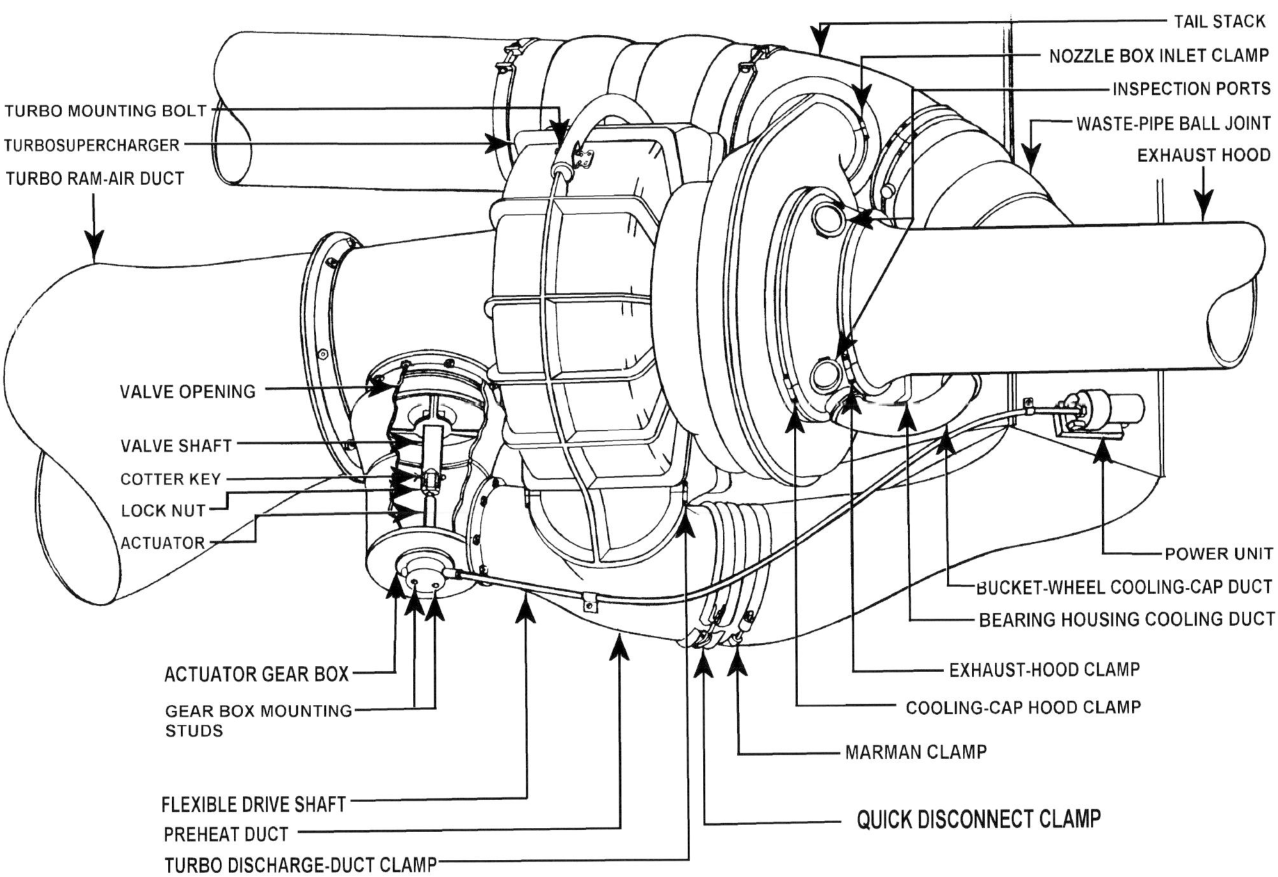

The problematic GE CH-7 turbosupercharger installed in Boeing B-50s. *(Erection & Maintenance Manual AN 01-20ELA-2. Revised February 5, 1948)*

During the worst of the CH-7's most problematic times, Boeing B-50s were flown with shrouds removed in order to offer better cooling. This in-flight photo shows such a B-50 with the shrouds removed. *(United Technologies Corp./Pratt & Whitney)*

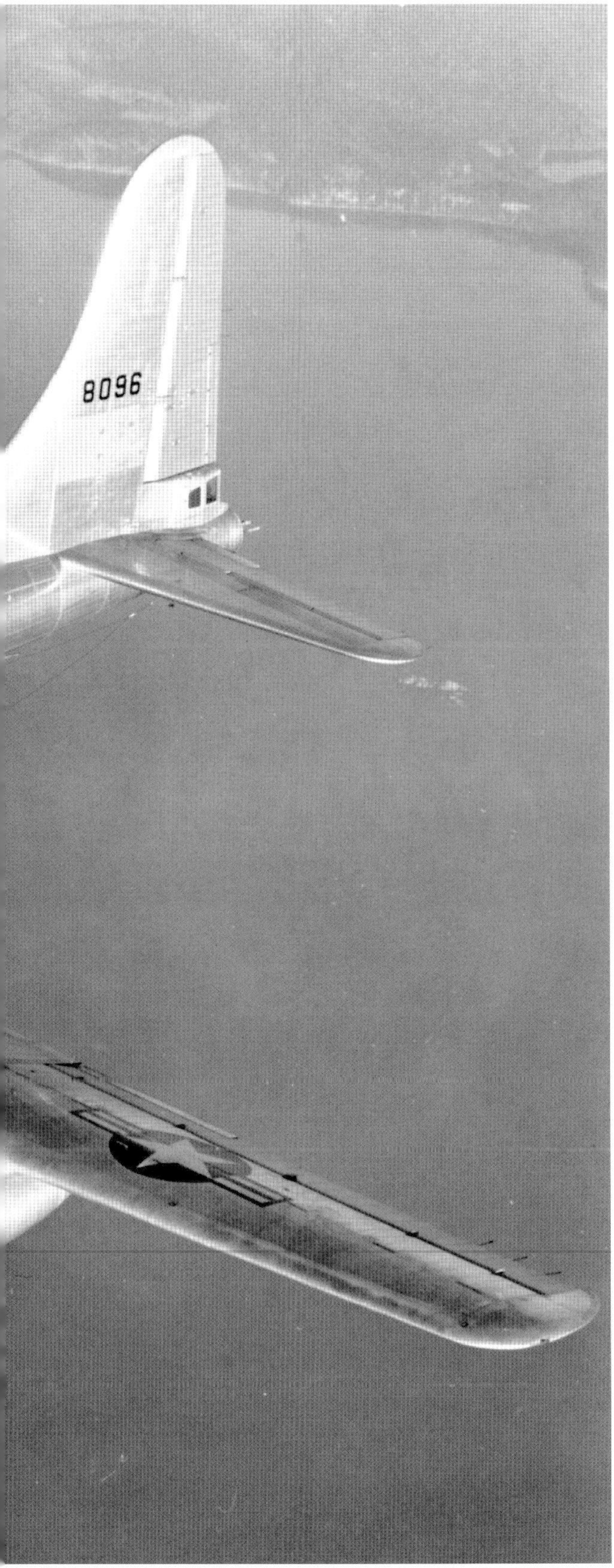

meantime a redesign to correct the CH-7A1's deficiencies was instigated. The new model, designated CH-7B1, entered production in March 1950. But it wasn't a simple case of swapping out CH-7A1s for CH-7B1s. A significant amount of re-engineering of the Boeing B-50 was required. Boeing furnished modification kits for all affected B-50s. The CH-7B1 corrected the turbine disk overheating problem and the nozzle was optimized for the R-4360. An integral fan in conjunction with a sealing plate fixed the disk overheating. CH-7B1s were now rated to run at 20,000 rpm with a 15-minute overspeed rating of 22,000 rpm.

These performance parameters would reestablish the B-50's 35,000-foot cruising altitude. Sounds good in theory, but unfortunately it turned out to be a classic case of jumping from the frying pan into the fire. Although all parties involved recognized that fundamental problems existed with the CH-7 and the definitive solution was to redesign the turbine buckets, this was not immediately done. Instead, a Band-Aid in the form of a modified CH-7A1 turbine was tried in order to expedite the retrofit of turbos in B-50s. Fortunately or unfortunately, depending on one's viewpoint, the B-50 was beset with other problems as well, the most significant being fuel cells. So in this case the CH-7B1 was not the "gate" for B-50 deployment. CH-7B1s suffered a variety of problems including poor quality in that they did not conform to the final production fix. As yet another Band-Aid, the shroud covers fitted to the B-50 for streamlining were removed in order to offer better cooling to the turbo. This configuration was the so-called "red dot" version. Even worse than the CH-7A1 turbosupercharger they were supposed to replace, the CH-7B1 was restricted to 30,000 feet. CH-7A1 red-dot turbos were further modified to green-dot configuration by increasing the clearances and sealing plate holes plugged.

The green-dot turbos were a considerable improvement over red dots. However, at their critical altitude of 35,000 feet, the margin of safety with respect to turbine RPM and temperature was not as wide as normally hoped for to assure reasonable turbo life. This was reflected in the attrition rate of CH-7B1 green-dot turbos, which in turn created shortages of parts and slowed down manufacture of new units. Problems with green-dot turbos included cracked cooling caps and bearing failures. Inevitably, things came to a head in December 1949 when B-50As, Bs, and Ds fitted with CH-7A1 turbos were grounded due to suspicion that turbosupercharger problems had contributed to a number of B-50 crashes. A flurry of activity kept the B-50 fleet more or less ready by setting up an "exchange pool" that exchanged red-dot turbos for green-dot turbos when they became available *(Ref. 4-1)*.

Of course, these serious problems did not occur in a vacuum. The highest levels of military brass were appraised of the B-50 situation. A conference was convened at Castle Air Force Base in December 1950 to discuss B-50 problems. Suppliers invited to this conference were Pratt & Whitney, General Electric, and Jack & Heintz. Everyone in attendance agreed that B-50 problems needed to be resolved "as quickly as humanly possible."

Green-dot turbos were further improved in 1950 by (i) stiffening the baffle support, (ii) development of an extended surface sealing plate in order to provide better turbine wheel cooling which in turn would allow an overspeed rating of 22,000 rpm, (iii) development of a configuration to pressurize seal space with cooled exhaust gas to, again, improve turbine wheel cooling, and (iv) development of an integral exhaust hood and cooling cap for improved durability and less warping. At the same time the aforementioned four improvements were being developed, investigations continued to look at different materials for the compressor's impeller. January 1, 1951, was set as the target date for design completion of the improved green-dot CH-7B1 turbos *(Ref. 4-1)*.

Problems with the CH-7 sound like a horror story, and for the folks involved with it during the late 1940s and early 1950s it undoubtedly was. However, the litany of problems GE experienced begs the question: How did the world's largest manufacturer of aircraft turbosuperchargers get mired in such a mess? No definitive answers have been forthcoming, although at times things got rather ugly between GE and the government. The government accused GE of cost gouging and not focusing on the problems at hand. For its part, GE claimed they were already overextended by trying to develop the CH-2 for the VDT program. And on top of all this, one can only sympathize with Air Force personnel who had to figure out what engineering change level turbos were fitted to a particular aircraft. It should also be borne in mind that each B-50 required four turbos. So trying to keep tabs on CH-7A1s, CH-71Bs, CH-7B1 red dot, and CH-7B1 green dot must have been the logistical nightmare from hell *(Ref. 4-1)*.

Although not mentioned in any of the official reports, at this time GE was hot and heavy into gas turbine development and they clearly saw this as the way of the future. So like the R-4360 program, in all likelihood, many of their top engineers were siphoned off to work on this more promising and profitable technology. The foregoing is conjecture, but the fact of the matter is the B-50 represented the best and most significant long-range strategic bomber the Air Force had in service at the time. Until the B-36 and later the B-47 and B-52 entered service, the B-50 had to shoulder the responsibility of responding to any Russian attack on the United States. When one thinks of the horrendous technical difficulties afflicting the B-50, it's a miracle the U.S. managed to pull it off. However, one has to assume that the Russians also had their share of technical difficulties to even the score.

CH-8

As previously described, the CH-8 is used in conjunction with a pair of CHM-2s for high altitude performance and specifically with VDT engines. The CH-8 is supplied with air that has already been compressed via the pair of two-stage compressors integral with the CHM-2s (see Figs. 4-3 and 4-4). This supercharged and heated air is

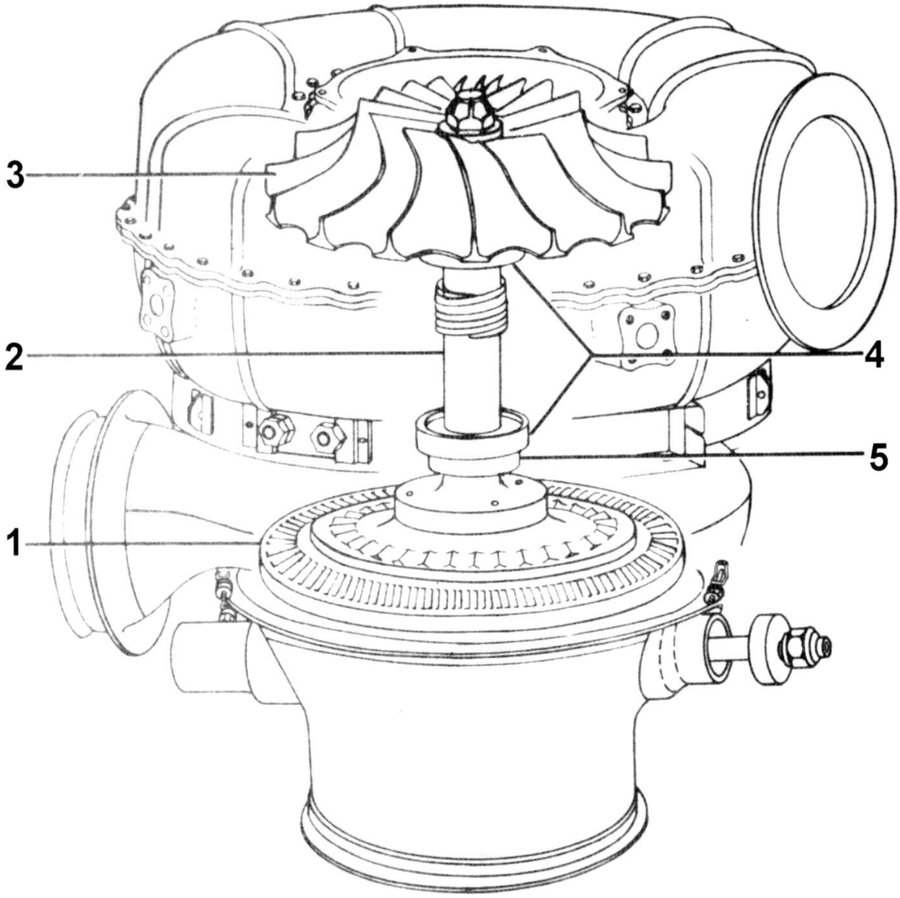

This line drawing shows the rotating assembly of the GE CH-8 turbosupercharger. 1. Bucket wheel; 2. Pump drive sleeve; 3. Impeller; 4. Bearings; 5. Oil deflectors. *(GEI-32056 Preliminary Operation and Service Instructions for Turbosuperchargers Models CHM-2C, CHM2-2D, CHM-2E, CH-8C, CH8-D. April 15, 1950)*

cooled via an air-to-air intercooler. The single CH-8 then further compresses the induction air, which then flows through another air-to-air cooler before entering the engine induction system. In order to preserve the turbine and offer some degree of longevity, a heat exchanger is fitted between the exhaust collector ring and the nozzle box. In this way the exhaust gas temperature is kept within a safe range of less than 1,650 degrees F. As previously described for the CHM-2, under most operating conditions part of the exhaust gas bypasses the turbine via an adjustable valve. Bypass volume is determined by supercharger requirements—the greater the supercharger demand, the less exhaust is bypassed *(Ref. 4-1 and 4-5).*

The CH-8 features a single-stage turbine and a single stage compressor.

CH-8 parameters:

Performance	pressure ratio; 2.5:1
Discharge	356 lbs/minute
Weight	300 lbs

CH-9

Designed for the B-36F, a CH-9 was designed to work in series with a pair of CHM-3s operating in parallel. In other words, it was another triple turbo setup. CH-9s were similar to the earlier BH-4. Fuel-injected R-4360-55s and R-4360-57s were the intended recipients of this turbosupercharger design. It was anticipated that all supercharging requirements would be arrived at through the triple turbosupercharger design. It was similar to a VDT setup except power control would be more conventional. That is, a throttle was on the induction side of the engine. It turned out to be yet another stillborn design. However, it wasn't for the lack of trying as GE tried to peddle CH-9s to the airlines and the Air Force. Full-page ads espousing the advantage of the CH-9 adorned aviation magazines in the late 1940s. Speed increases of 40 mph and 50 percent greater payloads were claimed. Claimed specifics for the CH-9 were impressive—350 pounds of air per minute, 75 in. Hg. at the carburetor deck for a weight of 300

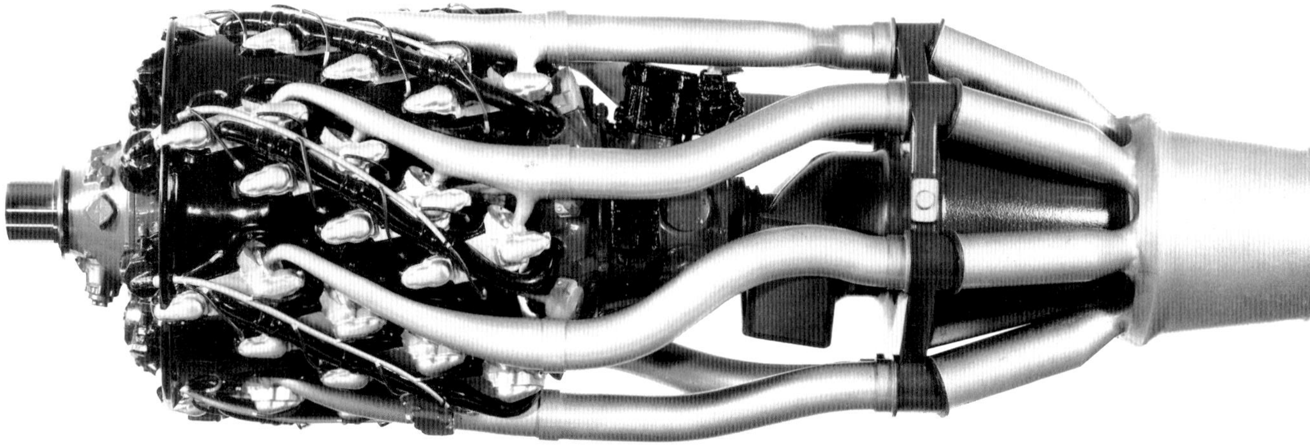

This concept is purely turbo compounded, that is, no turbosupercharging. The drive shaft joining the turbine to the rear of the R-4360 can be seen between the exhaust branches where it terminates at a reduction gearbox rear of the engine. It also appears that bleed air from the compressor feeds burner cans that in turn would offer more power to the turbine section plus additional jet thrust. *(United Technologies Corp./Pratt & Whitney)*

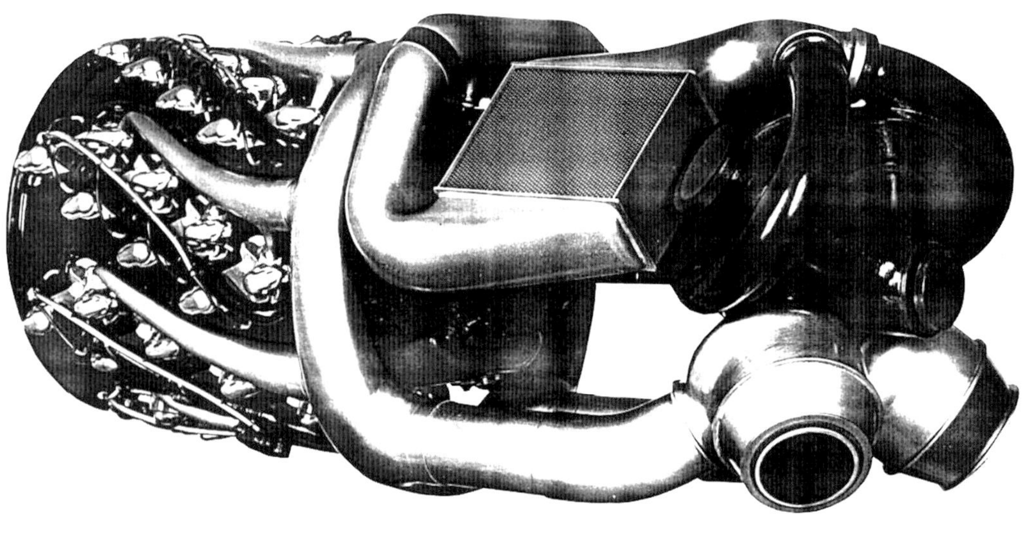

Two branches from the exhaust collector ring power two gas turbines. Interestingly, the compressors are not coupled directly to the turbine, but rather they appear to use a bevel gear-drive. An output shaft also feeds power back to the engine. The compressors discharge into a pair of rectangular air-to-air aftercoolers arranged in parallel. *(United Technologies Corp./Pratt & Whitney)*

pounds, and capable of providing cabin pressurization to an altitude of 30,000 feet. Maximum continuous exhaust temperature is a respectable 955 degrees C and maximum rated RPM is 18,000. Discharge temperature from the compressor is 233 degrees C, which, of course, is considerably reduced by the time it reaches the carburetor deck via an intercooler *(Ref. 4-1, 4-3 and 4-5).*

Paper Engines

Many wild and woolly schemes were cooked up for the R-4360. These schemes demonstrated the remarkable imagination of Pratt & Whitney engineers. Incorporation of turbosupercharging in combination with turbo compounding, multistage turbocharging, and various intercooler

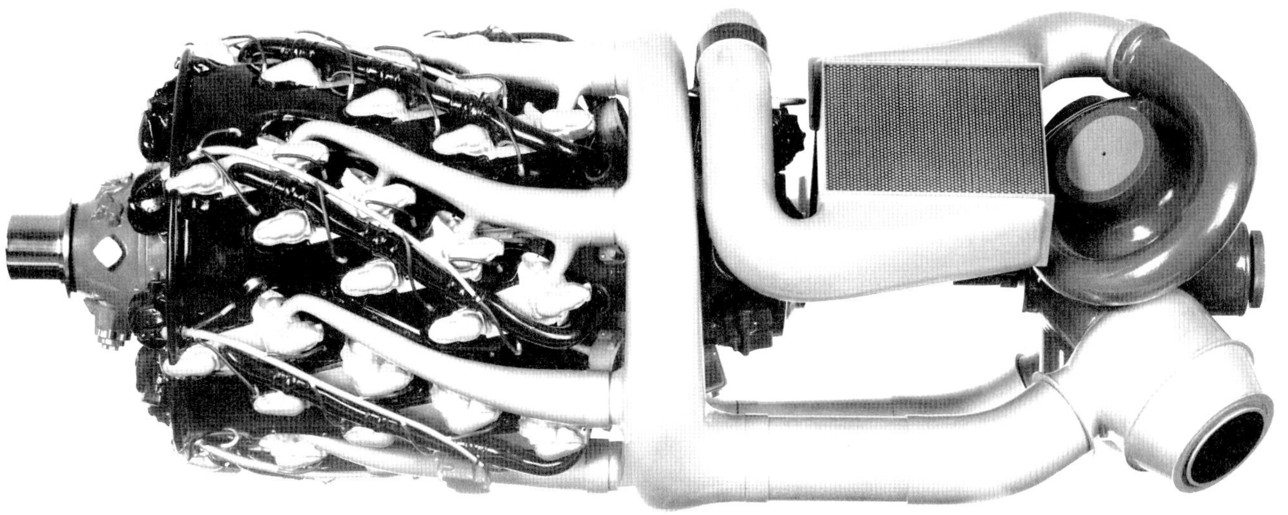

This illustration offers another perspective on the concept shown in the previous illustration. The gearbox can be just discerned between the turbines and compressors. The advantage with this setup is the fact that only one power feedback driveshaft is required. *(United Technologies Corp./Pratt & Whitney)*

schemes were all investigated and in many cases modeled. One scheme appears to have employed gas turbine-type burner cans discharging into a turbine. It appears to have used bleed air from the turbo to supply oxygen to the burner cans. It's doubtful if any of the following engines were ever built, even in prototype form. The accompanying illustrations demonstrate some of these wonderful concepts. The only engines to enter production featuring turbo compounding were some variations of the Wright R-3350. It represented another method to utilize exhaust gas energy. Turbo compounding simply uses exhaust energy to drive a gas turbine—similar to a turbosupercharger. But instead of driving a turbosupercharger's compressor, the turbo compound turbine feeds power back to the engine via a drive shaft and reduction gearing.

Variable Discharge Turbine (VDT)

Representing the last piston engine developed by Pratt & Whitney, the R-4360 also epitomized the most sophisticated development from this company. However, if the production version of the R-4360 incorporated the most advanced thinking, the VDT was downright radical. Standing for

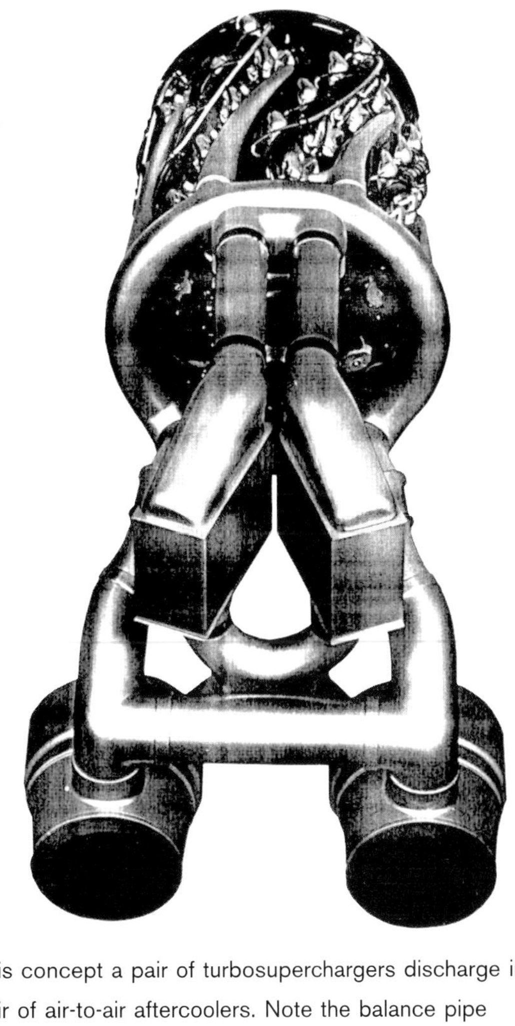

In this concept a pair of turbosuperchargers discharge into a pair of air-to-air aftercoolers. Note the balance pipe between the two turbines. Both systems operate in parallel. *(United Technologies Corp./Pratt & Whitney)*

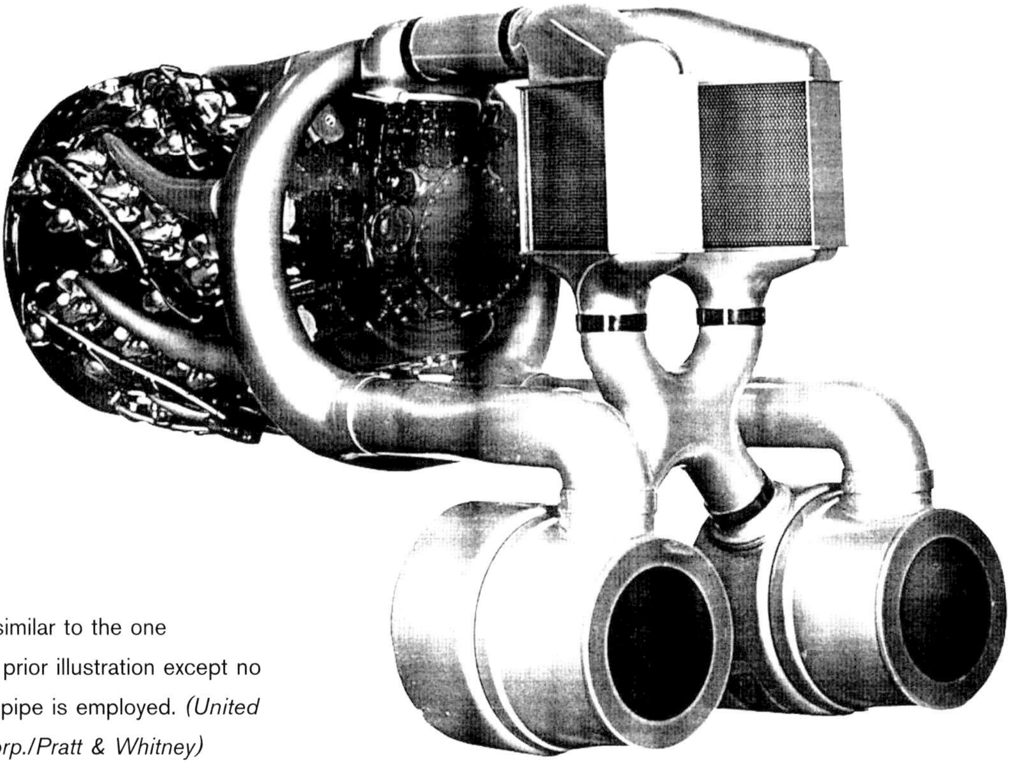

This concept is similar to the one described in the prior illustration except no exhaust balance pipe is employed. *(United Technologies Corp./Pratt & Whitney)*

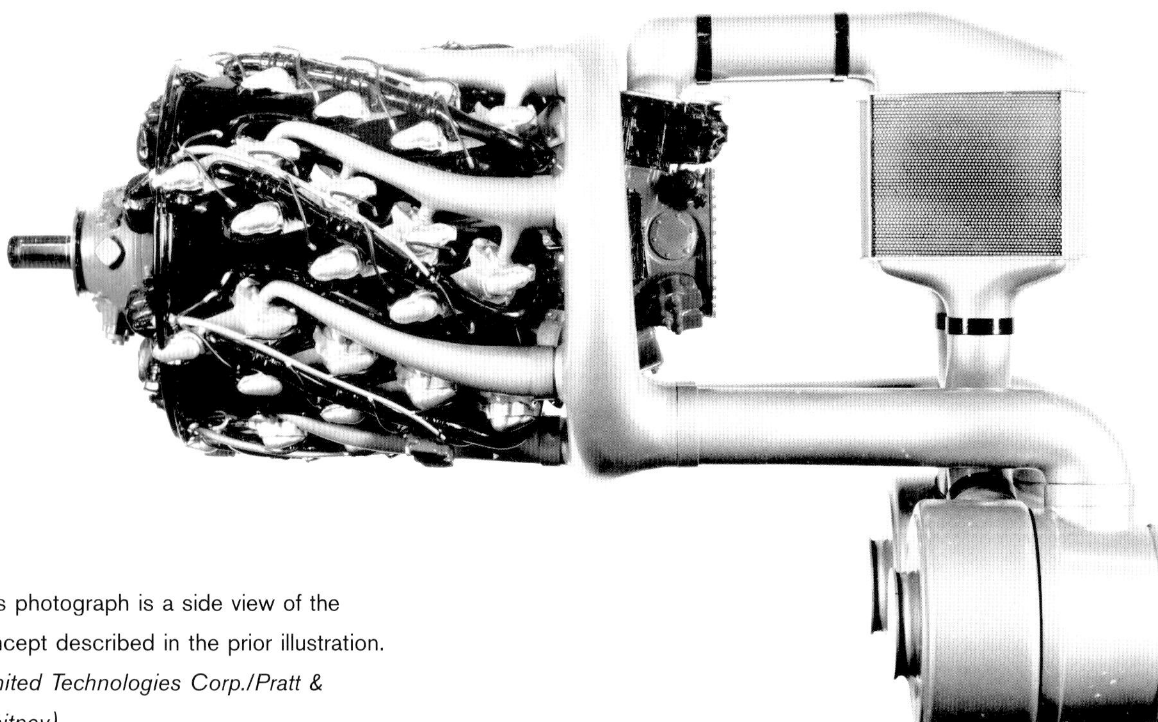

This photograph is a side view of the concept described in the prior illustration. *(United Technologies Corp./Pratt & Whitney)*

Variable Discharge Turbine, this engine re-wrote the rules for piston engine design. No engine before or since has incorporated such unusual and revolutionary thinking.

Harnessing exhaust gas energy has always been the goal of engine designers. From deceptively simple jet stacks to turbosupercharging to turbo compounding, the R-4360 VDT encompassed all these concepts—and lots more. A number of variations on the VDT theme were designed and in some instances built and actually flown.

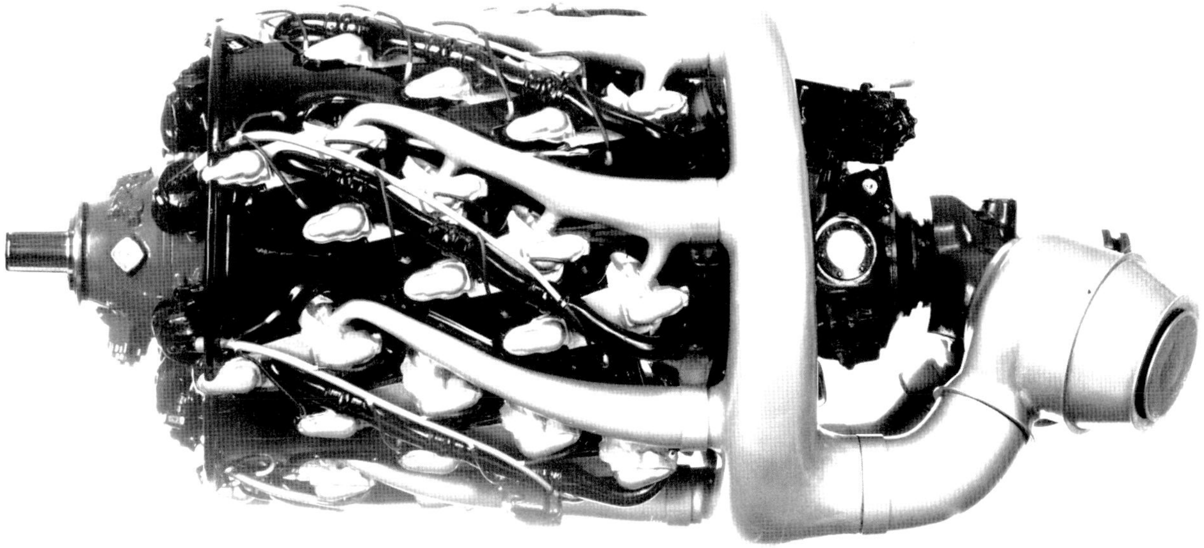

This is a pure turbo compound concept, meaning there is no turbosupercharger. Two turbines feed power into a bevel-type gearbox, which then transmits power to the rear of the engine and is then utilized to drive the supercharger. Although no figures have been forthcoming, it's probably fair to say that 400 to 500 hp could be recovered from the exhaust energy of the R-4360. *(United Technologies Corp./Pratt & Whitney)*

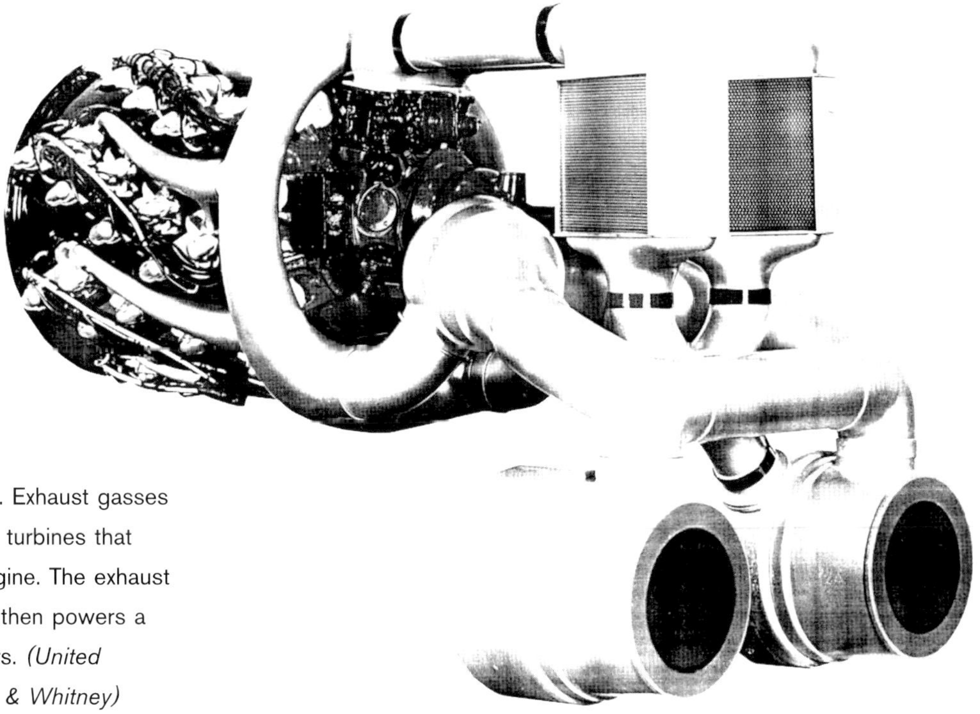

This is an interesting one. Exhaust gasses power a pair of feedback turbines that transmit power to the engine. The exhaust from this pair of turbines then powers a pair of turbosuperchargers. *(United Technologies Corp./Pratt & Whitney)*

It is taken for granted that all high-performance aircraft piston engines use some type of supercharging. It may be (i) single-stage, single-speed, (ii) single-stage, variable-speed, (iii) two- stage, single stage augmented by turbosupercharging, or combinations of these concepts. One common theme, however, was the use of at least one gear-driven supercharger. And there was a good reason for this. Beyond boosting manifold pressure, the supercharger and specifically its impeller offered a good means of ensuring even fuel/air mixture distribution. To put it in simple terms, the impeller,

SCHEMATIC DIAGRAM OF WASP MAJOR - VDT POWER PLANT

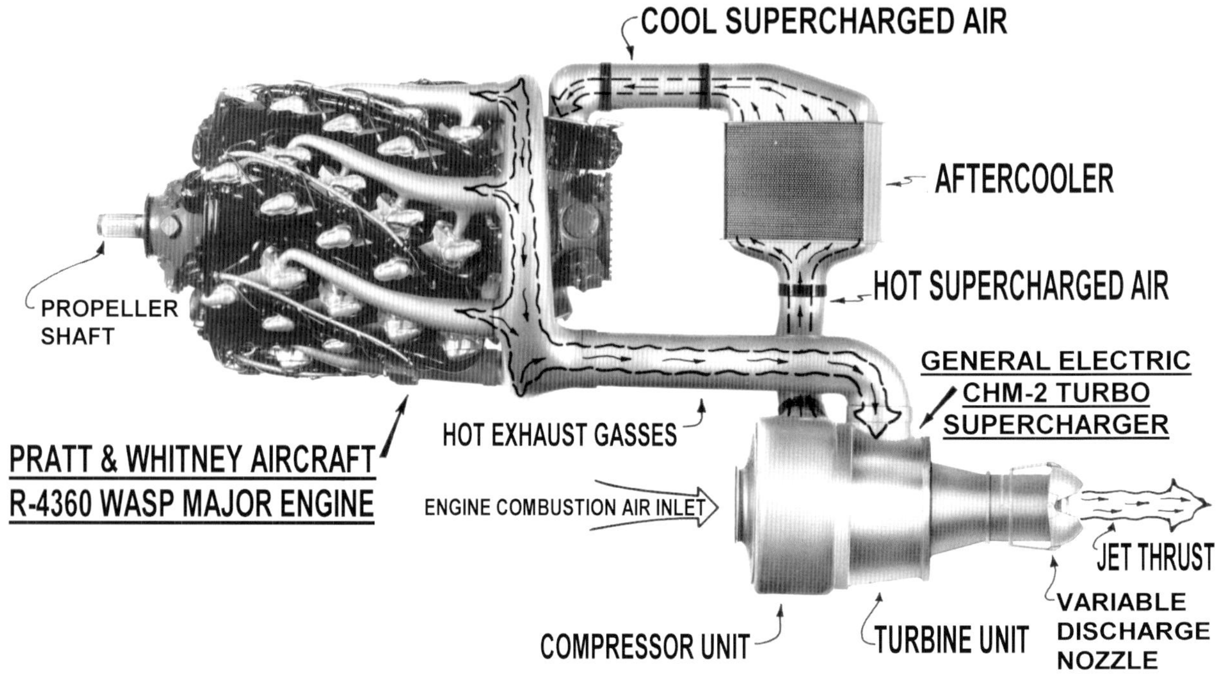

One of the most radical methods ever devised for taking advantage of exhaust energy was the Variable Discharge Turbine (VDT). Using the exhaust as a throttle and relying on turbosupercharging for all manifold boost requirements are just two of the innovative ideas incorporated into the VDT concept. *(United Technologies Corp./Pratt & Whitney)*

rotating with tip speeds approaching the speed of sound, "stirred up" the mixture and helped in the goal of getting a consistent and well-atomized mixture to each cylinder.

Of course, the success of this goal varied according to engine design, but at least that was the intent. Some low-power radials have been manufactured with no supercharger impeller. Invariably, these engines suffered from poor mixture distribution, which is why the majority of radials use some manner of supercharging even if it doesn't boost manifold pressure. The VDT departed from this methodology. Instead, this engine relied entirely on turbosupercharging for boost requirements. This now freed up several hundred additional horsepower to drive the propeller rather than drive the supercharger. In other words, where one would normally expect to see a supercharger, instead the VDT had a plenum that led to the seven intake manifolds, as would be the case for a "conventional" R-4360. If the usual injection-type carburetor had been used, either on the suction side or the discharge side of the turbosupercharger compressor, serious mixture distribution, problems would have arisen. In fact, the idea would have been impractical. The saving grace was the use of direct fuel injection. Although this still did not guarantee absolutely even mixture distribution it was infinitely superior to the use of a carburetor where fuel would have come out of solution resulting in lean and rich cylinders.

So far, the VDT sounds fairly conventional—turbosupercharged with direct fuel injection. However, things get radical in a hurry when one realizes that no conventional throttle was used—at least for anything beyond low power settings. Instead, power was regulated through the use of clamshell doors on the discharge side of the turbosupercharger. By closing off the clamshell doors, the mass flow of exhaust gas driving the turbosupercharger turbine was restricted. Therefore, in this instance, the turbine would rotate at a slower rate.

Consequently, the compressor side of the turbosupercharger would rotate at a slower rate, which results in a lower mass airflow to the engine. The master control unit of the direct fuel injection system senses the lower mass airflow rate and injects the appropriate amount of fuel into the cylinder for this decreased power setting. As the clamshells are opened for additional power, an increase in mass exhaust gas flow is allowed through the turbine, which in turn provides additional power to the compressor, which in turn provides a greater mass airflow to the engine *(Ref. 4-6 and 4-7).*

This greater mass airflow demands additional fuel to be injected into the cylinder in order to maintain the correct fuel/air ratio. And so it goes. Sounds simple, right? In reality, the controls for the entire system, especially the clamshells, posed a daunting challenge to the engineers. In addition to relieving the engine of drive requirements for a gear-driven supercharger, a substantial amount of residual thrust was available through the variable discharge nozzle. Depending on altitude and aircraft speed, this thrust could amount to over 500 pounds *(Ref. 4-8).* Some thought was also

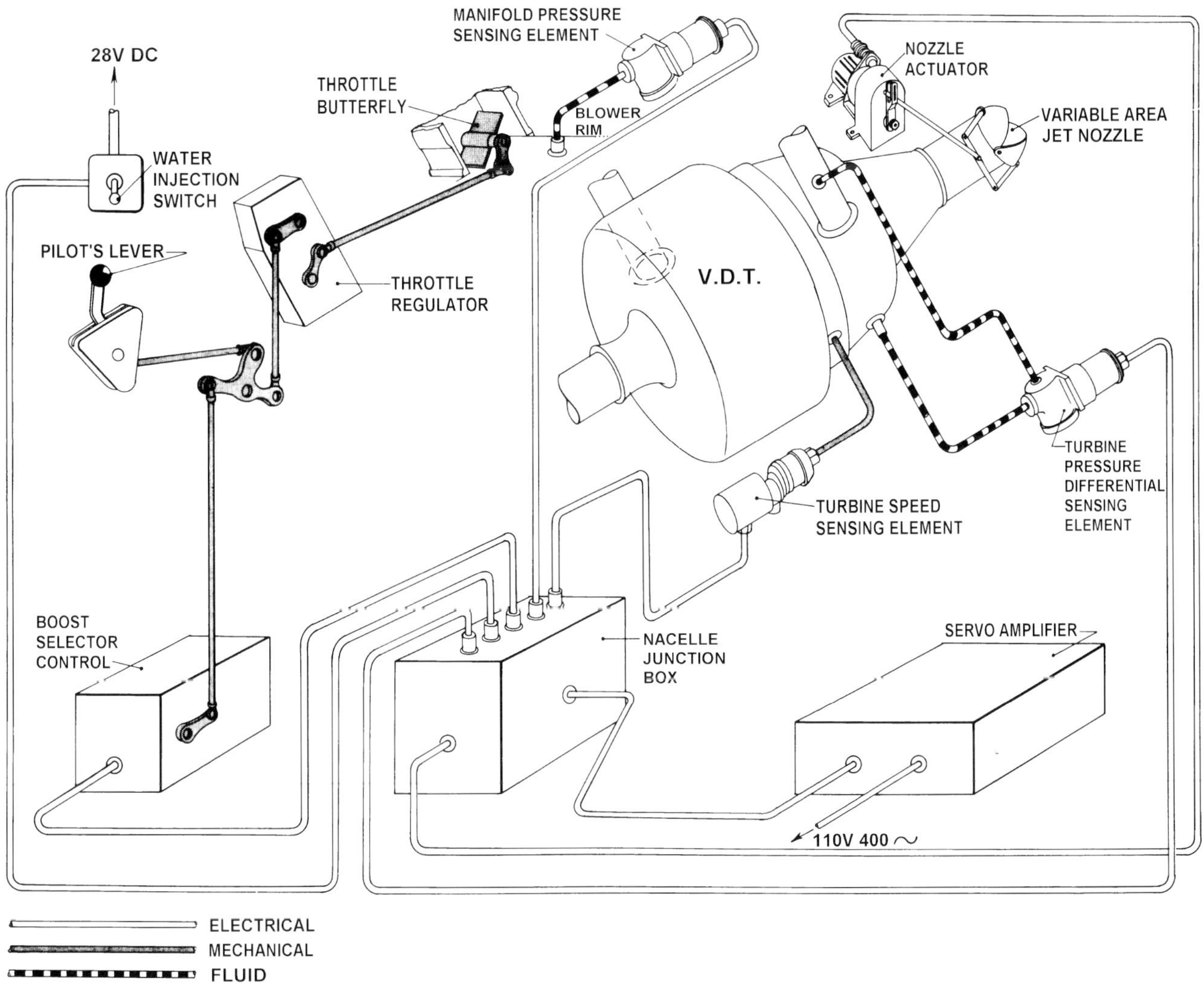

Although the basic concept of the VDT engine was relatively simple, finding a reliable method of throttling and controlling it was not. With today's solid-state electronic controls and FADEC (full authority digital engine control) it would be a simple task. However, those tools were not available in the late 1940s. A complex hydro-mechanical-electrical system was required, but GE, who was responsible for the design of this controlling mechanism, never managed to get one working reliably. *(United Technologies Corp./Pratt & Whitney)*

given to injecting raw fuel into the exhaust in order to offer an after-burning effect. Massive air-to-air aftercoolers were incorporated on the discharge side of the compressor to keep charge temperatures under control.

The foregoing describes the basic VDT theme; however, a number of variations were planned and in some cases actually manufactured and flown. But in all cases, every bit of exhaust energy is utilized, meaning no waste gate is used or needed. Pratt & Whitney made quarter-scale mock-ups of every imaginable variation on the VDT and turbo compounding theme. Some concepts drove a turbine, which fed power back to the crankshaft. The exhaust from the power recovery turbine then drove the turbosupercharger turbine, in series with the recovery turbine. Some VDTs used paired turbosuperchargers running in parallel. There was no limit to the imagination Pratt & Whitney engineers used for the VDT project. As can be ascertained from the foregoing, the term "VDT" was actually quite generic to the extent it covered a whole family of designs. The common theme was that of relying entirely on turbosupercharging for boost requirements; there were no gear driven or mechanically driven superchargers.

General Electric's Participation

As related previously, GE's participation in the VDT program was half hearted at best. Even though Pratt & Whitney was charged with overall responsibility for the VDT program, they relied heavily on GE's participation to furnish the CHM turbos and their associated controls. It should also be remembered that at the time GE was developing the two-stage CHM series of turbos. They were also inundated with production requirements and solving CH-7 problems—problems that all but brought Strategic Air Command (SAC) to its knees in 1950. Factor in GE's increasing focus on gas turbine technology and it's easy to see how they got behind. Exacerbating the situation was the fact that they never came up with a satisfactory control for the VDT. In retrospect, it was understandable. Even with today's FADEC technology it would be a daunting task to design such a control. Many parameters needed to be taken into account such as (i) balancing compressor discharge of the CHM with the CH turbos, (ii) ensuring the correct exhaust gas flow to the three turbosuperchargers, (iii) keeping intercooler air and aftercooler air at the appropriate temperature, and (iv) after taking into account these parameters, opening the variable discharge nozzle to the appropriate angle as determined by the foregoing inputs, and the power requirements demanded by the flight engineer/pilot. Little wonder that on the few occasions when a VDT engine flew in a modified B-50 during the summer of 1947, everything was controlled manually. Even so, as Phil Hopper, one of Pratt & Whitney's development engineers, later related it was a scary situation wondering if any of those turbos would overspeed or run away.

Applications

As a "paper" engine, the VDT showed great promise—so much so that several airframe manufacturers relied upon its performance potential. With the holy grail of one HP per cubic inch being realized, the VDT could have converted some mundane aircraft into aerial hot rods. Alas, for a number of reasons, it never happened.

Boeing B-50C/B-54

Going back to the B-29 as a starting point, Boeing developed the B-50 strategic bomber for the newly formed Strategic Air Command immediately after World War II. Ostensibly, the B-50 was simply a B-29 re-engined with the R-4360. In reality, the B-50 was a major redesign that entailed significant reengineering that, among other things, took advantage of newer and lighter materials. The USAF was ultimately planning on updating the B-50 design with VDT power. In this configuration it would have been designated B-54. On July 10, 1947, the USAF ordered the

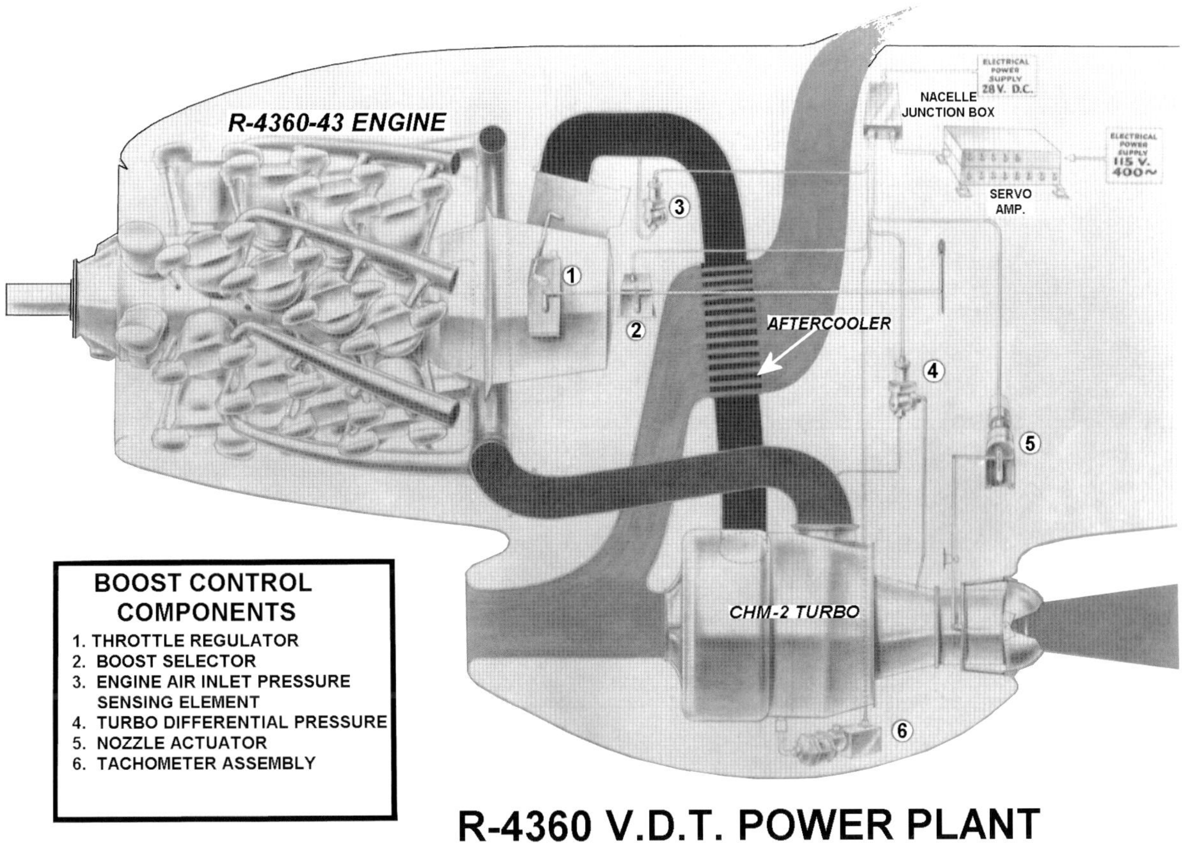

Artist's concept of how an R-4360-43 VDT engine may have looked in a tractor configuration such as a B-54. *(United Technologies Corp./Pratt & Whitney)*

modification of one B-50 powered by VDT engines. This aircraft was designated B-50C. As mentioned earlier, Phil Hopper, a Pratt & Whitney development engineer, went along for several flights in the modified B-50. Only one nacelle was modified for VDT power, the number three position or starboard inner. The following is how Phil recollected the VDT B-50 program *(Ref. 4-9)*.

"As I remember the B-50 aircraft we had at Experimental Flight Test (Pratt & Whitney) had the special nacelle (VDT nacelle) in the starboard inner position (#3 position). The axis of the CHM-2 turbosupercharger from G.E. lay parallel to the engine with the first stage inlet of the centrifugal compressor at the lower right-hand corner of the nacelle front face. The first installation did not look right, as the compressor inlet lip did not provide a clean entry to the inducer. Subsequently, the Experimental Hanger sheet metal shop reoperated the nacelle to open up the compressor inlet face.

"Since we did not have an automatic control (G.E. responsibility) for the turbo's compressor an exhaust valve was rigged up to control the volume of exhaust gas to the (CHM-2) turbine. A tachometer was set up to indicate CHM-2 turbosupercharger speed to the flight engineer's station in the cockpit. Along with it was a valve position control with a rheostat to adjust the exhaust valve. On takeoff I sat at the flight engineer's station to adjust the turbosupercharger speed according to scheduled limits.

"There were no significant problems during the flight testing at Rentschler Field starting in May 1947. We did shut down the unit in flight a couple of times due to over temps warning from the nacelle temp instrumentation but otherwise the testing went very smoothly during the summer. We never received a turbosupercharger control from G.E.

"Toward the end of the program on the 4360, a high altitude combination of three turbosuperchargers (2 CHM-2s and a CH-8) was assembled in the Installation Laboratory. It consisted of two two-stage turbosuperchargers (CHM-2s) in parallel and in series with a single stage turbosupercharger (CH-8). We were able to demonstrate the capability of Normal Rated power at a simulated altitude of 50,000 feet."

The parenthetical comments are those of the author. Phil Hopper's remarkable recollection jives exactly with recorded documentation. However, one can only admire a young and eager Phil Hopper in 1947 going for the ride of his life in the hopped-up Boeing B-50!

Republic RF-12 Rainbow

A victim of postwar budget cuts, the XF-12 never even got as far as having a VDT powerplant installed. Instead, the VDT powered XF-12 remained a design concept.

Genesis and Synopsis of VDT Development

During World War II, Pratt & Whitney had far more business than it could ever hope to handle. But then the hammer dropped at the cessation of hostilities and previously full order books evaporated overnight. This put Pratt & Whitney in the position of scrambling for all the business it could garner, both from the military and commercial markets. In February 1946, Pratt & Whitney proposed to the USAAF a $16,000,000 development program that would cover two projects: (i) improving the basic R-4360 design for higher horsepower ratings, modifying for use with direct fuel injection and for use with exhaust-driven gas turbines, and (ii) compounding the improved engine with exhaust-driven external superchargers or other auxiliary power units for further augmentation of power output and operation at higher altitudes. Pratt & Whitney envisioned a three-year development program to accomplish the goals proposed to the Air Force.

The first results of this new program were three R-4360-39s, later converted to true VDT engines in -43 guise. It was during the development of the -39 that the term "VDT" was coined. Pratt & Whitney issued the first VDT report in March 1947. Based on this early report, Consolidated Vultee (soon to be Convair) and Boeing were enthused at the estimated performance of the B-50C and B-36C. Pratt & Whitney were committed to successfully passing a 150-hour test of the R-4360-39 by June 30, 1948, plus a 250-hour test of the "C" series engine with direct fuel injection. Additionally, by June 30, 1949, Pratt & Whitney were committed to completing a 150-hour test on the R-4360-43 plus the manufacture of eight engines for flight testing. Other requirements included dynamometer calibration with high compression ratio pistons. Ambitious goals by anybody's standards.

Things started to come unraveled when the austere postwar defense budget interfered and it was declared that the 1948 budget was inadequate for all the aircraft and engine projects that were ongoing. From March to July 1947 two choices were mulled over: (i) improvement of the B-50 and B-36 along with development of the VDT engine, or (ii) financing the design phase of a new medium bomber. Initially, the AAF favored the latter. However, in May the AAF reversed its decision and decided to support the improved B-50 and B-36 along with the VDT program. The bean counters balanced the books by financing the modification of one B-50 and one B-36 and cutting back on the number of production aircraft for fiscal year 1948.

Pratt & Whitney revised its original proposal in May 1947 to cover the following: (i) conversion of the R-4360-39 to approved installation types, (ii) limited production of types for installation in the prototypes of the proposed modernized aircraft (B-50 and B-36), and (iii) quantity production of the VDT types for the proposed improved aircraft.

The following aircraft were earmarked for VDT power: 22 Republic F-12Cs (production

Original Time Line For VDT Development

(Table 4-1)

Contractual Authorization	Contract Number	Number of Engines	Deliveries to Begin	Deliveries To Be Completed
Development August 8, 1947	W33-038 ac-18653	2 R-4360-43 1 R-4360-51 (later converted to R-4360-43)		December 31, 1947 March 31, 1948 (November 1, 1948)
Development August 8, 1947	W33-038 ac-18653	150-hr. qualification test		December 31, 1948
Development August 8, 1947	W33-038 ac-18653	50-hr. flight test		One year after receipt of B-50
Prototype December 4, 1947	W33-038 ac-19032	8 R-4360-43 8 R-4360-51	March 1949 April 1949	July 1949 May 1949
Production December 31, 1947	NOa(s) 9511	14 R-4360-43	July 1949	
Production February 3, 1948	NOa(s) 8333	213 R-4360-51	September 1949	May, 1950
Production June 17, 1948	MIPR R-49-5N	186 R-4360-43	September 1949	August 1950

model to be designated R-12A), 43 Boeing B-50Cs (production model designated B-54), and 34 Convair B-36Cs. Nevertheless, the Republic F-12's 14 allocated engines were reallocated to the B-54 program. However, changes in the aircraft programs negated the possibility of carrying out the foregoing VDT program beyond the delivery of eight R-4360-43 VDT prototype engines. Neither the Convair B-36C nor the Boeing YB-50C reached flight test status. The program was therefore reduced to converting XR-4360-39s to R-4360-43s.

R-4360-43

An intensive two-year development program produced the R-4360-43 with eight of these engines being delivered by April 1949 for the B-50C program. Per the terms of contract W33-038 ac-18653, Pratt & Whitney converted two R-4360-39s to VDT (R-4360-43) configuration to power the Boeing B-50C. A key element of the VDT was the controls systems, particularly for the VDT clamshells. With their extensive experience in this field, General Electric was sub-contracted to perform this development work. In addition to the controls, General Electric was also responsible for development of the CHM-2 turbosupercharger. A government-owned B-50 was supplied to Pratt & Whitney for flight test of the VDT engine. One nacelle, number three, was modified by Boeing to accept the VDT engine. From the foregoing it can easily be seen how dependent Pratt & Whitney was on Boeing and General Electric, commitments that these two companies did not always live up to. Pratt & Whitney delivered two out of three R-4360-43 VDT engines to Boeing in October and December 1947. Boeing used these engines as mock-ups for the design of the YB-50C and the single B-50 with the modified number three nacelle to be used for flight testing by Pratt & Whitney. The third VDT engine was converted to a -43. As Pratt & Whitney progressed with development work in the test cell, engines were updated. At its peak, Pratt & Whitney employed 12 R-4360s for VDT development work.

Development work by Pratt & Whitney bore fruit not just for the VDT program, but also for all models of the R-4360. For instance, a modified and improved crankshaft was developed. Other improvements resulted in the -43 and basic data for higher ratings. Improvements in the -43 included a hydraulically damped flywheel in the crankshaft, cylinders with improved cooling characteristics, and improved exhaust system. Cooling was always an ongoing problem and in February/March 1948, Pratt & Whitney revised

R-4360-43 VDT engine. This photograph does not show the VDT turbosupercharger. With a great deal of difficulty, this engine managed to pass its 150-hour type test prior to cancellation of the entire VDT program. *(United Technologies Corp./Pratt & Whitney)*

downward its previous performance estimates for cooling requirements. This revelation resulted in a decrease in performance estimates for the YB-50C. Even so, the AMC (Air Material Command) believed the R-4360-43 cooled as well, if not better, than the B-29 or B-50A. However, being compared to the problematic R-3350 in the B-29 was not exactly a resounding endorsement. Reference is made to the development of a five-piece, built-up crankshaft in order to overcome crankshaft failures. It is not known if any engines were built with this crank configuration, but it certainly received serious consideration and probably got to the design stage. It would have to be assumed that had a built-up crankshaft been used, one-piece master rods would have replaced the former two-piece master rods. Tests were conducted with higher-compression-ratio pistons for improved fuel economy. In fact, it was the goal of Pratt & Whitney to reduce fuel consumption down to 0.37 pounds per HP per hour, which would put it in the same category as a diesel. To further put things in perspective, typical cruise power fuel consumption would be in the range of 0.45 pounds per HP per hour.

Pratt & Whitney made three attempts from November 1947 to August 1948 at completing an official model test with the -43. In December 1947, the -43 ran for 11.5 hours but then had to be torn down for redesign and fabrication of parts that failed. By May 1948, Pratt & Whitney had accumulated 150 hours of testing. But it was not smooth sailing, several more interruptions due to failures required more redesign and fabrication of new parts.

The next attempt at an official 150-qualification-hour test occurred in June 1948. The Air Material Command urged Pratt & Whitney to conduct the test in a more realistic environment—with all VDT components attached, driving a flight propeller, and driving all accessories that would be used in the B-50C. More delays followed with the non-availability of the General Electric CHM-2 turbosupercharger. GE came through with the all-important CHM-2 in early June 1948; however, Pratt & Whitney's woes were far from over. GE announced that it would need to make frequent inspections of the CHM-2 during the test. The test was started without the CHM-2 until its availability on June 8. However, the test only lasted 1.5 hours before the crankshaft failed.

Pratt & Whitney completed a 150-hour test in early July 1948 with the R-4360-43 fitted with a crankshaft flywheel damper. On July 6, 1949, Pratt & Whitney announced its intention of attempting another 150-hour test with an improved crankshaft design. But General Electric presented more flies in the ointment by claiming the CHM-2 was not ready for a 150-hour test; in fact it had not even completed a factory test. Chomping at the bit, Pratt & Whitney suggested going ahead without the GE CHM-2 turbosupercharger. Not surprisingly, the Powerplant Laboratory did not approve this idea. A number of concerns were expressed by the Powerplant Laboratory: The proposed cylinder head temperatures and engine ratings were not those guaranteed in the model specification, the B-50C engine-driven accessories were not being used, nor was the GE-furnished VDT power control. However, Pratt & Whitney would have preferred to use the B-50C flight equipment if it were available on time. But Pratt & Whitney had commenced testing and pointed out two reasons to conclude it quickly: (i) Pratt & Whitney was funding the test, and (ii) the contractor (Pratt & Whitney) could not go ahead with the production schedule until the bill of materials for the production engine had been approved by an acceptable qualification test on the development engines.

Pratt & Whitney got some relief from the AMC when they conceded that the development status of B-50C engine-driven flight accessories would delay the 150-hour qualification test. In order to avoid further delays to the production schedule, the AMC authorized Pratt & Whitney to continue testing with simulated loads for the accessories. In the meantime, the AMC asked Pratt & Whitney to submit a contract proposal for yet another 150-hour qualification test to be run when all the required equipment would be available. Pratt & Whitney finally completed the R-4360-43 150-hour qualification test on August 5, 1948. Even so, Pratt & Whitney was not out of the woods yet. The 150-hour test was completed with the understanding of the AMC that the complete VDT assembly, along with the B-50C propeller and other engine-driven accessories, would be given a 150-hour test at a later date.

A full complement of VDT components was used in the 50-hour flight test program. An Air-Force-furnished B-50 was delivered to Pratt & Whitney in East Hartford in March 1948. This B-50 was unique inasmuch as Boeing had a modified number three nacelle for the R-4360-43 VDT engine. As usual, General Electric held up the party. The CHM-2 turbosupercharger was not available until April. May 26, 1948, marked the first flight of a VDT engine in the modified Boeing B-50. Test work continued until December 1948. An additional five hours of testing was added to the contract making for a total of 55 hours of flight testing. This additional five hours was used for the purpose of vibration studies and testing of the Curtiss Electric C-6455-A propeller. Testing pointed up several key areas that required redesign or improvement. These included VDT components, particularly the troublesome GE CHM-2 turbosupercharger. The aircraft nacelles needed changes to improve compressor entrance airflow characteristics, better cooling for the turbine, and a change to the compressor housing material from magnesium to aluminum. These and other requirements were passed onto Boeing to incorporate into the prototype YB-50C and production B-54s.

In January 1949, the main participants in the B-50C/B-54 program, Curtiss Electric Propellers, General Electric, Boeing, and Pratt & Whitney convened for a conference. All agreed that further single-nacelle testing would be required. Negotiations were underway for an additional 95 hours of flight-testing in the modified B-50 when news came down that the B-54 program had been cancelled. As it turned out, the one nacelle in the modified B-50 would be the only time a VDT engine ever flew. In the meantime, Pratt & Whitney complied with their contractual obligations and delivered eight R-4360-43s ahead of schedule, the last one being delivered in April 1949 *(Ref. 4-1 and 4-10).*

R-4360-51

Unlike its R-4360-43 stablemate, the R-4360-51 VDT did not receive anything like the intensive development effort accorded the -43. In fact, the program was cancelled prior to any R-4360-51 engines being delivered to the Air Force. Much of this was due to the problematic B-36C, the intended recipient of the -51. Problems went far beyond engineering and development. The B-36C was also cursed with political problems. All production B-36s had pusher propellers; however, the B-36C was intended to use tractor propellers. In an effort to reduce engineering time, the R-4360 was mounted in the same place as the pusher versions. This required a long extension shaft to drive the propeller. The USAF and AMC agonized over the decision of whether or not to proceed with the VDT program for the B-36C (also termed B-36Z) and B-54. The debate raged from March until July 1947. The decision was made to go ahead by using production funds to support VDT development. Nevertheless, the AMC expressed concern regarding the YB-36C and its highly experimental VDT engines.

In May and June 1947, the key players—the AMC, Convair, General Electric, and Pratt & Whitney—met on numerous occasions to hammer out the configuration of the VDT engine, specifically the R-4360-51. Although based on the R-4360-43, the R-4360-51 had to accommodate the aforementioned driveshaft along with other modifications. Convair also had their work cut out for them by having to accommodate the extension driveshaft and its associated reduction gearing attached at the propeller end. Under contract number W33-038 ac-18653, one of three XR-4360-39 engines was converted to R-4360-51 status and, rather surprisingly, delivered without qualification testing.

On August 22, 1947, the AAF Aircraft and Weapons Board pulled the plug when they cancelled the B-36Z program with 100 aircraft on order. But all was not lost, not yet anyway. Convair countered with a proposal on September 4, 1947, to reinstate the B-36Z (a.k.a. B-36C) project by modifying the last 34 B-36s on contract with VDT engines. Convair proposed financing these 34 modified aircraft through cutting the number of aircraft on contract from 100 to 95. Furthermore, Convair agreed to complete these aircraft by June 1950, an extension of seven months. However, the revised time frame represented an acceleration of the B-36C schedule. Many skeptics doubted if Convair could meet the new B-36C schedule. The AMC was requested to furnish a B-36C time frame based on realistic engine development schedules. The Chief of Staff directed that members of the USAF Aircraft and Weapons Board evaluate Convair's new proposal and reconsider their decision to terminate the B-36C program. A revised B-36C schedule was sent by the AMC to USAF Headquarters on September 30, 1947 (USAF was created on September 18, 1947). All but one approved the revised contract and schedule in November 1947. That one holdout was Chief of Strategic Air Command. Even with this key holdout, on December 5, 1947, HQ USAF directed the AMC to proceed with the B-

36C. So from March 1947 to December 1947, decisions zigzagged back and forth on the B-36C. Of course, this is not the way to start a new program, particularly one as complex as the B-36C.

One result of these decision reversals was delay in initiating the program. This affected Pratt & Whitney to the extent that work was delayed from August until December 1947. Therefore, Pratt & Whitney did not commence conversion work on the R-4360-39 to R-4360-51 status until January 1948. By March 1948, the first R-4360-51 was assembled and ready for testing. Major problems surfaced almost immediately. In May 1948 an extension shaft coupling failed and, perhaps coincidentally, the B-36C program was cancelled (again). Additional problems from the zigging and zagging that occurred for eight months was that of delayed engine delivery. Convair's proposed accelerated B-36C production was optimistically based on availability of production engines by 1949. This was clearly out of the picture. On the day the USAF was formed, the AMC conferenced with Pratt & Whitney and Convair for the latest production schedule for the R-4360-51 and for all VDT type engines. Four VDT engines were promised for delivery by May 1949, albeit R-4360-43s for the B-54 because Pratt & Whitney had planned on the B-54 getting into production prior to the B-36C. No -51s would be available before July 1949. With this in mind the AMC, assuming a realistic B-36C time frame of September 30, 1947, with a three-month lead-time for the engines, established delivery of the first production aircraft in October 1949. Due to the ongoing delays and indecision, delivery of the first B-36C was pushed back to December 1949 and the last delivery to September 1950.

It was becoming increasingly apparent to all involved that interest in the B-36C was waning. Nevertheless, contractual commitments had to be met. The one driving incentive to continue with B-36C development was the promise of higher performance with VDT power. But even this promise began to evaporate after Pratt & Whitney's three abortive attempts at a 150-hour qualification test. Therefore, planning for the YB-36C, which was to be available in July 1949 and followed within a short five months, had to proceed on estimated and experimental values for the VDT engine.

In January 1948 Convair sent a statement to the AMC urgently requesting information for proceeding with the VDT installation in their aircraft, the B-36C. This issue was brought to a head on January 26, 1948, when Convair, Pratt & Whitney, and the AMC met to discuss the VDT installation requirements. As a carrot and stick tactic, the AMC threatened cancellation of the entire B-36C/VDT program if performance parameters were not met. Among many others, Pratt & Whitney were faced with three major problems: (i) delivery of a mock-up engine to Convair at a time when the final configuration had not been arrived at, (ii) development of a sophisticated two-speed cooling fan, and (iii) establishment of cooling curves for the VDT engine. The next three months only saw further confusion and delay as Pratt & Whitney struggled to arrive at viable solutions to innumerable problems.

On March 1, 1948, Pratt & Whitney agreed to furnish the requested engine mock-up within five months at no cost to the customer (the USAF). At the same time, Pratt & Whitney pointed out that it would have to be a mock-up of the latest design, as it existed at that time. Of course, with an ambitious program like this, it was almost assured that significant changes to the engine would occur. Even so, Pratt & Whitney agreed to furnish updated mock-ups to Convair as the engine developed. But Convair required an engine mock-up no later than June rather than the August time frame Pratt & Whitney had promised.

During the January 1948 conference between Pratt & Whitney, the AMC, and Convair, it was agreed that Pratt & Whitney would furnish the required two-speed cooling fan similar to that used on the R-4360-41, the engine used to power the B-36B, D, E, and Convair XC-99. Convair agreed to provide the cooling fan gear ratios and horsepower loading based on data supplied by Pratt & Whitney. Pratt & Whitney presented a contract proposal to the AMC on February 16, 1948, to address the cooling fan drive. On March 19, 1948, Consolidated put another fly in the ointment by releasing revised gear ratios for the cooling fan and its associated horsepower loading. Pratt & Whitney countered on April 6, 1948, by telling all parties that the revised cooling requirements would necessitate a totally new fan drive, which, as a consequence, would result in serious delays to the R-4360-51 and by default, the entire B-36C program. A remarkable 18 months was quoted by Pratt & Whitney to develop the new cooling fan.

By now it was apparent to everyone that cooling and developing a satisfactory solution was of paramount importance. During the January 1948 conference, Pratt & Whitney pointed out that determining the exact cooling requirements due to the ongoing development nature of the VDT engine would be difficult. It was suggested that GE and Convair jointly study with Pratt & Whitney cooling requirements and draw up a revised report on the engine's performance. Pratt & Whitney completed the study in February 1948, and published the results in March. This report included the latest VDT cooling characteristics and performance estimates for the R-4360-51. During April 12 and 13, the AMC, Convair, Curtiss Electric Propellers, GE, and Pratt & Whitney met yet again at Convair's facility in Fort Worth, Texas. The purpose was to go over the revised cooling and performance characteristics of the -51 and its effect on B-36C performance. As everyone feared, the results indicated yet another reduction in performance of the aircraft.

At this point, the AMC basically said enough is enough and recommended termination of the B-36C program. Things now moved quite rapidly and the chief of the Fort Worth procurement office advised Consolidated to slow down on factory operations in anticipation of the B-36C's termination. On or about April 21, 1948, the AMC asked HQ USAF to re-evaluate the B-36 aircraft in view of the latest developments and reduced performance estimates. Four recommendations were made by the AMC, none of which included continuation of the B-36C program (Ref. 4-1). Cancellation of the B-36C set off a firestorm, particularly in the Research and Engineering Division. The Deputy Chief of Staff, Material asked the Inspector General to make a complete investigation. Things had now gone from bad to downright ugly. Fortunately, cooler heads prevailed and the investigation was scaled down in early June to include just the VDT program and the information available at HQ USAF. Formal cancellation of the B-36C program came on June 25, 1948, at a conference in the office of the Secretary of the Air Force. The B-36 program was now revised with the authorization of 95 aircraft, of which 22 were B-36As and the remaining 73 were B-36Bs. As we shall see, significantly more B-36s were eventually procured instigating the "Admiral's Revolt" of Navy brass.

If any good could come out of the B-36C debacle, it was the fact that decision makers realized they had a tiger by the tail and the sooner it was concluded the better. This resulted in a relatively small expenditure by military standards. An offer was made by Pratt & Whitney to convert one R-4360-51 development engine into an R-4360-43 at no cost to the government. The Navy estimated that approximately $200,000 in termination fees would need to be paid to Pratt & Whitney for the changes made to convert the R-4360-41 to R-4360-51 and then back again to R-4360-41. Little wonder that Pratt & Whitney hardly knew if they were coming or going during this tortuous time period.

Plans for Further Development of the R-4360 "C" Series Engines

On April 13, 1948, the Powerplant Laboratory requested that Pratt & Whitney prepare a design study and contract for further development of the R-4360-43 for operation in a B-50C to operate at altitudes of 40,000 to 45,000 feet. Pratt & Whitney requested and were granted a postponement of the new study until contract W33-038 ac-18653 was complete. This contract covered the terms and conditions under which Pratt & Whitney would complete a successful 150-hour qualification test of the R-4360-43 VDT engine. Upon completion of the 150-hour qualification test, Pratt & Whitney presented a report in October 1948 to the AMC for further improvement of the R-4360. This report showed a potential 4,800 hp and operation in excess of 43,000 feet with the remarkably low specific fuel consumption of 0.375. A 12-month development program was proposed that would include the following: (i) conversions of two R-4360-43s, (ii) a 150-hour qualification test, and (iii) development testing of an Eclipse power control. A price of $3,920,000 was proposed plus an additional $3,000 per hour for flight testing. Even allowing for inflation, this was a really good deal for the government.

The wheels of the government bureaucracy turned slowly as they mulled over Pratt & Whitney's proposal from November 1948 to February 1949. At least one government employee, the Director of Research and Development, was sold on the idea. He felt the R-4360-43 had already exceeded expectations and this engine would be the principle powerplant for future heavy transport aircraft. Additional support came from the Director of Procurement and Industrial Planning.

Alas, in April 1949 the hammer dropped when HQ USAF directed cancellation of the B-54 program, but requested an additional B-36s be procured. To date, the AMC had invested approximately $12,500,000. This does not include funds Pratt & Whitney had invested at its own expense. The AMC had allocated a total of $22,000,000 for production versions of the R-4360-43 for B-54s. Cancellation costs for the engine came in at $7,500,000 plus $1,500,000 for the CHM-2 turbosupercharger. In all, the VDT program cost the U.S. taxpayers approximately $21,000,000, which by the standards of the late 1940s could not be regarded as chump change

After all the problems had been ironed out on the R-4360-43 and it passed its 150-hour qualification test, it could have been a very successful engine in the 4,500- to 4,800-hp category. However, the same could not be said of its stablemate, the R-4360-52 intended for the stillborn B-36C.

The intense development of the VDT versions of the R-4360 spawned a new series, the "C" series R-4360. In order to salvage some of the intense development, not to mention the costs involved, the AMC asked HQ USAF to authorize further development and procurement of an advanced "C" series R-4360 for use in B-36s. It was estimated, rather optimistically as it turned out, that up to 80 percent of the VDT program termination charges could be recovered. HQ USAF agreed but couched their final approval contingent upon more detailed justification, which came on July 18.

Two phases were proposed: (i) developing the engines and turbosuperchargers and flight testing them in a B-36 and a C-124, and (ii) procuring sufficient R-4360-53s for 36 B-36s. It was also specified that the start of the second phase was contingent upon acceptance of the first phase.

The first phase called out for R-4360 "C" series engines to be converted and developed to an R-4360-55 standard with low-compression pistons and VDT. The second phase called for the development of a VDT R-4360-57 with high-compression pistons. In conjunction with these tests, General Electric was to develop the CH-8 and CHM-2 turbosuperchargers. Unlike the B-36C, which was designed as a tractor, the R-4360-55 and -57s were arranged as pushers in a similar fashion to production B-36s. Two B-36s were earmarked for this development along with 24 engines and 24 turbosuperchargers. Convair didn't give up on the tractor configuration and to this extent, internally supercharged "C" series R-4360s (not VDT) were to be procured for testing in two prototype Douglas C-124s. A total of 16 engines were required for the C-124 development. A total cost of $12,500,000 was estimated for the first phase of this program.

The second phase called for the procurement of R-4360-53 engines for 36 B-36 applications. The second phase included modifications of the airframe along with procurement of propeller hubs. The total cost of the second phase was estimated at $31,251,680. Funds were found by applying this cost against funds already available for procurement of additional B-36 engines, currently unallocated funds, and credits available through reduction of termination costs associated with B-54 engines.

On July 20, 1949, HQ USAF approved these plans. By August 5, 1949, the Procurement Division had initiated procurement action with the Navy for the R-4360-53 engines in lieu of R-4360-41s previously contracted. With the Navy's concurrence, the USAF was able to contract directly with Pratt & Whitney for development of the prototype engines. Most of the engineering and development for these engines had already been completed with the development of the R-4360-43. Pratt & Whitney had delivered eight R-4360-43s per the terms of a prior contract. The USAF returned these eight engines to Pratt & Whitney for conversion to R-4360-55 status. A lead-time of 6½ months was estimated by Pratt & Whitney to complete this conversion work from the date of contract execution. Pratt & Whitney estimated 11½ months for delivery of new prototype (not converted) engines and make delivery of production engines within 12 months (Ref. 4-1).

Okay, so what happened? After all the development and expenditures, the only engines that saw production were the R-4360-41, R-4360-53, and the "C" series engines that later saw extensive use in the C-124 and B-36B, D, E, F, H, and J. Was it a wasted effort? Yes and no. Remember that at this time frame, late 1940s, gas turbines were rapidly surpassing piston engines both in performance and production quantity. Development of the piston engine was in a death spiral. The military wanted to get on with developing the gas turbine rather than spin their wheels on more and more sophisticated piston engines. Could the VDT have worked? No question that with the appropriate support both monetarily and engineering wise, the VDT could have been developed into a remarkable piston engine. This fact was born out by the successful completion by the R-4360-43 of its 150-hour qualification test.

Even though Republic's RF-12 Rainbow was dropped for VDT power quite early on in the program, performance estimates for this aircraft with VDT power were nothing short of remarkable. It would have been capable of over 500 mph at 40,000 feet. And don't forget, the Rainbow was a large, four-engined aircraft. As it was, some of the lessons learned in the VDT development program filtered through to production engines. Even so there were many avenues of development that were abandoned. If the VDT had progressed into production there seems little doubt that the crankshaft would have been redesigned as a five-piece built-up unit resulting in an improvement in weight reduction, better vibration characteristics because of lighter and stronger one-piece master rods, and a stouter lower end capable of withstanding higher BMEPs, horsepower, and higher RPM. In retrospect, no one can criticize the program. It was advancing the state-of-the-art of aero propulsion, but what the program originators did not realize was the fact that gas turbines would develop so rapidly. It took many people by surprise. Band-Aids such as the additional J47 jet pods on the B-36 and B-50 represented a transition period where gas turbines quickly displaced their piston brethren. Even though piston-powered aircraft were still on the USAF inventory as late as the 1980s, piston-powered aircraft were replaced for frontline duties by the late 1950s. By this time the gas turbine reigned supreme—it was no contest. In a way this is too bad, some of the engines on the drawing board in the late 1940s showed amazing ingenuity. However, the argument has been presented, and quite correctly, maintaining sophisticated piston engines would have been a nightmare.

As it was, B-36s were probably the most maintenance intensive aircraft the USAF had ever operated. As much as they oozed charisma, reality was they could be a monumental pain with relatively

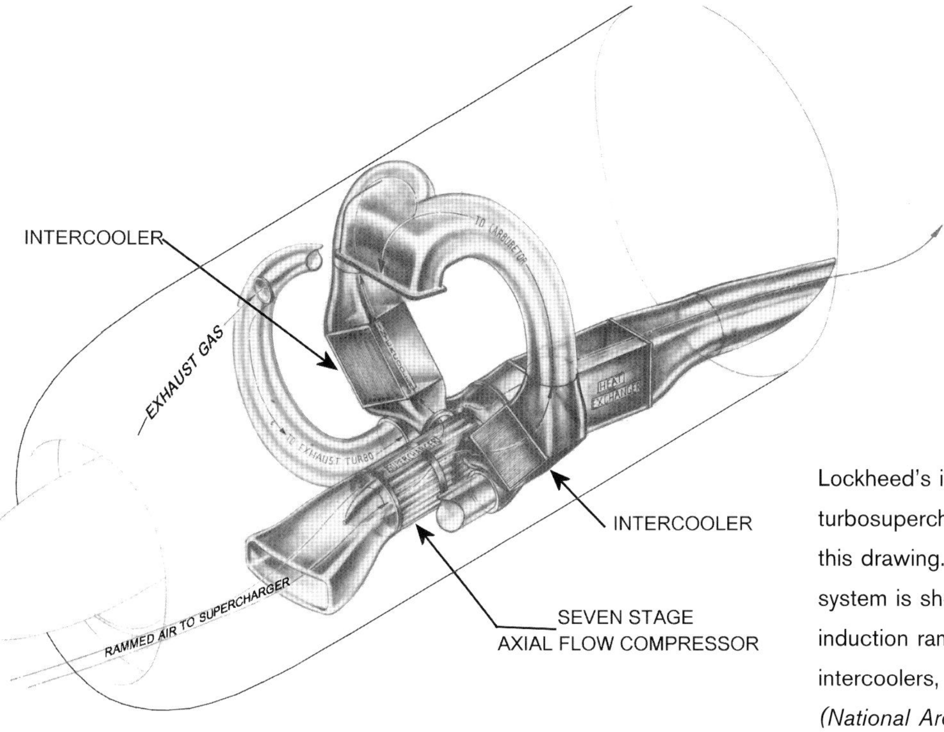

Lockheed's innovative axial flow turbosupercharger is illustrated in this drawing. The entire induction system is shown including the induction ram scoop, dual intercoolers, and exhaust system. (National Archives and Records Administration)

short times between overhauls—and that's even if the engine managed to reach its time between overhaul (TBO), which did not occur that often. On the other hand, the promised fuel economy and altitude performance of the VDT engine could have handed the USAF a wonderful powerplant for high-altitude photo reconnaissance. Of course, this was the mission for the Republic Rainbow. It was not until the advent of Lockheed's U-2 that the Rainbow's performance, when powered by a VDT, could have been matched.

Jet Stacks

When the exhaust valve opens, in the case of the R-4360, 70 degrees after top dead center, a significant amount of residual energy still resides within the cylinder. As the exhaust valve opens, this residual energy in the form of pressure exits the exhaust port at a supersonic velocity. When directed rearwards, exhaust energy is capable of imparting a significant forward thrust. This thrust can amount to well over 400 pounds.

Lockheed Constitution

Development of this large four-engined transport began in 1942. One overriding requirement for this aircraft was long-range economical cruise. During various phases of its development, four turbosuperchargers were looked at—a Lockheed axial flow design, a Turbo Engineering design, and two designs from General Electric *(Ref. 4-3)*.

Lockheed Turbosupercharger

Surprisingly, no designation was given to the innovative Lockheed turbosupercharger. It featured a seven-stage axial flow compressor driven by a modified GE Type B turbine wheel. This complex turbo was a two-shaft design. The first four stages of the compressor were driven directly from the turbine wheel. The last three stages were driven through a fluid coupling. This two-shaft arrangement allowed a wide range of operating parameters. It should be remembered that the Constitution was originally designed around R-4360-18s rated at 3,000 horsepower.

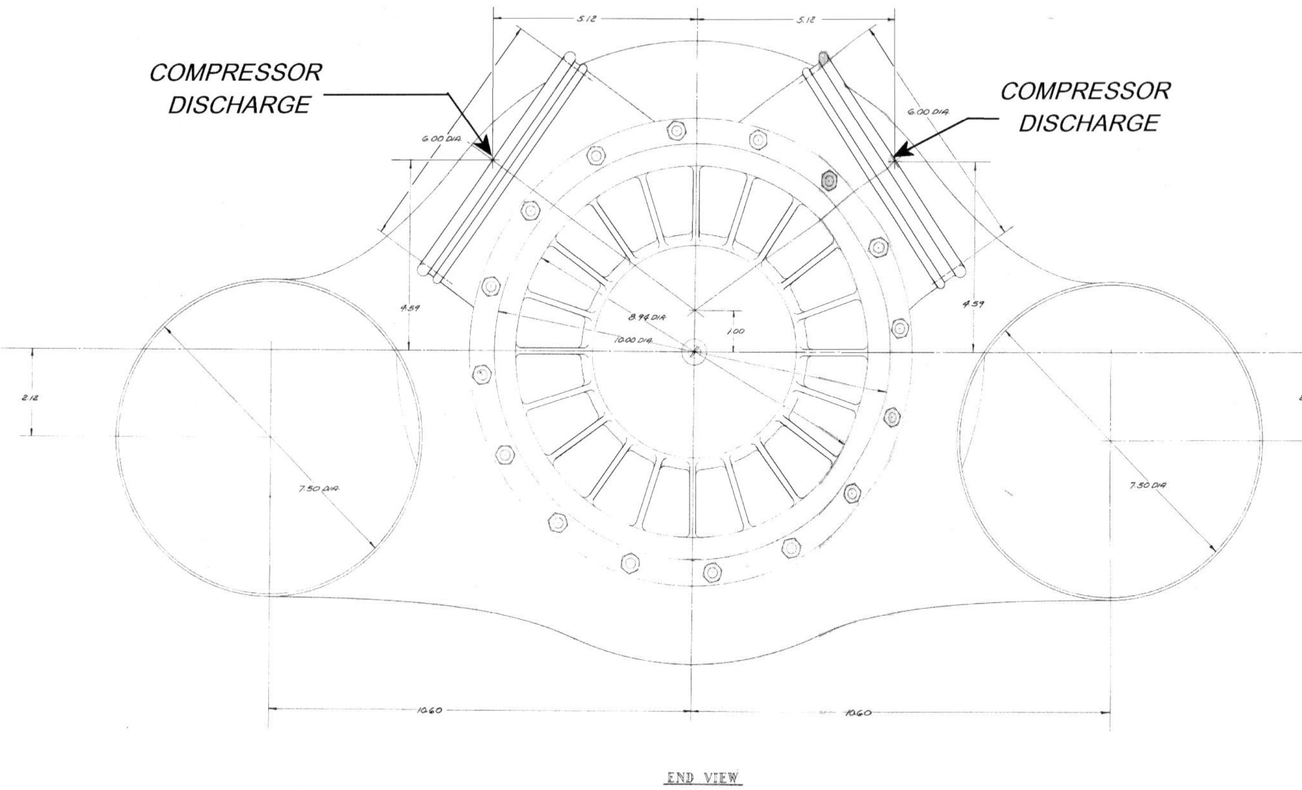

This end view of the Lockheed turbosupercharger illustrates its compact dimensions and dual discharges from the seven-stage axial flow compressor section. *(National Archives and Records Administration)*

And yet economy cruise demanded a power of 900 hp. A sophisticated speed control determined how fast the last three stages should rotate. Two discharges are provided for the seven-stage compressor. Each outlet feeds into a rectangular air-to-air intercooler and then to the carburetor deck. Compressor blades were cast and could be individually adjusted. Between each compressor stage was a ring of stator vanes. Modifications to the GE Type B turbine wheel included modifying the buckets from a pure impulse to a 50 percent reaction and modified nozzles. Although it was slightly heavier than its GE counterpart, most of that weight penalty was negated by the fact that the Lockheed compressor, being more efficient than centrifugal flow designs, could get by with a smaller intercooler. Of all the turbosuperchargers developed for the R-4360, Lockheed's was one of the more innovative. It's not clear if one was ever built *(Ref. 4-3).*

Turbo Engineering Company

Rudolph Birman was a brilliant Westinghouse turbo machinery engineer. He had innovative ideas on how turbosuperchargers should be designed, and the U.S. Navy recognized his talents. The Navy was always lukewarm on turbosuperchargers, preferring, instead of relying, on multi-stage gear-driven superchargers. Nevertheless, the Navy recognized the considerable strides turbosuperchargers were making powering Army Air Force aircraft. With this in mind, the Navy set Birman up with his own company, the Turbo Engineering Company. Although a number of excellent designs came from Birman, and in some cases were manufactured and test flown, none went into production. Same with his PT-15B-3 design for the Constitution. Very little information has survived on this turbosupercharger. From internal Lockheed reports, it is apparent that the PT-15B was not a suitable match for the

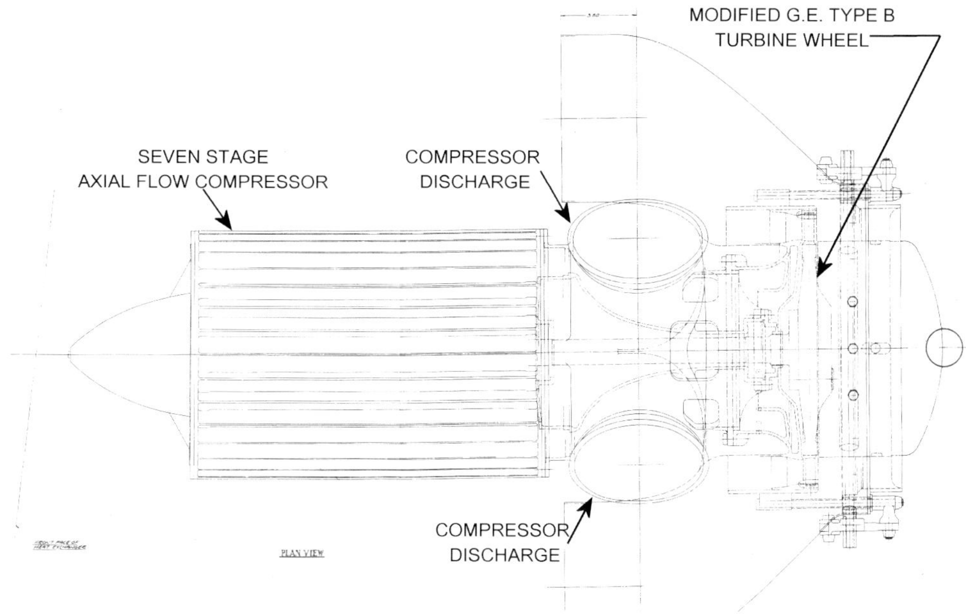

This top view of the Lockheed turbosupercharger illustrates the unique seven-stage axial flow compressor and GE derived turbine section. *(National Archives and Records Administration)*

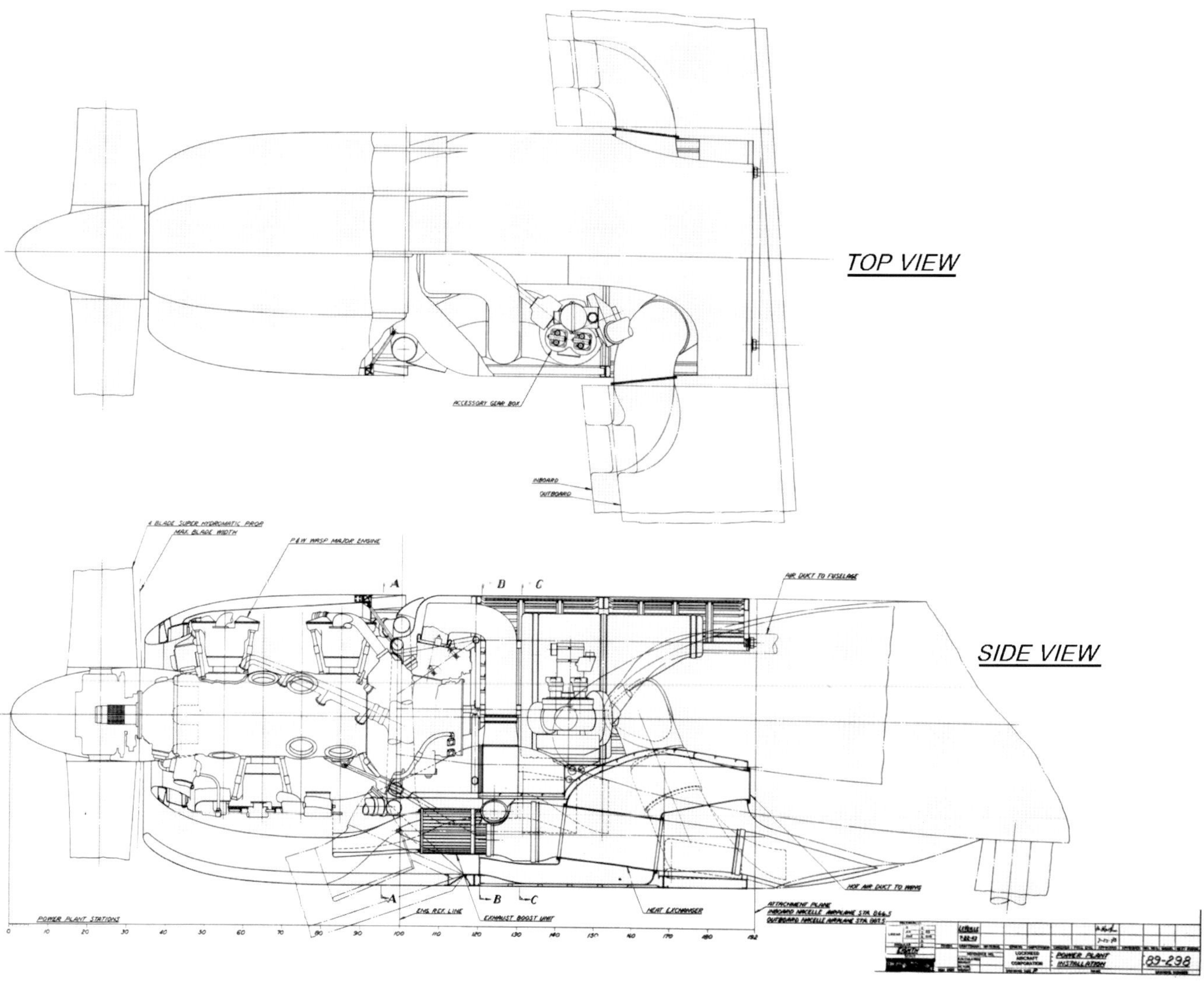

Lockheed developed a nice QEC (quick engine change) package designed around their axial flow turbosupercharger. These side and top views give a good perspective of how everything fit. *(National Archives and Records Administration)*

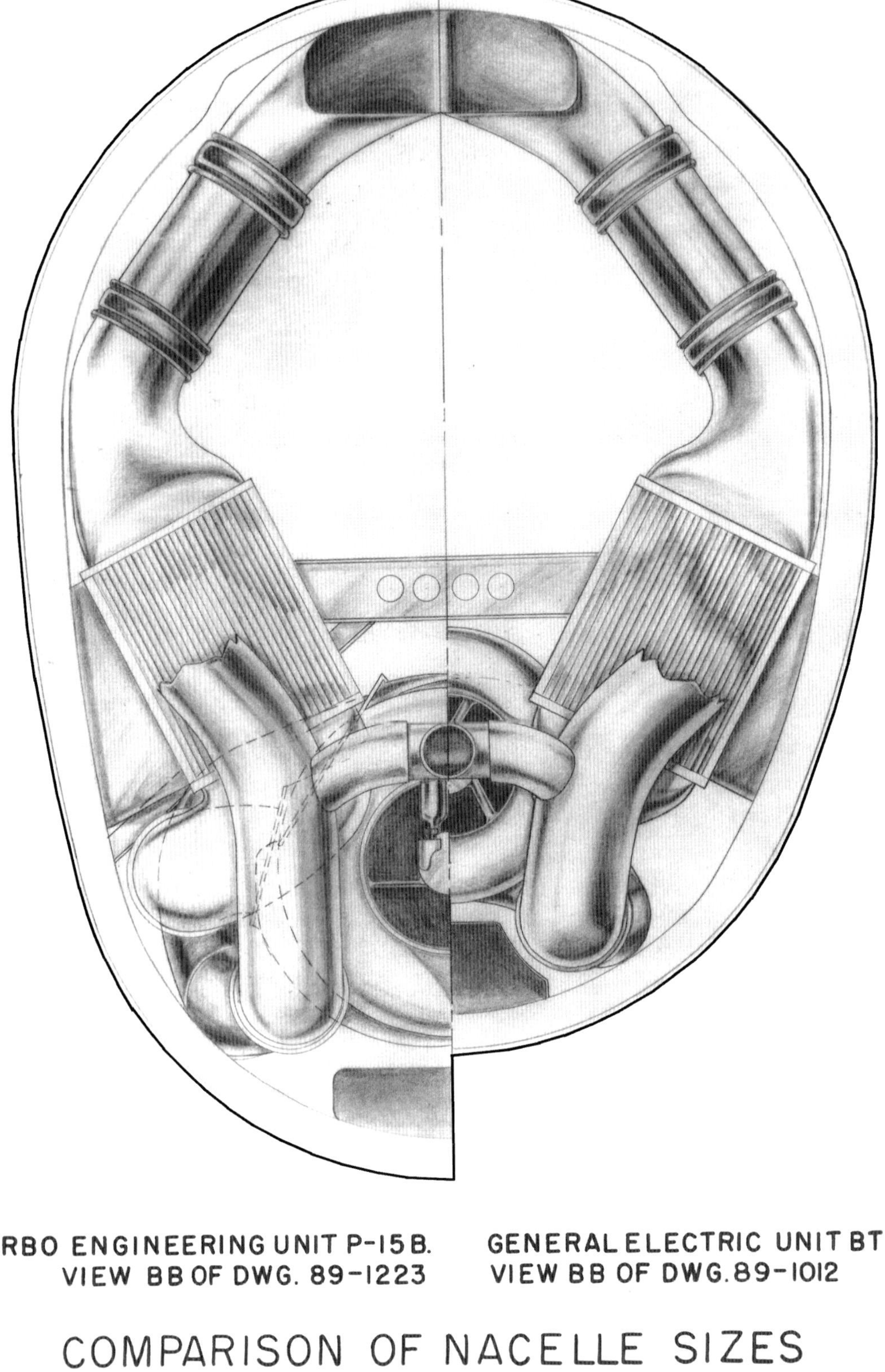

Although Rudolph Birman produced some excellent turbosupercharger designs for the Navy, his turbo for the Constitution was not a good fit, literally and figuratively. This cross section through the nacelle shows how much larger the Turbo Engineering unit would have been compared to a GE turbo. *(National Archives and Records Administration)*

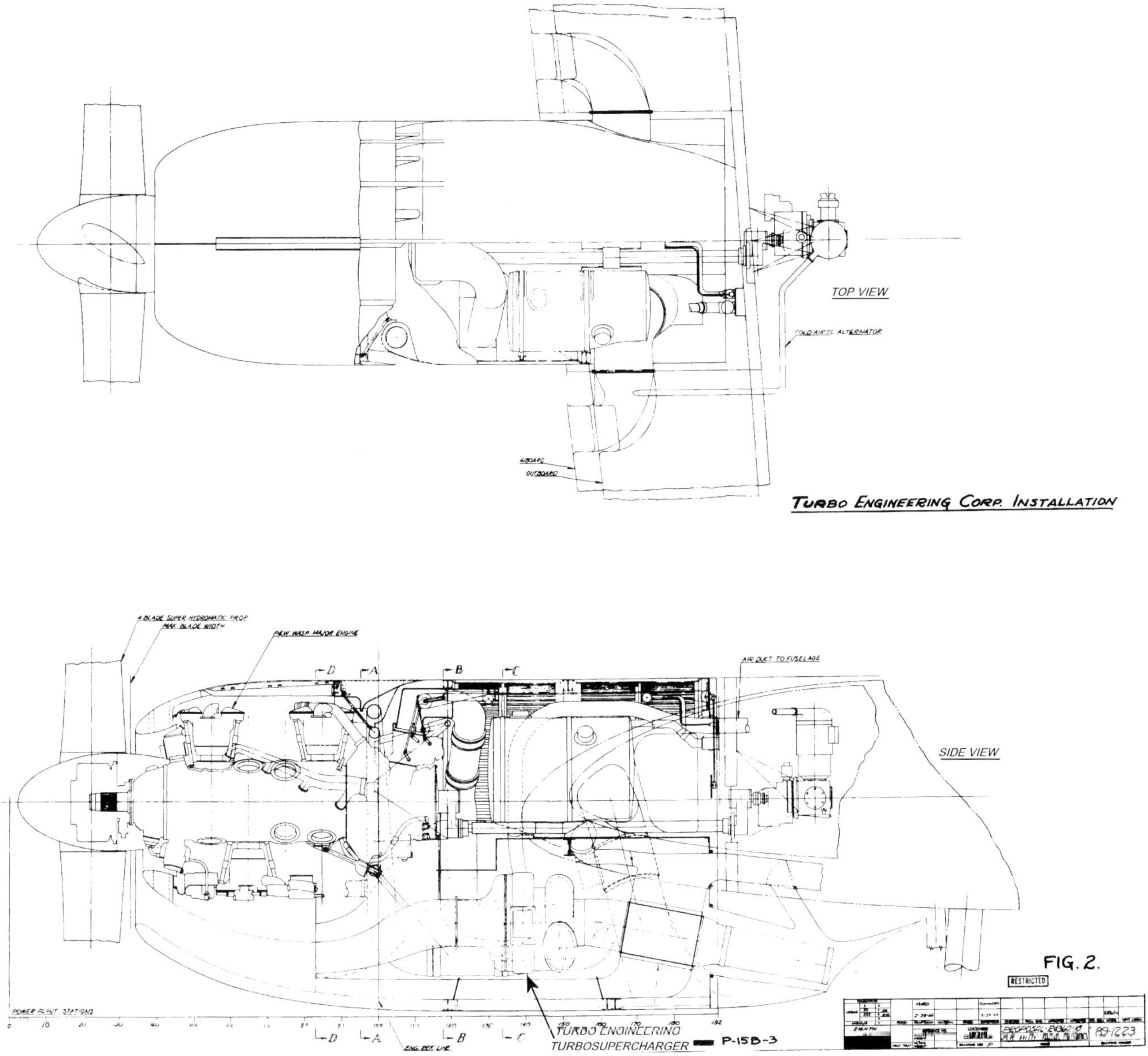

Top and side views of the Turbo Engineering Corp. installation. It quickly became apparent that this unit was optimized for high power, not low-power economy cruise. *(National Archives and Records Administration)*

Constitution. At least two factors contributed to this. The PT-15B was a far bulkier unit than the GE or Lockheed design, resulting in a larger nacelle. Worse yet, it was optimized for high power, not cruise powers of 900 to 1,300 hp. Consequently, the fuel consumption with this unit would have been unacceptably high. Adding to the PT-15B-3's woes was the fact that it was considerably heavier; in fact, a weight penalty of 1,200 pounds per aircraft would have resulted when compared to the GE unit. Not surprisingly, this turbosupercharger soon dropped out of the running *(Ref. 4-3).*

General Electric

According to Lockheed engineering reports, GE developed two turbosuperchargers for the Constitution—the B-133 and BH-3. It is quite possible that these two turbosuperchargers were one and the same. As with other GE turbo designs, it was

This illustration shows the GE installation. This is how the two production Constitutions were equipped. *(National Archives and Records Administration)*

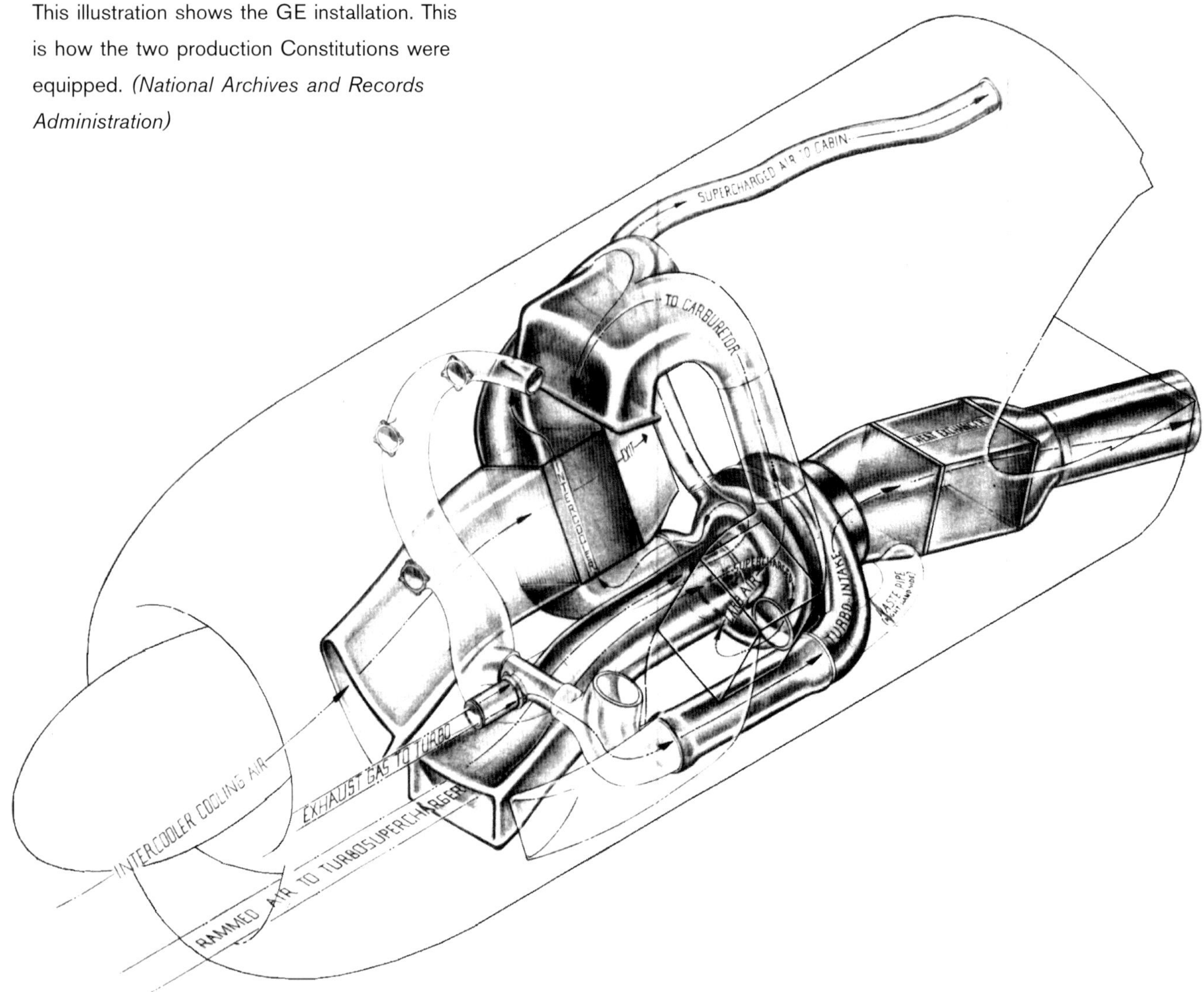

quite conventional. It featured a single-stage centrifugal compressor and single-stage turbine. Like the other installations, bleed air was tapped off each compressor for cabin pressurization. Each turbo had to be capable of providing 30 pounds per minute of bleed air. De-icing heat was provided via a heat exchanger built into the turbo exhaust. At the end of the day, GE won the business; however, it was not very lucrative. Only two Constitutions were built, which must have been a bitter disappointment to Lockheed, which had high hopes that the airlines would utilize it *(Ref. 4-3)*.

References

4-1 Miller, Dorothy L., Historian. *Case History (Selected) Turbosuperchargers*. Historical Office Executive Secretariat Air-Material Command Wright-Patterson Air Force Base. March 1951.

4-2 Contract W535 ac-38306, August 13, 1943.

4-3 Johnson, C. L., Report Number 4217. *Comparison of General Electric and Lockheed Turbosuperchargers*. Lockheed Aircraft Corporation, Burbank, California. June 22, 1943.

4-4 *Turbo-Supercharging*. Flight, February 8, 1951.

4-5 GEI-32056. *Preliminary Operation and Service Instructions for Turbosuperchargers*. April 15, 1950.

4-6 DeHaven, Ethel M., Historian. *Case History of The VDT Engine*. Historical Office Executive Secretariat Air-Material Command

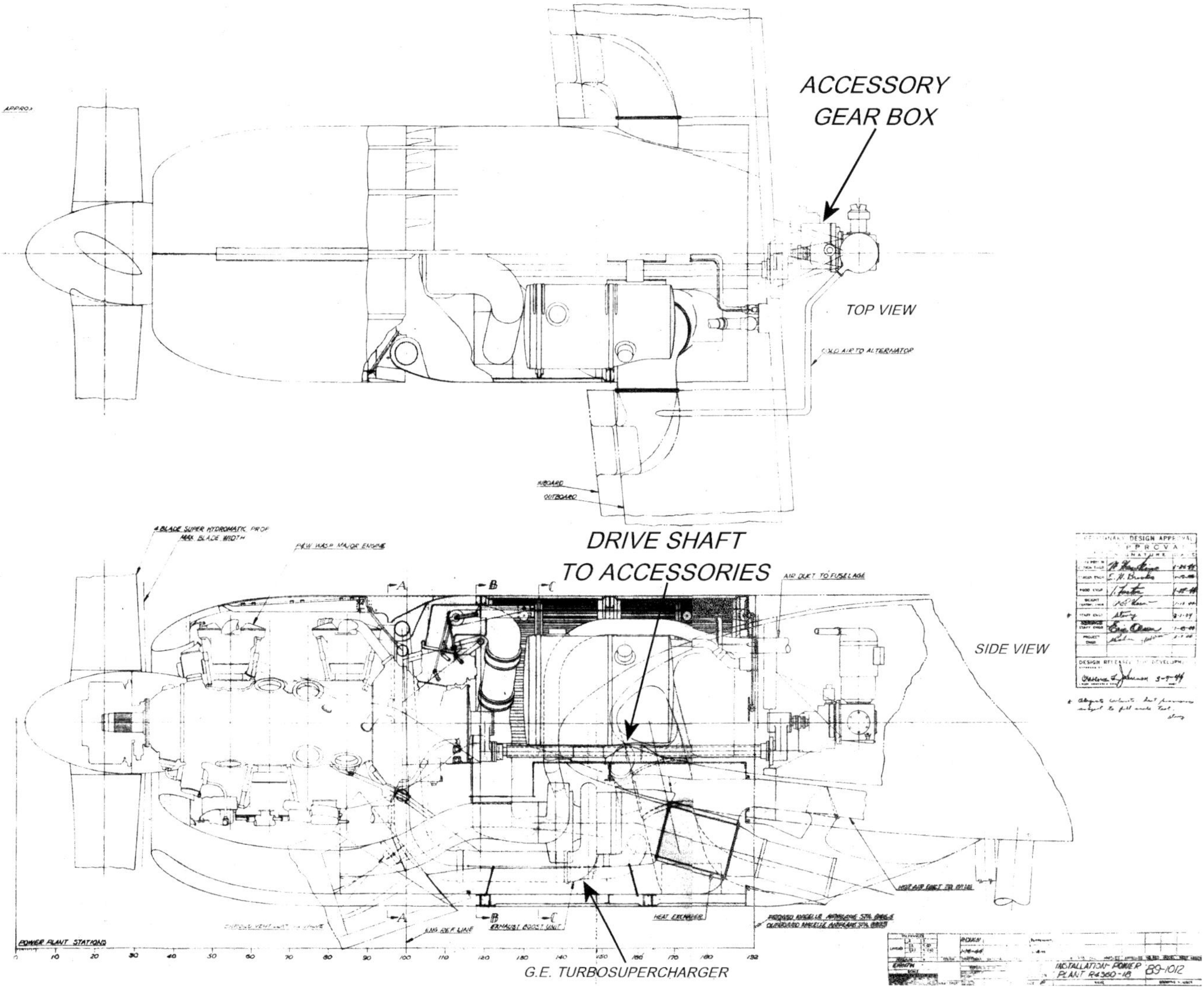

Top and side views of the GE installation. It was probably a wise decision that Lockheed made by going with GE. If Lockheed had pursued its own design, no doubt it would have run into major development issues with its innovative design. Note the driveshaft from the rear of the engine terminating at the accessory gearbox mounted in the wing structure. *(National Archives and Records Administration)*

Wright-Patterson Air Force Base. December 22, 1949.

4-7 *Combination Piston-Jet Engine Being Developed for Air Force*. National Military Establishment Department Of The Air Force. Washington 25, D.C. Thursday, October 7, 1948.

4-8 Mekley, W. O., Bulletin N. DF-8154. *The Variable Discharge Turbosupercharger (VDT)*. General Electric Company, Schenectady, N.Y., U.S.A. August 25, 1947.

4-9 Correspondence and interviews with Phil Hopper, P&W development test engineer, 2002.

4-10 Susag, M. P., E. J. Sceggel, John Johnson, Mary Bagga. *Estimated Performance of a Wasp Major R-4360-C3 Variable Discharge Turbo Supercharged Engine*. Pratt & Whitney Aircraft, East Hartford, Connecticut. March 1947.

CHAPTER FIVE

Model Types & Specifications

Although not manufactured in the huge numbers of the R-1830 or R-2800, R-4360 production amounted to nearly 16,000, which is still a significant figure. Additionally, it did not go through the profound design changes that the R-2800 went through. Indeed, the R-2800 was totally re-designed from the "B" series to the "C" series. Instead, the R-4360 went though an evolutionary improvement. Nevertheless, the last of the R-4360s manufactured shared very little with the early ones. Few, if any, parts were interchangeable. But it still looked like the same engine and all models shared the same basic design philosophies. After 1947, it was clear that Pratt & Whitney had to get on the gas turbine bandwagon or risk being left behind. As a jumpstart they manufactured Rolls-Royce Nenes under license. Consequently, less and less manpower along with engineering talent was available for R-4360 development. And herein lies the reason such promising projects as the VDT were allowed to die on the vine: Every manufacturer was enamored with the new-fangled gas turbine. Despite having a laboratory named after him, Andy Wilgoos, one of the major contributors to the R-4360's development, could never get excited about gas turbines, a sentiment this author can empathize with. A stainless-steel pipe that spits out kerosene just doesn't seem exciting. By the early 1950s, most if not all R-4360 production had been farmed out to Ford Motor Company. The last R-4360 was manufactured in 1955 *(Ref. 5-1).*

Pratt & Whitney Military Engine Designations

The following tables are divided into two sections: Commercial and Military. The U.S. Army Air Force (later U.S. Air Force) and Navy employed a numerical suffix to the military type designation (R-4360-25, etc.) to identify the complete engine model in accordance with the applicable model specification as amended by contract change orders.

Military designations are quite easy to follow. The first R-4360 was the -2 and the last was the -65. But like many engine programs, not all dash numbers were manufactured. In fact only a relative few saw mass production. The usual convention of assigning dash numbers was followed. That is, odd dash numbers were Air Force sponsored and even dash numbers were Navy sponsored. As with other engines, however, it was possible for a Navy-sponsored aircraft to be powered by an Air Force engine. That is, on occasion a Navy aircraft could be powered by an odd dash number engine and vice versa. The R-4360 is still, insultingly in the view of the author, referred to as the "corn cob." It's too bad that folks who wouldn't know a piston from a hole in the ground put down such a fine example of the engineer's art in this manner.

Pratt & Whitney Aircraft (Commercial) Engine Designations

Commercial engines followed Pratt & Whitney's designation using a complex alphanumeric system. Furthermore, officially they were not R-4360s, but rather Wasp Majors. Not too many people, however, referred to the R-4360 as a "Wasp Major." The following convention was not just reserved for R-4360s, but applied across the board for all Pratt & Whitney piston engines.

Engines are classified according to takeoff horsepower and/or type variations by a letter designation used in conjunction with the engine trade name. This letter designation indicating engine series is modified to facilitate spare parts identification and to indicate model variations as follows.

During the year 1940, in order to clarify the identification of engines equipped with various types of superchargers, the following symbols were adopted *(Ref. 5-2).*

S for single stage – single speed
VS for single stage – variable speed
2S for single stage – two speed
SS for two stage – two speed
TS for single stage – single speed with turbo

Above and next two pages: X-103. This proof of concept engine was cobbled together from off-the-shelf parts. Production engines differed significantly from this prototype.

Example

(Table 5-1)

Series	Letter Designation	Example
Unmodified	Letter Designation	C
Propeller reduction gearing	G	C-G
Supercharger - Numerical suffix to letter designation:		
Single stage - single speed	1 through 9	C1-G
Single stage - two speed	11 through 19	C11-G
Two stage - two speed	21 through 29	C21-G
Variations in major parts	Change in numerical suffix to letter designation	C12-G
Model:		
1.) Single stage - single speed rating.	Prefix S	SC2-G
2.) Single stage - two speed rating.	Prefix 2S	2SC12-G
3.) Two stage - two speed rating.	Prefix SS	SSC22-G
4.) Single stage - single speed rating with turbosupercharger.	Prefix TS	TSC2-G
5.) Variations in ratings	Prefix modified	2S1C12-G
6.) Propeller gear ratio	Suffix	2S1C12-G20:9

Complete model designation from example above: Wasp Major 2SC1C12-G20:9

Supercharging Designations (model prefix)

(Table 5-2)

Single stage - variable speed and turbosupercharging:	TVS
Single stage - single speed	A1 through A9
Single stage - variable speed:	A11 through A19
Two stage - variable speed:	A21 through A29
Multi stage - variable speed	A31 through A39

Propeller Reduction Gearing (model suffix)

(Table 5-3)

Propeller Reduction Gearing:	
Single rotation - single speed	-G
Single rotation - two speed	-G2
Dual rotation - single speed	-GD
Remote drive	-RG, -RG2, etc.

(Table 5-4)

As an example of how these designations would be used for the Wasp Major two-stage engine, the following examples are outlined.

Single rotation – single speed	Wasp Major SSC22-G
Single rotation – two speed	Wasp Major SSC22-G2
Dual rotation – single speed	Wasp Major SSC22-GD
Remote drives (extension shafts)	Wasp Major SSC22-RG

Sample Designations
Series A1-G
(Table 5-5)

Model Wasp Major TSA1-G	
TS	TS - Single stage, single speed and turbosupercharging
A1-G-A	(2800 hp) power section with single stage, single speed supercharger - single rotation, single speed reduction gearing.

Up to 1940 the Army and Navy adhered strictly to the rule of using odd dash numbers for the Army and even numbers for the Navy. However, this policy was not always strictly adhered to. In the interests of standardization, wherever both the Army and the Navy used an identical engine, the same dash number was used.

The designation of an engine was changed whenever an engine modification was made, which affected either performance or installation in an aircraft or other modifications that required identification.

Pratt & Whitney Aircraft designations identified (i) the engine series and (ii) the performance characteristics of the particular model as defined previously. They were applied for that purpose to development of stock engines. Thus for individual customers, such as the U.S. Navy or USAAF, requirements defined the equipment schedules covering such items as reduction gear ratio, supercharger parameters, accessory equipment, etc.

Pratt & Whitney Aircraft designations were also used to facilitate the identification of military-designated engines (for example, R-4360-24 would be a Wasp Major VSB11-G or variable speed supercharger, "B" series, and geared). In such cases, the specifications applicable to the military dash numbers defined the equipment schedules.

Pratt & Whitney Specification Forms
(Ref. 5-2)

Note: the following specification forms were never used on the R-4360, but are included here for reference purposes to identify Pratt & Whitney engines that did. An example would be the R-1340 AN-1.

Prefix	Indicating	Example
PW	Standard Commercial (PW) form	PW-100
A-	AN form - Coordination with Army	A-100
N-	AN form - Coordinating with Navy	N-100
AN-	AN form - Joint coordination with Army & Navy	AN-100
None	Army form prior to AN specifications (1939)	100

(Note: All engine photos in this chapter are courtesy of Pratt & Whitney archives.)

Commercial (Wasp Major) Versions of the R-4360

(Table 5-6) (Ref. 5-3)

Engine Model	**Type** **P&W** **Aircraft** Air Force Navy	**Wasp Major** **X Wasp**	**Type** **P&W** **Aircraft** Air Force Navy	**Wasp Major** **X Wasp**	**Type** **P&W** **Aircraft** Air Force Navy	**Wasp Major** **X Wasp**
Specification Number	PW-7023		PW-7024		PW-7025	
Engine Series	A Series		A Series		A Series	
Rating; Take-off	2800 hp @ 2600 rpm		2800 hp @ 2600 rpm		2800 hp @ 2600 rpm	
Rating; Military	2800 hp @ 2600 rpm @ 2000 feet 2300 hp @ 2600 rpm @ 13500 feet		2800 hp @ 2600 rpm @ 2000 feet 2600 hp @ 2600 rpm @ 13500 feet		2800 hp @ 2600 rpm @ 25000 feet with turbo 2200 hp @ 2600 rpm @ 25000 feet	
Rating; Normal	2400 hp @ 2400 rpm @ 4700 feet 2150 hp @ 2400 rpm @ 13400 feet		2400 hp @ 2400 rpm @ 4700 feet 2250 hp @ 2400 rpm @ 14500 feet		2400 hp @ 2400 rpm @ 25000 feet with turbo 2000 hp @ 2400 rpm @ 25800 feet	
Max. Continuous						
Cruise						
Fuel Grade	100 Octane		100 Octane		100 Octane	
Curves	T-753		T-745		T-755	
Weight, dry	3200 lbs		3450 lbs		3200 lbs less turbo	
Prop. Reduction Ratio	0.315:1		0.315:1		0.315:1	
Prop. Shaft Spline	70		70		70	
Compression Ratio						
Blower Ratio(s)						
Carburetor						
Magnetos						
Installation Drwg. No.						
Dimensions	Diameter: 51.75 Length:		Diameter: 51.75 Length:		Diameter: 51.75 Length:	
A.T.C. number						
Number Manufactured						
Applications			Vultee Model 85, modified Vultee Vengeance, crashed in late 1942. Vought Model VS-326			
Notes						

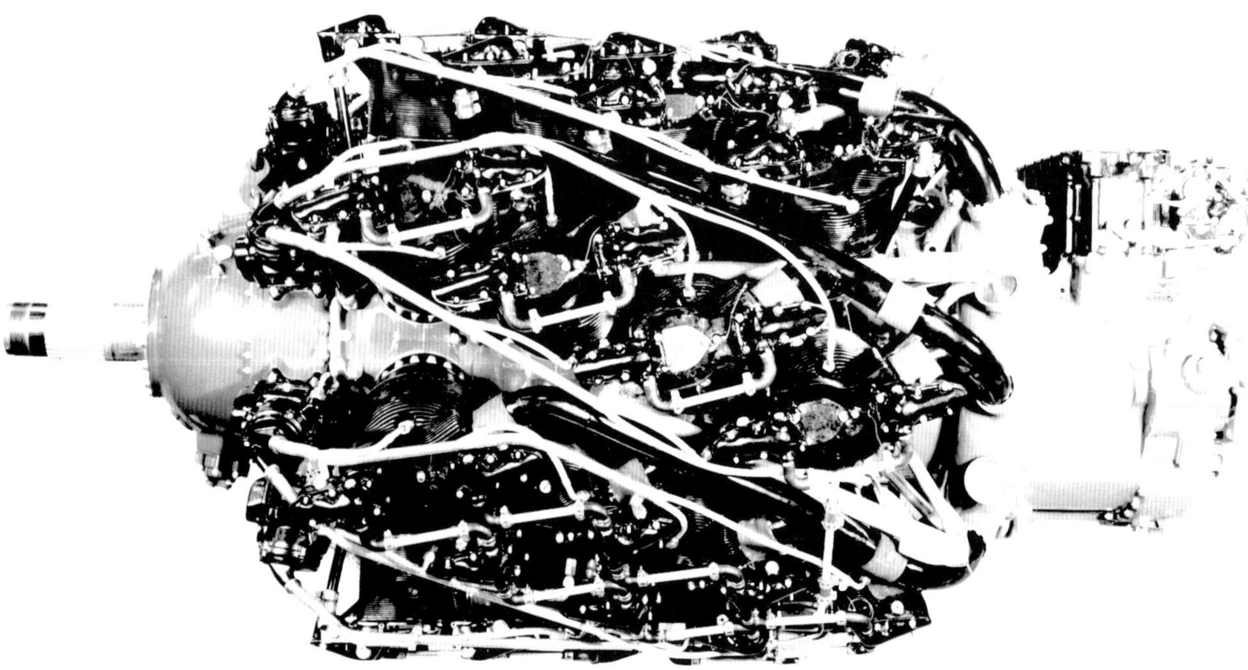

Starting to evolve into a recognizable R-4360, the "X Wasp" now had a fore and aft valve arrangement, however, the cylinder heads appear to be cast rather than forged.

This R-4360-2 is equivalent to a commercial TSB1-G. Now all the basic elements of production R-4360s–fore and aft valves, forged cylinder heads, and its unique accessory drive arrangement–are in place.

(Table 5-7) (Ref. 5-3)

Engine Model	**Type** **Wasp Major** **P&W** **TSB1-G** **Aircraft** Air Force Navy	**Type** **Wasp Major** **P&W** ***TSB2-G** **Aircraft** Air Force Navy	**Type** **Wasp Major** **P&W** **TSB3-G** **Aircraft** Air Force Navy
Specification Number	7031	7050	PW-7052
Engine Series	B Series	B Series	B Series
Rating; Take-off	3000 hp @ 2700 rpm	3500 hp @ 2700 rpm @ 500 feet wet 3250 hp @ 2700 rpm @ 1000 feet dry	3500 hp @ 2700 rpm @ 500 feet wet 3250 hp @ 2700 rpm @ 700 feet dry
Rating; Military	3000 hp @ 2700 rpm @ 1500 feet		
Rating; Normal	2500 hp @ 2550 rpm @ 6500 feet	2600 hp @ 2550 rpm @ 6500 feet	2650 hp @ 2550 rpm @ 5500 fee
Max. Continuous		2800 hp @ 2550 rpm @ 3500 feet stand by	2800 hp @ 2550 rpm @ 3500 feet
Cruise			
Fuel Grade	100/130 PN	115/145 PN	*108/135 PN
Curves	T-833	T-979	T-984 Inst. #7240
Weight, dry	3325 lbs	3470 lbs	3482 lbs
Prop. Reduction Ratio	0.381:1 or 0.425:1	0.375:1	0.375:1
Prop. Shaft Spline	60-A	60-A	60-A
Compression Ratio	7:1		6.7:1
Blower Ratio(s)	6.08:1		6.375:1
Carburetor	Optional	Optional	Bendix PR-100B3
Magnetos	Optional	Optional	D4RN-2
Installation Drwg. No.			97801
Dimensions	Diameter: 52.50 inches Length: 96.75 inches	Diameter: 53.50 inches Length: 96.75 inches	Diameter: 54.00 inches Length: 96.50 inches
A.T.C. number			247
Number Manufactured	4	None manufactured	391
Applications	*Vought F4U-1 WM		** Boeing Model 377 Republic RC-2
Notes	* #3 aircraft. This was a converted "Birdcage" F4U-1 built as a proof of concept aircraft for the follow-on Goodyear F2G (Ref. 5-4).	*Cancelled for TSB3-G	Incorporates manifold pressure actuated water (ADI) control switch. Power control not provided. *If unavailable, 115/145 was authorized. **TSB3-Gs converted to CB1, B6 and CB2.

Three-quarter right front view showing seven high-tension magnetos.

(Table 5-8) (Ref. 5-3)

Engine Model	Type P&W **Aircraft** Air Force Navy	**Wasp Major TSB3-G**	Type P&W **Aircraft** Air Force Navy	**Wasp Major *B-4**	Type P&W **Aircraft** Air Force Navy	**Wasp Major B-5**
Specification Number	PW-7052, Appendix A					
Engine Series	B Series		B Series		B Series	
Rating; Take-off	3500 hp @ 2700 rpm @ 500 feet wet 3250 hp @ 2700 rpm @ 700 feet dry				3250 hp @ 2700 rpm @ 500 feet wet 3000 hp @ 2700 rpm @ 1300 feet dry	
Rating; Military						
Rating; Normal	2650 hp @ 2550 rpm @ 5500 feet				2650 hp @ 2550 rpm @ 2800 feet	
Max. Continuous	2800 hp @ 2550 rpm @ 3500 feet				2650 hp @ 2550 rpm @ 2800 feet	
Cruise						
Fuel Grade	*108/135 PN				100/130 PN	
Curves	T-984 Inst. #7240				Inst. #7337	
Weight, dry	3482 lbs				3490 lbs	
Prop. Reduction Ratio	0.375:1				0.375:1	
Prop. Shaft Spline	60-A				60-A	
Compression Ratio	6.7:1				6.7:1	
Blower Ratio(s)	6.375:1				6.375:1	
Carburetor	Bendix PR-100B3				Bendix PR-100B3	
Magnetos	D4RN-2				D4RN-2	
Installation Drwg. No.	97801				97801	
Dimensions	Diameter: 54.0 inches Length: 96.5 inches				Diameter: 54.0 inches Length: 96.5 inches	
A.T.C. number	247				247	
Number Manufactured	391 (total)					
Applications	** Boeing Model 377 Model 10 & 19 ***Hughes HK-1				Boeing Model 377 Model 10 & 19	
Notes	Torquemeter supplied. No manifold pressure regulator. Power control not provided. Starter drive located at #2 pad. *If unavailable, 115/145 was authorized. **TSB3-Gs converted to CB1, B6, and CB2. *** Installed in 1951 but never flown with these engines (Ref. 5-5).		*No specification written Commercial counterpart to military R-4360-4, A spec. A-7063		Similar to Wasp Major TSB3-G except rated with 100/130 grade fuel. None manufactured.	

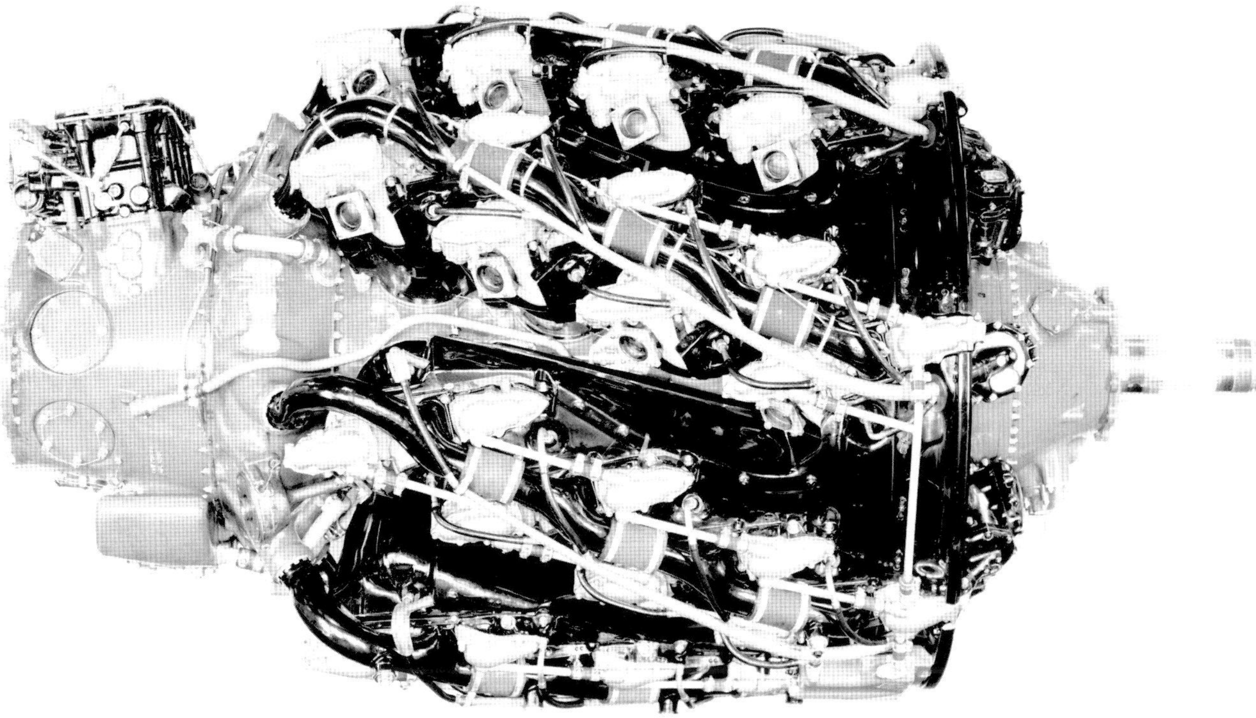

Right side view showing early rubber hose clamp type intake manifold.

(Table 5-9) (Ref. 5-3)

Engine Model	**Type** **P&W** **Aircraft** Air Force Navy	**Wasp Major** **B6**	**Type** **P&W** **Aircraft** Air Force Navy	**Wasp Major** ***B6C**	**Type** **P&W** **Aircraft** Air Force Navy	**Wasp Major** ***B-7**
Specification Number	7097					
Engine Series	B Series		B Series		B Series	
Rating; Take-off	3500 hp @ 2700 rpm @ 500 feet wet 3250 hp @ 2700 rpm @ 700 feet dry				3250 hp @ 2700 rpm @ 500 feet wet 3000 hp @ 2700 rpm @ 1300 feet dry	
Rating; Military						
Rating; Normal	2650 hp @ 2550 rpm @ 5500 feet				2650 hp @ 2550 rpm @ 2800 feet	
Max. Continuous Cruise	2800 hp @ 2550 rpm @ 3500 feet				2650 hp @ 2550 rpm @ 2800 feet	
Fuel Grade			108/135 PN		100/130 PN	
Curves	Inst. #16859			Inst. #17494		
Weight, dry	*3584 lbs				3584 lbs	
Prop. Reduction Ratio	0.375:1				0.375:1	
Prop. Shaft Spline	60-A				60-A	
Compression Ratio	6.7:1				6.7:1	
Blower Ratio(s)	6.375:1				6.375:1	
Carburetor	Bendix PR-100B3				Bendix PR-100B3	
Magnetos	S14RN-15 (Low tension)				D4RN-2	
Installation Drwg. No.	176601				97801	
Dimensions	Diameter: 55.0 inches Length: 96.5 inches				Diameter: 54.0 inches Length: 96.75 inches	
A.T.C. number	247				247	
Number Manufactured						
Applications	Boeing Model 377 Aero Spacelines B-377MG Mini Guppy Aero Spacelines B-377SG Super Guppy Aero Spacelines B-377PG Pregnant Guppy		Boeing Model 377			
Notes	Intermediate engine for conversion to Wasp Major TSB3-G to Wasp Major CB2. *Additional weights: Mount structure: 110lbs Torquemeter, Chandler Evans carburetor. optional: 19lbs TSB3-G rating with "C" power section and nose section with short rod "B" series cylinders		*Identification for Wasp Major B6 with "C" series cylinders and 4 bolt exhaust flange with no coupling. None manufactured.		* Identification for Wasp Major B6 with 100/130 fuel. None manufactured.	

R-4360 B-6. Engines are now starting to feature hooded baffles and low-tension ignitions, although for a time during the inevitable transition period engines were produced with and without hooded baffles, and they could also feature low- or high-tension ignitions.

(Table 5-10) (Ref. 5-3)

Engine Model	Type **P&W Aircraft** Air Force Navy	**Wasp Major *B7C**	Type **P&W Aircraft** Air Force Navy	**Wasp Major VSB11-G**	Type **P&W Aircraft** Air Force Navy	**Wasp Major VSSB21-G**
Specification Number			7029		7030	
Engine Series	B Series		B Series		B Series	
Rating; Take-off			3000 hp @ 2700 rpm @ 1500 feet		3000 hp @ 2700 rpm 3000 hp @ 2700 rpm @ 3000 feet	
Rating; Military					2400 hp @ 2700rpm @ 25000 feet	
Rating; Normal			2500 hp @ 2550 rpm @ 5000 feet 2200 hp @ 2550 rpm @ 14500 feet		2500 hp @ 2550 rpm @ 6000 feet 2200 hp @ 2550 rpm @ 25000 feet	
Max. Continuous						
Cruise						
Fuel Grade			100/130 PN		100/130 PN	
Curves			T-831 Inst. #1722 Ceco carburetor (Chandler Evans) Inst. #7318 Bendix carburetor		T-832	
Weight, dry			3498 lbs or 3490 lbs depending on nose case		3610 lbs	
Prop. Reduction Ratio			0.375:1 or 0.425:1		0.425:1 or 0.381:1	
Prop. Shaft Spline			60-A		60-A	
Compression Ratio			7:1		7:1	
Blower Ratio(s)			Variable speed, maximum: 7.52:1		6.08:1 & 5.75:1	
Carburetor			Bendix PR-100B1-2 Bendix PR-100B2-3 Bendix PR-100B3-3, -7		Optional	
Magnetos			D4RN-2		Optional	
Installation Drwg. No.			97801			
Dimensions			Diameter: 54.0 inches Length: 96.75 inches		Diameter: 52.50 inches Length: 109.07 inches	
A.T.C. number			247		Military	
Number Manufactured			25			
Applications			Aero-Sud-Est SE-2010			
Notes	*Identification for Wasp Major B7 with "C" series cylinders and 4 bolt exhaust flange with no coupling. None manufactured.		Variable speed impeller drive. Many features similar to TSB3-G. Engines converted by airports to Wasp Major B-13		None manufactured	

R-4360 VSB11G. High-tension ignition and hooded baffles.

(Table 5-11) (Ref. 5-3)

Engine Model	Type P&W Aircraft Air Force Navy — Wasp Major B12	Type P&W Aircraft Air Force Navy — Wasp Major B13	Type P&W Aircraft Air Force Navy — Wasp Major *B14
Specification Number	7059	7060	None
Engine Series	B Series	B Series	B Series
Rating; Take-off	3500 hp @ 2700 rpm @ 750 feet wet 3250 hp @ 2700 rpm @ 1500 feet dry	3500 hp @ 2700 rpm @ 750 feet wet 3250 hp @ 2700 rpm @ 1500 feet dry	3500 hp @ 2700 rpm @ 1000 feet wet 3250 hp @ 2700 rpm @ 1700 feet dry
Rating; Military	3500 hp @ 2700 rpm @ 750 feet wet 3250 hp @ 2700 rpm @ 1500 feet dry 2500 hp @ 2700 rpm @ 15400 feet dry		
Rating; Normal	2650 hp @ 2550 rpm @ 6000 feet 2300 hp @ 2550 rpm @ 16300 feet	2650 hp @ 2550 rpm @ 6000 feet 2300 hp @ 2550 rpm @ 16300 feet	2650 hp @ 2550 rpm @ 6500 feet 2000 hp @ 2550 rpm @ 22300 feet
Max. Continuous			
Cruise			
Fuel Grade	115/145 PN	115/145 PN	108/135 PN
Curves	T-1022 Inst. #7266-1, -2 (exhaust thrust)	T-1024 Inst. #7389	Inst. #17051
Weight, dry	*3505 lbs	*3535 lbs	3535 lbs
Prop. Reduction Ratio	0.425:1	0.425:1 or ** 0.375:1	0.425:1
Prop. Shaft Spline	60-A	60-A	60-A
Compression Ratio	6.7:1	6.7:1	6.7:1
Blower Ratio(s)	6.95:1 low & 9.07:1 high	6.95:1 low & 9.07:1 high	6.95:1 low & 9.07:1 high
Carburetor	Bendix PR-100B3 Optional	Bendix PR-100B3-11	Bendix PR-100B3-11
Magnetos	D4RN-2	D4RN 2	D4RN-2
Installation Drwg. No.	115101	132601	132601
Dimensions	Diameter: 54.00 inches Length: 101.76 inches	Diameter: 54.00 inches Length: 101.76 inches	Diameter: 54.00 inches Length: 101.76 inches
A.T.C. number	Military	247	247
Number Manufactured		46	
Applications		Aero-Sud-Est SE-2010	Aero-Sud-Est SE-2010
Notes	*Weight increases: Torquemeter: 20 lbs None manufactured	*Includes torquemeter *Weight increases: Constant water regulator: 8lbs Fireseal diaphragm: 8lbs **0.375:1 reduction gear: 8lbs Original Wasp Major B-13 converted from Wasp Major VSB11-G	*Identification for Wasp Major B13 rated with 108/135 PN fuel and revised high blower gear rating. None manufactured.

(Table 5-12) (Ref. 5-3)

Engine Model	Type P&W Aircraft Air Force Navy — Wasp Major TSC1-G	Type P&W Aircraft Air Force Navy — Wasp Major C2	Type P&W Aircraft Air Force Navy — Wasp Major C3
Specification Number	7045	7053	7068
Engine Series	C Series	C Series	C Series
Rating; Take-off	3300 hp @ 2800 rpm dry 3500hp@ 2800 rpm wet	3500 hp @ 2800 rpm dry 3800 hp @ 2800 rpm wet	4300 hp @ 2800 rpm plus 300 pounds of thrust
Rating; Military	3300 hp @ 2800 rpm @ 1500 feet	3500 hp @ 2800 rpm @ 500 feet dry 3800 hp @ 2800 rpm @ 1000 feet wet	4300 hp @ 2800 rpm @ 27500 feet plus 380 pounds of thrust
Rating; Normal	2650 hp @ 2600 rpm @ 5000 feet	2800 hp @ 2600rpm @ 5500 feet	3150 hp @ 2600 rpm @ 32500 feet plus 300 pounds of thrust
Max. Continuous			
Cruise			
Fuel Grade	100/130 PN	115/145 PN	115/145 PN
Curves	T-971	T-990	T-105
Weight, dry	3600 lbs	*3600 lbs	*3720 lbs
Prop. Reduction Ratio	0.333:1 or 0.425:1	0.375:1	0.375:1
Prop. Shaft Spline	60-A	60-A	60-A
Compression Ratio	6.7:1	7:1	7:1
Blower Ratio(s)	6.10:1		
Carburetor	Optional	Optional	Bendix PR-100-28-A3 (Fuel injection)
Magnetos	Optional	Optional	
Installation Drwg. No.	100001	100001	126801
Dimensions	Diameter: 53.50 inches Length: 98.41 inches	Diameter: 54.00 inches Length: 102.00 inches	Diameter: **61.00 inches Length: 103.3 inches
A.T.C. number	Military	Military	Military
Number Manufactured			
Applications			
Notes	Single speed version of VSC11-G None manufactured.	*Weight increases: Fuel injection equipment.: 90 lbs Torquemeter: 15 lbs Single speed version of C12. None manufactured.	*Weight includes torquemeter. **61.00 inches over exhaust header and 65.00 inches over exhaust collector. VDT engine with GE CHM-2 exhaust driven supercharger. None manufactured.

(Table 5-13) (Ref. 5-3)

Engine Model	Type P&W Aircraft Air Force Navy — Wasp Major C4	Type P&W Aircraft Air Force Navy — Wasp Major *C5	Type P&W Aircraft Air Force Navy — Wasp Major *C6
Specification Number	*7071		
Engine Series	C Series	C Series	C Series
Rating; Take-off	4800 hp @ 2800 rpm wet plus 300 pounds thrust 4300 hp @ 2800 rpm dry plus 245 pounds thrust	3800 hp @ 2800 rpm wet	
Rating; Military	4800 hp @ 2800 rpm @ 25000 feet wet plus 380 pounds thrust 4300 hp @ 2800 rpm @ 27000 feet dry plus 335 pounds thrust	3800 hp @ 2800 rpm wet 3500 hp @ 2800 rpm dry	
Rating; Normal	3500 hp @ 2600 rpm @ 30000 feet plus 300 pounds thrust	2800 hp @ 2600 rpm	
Max. Continuous			
Cruise			
Fuel Grade	**115/145 PN	115/145 PN	
Curves	T-1077		
Weight, dry	3820 lbs	**3950 lbs	*3720 lbs
Prop. Reduction Ratio		0.3125:1	0.375:1
Prop. Shaft Spline		70	
Compression Ratio		6.7:1	
Blower Ratio(s)		External	
Carburetor		Bendix 100-28-A3 (Fuel injection)	
Magnetos			
Installation Drwg. No.		149810	
Dimensions		Diameter: 55.00 inches Length: 103.5 inches less gearbox	
A.T.C. number	Military	Military	Military
Number Manufactured			
Applications			
Notes	*Specification limited to rating and weight. **With 3cc of Tetraethyl Lead. Intended as VDT with fuel injection.	*See bottom of page 223	*Made up as follows: 1. Internal supercharging 2. R-4360-20 rear section features 3. Pusher installation. None manufactured.

(Table 5-14) (Ref. 5-3)

Engine Model	Type **P&W Aircraft** Air Force Navy — **Wasp Major C7**	Type **P&W Aircraft** Air Force Navy — **Wasp Major C8**	Type **P&W Aircraft** Air Force Navy — **Wasp Major C9**
Specification Number		7077	7081
Engine Series	C Series	C Series	C Series
Rating; Take-off		4300 hp @ 2800 rpm	3800 hp @ 2800 rpm @ sea level wet 3600 hp @ 2800 rpm @ 2000 feet dry
Rating; Military		4300 hp @ 2800 rpm @ 5500 feet	3800 hp @ 2800 rpm @ sea level wet
Rating; Normal		3150 hp @ 2600 rpm @ 15500 feet	2800 hp @ 2600 rpm @ 5000 feet
Max. Continuous			
Cruise			
Fuel Grade		115/145 PN	115/145 PN
Curves		T-1105	T-1120
Weight, dry		*3720 lbs	*3840 lbs
Prop. Reduction Ratio		0.375:1	0.375:1
Prop. Shaft Spline		60-A	60-A
Compression Ratio	7.5:1	6.7:1	6.7:1
Blower Ratio(s)		External turbosupercharger	6.6:1
Carburetor		Bendix 100-28-A3 (Fuel injection)	Bendix 100-28-A3 (Fuel injection)
Magnetos		Scintilla S14RN-15 low tension	Scintilla S14RN-15 low tension
Installation Drwg. No.		126801	155401
Dimensions		**Diameter: 55.00 inches Length: 101.63 inches less gearbox	Diameter: 55.00 inches Length: 101.63 inches
A.T.C. number	Military	Military	Military
Number Manufactured			
Applications			
Notes	Similar to Wasp Major C5 except incorporates high compression ratio (7.5:1). Pusher installation. None manufactured	*Includes torquemeter. ** 65 inches over collector ring. Single stage supercharger. Similar to Wasp Major C5 except tractor and low compressor. None manufactured.	Single speed version of C14 *Includes torquemeter and fuel injection equipment. Engine weighs 140 lbs less with carburetor. Similar to Wasp Major C6 (R-4360-53) except tractor. None manufactured.

(Table 5-15) (Ref. 5-3)

Engine Model	Type **P&W Aircraft** Air Force Navy — **Wasp Major C-01**	Type **P&W Aircraft** Air Force Navy — **Wasp Major C-02**	Type **P&W Aircraft** Air Force Navy — **Wasp Major C-03**
Specification Number	*7082	7083	*7084
Engine Series	C Series	C Series	C Series
Rating; Take-off		4300 hp @ 2800 rpm	
Rating; Military		4300 hp @ 2800 rpm @ 6500 feet	
Rating; Normal		3500 hp @ 2600 rpm @ 14200 feet	
Max. Continuous			
Cruise			
Fuel Grade		115/145 PN	
Curves		T-1122	
Weight, dry		*3760 lbs	
Prop. Reduction Ratio		0.375:1	
Prop. Shaft Spline		60-A	
Compression Ratio	7.5:1	7.5:1	7.5:1
Blower Ratio(s)		6.6:1	
Carburetor		Bendix 100-28-A3 (Fuel injection)	
Magnetos		Scintilla S14RN-15 low tension	
Installation Drwg. No.		126801	155401
Dimensions		Diameter: 65.00 inches Length: 101.63 inches	
A.T.C. number		Military	
Number Manufactured			
Applications			
Notes	High compression ratio (7.5:1). Wasp Major C5 tractor *No specification written. None manufactured.	*Includes torquemeter. Similar to Wasp Major C8 except high compression ratio (7.5:1). None manufactured.	Similar to Wasp Major C7 (R-4360-57) except tractor. *No specification written. None manufactured.

Made up as follows: 1. Wasp Major C3 power section arranged for pusher installation, remotely mounted cooling fan drive gearbox power take-off exhaust driven GE superchargers; one CH-9 and two CHM-3s. 2. R-4360-43 fuel injection system 3. R-4360-51 nose 4. Wasp Major C3 low tension ignition system except spark plugs 5. R-4360-41 accessory drives. **Additional weights: Ex. system, including collector ring.: 300 lbs Engine mounts with no isolators: 112 lbs Remote gearbox and power take-off: 200lbs Boost control system: 80 lbs Weight includes torquemeter and fuel injection equipment. None manufactured.

(Table 5-16) (Ref. 5-3)

Engine Model	Type P&W **Wasp Major C-04**	Type P&W **Wasp Major C-05**	Type P&W **Wasp Major C-06**
	Aircraft Air Force Navy	**Aircraft** Air Force Navy	**Aircraft** Air Force Navy
Specification Number	7085	7095	No specification written
Engine Series	C Series	C Series	C Series
Rating; Take-off	3500 hp @ 2700 rpm @ 1000 feet	4000 hp @ 2800 rpm @ 7500 feet	4300 hp @ 2800 rpm @ 5500 feet
Rating; Military			
Rating; Normal	2800 hp @ 2550 rpm @ 3500 feet	2800 hp @ 2600 rpm @ 18700 feet	
Max. Continuous	2800 hp @ 2550 rpm @ 3500 feet	3000 hp @ 2600 rpm @ 17000 feet	2800 hp @ 2600 rpm @ 18700 feet 3150 hp @ 2600 rpm @ 15500 feet
Cruise			
Fuel Grade	108/135 PN	108/135 PN	115/145 PN
Curves	T-1142	T-1158	Inst. #17486
Weight, dry	3700 lbs	*3720 lbs	*3700 lbs
Prop. Reduction Ratio	0.375:1	0.375:1	0.4375:1 or 0.375:1
Prop. Shaft Spline	SAE #60-A	SAE #60-A	SAE #60-A
Compression Ratio	6.7:1	7.5:1	7.5:1
Blower Ratio(s)	6.6:1	External turbosupercharger	External turbosupercharger
Carburetor	Bendix PR-100B4	Bendix 100-28-A3 (Fuel injection)	Bendix 100-28-A3 (Fuel injection)
Magnetos	Scintilla S14RN-15 low tension	Scintilla S14RN-15 low tension	Scintilla S14RN-15 low tension
Installation Drwg. No.	160701	126801	Inst. 17479
Dimensions	Diameter: 55.00 inches Length: 102.00 inches	Diameter: 65.00 inches Length: 101.63 inches	Diameter: 55.00 inches Length: 101.63 inches
A.T.C. number	Military	Military	
Number Manufactured			
Applications			
Notes	"C" series engine for TSB3G replacement. This project dropped in favor of Wasp Major CB2	Rated with GE CH-9 turbosupercharger. *Includes torquemeter. None manufactured.	*Includes torquemeter. Generally similar to Wasp Major C-05 except ratings. *Additional weights: Engine mount structure less isolators...110 lbs GE CH-9 turbosupercharger: 300 lbs Honeywell control.: 20 lbs Exhaust system: 235 lbs None manufactured.

(Table 5-17) (Ref. 5-3)

Engine Model	Type P&W **Wasp Major VSSC21-G**	Type P&W **Wasp Major *VSC11-G**	Type P&W **Wasp Major C-12**
	Aircraft Air Force Navy	**Aircraft** Air Force Navy	**Aircraft** Air Force Navy
Specification Number	7046	7044	7054
Engine Series	C Series	C Series	C Series
Rating; Take-off	3300 hp @ 2800 rpm dry 3500 hp @ 2800 rpm wet	3300 hp @ 2800 rpm dry 3500 hp @ 2800 rpm wet	3500 hp @ 2800 rpm dry 3800 hp @ 2800 rpm wet
Rating; Military	3300 hp @ 2800 rpm @ 1500 feet 2800 hp @ 2800 rpm @ 23000 feet	3300 hp @ 2800 rpm @ 1500 feet 2650 hp @ 2800 rpm @ 16000 feet	3800 hp @ 2800 rpm @ 500 feet wet 3000 hp @ 2800 rpm @ 13000 feet wet 3500 hp @ 2800 rpm @ 1000 feet dry 2800 hp @ 2800 rpm @ 14500 feet dry
Rating; Normal	2650 hp @ 2600 rpm @ 5000 feet 2300 hp @ 2600 rpm @ 23000 feet	2650 hp @ 2600 rpm @ 5000 feet 2300 hp @ 2600 rpm @ 18500 feet	2800 hp @ 2600 rpm @ 5500 feet 2500 hp @ 2600 rpm @ 17000 feet
Max. Continuous			
Cruise			
Fuel Grade	100/130 PN	100/130 PN	115/145 PN
Curves	T-972	T-970	T-991
Weight, dry	3835 lbs	3650 lbs	*3650 lbs
Prop. Reduction Ratio	0.333:1 or 0.425:1	0.333:1 or 0.425:1	0.375:1
Prop. Shaft Spline	SAE #60-A	SAE #60-A	SAE #60-A
Compression Ratio	7.0:1	7.0:1	7.0:1
Blower Ratio(s)			
Carburetor	Optional	Optional	Optional
Magnetos	Optional	Optional	Optional
Installation Drwg. No.		100001	100001
Dimensions	Diameter: 53.50 inches Length: 112.50 inches	Diameter: 53.50 inches Length: 98.41 inches	Diameter: 54.00 inches Length: 102.00 inches
A.T.C. number			Military
Number Manufactured			
Applications			
Notes	Two-stage, variable speed supercharger. None manufactured.	*Was VSC1-G Variable speed supercharger. None manufactured	*Weight increases: Torquemeter: 20 lbs Fuel injection: 100 lbs Variable speed supercharger. None manufactured

(Table 5-18) (Ref. 5-3)

Engine Model	Type **P&W** **Aircraft** Air Force Navy — **Wasp Major C13**	Type **P&W** **Aircraft** Air Force Navy — **Wasp Major C14**	Type **P&W** **Aircraft** Air Force Navy — **Wasp Major C15**
Specification Number	7058	7078	
Engine Series	C Series	C Series	C Series
Rating; Take-off		3500 hp @ 2800 rpm dry 3800 hp @ 2800 rpm wet	3800 hp @ 2800 rpm @ sea level wet, or; 3600 hp @ 2800 rpm @ 2000 feet wet
Rating; Military	3500 hp @ 2800 rpm @ 1000 feet dry in low blower 3000 hp @ 2800 rpm @ 10000 feet dry in int. blower 2500 hp @ 2800 rpm @ 20000 feet dry in high blower 3800 hp @ 2800 rpm @ 500 feet wet in low blower	2800 hp @ 2600 rpm @ 5000 feet 2450 hp @ 2600 rpm @ 15000 feet	
Rating; Normal	2800 hp @ 2600 rpm @ 5500 feet in low blower 2500 hp @ 2600 rpm @ 13500 feet in int. blower 2250 hp @ 2600 rpm @ 21000 feet in high blower.		
Max. Continuous			
Cruise			
Fuel Grade	115/145 PN	115/145 PN	
Curves	T-1021	T-1107	
Weight, dry	*3705 lbs	*3730 lbs	
Prop. Reduction Ratio	0.375:1	0.375:1	
Prop. Shaft Spline	60-A		
Compression Ratio	6.7:1	6.7:1	
Blower Ratio(s)	Low blower: 5.66:1 Intermediate blower: 7.02:1 High blower.: 8.10:1		
Carburetor	Chandler Evans CECO 100CPB	Bendix PR-100B3	
Magnetos	Scintilla	Scintilla S14N-15 (low tension)	
Installation Drwg. No.	115401		
Dimensions	Diameter: 4.00 inches Length: 105.50 inches	Diameter: 55.00 inches Length: 112.50 inches	
A.T.C. number	Military	Military	
Number Manufactured			
Applications			
Notes	*Weight increases: Torquemeter: 15 lbs Fuel injection: 90 lbs Similar to C12 except three speed supercharger. None manufactured	* Includes torquemeter and fire seal diaphragm. Superseded by CB11 None manufactured.	Variable speed supercharger. Fuel injection. None manufactured.

(Table 5-19) (Ref. 5-3)

Engine Model	Type **P&W** **Aircraft** Air Force Navy — **Wasp Major CB1**	Type **P&W** **Aircraft** Air Force Navy — **Wasp Major CB2**	Type **P&W** **Aircraft** Air Force Navy — **Wasp Major CB3**
Specification Number		7086	
Engine Series	CB Series	CB Series	CB Series
Rating; Take-off		3500 hp @ 2700 rpm @ 500 feet wet 3250 hp @ 2700 rpm @ 700 feet dry	
Rating; Military			
Rating; Normal		2650 hp @ 2550 rpm @ 5500 feet	
Max. Continuous		2800 hp @ 2550 rpm @ 3500 feet	
Cruise			
Fuel Grade		108/135 PN	
Curves		T-1139 Inst. #16859	
Weight, dry		*3682 lbs	
Prop. Reduction Ratio		0.375:1	
Prop. Shaft Spline		60-A 00	
Compression Ratio		6.7:1	
Blower Ratio(s)		6.375:1	
Carburetor		Bendix PR-100B3	
Magnetos		Scintilla S14RN-15 low tension	
Installation Drwg. No.		166001	
Dimensions		Diameter: 55.00 inches Length: 96.50 inches	
A.T.C. number		247	
Number Manufactured		2	
Applications	Boeing Model 377	**Boeing Model 377 (C-97), Aero Spacelines B-377MG Mini Guppy, Aero Spacelines B-377SG Super Guppy, Aero Spacelines B-377PG Pregnant Guppy	
Notes	CB1 was converted from TSB3-G with "C" series cylinders for service test by Pan American. Two engines involved - none manufactured.	*Includes torquemeter, fire seal and water regulator. *Additional weight: Mount: 110 lbs **Original Wasp Major CB2 was converted Wasp Major TSB3-G and Wasp Major B6. Wasp Major CB2 incorporates "C" nose and powersection with a "B" series rear section. Single speed versions of Wasp Major CB11.	Single speed CB11. None manufactured.

(Table 5-20) (Ref. 5-3)

Engine Model	Type P&W **Wasp Major CB4** Aircraft Air Force Navy	Type P&W **Wasp Major *CB5** Aircraft Air Force Navy	Type P&W **Wasp Major CB11** Aircraft Air Force Navy
Specification Number		None	7087
Engine Series	CB Series	CB Series	CB Series
Rating; Take-off		3250 hp @ 2700 rpm @ 500 feet wet 3000 hp @ 2700 rpm @1300 feet dry	3800 hp @ 2800 rpm @ Sea level, wet 3400 hp @ 2800 rpm @ 2000 feet, dry
Rating; Military			3800 hp @ 2800 rpm @ sea level, wet in low blower. 3400 hp @ 2800 rpm @ 2000 feet, dry in low blower. 2450 hp @ 2550 rpm @ 15000 feet, dry in high blower.
Rating; Normal		2650 hp @ 2550 rpm @ 2800 feet	2800 hp @ 2600 rpm @ 5000 feet 2450 hp @ 2550 rpm @ 15000 feet
Max. Continuous		2650 hp @ 2550 rpm @ 2800 feet	
Cruise			
Fuel Grade		100/130 PN	115/145 PN
Curves		Inst. 17494	T-1144
Weight, dry			*3809 lbs
Prop. Reduction Ratio			0.375:1
Prop. Shaft Spline			SAE #60-A or SAE #70
Compression Ratio			6.7:1
Blower Ratio(s)			6.95:1 and 9.07:1
Carburetor			Bendix PR-100B4
Magnetos			Scintilla S14RN-15 low tension
Installation Drwg. No.			172801
Dimensions			Diameter: 55.00 inches Length: 102.00 inches
A.T.C. number			Military
Number Manufactured			
Applications		Boeing Model 377 for North West Airlines	
Notes	Single speed version of R-4360-59 & R-4360-61 None manufactured.	*Identification like CB2 except 100/130 PN fuel and Wasp Major B5 ratings. None manufactured.	Wasp Major C14 ratings with Wasp Major B13 rear section. *Additional weights: Engine mount: 110 lbs 70 spline prop. shaft: 40 lbs None manufactured.

(Table 5-21) (Ref. 5-3)

Engine Model	Type P&W **Wasp Major CB12** Aircraft Air Force Navy	Type P&W **Wasp Major CB13** Aircraft Air Force Navy
Specification Number		No specification written
Engine Series	CB Series	CB Series
Rating; Take-off		3600 hp @ 2800 rpm @ 2000 feet wet in low blower. 3250 hp @ 2800 rpm @ 3000 feet dry in low blower.
Rating; Military		
Rating; Normal		2650 hp @ 2600 rpm @ 6500 feet 2250 hp @ 2600 rpm @ 15500 feet
Max. Continuous		2800 hp @ 2600 rpm @ 5000 feet 2450 hp @ 2600 rpm @ 13500 feet
Cruise		1800 hp @ 2300 rpm @ 10000 feet 1700 hp @ 2300 rpm @ 18500 feet
Fuel Grade		115/145 PN
Curves		Inst. 17488
Weight, dry		*3811 lbs
Prop. Reduction Ratio		0.4375:1 or 0.375:1
Prop. Shaft Spline		
Compression Ratio		6.7:1
Blower Ratio(s)		6.95:1 and 8.0:1
Carburetor		Bendix PR-100B4
Magnetos		Scintilla S14RN-15 low tension
Installation Drwg. No.		17472
Dimensions		Diameter: 55.00 inches Length: 102.00 inches
A.T.C. number		
Number Manufactured		
Applications		
Notes	Wasp Major CB11 with R-4360-53 nose features. None manufactured.	Similar to Wasp Major CB11 except ratings. *Additional weight: Engine mount structure less isolators: 110 lbs None manufactured.

Military Versions of the R-4360

(Table 5-22) (Ref. 5-3)

Engine Model	Type R-4360 P&W Aircraft Air Force Navy -2	Type R-4360 P&W Aircraft Air Force Navy -2A	Type R-4360 P&W Aircraft Air Force Navy X-2
Specification Number	N-7027 applies	N-7039-A applies	
Engine Series	B Series	B Series	B Series
Rating; Take-off	3000 hp @ 2700 rpm	3000 hp @ 2700 rpm	3000 hp @ 2700 rpm
Rating; Military	3000 hp @ 2700 rpm @ 1500 feet 2400 hp @ 2700 rpm @ 13500 feet	3000 hp @ 2700 rpm @ 1500 feet 2400 hp @ 2700 rpm @ 13500 feet	3000 hp @ 2700 rpm @ 1500 feet 2400 hp @ 2700 rpm @ 13500 feet
Rating; Normal	2500 hp @ 2550 rpm @ 5000 feet. 2200 hp @ 2550 rpm @ 14500 feet	2500 hp @ 2550 rpm @ 5000 feet. 2200 hp @ 2550 rpm @ 14500 feet	2500 hp @ 2550 rpm @ 5000 feet 2200 hp @ 2550 rpm @ 14500 feet
Max. Continuous			
Cruise			
Fuel Grade			
Curves	Inst. #1694	Inst. #7273	
Weight, dry	3200 lbs	3450 lbs	3200 lbs less turbo
Prop. Reduction Ratio			0.400:1
Prop. Shaft Spline			
Compression Ratio			
Blower Ratio(s)			6.08:1
Carburetor		Chandler Evans CECO 100-CPB7-1	
Magnetos			
Installation Drwg. No.			
Dimensions			
A.T.C. number			
Number Manufactured			
Applications	Goodyear F2G-1 Goodyear F2G-2 Martin XBTM-1	Goodyear F2G-1 Goodyear F2G-2 Martin XBTM-1	
Notes	R-4360-2s were YR-4360-4 engines with Bendix PR-100 carburetor, changed to R-4360-2	R-4360-2s were YR-4360-4 engines with Chandler Evans CECO 100-CPB7-1 carburetor, changed to R-4360-2A	X-116 experimental Type Test Engine.

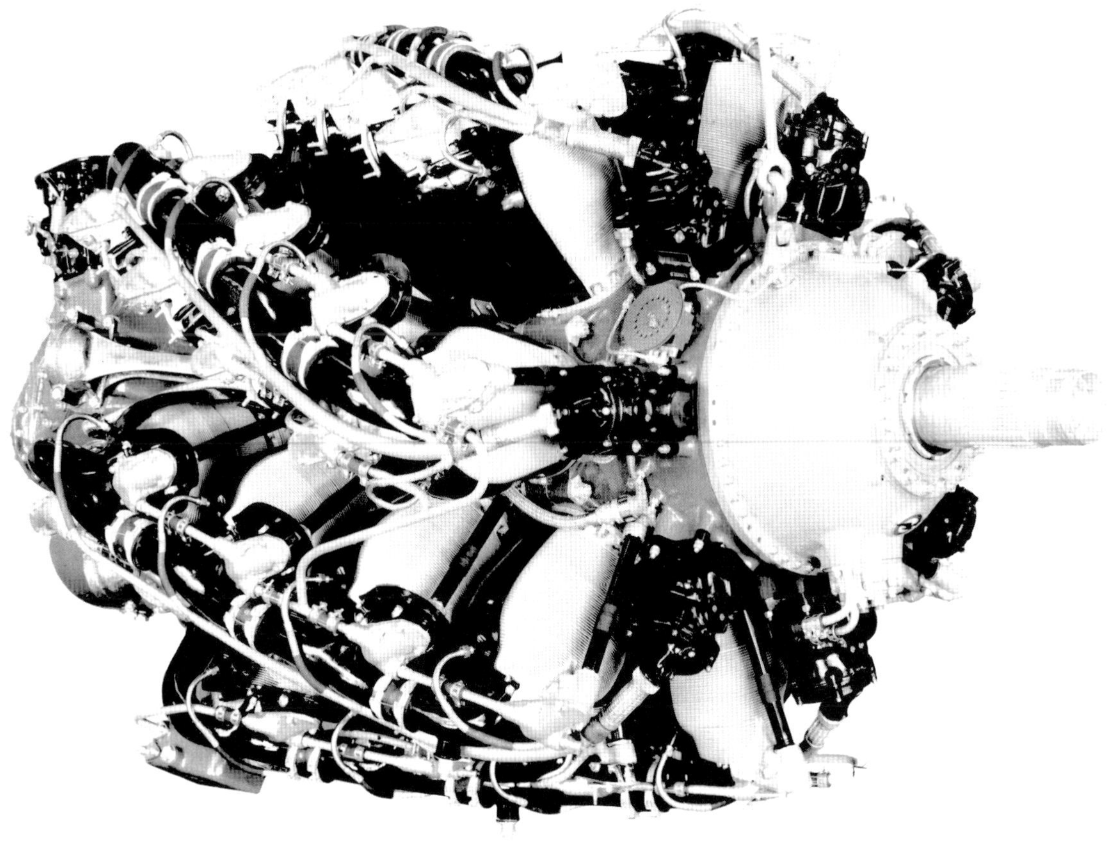

And overleaf: R-4360-2. First R-4360 to be shipped to a military customer.

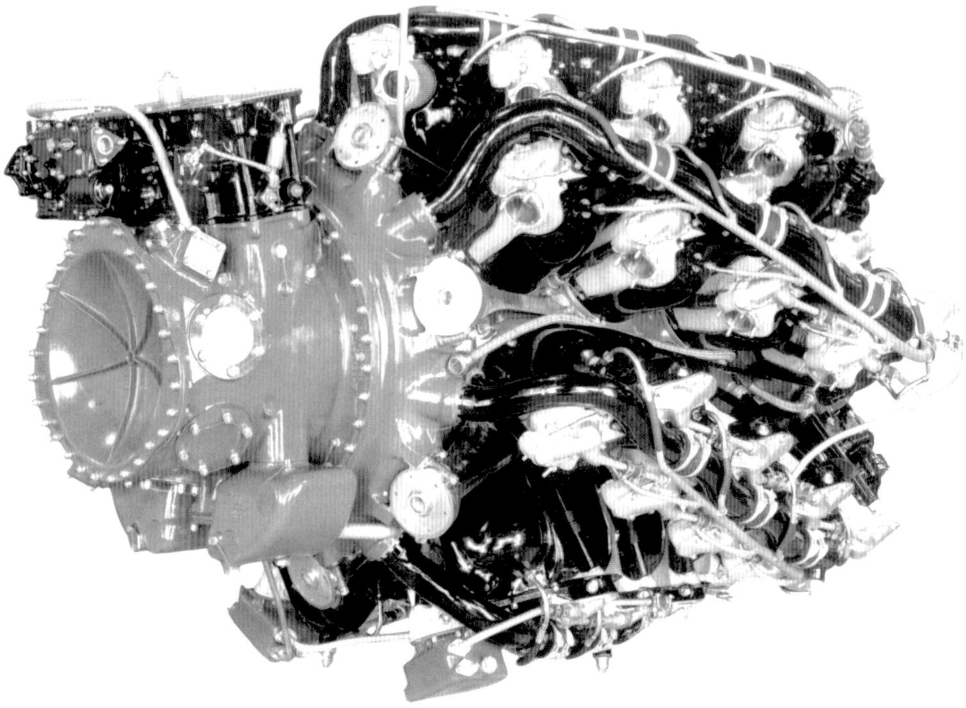

(Table 5-23) (Ref. 5-3)

Engine Model	**Type R-4360** P&W Aircraft **Air Force -3 (Semi production)** Navy	**Type R-4360** P&W Aircraft Air Force **Navy -4 (Semi production)**	**Type R-4360** P&W Aircraft Air Force **Navy -4 (*R-4360-4T)** ****** -4(*R-4360-4T)** **First production -4**
Specification Number	N-7027, Appendix A	N-7027	N-7039-C
Engine Series	B Series	B Series	B Series
Rating; Take-off	3000 hp @ 2700 rpm	3000 hp @ 2700 rpm	3000 hp @ 2700 rpm
Rating; Military	3000 hp @ 2700 rpm @ 1500 feet without turbosupercharger. 3000 hp @ 2700 rpm @ 40000 feet with turbosupercharger.	3000 hp @ 2700 rpm @ 1500 feet 2400 hp @ 2700 rpm @ 13500 feet	3000 hp @ 2700 rpm @ 1500 feet 2400 hp @ 2700 rpm @ 13500 feet
Rating; Normal	2500 hp @ 2550 rpm @ 5000 feet without turbosupercharger. 2500 hp @ 2550 rpm @ 40000 feet with turbosupercharger.	2500 hp @ 2550 rpm @ 5000 feet 2200 hp @ 2550 rpm @ 14500 feet	2500 hp @ 2550 rpm @ 5000 feet 2200 hp @ 2550 rpm @ 14500 feet
Max. Continuous			
Cruise			
Fuel Grade	100/130 PN	100/130 PN	100/130 PN
Curves	T-815	T-814	T-902 Inst. 7273
Weight, dry	*3475 lbs	*3325 lbs	*3400 lbs
Prop. Reduction Ratio	0.425:1	0.425:1	0.425:1
Prop. Shaft Spline	SAE #60 and #80 dual rotation	SAE #60- A	SAE #60- A
Compression Ratio	7.0:1	7.0:1	7.0:1
Blower Ratio(s)	6.08:1	Variable speed, maximum: 7.52:1	Variable speed, maximum: 7.52:1
Carburetor	Bendix PR-100-A3-1	Bendix PR-100-A3-1	Chandler Evans CECO 100-CPB7-1
Magnetos	DF-4RN-1	DF-4RN-1	D4RN-2
Installation Drwg. No.	70820	771207	86301
Dimensions	Diameter: 52.50 inches Length: 120.25 inches	Diameter: 52.50 inches Length: 96.75 inches	Diameter: 52.50 inches Length: 96.75 inches
A.T.C. number			
Number Manufactured		200	
Applications	Republic XP-72 dual rotation version. Curtiss XP-71 (none built)	Goodyear XF2G-1, Goodyear F2G-1 Martin XBTM-1, Martin JRM-2	Goodyear F2G-1, Martin AM-1 Martin XBTM-1, Martin JRM-2 ***Martin XP4M-1
Notes	Single stage, single speed. Supplied with fan drive. *Additional Weight: Torquemeter: 20 lbs	Single stage, variable speed. *Additional Weight: Torquemeter: 20 lbs	Single stage, variable speed. Supplied with Eclipse-Pioneer Type 1559 manifold pressure regulator. *R-4360-4T supplied with torquemeter. *Additional Weight: Torquemeter: 20 lbs ***With two GE I-40 jets. ****Sold commercially and to Navy and Air Force.

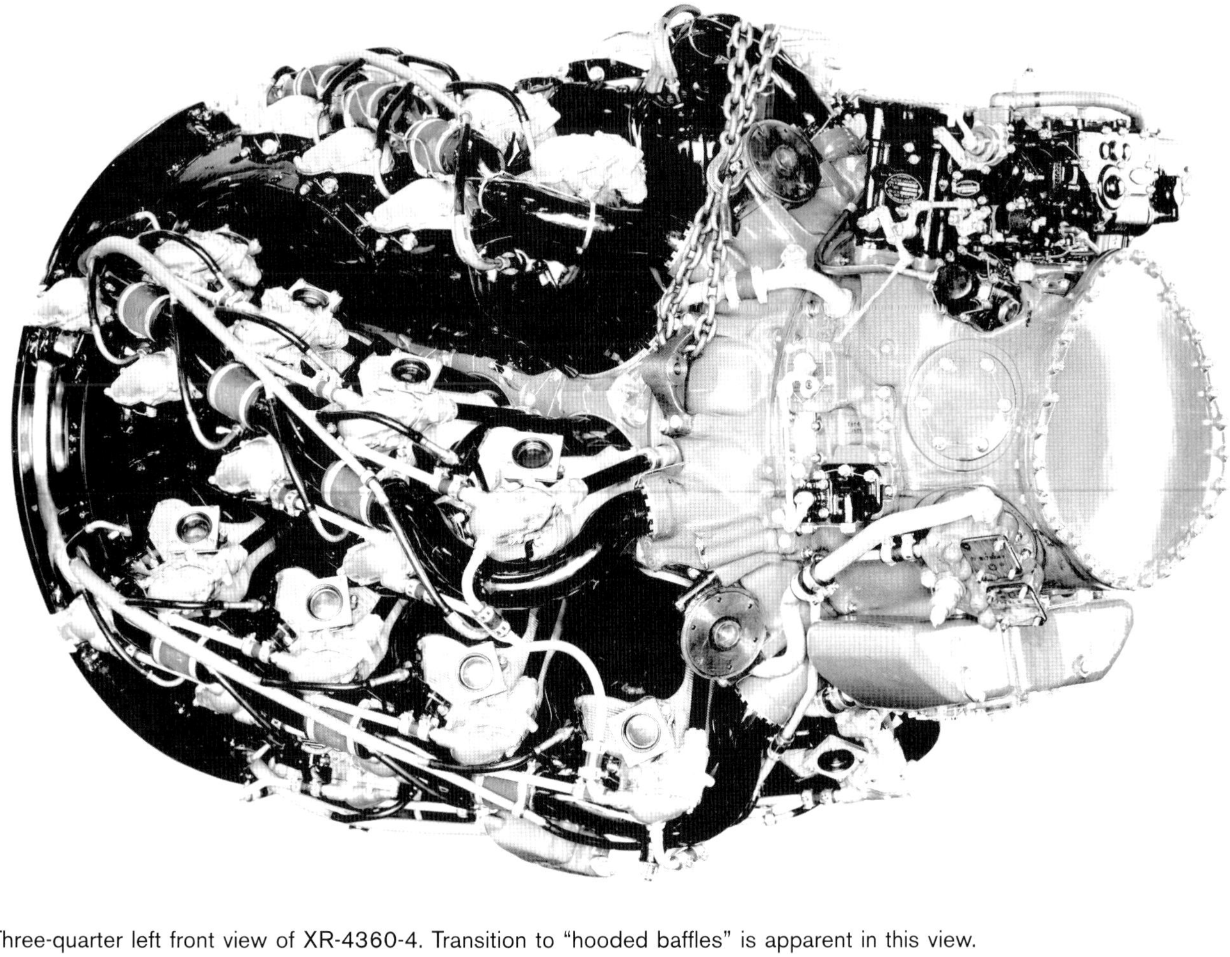

Three-quarter left front view of XR-4360-4. Transition to "hooded baffles" is apparent in this view.

Right side view of XR-4360-4. Bell-shaped nose case is notable in this shot.

(Table 5-24) (Ref. 5-3)

Engine Model	**Type** **R-4360** P&W Aircraft Air Force **Navy** **-4W (*R-4360-4T)**	**Type** **R-4360** P&W Aircraft **Air Force** ***-4A** Navy	**Type** **R-4360** P&W Aircraft **Air Force** **-5 (Semi production)** Navy
Specification Number	N-7039-C, Appendix A	N-7039-A, Appendix A	N-7027, Appendix B
Engine Series	B Series	B Series	B Series
Rating; Take-off	3000 hp @ 2700 rpm	3000 hp @ 2700 rpm	3000 hp @ 2700 rpm
Rating; Military	3000 hp @ 2700 rpm @ 1500 feet 2400 hp @ 2700 rpm @ 13500 feet	3000 hp @ 2700 rpm @ 1500 feet 2400 hp @ 2700 rpm @ 13500 feet	3000 hp @ 2700 rpm @ 1500 feet without turbosupercharger 3000 hp @ 2700 rpm @ 40000 feet with turbosupercharger
Rating; Normal	2500 hp @ 2550 rpm @ 5000 feet. 2200 hp @ 2550 rpm @ 14500 feet	2500 hp @ 2550 rpm @ 5000 feet. 2200 hp @ 2550 rpm @ 14500 feet	2500 hp @ 2550 rpm @ 5000 feet without turbosupercharger 2500 hp @ 2550 rpm @ 40000 feet with turbosupercharger
Max. Continuous			
Cruise			
Fuel Grade	100/130 PN	100/130 PN	100/130 PN
Curves	T-902 Inst. 7273	T-902 Inst. 1694	T-815
Weight, dry	**3413 lbs	**3390 lbs	*3605 lbs
Prop. Reduction Ratio	0.425:1	0.425:1	0.290:1 and 0.500:1 (two speed)
Prop. Shaft Spline	SAE #60- A	SAE #60- A	SAE #60- A
Compression Ratio	7.0:1	7.0:1	7.0:1
Blower Ratio(s)	Variable speed, maximum: 7.42:1	Variable speed, maximum: 7.52:1	6.06:1
Carburetor	Chandler Evans CECO 100-CPB7-1	Bendix PR-100-B3-3	Bendix PR-100-A3-1
Magnetos	D4RN-2	D4RN-2	DF-4RN-1
Installation Drwg. No.	86301	86301	70828
Dimensions	Diameter: 52.50 inches Length: 96.75 inches	Diameter: 52.50 inches Length: 96.75 inches	Diameter: 52.50 inches Length: 109.75 inches
A.T.C. number			
Number Manufactured		16	9
Applications	Martin AM-2 Martin AM-1 Martin AM-1Q	Hughes XF-11 *** Hughes HFB-1 (H4)	Convair XB-36
Notes	Single stage, variable speed. Includes water injection equipment (ADI) *R-4360-4T supplied with torquemeter. **Additional Weight: Torquemeter: 20 lbs	Single stage, variable speed supercharger. *Sold to Army. **Includes manifold pressure regulator Additional Weight: Torquemeter: 20 lbs ***Hughes exchanged 15 for 11 engines from Navy. Incorporated torquemeter nose.	Incorporates power take-off co-axial with crankshaft and cooling fan drive. Single stage, single speed. *Additional Weight: Torquemeter: 6 lbs Cooling fan on accessory drive: 103 lbs

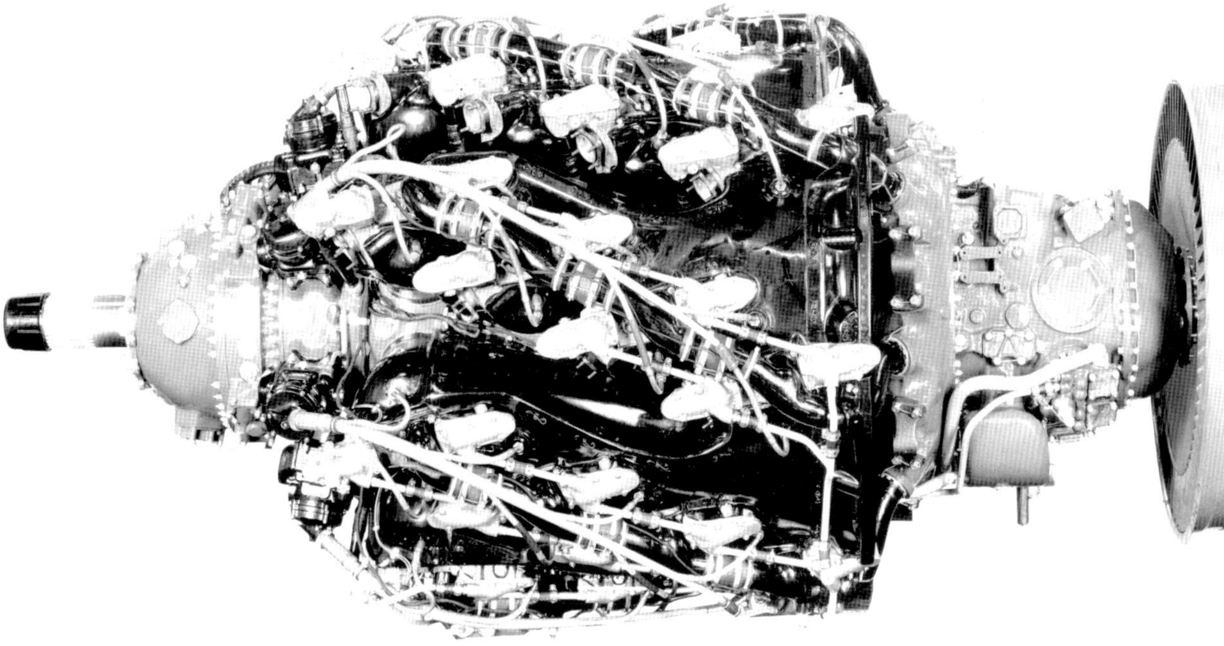

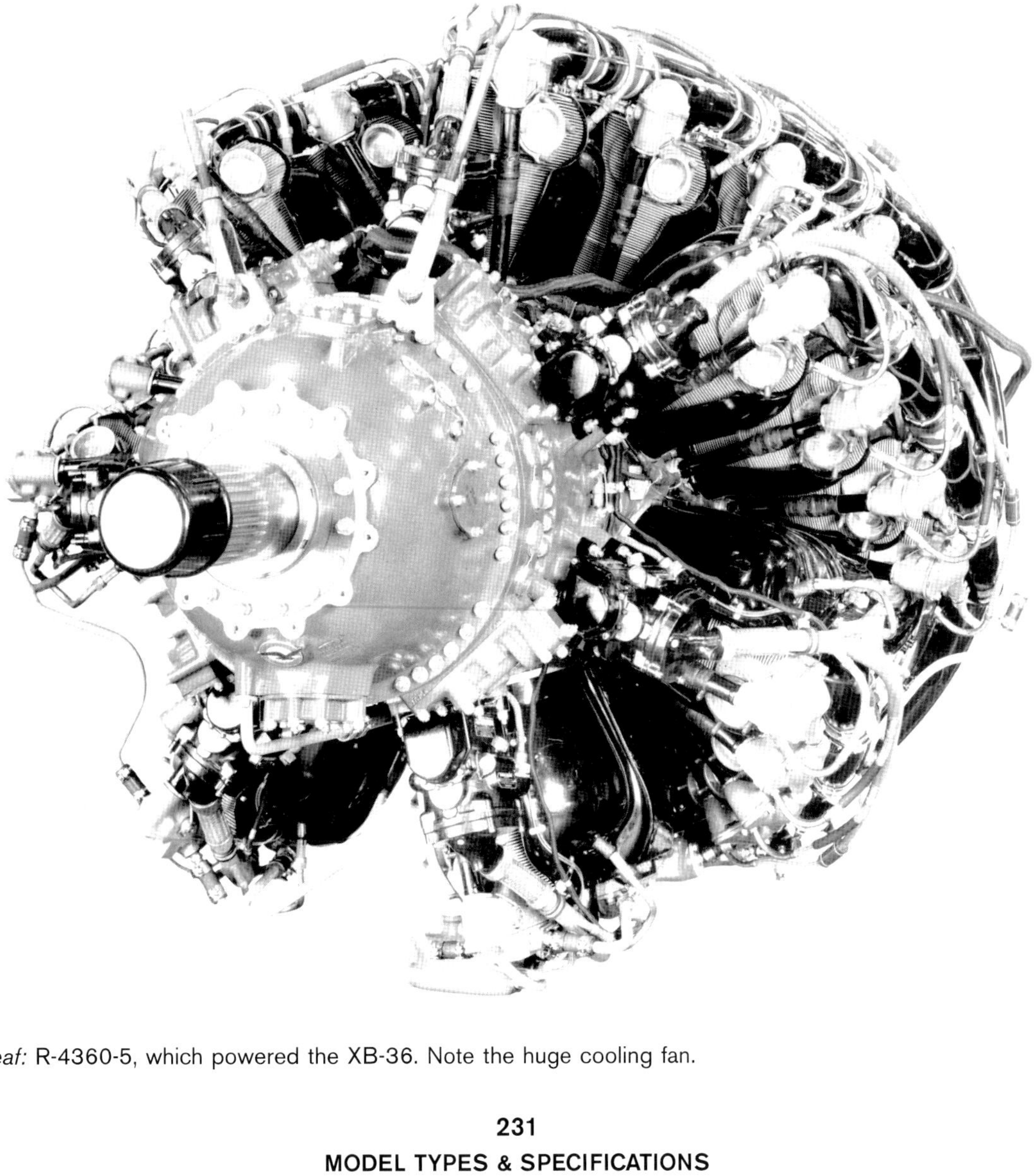

And overleaf: R-4360-5, which powered the XB-36. Note the huge cooling fan.

(Table 5-25) (Ref. 5-3)

Engine Model	**Type** **R-4360** P&W Aircraft Air Force **Navy** **-6 (semi production)**	**Type** **R-4360** P&W Aircraft **Air Force** **-7** Navy	**Type** **R-4360** P&W Aircraft Air Force **Navy** **-8 (Semi production)**
Specification Number	N-7028	N-7027, Appendix C1	N-7027, Appendix D
Engine Series	B Series	B Series	B Series
Rating; Take-off	3000 hp @ 2700 rpm	3000 hp @ 2700 rpm	3000 hp @ 2700 rpm
Rating; Military	3000 hp @ 2700 rpm @ 1500 feet 2400 hp @ 2700 rpm @ 25000 feet	3000 hp @ 2700 rpm @ 1500 feet without turbosupercharger 3000 hp @ 2700 rpm @ 40000 feet with turbosupercharger	3000 hp @ 2700 rpm @ 1500 feet 2400 hp @ 2700 rpm @ 13500 feet
Rating; Normal	2500 hp @ 2550 rpm @ 5000 feet 2200 hp @ 2550 rpm @ 25000 feet	2500 hp @ 2550 rpm @ 5000 feet without turbosupercharger 2500 hp @ 2550 rpm @ 40000 feet with turbosupercharger	2500 hp @ 2550 rpm @ 5000 feet 2200 hp @ 2550 rpm @ 14500 feet
Max. Continuous			
Cruise			
Fuel Grade	100/130 PN	100/130 PN	100/130 PN
Curves	T-812 Inst. #7223	T-812 Inst. #1691	T-814 Inst. #1694 (Bendix carburetor) Inst. #7273 (Chandler Evans carburetor)
Weight, dry	*3585 lbs	*3183 lbs	*3525 lbs
Prop. Reduction Ratio	0.425:1	0.381:1 remote drive	0.425:1
Prop. Shaft Spline	SAE #60- A	SAE #60- Aand SAE #80 dual rotation	SAE #50- and SAE #70 dual rotation
Compression Ratio	7.0:1	7.0:1	7.0:1
Blower Ratio(s)	6.08:1 maximum	6.08:1	7.52:1 maximum
Carburetor	Bendix PR-100-A3	Bendix PR-100-A3-1	Bendix PR-100-A3-1 or, Chandler Evans CECO 100-CPB7-1
Magnetos	DF-4RN-1	DF-4RN-1	DF-4RN-1
Installation Drwg. No.	57448	77301	86400
Dimensions	Diameter: 52.50 inches Length: 116.50 inches	Diameter: 52.50 inches Length: 87.00 inches for direct drive engine Length: 213.847 inches from engine center line to propeller thrust nut face	Diameter: 52.50 inches Length: 114.25 inches
A.T.C. number			
Number Manufactured		2	5
Applications		Northrop B-35 (Outboard engine)	Douglas TB2D-1
Notes	Two-stage, variable speed supercharger Additional Weights: *Torquemeter: 20 lbs None manufactured.	Single speed, single stage supercharger. Suitable for use with turbosupercharger. *Additional Weights: Torquemeter (available option): 20 lbs Cooling fan on accessory end: 179 lbs Extension shaft for outboard engine: 695 lbs	Single-stage, variable speed supercharger. Dual rotation. *Additional Weights: Torquemeter: 20 lbs

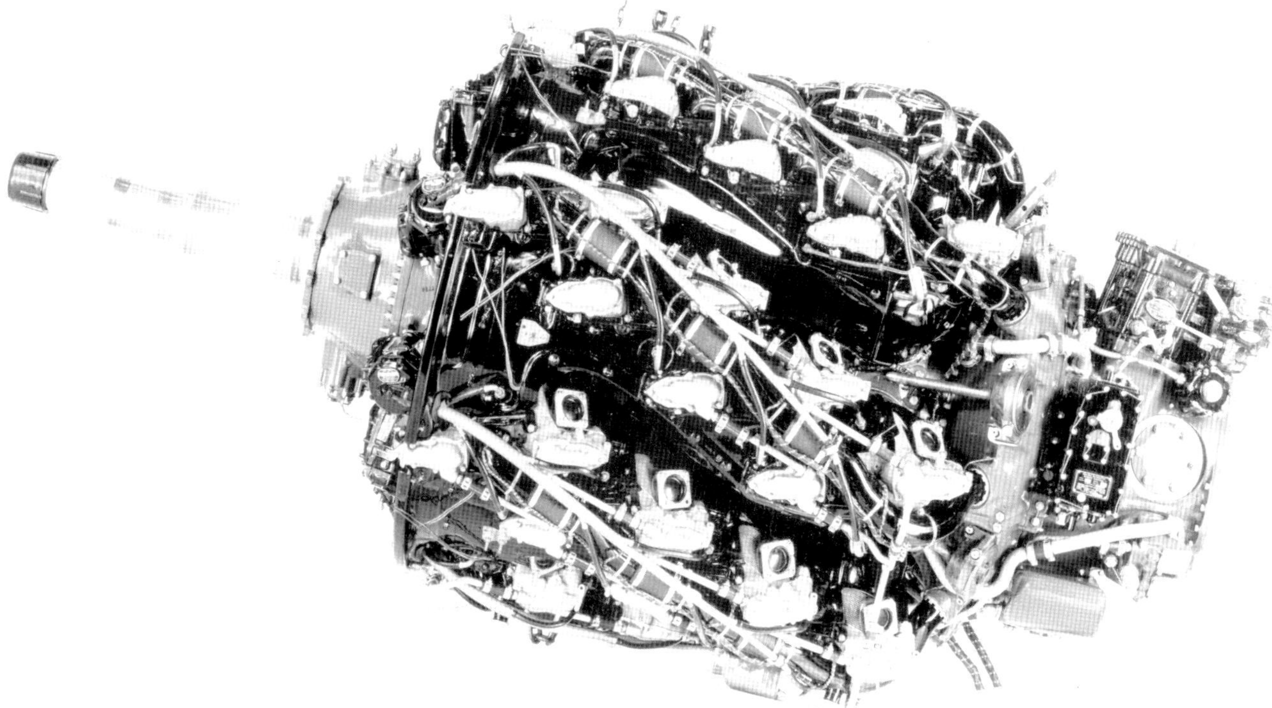

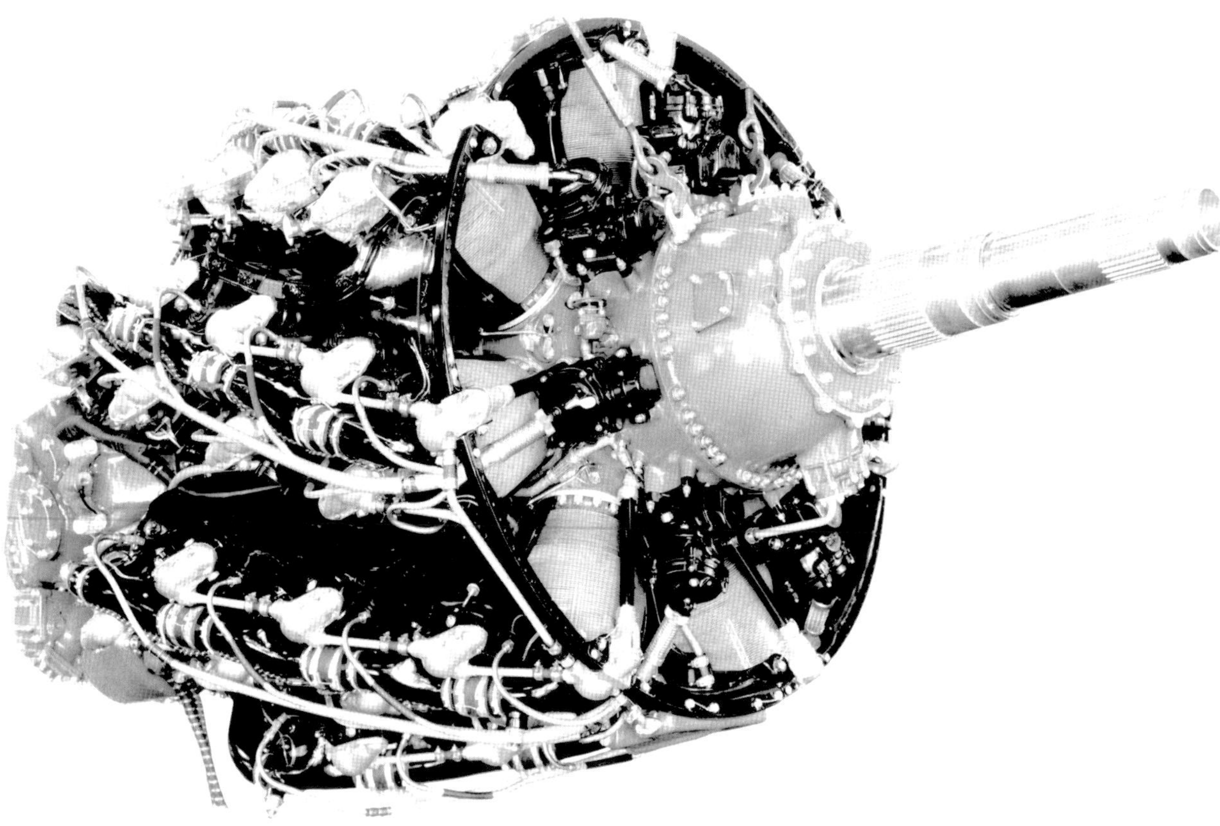

And on next two pages: R-4360-8. Note the dual-rotation propeller shafts. The black rectangular device bolted to the side of the supercharger is an Eclipse automatic engine (boost) controller. It had high-tension ignition and no hooded baffles.

(Table 5-26) (Ref. 5-3)

Engine Model	Type R-4360 P&W Aircraft Air Force **Navy** **-8A (semi production)**	Type R-4360 P&W Aircraft Air Force -7 **Navy** **-8 (production version of R-4360-8**	Type R-4360 P&W Aircraft Air Force **-9 (Semi production)** Navy
Specification Number	N-7027, Appendix F	N-7041A	N-7027, Appendix H
Engine Series	B Series	B Series	B Series
Rating; Take-off	3000 hp @ 2700 rpm	3000 hp @ 2700 rpm	3000 hp @ 2700 rpm
Rating; Military	3000 hp @ 2700 rpm @ 1500 feet 2400 hp @ 2700 rpm @ 13500 feet	3000 hp @ 2700 rpm @ 1500 feet 2400 hp @ 2700 rpm @ 13500 feet	3000 hp @ 2700 rpm @ 1500 feet 2400 hp @ 2700 rpm @ 13500 feet
Rating; Normal	2500 hp @ 2550 rpm @ 5000 feet 2200 hp @ 2550 rpm @ 14500 feet	2500 hp @ 2550 rpm @ 5000 feet 2200 hp @ 2550 rpm @ 14500 feet	2500 hp @ 2550 rpm @ 5000 feet 2200 hp @ 2550 rpm @ 14500 feet
Max. Continuous			
Cruise			
Fuel Grade	100/130 PN	100/130 PN	100/130 PN
Curves	T-814 Inst. #1694	T-902 Inst. #1694	T-814 Inst. #1694
Weight, dry	*3525 lbs	*3570 lbs	*3425 lbs
Prop. Reduction Ratio	0.425:1	0.425:1	**0.500:1 and 0.290:1
Prop. Shaft Spline	SAE #50- and SAE #70 dual rotation	SAE #60- and SAE #80 dual rotation	SAE #60-A
Compression Ratio	7.0:1	7.0:1	7.0:1
Blower Ratio(s)	7.52:1 maximum	7.52:1 maximum	7.52:1 maximum
Carburetor	Bendix PR-100-A3-1	Bendix PR-100-B2-3	Bendix PR-100-A3-1
Magnetos	DF-4RN-1	DF-4RN-2	DF-4RN-1
Installation Drwg. No.	76042	90301	80001
Dimensions	Diameter: 52.50 inches Length: 111.00 inches	Diameter: 52.50 inches Length: 114.25 inches	Diameter: 52.50 inches Length: 101.00 inches
A.T.C. number			
Number Manufactured	2		2
Applications		Douglas XTB2D-1 Douglas TB2D-1 Curtiss XBTC	Vultee XA-41 Convair A-41
Notes	Single-stage, variable speed supercharger *Additional Weights: Torquemeter: 20 lbs	Single-stage, variable speed supercharger *Additional Weights: Torquemeter: 20 lbs Weight includes manifold pressure regulator.	Single-stage, variable speed supercharger *Additional Weights: Torquemeter: 20 lbs **Two-speed nose with .290:1 ratio locked out.

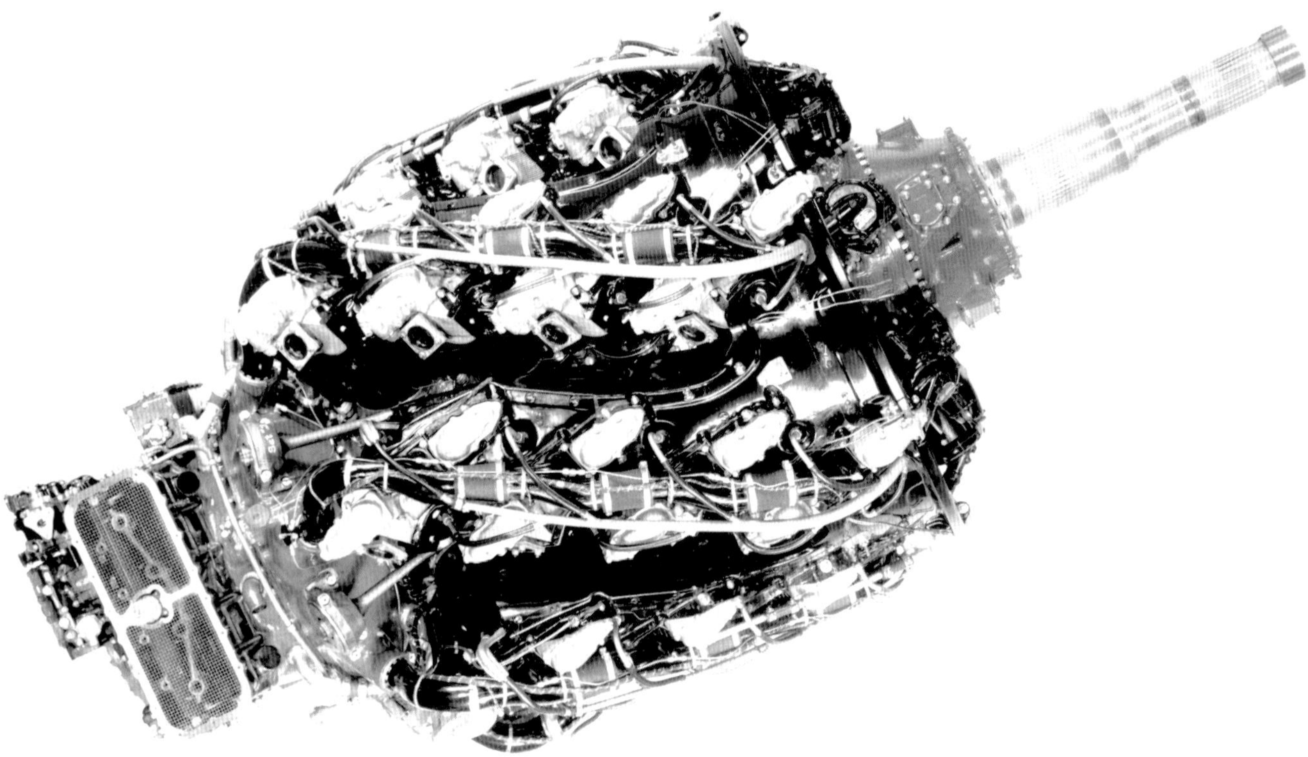

R-4360-8. This is a top view—note the immense size of the PR-100 carburetor.

(Table 5-27) (Ref. 5-3)

Engine Model	Type R-4360 P&W Aircraft **Air Force 9T** Navy	Type R-4360 P&W Aircraft Air Force **Navy -10 (semi production)**	Type R-4360 P&W Aircraft Air Force **Navy -10 (production version)**
Specification Number		N-7028, Appendix A	N-7043, Appendix A
Engine Series	B Series	B Series	B Series
Rating; Take-off	3000 hp @ 2700 rpm	3000 hp @ 2700 rpm	3000 hp @ 2700 rpm
Rating; Military	3000 hp @ 2700 rpm @ 1500 feet 2400 hp @ 2700 rpm @ 13500 feet	3000 hp @ 2700 rpm @ 1500 feet 2400 hp @ 2700 rpm @ 25000 feet	3000 hp @ 2700 rpm @ 1500 feet 2400 hp @ 2700 rpm @ 25000 feet
Rating; Normal	2500 hp @ 2550 rpm @ 5000 feet 2200 hp @ 2550 rpm @ 14500 feet	2500 hp @ 2550 rpm @ 5000 feet 2200 hp @ 2550 rpm @ 25000 feet	2500 hp @ 2550 rpm @ 5000 feet 2200 hp @ 2550 rpm @ 25000 feet
Max. Continuous			
Cruise			
Fuel Grade	100/130 PN	100/130 PN	100/130 PN
Curves	T-814 Inst. #1694	T-812	T-951 Inst. #1721
Weight, dry	3525 lbs	3785 lbs	3775 lbs
Prop. Reduction Ratio	0.381:1	0.381:1	0.381:1
Prop. Shaft Spline	SAE #60- and SAE #80 dual rotation	SAE #50- and SAE #70 dual rotation	SAE #50 and SAE #70 dual rotation
Compression Ratio	7.0:1	7.0:1	7.0:1
Blower Ratio(s)	7.52:1 maximum	6.08:1 maximum 5.75:1 minimum	6.08:1 maximum 5.75:1 minimum
Carburetor	Bendix PR-100-A3-1	Bendix PR-100-A3-1	Bendix PR-100-B2-3
Magnetos	DF-4RN-1	DF-4RN-1	D-4RN-2
Installation Drwg. No.	R-86400	70836	91701
Dimensions	Diameter: 52.500 inches Length: 113.996 inches	Diameter: 52.50 inches Length: 130.75 inches	Diameter: 52.50 inches Length: 123.50 inches
A.T.C. number			
Number Manufactured	1	6	
Applications		Boeing XF8B-1	
Notes	This engine was used to show general characteristics. Army test engine hence "T" designation. Like R-4360-8 except reduction gear ratio.	Two-stage, variable speed supercharger *Additional Weights: Torquemeter: 20 lbs	Two-stage, variable speed supercharger *Additional Weights: Torquemeter: 20 lbs Incorporates manifold pressure regulator.

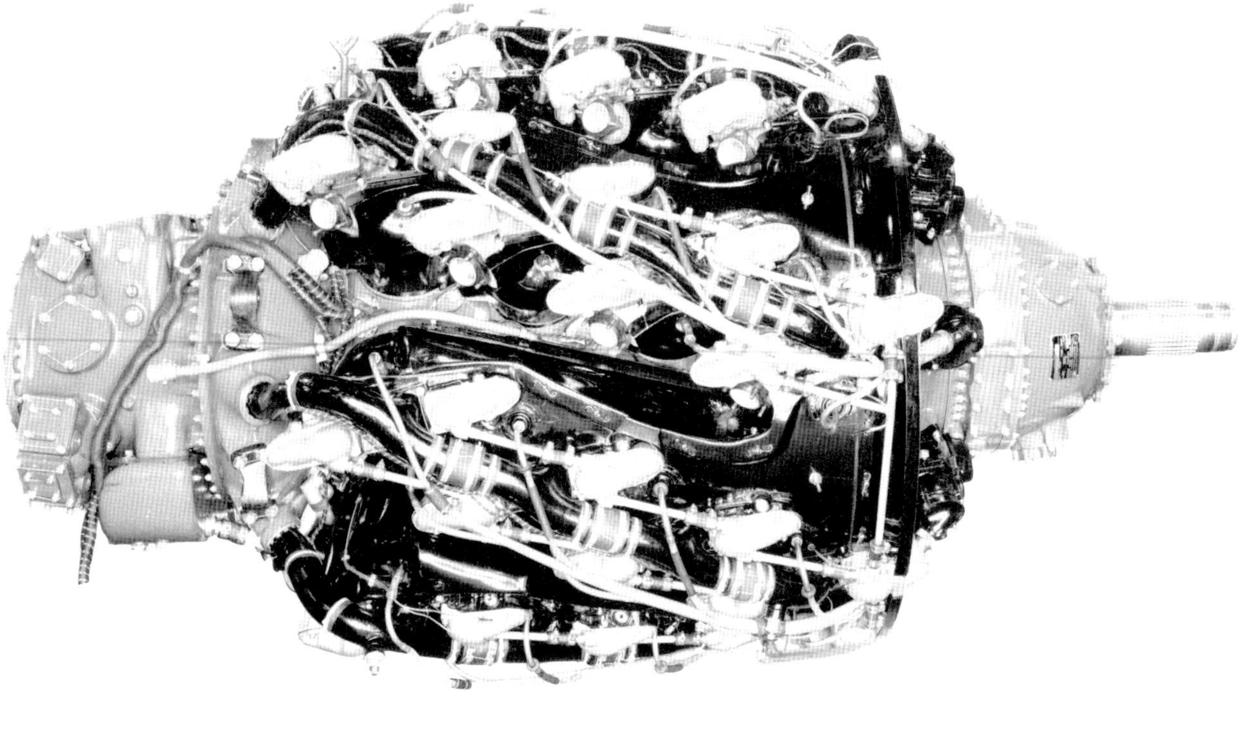

R-4360-9. Like many factory portraits, no carburetor is fitted.

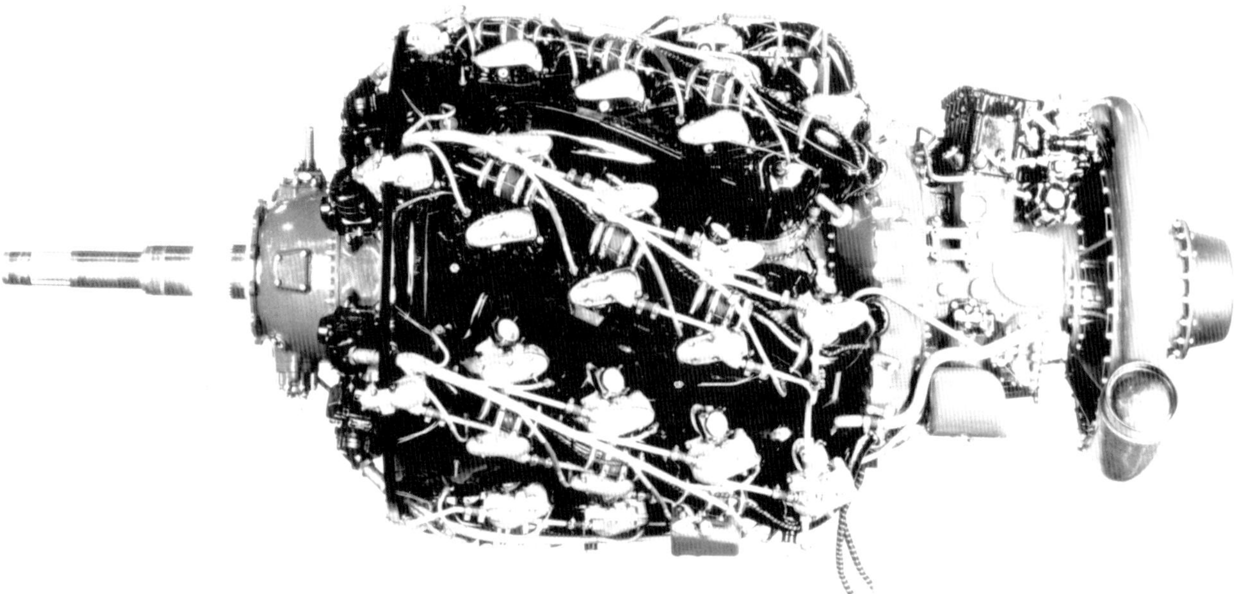

R-4360-10. It would appear that Boeing was sold on the idea of gear-driven two-stage supercharging, at least until the C-97/B-50 came along. Similar to the -33 that powered the Boeing XB-44, the R-4360-10 has an auxiliary supercharger driven off the rear end.

(Table 5-28) (Ref. 5-3)

Engine Model	**Type** **R-4360** P&W Aircraft **Air Force** **-11 (semi-production)** Navy	**Type** **R-4360** P&W Aircraft Air Force **Navy** **-12**	**Type** **R-4360** P&W Aircraft Air Force **Navy** **-12A**
Specification Number	N-7027, Appendix C2	N-7027, Appendix E	N-7027, Appendix G
Engine Series	B Series	B Series	B Series
Rating; Take-off	3000 hp @ 2700 rpm	3000 hp @ 2700 rpm	3000 hp @ 2700 rpm
Rating; Military	3000 hp @ 2700 rpm @ 1500 feet without turbosupercharger 2000 hp @ 2700 rpm @ 40000 feet with turbosupercharger	3000 hp @ 2700 rpm @ 1500 feet 2400 hp @ 2700 rpm @ 13500 feet	3000 hp @ 2700 rpm @ 1500 feet 2400 hp @ 2700 rpm @ 13500 feet
Rating; Normal	2500 hp @ 2550 rpm @ 5000 feet without turbosupercharger 2500 hp @ 2550 rpm @ 40000 feet with turbosupercharger	2500 hp @ 2550 rpm @ 5000 feet 2200 hp @ 2550 rpm @ 14500 feet	2500 hp @ 2550 rpm @ 5000 feet 2200 hp @ 2550 rpm @ 14500 feet
Max. Continuous			
Cruise			
Fuel Grade	100/130 PN	100/130 PN	100/130 PN
Curves	T-815 Inst. #1691	T-814	T-814
Weight, dry	*3183 lbs	*3690 lbs	*3690 lbs
Prop. Reduction Ratio	0.381:1 (remote drive)	0.390:1 and 0.500:1 - two speed reduction gearing.	0.390:1 and 0.500:1 - two speed reduction gearing.
Prop. Shaft Spline	SAE #60 and SAE #80 dual rotation	SAE #60 and SAE #80 dual rotation	SAE #50 and SAE #70 dual rotation
Compression Ratio	7.0:1	7.0:1	7.0:1
Blower Ratio(s)	6.08:1 minimum 7.52:1 maximum	Variable speed, 7.52:1 maximum	Variable speed, 7.52:1 maximum
Carburetor	Bendix PR-100-A3-1	Bendix PR-100-A3-1	Bendix PR-100-A3-1
Magnetos	DF-4RN-1	DF-4RN-1	D-4RN-1
Installation Drwg. No.	77301	R-70833	R-76045
DImensions	Diameter: 52.50 inches Length: 87.00 inches for direct-drive engine Length: 326.40 inches from engine center line to propeller thrust nut face	Diameter: 52.50 inches Length: 119.25 inches	Diameter: 52.50 inches Length: 119.25 inches
A.T.C. number			
Number Manufactured	1		
Applications	Northrop B-35 (Inboard engine)		
Notes	Single-stage, variable-speed supercharger. Suitable for turbosupercharger. *Additional Weights: Torquemeter: 20 lbs Extension shaft for inboard engine: 830 lbs Cooling fan on accessory drive end: 179 lbs	Single-stage, variable speed supercharger *Additional Weights: Torquemeter: 6 lbs None manufactured.	Single-stage, variable speed supercharger *Additional Weights: Torquemeter: 6 lbs None manufactured.

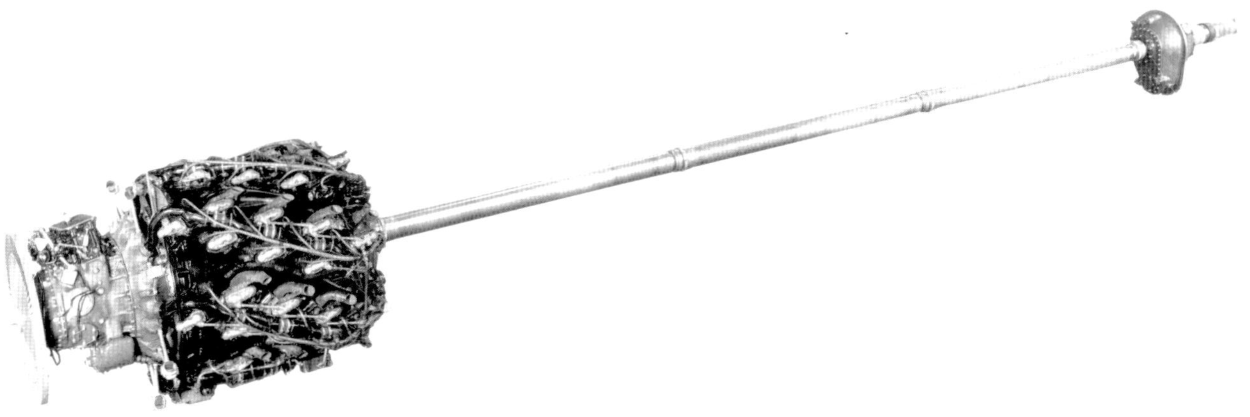

R-4360-11. The unique aspect of the -11 was the fact that it was direct drive. Built for the XB-35, propeller reduction gearing was mounted in the end of the extension shaft. Basically the same power section and rear end as the R-4360-5 installed in the XB-36. Dash 11s were the inboard engines and drove contra-rotating propellers. Note the cooling fan attached to the rear case. This large axial flow fan discharged cooling air over the cylinders, meaning cooling flow is from left to right in this view. Wide slots in the wing leading edge ducted air to the fan. Leading-edge ducts also fed ram air to the Bendix PR-100-A3-1 carburetor.

(Table 5-29) (Ref. 5-3)

Engine Model	**Type** **R-4360** P&W Aircraft **Air Force** **-13 (semi-production)** Navy	**Type** **R-4360** P&W Aircraft Air Force **Navy** **-14 (Production version of R-4360-8A**	**Type** **R-4360** P&W Aircraft **Air Force** **-15 (semi prod.)** Navy -12A
Specification Number	N-7028, Appendix C	N-7037	N-7027, Appendix K
Engine Series	B Series	B Series	B Series
Rating; Take-off	3000 hp @ 2700 rpm	3000 hp @ 2700 rpm	3000 hp @ 2700 rpm
Rating; Military	3000 hp @ 2700 rpm @ 1500 feet 2400 hp @ 2700 rpm @ 25000 feet	3000 hp @ 2700 rpm @ 1500 feet 2400 hp @ 2700 rpm @ 13500 feet	3000 hp @ 2700 rpm @ 1500 feet 2400 hp @ 2700 rpm @ 13500 feet
Rating; Normal	2500 hp @ 2550 rpm @ 5000 feet 2200 hp @ 2550 rpm @ 25000 feet	2500 hp @ 2550 rpm @ 5000 feet 2200 hp @ 2550 rpm @ 14500 feet	2500 hp @ 2550 rpm @ 5000 feet 2200 hp @ 2550 rpm @ 14500 feet
Max. Continuous			
Cruise			
Fuel Grade	100/130 PN	100/130 PN	100/130 PN
Curves	T-812 Inst. #1721	T-902 Inst: #1694 with Bendix carburetor Inst: #7273 with Chandler Evans carburetor.	T-814 Inst: #1694
Weight, dry	3685 lbs	*3584 lbs	3325lbs
Prop. Reduction Ratio	0.425:1	0.425:1	0.381:1
Prop. Shaft Spline	SAE #60-A	SAE #50- and SAE #70 dual rotation	SAE #60
Compression Ratio	7.0:1	7.0:1	7.0:1
Blower Ratio(s)	Variable speed: 6.08:1 maximum 5.75:1 minimum	Variable speed, 7.52:1 maximum	Variable speed, 7.52:1 maximum
Carburetor	Bendix PR-100-A3-1	Bendix PR-100-B2-3 Chandler Evans CECO 100-CPB7-1	Bendix PR-100-A3-1
Magnetos	DF-4RN-1	DF-4RN-1	D-4RN-1
Installation Drwg. No.	57448	87001	71207
Dimensions	Diameter: 52.00 inches Length: *50.25 inches Length: ** 100.00 inches Length *** 216.37 inches	Diameter: 52.50 inches Length: 111.00 inches	Diameter: 52.50 inches Length: 96.75 inches
A.T.C. number			
Number Manufactured	3	2	1
Applications	Republic XP-72	**Curtiss XBTC-2	Douglas C-74
Notes	Two-stage, variable speed supercharger. *Power section and nose case. **Without remotely mounted supercharger. ***Includes remotely mounted supercharger. R-4360-14 incorporates R-4360-19 dual rotation nose case.	Single-stage, variable speed supercharger *Additional Weights: Torquemeter: 20 lbs *Weight includes manifold pressure regulator. **First flight July 19, 1945, at Columbus.	Single-stage, variable speed supercharger.

R-4360-13. The following six photographs were taken in September 1943.

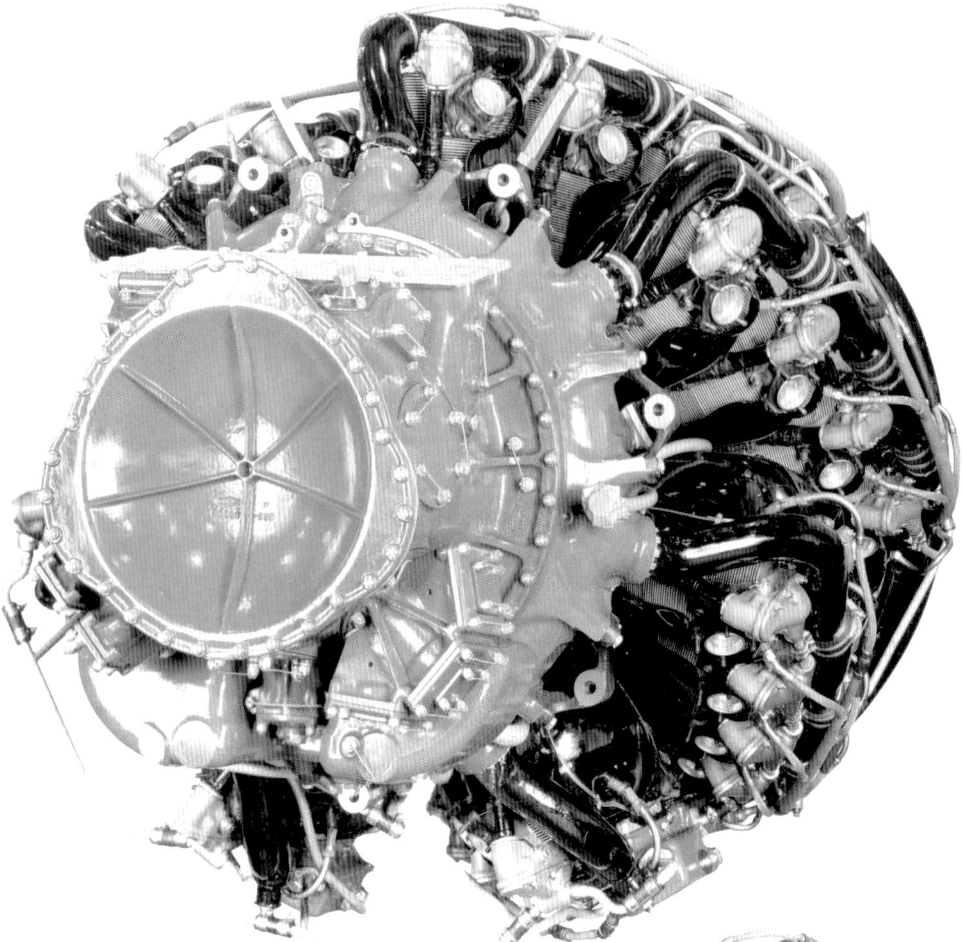

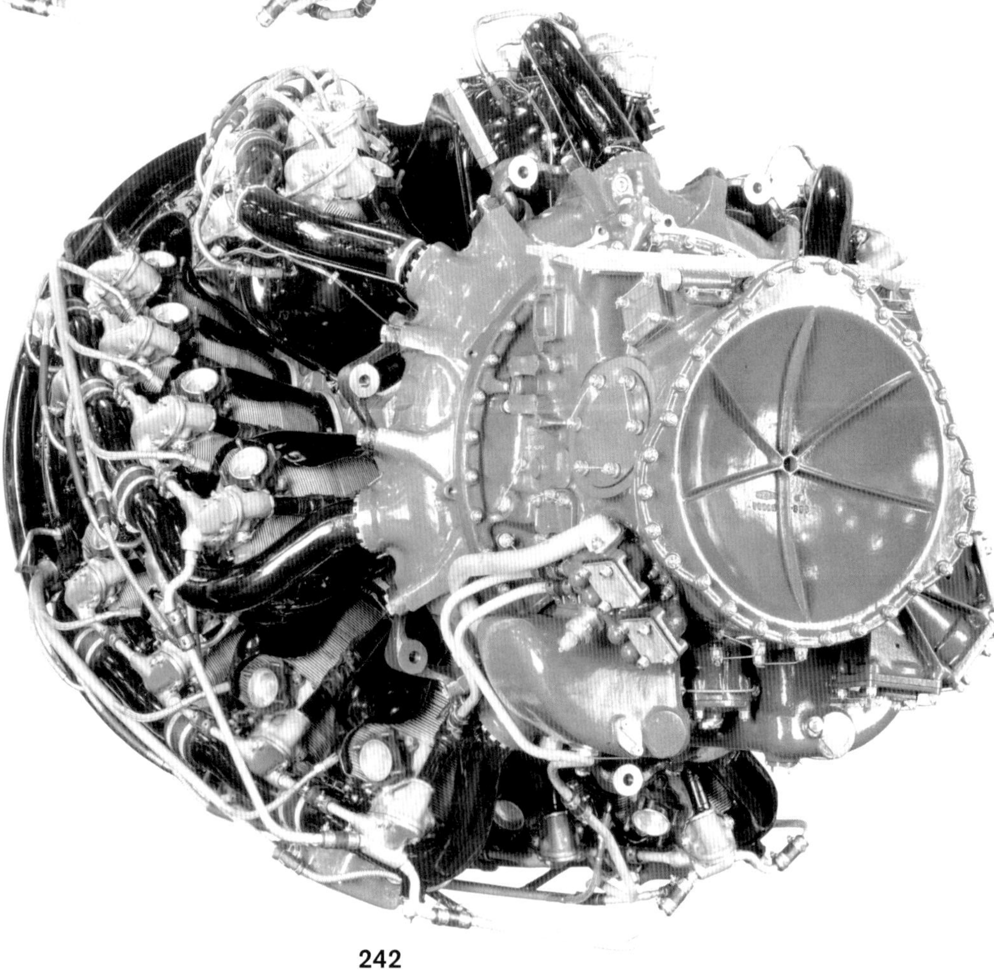

(Table 5-30) (Ref. 5-3)

Engine Model	Type R-4360 P&W Aircraft **Air Force** -17 (production version of R-4360-7) Navy	Type R-4360 P&W Aircraft Air Force **Navy** -18 (production) No semi production engines built.	Type R-4360 P&W Aircraft **Air Force** -19 (production version of R-4360-13) Navy
Specification Number	N-7032-D, Appendix A	N-7038	N-7033
Engine Series	B Series	B Series	B Series
Rating; Take-off	3000 hp @ 2700 rpm	3000 hp @ 2700 rpm	3000 hp @ 2700 rpm
Rating; Military	3000 hp @ 2700 rpm @ 1500 feet without turbosupercharger 3000 hp @ 2700 rpm @ 40000 feet with turbosupercharger	*3000 hp @ 2700 rpm @ 30.0 in. Hg. back pressure	3000 hp @ 2700 rpm @ 3000 feet 2400 hp @ 2700 rpm @ 25000 feet
Rating; Normal	2500 hp @ 2550 rpm @ 5000 feet without turbosupercharger 2500 hp @ 2550 rpm @ 40000 feet with turbosupercharger	*2500 hp @ 2550 rpm @ 27.5 in. Hg. back pressure	2500 hp @ 2550 rpm @ 6000 feet 2200 hp @ 2550 rpm @ 25000 feet
Max. Continuous			
Cruise			
Fuel Grade	100/130 PN	100/130 PN	100/130 PN
Curves	T-869 Inst. #1691	T-889 Inst. #7278	T-875
Weight, dry	*3308 lbs	**3355 lbs	*3889 lbs
Prop. Reduction Ratio	0.381:1 remote drive	0.381:1	0.425:1
Prop. Shaft Spline	SAE #60A and SAE #80 dual rotation.	SAE #60-A	SAE #50 and SAE #70 dual rotation.
Compression Ratio	7.0:1	7.0:1	7.0:1
Blower Ratio(s)	6.08:1	6.08:1	Variable speed: 6.08:1 maximum 5.75:1 minimum
Carburetor	Bendix PR-100-B2-3	Chandler Evans CECO 100-CPB7-1	Bendix PR-100-B-1
Magnetos	D4RN-2	D4RN-1	DF-4RN-2
Installation Drwg. No.	79201	87601	R-88020
Dimensions	Diameter: 52.50 inches Length: 87.00 inches for direct drive engine Length: 213.85 inches from engine centerline to propeller thrust nut face	Diameter: 52.00 inches Length: 96.75 inches Length: **230.75 inches	Diameter: 52.50 inches Length: 114.25 direct drive
A.T.C. number			
Number Manufactured	28	13	
Applications	Northrop B-35 (Outboard engine) Northrop XB-35 (Outboard engine) Northrop YB-35 (Outboard engine)	Lockheed XR60-1 Constitution. ***Lockheed R6V-1 Model 89	Planned for Republic P-72, never built.
Notes	*Additional weights: Extension shaft and coupling asm: 185 lbs Reduction gear housing asm. and power take-off, propeller governor drives, and torquemeter: 556 lbs	Single-stage, single speed supercharger. Suitable for turbosupercharging. GEModel BH-3 turbosupercharger designed around this engine. Turbo Engineering Company developed PT 15B 3 turbo, did not see production. Lockheed also developed an axial flow turbo, again, did not see production. Power take-off drive furnished on accessory end and co-axial with crankshaft. *30 in. Hg. back pressure and 27.5 in. Hg. back pressure based on 100 degrees F carburetor air and 31 in. Hg. back pressure. **Includes torquemeter. ***Replaced by R-4360-22W.	Two-stage, variable speed supercharger. Power take-off drive furnished on accessory end and co-axial with crankshaft. *Weight includes remote auxiliary stage supercharger and supercharger pressure regulator. **Includes remotely mounted super charger. None manufactured.

The R-4360-13 could also handle an auxiliary supercharger. This application was built for the Republic XP-72. It's unclear if the XP-72 ever flew with this supercharger.

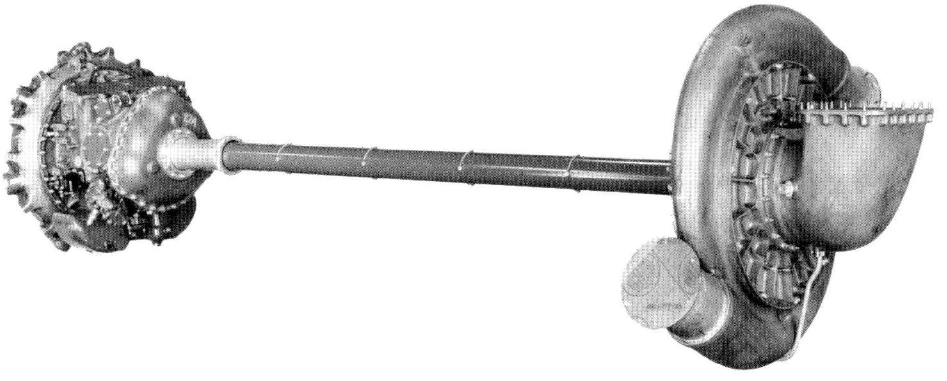

R-4360-14. Another dual-rotation setup.

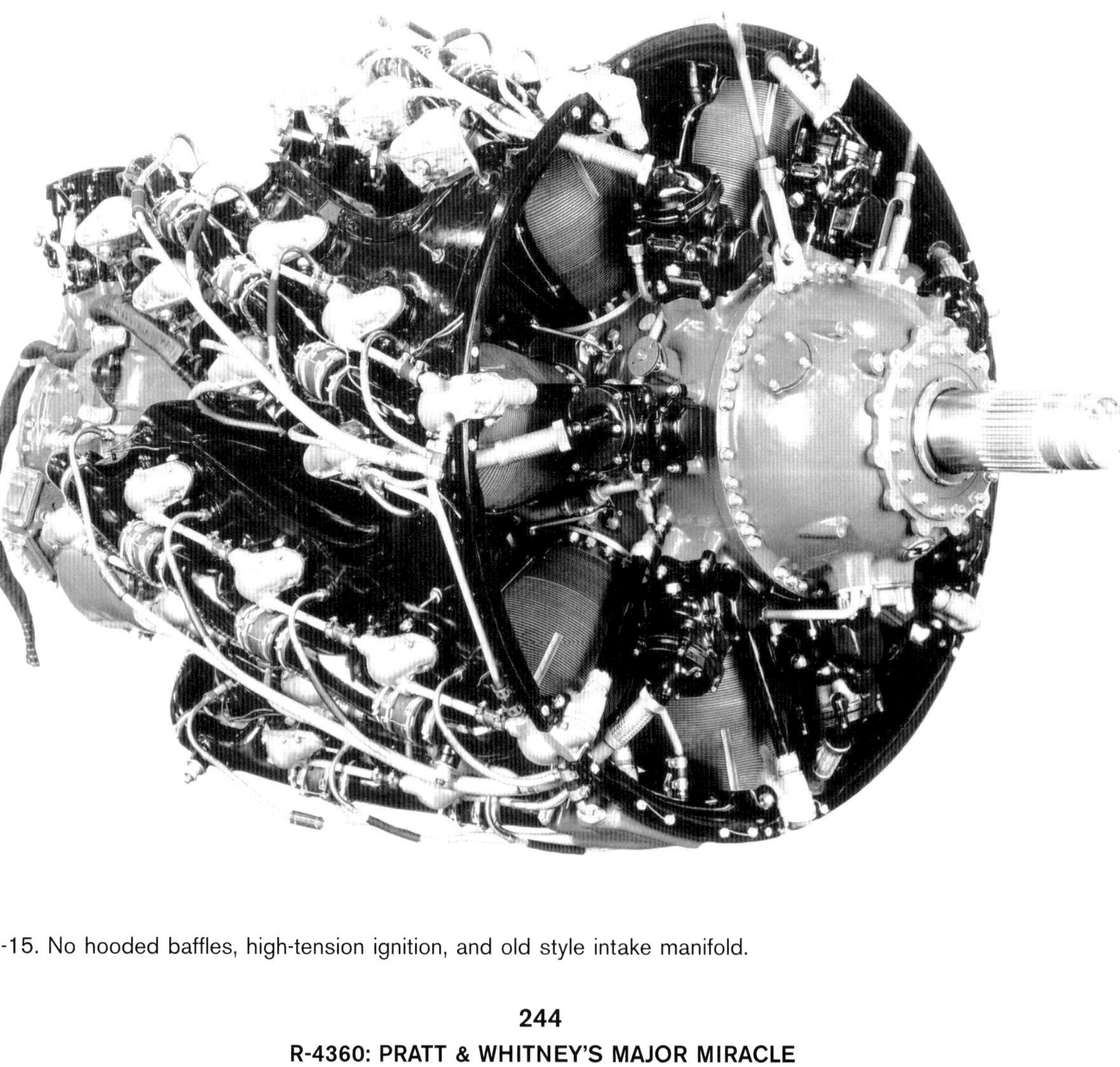

R-4360-15. No hooded baffles, high-tension ignition, and old style intake manifold.

(Table 5-31) (Ref. 5-3)

Engine Model	Type R-4360 P&W Aircraft **Air Force** *-20, -20A **Navy** *-20W, -20WA (Appendix A)	Type R-4360 P&W Aircraft **Air Force** -20B, -20C, -20WB, **Navy** -20WC	Type R-4360 P&W Aircraft **Air Force** -21 (Production Navy version of R-4360-11)
Specification Number	N-7056-F		N-7032, Appendix D
Engine Series	B Series	B Series	B Series
Rating; Take-off	3500 hp @ 2700 rpm @ 750 feet wet (-20W) 3250 hp @ 2700 rpm @ sea level dry	3500 hp @ 2700 rpm @ 750 feet wet (-20W) 3250 hp @ 2700 rpm @ sea level dry	3000 hp @ 2700 rpm
Rating; Military	3500 hp @ 2700 rpm @ 750 feet wet (-20W) 3250 hp @ 2700 rpm @ 1500 feet dry 2500 hp @ 2700 rpm @ 17000 feet dry	3500 hp @ 2700 rpm @ 750 feet wet (-20W) 3250 hp @ 2700 rpm @ 1500 feet dry 2500 hp @ 2700 rpm @ 17000 feet dry	3000 hp @ 2700 rpm @ 1500 feet without turbosupercharger 3000 hp @ 2700 rpm @ 40000 feet with turbosupercharger
Rating; Normal	2650 hp @ 2550 rpm @ 6000 feet 2300 hp @2550 rpm @ 18000 feet	2650 hp @ 2550 rpm @ 6000 feet 2300 hp @2550 rpm @ 18000 feet	2500 hp @ 2550 rpm @ 5000 feet without turbosupercharger 2500 hp @ 2550 rpm @ 40000 feet with turbosupercharger
Max. Continuous			
Cruise			
Fuel Grade	115/145 PN		100/130 PN
Curves	T-1013 Inst. #7264 not in specification. T-1013 in specification		T-869 Inst. #1691
Weight, dry	**3540 lbs		*3308 lbs
Prop. Reduction Ratio	0.425:1		0.381:1 (remote drive)
Prop. Shaft Spline	SAE #60-A		SAE #60- and SAE #80 dual rotation
Compression Ratio	6.7:1		7.0:1
Blower Ratio(s)	6.95:1 low blower 9.07:1 high blower		6.08:1
Carburetor	Bendix PR-100-B4		Bendix PR-100-B2-3
Magnetos	D4RN-2		D4RN-2
Installation Drwg. No.	99001		79201
Dimensions	Diameter: 54.00 inches Length: 102.00 inches		Diameter: 52.50 inches Length: 87.00 inches for direct drive engine Length: 326.40 inches from engine centerline to propeller thrust nut face
A.T.C. number			
Number Manufactured	275....................-20 583..................-20W 2,053.................-20WA		28
Applications	Douglas C-124A (-A wet) Fairchild R4Q-1 (-A wet) Fairchild C-119B (-A wet) Fairchild C-119C (-A wet) Fairchild C-120 Fairchild XC-120 (-A, dry, i.e., no ADI) ***Martin P4M-1 (-A, dry, i.e., no ADI) Martin XP4M-1 (-A, dry, i.e., no ADI)		Northrop B-35 (Inboard engine) Northrop XB-35 (Inboard engine) Northrop YB-35 (Inboard engine)
Notes	Single-stage, variable speed. *"A" denotes long connecting rods. *"W" denotes wet (ADI) rating. **Weight increases: Torquemeter: 20 lbs (-20A & -20WA) ADI equipment: 13 lbs (-20WA) Fire seal diaphragm 8.4 lbs (-20A & -20WA) ***With GE I-40 jets	Similar to -20, -20A, -20W, -20WA except no provision for hydraulic couplings and no power connection for Navy.	Single-stage, single speed supercharger. Suitable for turbosupercharger. *Additional Weights: Extension shaft and coupling asm: 830 lbs Reduction gear housing asm. and power takeoff, propeller governor drives, and torquemeter: 556 lbs

MODEL COMPARISON OF THE -20 AND -35 ENGINES

(Table 5-32)

FEATURE	-20W	-20WA	-20WB	-20WC	-35	-35B
Reduction Gearing	0.425:1	0.425:1	0.425:1	0.425:1	0.375:1	0.375:1
Long Rods	No	Yes	No	Yes	No	No
Spacer Case	Yes	Yes	Yes	Yes	No	No
Supercharger Drive	Fluid	Fluid	Rigid	Rigid	Rigid	Rigid
Water Regulator	Automatic	Automatic	Automatic	Automatic	Automatic	Automatic
Automatic Engine Control	No	No	No	No	Yes	Yes

(Table 5-33) (Ref. 5-3)

Engine Model	Type R-4360 P&W Aircraft Air Force **Navy** ***-22W**	Type R-4360 P&W Aircraft Air Force **Navy** **-24**	Type R-4360 P&W Aircraft **Air Force** **-25 (Production version of R-4360-5)** Navy
Specification Number	N-7066	N-7039-C, Appendix B	N-7035, Appendix F
Engine Series	B Series	B Series	B Series
Rating; Take-off	3500 hp @ 2700 rpm @ sea level wet 3250 hp @ 2700 rpm @ sea level dry	3000 hp @ 2700 rpm	3000 hp @ 2700 rpm
Rating; Military	3500 hp @ 2700 rpm @ 500 feet dry	3000 hp @ 2700 rpm @ 1500 feet 2500 hp @ 2700 rpm @ 13500 feet	3000 hp @ 2700 rpm @ 1500 feet without turbosupercharger 3000 hp @ 2700 rpm @ 40000 feet with turbosupercharger
Rating; Normal	2650 hp @ 2550 rpm @ 5500 feet	2500 hp @ 2550 rpm @ 5000 feet 2500 hp @ 2550 rpm @ 14500 feet	2500 hp @ 2550 rpm @ 5000 feet without turbosupercharger 2500 hp @ 2550 rpm @ 40000 feet with turbosupercharger
Max. Continuous			
Cruise			
Fuel Grade	115/145 PN	100/130 PN	100/130 PN
Curves	T-1052 Inst. #7338	T-902	T-869 Inst. #1691
Weight, dry	*3490 lbs	*3411 lbs	*3483 lbs
Prop. Reduction Ratio	0.375:1	0.375:1	0.381:1
Prop. Shaft Spline	SAE #60-A	SAE #60-A	SAE #60-A
Compression Ratio	6.7:1	7.0:1	7.0:1
Blower Ratio(s)	6.375:1 low blower	7.52:1 maximum	6.06:1
Carburetor	Bendix PR-100-B3	Chandler Evans CECO 100-CPB7-1	Bendix PR-100-B2-3
Magnetos	D4RN-2	D4RN-2	D4RN-2
Installation Drwg. No.	125901	86301	R-80901
Dimensions	Diameter: 54.00 inches Length: 96.75 inches	Diameter: 52.50 inches Length: 96.75 inches	Diameter: 52.50 inches Length: 109.50 inches
A.T.C. number			
Number Manufactured	24		221
Applications	**Lockheed R6V-1 (Constitution)	Martin JRM-2	**Convair B-36A Model 37 ***Convair XC-99 Model 37
Notes	*Weight increases: Fire seal diaphragm....9.7 lbs Three-way adapter on #5 accessory pad...14 lbs *Weight includes torquemeter. **Only two aircraft manufactured: #1 flown late 1946 #2 flown June 1948 **R6V-1 was originally powered by R-4360-18s. R-4360-22W similar to R-4360-35 except axial power take-off and three-way adapter on #5 accessory pad. See R-4360-18 for turbo information.	Single-stage, variable speed supercharger *Weight increases: Torquemeter: 20 lbs *Weight includes manifold pressure regulator. Similar to and converted from R-4360-4 except R-4360-24 incorporates .375:1 reduction gearing. Sold separately. None manufactured.	*Includes torquemeter. **First flight; August 8, 1946, at Convair's facility in Fort Worth, Texas. ***First flight; November 23, 1947. Single stage, single speed supercharger. Suitable for turbosupercharging. Provision for cooling fan on accessory end. Power take-off co-axial with fan drive.

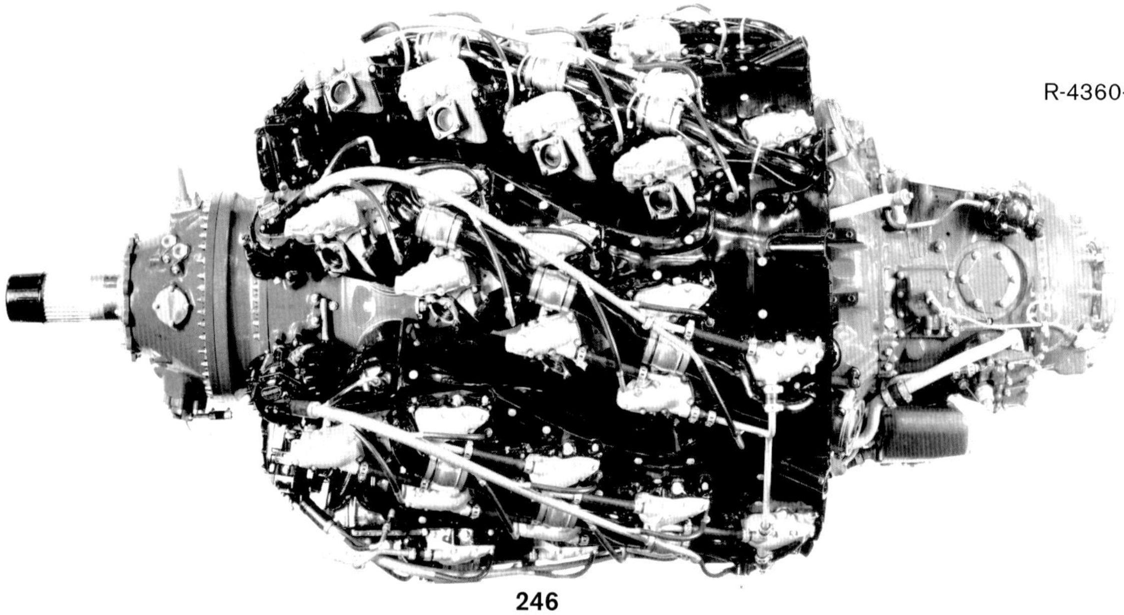

R-4360-25

R-4360-15

R-4360-15

R-4360-25

R-4360-25

(Table 5-34) (Ref. 5-3)

Engine Model	Type R-4360 P&W Aircraft **Air Force** -27 (Production version of R-4360-15 Navy	Type R-4360 P&W Aircraft **Air Force** -29 (semi-production) Navy	Type R-4360 P&W Aircraft **Air Force** ****-31 (Production version) Navy (No semi-production)
Specification Number	A-7036-D	N-7028, Appendix B	A-7040-D
Engine Series	B Series	B Series	B Series
Rating; Take-off	3500 hp @ 2700 rpm	3000 hp @ 2700 rpm	3000 hp @ 2700 rpm
Rating; Military	3000 hp @ 2700 rpm @ sea level to 1500 feet 2400 hp @ 2700rpm @ 13500 feet	3000 hp @ 2700 rpm @ 1500 feet 2400 hp @ 2700 rpm @ 25000 feet	3000 hp @ 2700 rpm @ 1500 feet without turbosupercharger 3000 hp @ 2700 rpm @ 40000 feet with turbosupercharger
Rating; Normal	2500 hp @ 2550 rpm @ sea level to 5000 feet 2500 hp @ 2550 rpm @ 14500 feet	2500 hp @ 2550 rpm @ 5000 feet	2500 hp @ 2550 rpm @ 5000 feet without turbosupercharger 2500 hp @ 2550 rpm @ 40000 feet with turbosupercharger
Max. Continuous			
Cruise			
Fuel Grade	100/130PN	100/130 PN	100/130 PN
Curves	T-886 Inst. #1694	T-812	T-869 Inst. #1691
Weight, dry	*3404 lbs	3585 lbs	*3506 lbs
Prop. Reduction Ratio	0.381:1	0.381:1	0.381:1
Prop. Shaft Spline	SAE #60-A	SAE #60-A	SAE #60- and SAE #80 dual rotation
Compression Ratio	7.0:1	7.0:1	7.0:1
Blower Ratio(s)	7.52:1 maximum	6.06:1 maximum 5.75:1 minimum	6.06:1
Carburetor	Bendix PR-100-B3-3	Bendix PR-100-A3-1	Bendix PR-100-B3-3
Magnetos	D-4RN-2	DF-4RN-1	D-4RN-2
Installation Drwg. No.	81001	57448	89601
Dimensions	Diameter: 52.50 inches Length: 96.75 inches	Diameter: 52.50 inches Length: 116.50 inches	Diameter: 52.50 inches Length: 114.25 inches
A.T.C. number			
Number Manufactured	84		29
Applications	**Douglas C-74 (DC-7) **Douglas XC-74 (DC-7)	Planned for XB-44, re-engined version of B-29	**Hughes XF-11 ***Republic XF-12
Notes	Single-stage, variable speed supercharger. *Weight increases: Torquemeter: 20 lbs *Weight includes manifold pressure regulator. ** R-4360-27 engines later replaced by R-4360-49s.	Two-stage, variable-speed supercharger. None manufactured. Engine cancelled in favor of R-4360-33.	*Weight increases: Torquemeter...20 lbs **First aircraft crashed on first flight at Culver City, California, July 7, 1946, with Howard Hughes at controls. Second aircraft flew April 5, 1947, again with Hughes at the controls. ***First flew February 4, 1946, at Republic's facility at Farmingdale, N.Y. ****Changed to R-4360-37 single rotation by Army. Single stage, single speed supercharger. Suitable for turbosupercharging.

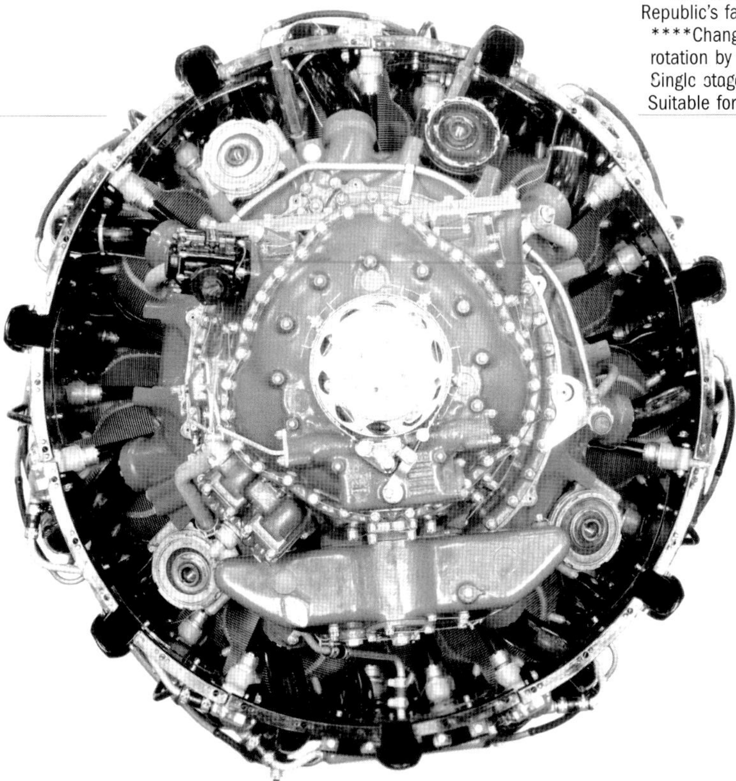

R-4360-25. This rear shot offers a nice view of the fan brake mechanism.

R-4360-25. The ring of studs around the fan brake is for attaching the cooling fan. The bulged out rear cover houses the fluid couplings that drive the fan.

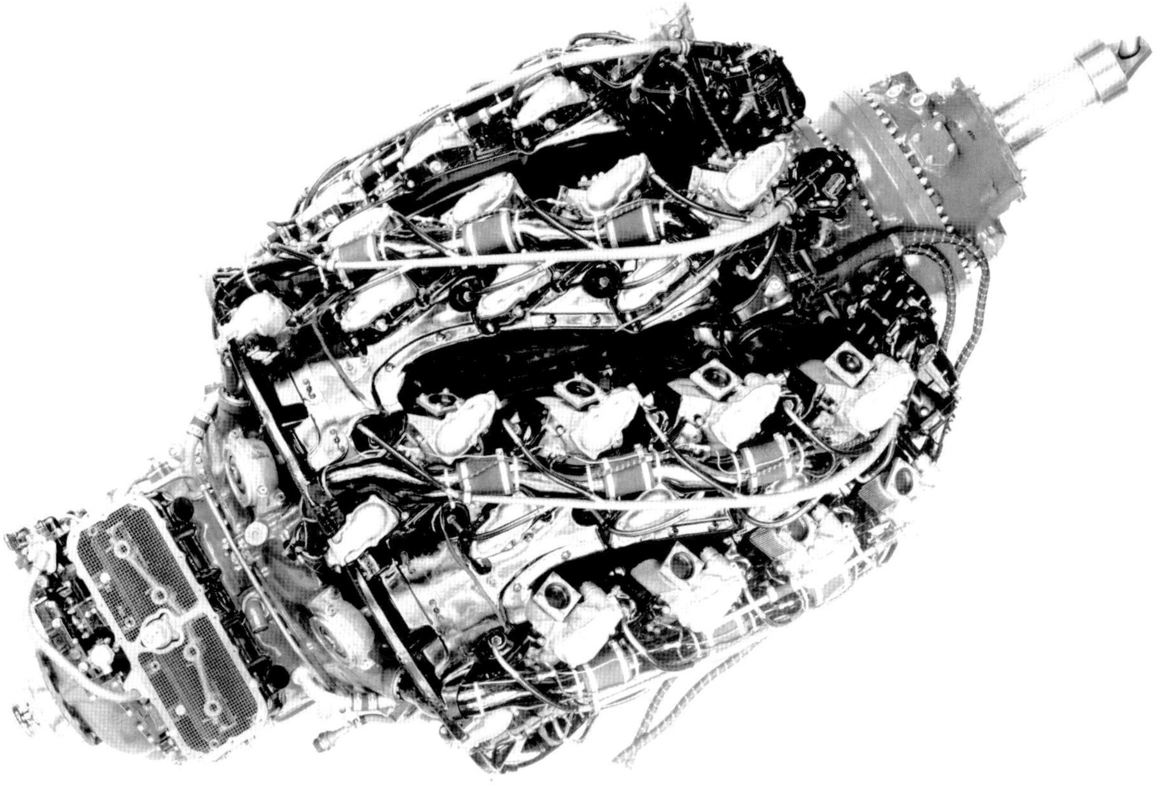

R-4360-25. Top view.

R-4360-25. Bottom view.

Montage of views continued on next three pages showing the R-4360-27. This 3,000-hp variant powered the Douglas C-74. It still retains the early high-tension ignition, rubber hose intake manifold joints, and no hooded baffles–yet.

(Table 5-35) (Ref. 5-3)

Engine Model	Type R-4360 P&W Aircraft **Air Force -33 (Production version** Navy **of R-4360-29**	Type R-4360 P&W Aircraft **Air Force **-35, -35B** Navy	Type R-4360 P&W Aircraft **Air Force *-35A, -35C** Navy
Specification Number	A-7042-A	A-7051-F (Applies to R-4360-35)	A-7051-F (Applies to R-4360-35A)
Engine Series	B Series	B Series	B Series
Rating; Take-off	3000 hp @ 2700 rpm	3500 hp @ 2700 rpm wet 3250 hp @ 2700 rpm dry	3500 hp @ 2700 rpm wet 3250 hp @ 2700 rpm dry
Rating; Military	3000 hp @ 2700 rpm @ 1500 feet 2400 hp @ 2700 rpm @ 25000 feet	3500 hp @ 2700 rpm @ 500 feet wet	3500 hp @ 2700 rpm @ 500 feet wet
Rating; Normal	2500 hp @ 2550 rpm @ 5000 feet 2200 hp @ 2550 rpm @ 25000 feet	2650 hp @ 2550 rpm @ 5500 feet	2650 hp @ 2550 rpm @ 5500 feet
Max. Continuous			
Cruise			
Fuel Grade	100/130 PN	115/145 PN	115/145 PN
Curves	T-951 Inst. #1721	T-983 Inst. #7223	T-983 Inst. #7223
Weight, dry	*3595 lbs	*3490 lbs	3490 lbs
Prop. Reduction Ratio	0.381:1	0.375:1	0.375:1
Prop. Shaft Spline	SAE #60-A	SAE #60-A	SAE #60-A
Compression Ratio	7.0:1	6.7:1	6.7:1
Blower Ratio(s)	6.06:1 maximum 6.375:1	6.375:1	5.75:1 minimum
Carburetor	Bendix PR-100-B3-3	Bendix PR-100-B3-4	Bendix PR-100-B3-4
Magnetos	D-4RN-2	D-4RN-2	D-4RN-2
Installation Drwg. No.	91101	96501	96501
Dimensions	Diameter: 52.50 inches Length: 109.25 inches	Diameter: 54.00 inches Length: 96.75 inches	Diameter: 54.00 inches Length: 96.75 inches
A.T.C. number			
Number Manufactured	6	-35: 1,931 -35B: 661	-35A: 382 -35C: 598
Applications	Boeing XB-44, converted B-29	Boeing TB-50A, D, H Boeing B-50A, B, D Fairchild XC-119A Republic XF-12 (Planned)	**Boeing C-97A, C **Boeing KC-97E Boeing YC-97A Boeing YC-97B Aero Spacelines B-377MG Mini Guppy Aero Spacelines B-377SG Super Guppy Aero Spacelines B-377PG Pregnant Guppy Douglas XC-124A
Notes	Two-stage, variable speed supercharger. *Weight increases: Torquemeter....20 lbs	*Weight increases: Manifold pressure regulator...12 lbs *Weight includes torquemeter. **R-4360-35B features long connecting rods.	Same as R-4360-35 and R-4360-35B except no manifold pressure regulator. *R-4360-35C features long connecting rods. **These aircraft converted to R-4360-65 power.

R-4360-33

Montage on this and next three pages of R-4360-33. This two-stage engine powered the Boeing XB-44–a B-29 converted to R-4360 power. It could be argued that the B-50 should have had this powerplant instead of the -35. Still, that's pure conjecture.

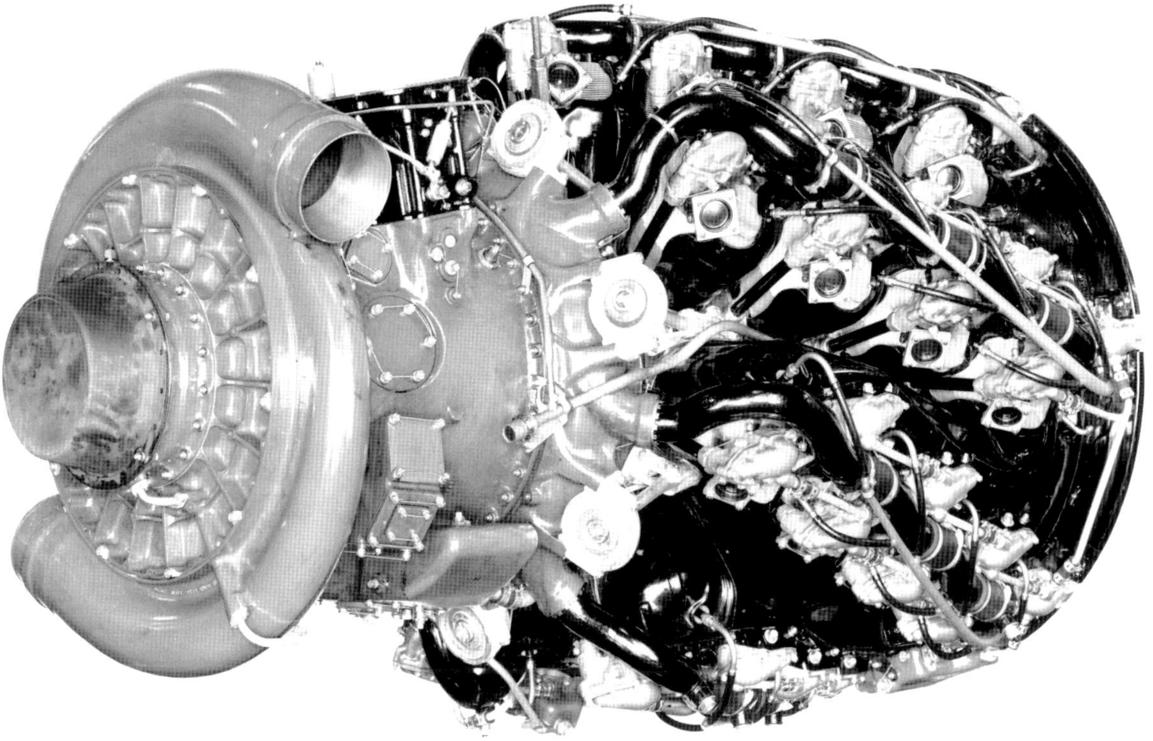

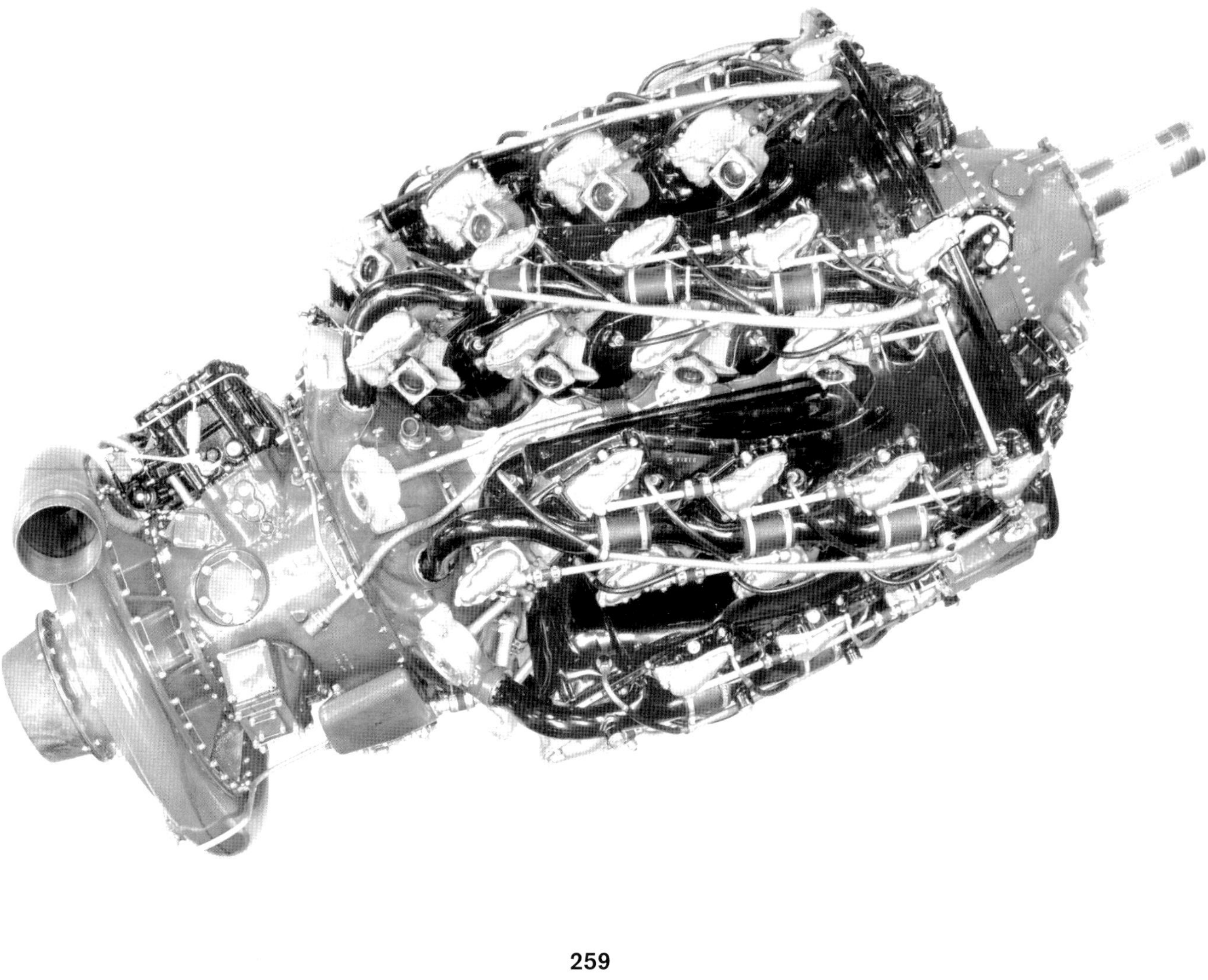

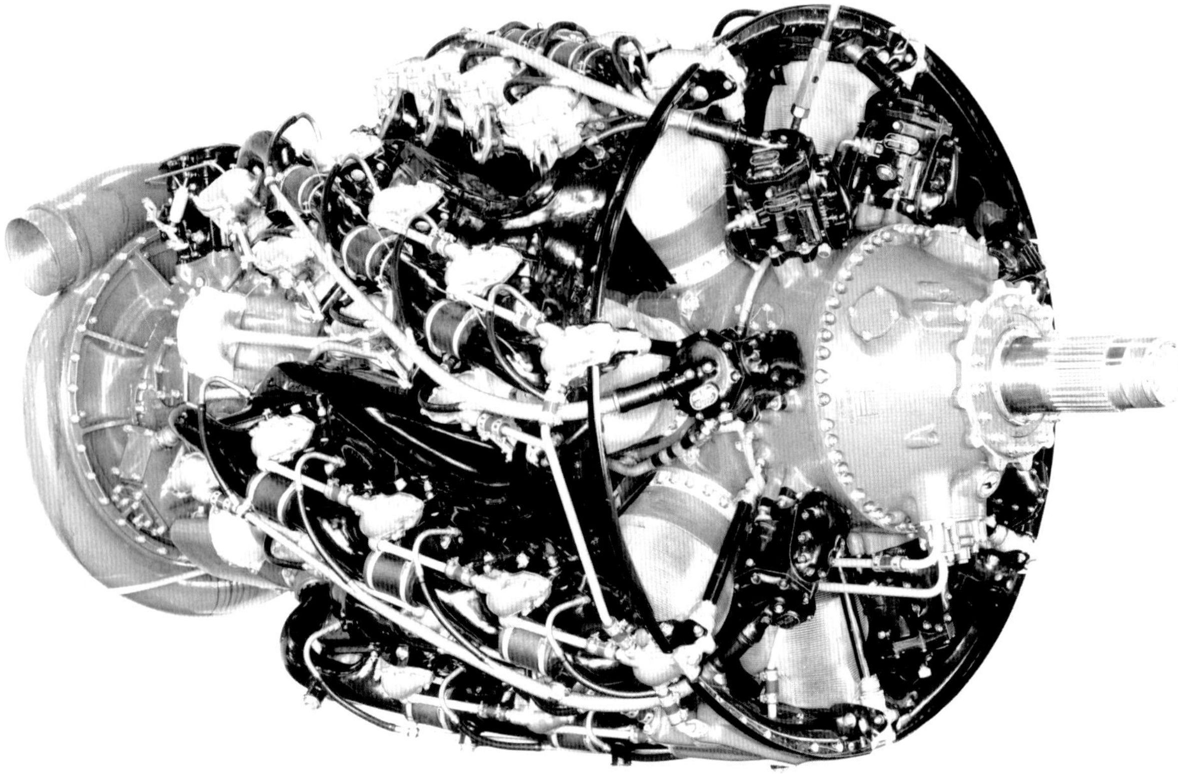

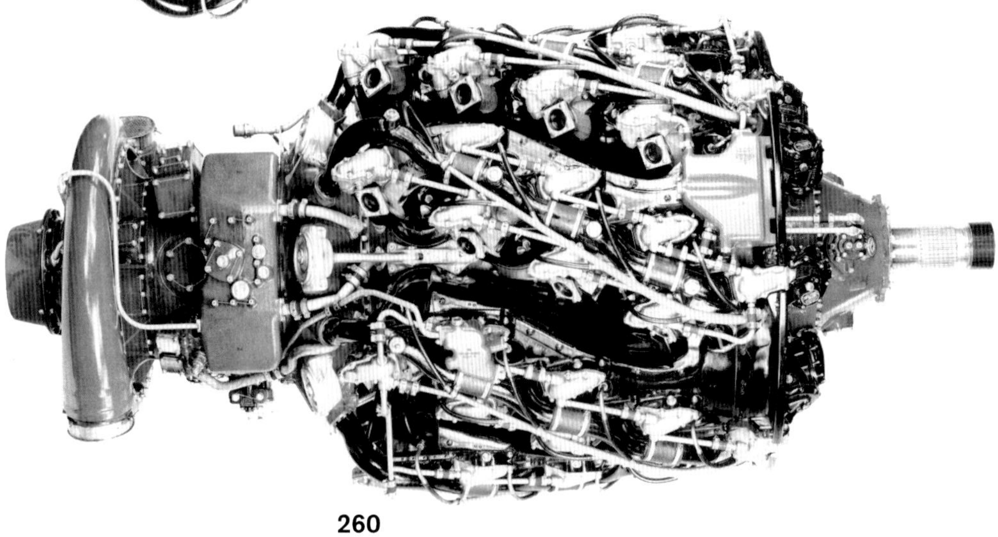

(Table 5-36) (Ref. 5-3)

Engine Model	**Type** **R-4360** P&W Aircraft **Air Force** ***-37** Navy **of R-4360-29**	**Type** **R-4360** P&W Aircraft **Air Force** **-39** Navy	**Type** **R-4360** P&W Aircraft **Air Force** **-41** Navy
Specification Number	A-7040-D	A-7055-A	A-7061-B
Engine Series	B Series	C Series	B Series
Rating; Take-off	3000 hp @ 2700 rpm	3800 hp @ 2800 rpm wet 3500 hp@ 2800 rpm dry	3000 hp @ 2700 rpm
Rating; Military	3000 hp @ 2700 rpm @ 1500 feet without turbosupercharger 3000 hp @ 2700 rpm @ 40000 feet with turbosupercharger	3800 hp @ 2800 rpm @ 29.4 in. Hg. back pressure wet 3500 hp @ 2800 rpm @ 29.9 in. Hg. back pressure dry	3000 hp @ 2700 rpm @ 1500 feet without turbosupercharger 3000 hp @ 2700 rpm @ 40000 feet with turbosupercharger
Rating; Normal	2500 hp @ 2550 rpm @ 5000 feet without turbosupercharger 2500 hp @ 2550 rpm @ 40000 feet with turbosupercharger	2800 hp @ 2600 rpm @ 27.0 in.Hg. back pressure	2500 hp @ 2550 rpm @ 5000 feet without turbosupercharger 2500 hp @ 2550 rpm @ 40000 feet with turbosupercharger
Max. Continuous			
Cruise			
Fuel Grade	100/130 PN	115/145 PN	100/130 PN
Curves	T-869 Inst. #1691		T-1038
Weight, dry	**3346 lbs	*3730 lbs	*3567 lbs
Prop. Reduction Ratio	0.425:1	0.375:1	0.375:1
Prop. Shaft Spline	SAE #60-A	SAE #60-A	SAE #60-A
Compression Ratio	7.0:1	6.7:1	6.7:1
Blower Ratio(s)	6.08:1		6.375:1
Carburetor	Bendix PR-100-B3	Bendix PR-100-28-A1	Bendix PR-100-B3
Magnetos	D-4RN-2		D-4RN-2
Installation Drwg. No.	89601	101601	117901
Dimensions	Diameter: 52.50 inches Length: 96.75 inches	Diameter: 52.50 inches Length: 102.00 inches	Diameter: 53.50 inches Length: 109.75 inches
A.T.C. number			
Number Manufactured			1
Applications	XF-11 Republic XF-12 Republic XR-12		Convair B-36B
Notes	*Sold as R-4360-31, no R-4360-37s None manufactured. **Includes torquemeter.	Single-stage, single speed supercharger *Includes torquemeter and fuel injection. 2 or 3 engines built for experimental purposes, later converted to R-4360-43 Additional weight: Engine mounts...125 lbs	Pusher configuration. Single-stage, single speed supercharger *Includes torquemeter and water injection (ADI) equipment. Incorporates R-4360-25 and R-4360-35 features. One experimental engine #P-164

Montage on next five pages of the R-4360-35. Progress in the development process is evident via the hooded baffles; a feature that kept ignition harnesses from frying. This B-50 engine was beset with turbosupercharger problems.

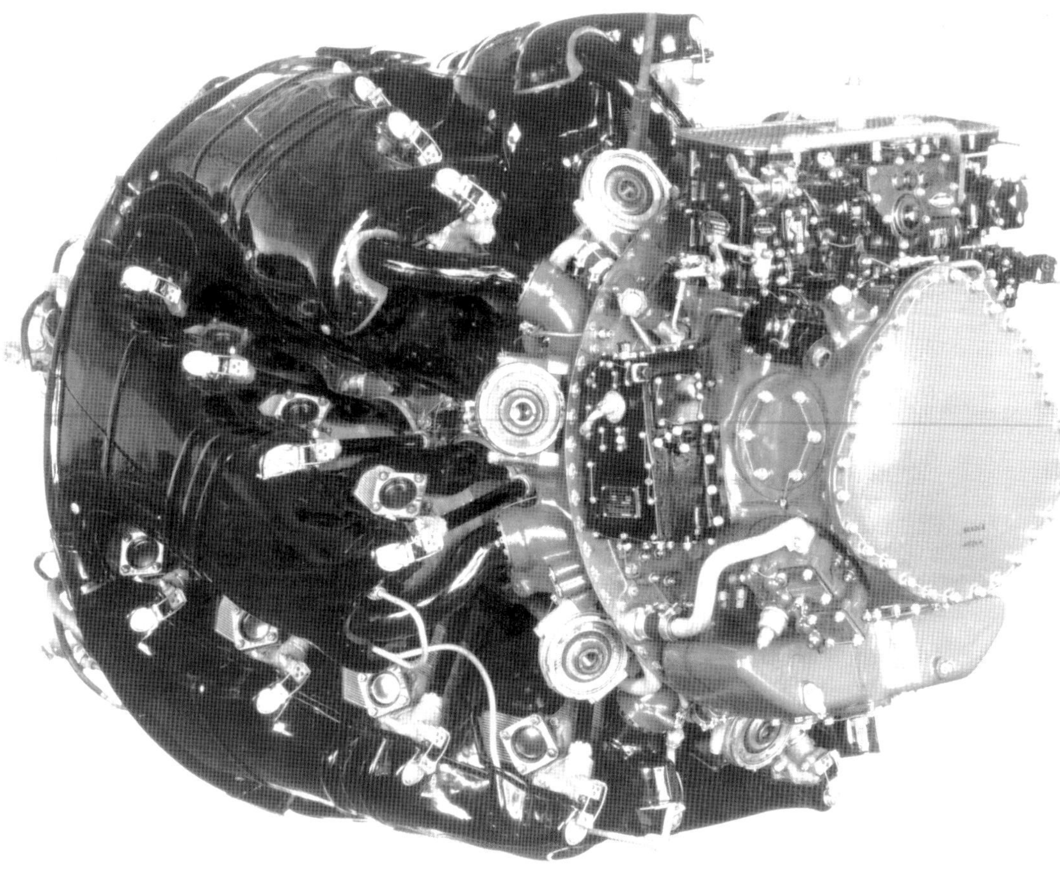

(Table 5-37) (Ref. 5-3)

Engine Model	Type R-4360 P&W Aircraft **Air Force** ****-4!, -41A** Navy	Type R-4360 P&W Aircraft **Air Force** **-43** Navy	Type R-4360 P&W Aircraft **Air Force** **-45** Navy
Specification Number	A-7063-D1 (R-4360-41)	A-7065-F	N-7032-D, Appendix B
Engine Series	B Series	C Series	B Series
Rating; Take-off	3500 hp @ 2700 rpm @ sea level wet 3250 hp @ 2000 rpm @ sea level dry	4300 hp @ 2800 rpm	3000 hp @ 2700 rpm
Rating; Military	3500 hp @ 2700 rpm @ 500 feet wet	4300 hp @ 2800 rpm	3000 hp @ 2700 rpm @ 1500 feet without turbosupercharger 3000 hp @ 2700 rpm @ 40000 feet with turbosupercharger
Rating; Normal	2650 hp @ 2550 rpm @ 5500 feet	3150 hp @ 2600 rpm	2500 hp @ 2550 rpm @ 5000 feet without turbosupercharger 2500 hp @ 2550 rpm @ 40000 feet with turbosupercharger
Max. Continuous			
Cruise			
Fuel Grade	115/145 PN	115/145 PN	100/130 PN
Curves	T-1039 Inst. #7311	T-1088	T-869 Inst. #1691
Weight, dry	*3567 lbs	*3720 lbs	*3308 lbs
Prop. Reduction Ratio	0.375:1	0.375:1	0.425:1 remote drive
Prop. Shaft Spline	SAE #60-A	SAE #60-A	SAE #60-A
Compression Ratio	6.7:1	6.7:1	7.0:1
Blower Ratio(s)	6.375:1	External turbosupercharger, no engine driven blower.	6.08:1
Carburetor	Bendix PR-100-B3	Bendix 100-28-A2 (prototype) Bendix 100-28-A3 (production)	Bendix PR-100-B2-3
Magnetos	D-4RN-2	High or low tension ignition	D4RN-2
Installation Drwg. No.	117901	121001 (prototype) 136601 (production)	121901
Dimensions	Diameter: 54.00 inches Length: 109.75 inches	Diameter: 55.00 inches Length: 103.50 inches	Diameter: 52.50 inches Length: 87.00 inches for direct drive engine. Length: 213.85 inches from engine centerline to propeller thrust nut face.
A.T.C. number			
Number Manufactured	-41......................947 -41A...................269	8	13
Applications	Convair B-36B, D, E Convair RB-36 Convair XC-99	**Boeing YB-50C **Boeing B-54A **Boeing RB-54A	Northrop B-35 (Outboard engine) Northrop XB-35 (Outboard engine) Northrop YB-35 (Outboard engine)
Notes	Pusher configuration. Single-stage, single-speed supercharger. *Includes torquemeter and water injection (ADI) equipment. Similar to Specification Number A-7061B-41 except for increased ratings. **R-4360-41A has longer connecting rods.	Tractor VDT type configuration for use with GE CHM-2 turbosupercharger. *Additional weights: Engine mounts: 105 lbs Boost control system: 50 lbs Exhaust system: 294 lbs Fuel filter parts: 6 lbs *Includes torquemeter. **These aircraft cancelled along with all other VDT projects.	*Additional weights: Remote assembly complete with reduction gear housing asm. and power take-off, propeller governor drives, and torquemeter...376 lbs Similar to R-4360-17 except for single rotation and 0.425:1 reduction gearing.

(Table 5-38) (Ref. 5-3)

Engine Model	Type R-4360 P&W Aircraft **Air Force** **-47** Navy	Type R-4360 P&W Aircraft **Air Force** ***-49, -49A** Navy	Type R-4360 P&W Aircraft **Air Force** **-51** Navy
Specification Number	N-7032-D, Appendix C	None	A-7067C
Engine Series	B Series	B Series	C Series
Rating; Take-off	3000 hp @ 2700 rpm	3500 hp @ 2700 rpm wet 3250 hp @ 2700 rpm dry	*4300 hp @ 2800 rpm wet 4000 hp @ 2800 rpm dry
Rating; Military	3000 hp @ 2700 rpm @ 1500 feet without turbosupercharger 3000 hp @ 2700 rpm @ 40000 feet with turbosupercharger	3500 hp @ 2700 rpm @ 500 feet wet	4300 hp @ 2800 rpm wet 4000 hp @ 2800 rpm dry
Rating; Normal	2500 hp @ 2550 rpm @ 5000 feet without turbosupercharger 2500 hp @ 2550 rpm @ 40000 feet with turbosupercharger	2650 hp @ 2550 rpm @ 5500 feet	3100 hp @ 2600 rpm
Max. Continuous			
Cruise			

continued on page 267

Three views of the R-4360-43, one of the VDT variants. This fascinating engine almost made it into production before cancellation.

Table 38 continued from page 265

Fuel Grade	100/130 PN		115/145 PN
Curves	T-869 Inst. #1691	T-983	Not released Inst. #7223
Weight, dry	*3308 lbs	3490 lbs	**4020 lbs
Prop. Reduction Ratio	0.425:1 remote drive	0.375:1	0.3125:1
Prop. Shaft Spline	SAE #60-A	SAE #60-A	SAE #70
Compression Ratio	7.0:1	6.7:1	6.7:1
Blower Ratio(s)	6.08:1	6.375:1	***External turbosupercharger, no engine driven blower.
Carburetor	Bendix PR-100-B2-3	Bendix PR-100-B3-6	Bendix 100-28-A3 (production)
Magnetos	D-4RN-2	D-4RN-2	Low tension or alternatively high tension ignition.
Installation Drwg. No.	121901	96501	126001
Dimensions	Diameter: 52.50 inches Length: 87.00 inches for direct drive engine Length: 326.40 inches from engine centerline to propeller thrust nut face	Diameter: 54.00 inches Length: 96.75 inches	Diameter: 55.00 inches Length: 36.00 inches (reduction gear unit) Length: 91.00 inches (direct drive)
A.T.C. number			
Number Manufactured	13		
Applications	Northrop B-35 (Inboard engine) Northrop XB-35 (Inboard engine) Northrop YB-35 (Inboard engine)	Douglas C-74	Convair B-36C study for tractor installation
Notes	*Additional weights: Remote assembly complete with reduction gear housing asm. and power take-off, propeller governor drives, and torquemeter..376 lbs Similar to R-4360-21 except for single rotation and .425:1 reduction gearing.	*Converted from R-4360-35 in field. Basically similar to R-4360-35 for C-74 installation except -49 incorporates 3-way adapter on #2 accessory pad and Bendix PR-100-B3-6 carburetor. R-4360-49A features long connecting rods.	*At critical altitude **Additional weight: Engine mounts without isolators: 94 lbs Exhaust system boost control: 50 lbs Extension shaft asm. "A": 380 lbs Extension shaft asm. "B": 297 lbs Extension shaft asm. "C": 266 lbs **Weight includes torquemeter. ***Tractor VDT type configuration for use with GE CHM-2 turbosupercharger. Similar to R-4360-43 except for remotely mounted reduction gear. None manufactured.

(Table 5-39) (Ref. 5-3)

Engine Model	**Type** **R-4360** P&W Aircraft **Air Force** ****-53** Navy	**Type** **R-4360** P&W Aircraft **Air Force** **-55** Navy	**Type** **R-4360** P&W Aircraft **Air Force** **-57** Navy
Specification Number	A-7076J	A-7074-D	A-7075
Engine Series	C Series	C Series	C Series
Rating; Take-off	3800 hp @ 2800 rpm wet	4300 hp @ 2800 rpm	4300 hp @ 2800 rpm
Rating; Military	3800 hp @ 2800 rpm wet 3500 hp @ 2800 rpm dry	4300 hp @ 2800 rpm	4300 hp @ 2800 rpm
Rating; Normal	2800 hp @ 2600 rpm	3150 hp @ 2600 rpm	3500 hp @ 2600 rpm
Max. Continuous			
Cruise			
Fuel Grade	115/145 PN	115/145 PN	115/145 PN
Curves	T-1124 Inst. #14882 (not in T-1124 spec.)	T-1132	
Weight, dry	*4040 lbs	*3892 lbs	*3892 lbs
Prop. Reduction Ratio	0.375:1	0.375:1	0.375:1
Prop. Shaft Spline	SAE #70-	SAE #70	SAE #70
Compression Ratio	6.7:1	6.7:1	7.5:1
Blower Ratio(s)	6.6:1	External turbosupercharger, no engine driven blower.	External turbosupercharger, no engine driven blower.
Carburetor	Bendix PR-100-28-A5	Bendix 100-28-A2 (prototype) Bendix 100-28-A3 (production)	Bendix 100-28-A3
Magnetos	S14RN-15 (low tension)	Low tension ignition	Low tension ignition
Installation Drwg. No.	154101	151001	148001
Dimensions	Diameter: *55.00 inches Length: 117.00 inches	Diameter: 55.00 inches (does not include exhaust system) Length: 142.50 inches	Diameter: 55.00 inches (does not include exhaust system) Length: 140.251 inches
A.T.C. number			
Number Manufactured	2,240	16	
Applications	Convair B-36D, E, F, H, J Convair RB-36D, E, F, H		
Notes	Pusher configuration. Single-stage, single speed *Includes deflector torquemeter, ignition system, priming system, Bendix K1 water (ADI) regulator. Additional weight: Exhaust system including collector ring: 310 lbs Engine mount (no vibration isolators): 117 lbs Similar to R-4360-41 except for "C" series features. Also manufactured by Ford Motor Company.	Similar to R-4360-43. Pusher VDT type configuration for use with two GE CHM-3s operating in parallel and a CH-9 turbosupercharger in series with the CHM-3s. Provision for driving remotely mounted cooling fan and power take-off *Additional weight: Engine mounts (no isolators): 112 lbs Boost control system: 80 lbs Exhaust system including collector ring: 300 lbs Remote gearbox and power take-off: 200 lbs *Includes torquemeter and fuel injection equipment.	Similar to R-4360-55 except high (7.5:1) compression and ratings. Pusher VDT type configuration for use with turbosupercharger. Provision for driving remotely mounted cooling fan and power take-off *Additional weight: Engine mounts (no isolators): 117 lbs Boost control system: 55 lbs Exhaust system including collector ring: 310 lbs Remote gearbox and power take-off: 220 lbs *Includes torquemeter and fuel injection equipment. None manufactured.

With this version, Pratt & Whitney had reached the definitive production version; fuel injection, hooded baffles, low-tension ignition, and rated at 3,800 hp. R-4360-53s powered B-36s.

R-4360-53

R-4360-53

R-4360-53

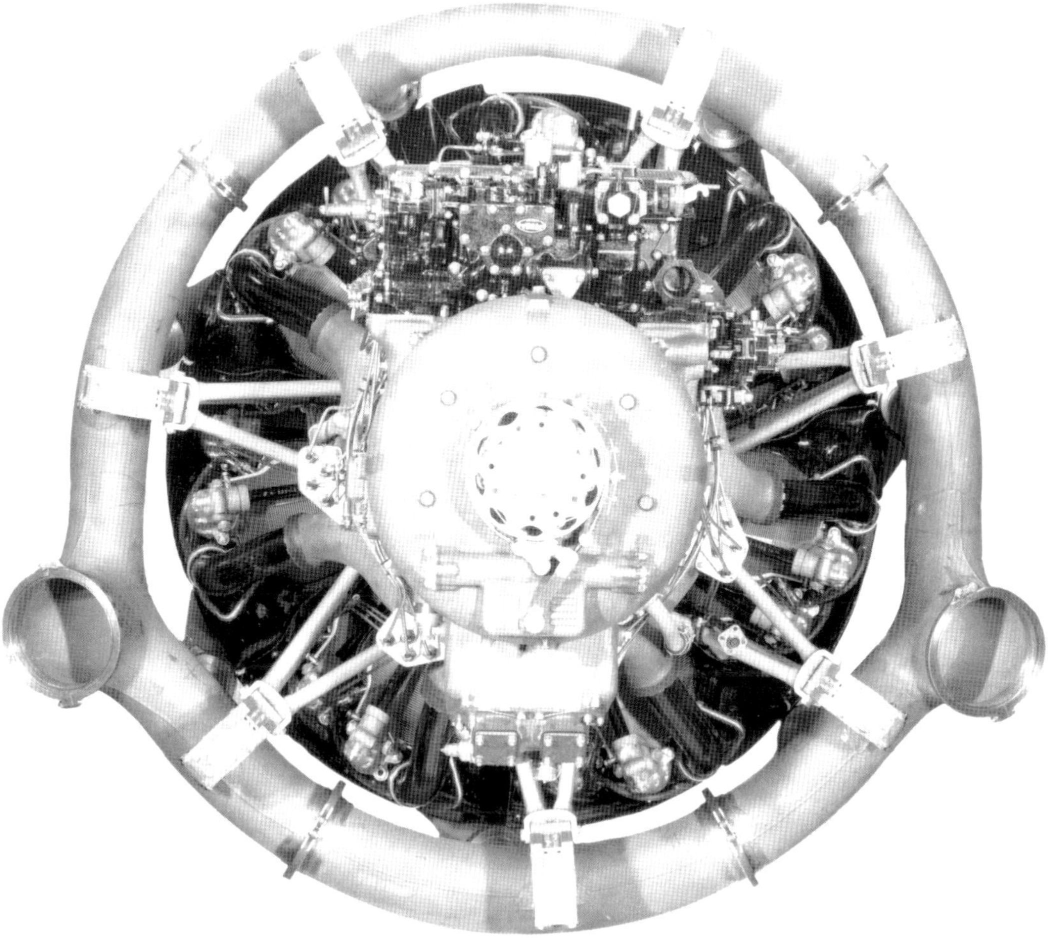

R-4360-53

R-4360-57. Intended for the B-36C, this 4,300-hp VDT engine never entered production.

R-4360-57. The device off to the left of this photograph is a remotely mounted cooling fan drive.

R-4360-57 fan drive close up.

(Table 5-40) (Ref. 5-3)

Engine Model	Type R-4360 P&W Aircraft **Air Force** **-59 Navy	Type R-4360 P&W Aircraft **Air Force** -59 Navy	Type R-4360 P&W Aircraft **Air Force** *-59A Navy
Specification Number	A-7080	A-7091-E	A-7096
Engine Series	CB Series	CB Series	CB Series
Rating; Take-off	3800 hp @ 2800 rpm @ sea level wet	3500 hp @ 2700 rpm wet 3500 hp @ 2800 rpm wet. Alternative	3500 hp @ 2700 rpm wet 3500 hp @ 2800 rpm wet. Alternative
Rating; Military		3500 hp @ 2700 rpm 3250 hp @ 2700 rpm dry	3500 hp @ 2700 rpm 3250 hp @ 2700 rpm dry
Rating; Normal	2800 hp @ 2600 rpm @ 5000 feet 2300 hp @ 2600 rpm @ 18000 feet	2650 hp @ 2550 rpm	2650 hp @ 2550 rpm
Max. Continuous			
Cruise			
Fuel Grade	115/145 PN	115/145 PN	115/145 PN
Curves		T-1145	T-1145
Weight, dry	*No weight was ever established for this engine	*3689 lbs	
Prop. Reduction Ratio	0.375:1	0.375:1	0.375:1
Prop. Shaft Spline	SAE #70	SAE #60-A	SAE #60-A
Compression Ratio	6.7:1	6.7:1	6.7:1
Blower Ratio(s)		6.375:1	6.375:1
Carburetor	Bendix 100-28-A3	Chandler Evans CECO 100CPB-9	Chandler Evans CECO 100CPB-9
Magnetos		Scintilla S14RN-15 (Low tension)	Scintilla S14RN-15 (Low tension)
Installation Drwg. No.	155501	172701	
Dimensions	Diameter: 55.00 inches Length: 102.00 inches	Diameter: 55.00 inches Length: 96.50 inches	Diameter: 55.00 inches Length: 96.50 inches
A.T.C. number			
Number Manufactured		8	
Applications	Douglas C-124 study	Boeing C-97D Aero Spacelines B-377MG Mini Guppy Aero Spacelines B-377SG Super Guppy Aero Spacelines B-377PG Pregnant Guppy	
Notes	*Includes fuel injection equipment and torquemeter **This engine was based on Wasp Major C14 but dropped by AMC for basic specification A-7091 around Wasp Major CB	*Additional weight: Exhaust system with collector ring: 35 lbs Engine mounts: 110 lbs Fluid power pump adapter (optional): 5 lbs *Weight includes torquemeter. "C" series nose section and power section. "B" series rear section. Also manufactured by Ford Motor Company.	*Redesignated -61. Similar to R-4360-59 except -59A has manifold pressure regulator.

Montage of the R-4360-59. This is probably the most common of the surviving R-4360s. Has all the later developments including hooded baffles, low-tension ignition, and rated at 3,500 hp. -59s powered Boeing KC-97s.

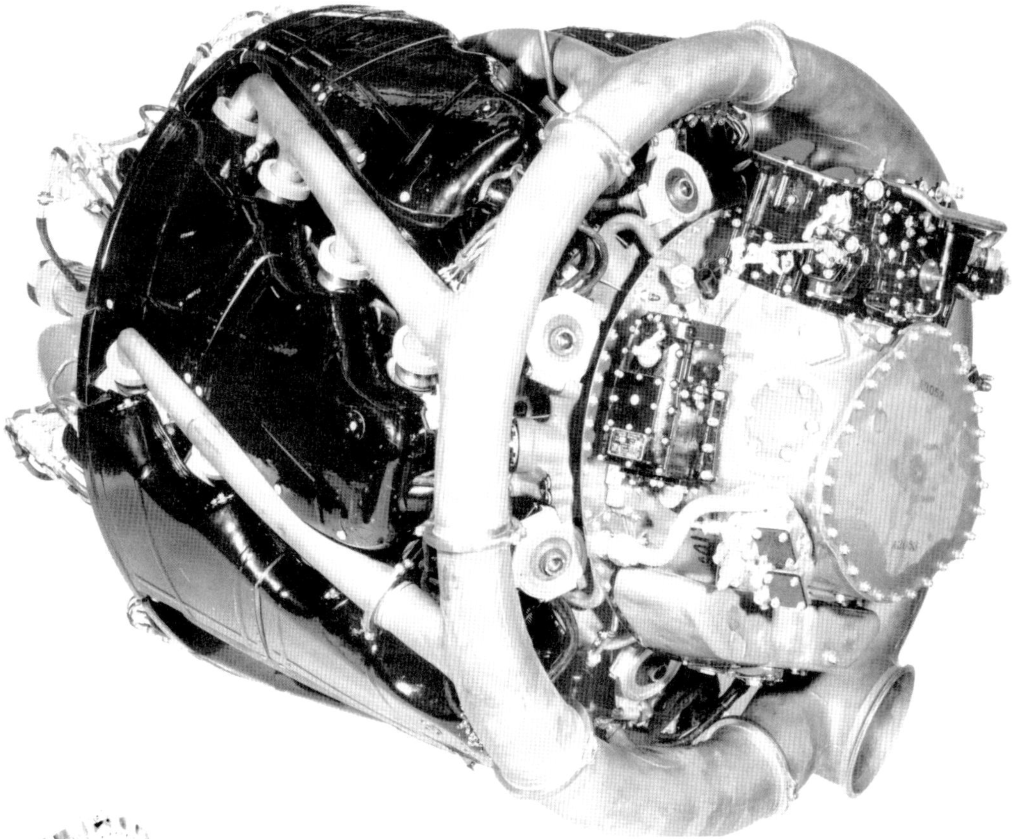

(Table 5-41) (Ref. 5-3)

Engine Model	**Type** R-4360 P&W Aircraft **Air Force** *-59B Navy	**Type** R-4360 P&W Aircraft **Air Force** *-61 Navy	**Type** R-4360 P&W Aircraft **Air Force** **-63 Navy
Specification Number	A-7091-G	A-7096-G	A-7092-G
Engine Series	CB Series	CB Series	CB Series
Rating; Take-off	3500 hp @ 2700 rpm wet 3500 hp @ 2800 rpm wet. Alternate	3500 hp @ 2700 rpm wet 3500 hp @ 2800 rpm wet. Alternate	3800 hp @ 2800 rpm @ sea level wet 3800 hp @ 2800 rpm @ 2000 feet wet 3400 hp @ 2800 rpm @ 2000 feet dry
Rating; Military	3500 hp @ 2700 rpm wet 3250 hp @ 2700 rpm dry	3500 hp @ 2700 rpm wet 3250 hp @ 2700 rpm dry	3800 hp @ 2800 rpm @ sea level wet 3400 hp @ 2800 rpm @ 2000 feet dry 2450 hp @ 2550 rpm @ 15000 feet dry
Rating; Normal	2650 hp @ 2550 rpm	2650 hp @ 2550 rpm	2800 hp @ 2600 rpm @ 5000 feet 2450 hp @ 2550 rpm @ 15000 feet
Max. Continuous			
Cruise			
Fuel Grade	115/145 PN	115/145 PN	115/145 PN
Curves	T-1145 Inst. #16987	T-1145	T-1146 Inst. #16770
Weight, dry	**3689 lbs	**3701 lbs	*3811 lbs
Prop. Reduction Ratio	0.375:1	0.375:1	0.375:1
Prop. Shaft Spline	SAE #60-A	SAE #60-A	SAE #70
Compression Ratio	6.7:1	6.7:1	6.7:1
Blower Ratio(s)	6.375:1	6.375:1	6.95:1 minimum 9.07:1 maximum
Carburetor	Bendix PR-100B3-4	Chandler Evans CECO 100CPB-9	Chandler Evans CECO 100CPB-9
Magnetos	Scintilla S14RN-15 (Low tension)	Scintilla S14RN-15 (Low tension)	Scintilla S14RN-15 (Low tension)
Installation Drwg. No.	172701	179601	172901
Dimensions	Diameter: 55.00 inches Length: 96.50 inches	Diameter: 55.00 inches Length: 96.50 inches	Diameter: 55.00 inches Length: 103.75 inches
A.T.C. number			
Number Manufactured	4,260	6	6
Applications	Boeing KC-97F Boeing KC-97G Aero Spacelines B-377MG Mini Guppy Aero Spacelines B-377SG Super Guppy Aero Spacelines B-377PG Pregnant Guppy		
Notes	*-59Bs were also manufactured by Ford Motor Company. R-4360-59Bs were developed from 125 R-4360-59s converted with Bendix carburetor replacing the -59's Chandler Evans carburetor. **Weight includes torquemeter Additional weights: Exhaust system with collector ring. ..235 lbs Engine mounts, less isolators.110 lbs Fluid power pump adapter: 5 lbs "C" series nose section and power section. "B" series rear section.	*-61s were originally -59As. Six manufactured, two sent to Ford Motor Company for test stand engines. Four were made into cutaway display engines. **Additional weights: Exhaust system with collector ring..294 lbs Engine mounts, less isolators: 110 lbs Weight includes torquemeter and manifold pressure regulator. Also manufactured by Ford Motor Company.	*Includes torquemeter. *Additional weight: Engine mounts, less isolators: 110 lbs Fluid power pump adapter (optional): 5 lbs Also manufactured by Ford Motor Company.

This montage (next four pages) shows the R-4360-63. Surprisingly, along with some B-36 variants, this engine was rated at the highest power of all production R-4360s and yet was installed in the ubiquitous Douglas C-124.

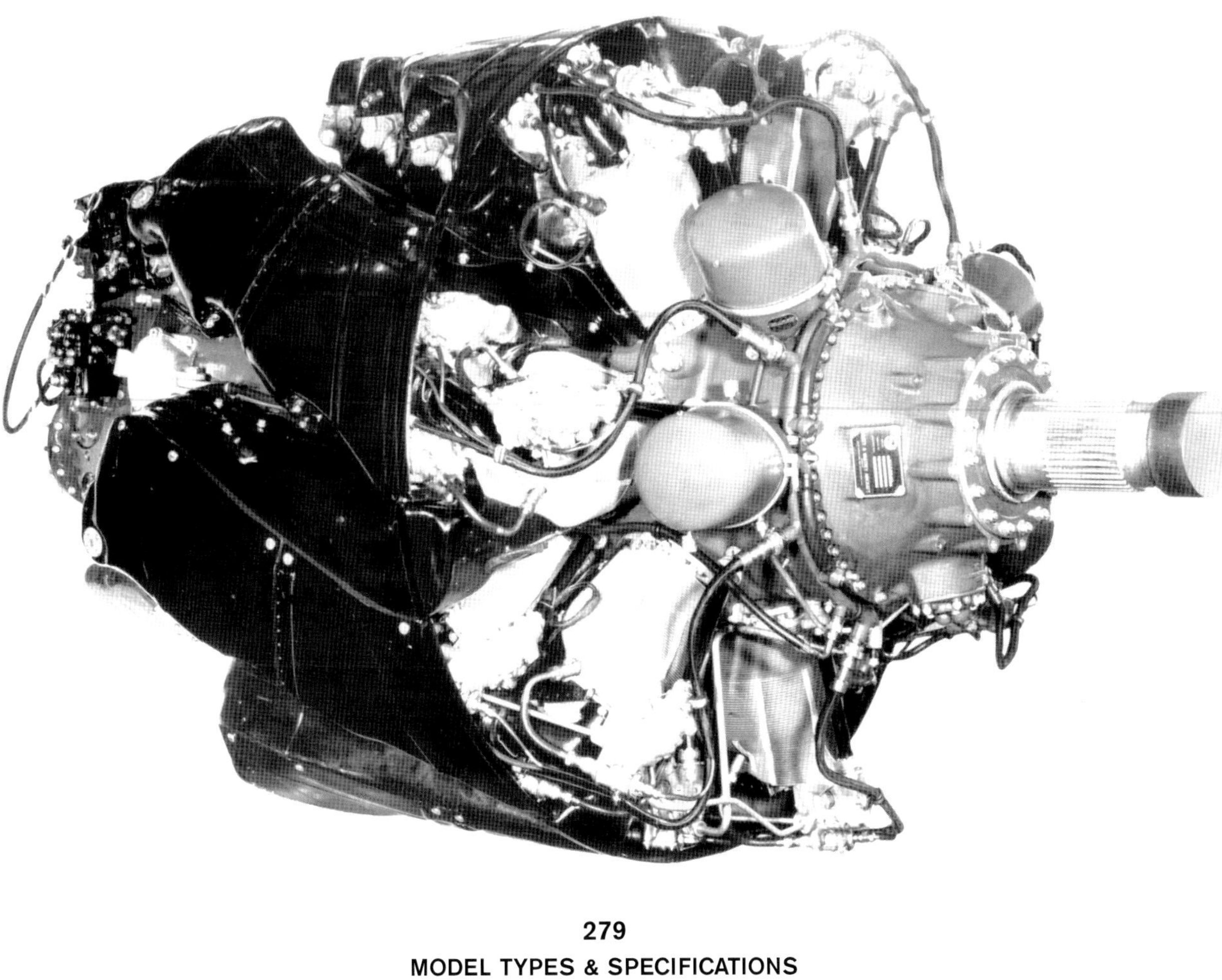

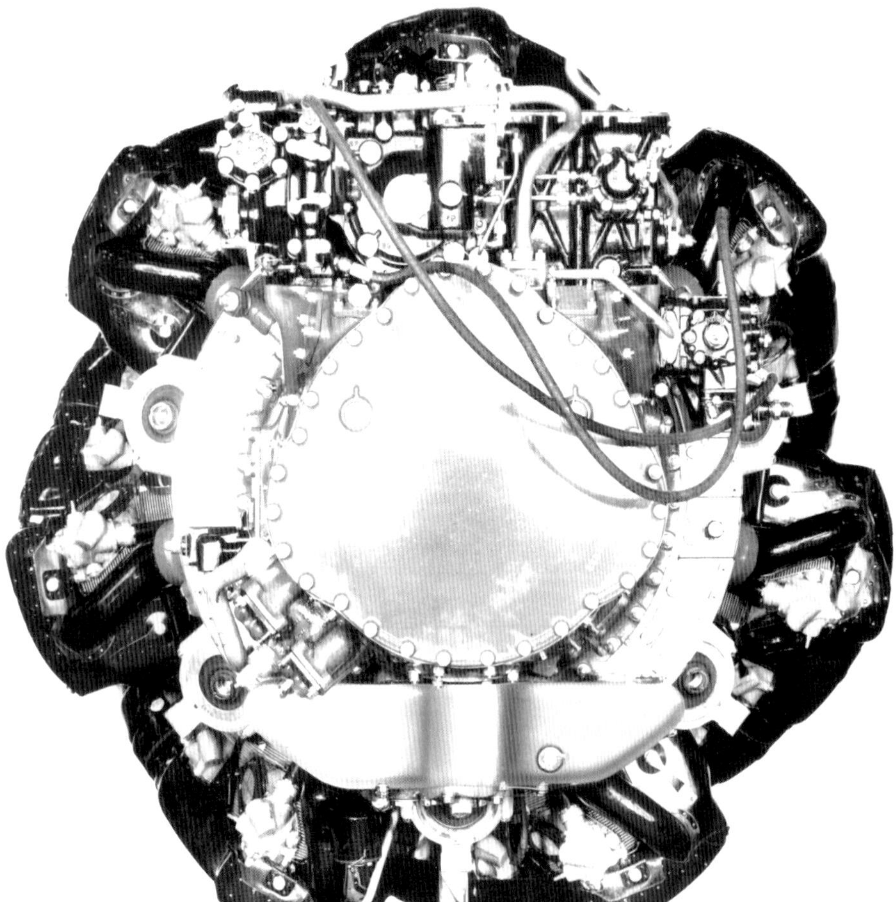

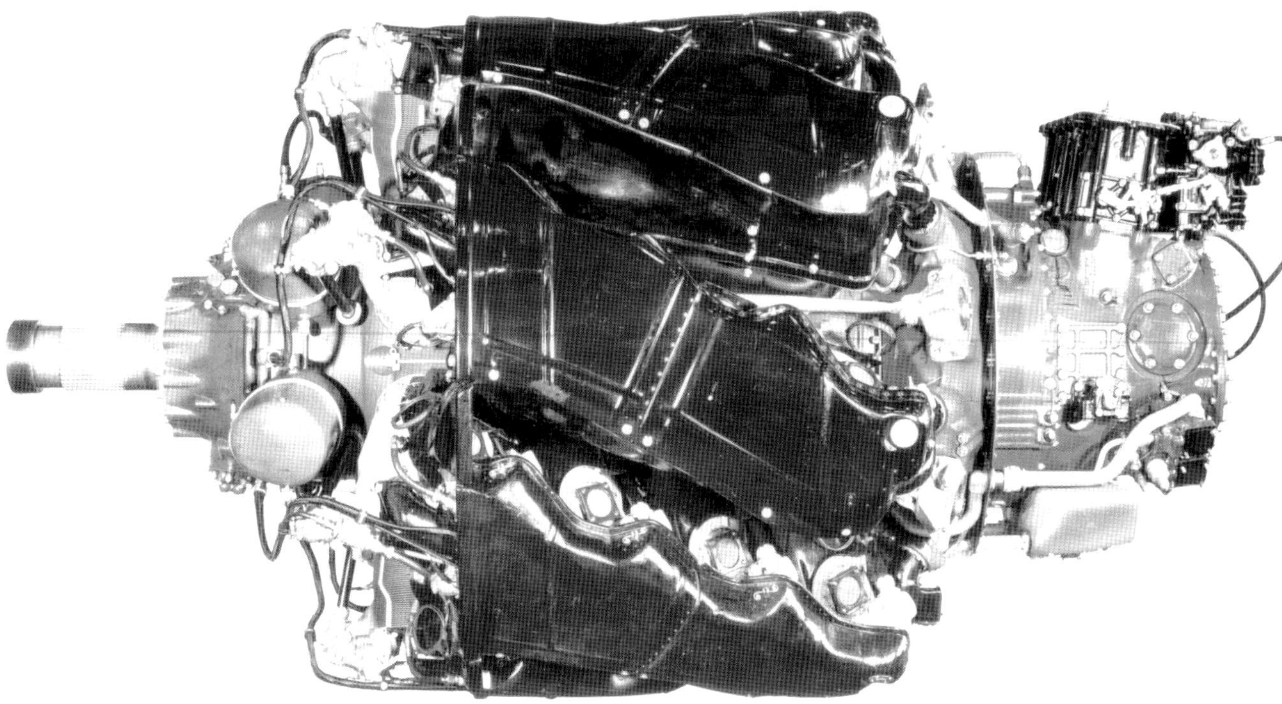

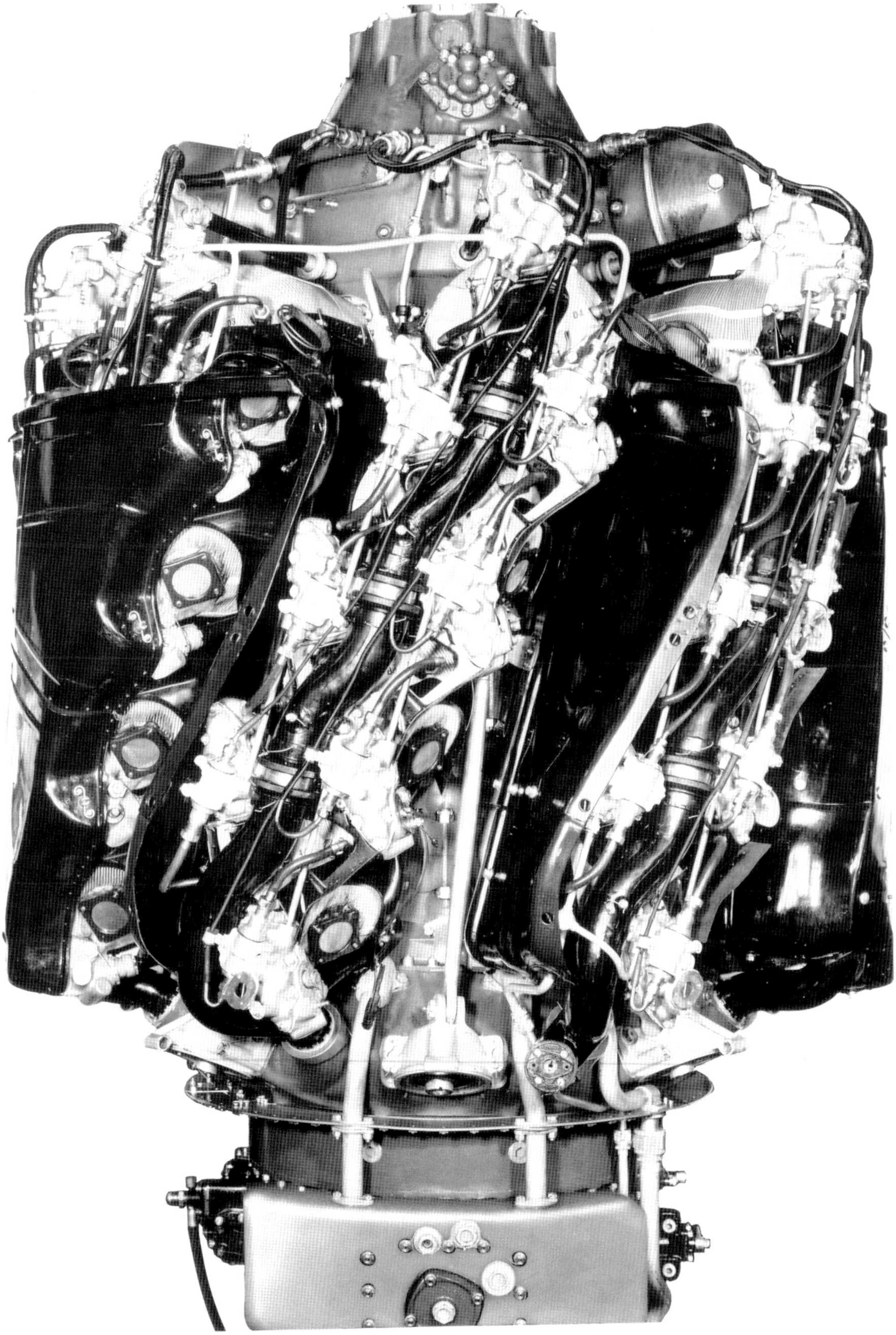

R-4360-63. Another bottom view, however, this one shows two hooded baffles removed exposing the cylinder heads and ignition boost coils.

(Table 5-42) (Ref. 5-3)

Engine Model	Type R-4360 P&W Aircraft Air Force *-63A Navy	Type R-4360 P&W Aircraft Air Force *-65 Navy	Type R-4360 P&W Aircraft Air Force Navy
Specification Number			
Engine Series	CB Series	CB Series	
Rating; Take-off	3800 hp @ 2800 rpm @ sea level wet 3800 hp @ 2800 rpm @ 2000 feet wet 3400 hp @ 2800 rpm @ 2000 feet dry	3500 hp @ 2700 rpm wet 3250 hp @ 2700 rpm dry	
Rating; Military	3800 hp @ 2800 rpm @ sea level wet 3400 hp @ 2800 rpm @ 2000 feet dry	3500 hp @ 2700 rpm @ 500 feet wet	
Rating; Normal	2800 hp @ 2600 rpm @ 5000 feet 2450 hp @ 2550 rpm @ 15000 feet	2650 hp @ 2550 rpm @ 5500 feet	
Max. Continuous			
Cruise			
Fuel Grade	115/145 PN	115/145 PN	
Curves	Inst. #16808		
Weight, dry		3490 lbs	
Prop. Reduction Ratio	0.375:1	0.375:1	
Prop. Shaft Spline	SAE #70	SAE #60-A	
Compression Ratio	6.7:1	6.7:1	
Blower Ratio(s)		6.375:1	
Carburetor	Bendix PR-100B4-12	Bendix PR-100-B3-4	
Magnetos	Scintilla S14RN-15 (Low tension)	D-4RN-2	
Installation Drwg. No.	155501	96501	
Dimensions	Diameter: 55.00 inches Length: 102.00 inches	Diameter: 54.00 inches Length: 96.75 inches	
A.T.C. number			
Number Manufactured			
Applications	Douglas C-124C	Boeing C-97A, C Boeing KC-97E Aero Spacelines B-377MG Mini Guppy Aero Spacelines B-377SG Super Guppy Aero Spacelines B-377PG Pregnant Guppy	
Notes	Similar to R-4360-63 except -63A uses Bendix PR-100B4-12 carburetor. Weight includes fuel injection equipment and torquemeter. Additional weight: Engine mounts: 110 lbs Fluid power pump adapter (optional): 5lbs Also manufactured by Ford Motor Company and designated by engineering change.	*R-4360-65s were R-4360-35A and R-4360-35Cs converted to Wasp Major B6 standard. -65 features .964:1 propeller governor ratio Wasp Major B6 features 0.961:1 propeller governor ratio.	

Front view of the R-4360-65 illustrating its small frontal area and low-tension ignition.

R-4360-65. In this factory portrait, note that the rear half of a hooded baffle has been removed.

Rear view of the R-4360-65. Note the seven mounts used to support the engine in the airframe. This significant design challenge was a major accomplishment.

R-4360-65. It would appear that Pratt & Whitney used the same engine for this series of portraits for the R-4360-65. Note the same rear half hooded baffle is missing as that in the image on top of page 285.

R-4360-65. A view from the other side reveals another hooded baffle removed; this time the front and rear halves.

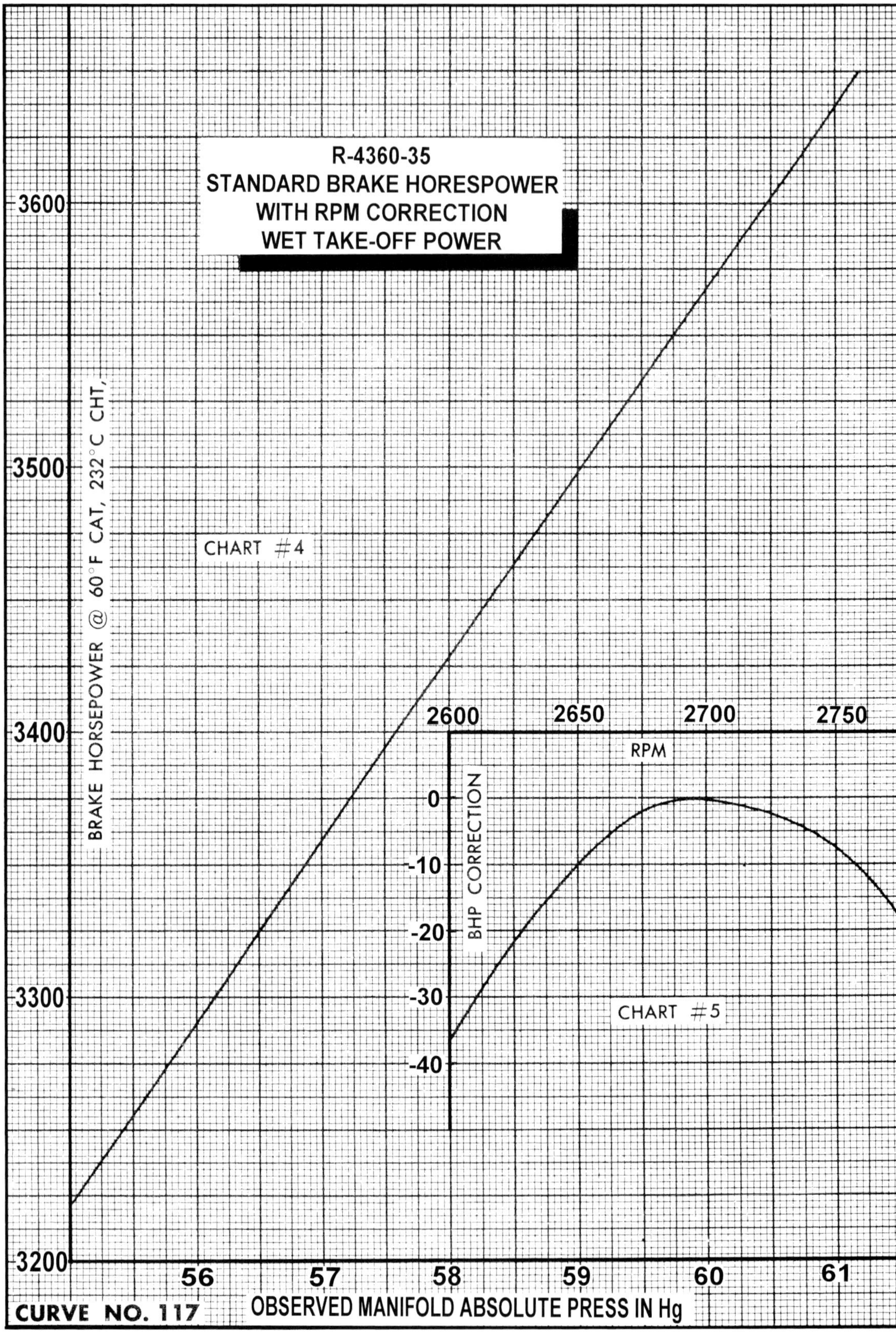

Each variation on the R-4360 was put through a rigorous testing program that culminated in the development of a power curve. Power was measured at all manifold pressures up to its maximum and at all RPMs. This is a typical power curve, in this instance a "B" series R-4360 rated at 3,500 hp at 2,700 rpm. From the chart it can be seen that 59 in. Hg. manifold pressure will result in this rating.

CROSS REFERENCE OF COMMERCIAL (WASP MAJOR), NAVY AND AIR FORCE R-4360s

(Table 5-43)

AIR FORCE	NAVY	WASP MAJOR (COMMERCIAL)
	R-4360-2	TSB1-G
	R-4360-2A	TSB1-G
	R-4360X-2	TSB1-G
R-4360-3		TSB1P-G
R-4360-4	R-4360-4, R-4360-4T	VSB11-G
	R-4360-4W,	
	R-4360-4T	VSB11-G
R-4360-5		TSB1P-G
	R-4360-6	VSSB21-G
R-4360-7		TSB1P-RGD
	R-4360-8	VSB11-GD
	R-4360-8A	VSB11-GD
R-4360-9		VSB11-G2
R-4360-9T		VSB11GD
	R-4360-10	VSSB21-GD
R-4360-11		TSB1P-RGD
	R-4360-12	VSSB11-G2D
	R-4360-12A	VSSB11-G2D
R-4360-13		VSSB21-G
	R-4360-14	VSSB11GD
R-4360-15		VSSB11-G
R-4360-17		TSB1P-RGD
	R-4360-18	TSB1-G
R-4360-19		VSSB21-GD
	R-4360-20,	B12
	R-4360-20A	
	R-4360-20W	
	R-4360-20WA	
	R-4360-20B,	B12
	R-4360-20C,	
	R-4360-20WB,	
	R-4360-20WC	
R-4360-21		TSB1P-RGD
	R-4360-22W	TSB3-G
	R-4360-24	VSB11-G
R-4360-25		TSB1P-G
R-4360-27		VSB11-G
R-4360-29		VSSB21-G
R-4360-31		TSB1-GD
R-4360-33		VSSB21-G
R-4360-35,		
R-4360-35B		TSB3-G
R-4360-35A,		
R-4360-35C		TSB3-G
R-4360-37		TSB1-G
R-4360-39		C2
R-4360-41,		
R-4360-41A		B4
R-4360-43		C3
R-4360-45		TSB1P-RG
R-4360 47		TSB1P-RG
-49, -49A		TSB3-G
R-4360-51		C3
R-4360-53		C6
R-4360-55		C5
R-4360-57		C7
R-4360-59		CB4
R-4360-59A		CB4
R-4360-59B		CB4
R-4360-61		CB4
R-4360-63		CB11
R-4360-63A		CB11
R-4360-65		B6

Summary of R-4360/Wasp Major Production:

(Table 5-44)

ENGINE MODEL (NAVY)	ENGINE MODEL (USAF/ARMY)	ENGINE MODEL (COMMERCIAL)	QUANTITY (NAVY)	QUANTITY (USAF/ARMY)	QUANTITY (COMMERCIAL)	QUANTITY (TOTAL)
		TSB-1G			4	4
-4	-4	-4	197	1	2	200
	-4A			16		16
-4W			130			130
		TSB-11G				
	-5			8		8
	-7			2		2
-8			5			5
-8A			2			2
	-9			2		2
		VSB-11G			25	25
		B13			46	46
	-9T			1		1
-10		-10	2		4	6
	-11			1		1
	-13			3		3
-14			2			2
	-15			1		1
	-17			28		28
-18			13			13
-20	-20		78	197		275
-20W	20W		37	546		583
-20WA	-20WA		155	1,948		2,103
	-21			28		28
-22W			24			24
	-25			221		221
	-27			84		84
		TSB-3G			391	391
	-31	-31		27	1	28
	-33			6		6
	-35	-35		1,929	2	1,931
	-35A			382		382
	-35B			661		661
	-35C			598		598
	-41			947		947
	-41A			269		269
	-43			8		8
	-45			13		13
	-47			13		13
	-53			2,240		2,240
	-55			16		16
	-59			8		8
	-59B	-59B		4,260	2	4,262
	-61			6		6
	-63			6		6
		CB2			2	2
GRAND TOTAL:						**15,600**

Production Totals By Year

(Table 5-45)

YEAR	NAVY	ARMY/USAF	COMMERCIAL	TOTALS BY YEAR
1943	2	6	3	11
1944	11	11	5	27
1945	66	45	3	114
1946	142	238	7	387
1947	68	722	195	985
1948	21	984	215	1,220
1949	144	1425	7	1,576
1950	36	1565	12	1,613
1951	102	2227	5	2,334
1952	53	2750	26	2,829
1953	0	2258	0	2,258
1954	0	1608	1	1609
1955	0	637	0	637
SUB TOTALS:	645	14476	479	
GRAND TOTAL:				**15,600**

R-4360-65. This view shows the same hooded baffle (front and rear halves) removed.

References

5-1 Inter-office correspondence from W. G. Gwinn to Mr. E. H. Benham recognizing end of R-4360 production. June 23, 1955.

5-2 Symposium Lecture #2 General Pratt & Whitney Engine Specifications.

5-3 Index of Wasp Major & R-4360 Designated Engines. Revised July 2, 1956.

5-4 Horwath, A. S., and R. N. Wallace. *Installation And Test of an R-4360 Engine In The F4U-1 (WM) Airplane*. Report No. PWA. Inst. 146. November 15, 1944.

5-5 Gunston, Bill. *Giants of the Sky*. Patrick Stephens Limited. 1991.

CHAPTER SIX

Carburetors & Fuel Injection

The basic requirements of a carburetor sound pretty simple: Supply the engine with a correct fuel/air mixture. But when we take a harder look, things get complicated—in a hurry. The "correct fuel/air mixture" needs to be supplied to the cylinders under all conditions. This includes all throttle positions, all load conditions, at all altitudes, at a wide range of temperature and humidity, and for the complete RPM range.

Fundamental Requirements of a Fuel System

Ideally, fuel enters the cylinder finely dissolved in air—as a "fog"—mixed with the correct amount of air. Liquid fuel in the form of droplets does not burn completely because each droplet must dissolve its surface into the air. For the reasons stated, fuel less than finely atomized burns slower, if at all, at least in the cylinder. The ideal scenario is to burn the fuel/air mixture as rapidly as possible in a controlled fashion inside the cylinder. Under some conditions, combustion occurs spontaneously. The more common cause for this is detonation, which has already been described.

The appropriate fuel/air mixture needs to enter the cylinder. This mixture changes with operating conditions. For instance, under takeoff and maximum power conditions, a rich mixture is required. Cruise conditions demand a relatively lean mixture and when operating with ADI (anti-detonation injection) the mixture requires derichment. Advancing the throttle requires a temporary enrichment, or accelerator pump. And to shut the engine down, fuel to the engine needs to be cut off. Shutting down an aircraft engine is accomplished through starving it of fuel rather than killing the ignition, as would be the case for an automobile. Fuel/air mixtures need to be maintained at the same level regardless of changing atmospheric conditions. This is due to altitude, temperature, or humidity. Provisions should also be provided to allow for manual adjustment of the mixture.

At high power settings (3,500 hp and above) upwards of 14 tons of air an hour is consumed by the R-4360. Airflow, in pounds per hour divided by 7.3, gives an approximation of horsepower. Likewise, under the same operating conditions just described, over one ton of fuel per hour is consumed. Clearly, the R-4360's fuel system has to deal with immense mass flows of fuel and air and at the same time meter this mixture accurately.

Two companies supplied fuel systems for the R-4360: Bendix/Stromberg and Chandler Evans (CECO). With later developments of the R-4360, Bendix developed a direct fuel injection system.

So Who's on First?

Clearly, the carburetor or fuel injection system is an integral and key component of the engine's fuel system. But who takes responsibility for what?

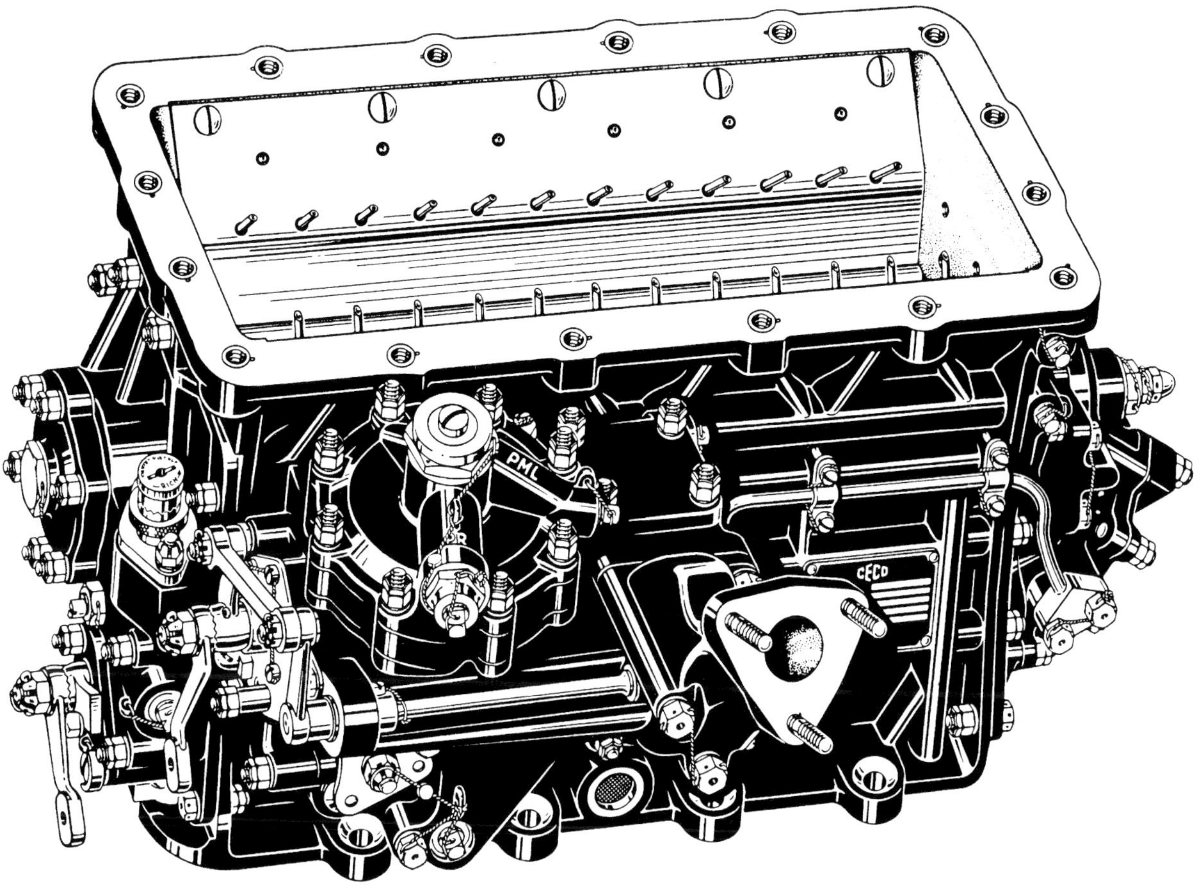

Overall view of the Chandler Evans 100-CPB9 carburetor. Note the "cleaner" intake system compared to its Bendix counterpart. No boost venturi or its associated support structure is in sight. *(Handbook of Overhaul Instructions. Chandler Evans Carburetor Model 100-CPB9)*

For instance, Bendix or CECO manufactures the carburetor, but who makes the fuel feed valve? One would think the carburetor manufacturer would. However, that is not the case. Surprisingly, Pratt & Whitney was charged with design of the fuel feed valve. It was the same scenario with the water regulator. Again, this is a key component that has to be integrated with the carburetion system. But it was Pratt & Whitney's responsibility to design the water regulator. In an unrelated but similar scenario, the same situation occurs with exhaust systems. The airframe manufacturer takes responsibility for the design and manufacture of this all important engine component. The same situation occurs with the design of the cowl and engine mount. Although in the case of the R-4360, Pratt & Whitney realized that its engines were getting blamed for poor installation design regarding cowls that did not offer adequate cooling and compromised exhaust systems.

It was not just Pratt & Whitney that was placed in this difficult position. Rolls-Royce, for example, had its engines compromised through poor installation designs. As with many other concepts, Rolls-Royce pioneered the idea of the engine manufacturer taking responsibility for the entire QEC (quick engine change) design. This includes the cooling system, engine mount, and exhaust system—all integral and key components of the overall engine installation. In this way, the engine manufacturer took total responsibility for the entire powerplant package. The U.S. was behind the British in this respect. Airframe manufacturers would contract out engine installation design and manufacturing to experts such as the Rohr Corporation, which specialized in QEC designs. The engine manufacturer was kept out of the picture. Although this arrangement worked well for most installations, as engines became more sophisticated—and consequently more critical regarding their cooling, exhaust, and fuel system requirements—the engine manufacturers were forced into taking a more active role in engine installation. For Pratt & Whitney, this trend surfaced with the R-4360.

Getting back to the fuel system, the airframe manufacturer was responsible for providing fuel at a pressure of 20 psi at the carburetor. Other requirements went along with this requirement such as filtration and appropriate flow rates. Once the fuel reached the carburetor, Bendix or CECO took responsibility for metering fuel according to mass airflow. Metered fuel is discharged from the carburetor and this is where the carburetor manufacturer's responsibility ends. Pratt & Whitney then took responsibility for the fuel feed valve. The foregoing begs the question: Why did the carburetor manufacturer not take responsibility for the entire fuel system from the airframe fuel tank to the methodology of entering fuel into the engine? In the case of direct fuel injection, Bendix did design everything from fuel entering the master control unit to the fuel being injected into the cylinders. But this was not the norm.

Chandler Evans 100-CPB-9

Today, it's almost impossible to find an aircraft fitted with a Chandler Evans carburetor—R-4360 or any other application. And yet the U.S. Navy expressed an opinion felt the CECO carburetor was superior to the Bendix *(Ref. 6-1)*.

This carburetor operates in a similar fashion to Bendix injection carburetors. In simple terms, the carburetor measures mass airflow and meters the appropriate amount of fuel depending on what the mass airflow is. And like its Bendix counterpart, the 100-CPB-9 offers 100 square inches of inlet area.

Airflow to the engine is regulated via a pair of rectangular throttle plates. The venturi and impact tubes measure this airflow. A boost venturi mounted in the middle of the air stream above the throttle plates is used to amplify airflow measurement. Air passing through the boost venturi and compared to the impact tubes pressure is in direct proportion to the air consumed by the engine.

Airflow measured by venturi suction is used to control fuel flow to the engine via the action of the pressure regulator assembly. This assembly consists of a valve operated by three diaphragms in series.

To increase power from the engine, the throttles are opened. This causes venturi suction to increase, which in turn causes the pressure regulator valve to open. This sequence of events causes an increase in fuel flow to the regulator chamber until fuel pressure against the regulator diaphragm balances the metered suction differential on the air diaphragm and discharges fuel pressure on the fuel balance diaphragms. Under all operating conditions, the ratio of fuel forces to air forces remains constant. In this way, any change in airflow to the engine results in a proportionate change in fuel flow through the carburetor.

To compensate for a decrease in reduced mass airflow as altitude and/or temperature are increased, an automatic mixture control is provided. It consists of an altitude valve operated by a nitrogen-filled brass bellows placed in series with the boost venturi. A decrease in air density results in the bellows expanding, which moves the altitude valve further onto its seat, thus leaning the mixture fed to the fuel feed valve.

100-CPB-9 carburetors are provided with five jets and a "B" valve to accommodate varying mixture requirements demanded by changing engine operating conditions. These five jets and the "B" valve operate in conjunction with the manual control valve position as follows.

a.) Manual Lean

In this position Port 1 is fully open, permitting full flow through the "A" jet. Port 2 is cracked open, permitting partial flow through the "Y2"

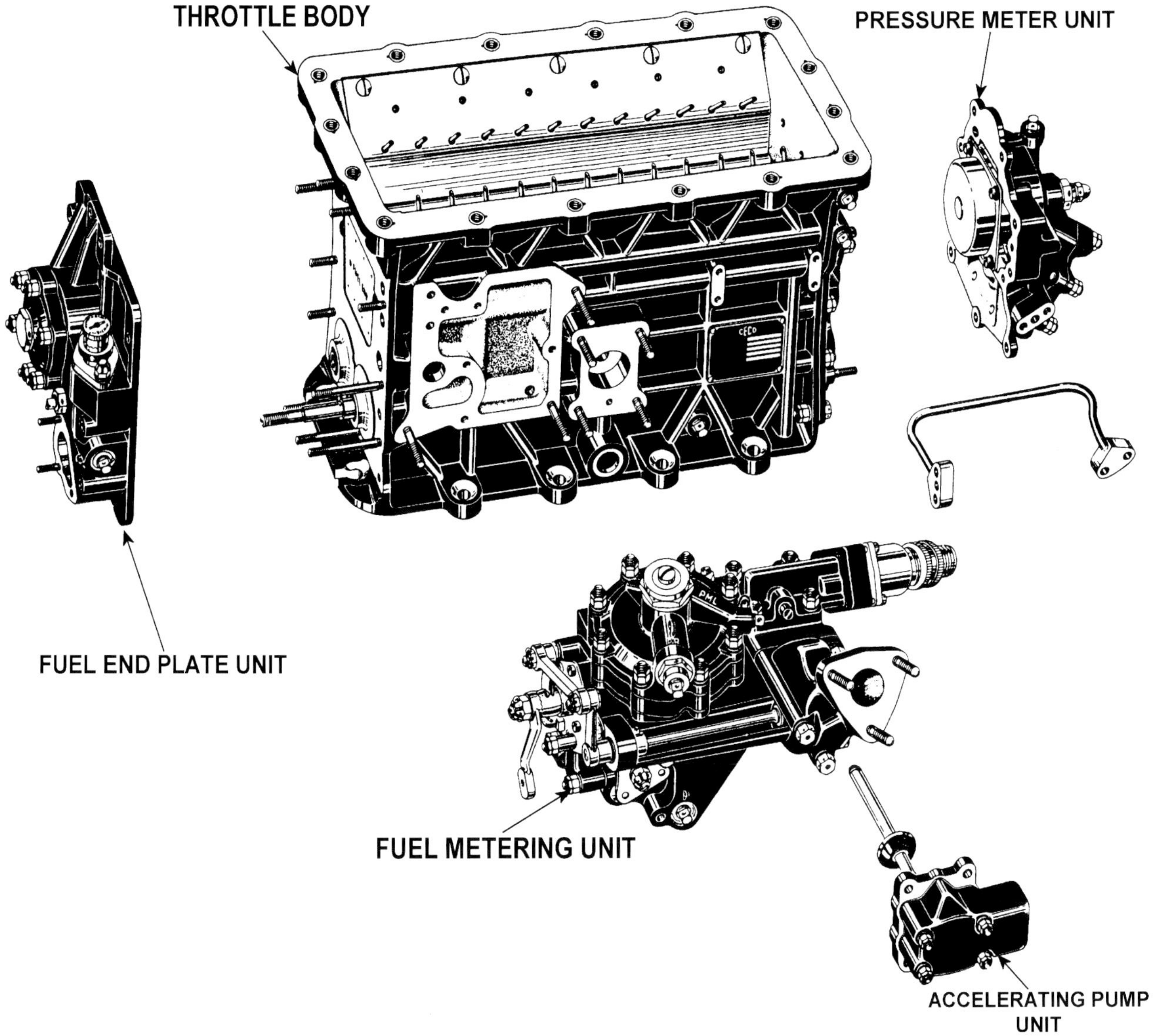

This illustrates nicely the major sub-assemblies for the CECO carburetor. *(Handbook of Overhaul Instructions. Chandler Evans Carburetor Model 100-CPB9)*

jet. Bleed "E" is open, permitting pressure regulator pressure to be applied to the "B" valve diaphragm, which is actuated when this pressure reaches a sufficient force.

b.) Auto Lean

Both Port 1 and Port 2 are fully open in auto lean, which permits full flow through the "A" and Y2" jets. Bleed "E" remains open.

c.) Auto-Rich

Ports 1, 2, 3, and 4 plus Bleed "E" are fully open in auto-rich. Ports 1, 2, and 4 permit full fuel flow through jets "A," "Y2," and "C." Opening of Port 4 causes additional regulated fuel pressure to be applied to the "B" valve diaphragm. When open, the "B" valve discharges fuel from Port 1 to join flow from Port 3. This combined flow passes to discharge through the limiting jet "D."

d.) Derichment

When water pressure, such as ADI, is applied to the derichment valve, valve "M" opens and permits fuel flow from Port 1 through jet "DE."

Under engine idling conditions, the desired mixture varies from that controlled by the

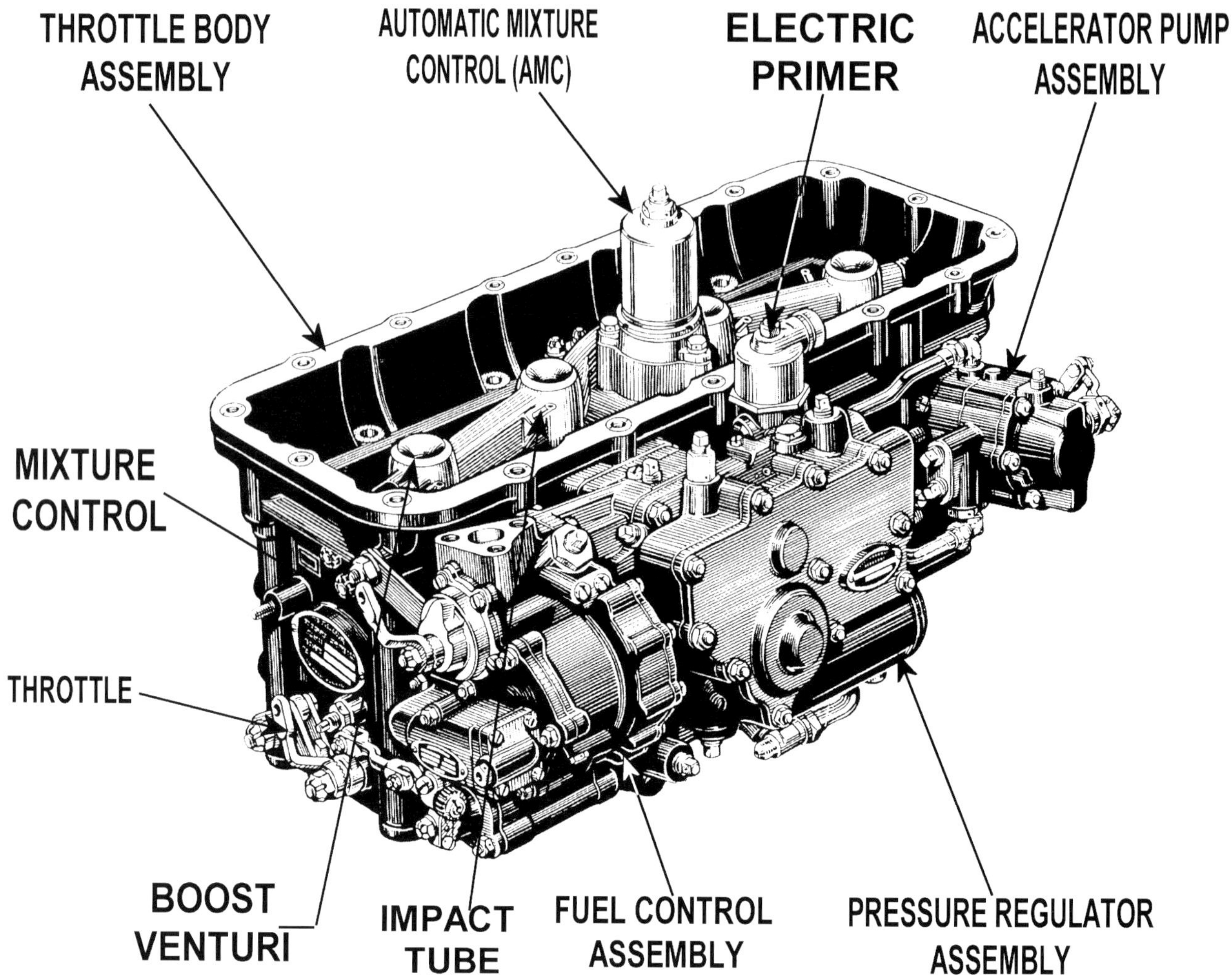

Overall view of the Bendix PR-100 carburetor. Note the boost four venturis located in the throttle-body intake. Comparing the air pressure difference between the boost venturis and the impact tubes regulated fuel flow to the engine. *(Illustrated Parts Breakdown, Injection Carburetor, Model PR-100 B4 Used on R-4360)*

metering system. A throttle-operated idle valve is provided in the fuel discharge passage for control of the mixture under idling conditions. An adjustment on the linkage between the throttle and idle valve provides the means for obtaining the correct mixture.

As with any piston engine, a rapid acceleration causes a slow, relative response of the fuel metering system. To compensate for this momentary delay, an acceleration pump is linked to the throttle so that sudden throttle openings force the opening of the acceleration valve and raw fuel is momentarily sprayed into the manifold suction via a pair of nozzles.

Bendix/Stromberg

By far the most numerous of all carburetors fitted to R-4360s were the PR-100 series made by Bendix/Stromberg.

Not unnaturally, the Bendix PR-100 operates in a similar fashion to the Chandler Evans. In very simple terms, it measures mass airflow and meters the appropriate amount of fuel based on this measurement. Airflow, and by default, power, is controlled by a pair of rectangular throttle plates. Fuel flow is metered through fixed jets proportionate to mass airflow. This mass airflow is controlled, measured, and converted into a working

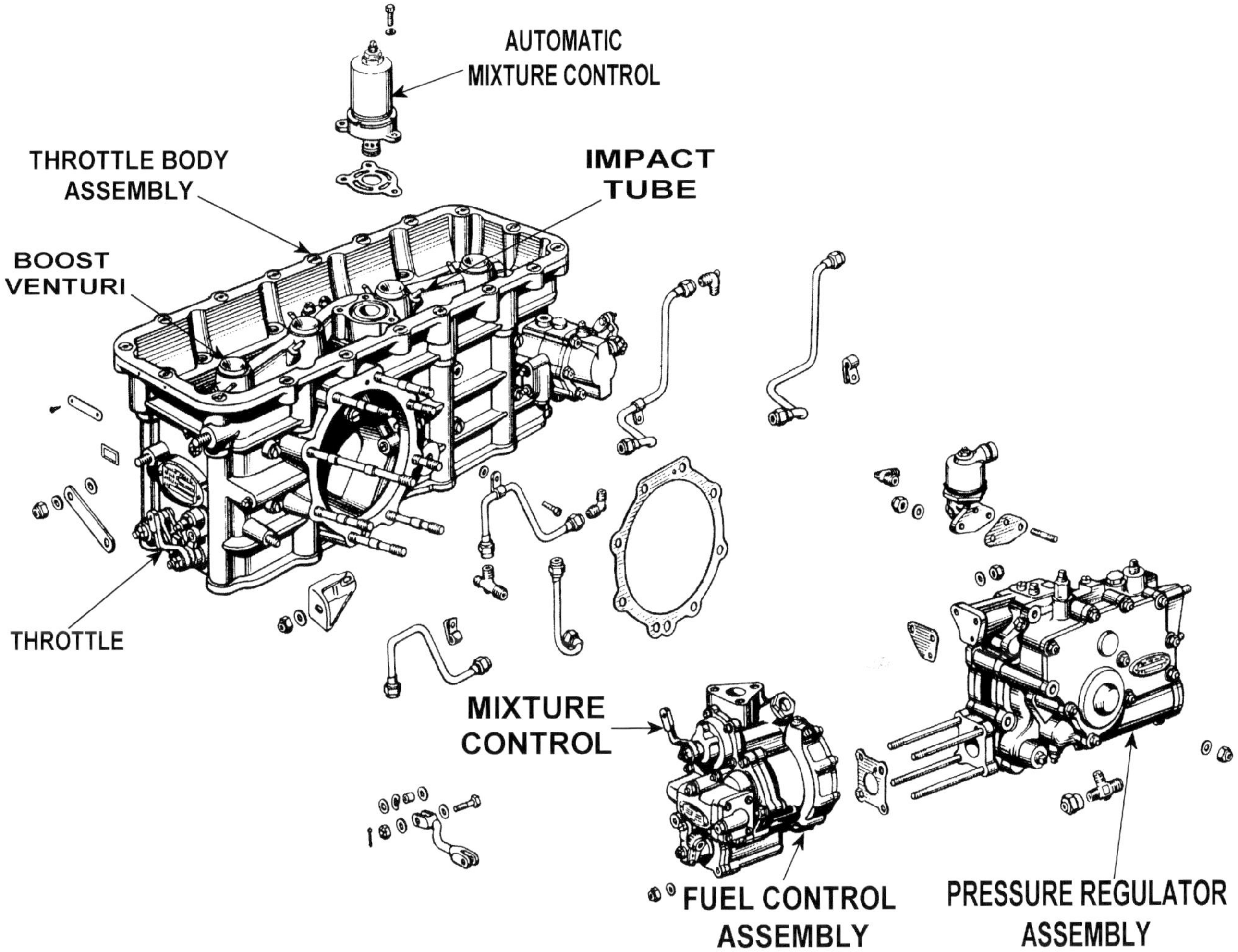

Major sub-assemblies of the PR-100 broken out. *(Illustrated Parts Breakdown, Injection Carburetor, Model PR-100 B4 Used on R-4360)*

force to supply fuel. Consequently, any increase in mass airflow provides an increase in fuel flow to the engine. By sensing changes in air pressure and temperature, automatic adjustment of the air metering force provides accurate fuel metering during all conditions of engine operation. The fact that fuel is discharged downstream from the throttle valves negates the possibility of ice formation due to fuel evaporation within the carburetor. Additionally, positive fuel metering is assured even during maneuvers and aerobatics. A manual means for selecting fuel requirements provided for under all operating conditions regardless of engine speed, propeller load, or throttle position. The PR-100 is made up from the following sub-assemblies:

I. Automatic mixture control
II. Fuel control unit
III. Electric primer unit
IV. Regulator unit
V. Throttle body unit

I. Automatic Mixture Control

This unit regulates impact air pressure to the regulator unit. Like the Chandler Evans, it uses a sealed metallic bellows containing a measured amount of inert oil and nitrogen. It senses air pressure changes due to altitude changes or atmospheric changes. Bellows movement positions a contoured needle to predetermined points, thereby regulating impact air pressure to compensate for air density changes.

II. Fuel Control Unit

Metering jets, a power enrichment valve, an idle metering valve, and the manual mixture control are contained within the fuel control unit. It attaches directly to the regulator unit from which it receives unmetered fuel. Fuel metering throughout the idle range is accomplished via the idle metering valve, which is mechanically connected to the throttle shaft. The diaphragm-operated power enrichment valve provides fuel enrichment in proportion to mass airflow through the throttle body. During water injection (ADI), a derichment valve comes into operation. This reduces fuel flow. Manual means for selecting the appropriate fuel flow for various engine operations is provided by the "rich," "normal," and "idle cut-off" positions of the manual mixture control valve.

III. Primer Unit

A richer than normal mixture is required for starting purposes. To accomplish the priming requirement, the carburetor is provided an electric primer unit, which is simply a solenoid actuated valve. This unit relies on engine pump pressure to supply raw fuel to a pair of discharge nozzles downstream from the carburetor.

IV. Regulator Unit

Air and fuel differential pressure operate the regulator unit. Regulated impact air pressure and boost venturi suction establish the required air pressure differential (metering suction). Unmetered and metered fuel creates a corresponding fuel differential pressure (metering head). An air diaphragm, fuel diaphragm, and sealing diaphragms—mechanically connected by a single stem to the poppet valve—provide a means for balancing the differential pressures (fuel and air).

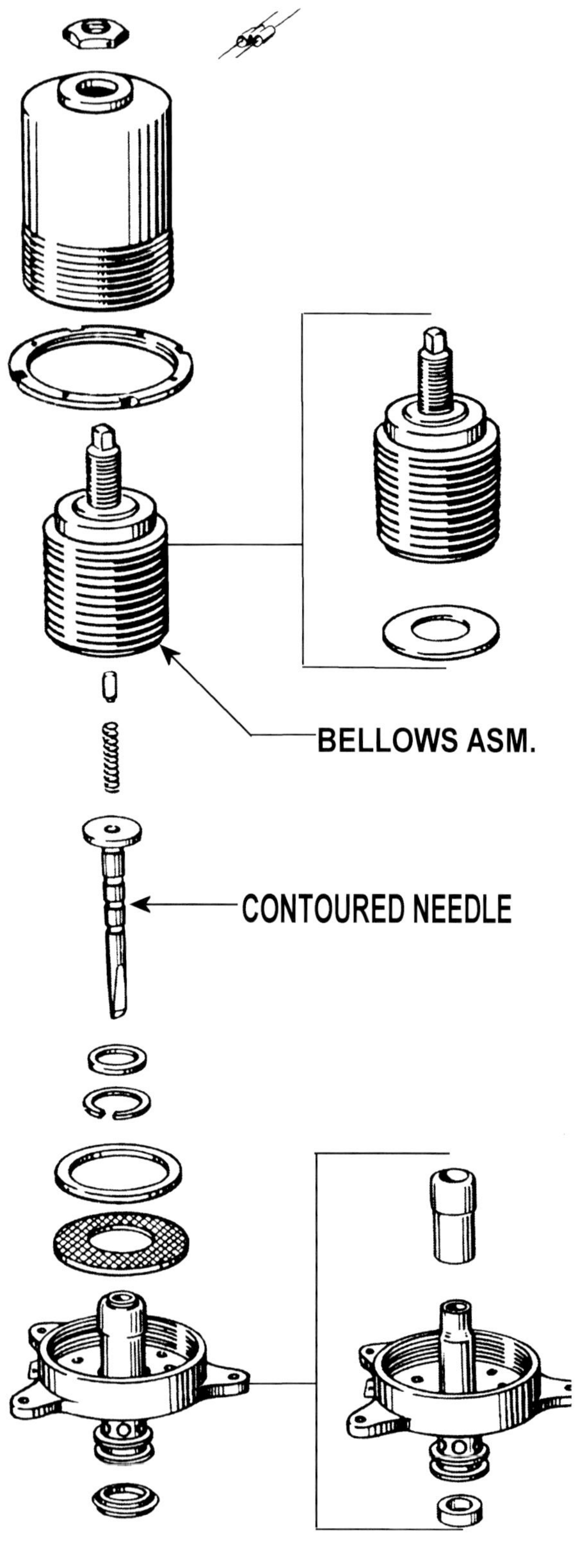

Sensing air pressure changes, the automatic mixture control accordingly adjusts the mixture entering the engine and thus compensates for changes in altitude. *(Illustrated Parts Breakdown, Injection Carburetor, Model PR-100 B4 Used on R-4360)*

Consequently, the regulator automatically supplies fuel to the fuel control in proportion to the mass airflow through the throttle body. To prevent vapor from affecting accurate metering, two vapor separators are provided—one in each top of chamber "D" and "E." A fuel strainer located at the fuel inlet protects the function of the carburetor from foreign particles. A bypass valve incorporated in the strainer allows fuel flow in the event ice crystals carried in suspension in extremely cold fuel clog it.

V. Throttle Body Unit

The throttle body unit controls mass airflow to the engine by use of a pair of long rectangular throttle valves. Measurement of mass airflow is accomplished through sensing pressure differential. Air flows tapped off at the impact tubes and boost venturis create the necessary pressure differential for mass airflow measurement. These pressures (impact tube and boost venturi), when applied to either side of the air diaphragm in the regulator unit, provide air metering force (metering suction). Metering suction is a unit of measurement of mass airflow to the engine. Corrections for air pressure and temperature are accommodated by the automatic mixture control. Flanges on the throttle body unit provide mounting points for the following sub-assemblies: (i) the automatic mixture control and regulator unit, (ii) internal channels for metering suction pressure, and (iii) an acceleration pump.

Principles of Operation

In simple terms, the function of the carburetor can be compared to that of a metering orifice. The only way to control fuel flow through an orifice is by one of two methods: Change the orifice area or change fuel differential pressure across the orifice. Consequently, a carburetor must have the ability to alter the orifice area and/or differential pressure.

The pair of throttle valves regulates the air being pumped through the throttle body. As induction-air flows through the throttle body, it is divided into the following categories: (i) impact pressure, (ii) boost venturi suction, and (iii) pressure below the throttle plates. The group of impact tubes spaced on the boost hanger sense an average impact pressure and by internal channels supply this pressure to chamber "A" of the regulator.

Boost venturis are located in the boost venturi hanger in order to utilize the lowest pressure area created by the main venturi. Boost venturi pressure is channeled to chamber "B" of the regulator.

A flexible rubber air diaphragm divides chamber "A" (impact tube pressure) and chamber "B" (boost venturi suction) of the regulator unit. Under the influence of the pressure differential between impact tube pressure and boost venturi suction, this diaphragm moves in relation to the differential pressure. The result of "A" chamber pressure minus "B" chamber pressure is called "metering suction." Chamber "D," unmetered fuel pressure, and chamber "C," metered fuel pressure, are separated by another flexible rubber fuel diaphragm. The fuel diaphragm moves in relation to the difference between the two fuel pressures. The result of "D" chamber pressure minus "C" chamber pressure is referred to as the "metering head." A single stem to the poppet (fuel) valve connects the air and fuel diaphragms.

Chamber "E" of the regulator unit houses a fuel strainer, vapor separator, and fuel at fuel pump pressure. The poppet valve is located between chambers "E" and "D." When chamber "D" fills with fuel, the vapor separator in chamber "D" eliminates any vapor not discharged by the vapor separator in chamber "E." Fuel flow from chamber "D" (unmetered fuel pressure) is accompanied by a pressure drop in chamber "C" (metered fuel pressure) through the metering jets in the fuel control. The fuel discharge nozzle (interestingly, not furnished by Bendix) provides a constant backpressure. Metered fuel pressure does not vary any appreciable amount during engine operation, although fuel flow increases in proportion to engine demand. During operation, "B" chamber pressure (boost venturi suction) is usually one third that of chamber "A" pressure (impact tube pressure).

Metering suction, from the air-side of the regulator, controls the metering head in the fuel section. Metering suction (air pressure differential) and metering head (fuel pressure differential) always tend to stabilize the poppet valve. When metering suction increases, the air diaphragm moves to the right, thus opening the poppet valve. This action increases fuel flow into chamber "D." This increased fuel pressure in chamber "D" increases the drop of metering pressure (metering head) across the jets in the fuel control unit. When the metering suction decreases, the metering head, being momentarily greater than metering suction, moves the fuel diaphragm to the left, thus decreasing the supply of fuel to chamber "D." The resulting lower pressure in chamber "D" lowers the metering head, again stabilizing the poppet valve. Balance of metering suction and metering head provides a stable poppet valve.

A sealing diaphragm between fuel chamber "C" and air chamber "B" allows adequate poppet valve stem movement and provides an effective fuel to air seal. If it were left like this, an unbalanced metering force would exist, tending to close the poppet valve. To compensate, a balance diaphragm of identical size is incorporated in the front body section of the regulator unit. One side of the diaphragm is exposed to unmetered fuel pressure "D" by internal galleries, the other to regulated impact tube air pressure "A." Differential pressure applied across the sealing and balance diaphragms results in equal and opposing forces. As long as the forces acting upon the diaphragms attached to the poppet valve are equal, the poppet valve remains in balance.

Airflow in the low RPM range of the engine is not sufficient to produce the required air metering force. Provision is made for a constant head idle spring designed to maintain an adequate fuel metering head during low RPM range. Located between the two telescoping parts of the regulator stem, the constant head idle spring opens the poppet valve until metering suction is equal to, or greater than, spring force. Beyond the low RPM range, metering suction overwhelms spring tension and the telescoping parts of the poppet valve stem make solid contact and function as an integral unit.

When operating in the engine idle range, proper air distribution is not provided for all 28 cylinders. Under these conditions, a richer mixture is demanded for smoother idle. Through a suitable linkage, the throttle lever is operated by the idle valve, which is located in the fuel control unit. The idle valve only provides fuel metering during the first few degrees of throttle opening. An external adjustment is provided in order to tweak idle mixture while the engine is running. When throttle movement passes the idle range, the idle valve no longer meters fuel but supplies unmetered fuel to the metering jet chamber.

Due to the immense variation in atmospheric conditions encountered by R-4360 operations (anything from sea level to 45,000 feet), large variations in fuel/air ratios need to be accommodated. Metering jets and other variables within the fuel control unit may be selected to provide fuel requirements of the engine under all conditions. Jet sizes and the enrichment valve rate are predetermined according to these requirements. The manual mixture control provides "rich," "normal," and "idle cut-off" positions. The selection of one of the three aforementioned positions determines which valve and jet or jets are metering fuel.

The idle valve meters fuel in either "rich" or "normal" positions during the idle range. The idle valve has completed its metering phase at a throttle opening of approximately 10 degrees from the closed position.

When the throttle position exceeds approximately 10 degrees and the manual mixture control is in the "rich" position, the auto-rich metering jet meters fuel in conjunction with the auto lean metering jet. An additional increase in fuel flow is required for engine cooling when entering the power range. The power enrichment valve provides the necessary fuel flow. Opening of the power enrichment valve can be set to a

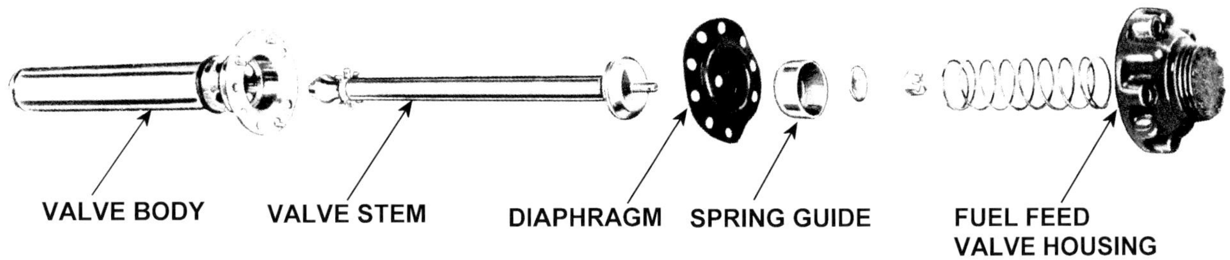

Metered fuel is fed into the engine via the fuel feed valve–shown dissembled here. The fuel feed valve is mounted on the rear section and injects fuel into the slinger ring. *(Overhaul Instructions USAF Models R-4360-20W, -20WA, -20WB, -20WC, -35, and -35B Aircraft Engines. December 15, 1961)*

predetermined point by adjusting the tension of the power enrichment valve spring. When unmetered fuel pressure in the "D" chamber exceeds that of the combined metered fuel pressure in chamber "C" plus the power enrichment spring pressure, the power enrichment valve is forced open. Fuel metering is then accomplished by the auto lean metering jet, auto-rich metering jet, and the power enrichment valve. The power enrichment valve opens to the point where, combined with the metering area of the rich metering jet, its metering area is greater than that of the power enrichment, derichment, and auto lean metering jets. It then takes over fuel metering. Maximum fuel flow is determined by the size of the power enrichment, derichment, and auto lean metering jets. Water pressure (when using ADI) operates the derichment valve via a diaphragm exposed to water pressure on one side and fuel pressure on the other. When ADI is in operation, water pressure closes the derichment valve, thus stopping fuel flow through the derichment jet.

The "normal" manual mixture control position is typically only used during the cruise range of power settings. By closing off fuel flow from the auto-rich metering jet, the engine operates at a more economical fuel/air ratio. Under cruise conditions with the mixture set in the "normal" position, the auto-lean metering jet meters fuel.

The pilot/flight engineer sets the "rich" and "normal" positions. Both positions have predetermined limits. The richer position provides the best power fuel/air ratio. "Normal" or leaner position provides a fuel/air ratio advantageous for use under normal cruise conditions for extended range.

"Idle cut-off" is used for stopping the engine. When this position is selected by the pilot/flight engineer, all fuel channels—including the channel that supplies metered fuel to the "C" chamber of the regulator—are closed, thus starving the engine of fuel.

A metered fuel vent restriction is employed in order to maintain "C" chamber pressure in the regulator unit. Its size provides no appreciable effect on fuel flow. When all the fuel pressure is cut off by closing the manual mixture valve, the vent ceases to function.

Metered fuel leaves the fuel control unit through a transfer tube and is discharged into the engine induction system by a fuel feed valve. The fuel feed valve is a Pratt & Whitney part, not a Bendix or Chandler Evans part. When fuel pressure exceeds that of the fuel feed valve discharge nozzle spring, fuel is injected into the slinger ring and consequently into the engine's induction system. A rubber diaphragm built into the fuel feed valve is vented to impact tube pressure in order to provide instant engine acceleration.

The automatic mixture control (AMC) maintains the correct fuel/air ratio under all operating conditions including changes in altitude, air density, and temperature. A pair of mixture control bleeds permits an adjusted airflow, via the impact pressure tubes, through the automatic mixture control orifice. One bleed is located in the

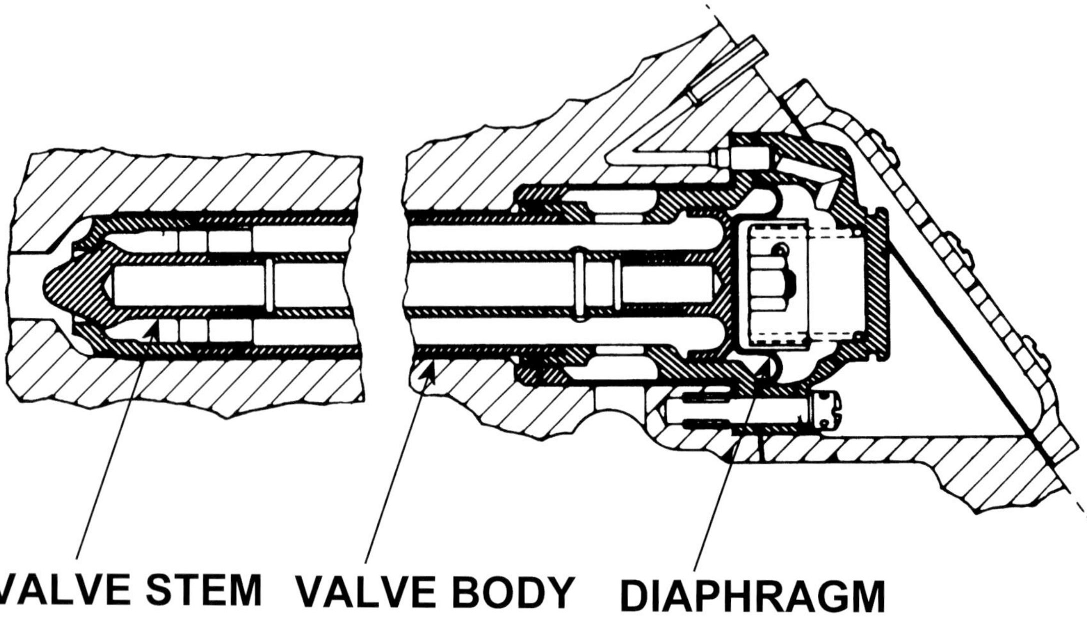

Cross section of the fuel feed valve. *(Overhaul Instructions USAF Models R-4360-20W, -20WA, -20WB, -20WC, -35, and -35B Aircraft Engines. December 15, 1961)*

regulator unit between "A" and "B" air chambers and the other is in the boost venturi hanger between the channels leading to air chambers "A" and "B." The AMC performs its function by using a contoured needle that controls the size of an orifice sealed with a bellows. Response of the bellows to changes in air density regulates impact pressure in air chamber "A." Metering suction, when corrected by the AMC, then becomes a measure of mass airflow. Being a critical component, each AMC is carefully calibrated then sealed with a lead seal. Each needle is contoured and sized for each application and dash number of R-4360. Back-venting the AMC to boost venturi suction eliminates possible contamination and provides a faster response to air density changes. Located in the back venting channel, the bleed hole provides a means of regulating pressure on the bellows.

Most models of the PR-100 use an electric primer valve assembly to supply raw, unmetered fuel to the priming nozzles. The electric primer valve is normally closed by spring action. Energizing the coil overcomes the spring and opens the valve, thus allowing priming fuel into the intake throat under the influence of boost pump pressure.

As with any piston engine, some type of temporary enrichment is required when the engine is accelerated. This is required due to the inherent lag in fuel flow during sudden acceleration of the engine. Throttle operated, the acceleration pump consists of a piston operating in a cylinder. An inlet check valve allows fuel to enter the pump cylinder when the piston is retarded. As the throttle is advanced, the piston travels into the cylinder and dischares fuel through a fuel transfer tube to the acceleration pump discharge nozzles.

Fuel Injection

Often touted as the Holy Grail of fuel systems, direct fuel injection did not see much usage by aircraft engine manufacturers—at least in the United States. The Wright R-3350 and Pratt & Whitney R-4360 were the only large U.S. engines that went into mass production to use direct fuel injection. And of course, not all R-3350s and R-4360s were fuel injected. Germany embraced direct fuel injection for all of its front line aircraft engines during World War II. Used with

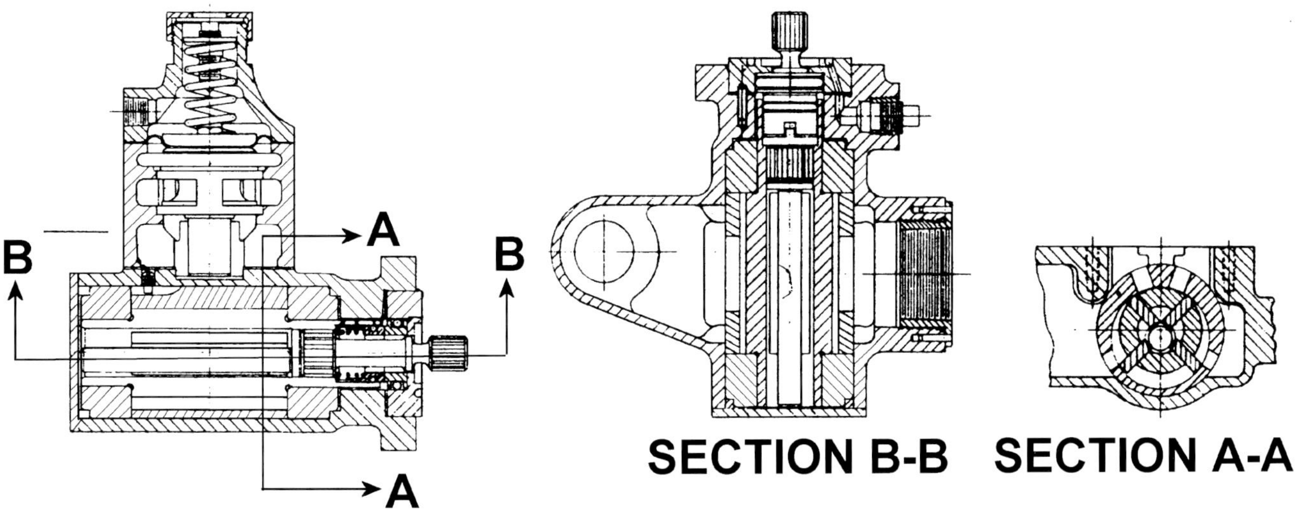

The correct fuel pressure needs to be supplied to the carburetor or fuel injection pump. In both cases a vane type pump is employed. *(Preliminary Instructions on Stromberg 100-28-A3 Direct Fuel Injection System. Courtesy of Pete Law)*

sophisticated controls, German direct fuel injection was very successful. Nevertheless, restorers rebuilding World War II German engines to airworthy condition would probably prefer a carburetor due to its (relative) simplicity.

Before delving into the intricacies of fuel injection, it should be understood what is meant by this over-used and generic term. Even the PR-100 carburetor is referred to as an "injection carburetor." Fuel injection can come in various flavors: Indirect, direct, throttle body, port injection, etc. The system described here is direct injection. In other words, a metered amount of fuel is injected directly into the combustion chamber during each cycle.

A total of 14 R-4360 dash numbers were specified with Bendix fuel injection. However, only a handful of these engines actually saw series production. The majority, if not all, fuel injected R-4360s were installed in Convair B-36s.

The U.S. Navy contracted development of a nine-cylinder injection pump for a Wright R-1820 in 1941. This initial development paved the way for follow-on developments, particularly for the R-3350 and later the R-4360. The R-3350 setup was essentially a doubled-up version of the nine-cylinder R-1820 unit. One of the driving requirements for fuel injection was even mixture distribution. As engines became larger, more complex, and in some cases with a convoluted intake manifold design, it was thought direct fuel injection would alleviate this problem. The situation became so serious that early carbureted Wright R-3350s exhibited huge variations in cylinder temperature. Much of this variation was attributable to uneven mixture distribution in the intake manifold. Even so, fuel injection could not compensate for uneven airflow within the intake system, so at best it was a partial solution to the problem.

Why Reinvent the Wheel?

Many components from the existing PR-100 series carburetor were employed in the 100-28 fuel injection system. All the functions of the carburetor were retained to read mass airflow. But instead of metering fuel, mass airflow readings were used to control the injection pump.

General Description

Three major units comprised the Stromberg 100-28-A series of fuel injection systems: (i) a vane-type fuel pump, (ii) a master control unit, and (iii) a 28-cylinder direct injection pump. All three of these major sub-assemblies are combined in one unit. This unit fits in the space normally

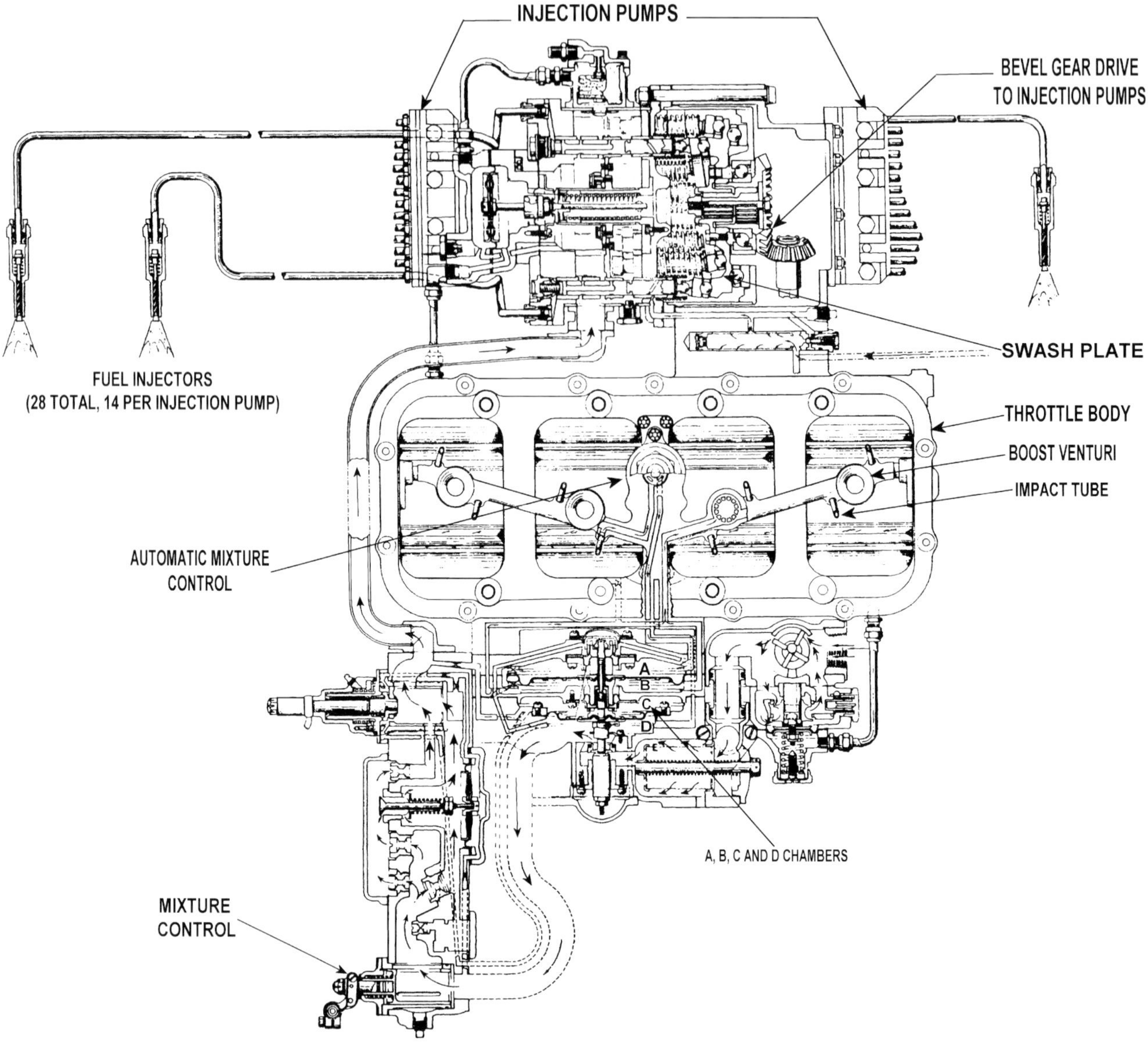

As with the PR-100 carburetor, the 100-28 fuel injection system reads mass airflow in a similar fashion via impact tubes and boost venturis. Known as the "master control unit," the mass airflow signal is sent to the injection pumps, which then determine how much fuel is injected into the engine. This line drawing shows the basic circuitry. *(Preliminary Instructions on Stromberg 100-28-A3 Direct Fuel Injection System. Courtesy of Pete Law)*

occupied by the PR-100 carburetor. The accompanying color figures show the fuel circuitry. Top and side views show an injection system fitted to an R-4360 below. Front and rear views of the master control unit are also illustrated.

Drive requirements for the vane-type pump and injection pump were derived from a splined drive extending upwards through the carburetor-mounting flange. Shafts and bevel gears supply the appropriate drive requirements located within the throttle body casting of the master control unit.

Each cylinder has an injection nozzle located in the cylinder head for injection directly into the combustion chamber. Fuel is injected at 500 psi during the intake stroke. High-pressure fuel is conducted from the injection pump through stainless-steel injection lines to the upper fuel distribution blocks, which are attached to the master control unit. Fuel pressurized to 500 psi is then transferred to passages within these blocks,

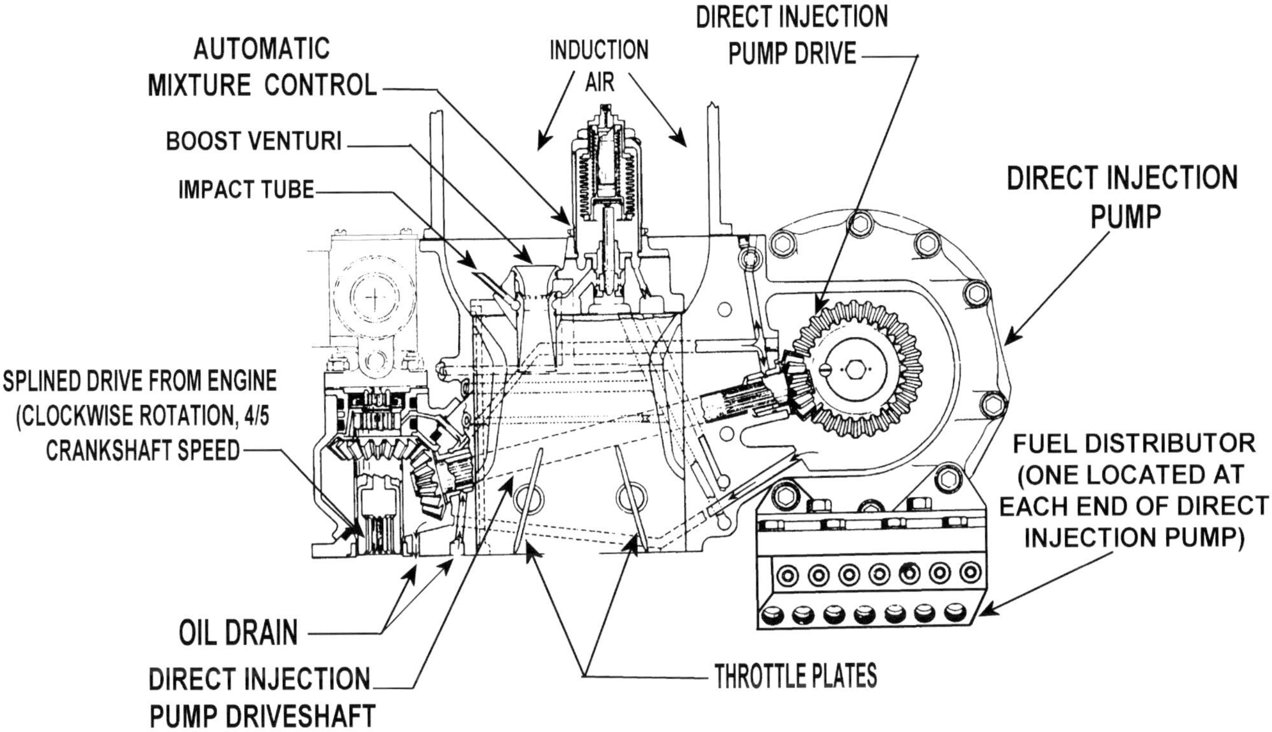

A positive displacement swash plate pump is used to feed the 28 individual injectors. The angle of the swash plate determines the stroke of the injection pump plungers and consequently how much fuel is injected. A tower shaft and bevel gears provide drive to the injection pump. It's somewhat reminiscent of a Rolls-Royce Merlin cam drive. *(Preliminary Instructions on Stromberg 100-28-A3 Direct Fuel Injection System. Courtesy of Pete Law)*

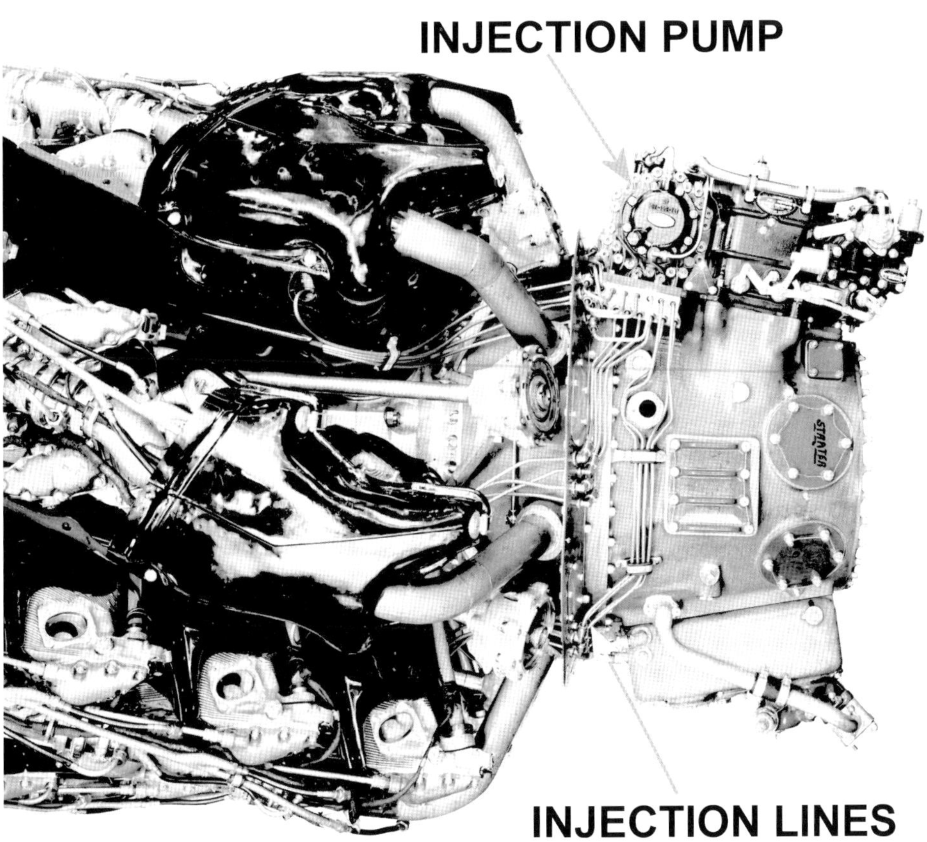

This side view shows one of the two injection pumps and some of the 28 injection lines. *(Courtesy of Pratt & Whitney)*

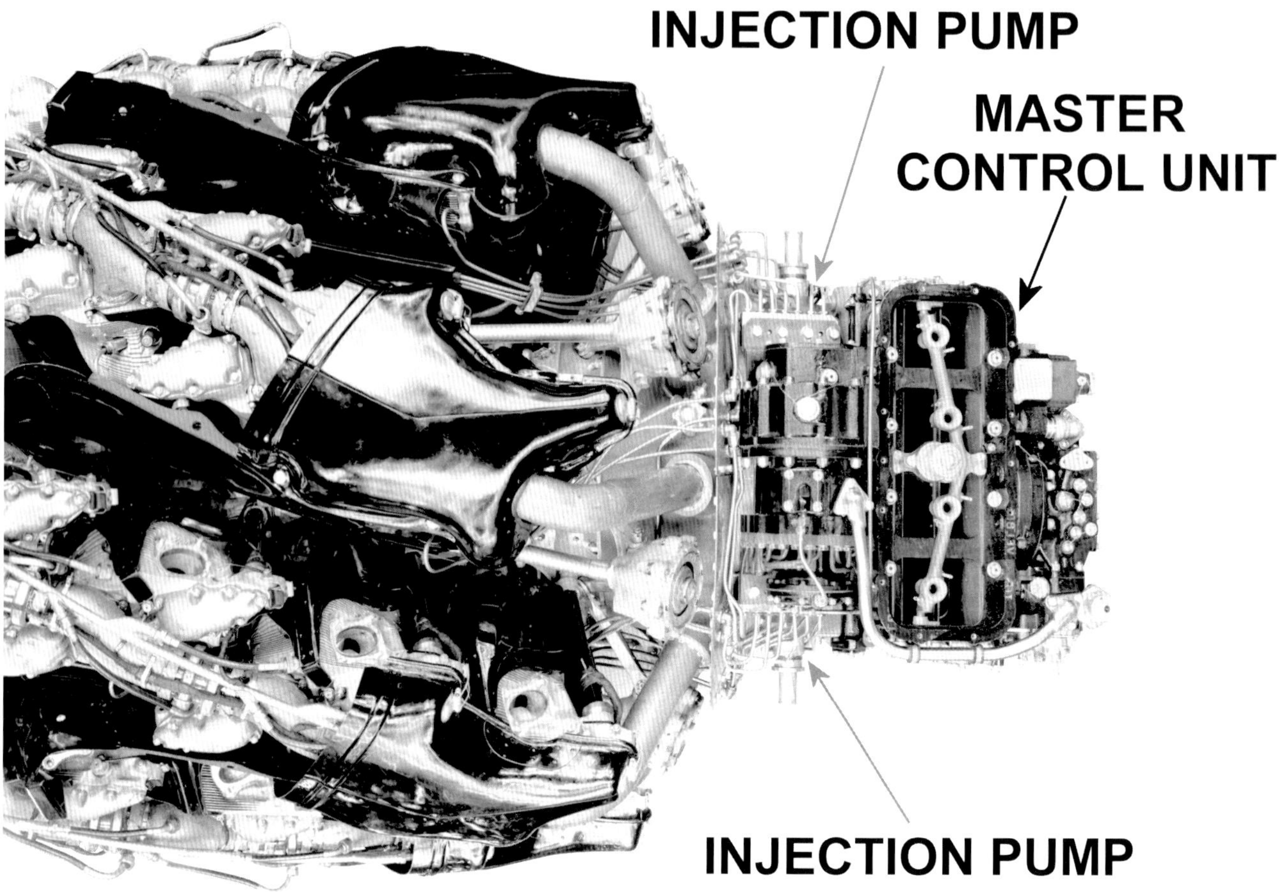

This top view shows the master control unit. It looks and operates in a similar fashion to the PR-100 carburetor. The dual injection pumps are also shown to good advantage. *(Courtesy of Pratt & Whitney)*

which in turn are mated with similar passages in the lower fuel distribution blocks. High-pressure fuel is conducted from the lower distribution blocks to the individual discharge nozzles through stainless-steel injection lines.

Operating in a similar fashion to the PR-100 carburetor, the manual mixture control has—like its carburetor sibling—three positions, which are "idle cut-off," "normal," and "rich."

Mounted on the master control unit and driven at 4/5-crankshaft speed, the vane-type pump provides a range of controlled fuel inlet pressures. These pressures vary with carburetor air inlet pressures by virtue of venting the relief valve diaphragm to the carburetor air entrance. Conditioned fuel is supplied to the vane pump by the aircraft's electric boost pump. Fuel is then delivered to the master control unit at the desired pressures. Provision is made that in the event of pump failure, fuel bypasses through the pump with a sufficient quantity to maintain engine operation.

Metered fuel is supplied to the 28-cylinder direct injection pump through a filter via the fuel control unit of the master control. This fuel is distributed to the high-pressure (500 psi) injection nozzles.

The 28 plungers of the injection pump are arranged in two concentric circles around the axis of the pump driveshaft. Plungers are operated by a rotating wobble, or swash, plate having two concentric lifts, each of which operates a circle of 14 plungers. Shoes at the bases of the plunger tappets ride on the surfaces of two rolling element thrust bearing plates. These plates remain stationary in relation to the wobble plate surface, which transmit the cam-like motion to the plungers. Each plunger is equipped with a return spring in

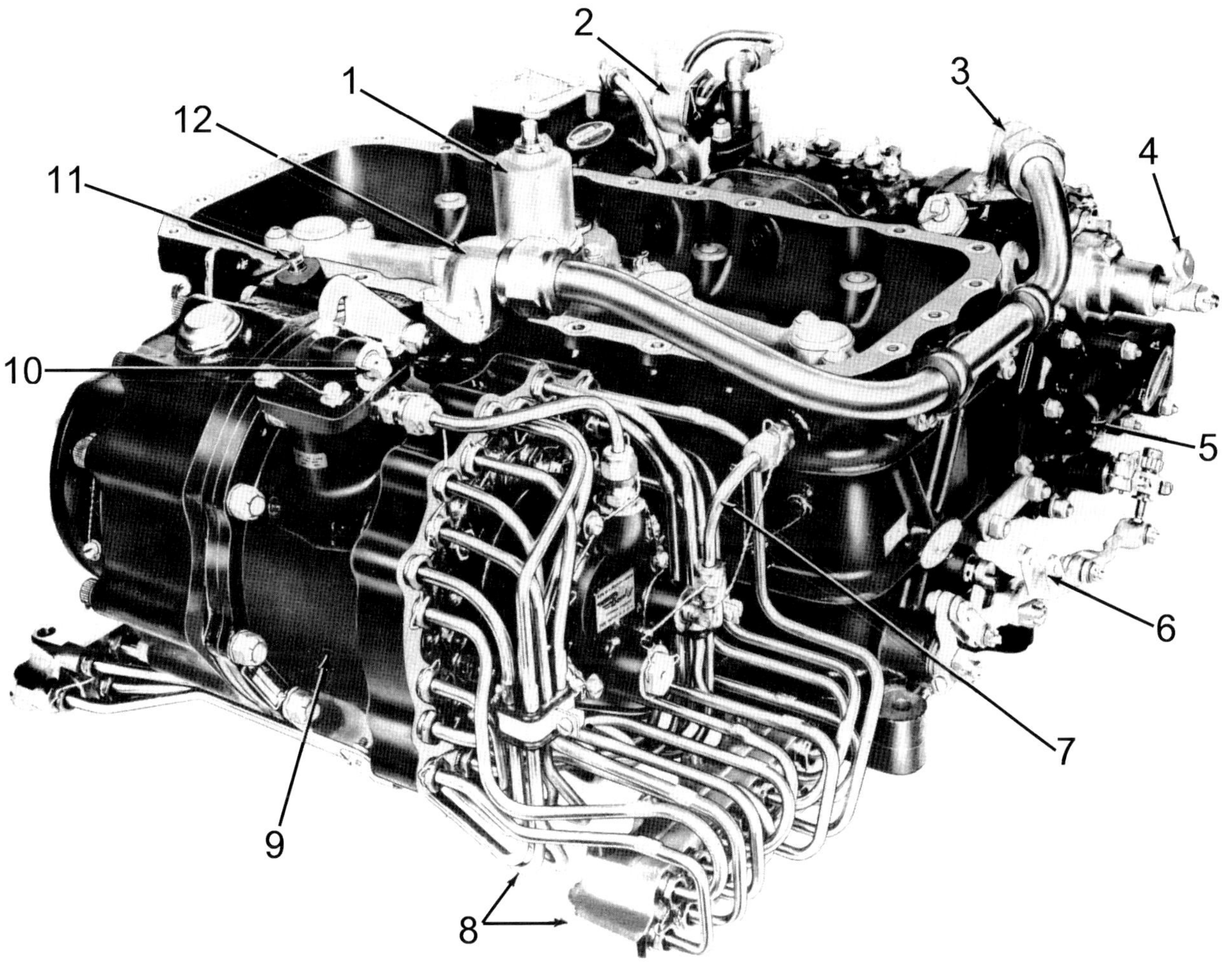

Bendix 100-28-A5 Direct Fuel Injection System (Front three-quarter view). 1. Unit Assembly-Automatic Mixture Control; 2. Unit Assembly/Electric Primer; 3. Fuel Outlet-Master Control; 4. Lever-Manual Mixture Control; 5. Valve-Derichment; 6. Lever-Throttle; 7. Vent to Top Deck, 8. Distributor Assembly-Fuel; 9. Pump Assembly-Direct Fuel Injection; 10. Vent Connection-Pump Vapor; 11. Pressure Connection-Oil to Injection Pump; 12. Fuel Inlet-Master Control to Injection Pump. *(Preliminary Instructions on Stromberg 100-28-A3 Direct Fuel Injection System. Courtesy of Pete Law)*

order to keep the plunger shoe in constant contact with the wobble plate contour, thereby returning plungers after each compression stroke and at the same time getting primed with fuel for the next cycle.

Lubrication of the vane pump, injection pump, and the driveshaft with its associated gears is accomplished through internal oil galleries. Engine oil at engine pump pressure is supplied to these parts with provision for drainback to the engine to be picked up by the scavenge system.

Cleanliness is essential in a system such as this. A dedicated oil filter is installed in the oil gallery leading into the injection pump body for this reason. An auxiliary fuel filter is mounted in the fuel transfer line located between the master control and injection pump. Magnetic chip detectors are located in the low-pressure fuel chambers of the direct injection pump.

Vapor is eliminated from the fuel section of the master control through two vapor vent valves. Vent lines from these valves are routed to the aircraft fuel tanks. A slide type vapor vent line is also located in the low-pressure fuel chamber of the injection pump and is, again, routed to the aircraft fuel tanks in a similar fashion.

So How Does Fuel Injection Work?

So far we have looked at the design details of the 100-28-A fuel injection system but haven't gone into great detail on how it actually accomplishes accurate fuel distribution to all 28 cylinders.

The master control is similar to the PR-100 series of carburetors, which has already been described in detail. Fuel is delivered to the vane pump of the master control by the aircraft's electric boost pump. The vane pump raises fuel pressure to the amount necessary to properly fulfill metering requirements of the master control. A pressure relief valve, integral with the vane pump, is controlled by carburetor air inlet pressure. A bypass valve incorporated in the vane pump allows fuel to be supplied to the injection unit at boost pump pressure in the event of vane pump failure.

Metered fuel from the fuel control unit of the master control is piped directly to the injection pump after being filtered through a secondary fuel filter.

A single bypass control carriage located inside the injection pump body holds and positions 28 individual bypass sleeves. Each of the 28 pump plungers operates inside one of these bypass sleeves. Therefore, control of the amount of fuel delivered by the plungers is dictated by the position of the sleeves, as a unit, which in turn varies the effective plunger strokes. The plungers have a constant length stroke, as controlled by the wobble plate contour, while the effective length of plunger stroke is controlled by the position of the bypass sleeves. "End of injection" of the plungers is a constant factor predetermined in the inherent design of the injection pump.

When in the retracted position, fuel in the low-pressure chamber of the injection pump enters into the open ports of the plunger. As the plungers move on their individual compression strokes under the influence of the wobble plate, the center ports in the plungers are covered by the plunger bushing, followed in sequence by the closing of the lower port by the bypass sleeve. At this time, the actual pumping begins. Under constant power conditions, the bypasses are held stationary in relation to the movement of the plungers except when their position is changed by the action of the bypass control plate for change in throttle setting.

As the upward movement of the plunger builds up sufficient pressure to overcome the force of the check valve spring, the check valve opens. This allows fuel to flow into the individual injection lines. Further upward movement of the plunger compresses fuel in the injection lines to 500 psi. At this point the injector, located in each cylinder head, opens and atomized fuel is sprayed into the cylinder. Pumping action continues until the center port in the plunger is opened by the annulus and port in the bypass bushing. High-pressure fuel then escapes back into the low-pressure area of the pump body. This relieves the pressure on the check valve, which closes, stopping fuel flow into the engine. In this way, fuel is trapped in the injection line. This prevents the line from being evacuated at the next stroke of the plunger. The point at which the center port in the plunger is opened by the annulus and port in the plunger bushing is termed "end of injection." It is a fixed, predetermined position.

The position of the plunger-bypasses determines the length of effective plunger stroke. The length of the effective stroke increases as the bypass sleeves are moved further down the plungers toward the pump wobble plate. Position of the bypasses is synchronized by means of the bypass control plate, which holds them in the same relative position on the plungers. The control shaft is connected to the pump control diaphragm and control spring by the bypass control plate. Metered fuel pressure is applied to one side of this control diaphragm and carburetor air inlet pressure is applied to the other side. The resulting force, metered fuel pressure minus carburetor air inlet pressure, overcomes the force of the bypass control spring. This force tends to move the bypass sleeve into the "idle cut-off" position. It properly locates the bypasses so that plungers pump the total volume of fuel delivered by the master control.

As the throttle valves in the master control are opened wider, metered fuel pressure applied to the pump control diaphragm increases more than the carburetor air inlet pressure increases. This results in increasing the differential pressure across the control diaphragm. This increased pressure differential imposed on the control diaphragm causes the diaphragm to move, compressing the bypass control spring and moving the bypass control plate in a direction away from "idle cut-off." This action increases the effective length of plunger strokes. Longer plunger strokes result in greater fuel delivery by each plunger to the injection nozzle.

When the pilot/flight engineer places the mixture control in "idle cut-off," fuel flow to the injection pump is cut off, consequently reducing the fuel pressure on the pump control diaphragm. The bypass control spring then moves the bypasses to a position such that even though the plungers are traveling their full stroke, the lower ports of the plungers are never covered. In this way, no fuel can be compressed sufficiently to open the check valves and no fuel is delivered to the engine. The engine quits as a result.

A similar scenario occurs if the fuel supply is exhausted from the aircraft fuel tank(s) while the aircraft is in flight. The bypasses move to the "idle cut-off" position, stop fuel flow to the engine, and leave the pump bodies full of fuel. In the event the pump is rotated due to a windmilling prop, no damage occurs to the pumping mechanism through lack of lubrication. It should be realized that fuel acts as a lubricant for the pump plungers. An additional bonus is the fact that fuel is always in the pump. This makes it possible to ensure a fast restart as soon as the fuel supply is re-established.

Overall, the R-4360 fuel injection system worked remarkably well considering it was conceived in World War II as a fast expedient to overcome the operating woes of R-3350s powering the Boeing B-29. However, the U.S. Navy wrote a scathing report on the value-add, or lack thereof, of direct fuel injection. Among their findings, the Navy claimed that with direct fuel injection there was a loss in altitude performance of the fuel injection engine compared to the carburetor engine. They also claimed that cylinder head temperature spread obtained with the fuel injection engine was no less than that obtained with the carburetor engine. The Navy followed through with their findings by not procuring aircraft powered by fuel-injected engines. A good example of this was the Wright R-3350 turbo-compound (TC). It is often assumed that all R-3350 TC engines were fuel injected. While this was true for the vast majority produced, some R-3350 TCs were manufactured with Bendix injection carburetors for Navy applications. And interestingly, the carbureted engines were rated at the highest power for this class of engine.

A Word About Water Injection, ADI, Wet and Dry Operations

Wet, water injection, and ADI all mean the same thing, although the term "water injection" is a misnomer because the mixture is actually 50/50 water/alcohol. Of course, dry means that the engine is operating on gasoline with no ADI—again a slight misnomer, but I think you get the idea. Alcohol is added to the mix to serve primarily as an anti-freeze.

So How Can an Engine Run on Water?

If an engine could be persuaded to run on water, it would be akin to perpetual motion. However, under certain circumstances, the use of water allows the engine to produce more power, not because the water acts as fuel but rather because it allows the fuel being burned in the engine to be used more efficiently. Due to the supercharger boosting manifold pressure to above atmospheric, the R-4360 spends most of its operational life running at a positive boost. Of course, this now gets us into the basic gas laws. In this case a mixture of air and fuel, or air in the case of a direct fuel injected engine, undergoes a rise in temperature.

That temperature rise can be considerable depending on the amount of boost, somewhere in the order of 200 degrees F at takeoff power. Under normal combustion conditions, the compressed fuel/air mixture in the cylinder is ignited by the spark plug. This event typically occurs at approximately 20 to 30 degrees before top dead center.

This ignition source creates a so-called flame front. In other words, the mixture burns in a very controlled, albeit rapid, fashion. As the flame front advances across the cylinder, the unburned fuel/air mixture is further heated due to the effects of radiated heat from the flame front and increasing pressure within the cylinder. Ideally, combustion is complete before the remaining unburned fuel/air mixture can ignite spontaneously. However, ideal conditions do not always prevail. A condition known as detonation can take over the normal combustion process. Under conditions of detonation the flame front is established, as with normal combustion, by the spark plug igniting the fuel/air mixture. As the flame front advances burning the fuel/air mixture in the normal fashion, the unburned fuel/air mixture undergoes considerable heating. If conditions are right, the temperature of the unburned fuel/air mixture reaches a point where it self ignites and burns in an uncontrolled fashion.

This uncontrolled burning causes a dramatic pressure spike that, if left to its own device, causes considerable mechanical damage and reduced power output. If the temperature of the unburned fuel/air mixture can be kept below that of self-ignition, all is well. This is where water comes to the rescue. Two things are accomplished by the use of water. First, it reduces the temperature of the incoming charge into the cylinder. By spraying water into the supercharger impeller along with the fuel/air mixture, a dramatic reduction of charge temperature results due to evaporation of the water. Furthermore, when the fuel/air mixture is ignited within the cylinder, the temperature of the flame front is reduced due to the influence of the water. Of course, conditions favorable to detonation only occur during high power settings such as takeoff. It is at these times, such as high power operation, that water comes into its own. If the engine is operated at, for instance, cruise conditions, the use of water accomplishes very little. Another way of looking at the advantages of water is to think of it as a fuel performance number enhancer. On a highly supercharged engine, it only takes a few seconds of running in the detonation range to wreak havoc resulting in melted piston crowns and damaged combustion chambers. For an R-4360, the maximum safe manifold pressure with 115/145 fuel was approximately 55 in. Hg. With the application of ADI, manifold pressure can be increased to 59 in. Hg., resulting in a 500 hp gain. In summary, it should be emphasized that the use of ADI does not in and of itself increase power, but rather allows the use of a higher manifold pressure than would be possible without the use of water. This higher manifold pressure consequently increases the power of the engine.

Water Injection, a Historical Perspective

Engine development pioneers such as England's Sir Harry Ricardo had experimented with water injection in the early 1900s. The U.S. Navy tested a World War I Liberty engine with water injection in 1926. Their findings were similar to that of Ricardo's. They concluded:

Although the Navy had determined, during the course of its laboratory tests of a Liberty engine in 1926, that the use of water would provide a substantial increase in the maximum power output of an aircraft engine due to the cooling effects and the tendency to act as a detonation suppresser, the investigation was not placed on high priority at that particular time for a number of reasons:

(a). It was appreciated that the gain with water injection would depend upon the particular engine for which it was intended, inasmuch as the quantities of permissible water flow would be linked directly to the structural strength and com-

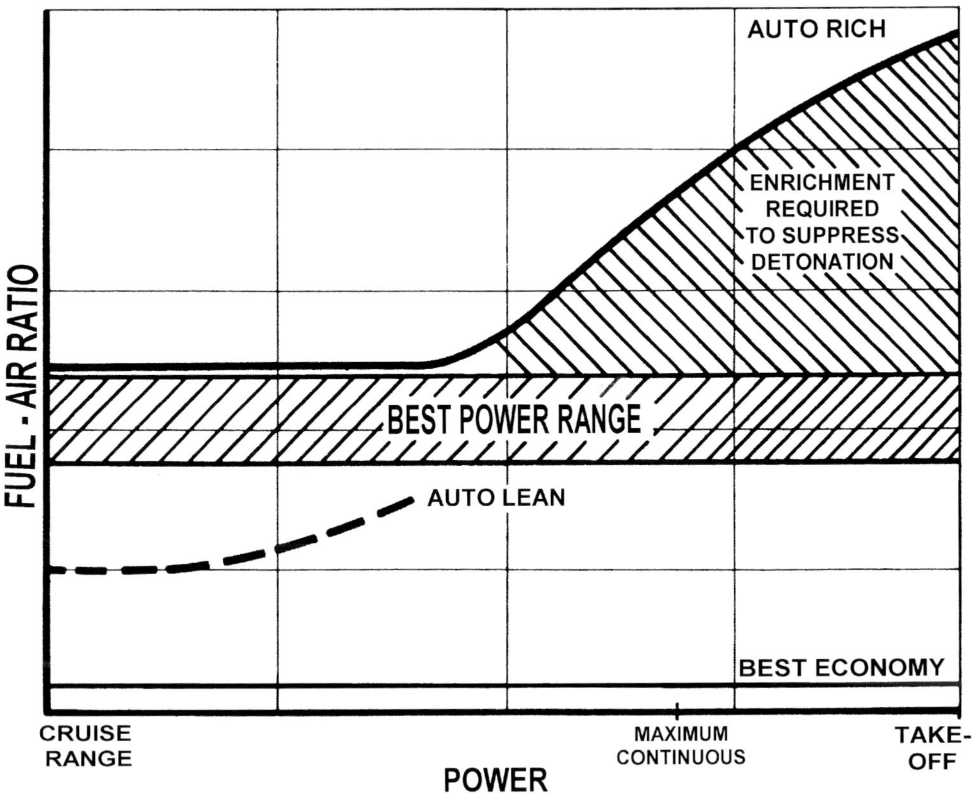

Without ADI (water injection), the fuel/air mixture needs to be richened by a considerable amount as power levels approach the detonation region. With the application of ADI, this enrichment is not necessary and in fact could be harmful. This graph shows the degree of derichment applied when the engine operates in what would normally be the detonation range if it were not for the application of ADI. *(Courtesy of Pete Law)*

bustion characteristics of the particular engine model or design in question.

(b). It was known that the determination of the best water flows for any engines could be quickly established when the need for higher powers arose or was anticipated.

(c). It was known that the use of ratings higher than those guaranteed by the engine manufacturers as safe for engine operation was, in general, undesirable, because the decrease in engine durability and reliability over balanced any gain to be expected from the higher powers.

The greatest deterrent at the time to the use of the higher powers, which might be obtained with water injection, had the need for such powers existed, was the fact that the only propellers available were of the "fixed-pitch" type. This propeller could not absorb large increases in power without seriously overspeeding. Such overspeeding would cause structural failure in some instances and, at the higher rotational speeds, the propellers would become less efficient in their absorption of power.

Reading the foregoing is a rather scary thought. The Navy admitted that water injection would increase power, yet refused to pursue it at the time (1926) on the grounds that it would over stress existing engine designs; propellers were not designed to absorb the additional power, etc. Fortunately, this "head-in-the-sand" viewpoint did not last too long. By World War II, the Navy, along with the Army Air Force, was heavily promoting the use of water injection.

Description of the Water System

Design of a water system involved more than a few "design challenges." The correct amount of water has to be introduced into the engine under the correct operating conditions. Also, in the event the aircraft water supply runs out, safety features are required to ensure that a "wet" power setting is not maintained under "dry" conditions. When water flows into the engine, the fuel needs to be leaned out in order to keep the liquid inflow to the engine at roughly the same value regardless of whether the engine is running "wet" or "dry." This requirement is accomplished via a

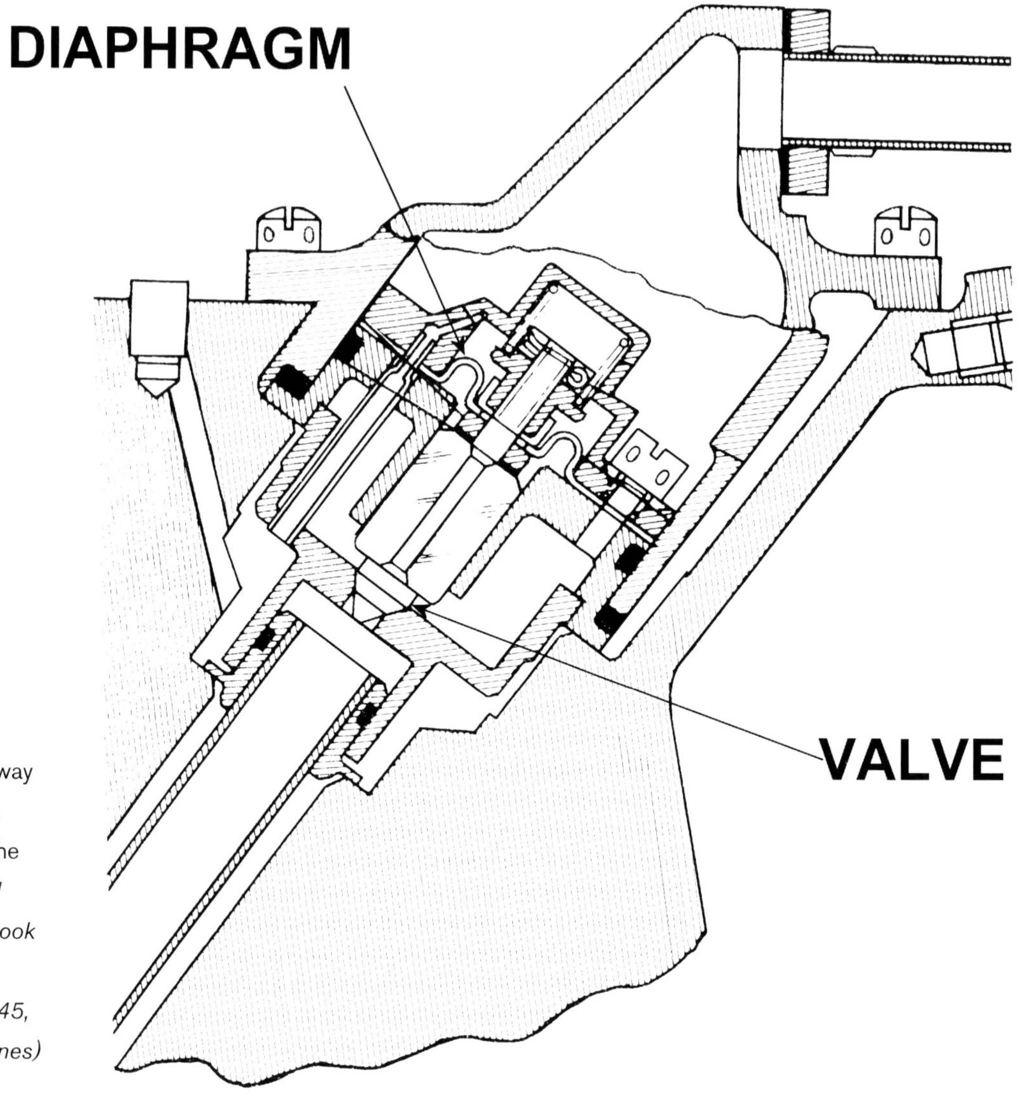

The water feed valve operates in a similar way to the fuel feed valve. Water pressure lifts the valve off its seat. *(AN 02A-10HC-2. Handbook Service Instructions Models R-4360-17, -45, and -47 Aircraft Engines)*

so-called "derichment" valve mounted on the carburetor. And as if the foregoing requirements were not enough, severe corrosion from the ADI mix had to be accounted for. The accompanying graph shows the additional fuel required as the engine starts to operate in the detonation range. This graph also shows why derichment is necessary.

Components

The ADI system is made up from the following sub components: a water tank (typically holds 20 to 50 gallons per engine), a pump with built-in pressure regulator, water supply line, water regulator (meters the correct amount of water to the engine), carburetor derichment valve to reduce fuel flow during ADI operation, and a cockpit control switch to actuate the system. In addition to the basic requirements described, other controls are sometimes used such as safety devices to prevent operation of the water system without the engine running. If the water system was operated without the engine running, serious consequences could and did occur. These included hydrolocking due to water entering the lower cylinders and the ever-present corrosion problem.

Some systems mixed the ADI fluid with the fuel at the fuel feed valve. Other systems had a separate water feed valve.

Description of the Water Regulator

The heart of any water system is the regulator. In the case of the R-4360, it is bolted to the side of the accessory drive case. Two basic types of water regulators are used—constant flow and variable flow. As these descriptions imply, the constant flow type only allows for one flow rate. This means that it has to be set at a flow rate to accommodate the highest power envisioned for the engine. At less than maximum power, water flow is "rich." The variable flow regulator, on the other hand, allows for varying water flow rates according to power setting.

The variable flow regulator requires the use of a two-position solenoid valve, used primarily as a selector valve. Acting as an on/off switch, the solenoid valve determines whether or not water injection is to be used. When water injection is used, the water pump is turned on and the solenoid valve is actuated to move the plunger to the right as shown in the schematic drawing. With the solenoid valve actuated, water is free to enter the regulator. This is because low pressure resulting from low blower throat suction no longer acts on the poppet valve diaphragm. But water does not enter the engine until sufficient metering force is developed to permit water pressure and open the spring-loaded check valve located immediately downstream from the metering pressure control valve.

When normal rated power is achieved, the spring-loaded check valve opens, giving a rapid rise in water/air ratio as power is increased. At the onset of water flow through the main water jet, the pressure differential is transmitted to a diaphragm-controlled spring-loaded valve which, when opened, allows water pressure to build up in the derichment valve. This water pressure closes the derichment valve and consequently reduces fuel flow through the carburetor to a predetermined setting for best power mixture ratio, rather than "full rich."

When the check valve has opened, sufficient pressure differential is developed across the spring-loaded enrichment valve diaphragm to permit the enrichment valve to start opening. As power is further increased, the water/air ratio is increased correspondingly until the water enrichment valve is wide open. At this point the water/air ratio levels off at a constant value. This enriching action, or increase in water/air ratio, corresponds to the action of the enrichment valve and jet in the carburetor.

Finally, as with most other engines fitted with ADI, an extra 23 percent improvement in takeoff power was typically available for a relatively small increase in weight. Despite the maintenance requirements due to corrosion and overhauling water regulators, the payback was well worth the trouble. And this was true regardless of whether it involved military or commercial operations.

References

6-1 *U.S. Navy Progress 1935 – 1945 Vol. II – Powerplant Design.* Unpublished Navy report compiled by Dan Whitney.

6-2 *Handbook of Overhaul Instructions. Chandler Evans Carburetor Model 100-CPB-9.* Navy AN-03-BE-2. November 20, 1951.

6-3 *Parts Catalogue. Chandler Evans Carburetor Model 100-CPB-9.* Navy AN-03-BE-3. November 20, 1951.

6-4 *Handbook of Overhaul Instructions. Direct Fuel Injection Systems Models 100-28-A4, 100-28-A5.* Bendix Products Division, South Bend, Indiana, U.S.A. November 1, 1953.

6-5 *Preliminary Instructions on Stromberg 100-28-A3 Direct Fuel Injection System.* Bendix Aviation Corporation. South Bend, Indiana. Form 10-1167. December 6, 1948.

6-6 *Illustrated Parts Breakdown, Injection Carburetor, Model PR-100 B4 Used on R-4360.* Bendix Products, South Bend, Indiana. June 15, 1951.

6-7 *Water Injection Systems For Pratt & Whitney Aircraft Reciprocating Engines.* Report No. PWA. Inst. 92C. July 15, 1954. Pratt & Whitney Aircraft, East Hartford, Connecticut.

CHAPTER SEVEN

Military Applications

From one perspective it could be argued that the R-4360 was a day late and a dollar short. It was desperately needed for World War II, but despite the valiant efforts of Pratt & Whitney engineers, it didn't quite achieve this goal. Considering the complexity and new ground being pioneered, it's a miracle that so much was accomplished in such a short time period. Even so, the R-4360's heyday was late 1940s to early 1950s. During this time frame gas turbine powered aircraft were developed. These aircraft could outstrip any piston engine regardless of how good it was. But some niches still remained for the R-4360, such as strategic bombing and transport aircraft. Until the worst of the development problems associated with the Boeing B-47 could be ironed out, aircraft such as the Boeing B-50 and B-36 held the line against the Red Peril. But it wasn't long before the B-50 was shuffled off to secondary roles such as aerial refueling and weather reconnaissance. Notwithstanding its size, the B-50 was redesignated as a medium bomber in the early 1950s. However, many aircraft didn't even get to this stage. Excellent aircraft such as the Douglas XTB2D-1 and Boeing XF8B-1 were built as prototypes then dropped like a hot potato as the military realized they had to deal with a paradigm shift from piston power to gas turbine power. Despite being an engineering marvel, the B-36 was outstripped by the infinitely more capable B-52. Consequently, Convair's Peacemaker was the last of the large piston powered strategic bomber developments, even though it soldiered on through the 1950s. The only role that survived the 1950s for the R-4360 was that of transport and aerial refueling. Douglas C-124s and Boeing C/KC-97s earned their keep well into the 1970s, and KB-50s were phased out in the early 1960s.

The table opposite summarizes military versions of the R-4360 and what they powered *(Ref. 5-3)*.

Aircraft are described in the order they appear in the table.

Goodyear XF2G-1, F2G-1, F2G-2

One of the finest Navy/Marine aircraft to come out of World War II was the famous F4U Corsair. Although a Vought design, Corsairs were also manufactured under license by Goodyear and Brewster. Normally powered by various iterations of the R-2800, Pratt & Whitney felt that an R-4360 could be shoehorned into the Corsair without too much difficulty. As hot rodders know only too well, too much power is just enough. And so it was with the "Super" Corsair. Despite many opinions to the contrary, it's difficult to over-power an aircraft. Adding urgency to the program was the ever-increasing menace of Kamikaze attacks on U.S. Navy ships. A high-performance, low-altitude aircraft was desperately needed to ward off this increasing threat.

(Table 7-1)

AIR FORCE	NAVY	APPLICATIONS
	R-4360-2	Goodyear F2G-1 Goodyear F2G-2 Martin XBTM-1
	R-4360-2A	Goodyear F2G-1 Goodyear F2G-2 Martin XBTM-1
	R-4360X-2	None
R-4360-3		Republic XP-72 (dual rotation) Curtiss XP-71 (none built)
R-4360-4	R-4360-4, R-4360-4T	Goodyear XF2G-1, F2G-1 Martin XBTM-1 Martin JRM-2
	R-4360-4W,	Goodyear F2G-1
	R-4360-4T	Martin AM-1, AM-2, AM-1Q Martin XBTM-1 Martin JRM-2 Martin XP4M-1
R-4360-4A		Hughes XF-11 Hughes HFB-1 (H4) Hercules "Spruce Goose"
R-4360-5		XB-36
	R-4360-6	None
R-4360-7		Northrop B-35 (Outboard engine)
	R-4360-8 (Semi-production)	Douglas TB2D-1 Skypirate
	R-4360-8A	None
	R-4360-8 (Production)	Douglas XTB2D-1, TB2D-1 Skypirate
R-4360-9		Vultee XA-41 Convair A-41
R-4360-9T		None
	R-4360-10 (Semi-production)	Boeing XF8B-1
	R-4360-10 (Production)	None
R-4360-11		Northrop B-35 (Inboard engine)
	R-4360-12	None
	R-4360-12A	None
R-4360-13		Republic XP-72 (single rotation)
	R-4360-14	Curtiss XBTC-2
R-4360-15		Douglas C-74
R-4360-17		Northrop B-35 (Outboard engine) Northrop XB-35 (Outboard engine) Northrop YB-35 (Outboard engine)
	R-4360-18	Lockheed XR60-1 (Constitution) Lockheed R6V-1 Model 89
R-4360-19		Planned for Republic P-72 – never built
	R-4360-20, R-4360-20A, R-4360-20W, R-4360-20WA	Douglas C-124, -124A Fairchild R4Q-1 Fairchild C-119B, C Fairchild C-120 Martin XP4M-1, P4M-1
	R-4360-20B, R-4360-20C, R-4360-20WB, R-4360-20WC	None
R 4360 21		Northrop B-35 (Inboard engine) Northrop XB-35 (Inboard engine) Northrop YB-35 (Inboard engine)
	R-4360-22W	Lockheed R6V-1 (Constitution)
R-4360-23		Curtiss XP-71 (Turbosupercharged and dual rotation. None built).
	R-4360-24	Martin JRM-2
R-4360-25		Convair B-36A Convair XC-99 Convair Model 37 (Never built)
R-4360-27		Douglas C-74 Douglas XC-74
R-4360-29		Planned for Boeing XB-44, re-engined version of B-29
R-4360-31		Hughes XF-11 Republic XF-12
R-4360-33		Boeing XB-44, converted B-29
R-4360-35, R-4360-35B		Boeing TB-50A, B, D, H Boeing B-50A, B, D Fairchild XC-119A Republic XF-12 (Planned)
R-4360-35A, R-4360-35C		Boeing C-97A, C-97C Boeing KC-97E, Boeing YC-97A, YC-97B, Douglas XC-124A

continued next page

(Table 7-1 continued)

AIR FORCE	NAVY	APPLICATIONS
R-4360-37		Hughes XF-11 Republic XF-12, XR-12
R-4360-39		None
R-4360-41,		Convair B-36A, B, D, E, RB-36
R-4360-41A		Convair XC-99
R-4360-43		Boeing YB-50C, Boeing B-54A, RB-54A
R-4360-45		XB-35, B-35, YB-35 (Outboard engine)
R-4360-47		XB-35, B-35, YB-35 (Inboard engine)
-49, -49A		Douglas C-74
R-4360-51		Convair B-36C study for tractor installation
R-4360-53		Convair B-36D, E, F, H, J Convair RB-36D, E, F, H
R-4360-55		None
R-4360-57		None
R-4360-59		Douglas C-124 study
R-4360-59A		Boeing C-97D
R-4360-59B		Boeing KC-97F, KC-97G
R-4360-61		None
R-4360-63		None
R-4360-63A		Douglas C-124C
R-4360-65		Boeing C-97A, C Boeing KC-97E

Pratt & Whitney made a proposal to the Bureau of Aeronautics in early 1943 that an R-4360 be installed in an F4U-1 airframe *(Ref. 5-4).* At this time, installing a 3,000-hp engine in a single-engine fighter was unknown territory. Concerns included stability, particularly with a single-rotation propeller. In other words, a dual-rotation or contra-rotating propeller would have eliminated one of the key concerns, that of torque reaction, particularly at low speeds and more especially so at takeoff.

On July 17, 1943, a standard production F4U-1 powered by a Pratt & Whitney R-2800-8 was delivered to Pratt & Whitney's East Hartford facility. First order of business upon delivery of the Corsair was to remove the R-2800 QEC. The proposed R-4360 engine was a semi-production TSB1-G, manufacturer's number P-3. The Hamilton Standard propeller was delivered on August 26, interestingly with feathering capability, a feature one does not normally associate with single-engine aircraft. Ground runs were initiated on September 6 and the first flight quickly followed on September 12, just 4½ months after the contract with the Bureau of Aeronautics had been executed. Powering the aircraft with the TSB1-G was an obvious choice because it featured a lighter and simpler single-stage, variable speed supercharger. This engine

met the requirement of high power at low altitude without the weight or complexity of a two-stage, intercooled installation. A single stroke eliminated all the ductwork and intercoolers of a two-stage installation. Designated F4U-1 (WM—presumably the WM stood for "Wasp Major"), the aircraft was an early production F4U with the "birdcage" style canopy that identified these initial production Corsairs. Rather then re-invent the wheel, the standard F4U-1 cowl was used with a 12-inch extension.

An interesting feature of the TSB1-G engine was its co-axial power take-off at the rear. It's conceivable that Pratt & Whitney also had ideas of installing an auxiliary blower. Had this been done, it would have been necessary to mount the blower in the rear fuselage in a similar fashion to the Republic XP-72; however, the foregoing is pure conjecture on the author's part. Nevertheless, it would have been doable. During its development, several ram recovery scoops were tried out, including a so-called "flush" scoop and a conventional ram-air scoop, which was the configuration chosen by Goodyear for the production aircraft.

continued on page 320

Opposite: This photo offers a good comparison of a standard F4U-1 powered by an R-2800-8 and the same aircraft fitted with an R-4360. Pratt & Whitney performed this modification in 1943. *(Courtesy of Pratt & Whitney)*

Above: A very early standard production F4U-1 was used for the R-4360 conversion. The so-called "Birdcage" canopy is an immediate identifier. *(Courtesy of Pratt & Whitney)* *Right:* Even though it was a proof-of-concept aircraft, the F4U WM was a nicely done conversion. The ring around the engine near the man's right hand represented the joining line between the original F4U-1 cowl and the extension for the R-4360. *(Courtesy of Pratt & Whitney)*

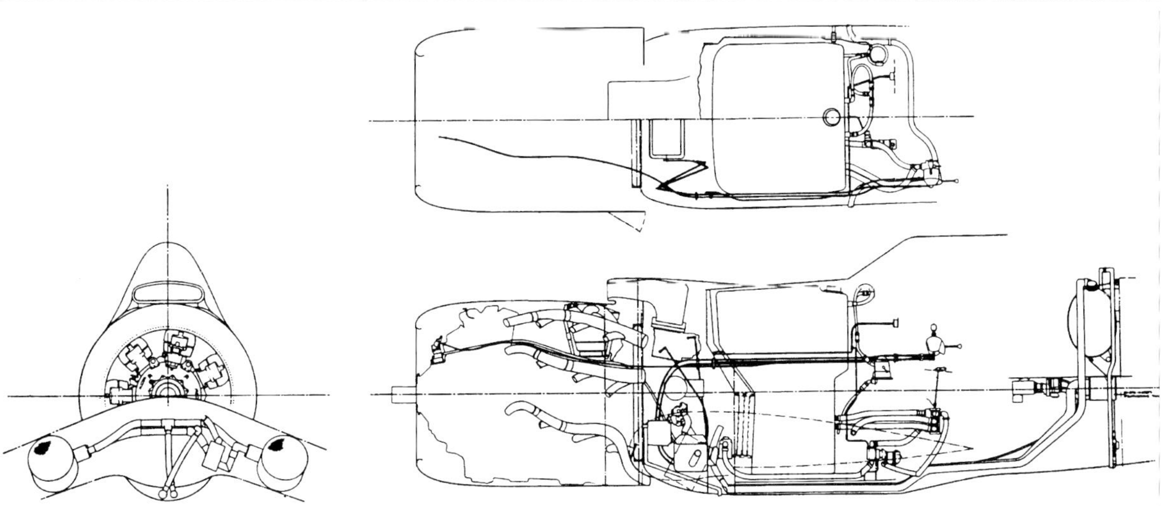

This is a copy of the original engineering installation drawing for the F4U-1 WM. A simple manifolded exhaust system rather than a siamesed system is notable. *(Courtesy of Pratt & Whitney)*

(Table 7-2)

Component	Weight for R-2800 Powered Aircraft	Weight for R-4360 Powered Aircraft	Difference
Airframe Group:			
Wing Group	2122	2122	0
Tail Group	164	164	0
Fuselage Group	755	755	0
Landing Gear	676	676	0
Engine Section Group:			
Engine mount	118	126	+8
Vibration mounts	39	128	+89
Cowl & cowl flaps	132	150	+18
Misc.	14	14	0
Powerplant (QEC):			
Engine	2459	3325	+866
Exhaust system	75	74	-1
Intercooler & supports	95	0	-95
Diaphragm	29	34	+5
Ducts	72	30	-42
Misc. accessories	8	8	0
Controls	27	27	0
Propeller installation	494	694	+200
Starting system	61	57	-4
Oil coolers	95	128	+133
Oil tank	21	28	+7
Piping, fittings and misc.	26	90	+64
Fuel system	272	272	0
Fixed equipment group	941	941	0
TOTAL WEIGHT DIFFERENCE:			**1,248 POUNDS**

continued from page 317

Table 7-2 shows comparative weights of the F4U-1 (WM) compared to a standard F4U-1 *(Ref. 5-4)*..

Goodyear is often credited with developing the R-4360-powered version of the Corsair. In fact, it was Pratt & Whitney who did all of the up front design work to install this huge engine into an airframe it was never intended for. Goodyear implemented a number of airframe modifications, the most noticeable being the bubble canopy in place of the turtle deck. This idea was first tried on an FG-1 (Goodyear's designation for the Vought-built F4U-1) by using a P-47D bubble canopy and modifying the rear fuselage *(Ref. 7-1)*. Built in two versions, F2G-1 and F2G-2, the primary difference was in the way the wings folded. On -1s they were folded by hand and -2s used hydraulic actuation for wing folding, thus making the -2 more suitable for carrier operation. In addition, -2s were also fitted with arresting gear. Elimination of the turtle deck plus the considerable increase in power necessitated a larger and taller vertical stabilizer.

With the massive cutbacks and contract cancellations at the end of World War II, the F2G was an early victim of these measures. Consequently, only 17 were manufactured, including prototypes. This number was made up from seven XF-2G-1s, five F2G-1s and five F2G-2s *(Ref. 7-2)*..

Interestingly, the main claim to fame for this formidable aircraft was its use in Thompson Trophy racing in the late 1940s.

Martin Mauler XBTM-1, Martin AM-2, Martin AM-1, Martin AM-1Q

Also and aptly known as Mauler, Martin's awesome AM-1 was another R-4360-powered postwar aircraft that saw very limited production and service. Initially known as the XBTM-1, the Mauler was developed at the request of the Navy in 1943 for a follow-on aircraft to the then current TBF/TBM, SB2C torpedo bombers. Assisting with the rapid development of this aircraft

continued on page 324

Front view of the F4U-1 WM showing the massive four-blade Hamilton Standard propeller. Also of note in this photograph is the small size of the leading-edge intakes. Unlike the standard F4U, the WM only used the wing intakes for oil cooling. The F4U-1 also used them for induction air and intercooler air, so consequently they were quite a bit larger. *(Courtesy of Pratt & Whitney)*

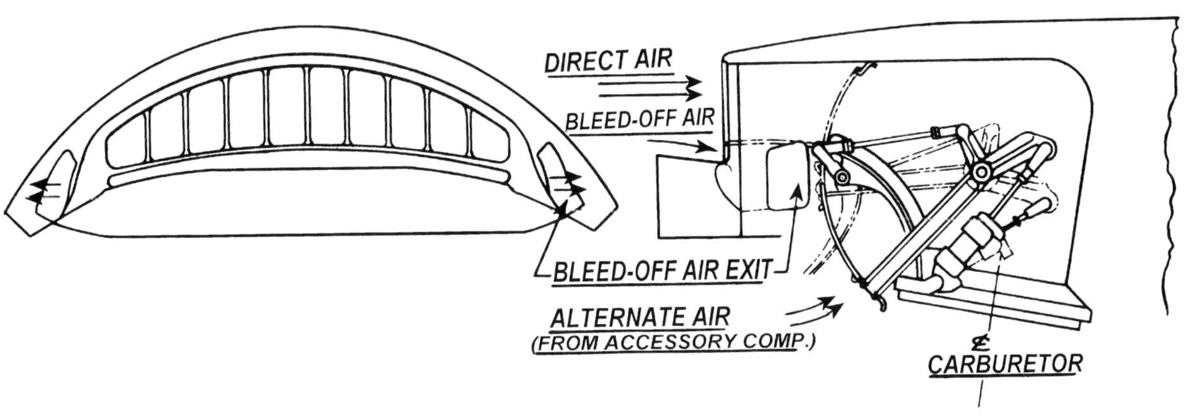

This is the induction scoop used for the production F2G. Note the boundary layer and bleed-off air exits. Clearly, a lot of engineering time was spent on this component and yet when F2Gs were raced in the immediate postwar years, the racers replaced this sophisticated scoop with a much simpler one that added even more power. *(Courtesy of Pratt & Whitney)*

Goodyear F2G Parameters *(Ref. 7-1)*:

Wing span	41 feet
Length	33 feet, 9 inches
Height	16 feet, 1 inch
Engine	R-4360-2, -2A, -4, -4T, and -4W
Max. speed (without ADI)	428 mph at 16,500 feet
Max. speed (with ADI)	450 mph at 16,500 feet
Max. rate of climb	7,000 feet per minute
Range (with aux. drop tanks)	2,500 miles

Above: This was the final production item based on the pioneering F4U WM–Goodyear's F2G. *(Courtesy of National Archives & Records Administration) Inset:* Several styles of induction scoops were tried in the F4U WM including a so-called flush scoop and a raised scoop (shown). An efficient ram recovery scoop could add a considerable amount of power. At 200 mph at sea level it amounts to 5 percent or about 150 hp. As speed increases, the power increase goes up dramatically. *(Courtesy of Pratt & Whitney)*

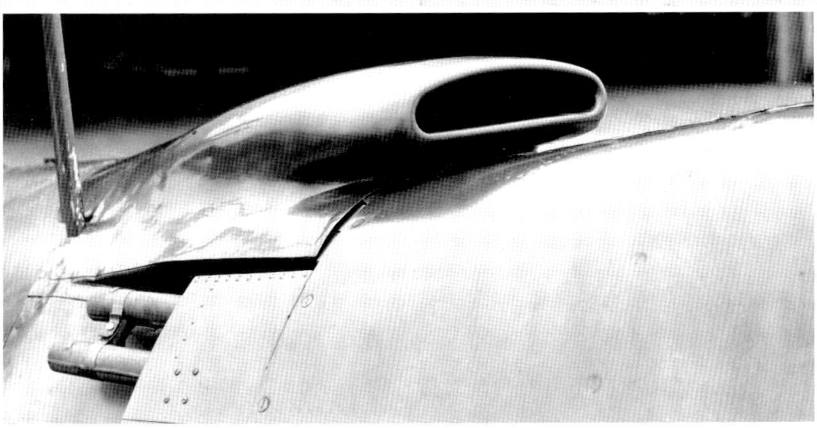

Above: Prototype for the Martin Mauler–the XBTM-1. This huge single-engine Navy aircraft was only produced in limited numbers. *(Courtesy of National Archives & Records Administration) Inset:* In this often-published photograph, the weight lifting capability of the Mauler is all too apparent. It's seen here with three 2,200-pound torpedoes, 800 rounds of ammunition, and 12 rockets. *(Courtesy of National Archives & Records Administration)*

continued from page 320
was the fact that B-26 Marauder production was winding down consequently freeing up a large number of engineers. First flight occurred in August 1944, at which time an order for 750 aircraft was placed. At the same time, the designation XBTM-1 was changed to AM-1 Mauler. Considering the austere financial environment following the cessation of hostilities, it's surprising that 149 Maulers were delivered. First entering service in 1948, it quickly became an orphan to the extent it was transferred to Navy Reserve squadrons following a brief and less than stellar performance aboard carriers.

Subjecting such a huge and heavy aircraft to carrier landings and arresting loads was more than the AM-1 could tolerate. Instances occurred in which the entire rear fuselage was ripped off. Fixes were incorporated, such as installing a roller on the arresting hook in order to reduce side loads.

AM-1 Parameters

Type	single-seat carrier based attack air craft (AM-1) or two seat electronic counter-measures (AM-1Q)
Crew	one (AM-1), two (AM-1Q)
Armament	four 20-mm cannons and 4,500 pounds of bombs, rockets, or torpedoes
Length	41 feet, 2 inches
Wingspan	50 feet, 1⅜ inches
Height	16 feet, 10 inches
Wing area	496 square feet
Empty weight	14,500 pounds
Max. T/O weight	23,386 pounds
Engine	(AM-1 & XTBM-1): R-4360-4, R-4360-4T, R-4360-4W
Range	1,800 miles (XTBM-1) 1,324 miles (AM-1)
Rate of climb	2,780 ft/min (XTBM-1) 3,130 ft/min (AM-1)
Cruise speed	189 mph
Max. speed	367 mph at 11,600 feet
Service ceiling	27,000 feet (XTBM-1) 25,630 feet (AM-1)

This in-flight photo of a production version AM-1 Mauler looks very similar to its prototype sibling. *(Courtesy of National Archives & Records Administration)*

Tail wheel assemblies tended to give trouble as well, oftentimes collapsing under landing loads. Exacerbating the AM-1's performance woes was the fact it was a heavy aircraft on the controls—not a desirable feature for a carrier-based aircraft. Its carrier days ended in October 1950 and it was totally withdrawn by 1953. Notwithstanding the Mauler's shortcomings, it was capable of hauling prodigious loads aloft. One famous instance, photographed in April 1949, shows the Mauler with three 2,200-pound torpedoes, 800 rounds of ammunition, and 12 rockets. Eighteen AM-1s were converted to the then new electronic counter measures aircraft and re-designated AM-1Q. This necessitated an additional crew space for the electronics operator *(Ref. 7-3, 7-4, and 7-5).*

The Mauler was a conventional design featuring tail dragger gear. Likewise with the QEC design,

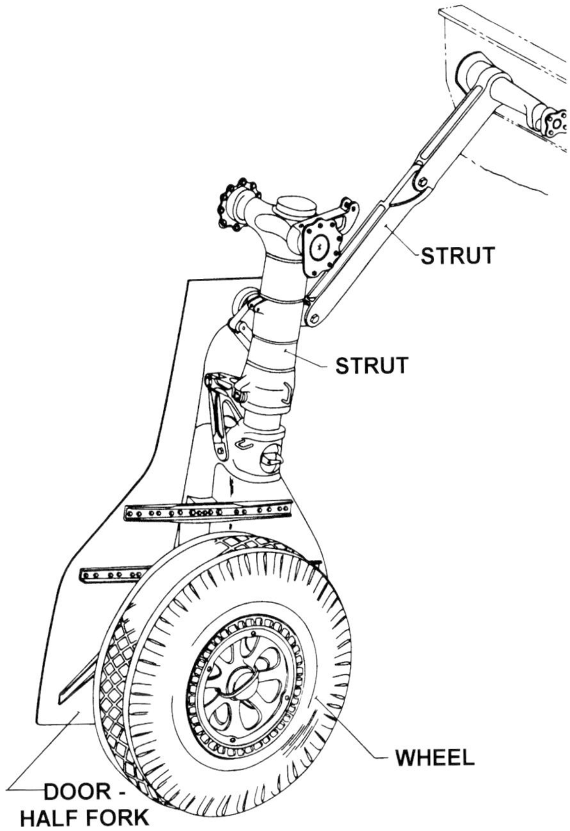

Above: Main gear for AM-1. It retracted to the rear in a similar fashion to a Grumman F6F Hellcat and its contemporary, the Douglas AD Skyraider. As it retracted it also rotated 90 degrees to lie flat in the wing. *(AN 01-35EF-2 Martin Mauler Erection & Maintenance Manual)*

Below: Tail wheel assembly for the AM-1. During carrier operations, a number of failures occurred with this assembly. One can only imagine the loads imposed upon the rear of the aircraft when snatching a cable during carrier operations. *(AN 01-35EF-2 Martin Mauler Erection & Maintenance Manual)*

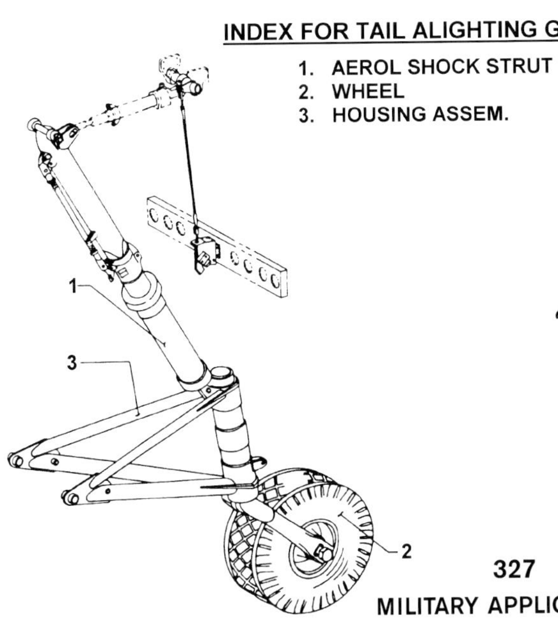

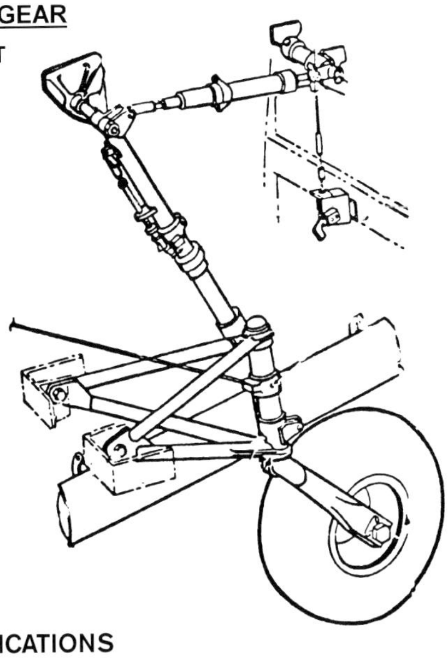

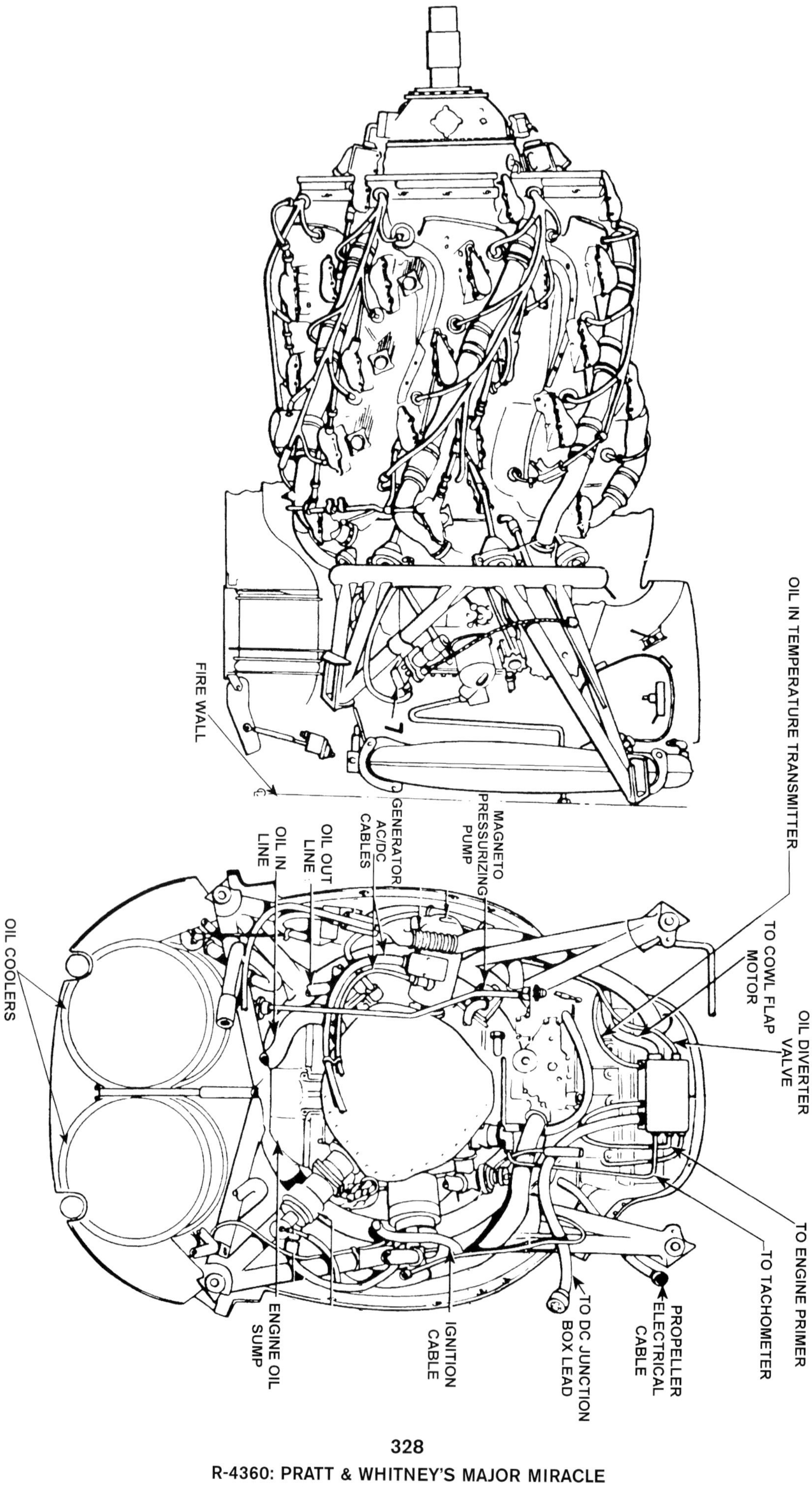

OIL IN TEMPERATURE TRANSMITTER
FIRE WALL
MAGNETO PRESSURIZING PUMP
GENERATOR AC/DC CABLES
OIL OUT LINE
OIL IN LINE
OIL COOLERS
TO COWL FLAP MOTOR
OIL DIVERTER VALVE
TO ENGINE PRIMER
TO TACHOMETER
PROPELLER ELECTRICAL CABLE
TO DC JUNCTION BOX LEAD
IGNITION CABLE
ENGINE OIL SUMP

Opposite: AM-1 QEC showing the R-4360-4W, engine mount, and oil coolers. Note these illustrations are drawn in (conventional) third angle projection, therefore the view on the right shows the rear of the engine. *(AN 01-35EF-2 Martin Mauler Erection & Maintenance Manual) Below:* Engine mount and cowl for AM-1. Large scoop on lower half of nose bowl provides cooling air to the dual-circular oil coolers. Induction ram-air scoop is shown on upper right. *(AN 01-35EF-2 Martin Mauler Erection & Maintenance Manual)*

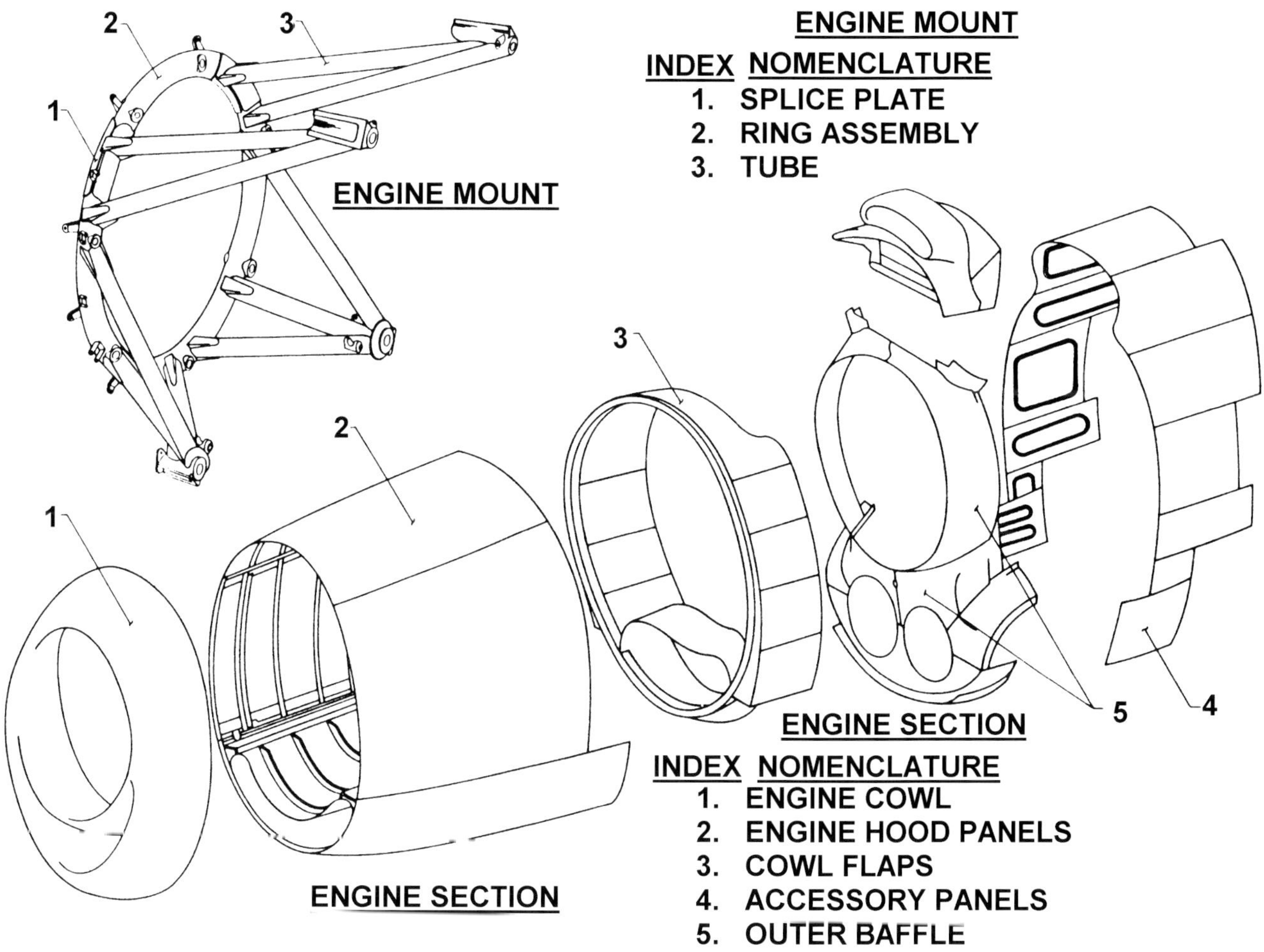

no innovations were incorporated. Two circular oil coolers controlled oil temperature. It's interesting to note the similarities between the Mauler's QEC design and that of its larger twin-engined sibling, the Mercator. From available documentation and drawings, they look identical. And that's not surprising considering it was probably the same design team that worked on both engine installations and both aircraft were powered by the -4W.

Republic XP-72 (dual rotation and single rotation)

In a similar fashion to the R-4360-powered Corsair, Republic Aviation engineers understandably figured that another 1,000 hp, give or take a few, would enhance the already impressive performance of the P-47. It's probably not too far from the truth to call the XP-72 a P-47 on steroids. The XP-72 was developed in parallel with the more

Above: Left side view of the first of two XP-72 prototypes. In simplistic terms, it's an R-4360-powered version of a Republic P-47. *(Courtesy of Pratt & Whitney) Below:* Right side view of the XP-72. Its P-47 heritage is evident in this shot. Note the clean lines of the cowl in contrast to a stock P-47 with its deep chin scoops. *(Courtesy of Pratt & Whitney)*

radical Republic XP-69 powered by the equally radical 42-cylinder Wright Tornado. When it became apparent that the engineering challenges facing the Republic in both installing the Tornado and Wright and in actually developing this highly complex engine were overwhelming, the logical decision was made to drop XP-69 development and concentrate on the XP-72 project. Two prototypes were ordered: 43-36598 and 43-36599. The first one, 36598, was fitted with a single-rotation four-blade propeller. The second, 36599, was fitted with a dual-rotation, six-blade prop.

Unfortunately, the second prototype (the one fitted with the dual rotation prop) had a short career. During a high-altitude test flight, fire broke out at 32,000 feet. When the pilot opened the canopy to bail out, the fire extinguished itself due to the lack of oxygen at this altitude so he decided to stay with the aircraft *(Ref. 7-6)*. Even though 32,000 feet would seem to be a good margin to plan an emergency descent, the drag was enormous with a massive six-blade contra-rotating prop up front—effectively an immense air brake. A gear-up landing was performed; however, the ensuing damage was deemed too excessive for economical repair. It's also likely that the Army Air Force had already pinned its hopes on improved versions of the P-47. This was a common theme during World War II: The best aircraft didn't always get into production; more practical decisions had to be made instead. If an incremental improvement in performance resulted in a delay in getting into production, then this idea would typically be dropped. And so it was even with potentially more capable aircraft such as the XP-72.

It's clearly apparent from photographs of the XP-72 that its heritage is derived from the P-47.

This photograph offers a good view of the belly scoop that feeds induction air plus cooling air to the intercooler and oil cooler. Interestingly the one-off XP-47J employed a similar looking scoop. *(Courtesy of Pratt & Whitney)*

Like its sibling, the XP-72 was designed for high-altitude operation. To achieve this goal, P-47's were fitted with a "C" series General Electric turbosupercharger. The XP-72 tackled this issue a little differently. Rather than relying on a turbosupercharger, Pratt & Whitney developed a remotely mounted auxiliary supercharger behind the pilot. Its location was in a similar position to the P-47's turbosupercharger. A driveshaft emanating from the rear of the R-4360 provided power to the auxiliary supercharger. For additional flexibility, the auxiliary supercharger featured a fluid drive, thus offering infinitely variable speed between its minimum and maximum speeds. Two discharges from the auxiliary blower, one at the four o'clock position and the other at the ten o'clock position, were ducted into a pair of air-to-air intercoolers, again in a similar way to the P-47. The now compressed and cooled air was fed into the PR-100 carburetor to be further compressed by the main engine stage supercharger. A cooling fan assisted with heat rejection of the tightly cowled R-4360 *(Ref. 7-7)*. The entire QEC was a clean and tight package, which represented one of the better R-4360 installations. Most of the airframe followed established P-47 practice including the landing gear.

continued on page 337

XP-72 Parameters

Type	single seat fighter
Crew	1
Armament	six 50-caliber machine guns
Length	36 feet, 7 13/16 inches (contra-prop version)
Wingspan	40 feet, 11 7/16 inches
Wing area	300 square feet
Height	14 feet, 6 inches
Empty weight	10,965 pounds
Max. T/O weight	14,760 pounds
Engine	R-4360-3 (dual rotation), R-4360-13 (single rotation)
Rate of climb	20,000 feet in 5 minutes
Max. speed	480 mph at S/L
Fuel	370 gallons internal
Range	1,200 miles

This depiction (opposite) of the XP-72's induction system is deceiving. For clarity, the fuselage has been shortened in this drawing. The pilot sits between the auxiliary blower (B) and the firewall. A drive shaft connects the engine to the auxiliary blower (not shown). Induction air enters the main duct under the fuselage (D) and flows into the auxiliary blower. Two discharges on the auxiliary blower located at the four o'clock and ten o'clock positions dump high-pressure and heated air into the air-to-air intercooler (E). Cooled induction air then travels to the carburetor (A). Cooling air for the intercooler also enters via duct D and is dumped over board at F. *(Republic XP-72/P-72 Erection & Maintenance Manual)*

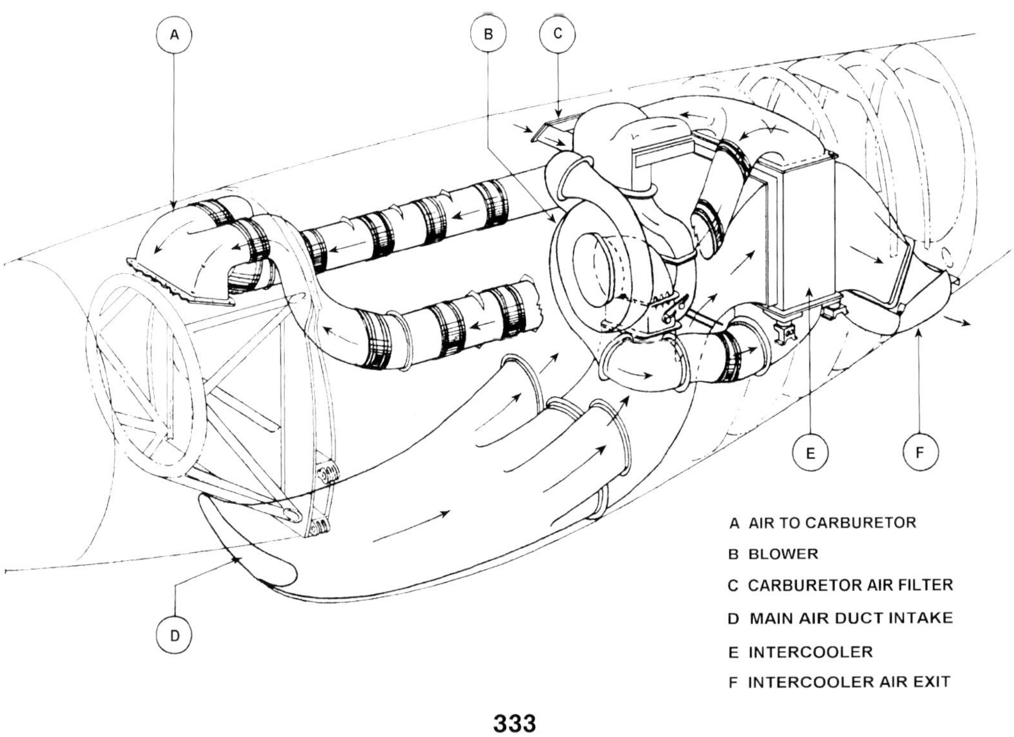
A
B
C
D
E
F
A AIR TO CARBURETOR
B BLOWER
C CARBURETOR AIR FILTER
D MAIN AIR DUCT INTAKE
E INTERCOOLER
F INTERCOOLER AIR EXIT

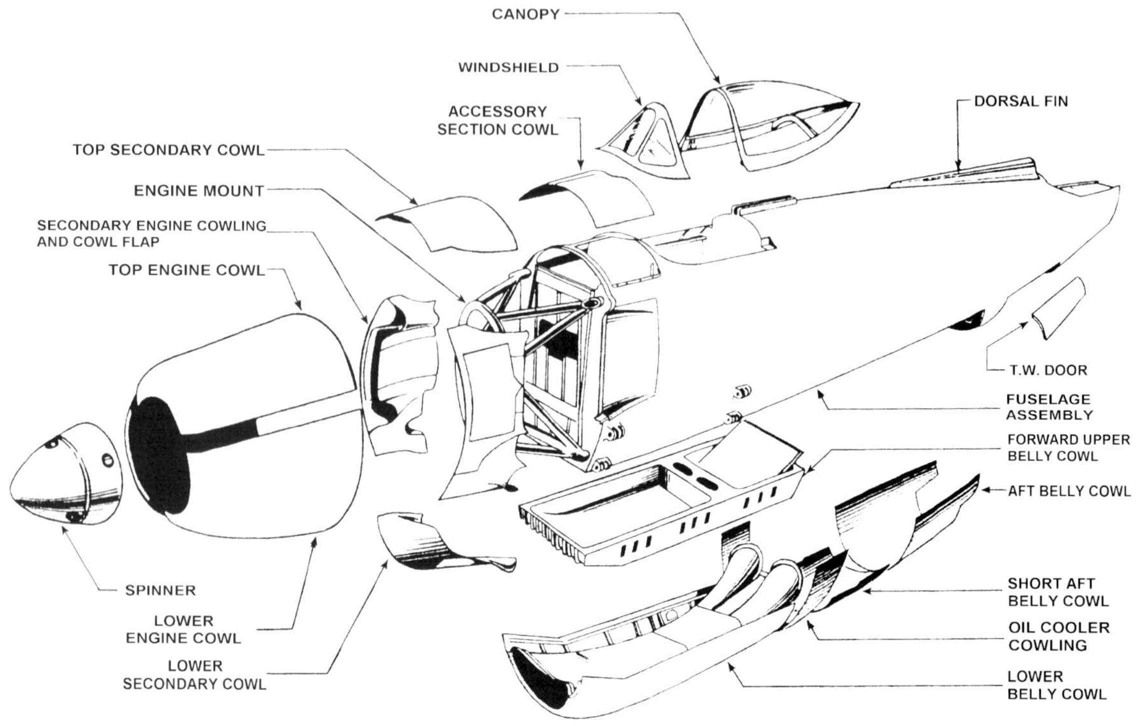

Fuselage structure of the XP-72 showing how the belly scoop feeds air to the intercooler and oil cooler. Also of note is the cowl construction and engine mount. *(Republic XP-72/P-72 Erection & Maintenance Manual)*

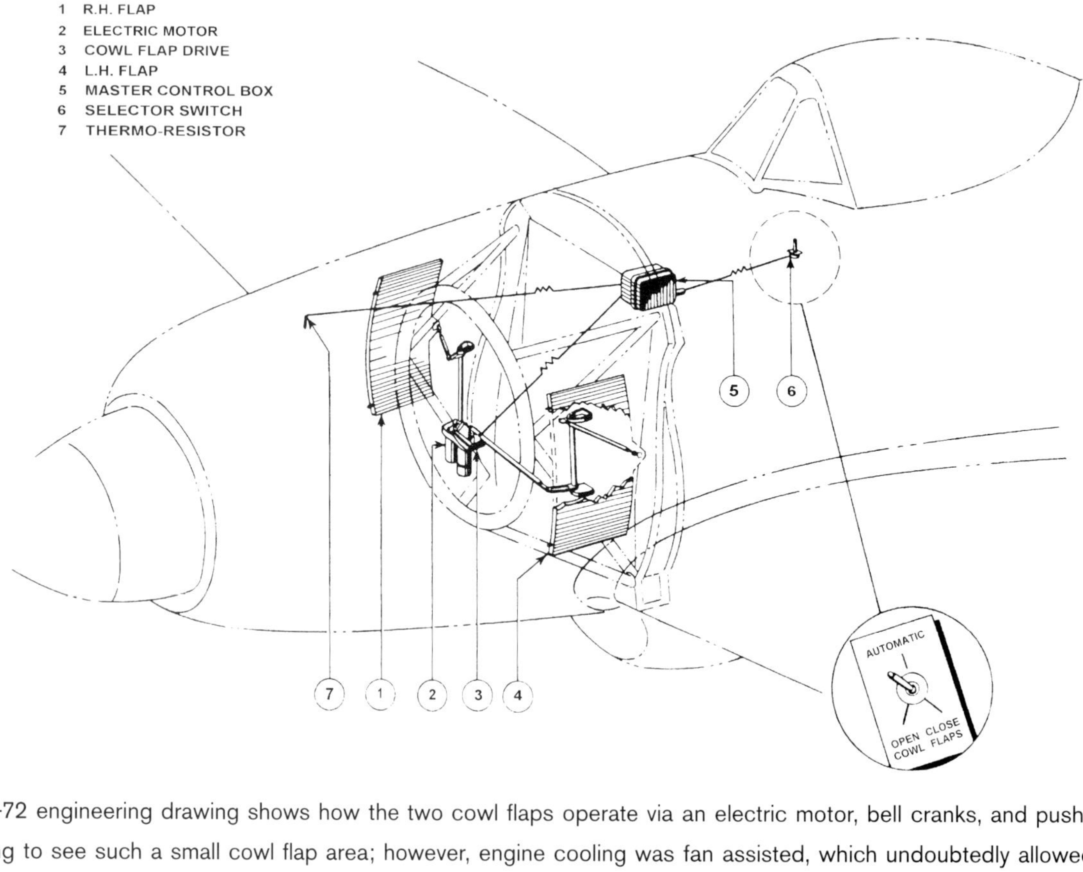

This XP-72 engineering drawing shows how the two cowl flaps operate via an electric motor, bell cranks, and pushrods. It's surprising to see such a small cowl flap area; however, engine cooling was fan assisted, which undoubtedly allowed Republic engineers to downsize the cowl flaps. *(Republic XP-72/P-72 Erection & Maintenance Manual)*

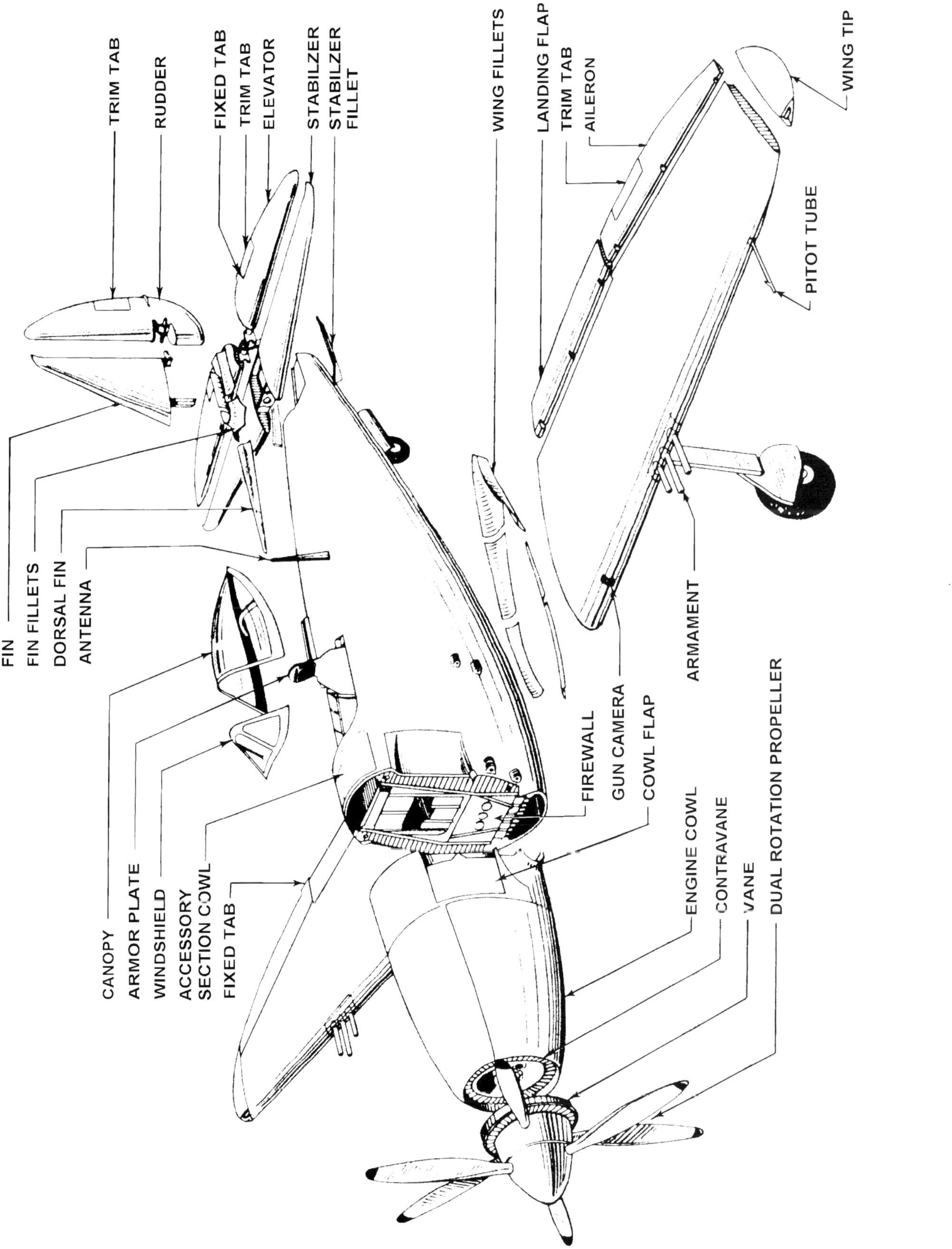

Exploded line drawing view of the XP-72 showing all major sub-assemblies. This view shows the dual rotation version. *(Republic XP-72/P-72 Erection & Maintenance Manual)*

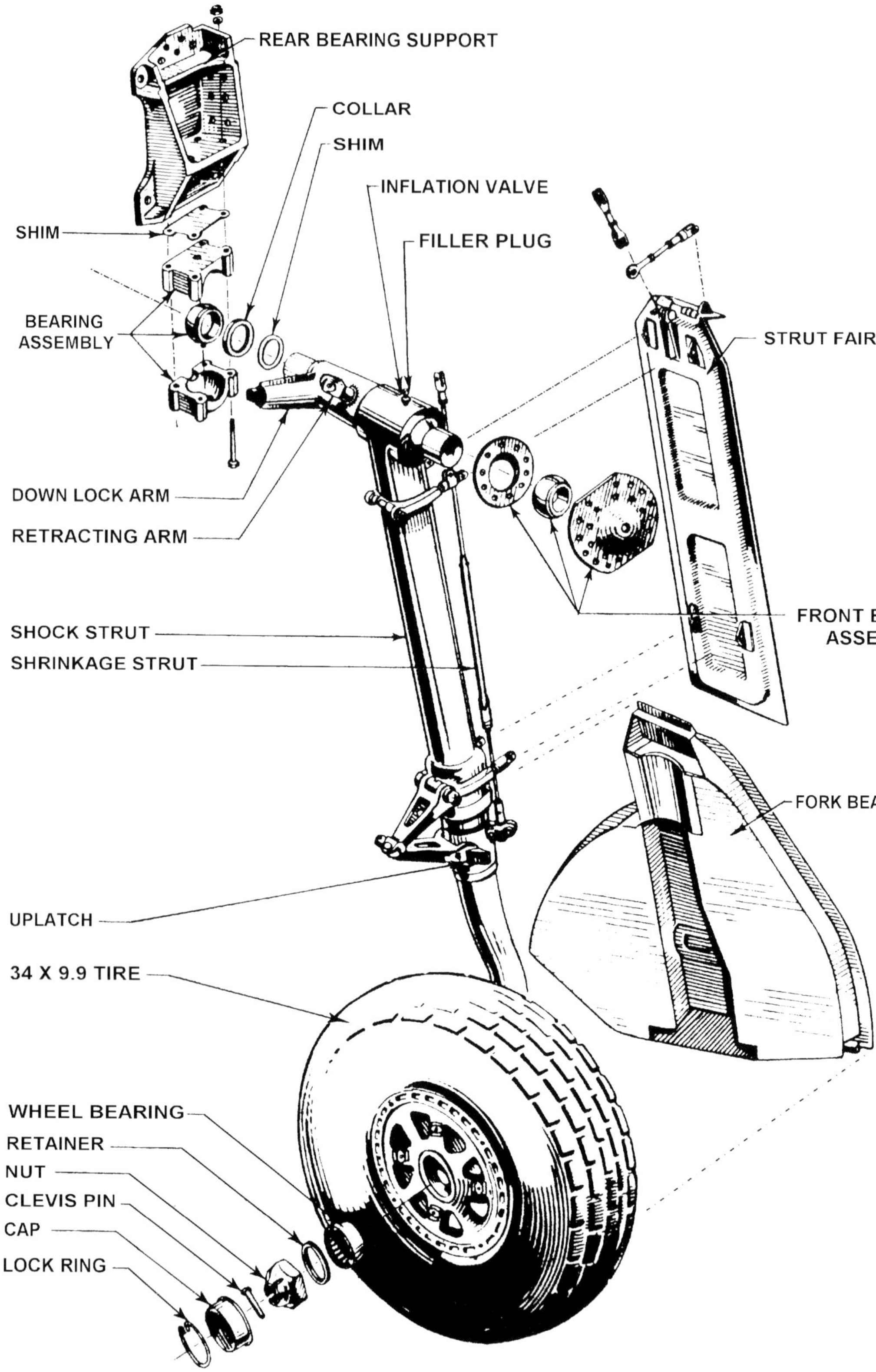

No surprises with the main gear–it's essentially the same as that used on the P-47. Gross weights in excess of 20,000 pounds were sometimes used in heavily loaded P-47s. Therefore, the XP-72's weight of 14,760 pounds was well within the capability of the P-47's gear. As so often happens, if the XP-72 had entered combat, no doubt that gross weight would have been considerably increased, possibly requiring a beef-up of the gear. *(Republic XP-72/P-72 Erection & Maintenance Manual)*

Martin JRM-2 Mars

The genealogy of the Mars goes back to 1935 when the U.S. Navy granted Martin a contract to build a large flying boat bomber designated XPB2M-1R. It got off to a bad start when an engine fire almost destroyed it on December 5, 1941, before it even flew. With the fire damage repaired, the XPB2M-1R flew on January 23, 1942, powered by four Wright R-3350s. By this time the Navy realized that a large and slow flying boat was not the most suitable aircraft to use for strategic bombing, so all the offensive equipment was stripped out and it was converted to a transport. It flew in this configuration throughout the war. Even though only one XPB2M-1R was built, the Navy was encouraged enough to sponsor a follow-on design. Based upon the XPB2M-1R, Martin produced the improved JRM-1 Mars still powered by four Wright R-3350s. The most obvious visual clue the JRM-1 gave compared to its predecessor was the use of a single vertical stabilizer rather than the twin tails of the XPB2M-1R. Four JRM-1s and one JRM-2 were built. The primary difference was the use of R-4360 power for the single JRM-2. And even this aircraft was later converted to Wright power later in its life, presumably for parts inventory considerations. A total of six Mars aircraft were built and each was christened with a name from a Pacific island. The first, *Hawaii Mars*, had a short career when it was destroyed in an accident in Chesapeake Bay in August 1945. However, the name would live on as another Mars was christened *Hawaii Mars*. The others were: *Philippine Mars*, *Marianas Mars*, *Marshall Mars*, and *Caroline Mars*, the sole R-4360-powered JRM-2.

After giving excellent service to the Navy, they came close to being scrapped until an enterprising Canadian, Dan McIvor, purchased the four survivors with the intent of using them for water bombing. Marianas Mars was soon lost in an accident and Caroline Mars (the JRM-2) was destroyed by a winter storm. Despite the initial problems, the two remaining aircraft continued in service for over 40 years—albeit powered by Wright R-3350s *(Ref. 7-1, 7-8, and 7-9).*

JRM-2 Mars Parameters:

Number built	one JRM-2
Engine	R-4360-4, later converted to Wright R-3350 power
Wingspan	200 feet
Length	117.25 feet
Height	38.4 feet
Empty Weight	75,573 pounds
Max. loaded weight	165,000 pounds
Maximum speed	221 mph
Service ceiling	14,600 feet
Range	4,945 miles

Hughes XF-11

Howard Hughes personified the term "enigmatic." Everything he did was in strict secrecy, adding yet more spice to his persona. Yet from the fertile mind of this talented individual came some of the most beautiful and graceful aircraft to slip the surly bonds of earth. The XF-11 was at the forefront of the old adage "form follows function." Its graceful fuselage gondola and twin booms gave it an appearance of going 500 mph when sitting on the ramp. As they say, if it looks good, then it should be good. The XF-11's lineage went back to the D-2A, a similarly configured twin-boom aircraft powered by a pair of turbosupercharged Pratt & Whitney R-2800-49s. Only one D-2A was built and this aircraft flew on just two occasions. Not long after its second flight, the aircraft was destroyed in a suspicious hanger fire *(Ref. 7-10).*

Designed as a purpose-built photo reconnaissance aircraft, like many Hughes projects, it was built as a prototype only with just two aircraft seeing the light of day. During World War II, aircraft not specifically designed for the job, usually modified fighters such as P-38s and P-51s, performed photo reconnaissance. It became apparent that photo reconnaissance was a key mission that required dedicated aircraft designed and built

continued on page 340

Nice overall view of the Martin JRM-2 Mars. Note the "V" struts supporting the wing floats. *(Courtesy of National Archives & Records Administration)*

The same aircraft photographed from the right. Martin produced the most elegant large flying boats–an era that will never be repeated. *(Courtesy of National Archives & Records Administration)*

824

824

The graceful lines of the XF-11 are shown to advantage in this shot. What a shame that the bureaucratic vandals had the sole survivor scrapped. This photo reconnaissance aircraft would have been a stellar performer if it had entered service. As it was, this role was performed by B-29s and B-50s, both of which were ill-suited for this key role. *(Courtesy of Warren Bodie)*

continued from page 337
from scratch with this purpose in mind. High altitude operation is a prerequisite for photo recon missions and one of the best ways to achieve this requirement is through turbosupercharging. In the case of the XF-11, a pair of General Electric BH-1 turbosuperchargers operating in parallel were used to augment each engine's gear driven supercharger. This arrangement allowed the R-4360-37s to maintain 3,000 hp up to an amazing 30,000 feet.

As an additional benefit, one of the two turbosuperchargers could be shut down during cruise conditions. Therefore the remaining operating turbo would be running at its peak performance. The inboard turbo in each engine installation had a "cruise" valve in the main exhaust pipe feeding that turbo. Shutting this valve off, and conse-

quently starving the turbo of its exhaust energy, shut down the turbo. The dual BH-1 turbosuperchargers were vertically mounted in the tail boom. Exhaust was discharged on both sides of the boom, one from each turbo, via an elegantly designed and streamlined stack. Discharged hot and compressed air from the dual BH-1's compressor sections went through a single rectangular air-to-air intercooler before entering the engine at the carburetor deck. Cooling air for the intercooler came from ducting designed into the nacelle scoop. After cooling the turbo discharge air, cooling air was dumped overboard above the nacelle. The weight of the entire engine installation came to 6,043 pounds, less oil and fuel systems. This weight would include the engine, twin turbosuperchargers, engine mount, and engine driven accessories. Oil was contained in tanks mounted in the tail boom under each wing. Each tank's capacity was 77 gallons with an oil capacity of 70 gallons. Oil cooling was accomplished via a round brass oil cooler mounted at the rear of the nacelle scoop. Each engine was mounted on a conventional chrome-moly tubular mount *(Ref. 7-11)*.

Operating in the stratosphere makes pressurization a highly desirable feature. By bleeding air

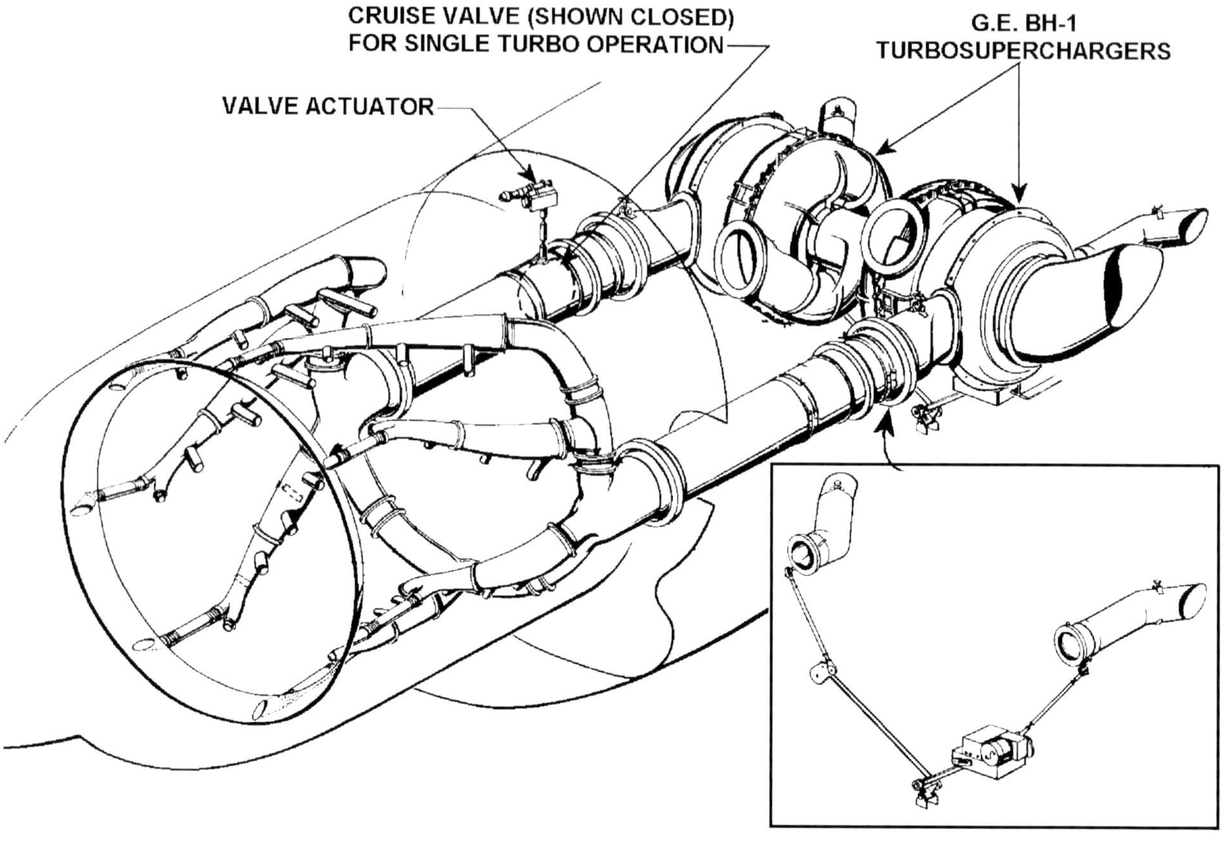

This restored engineering drawing shows how the pair of GE BH-1 turbosuperchargers were mounted vertically. It also shows the collector ring exhaust system to advantage. *(Courtesy of Pratt & Whitney)*

from the BH-1 turbosuperchargers, the required pressurization was achieved. A 10,500-foot cabin could be maintained to 33,500 feet, the critical altitude of the engines. Spoilers, in a way similar to the Northrop P-61 Black Widow, augmented roll control. In other words, if a turn to the right is initiated, the right aileron rises and a spoiler on the right wing also deploys, thus spoiling some of the lift generated by that wing. Unlike its predecessor, the D-2A, which was manufactured using the Duramold process, the XF-11 was a conventional all metal monocoque design.

The flight crew of two was housed on the central nacelle along with the camera equipment, which included eight cameras, the front three forming a trimetrogon group *(Ref. 7-3)*.

A typical trait of Howard Hughes was the fact that he liked (insisted on?) performing the maiden flight of his aircraft. This philosophy was to cost Hughes dearly. The first XF-11 prototype was fitted with Hamilton Standard Super Hydromatic contra-rotating propellers. An hour and ten minutes into an early test flight on July 7, 1946, something went horribly wrong. The right prop malfunctioned and was reported to have gone into reverse pitch. Although the "reverse pitch" statement has been made many times, the reality may have been somewhat different. In a recently discovered U.S. Army Air Force report, it was discovered that the propeller actually lost its oil due to a faulty seal. This being the case, the prop would go into fine pitch due to centrifugal turning force rather than reverse pitch. Even so, the net effect would be the same—a powerful asymmetric yaw. With such a violent asymmetric force acting on the aircraft there was no way Hughes could control it. Inevitably, the XF-11 went into a flat spin and crashed in a Beverly Hills neighborhood. Cause of the crash was subsequently determined to be a failed seal, which allowed total

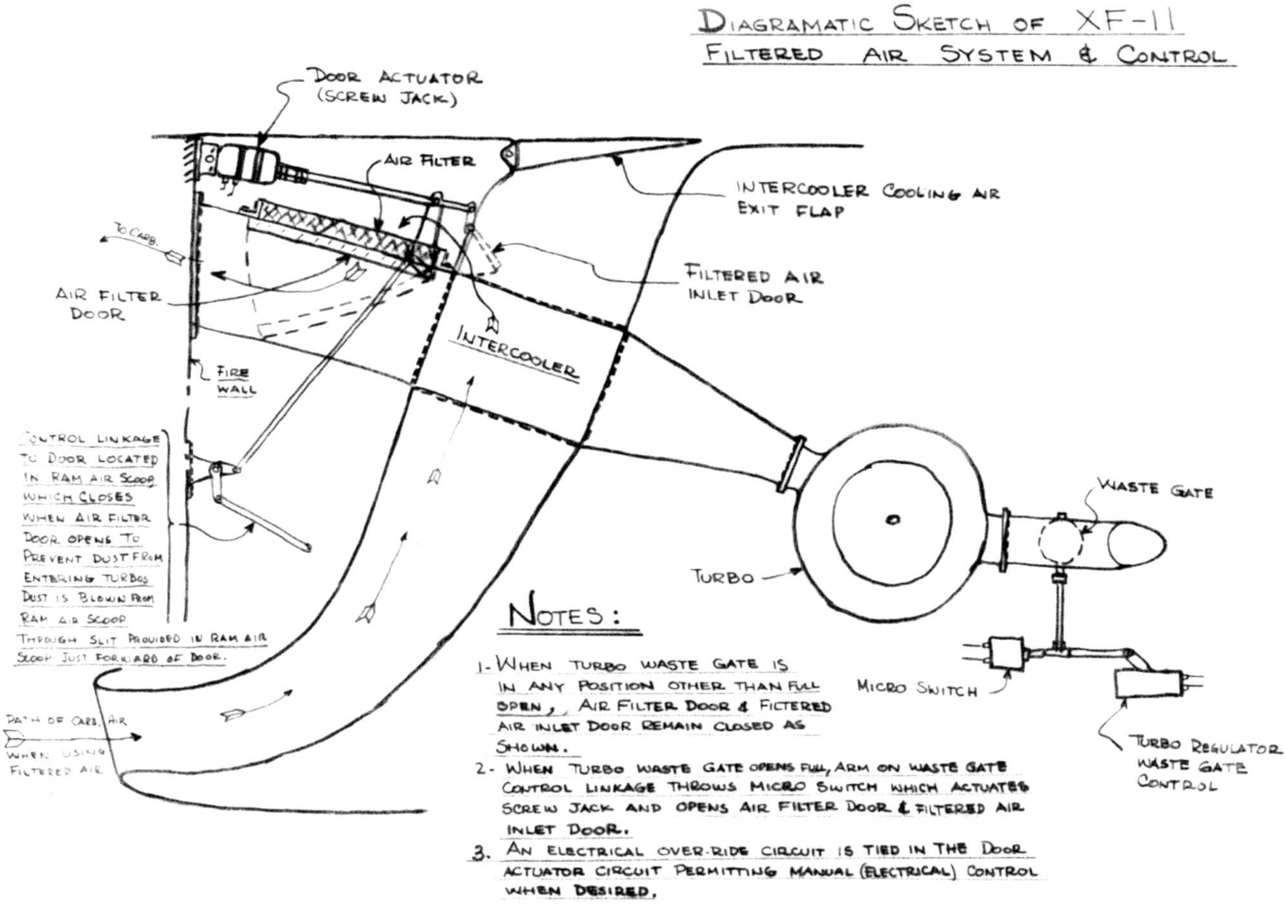

This XF-11 engineering sketch shows how cooling air for the intercooler entered via the chin scoop, flowed through the air-to-air intercooler, and exited over the top of the nacelle. *(Courtesy of Pratt & Whitney)*

fluid loss. It was this loss of fluid that caused the rear propeller to go into fine pitch. Being a highly experimental and complex propeller, it had not passed any type tests and had not even been tested by Wright Field, so it's not surprising that problems occurred *(Ref. 7-12).*

The ensuing crash almost killed Hughes, but fortunately didn't hurt anyone on the ground, even though it went down in a residential neighborhood. Consequences of his hospital treatment were to last for the rest of his life. One of the devastating after-effects of this hospital stay was his addiction to morphine. If this addiction contributed to his bizarre behavior in his later life, then it was truly a double tragedy *(Ref. 7-13).*

The second prototype was fitted with single-rotation Curtiss Electric propellers. This aircraft was flown to Eglin Air Force Base in Florida for further testing. Even though the XF-11 appeared to be a superlative aircraft, it got caught up in the defense cuts of the late 1940s. In its stead, the Air Force converted Boeing B-50s (RB-50s) for the role envisioned for the XF-11. However, Hughes' arcane personality possibly contributed to the XF-11's cancellation. Consequently, the last fixed-wing aircraft developed by Hughes was yet another non-starter.

Hughes XF-11 Parameters (Ref. 7-3):

Wingspan	101 feet, 4 inches
Length	65 feet, 5 inches
Height	23 feet, 3 inches
Empty weight	39,278 pounds
Max. weight	58,059 pounds
Useful load	18,781 pounds
Internal fuel	2,105 gallons
Max. speed	450 mph at 33,000 feet
Max. speed at sea level	295 mph
Rate of climb	17.4 minutes to 33,000 feet
Max range	5,000 miles
Ceiling:	42,000 feet
Engines	R-4360-4A (single rotation), R-4360-37 (single rotation), R4360-31 (dual rotation)
Number built	2

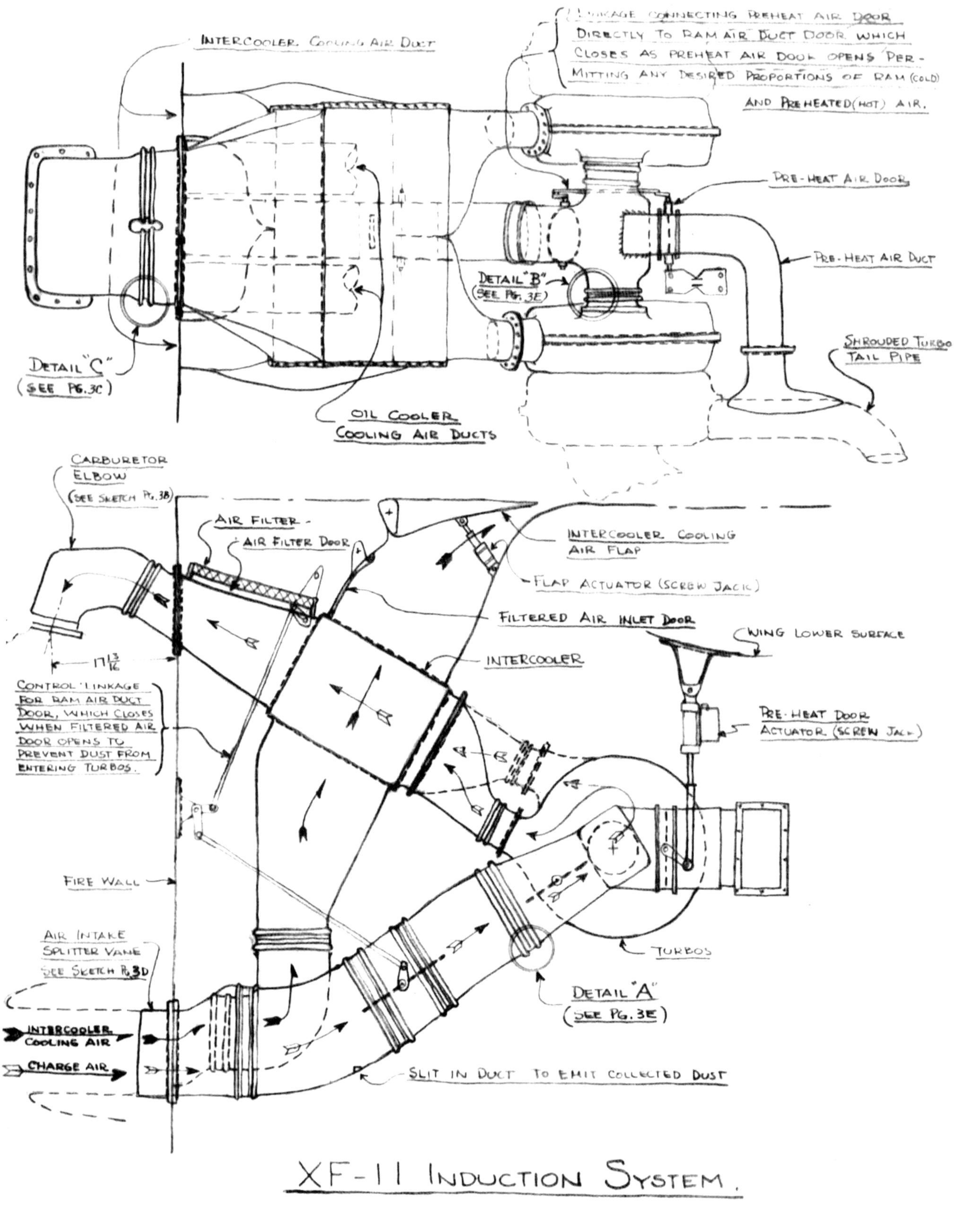

This engineering sketch shows the induction system. It's quite probable that the two prototype aircraft were constructed from sketches such as this. *(Courtesy of Pratt & Whitney)*

Hughes HFB-1, H-4, HK-1 Hercules "Spruce Goose"

It must have been a source of great irritation for Howard Hughes to hear his H-4 being referred to, rather derisively, as the "Spruce Goose." Unfortunately, that's the appellation that has gone down in history for the largest aircraft ever built. This eight-engined behemoth personified grace and beauty with its size dwarfing the nacelles that housed the huge R-4360-4As. Despite its Spruce Goose moniker, the primary material used in its construction was birch. Typical of Hughes, who had an eye for quality, workmanship was nothing short of superb. This was not surprising considering that the finest cabinetmakers were employed during its construction.

Variously known as the HFB (Hughes Flying Boat), HK-1 (Hughes-Kaiser number one), and H-4 (Hughes design number four), it is invariably known as Spruce Goose *(Ref. 7-14).* In the dark and desperate days of 1942, the Allies' fortunes of World War II were at their nadir. Hitler was rampaging through Europe and had invaded the Soviet Union. Japan launched a devastating attack on Pearl Harbor and it appeared nothing could stop the Axis Juggernaut. Making the situation even worse, if that was even possible, was the fact that Donitz's U-Boat wolf packs were close to bringing England to its knees. Ships were being sunk at a faster rate than they could be built—England was starving for food and matériel. In July 1942 alone, over 800,000 tons of Allied shipping was sunk.

Henry Kaiser came to the rescue by introducing the Liberty ship. Even by World War II standards, the Liberty ship was an outdated design going back to the turn of the century. However, this was exactly what was needed, nothing fancy and high tech, just a means of getting matériel across the Atlantic. Its triple-expansion steam engine was a throwback to the 1890s, but it got the job done. Kaiser's production methods were so efficient that Liberty ships were often built in less than a week. This astounding accomplishment was achieved through pre-fabrication of all key components and by simply using the shipyards as a final assembly line. However, even Kaiser realized that despite the prodigious output of his famed Liberty ship, they were still being sunk at a faster rate than they could be built. This prompted Kaiser to think of an alternate means of transportation. It appeared to him that a massive cargo plane would fit the bill. In retrospect, it could be argued that if the massive cargo plane envisioned by Kaiser had come to fruition, they would have been shot out of the sky at a faster rate than Liberty ships were being sunk. The Germans found this out when they embarked upon a similar program with the massive six-engined Messerschmitt 323 Gigant. Used primarily to re-supply Romel's Afrika Corps in North Africa, most of the 323s were shot down by Allied fighters roaming the Mediterranean.

Kaiser's original proposal for a flying freighter was a twin-hulled flying boat designed to carry a Sherman tank or several hundred troops. Initially rebuffed by the U.S. government due in part to his lack of experience with things aeronautical, Kaiser sought a partnership that would lend credence to his plan. That partner turned out to be Howard Hughes. Even though Hughes' forte was fighters and bombers, he soon warmed to the idea of a large aerial transport. One can only conjecture why Hughes took on this project, but it's probably a safe bet to say the monumental challenge piqued his interest. Another "kicker" associated with this mammoth undertaking was the fact that no strategic materials could be used, meaning no extensive use of aluminum. This left just one alternative—wood. Construction-wise, the aircraft was a conventional wood monocoque using birch ply skins as the primary load-bearing structure using the Fairchild-developed "Duramold" process—a process Hughes was familiar with from his prior D-2A. Unlike many other behemoth aircraft of the era, the HK-1 was a graceful and beautiful design. All the statistics associated with this aircraft can only be described as mind boggling. Its

One of the few photographs taken of the HK-1 airborne. Despite its huge size, it personifies elegance. Amazing to think that Hughes was essentially the sole pilot when this photo was taken. *(Courtesy of The Evergreen Aviation Museum)*

maximum gross weight of 400,000 pounds was in a different league compared to other aircraft, the 320-foot wingspan was more than twice as large as a Boeing B-29, which itself was no small aircraft, and so it was with all other dimensions and weights of this engineering marvel.

Construction commenced in 1942 with the building of the required jigs and special building for final assembly in Culver City, California. It soon became apparent that the HK-1 would play no part in World War II. Even though 1942 was a disastrous year, the Allies soon developed a countermeasure to the U-boat menace. Hughes, being the consummate salesman, convinced the government to continue funding of the aircraft under the auspices of a research project. Even this argument wore thin when cost over-runs and delays put the HK-1 behind schedule. Things came to a head in August 1947 when Senate hearings were called to investigate the project. In the

meantime, work continued although every step of the way created a daunting task. One of the more difficult hurdles to be overcome was how to transport the aircraft from its Culver City manufacturing plant to Long Beach—a distance of 28 miles. Ingenuity, as always, played a big part in getting the job done. The largest components were the fuselage and two wing assemblies, which represented the most intimidating challenge for the 28-mile trip. However, thanks to the cooperation of the communities on the route, telephone poles and anything else that posed interference were dealt with. A purpose-built graving dock was built on Long Beach for final assembly. At long last, the boat was launched on November 1, 1947, in the midst of acrimonious Senate hearings. Amazingly, on the next day, November 2, 1947, with Hughes at the controls, the HK-1 made its first and only flight. The co-pilot for that flight was not even a pilot, but rather an engineer who had intimate knowledge of the complex hydraulic system, so the choice may have been a good one. Hughes took along a crew of 18. A minimum crew of 11 people was required to fly this formidable aircraft made up of the following. *(Ref. 7-15)*

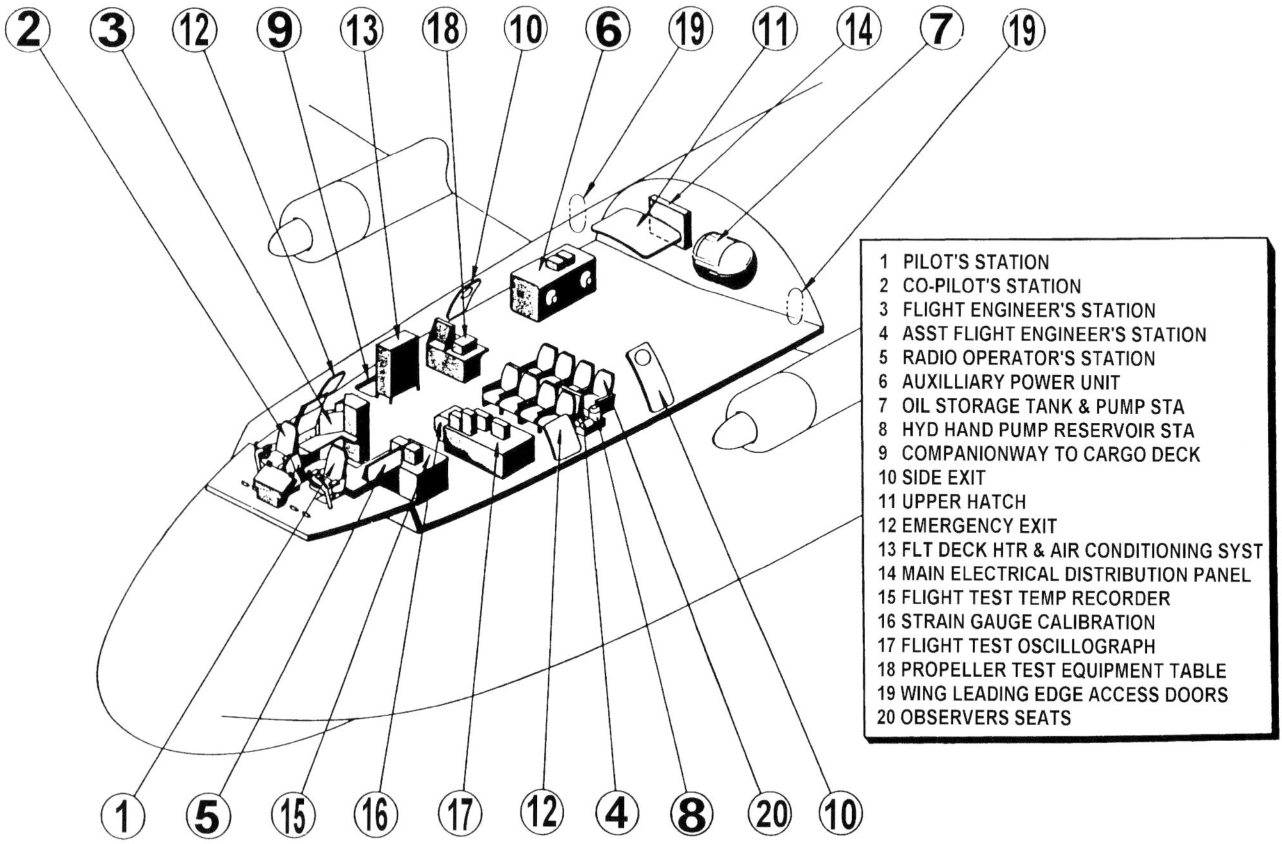

With a minimum crew requirement of 11, the HK-1's flight deck was huge. Looking more like a luxurious hotel lobby than the flight deck of an aircraft, the HK-1 was certainly not lacking in creature comforts. *(Hughes Flying Boat Manual. Courtesy of The Evergreen Aviation Museum)*

HK-1 Crew Requirements

I. Pilot
II. Co-pilot
III. First flight engineer
IV. Second flight engineer
V. Assistant flight engineer
VI. Radio operator
VII. Five additional crewmembers to operate the following stations:

A. One electrician to assist radio operator and maintain electrical equipment as directed by flight engineer and crew chief.
B. Two engine mechanics on flight deck to operate APU (auxiliary power unit) and assist in engine maintenance as directed by flight engineer and crew chief.
C. Two hydraulic mechanics located on cargo deck for operation of auxiliary hydraulic station and main hydraulic systems as directed by flight engineer and crew chief.

Reading the flight manual for the HK-1 it would appear that a Ph.D. in engineering was required. The following gauges are just some that had to be monitored by the flight engineer(s):

I. Flight Engineer's Engine Panel

A. Eight oil pressure gauges
B. Eight oil pressure low, warning lights
C. Eight oil dilution switches (inoperative)
D. Eight oil temperature gauges
E. Eight oil flap open, switches
F. Eight oil flap position indicators
G. Eight oil flap control switches
H. Eight carburetor air temperature gauges with

indicator light and control switch for each gauge

I. Carburetor air lights test switch
J. Eight fuel pressure gauges
K. Eight fuel pressure low, warning lights
L. Eight fuel flow gauges
M. Free air temperature gauge
N. Eight cylinder head temperature gauges with high temperature warning light for each gauge
O. Eight mixture control levers
P. Mixture lights test switch
Q. Pressure type altimeter
R. Eight cowl flap position indicators and individual switches
S. Master cowl flap switch
T. Eight-day 24-hour dial clock
U. Eight manifold pressure gauges
V. Propeller synchroscope indicator and sensitivity selector switch
W. Propeller synchroscope master engine selector switch
X. Eight tachometers
Y. Eight BMEP gauges

II. Flight Engineer's Oil System Control Panel
This panel is located at the top of the flight engineer's center instrument panel. It contains all controls and instruments relative to the engine oil supply and the transfer of oil from the oil reservoir to the nacelle oil tanks as follows:

A. Eight firewall oil valve switches
B. Eight oil level low, warning lights
C. Eight nacelle oil tank quantity gauges
D. Eight oil transfer selector switches
E. Oil transfer pump on indicator light
F. Reserve oil tank quantity gauge

III. Flight Engineer's Fuel System Control Panel
This panel is located on the flight engineer's center instrument panel just below the oil system panel. It contains all controls and instruments required to load fuel aboard the aircraft, transfer fuel to the wing tanks, and supply fuel to the engines both from the wing tanks and the emergency system. The panel is made up of the following instruments, switches, and indicator lights:

A. Eight firewall fuel valve switches
B. Four booster pump switches
C. Four fuel quantity gauges and warning lights
D. Two fuel tank selector switches
E. Two transfer pump switches and indicator lights
F. Fuel pump scavenge switch and indicator light
G. Several emergency pump selector switches and indicator lights
H. Several hull tank selector switches and lights
I. Two fuel loading port selector switches and lights.
J. Indicator selector switch

IV. Flight Engineer's Electrical Panel
This portion of the flight engineer's station is located aft of the flight engineer's seat and faces forward. This panel contains all controls and instruments required to operate the electrical system under normal and emergency conditions. The panel is made up of the following sub-panels with the necessary instruments, switches, and indicator lights to accomplish operation of the various systems concerned:

A. Generator control panel
B. Dynamotor control panel
C. Auto pilot, auto flying tab, and inverter control panel
D. Pneumatic system control panel
E. Fuel tank CO_2 and ventilator control panel
F. APU control panel

Aircraft of this era typically had manual controls. In other words, the pilot's and co-pilot's controls were attached to aircraft control surfaces via cables and pulleys. This would have been impossible in the case of the HK-1 due to the

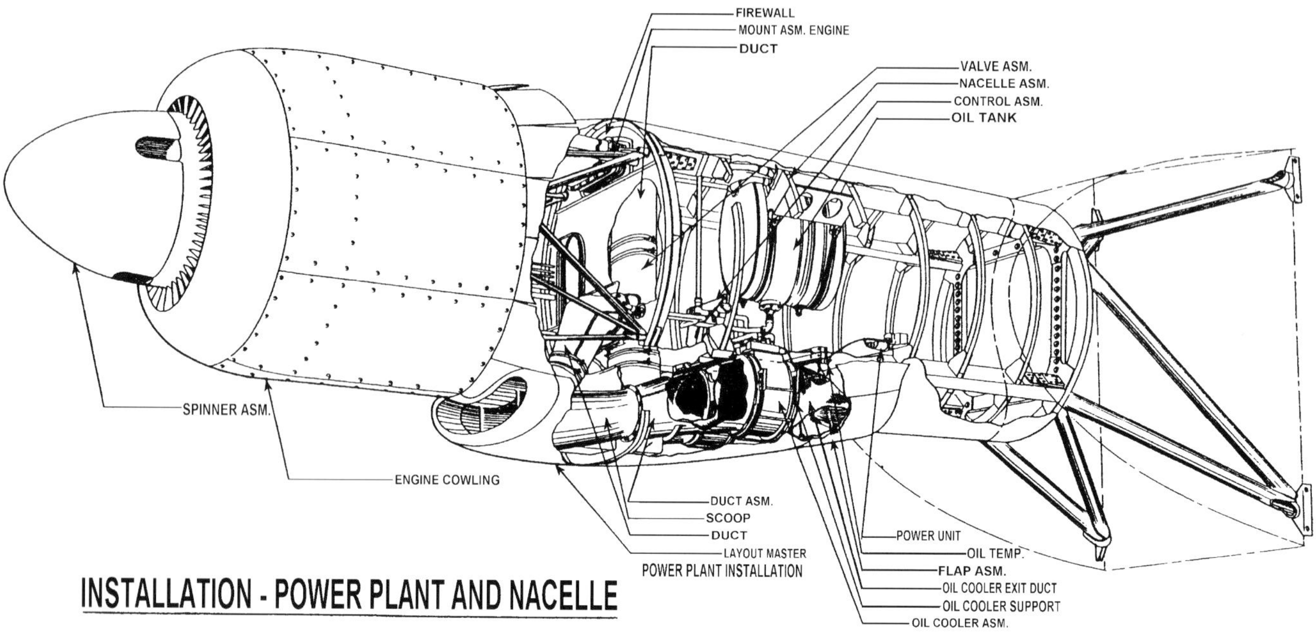

Complete nacelle and QEC. It's interesting to see in this view that the engine appears to have a fan to augment cooling. This feature was dropped when the aircraft was built. The lower duct that feeds induction air to the engine and cooling air to the pair of oil coolers is shown to good advantage in this drawing. Cantilevering a 5,000-pound plus QEC over 11 feet in front of a wooden wing with this material also used for the nacelle exemplified the engineer's art. *(Engineering drawing for HK-1. Courtesy of The Evergreen Aviation Museum)*

size, weight, and air loads on the flying surfaces. Therefore, the most practical alternative was the use of hydraulic assist. A dual backup was used in the event a hydraulic failure occurred. Either system could fly the aircraft. In the unlikely event that both independent hydraulic systems failed, a third system that used hydraulically assisted servo tabs could fly the aircraft. In a similar way, it would have been difficult at best to use conventional cables and pulleys to operate the engine throttles. Instead, in an early iteration of fly by wire, electric servomotors were used. Again, a backup was designed with an "inching" system operated by a switch. The electrical power requirements were, again, incredible. Three separate electrical power requirements demanded 120-volt DC, 24-volt DC, and 120-volt AC. The 120 DC came from two sources: (i) engines 1, 3, and 7 drove 15KW 120 VAC generators, (ii) two Franklin four-cylinder aircraft engines of 115 hp each drove 120 VAC generators. Two storage battery systems were installed in two banks of nine 12-volt DC batteries (18 in all) and hooked up in series providing 104 volts. A second battery system of two 12-volt batteries provided 24 volts to start the APUs. In addition to the foregoing electrical power requirements, provision was also made for receptacles for external power.

A major design challenge was how the QECs, which weighed over 5,000 pounds and cantilevered over 10 feet in front of the wing leading edge, should be designed. Not only did this structure need to support the considerable weights of the QEC, but it also had to handle the enormous gyroscopic loads from the massive 17 feet, 2-inch diameter four-blade propeller, G loads, and gust loads. And all this had to be handled by a lightweight wooden structure. The circular nacelle was made from a laminated wooden monocoque with four massive wooden longerons that pick up the four engine mounting points. It was essentially a large tube with the nacelle

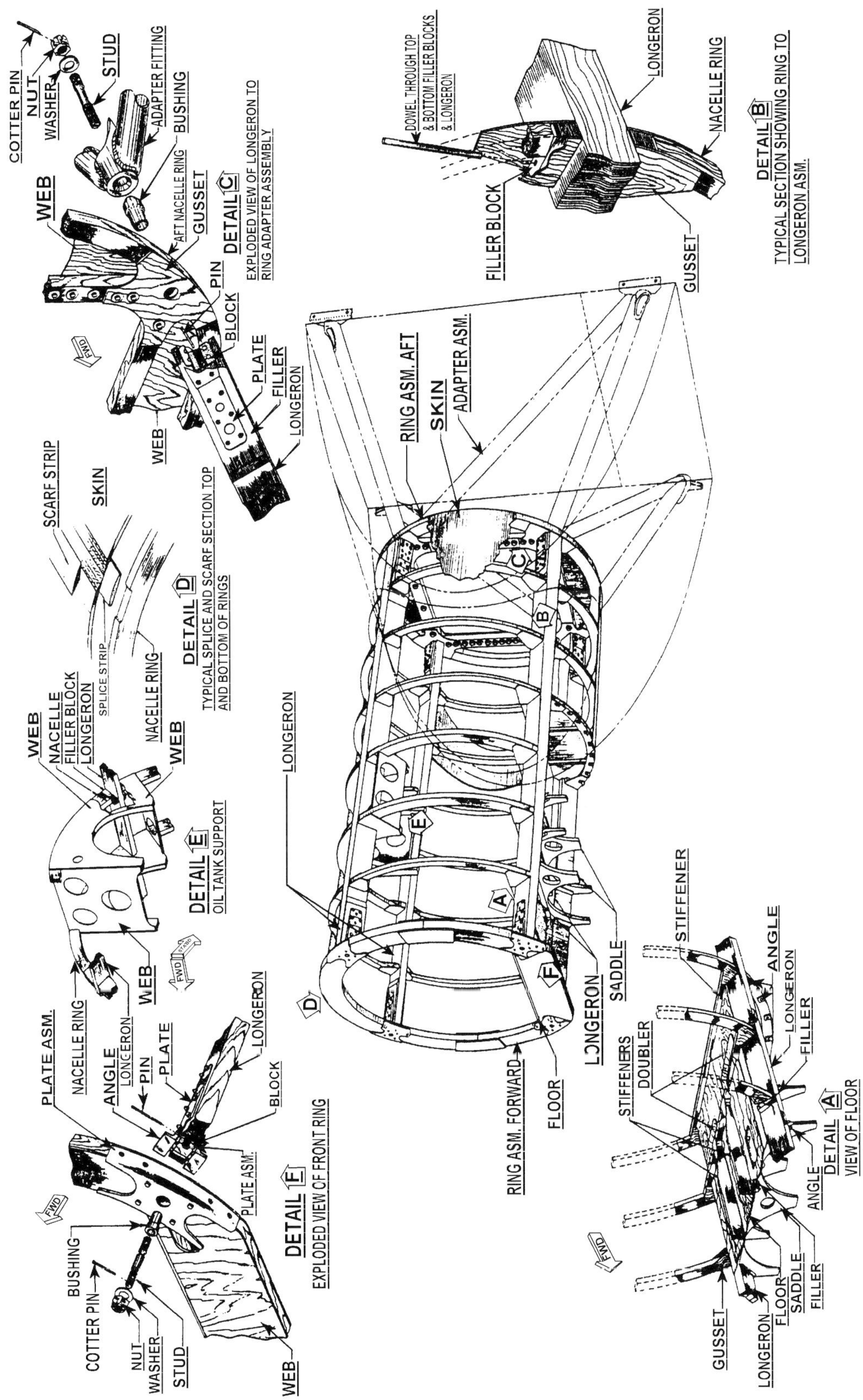

Skeletal structure of the HK-1 nacelle. The chrome-moly tubular frame bolted to the massive wing spar, which attached to the nacelle monocoque structure made entirely from wood. Four wooden longerons tie into the wing support structure and the engine mount. *(Engineering drawing for HK-1. Courtesy of The Evergreen Aviation Museum)*

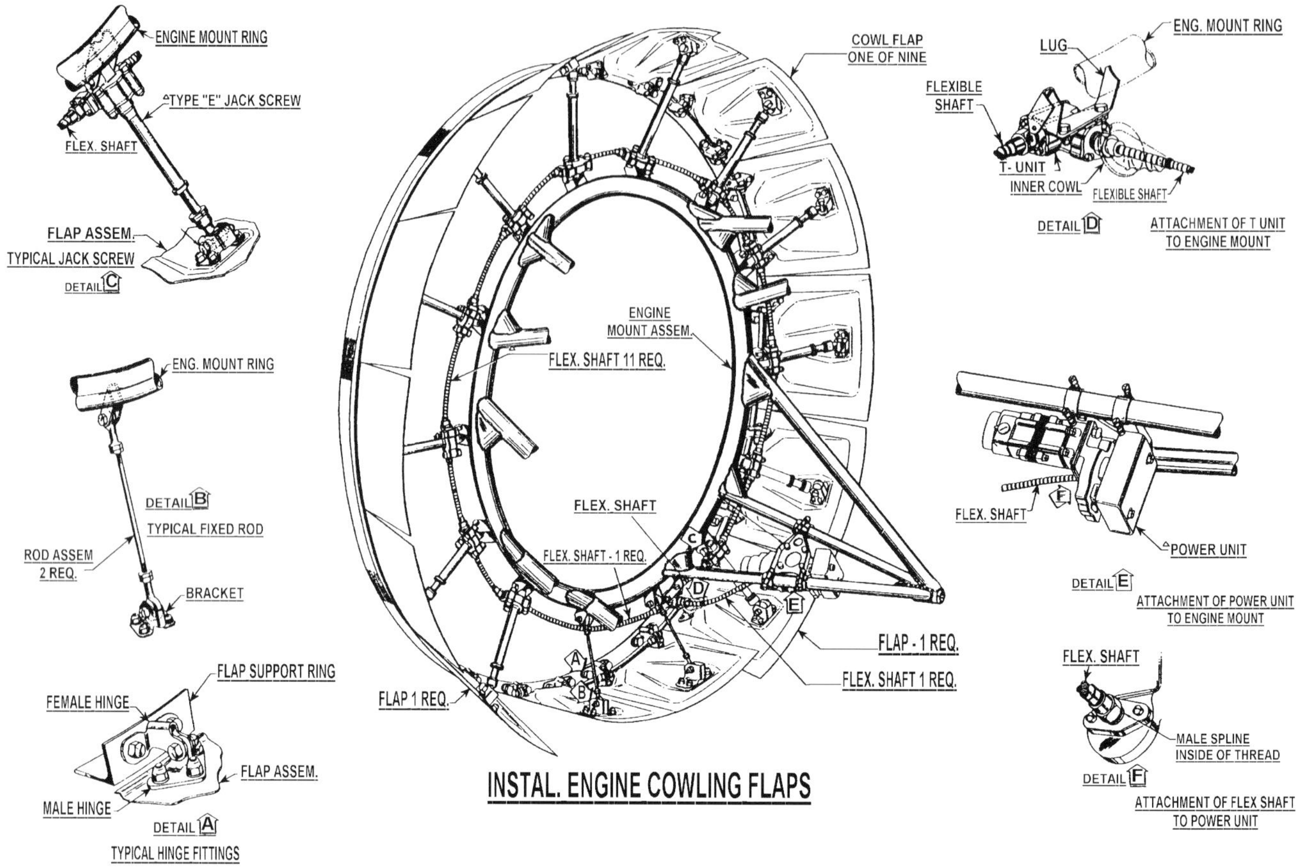

Above: Conventional design of the HK-1 cowl flap assembly and its electrically actuated screw jacks. *(Engineering drawing for HK-1. Courtesy of The Evergreen Aviation Museum)* *Opposite top:* Feeding eight thirsty R-4360s for 3,500 miles with up to 14,000 gallons of fuel demanded a complex fuel system. Considering the demands placed upon the fuel system, the HK-1's design was elegantly simple. *(Hughes Flying Boat Manual. Courtesy of The Evergreen Aviation Museum)* *Opposite bottom:* Extensive emergency equipment list for the HK-1. *(Hughes Flying Boat Manual. Courtesy of The Evergreen Aviation Museum)*

terminated behind the wing leading edge. A chrome-moly tubular structure was bolted to the huge wing spar that attached to the nacelle. Overall, it was a clever and wonderful piece of structural engineering *(Ref. 7-16 and 7-17).* Under the nacelle, a massive air scoop served two purposes. It provided cooling air for the pair of circular oil coolers. A ram-air scoop, integral with the duct, provided induction air. The duct was divided into three compartments. The center compartment fed induction air and the two outer compartments fed cooling air to the dual circular oil coolers. The R-4360-4As powering the HK-1 used downdraft induction; therefore, induction air was routed vertically, via the duct, inside the nacelle. The duct then redirects the air forward to discharge into the R-4360-4As PR-100 carburetor. A 31-gallon tank located in each nacelle provided oil.

Normally, 31 gallons would be totally inadequate for a long-range R-4360-powered aircraft. The HK-1 accommodated this requirement by having an auxiliary 275-gallon tank that fed each nacelle tank automatically. Or the flight engineer(s) could manually override the automatic system to replenish nacelle tanks from the auxiliary tank. All pumping requirements were performed electrically. Cooling the R-4360, particularly one installed in a relatively slow and heavily loaded transport such as the HK-1, offers another challenge. In order to assure the appropriate mass airflow through the cowl, 12 massive cowl flaps are employed, each one with an area

PRELIMINARY

FUEL SYSTEM DIAGRAM HK-1

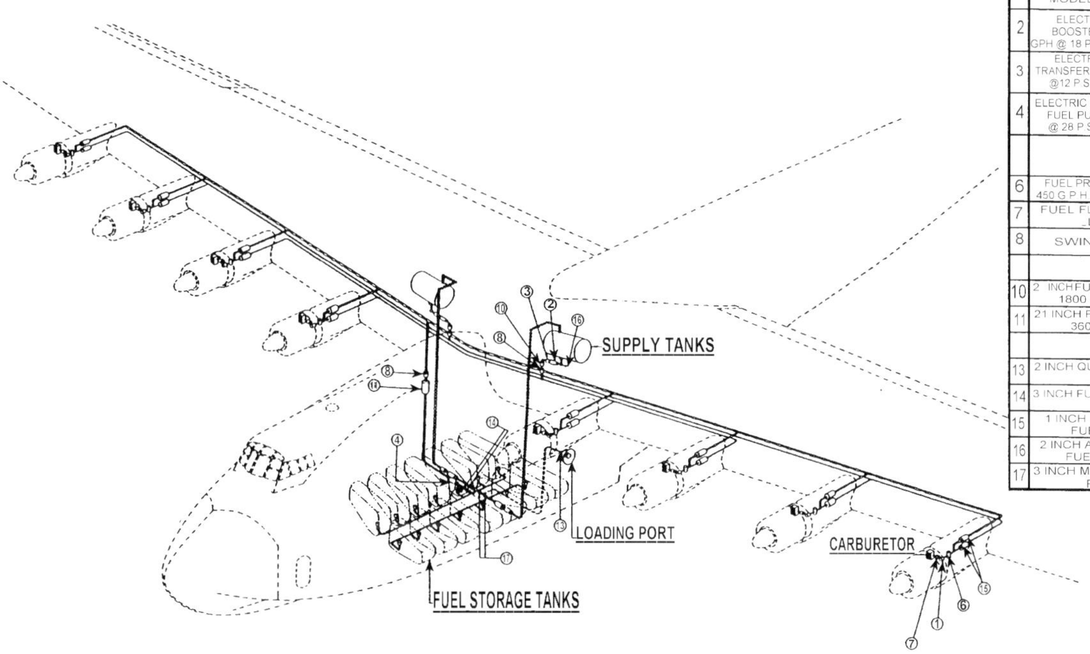

COMPONENTS			
ITEM	DESCRIPTION	MFG.	REQ.
1	ENGINE DRIVEN FUEL PUMP MODEL 2P-248-GB-TYPE G-10	PESCO	8
2	ELECTRIC MOTORDRIVEN FUEL BOOSTER PUMP - CAPACITY 1800 GPH @ 18 P.S.I. DISCHARGE PORESSURE	PESCO	2
3	ELECTRIC MOTOR DRIVEN FUEL TRANSFER PUMP CAPACITY 1125 G.P.H. @12 P.S.I. DISCHARGE PRESSURE	PESCO	2
4	ELECTRIC MOTOR DRIVEN EMERGENCY FUEL PUMP - CAPACITY 3600 G.P.H. @ 28 P.S.I. DISCHARGE PRESSURE	PESCO	1
6	FUEL PRESSURE REGULATOR 450 G.P.H. @ 15-17 P.S.I. CHARGE		8
7	FUEL FLOWMETER 200-2700 LB/HR RANGE		8
8	SWING CHECK VALVE	PARKER APPLIANCE CO	4
10	2 INCHFUEL STRAINER 1800 G.P.H. CAPACITY	PUROLATOR HUGHES	2
11	21 INCH FUEL STRAINER 360 GP.H. CAPACITY	PUROLATOR HUGHES	1
13	2 INCH QUICK DISCONNECT FUEL LINE FITTIGN WITH VALVE		1
14	3 INCH FUEL SELECTOR VALVE MANUALLY ACTUATED	PARKER APPLIANCE CO	2
15	1 INCH AIR ACTUATED FUEL SHUT-OFF VALVE	HUGHES AIRCRAFT CO	16
16	2 INCH AIR ACTUATED FUEL SHUT-OFF VALVE	HUGHES AIRCRAFT CO	2
17	3 INCH MANUALLY ACTUATED FUEL SHUT-OFF VALVE	PARKER APPLIANCE CO	2

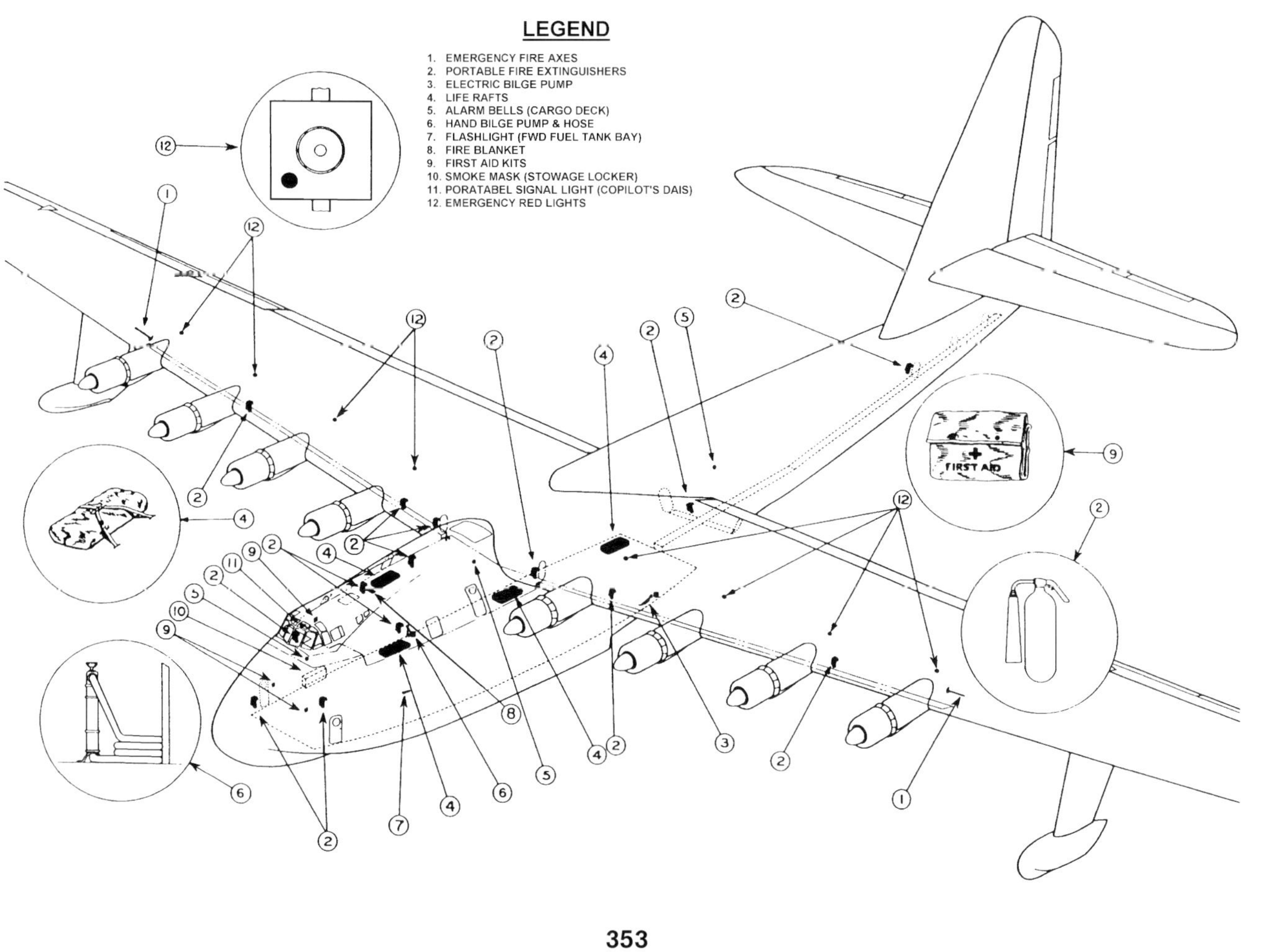

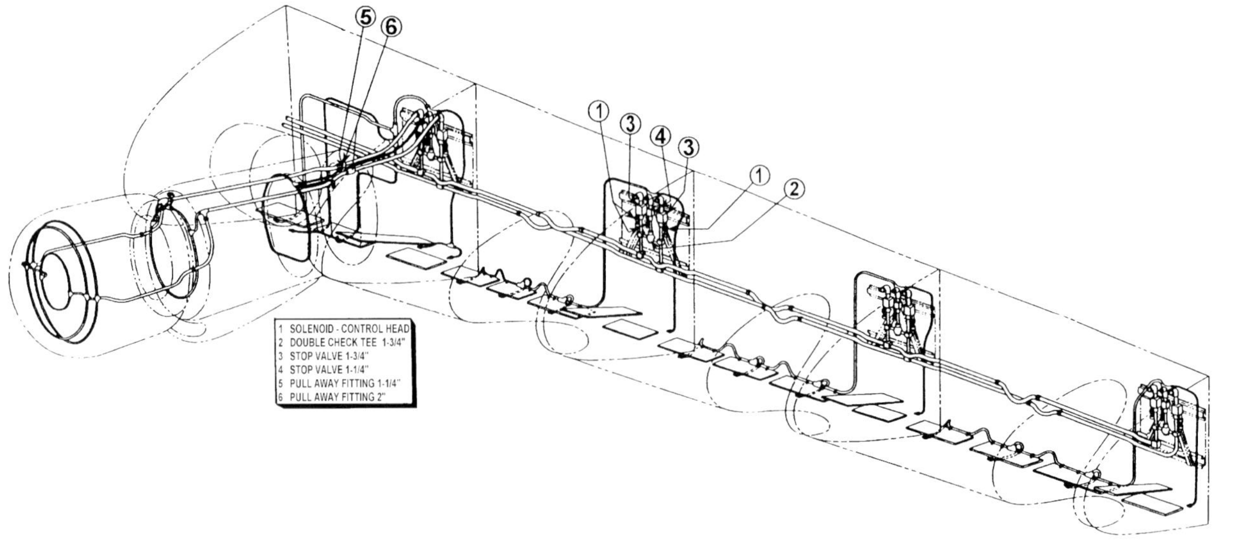

Fire is an ongoing hazard with aircraft, particularly with one made from wood. To combat this menace, a complex CO_2 fire extinguishing system was employed. *(Hughes Flying Boat Manual. Courtesy of The Evergreen Aviation Museum)*

of approximately one square foot. Cowl flap actuation comes via conventional screw jacks from an electric motor. Cooling-air flows past a conventional ribbed, stainless-steel dish pan, which then discharges the now heated air over board through the cowl flaps. Rather surprisingly, considering the degree of automation designed into the HK-1, the pilot/flight engineer regulates the cowl flaps. Keeping tabs on cylinder head temperatures for eight temperamental R-4360s must have been a real chore. Fourteen thousand gallons of fuel kept the eight R-4360s satiated via a complex fuel system. Safety features were abundant in the HK-1, including a sophisticated CO2 fire extinguishing system.

The foregoing is an overly brief look at the technical side of the HK-1. However, even from this snapshot, it can be seen that this aircraft advanced the state of the art in many areas. Even though the term was not used at the time, fly-by-wire technology was employed. Additionally, a very advanced hydraulic system for powered controls and a sophisticated servo system for engine power management are just a few of the sophisticated technologies employed. Little wonder that Hughes would be upset to hear this complex and sophisticated aircraft being derided as the Spruce Goose.

After its one and only flight on November 2, 1947, the aircraft was kept in a state of readiness. In fact, Hughes even had a number of changes incorporated into it. Additionally, the R-4360-4As were replaced in 1951 with commercial Wasp Major TSB-3Gs rated at 3,500 hp *(Ref. 5-5)*. In the meantime, the Senate hearings fizzled out and Hughes moved on to pursue different projects. Even so, the HK-1 always had a place in Hughes' heart, so rather than scrap it, a fate that befell many other one-off prototypes, he had it stored at

Hughes HK-1 Parameters:

Wingspan	320 feet
Tail span	113 feet, 6 inches
Wing area	11,430 square feet
Maximum wing thickness	11 feet, 6 inches
Length	218 feet, 6¼ inches
Height	79 feet, 3⅜ inches
Fuselage height	30 feet
Payload	180,000 pounds
Max. weight	400,000 pounds
Internal fuel	14,000 gallons
Max. speed	218 mph
Cruise speed	175 mph
Landing speed	78 mph
Max. range	3,500 miles
Service ceiling	24,000 feet (300,000-pound gross weight)
Rate of climb	700 fpm to 1,000 fpm depending on weight
Engines	Eight R-4360-4A, single stage, variable speed. Replaced in 1951 with commercial Wasp Major TSB-3Gs
Propellers	Hamilton Standard 24F60-35, 17-feet, 2-inch diameter*
Number built	1, order for 2 additional prototypes cancelled

* Engine #4 used a 16-feet, 2-inch diameter propeller.

considerable expense in a hanger at Long Beach. Upon his death in 1976, the fate of the HK-1 was in question. Fortunately, the Wrather Corporation rescued it and housed it in a geodesic dome next to the Queen Mary at Long Beach Harbor. But it was not all over yet. In the 1990s the Wrather Corporation wanted to use the valuable property that housed the HK-1 for more profitable enterprises. Again, the aircraft was at risk of being broken up. At one time serious thought was given to breaking it up and donating the parts to various museums. Of course, this would have been a tragedy. Once more rescue came, this time from The Evergreen Aviation Museum in McMinnville, Oregon. One can only imagine the logistics of transporting such a huge aircraft from Southern California to Oregon. Nevertheless, the impossible was pulled off and the HK-1 is once again securely housed in a museum.

It's interesting to note the takeoff time, distance and rate of climb illustrated below *(Ref. 7-15).*

TAKEOFF TIME AND DISTANCE

2700 RPM 202 BMEP CG 29.5%

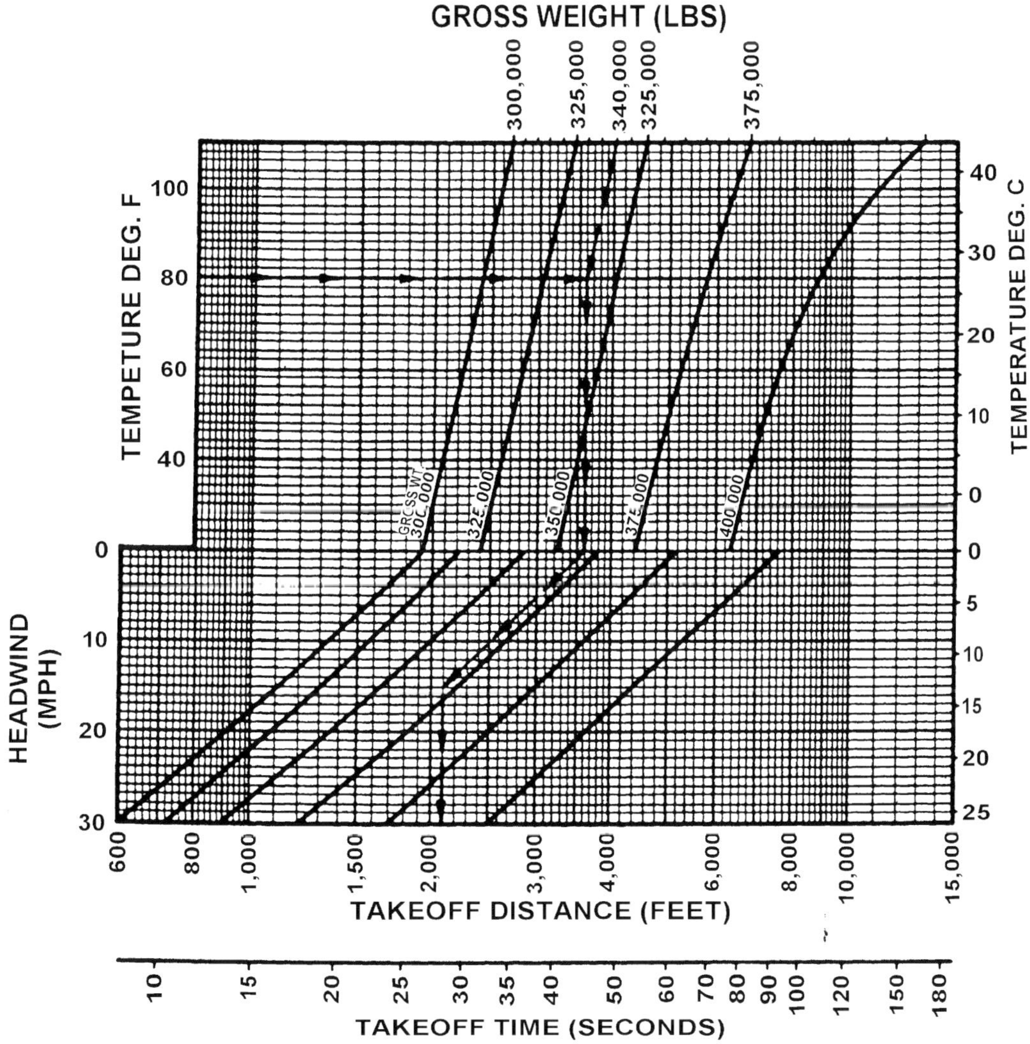

It's interesting to note that for takeoff, 20 degrees of flap defection is called for in the chart. It has been reported that Hughes used 15 degrees of flap for his flight. *(Hughes Flying Boat Manual. Courtesy of The Evergreen Aviation Museum)*

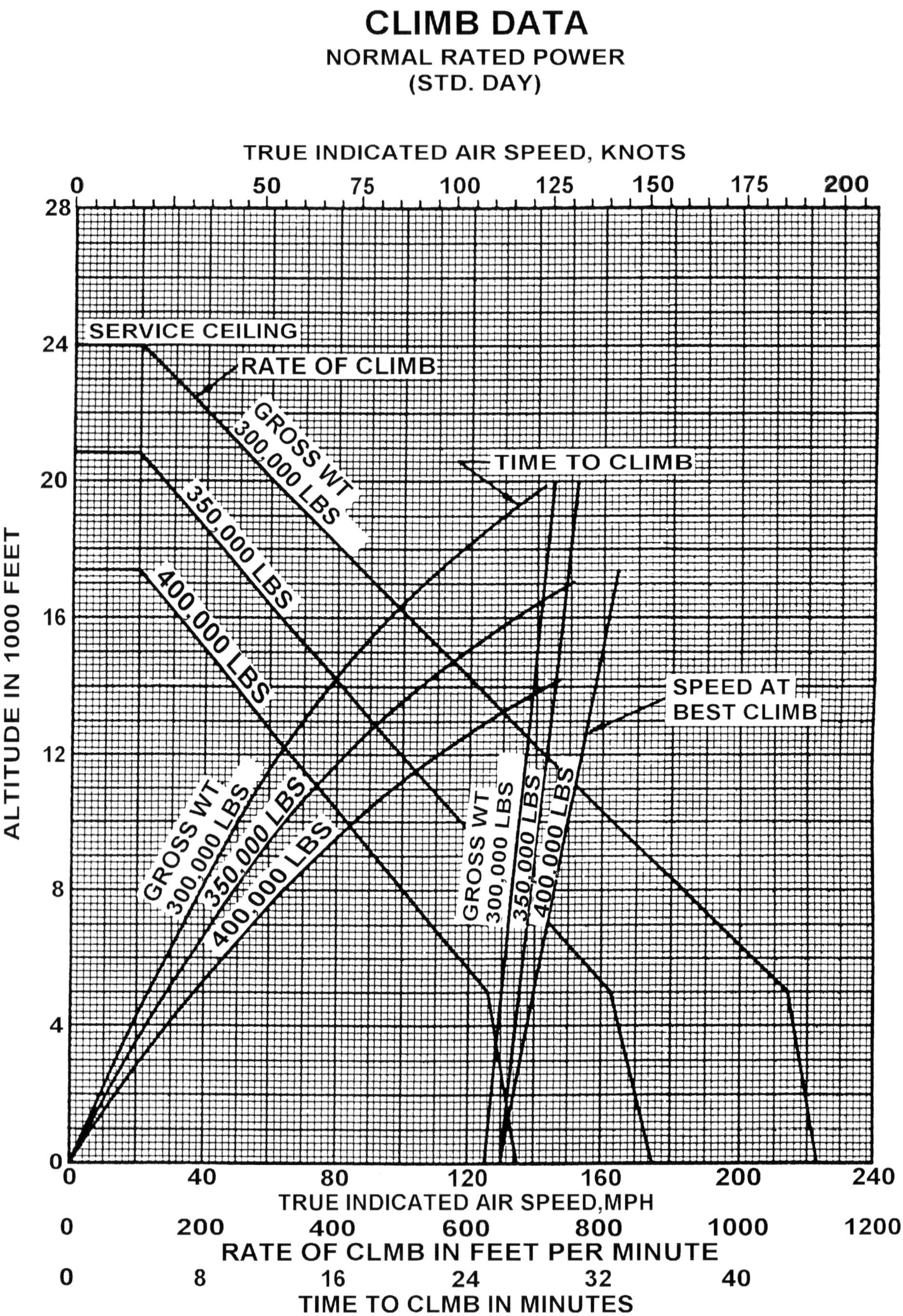

NOTE:- PRELIMINARY BASED UPON MODEL TEST DATA

The Enigmatic Howard Hughes

By any stretch of the imagination, Howard Hughes and his aircraft company produced some of the most beautiful and well-crafted aircraft ever. Yet none of his aircraft saw series production. Why was this? The answers are not easy or quick to arrive at. Like Hughes himself, the reasons are complex. Although he comes across as genial, easy going, and down to earth when one views old newsreel footage of him testifying in front of various committees, reality may have been somewhat different. He obviously rubbed the military brass the wrong way. On numerous occasions, damning internal reports castigated Hughes and his difficult demeanor. This is too bad because no one would doubt that, regardless of his egocentricity, he was a brilliant individual. Typical of the memos circulating about Hughes is the one below copied from a 1947 report on his testimony regarding a twin-engined fighter aircraft he proposed. If only half of this report is true, it casts the military brass and Hughes in a poor light—no one came out a winner. He claims that Lockheed got the idea for the P-38 from him. It further goes on to say that his H-1 was turned down as a military fighter. The H-1 was, of course, the purpose-built racer that Hughes had custom made in the 1930s which broke a number of records including the landplane speed record of 352.46 mph on September 12, 1935. Powered by a Pratt & Whitney R-1535, this engine would have been inadequate for military use. However, it would have been relatively easy to fit the larger and more powerful R-1830. If the foregoing is true, then it would appear that the Army Air Corps missed out on an excellent design that could have given the U.S. an advantage over Japanese Zeros in the early months of World War II. Instead, the U.S. had to make do with inferior designs such as the Curtiss P-36, P-40, and Seversky P-35. It seems that Hughes did not understand a sad fact of life, which is politics are key to securing military contracts. Simply having the finest mousetrap was not sufficient—then or now.

The report below is copied in its entirety including typos and spelling errors *(Ref. 7-12).*

8 October, 1947

Testimony of Mr. Howard Hughes regarding his Pursuit Aircraft Designs.

Chief of Staff, United States Air Force
Washington 25, D.C.
Attn: Director of Public Information
Legislative and Liaison Division

THRU: Chief, Research & Engineering Division
Office, Asst. Chief of Air Staff, 4

1. *It is considered desirable to call attention to certain erroneous or misleading statements contained in the testimony of Mr. Howard Hughes on 9 August, 1947 before Special Committee to investigate the National Defense Program. These statements are included in the inclosed extract of the stenographic transcript of the hearings (Inclosure 1 (sic)).*

2. *Speaking of a high speed airplane which he "finished around 1935", he stated "I tried to sell that airplane to the Army, but they turned it down because at that time the Army did not think a cantilever monoplane was proper for (a) pursuit ship. Our pursuits at that time were biplanes, or wire braced monoplanes, and they did not think that a cantilevered monoplane . . . was suitable for a pursuit ship". In this connection, attention is invited to the inclosed photographs of the YPB-2A airplane, delivered in January 1934, (Inclosure 2), and the YP-29 airplane delivered in August 1934, (Inclosure 3), both of which were cantilever*

For a typical gross weight of 350,000 pounds, rate of climb would have been a respectable 700 to 750 feet per minute. This was comparable to the contemporary B-36. *(Hughes Flying Boat Manual. Courtesy of The Evergreen Aviation Museum) monoplanes. It is also desired to draw attention to the inclosed extract from U.S. Army Specification No. 98-605 dated 22 January 1936 for "Airplane, Mono-place Pursuit", (Inclosure 4), in which it will be noted that "an all-metal monocoque fuselage, cantilever wings and tail surfaces are desired". This specification was cited in the Circular Proposal for mono-place pursuit airplanes, issued 12 March 1936, which resulted in the procurement of the Seversky YP-35 and Curtiss YP-36 aircraft, both of which were cantilever monoplanes. In a letter dated 30 March 1936, (Inclosure 5), the Hughes Aircraft Company stated that it was entering bids on any current Pursuit Circular Proposal.*

3. *In his testimony, Mr. Hughes further stated: "I submitted this plane to the Army; it was a cantilever monoplane pursuit ship. It was turned down, ostensibly for that reason . . . and I was asked to submit a design for an interceptor . . . and I went to work on it, and with some considerable fear I made a two-engine design. Now, at this time, nobody, I am sure, had considered really seriously a two-engine interceptor of pursuit plane, because nobody thought of one pilot flying an airplane with two engines." In this connection, in the letter of 30 March 1936 (Inclosure 5), previously referred to, the Hughes Aircraft Company requested recent design studies of pursuit types of airplanes, particularly any multi-engined studies which might be available. By letter dated 22 April 1936 (Inclosure 6), the Hughes Aircraft Company was furnished 8 eight design studies, including drawing P35K535 dated 24 November 1934, (Inclosure 7), which covers a twin-engine pursuit airplane.*

4. *Mr. Hughes further stated that ". . . when the design was finished, the Army informed me that it would be necessary for me to hold that design, and sit on it for a period of months, while they allowed Lockheed Company to make a similar design, and then the two would have to be evaluated in a competition". The inclosed copy of telephone conversation on 31 July 1936 between Brigadier General A.W. Robins, then Chief of the Material Division, and Mr. Alexander, Manager of the Hughes Aircraft Company, (Inclosure 8), indicates that development funds had been impounded, there were no funds then available for the Hughes airplane. Subsequently, when funds became available, and in accordance with established procedures for securing competitive designs, letters were dispatched on 8 January 1937 to the Howard Hughes Aircraft Company and four other aircraft companies, (Inclosure 9), announcing a design competition for an experimental two-engine pursuit aircraft, and inviting requests for the specifications and method of evaluation. By letter dated 23 February 1937, (Inclosure 10), the Hughes Aircraft Company was furnished a copy of Air Corps Specification X-608, dated 19 February 1937, covering the requirements of the Air Corps for a two-engine, interceptor pursuit type airplane.*

5. *By letter dated 16 April 1937, (Inclosure 11), the Hughes Aircraft Company submitted two-engine interceptor pursuit designs; one, the H-1 (Inclosure 12) being a single place design, and the other, the H-2 (Inclosure 13) being a two-place design. It should be noted that these designs are not radically different from the design prepared by the Material Division in 1934 and furnished (to) the Hughes Aircraft Company in April, 1936, although they do provide for the use of Allison V-1710 engines instead of the Pratt & Whitney R-1535-5 engines.*

6. *Mr. Hughes stated ". . . so by a strange coincidence, Lockheed also submitted a two-engine interceptor design, and I may say that at that time, I am sure that nobody had seriously considered an interceptor with two*

engines. . ." In view of the fact that Lockheed Aircraft Company was one of the five airplane manufacturers which had been invited on 8 January 1937 to participate in the design competition for experimental two-engine pursuit airplanes, it does not appear "a strange coincidence" that the design submitted by Lockheed was also a two-engine interceptor as called for in the specifications.

7. *In discussing the fact that his design was not accepted, Mr. Hughes stated, "I would like to say that there was considerable* [sic] *to that situation which did not meet the eye", and went on to say that, "I heard from some of the people who were on the evaluation board that my design was liked better but that the Army thought maybe Lockheed with a little bigger plant . . . might be able to get more airplanes out or something of that kind". Attention is invited to the inclosed copy of letter dated 13 May 1937, (Inclosure 14), transmitting the report of the Technical Sub-Committee on the evaluation of the two-engine interceptor design competition, together with inclosures. These indicate that, in accordance with the prescribed method of evaluation, the Lockheed design was awarded a figure of merit of 65.0 percent and the Hughes design received only 47.3 percent. A review of the points awarded for various factors indicated that this difference is primarily attributable to the greater high speed and shorter time of climb which could be expected from the Lockheed design.*

FOR THE COMMANDING GENERAL:
S. R. BRENTNALL
Brigadier General, U.S.A.
Asst. Deputy Commanding General
Research & Development,
Procurement & Industrial Mobilization, T-3

Incls. 1-14 as noted

Declassified by the author at the National Archives And Records Administration, November 6, 2002.

For XB-35 development, Jack Northrop had four ⅓-scale proof-of-concept flying wings built designated N-9M. *(Courtesy of National Archives & Records Administration)*

Northrop XB-35, B-35, YB-35

Always bucking trends, John K. Northrop, or "Jack" as he was usually referred to, reveled in designing unconventional aircraft. After working for several of the larger aircraft companies such as Lockheed and Douglas, Northrop started his own aircraft business in 1932. Hard to imagine anyone embarking upon a new business venture in the depths of the Depression, but Northrop did and furthermore, pulled it off. But its success was due in no small part to funding by Douglas. Of course, with an arrangement such as this, strings were attached that Northrop eventually found unacceptable. So in 1939, Northrop formed yet another company bearing his name: Northrop Aircraft, Inc.

Although Northrop did not invent the flying wing (the Dunne flying wing of 1913 is often credited with that accomplishment), it's probably fair to say that Northrop devoted more time to its development than anyone else. Reducing an aircraft down to its lowest common denominator—a flying wing—sounds good in theory, but in practice tremendous hurdles need to be overcome; hurdles that were all but insurmountable in the 1940s. With today's computer technology, fly by wire, and better understanding of control issues, flying wings are a viable design. And as if to prove this point, the B-2 stealth bomber employs similar dimensions to the B-35. But if it were not for computer-operated, fly-by-wire controls, the B-2 would be impossible to fly by a human.

Seeing war clouds brewing on the horizon, the USAAF issued a requirement for a bomber to carry a 10,000-pound load 10,000 miles. Even though two companies submitted proposals and built aircraft, this ambitious goal was never achieved, at least not with piston powered bombers. Northrop submitted a proposal that would eventually result in the B-35 flying wing. The other contender for the 10,000-mile, 10,000-pound load requirement was the Convair B-36 and more of that anon.

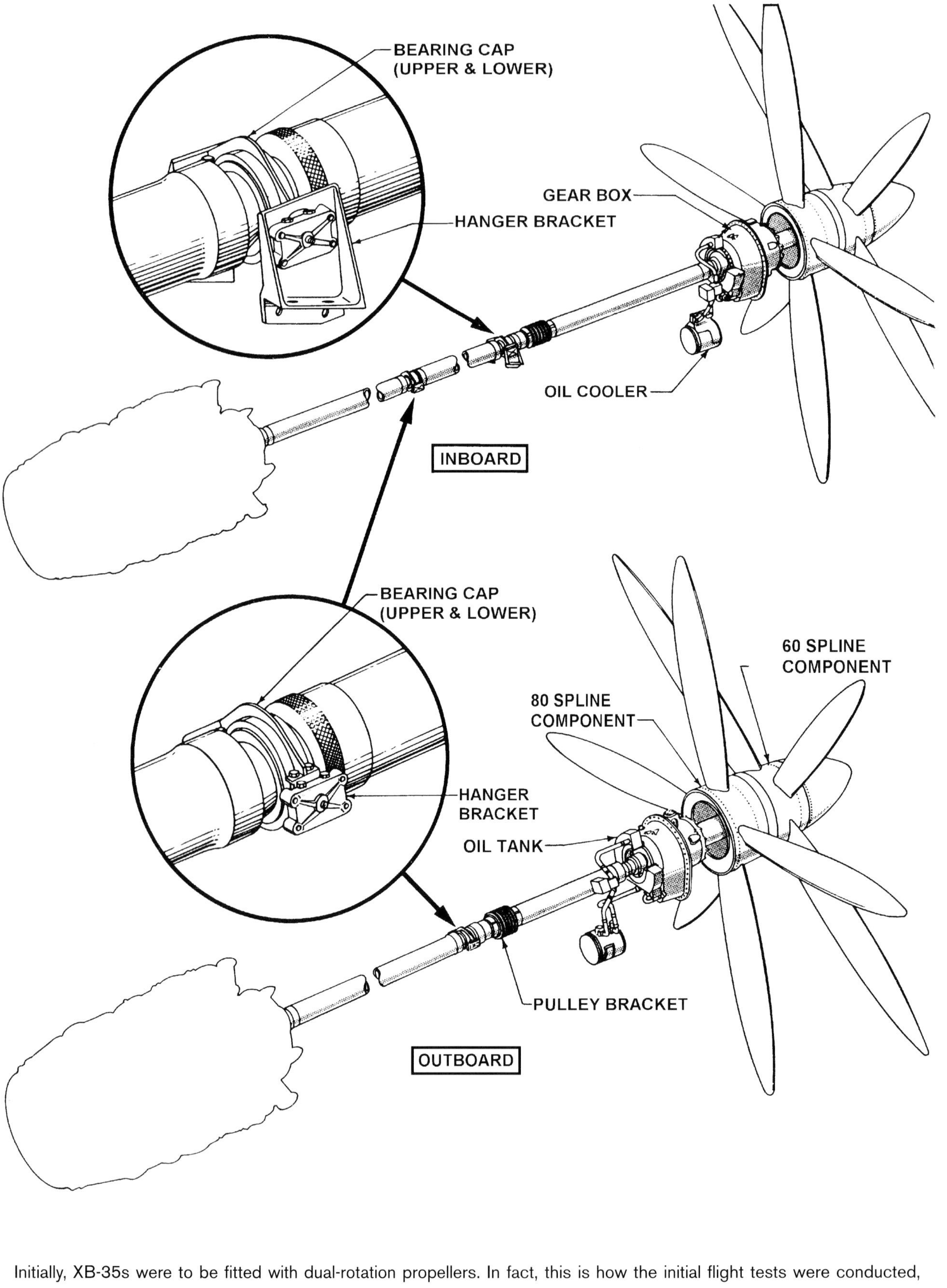

Initially, XB-35s were to be fitted with dual-rotation propellers. In fact, this is how the initial flight tests were conducted, that is until insurmountable vibration problems nixed that idea. Note the extension shafts from the engine to the propeller-reduction gearbox. *(Courtesy of National Archives & Records Administration)*

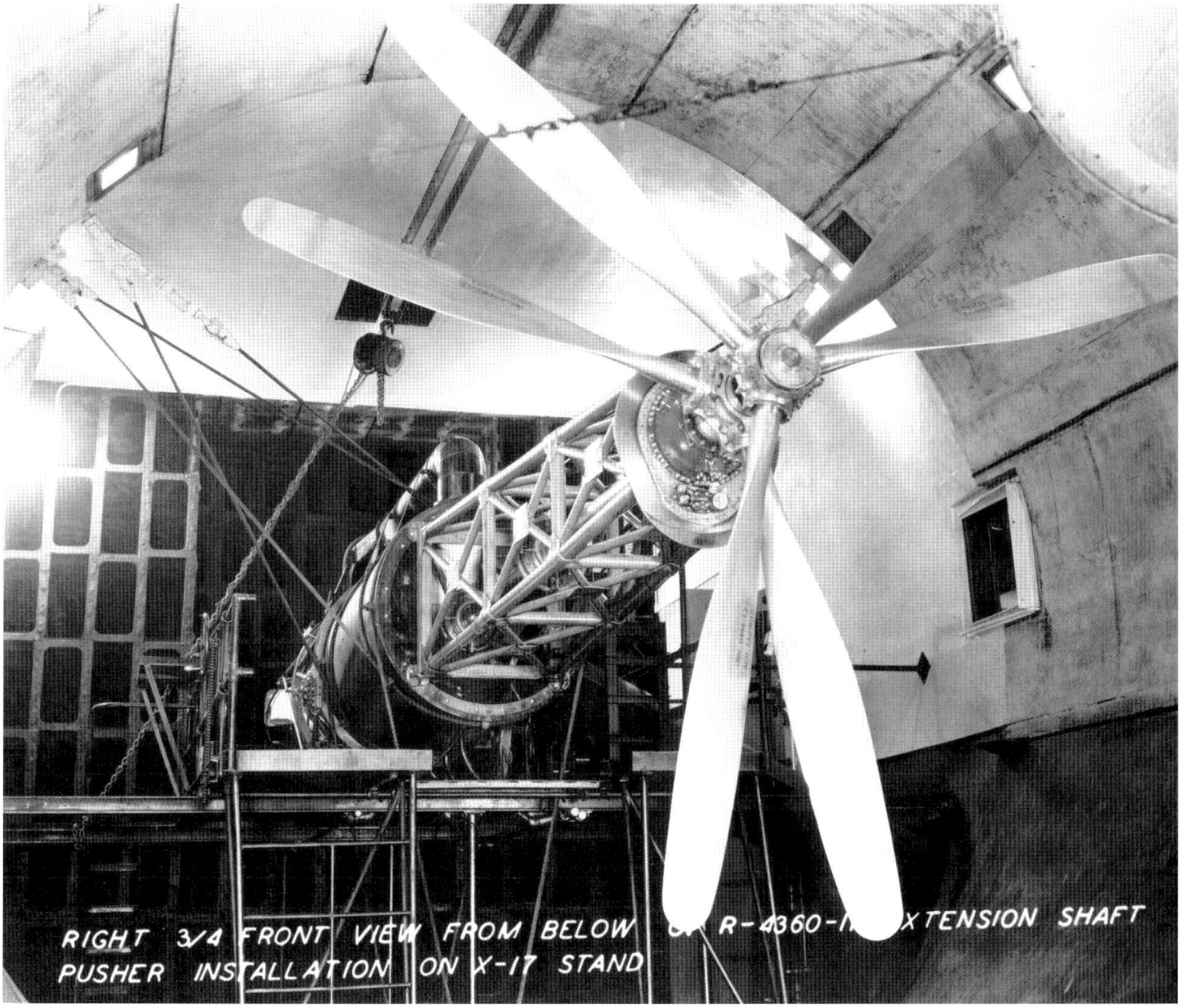

Under test at Pratt & Whitney's facility in East Hartford, the dual-rotation propeller drive performed flawlessly. However, when installed in the XB-35, severe vibration problems arose. *(Courtesy of Pratt & Whitney)*

As a proof of concept, several test aircraft were built to evaluate the flying wing and its control issues. With no vertical stabilizer, yaw control about the vertical axis was a difficulty that required a solution. Northrop devised a system of split flaps on the trailing edge at each wing tip. During straight and level flight, if the spit flaps were deployed on one wingtip, a yawing action would result. In other words, it was the same result as if the rudder were deflected on a conventional aircraft. The increased drag caused by this deflection would give the necessary yaw control about the vertical axis. For that reason the split flaps were connected to the rudder pedals, although the term "rudder pedal" now became somewhat of a misnomer. A more accurate term would be yaw control pedal. As an added benefit to the wing trailing edge split flap for yaw control, a spoiling action also resulted, thus the aircraft would bank into the turn. By depressing both "rudder" pedals at the same time, a rapid rate of descent could be achieved, giving excellent glide path control.

Elevators mounted on hinged surfaces of the horizontal stabilizer gives pitch control about the horizontal axis for a conventional aircraft. For the Northrop flying wings, pitch control was achieved by elevons located inboard of the wingtip split-flaps. When both were deflected in the same direction, the required pitch control was achieved. By deflecting the elevons differentially, roll control about the longitudinal axis was achieved, consequently giving the same result as the ailerons on a conventional aircraft. Flaps mounted inboard of the elevons on the trailing edge were used for landing only *(Ref. 7-1).*

Northrop's first pure flying wing design was designated N-1M powered by a pair of 65-hp Lycomings. Its first flight occurred in 1940 at Muroc Army Air Base, as it was known then. Based on the preliminary findings of the N-1M, Northrop designed what would eventually be the B-35. Design work was initiated in 1942. Four, ⅓-size test aircraft designated N-9M were built to evaluate the B-35's performance and handling characteristics *(Ref. 7-1).* It must have come as a rude awakening to realize that, based on N-9M testing, the projected 10,000-mile range with a 10,000-pound bomb load was now down to 8,400 miles. Nevertheless, the B-35 project proceeded and in June 1943 a contract was executed for the delivery of 200 aircraft. Like many projects started in 1941 or later, the B-35 had no hope of making a contribution to the war effort. In fact, Northrop's P-61 was one of the few aircraft designed, built, and put into service during World War II. The B-35 was a large aircraft with a wingspan of 172 feet and a maximum gross weight of 209,000 pounds. The engines were buried in the wing and drove the propellers via long extension shafts. A propeller reduction gearbox was fitted at the end of the extension shaft. This made perfect sense in that the engine was direct drive; therefore, the driveline did not have to deal with the torque multiplication of reduction gearing. Due to the differing length of the driveshafts, depending on whether it was an outboard or inboard engine, two different dash numbers of R-4360 were required for each aircraft. The inboard driveshaft was almost 9½ feet longer than that for the outboard engine.

XB-35s were initially powered by R-4360-7 (outboard) and R-4360-11 (inboard) engines driving contra-rotating propellers. Problems with severe vibration immediately surfaced. All past accounts have put the blame for this vibration squarely at the doorstep of Pratt & Whitney. Reality is different. Pratt & Whitney did extensive testing of the B-35 driveshaft arrangement in East Hartford and everything—engine, drive shaft, propellers, and reduction gearbox—all behaved perfectly. But then everything in the aircraft went dreadfully wrong. Being a pusher arrangement, the propellers were exposed to severe turbulence swirling off the trailing edge of the wing due to low pressure on the upper surface and high pressure on the lower surface. This would excite the propellers into vibration. Two additional factors further exacerbated the problem: (i) the fact that the wing trailing edge was swept back; therefore each side of the propeller disk saw very different degrees of turbulence, and (ii) the propeller centerline was considerably higher than the trailing edge. As a temporary fix, single rotation four-blade propellers replaced the contra-rotating propellers. It still vibrated. In retrospect, it was unfair to blame Pratt & Whitney, as they had a good design. But regardless of how good it was, expecting the propellers to not vibrate in a pusher arrangement, and especially one installed in the B-35, was simply expecting the impossible. As an aside, Convair had the same problem for similar reasons with the B-36. In the case of the B-36, the problem was literally fixed through brute force. The entire wing trailing edge was over-designed and beefed up in order to attenuate propeller vibration. Also, the B-36's propeller disks lay parallel with the wing trailing edge. Although the pusher arrangement sounds like a good idea, in reality it has some serious flaws. That's why the vast majority of propeller-driven aircraft are a tractor configuration.

Getting back to the XB-35 tale. As the Martin B-26 Marauder production schedule wound down, they were sub-contracted to take on B-35 work. Martin was tasked with the structural design of the wing and engine installation. Problem was, many Martin engineers were

The B-35 was fundamentally flawed. Possibly contributing to its problems was the fact that the propeller centerline was considerably higher than the wing trailing edge. Therefore, with low-pressure air swirling off the top of the wing and higher-pressure air coming off the bottom of the wing, severe turbulence hit the propeller disc. This would result in severe vibration and yet historians have never mentioned this. Instead the typical comment is: "The flawed R-4360 installation." *(Courtesy of National Archives & Records Administration)*

drafted into the services, leaving them short handed. The Otis Elevator Company was called upon to assist Martin with its design obligations. However, salvation arrived when it was obvious that B-35s would have no effect on the outcome of the war. Consequently, the Martin contract was cancelled. Construction of the XB-35 started in early 1943 at Northrop's Hawthorne, California, facility. Northrop kept plugging away after the war and the first XB-35 made its maiden flight on June 25, 1946, coincidently within a couple of months of its rival

This photograph of the XB-35 was taken April 30, 1946, shortly before its first flight. Of interest in this photo is the fact that the #1 engine is fitted with a pair of three-blade dual rotation propellers. The others are four-blade dual rotation. *(Courtesy of National Archives & Records Administration)*

the Convair B-36. Although the first flight went off without a hitch, vibration problems soon surfaced as a serious and, as it turned out, insurmountable issue. Plans were soon afoot to retrofit the B-35 fleet with jet engines. This modification was performed by installing eight Allison J33 jet engines and redesignating the aircraft XB-49. In an effort to at least salvage something from the B-35 program, R-4360-17s and R-4360-11s were converted to R-4360s for other uses.

Notwithstanding the foregoing, historians have often criticized the engine and its installation. In fact, the entire aircraft was fundamentally flawed. Of the 15 built only five ever flew, the remainder being scrapped before they had a chance to fly. Even worse was the incredible effort put forth by Pratt & Whitney and yet it was they who have borne the brunt of unjustified criticism over the years. Despite the fact that only a handful ever got to fly, an amazing five different engines were developed for this aircraft. All things considered, the B-35 program was a terrible waste of resources that accomplished nothing. Furthermore, it's probably a good thing that only five actually got to fly; this dog of an aircraft was a death trap due to its very limited CG range, marginal longitudinal stability, and other serious control problems.

Powered by a similar setup to that employed by the XB-36 (fan cooled, pusher with dual turbosuperchargers for each engine), the engineering challenges facing Northrop were daunting. Eight turbosuperchargers, four massive air-to-air intercoolers, engine oil coolers, induction ram air, and a complex exhaust system had to be mounted in an efficient and lightweight manner. As if the above mentioned wasn't difficult enough, additional challenges facing Northrop engineers were mounting of the extension shaft and propeller reduction gear with their attendant oil tanks, oil coolers, etc. It would have been considerably easier to simply mount the propeller reduction gearing on the engine in a conventional fashion. The problem with that arrangement would have been an excessively heavy driveshaft due to the torque multiplication. With an excess of 3,000 hp being transmitted through the reduction gearbox, a considerable amount of heat is generated. Even with an efficiency of 95 percent, that means 5 percent or 150 hp in heat needs to be rejected and the majority of this heat went into the gearbox oil cooler. As an example of the desperation facing both Northrop and Pratt & Whitney engineers, lubricating the shaft couplings could only be described as bizarre. The shaft couplings had

Engineering Study of the R-4360 Installation in the XB-35

(Ref. 7-18)

While the XB-35 has often been justifiably criticized as an unsafe aircraft, the engineering concepts embodied within its design, particularly its powerplant installation, are worthy of praise and a closer look.

Another view of the XB-35 also taken on April 30, 1946. Note the huge leading edge air intake that provided cooling air for the engine, oil cooler, and intercooler. It also provided induction air. *(Courtesy of National Archives & Records Administration)*

a paper Dixie cup filled with 600-W oil. After assembly, the cup was tipped over, thus liberating its oil into the coupling. Even today, surviving Pratt & Whitney engineers enjoy a chuckle over the Dixie cup lubrication system. Very rarely do you see an accessory on a military aircraft driven by a rubber V-belt. And yet this is what Northrop engineers resorted to for the hydraulic pumps. One would have to assume that this rather primitive automotive technology would have been replaced by a more substantial gear drive for production aircraft. But maybe not.

Engine Cooling

In a similar fashion to the B-36, the R-4360s buried within the wing of the B-35 needed augmented cooling via a sophisticated, variable speed fan to keep engine temperatures under control. Ductwork from the wing leading edge fed air to the fan, which then discharged over the engine. A sealed shroud encasing the engine ensured that all the output from the fan went to cooling the engine.

The hydraulically driven engine fan was secured to the accessory drive case at the rear end of the engine. However, rear in the context of the

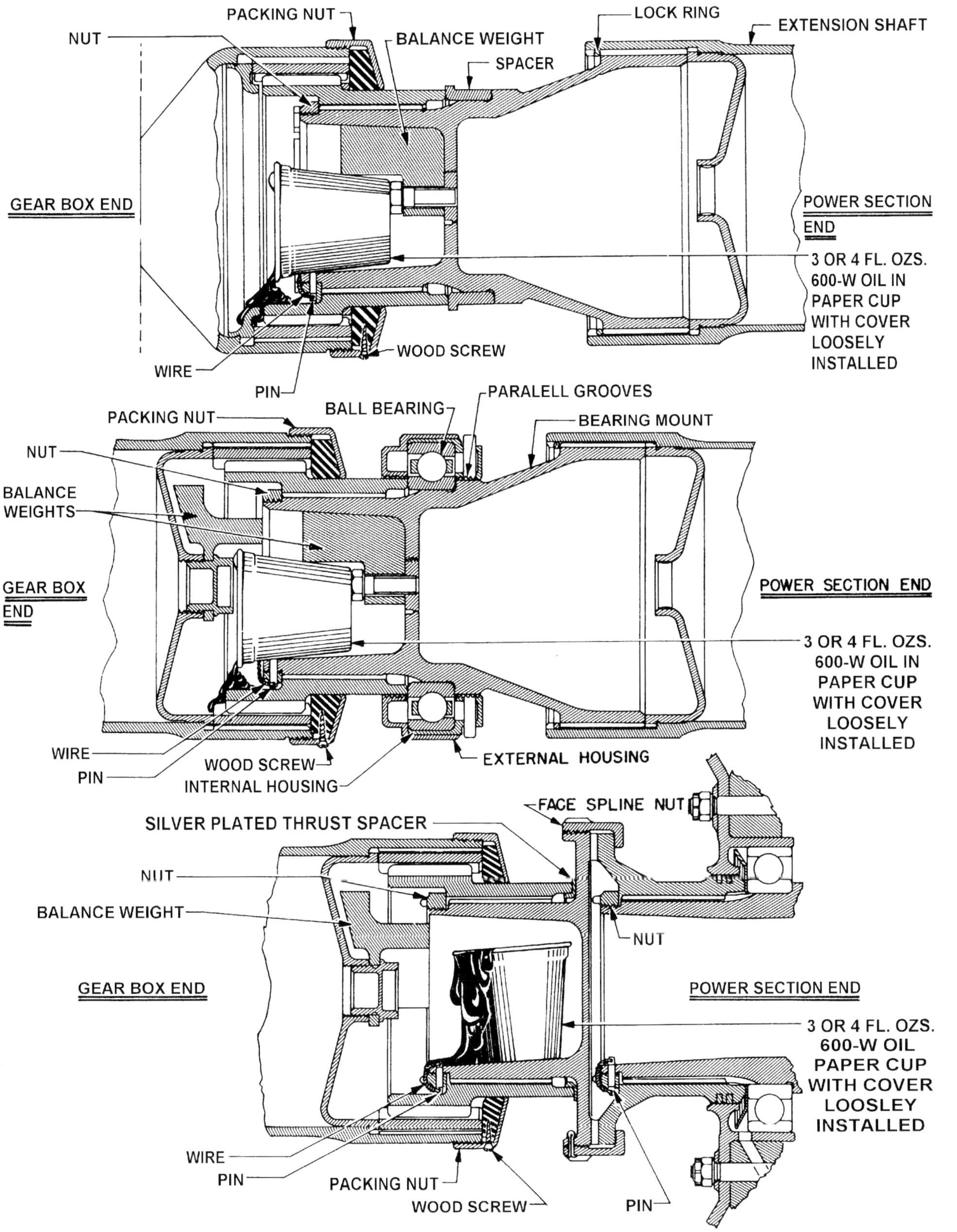

A bizarre lubrication system was installed for the B-35's intermediate propeller drive shaft bearings. Paper Dixie cups filled with 600-weight oil were inserted in the bearing housing and then tipped over to spill the oil in the appropriate place. Surviving Pratt & Whitney engineers still get a chuckle over this. *(Erection & Maintenance Manual for the XB-35. Courtesy of National Air & Space Museum)*

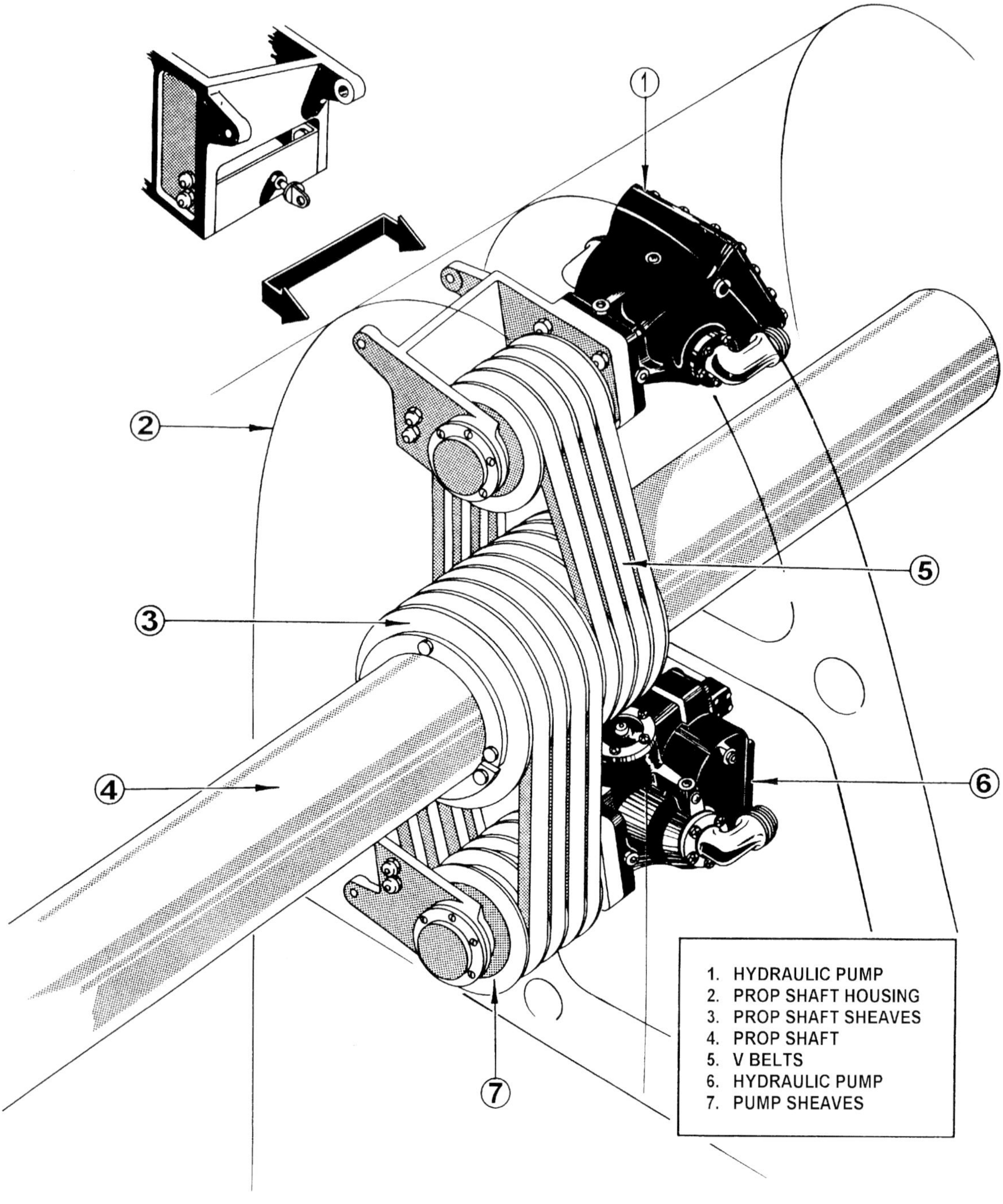

Rarely are "V" belts employed in frontline military aircraft, especially for a critical drive requirement such as for a hydraulic pump. And yet this is the method Northrop engineers used to drive the hydraulic pumps. *(Erection & Maintenance Manual for the XB-35. Courtesy of National Air & Space Museum)*

B-35 meant the end facing the wing leading edge. An adjacent selector valve (fan speed control) controlled it manually in accordance with cooling demands by the flight engineer. The fan drive low-ratio gear was bolted to the rear face of the fan drive high-ratio gear. This assembly was splined onto the fan driveshaft. The hub of the high-ratio gear sat against the thrust bearing inner race and was retained in place with a spanner nut. The low-ratio gear meshed with the low-ratio hydraulic coupling fan intermediate drive pinion and the high-ratio gear meshed with the two-high ratio coupling fan intermediate drive pinions. The cooling fan was bolted to the face of the flanged fan drive shaft.

A hydraulic valve actuated by a 28-volt DC reversible motor controlled the engine fan speed. This was mounted on the right side of the accessory section of the engine and wired through the engine disconnect box to the flight engineer's junction panel, which was located under the engineer's floor. The hydraulic valve controlled the fluid clutches in the engine. Fan speed control, located on the flight engineer's upper electrical control panel, was via a double throw momentary toggle switch that was normally in an "off" position. Fan speed could be increased or decreased by the two "on" positions designated "warmer" and "cooler." When the switch was placed in either of these positions, the circuit to the motor was closed. This caused the motor to rotate in the appropriate direction to respectively decrease or increase the fan speed. It should be realized that the cooling fan was capable of a prodigious output resulting in a significant amount of parasitic power loss. Leading-edge scoops supplied cooling air via ducts to the fan. These leading-edge scoops also provided cooling air for the oil coolers.

Turbosupercharger Installation

Two GE CH-1 tubosuperchargers augmented each engine-driven, single-stage, single-speed supercharger. The pair of CH-1s operated in parallel. With this arrangement, one turbo could be shut down during cruise conditions. In this way, a single turbo would operate far more efficiently than a pair running in parallel. Part of the rationale for this is the fact that the waste gate would be almost, if not completely, closed with a single

Like its XB-36 contemporary, the XB-35 used fan-assisted cooling. This photo shows the contra-vane. The fan, not viewable in this shot, is directly behind the contra-vane. The fan's disc and support shaft are shown with the tachometer generator for measuring fan speed. *(Courtesy of Pratt & Whitney)*

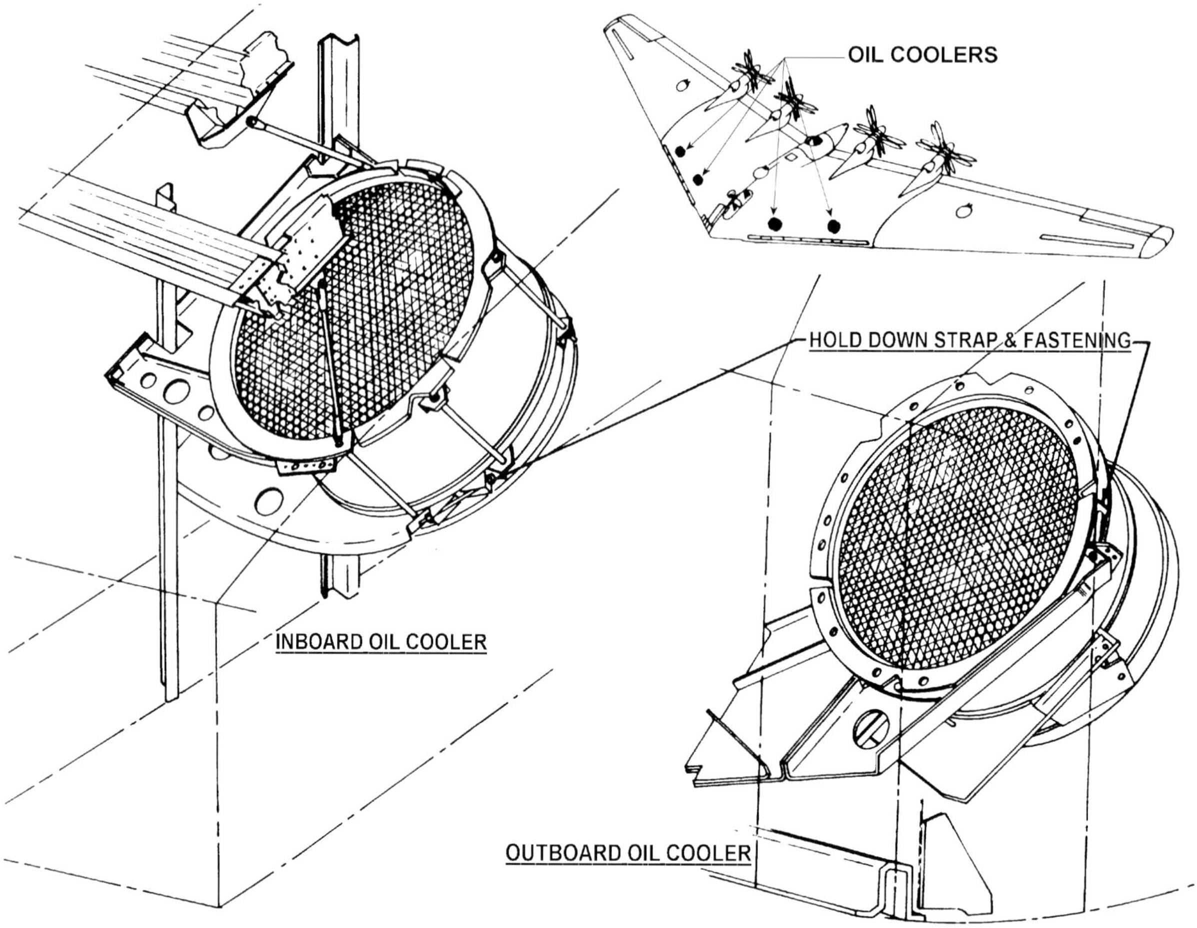

A single large circular oil cooler, mounted at an acute angle, was used for each engine. *(Erection & Maintenance Manual for the XB-35. Courtesy of National Air & Space Museum)*

turbo in operation. Whereas if both turbos were in operation, the exhaust waste gate would open, thus wasting some of the exhaust energy. At the critical altitude of single-turbo operation, the other turbo would be brought online. Additionally, in the event emergency power was required, the other turbo would be brought online. A simple butterfly valve closed off the engine's exhaust to one of the turbos when it was not needed. As with any supercharging arrangement, considerable heat was imparted to the induction air, which of course is anathema for detonation resistance. Excess heat was eliminated via air-to-air heat exchangers.

An engineering challenge faced Northrop engineers with respect to mounting a pair of R-4360s and four massive CH-1 GE turbosuperchargers in each side. The solution was to mount four turbos between the engines on each wing. Each turbo was mounted at approximately 45 degrees with the compressor facing forward (towards the wing leading edge). It should be noted that the B-35's contemporary, the B-36, shared a similar powerplant arrangement, but solutions to the engineering challenges made for an interesting contrast—different, but not necessarily better. Installing the CH-1s could be quite a chore. Mechanics had to insert a crane into the leading-edge air intakes. If the B-35 had entered service, one could just imagine the hanger rash these critical air intakes would have suffered.

Remote Propeller Reduction Gearboxes

Mounted on the trailing edge of the wing, the heavy and bulky reduction gearboxes presented a significant design challenge to the Northrop

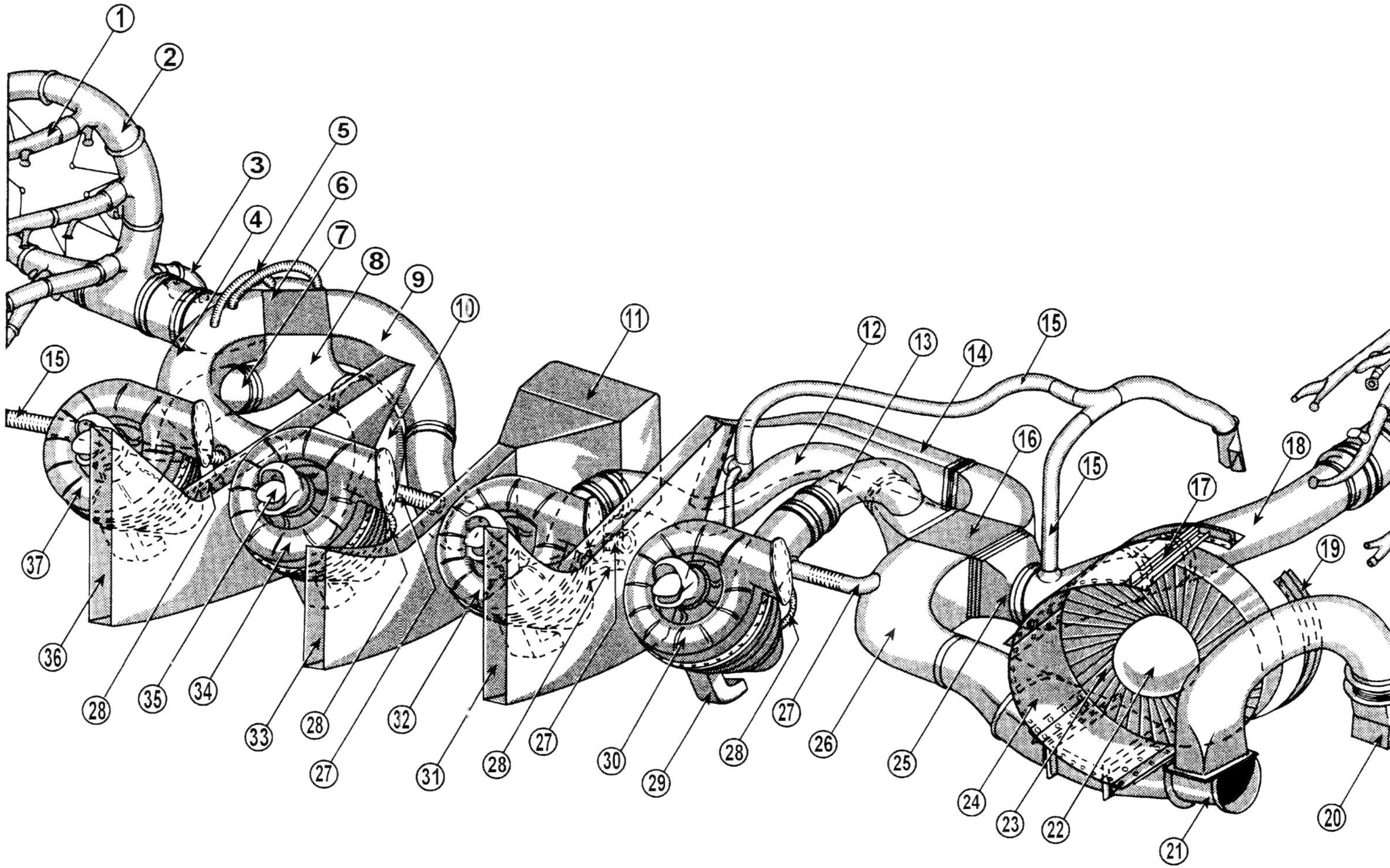

Each engine for the XB-35 used a pair of turbosuperchargers. The four turbos for the two engines on a wing were clustered together. 1. Exhaust Pipe Header; 2. Collector Ring Assembly; 3. Overboard Valve (inboard); 4. Inboard Anti-Icing Cold Air Duct; 5. Flex Cooling Ducts; 6. Inboard Heat Exchanger; 7. No. 4 Turbo Exhaust Pipe; 8. Inboard Turbo Exhaust Chamber; 9. Inboard Anti-Icing Hot Air Duct; 10. No. 3 Turbo Exhaust Pipe; 11. Cabin Heat Exchanger and Waste Gate; 12. No. 2 Turbo Exhaust Pipe; 13. No. 1 Turbo Exhaust Pipe; 14. Outboard Anti-Icing Cold Air Duct; 15. Ground Cooling Ducts; 16. Outboard Heat Exchanger; 17. Diffuser Stringer Assembly; 18. Exhaust Tailpipe Assembly; 19, Diffuser Frame and Aft Skin Assembly; 20. Anti-Icing Waste Gate; 21. Anti-Icing Valve to Outer Wing; 22. Diffuser Cap; 23. Outboard Engine Diffuser; 24. Duct Doors (3); 25. Tailpipe Transition Duct; 26. Outboard Anti-Icing Air Duct; 27. Cooling Air Junction; 28. Turbo Bearing Cooling Air Ducts; 29. Turbo Waste Gate; 30. No. 1 Turbo; 31. Outboard Ram Air Duct; 32. No. 2 Turbo; 33. Anti-Icing Rib Duct; 34. No. 3 Turbo; 35. Turbo Air Intake Scoop; 36. Inboard Ram Air Duct; 37; No. 4 Turbo. *(Erection & Maintenance Manual for the XB-35. Courtesy of National Air & Space Museum)*

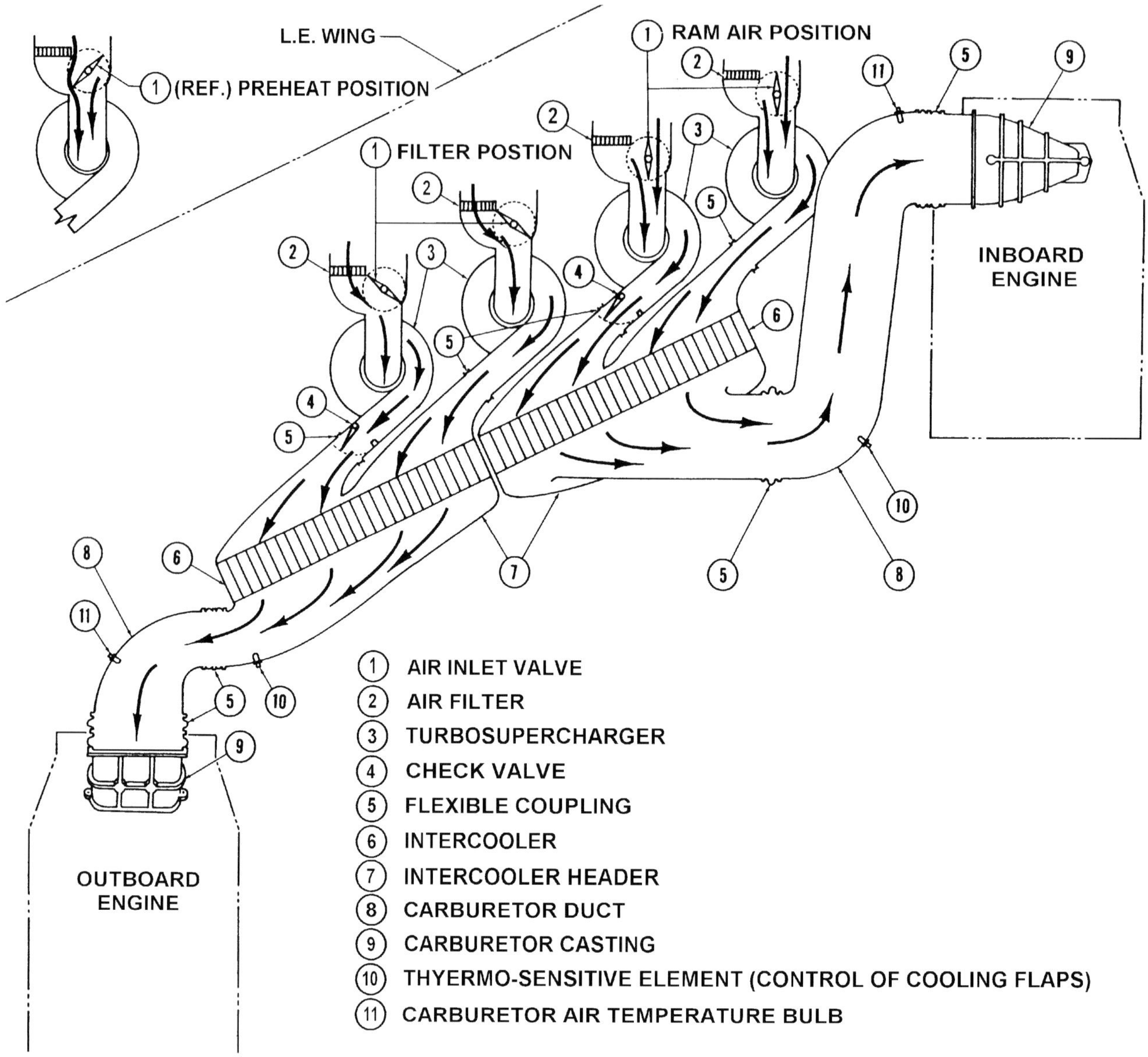

Four turbosuperchargers and a pair of air-to-air intercoolers were located between the inboard and outboard engines making for a nice tidy installation. *(Erection & Maintenance Manual for the XB-35. Courtesy of National Air & Space Museum)*

engineers. As related previously, two styles were employed—single rotation and dual rotation. Internally, they were similar to the reduction gears employed on R-4360s with integral gearing. The primary difference was the requirement for additional pressure and scavenge pumps for the lubrication system. An oil tank, oil cooler plus ducting, and an air scoop for the cooler were required. The dual-rotation gearbox used a pinion and reduction gear. A Farmen type epicyclic gear was used to give opposite rotation of the additional propeller shaft. This resulted in the driveshaft being somewhat lower than the pair of co-axial propeller shafts. With the single-rotation gearbox, the driveshaft and propeller shaft were co-axial. Therefore, one would have to assume that the single-rotation gearbox was mounted higher in order to maintain the same thrust line. This small and seemingly insignificant detail illustrates the engineering challenges facing Northrop when the decision was made to go from dual- to single-rotation propellers.

Exhaust System

In addition to conveying exhaust gasses from the engine, the engine exhaust system was utilized to provide motive power for the turbosuperchargers,

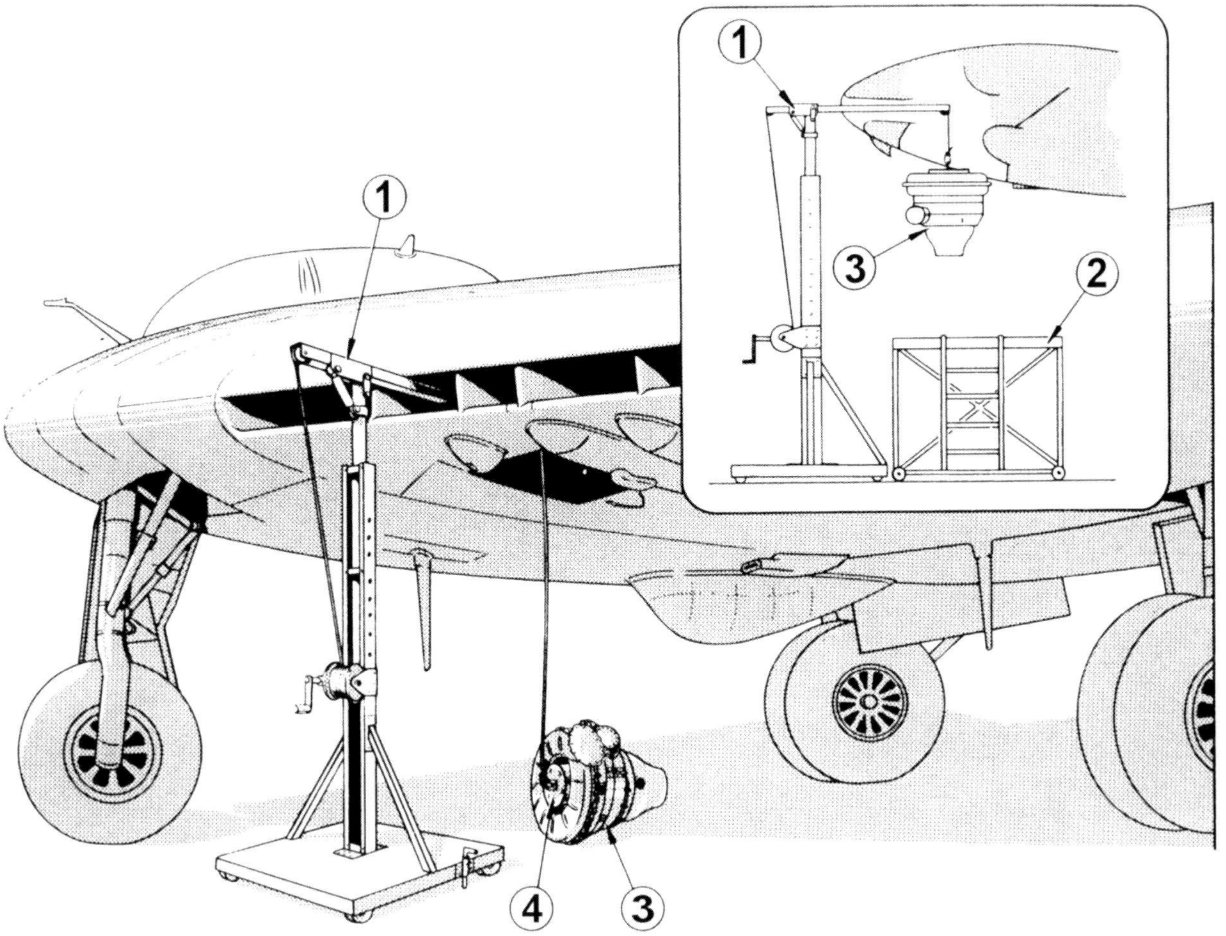

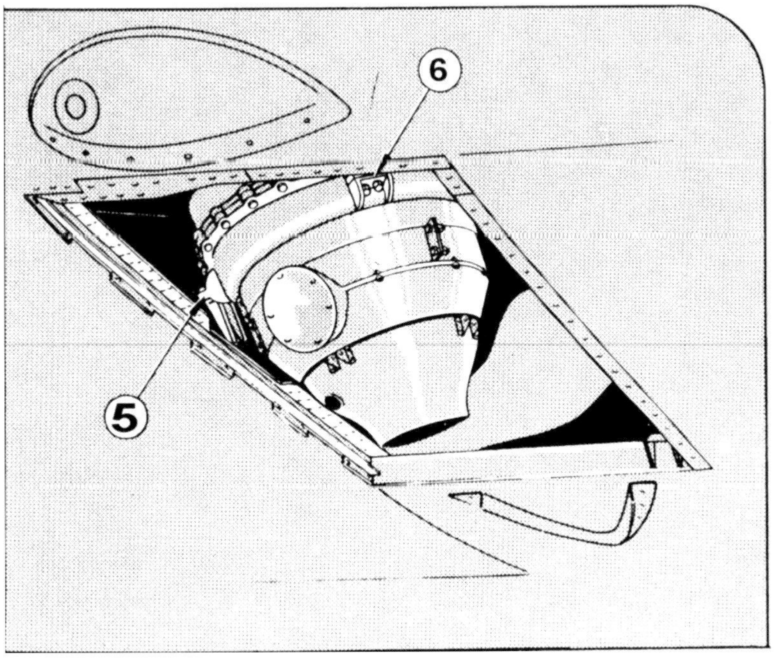

1. TURBO HOIST
2. MECHANICS' STAND
3. TURBOSUPERCHARGER
4. SPECIAL HOIST PLATE
5. SIDE HANGER
6. FRONT HANGER

One can only speculate how much hanger rash the critical leading-edge ducts would have suffered had the B-35 entered service. Poking a crane jib through the leading-edge intakes in order to change out a turbo was asking for trouble. *(Erection & Maintenance Manual for the XB-35. Courtesy of National Air & Space Museum)*

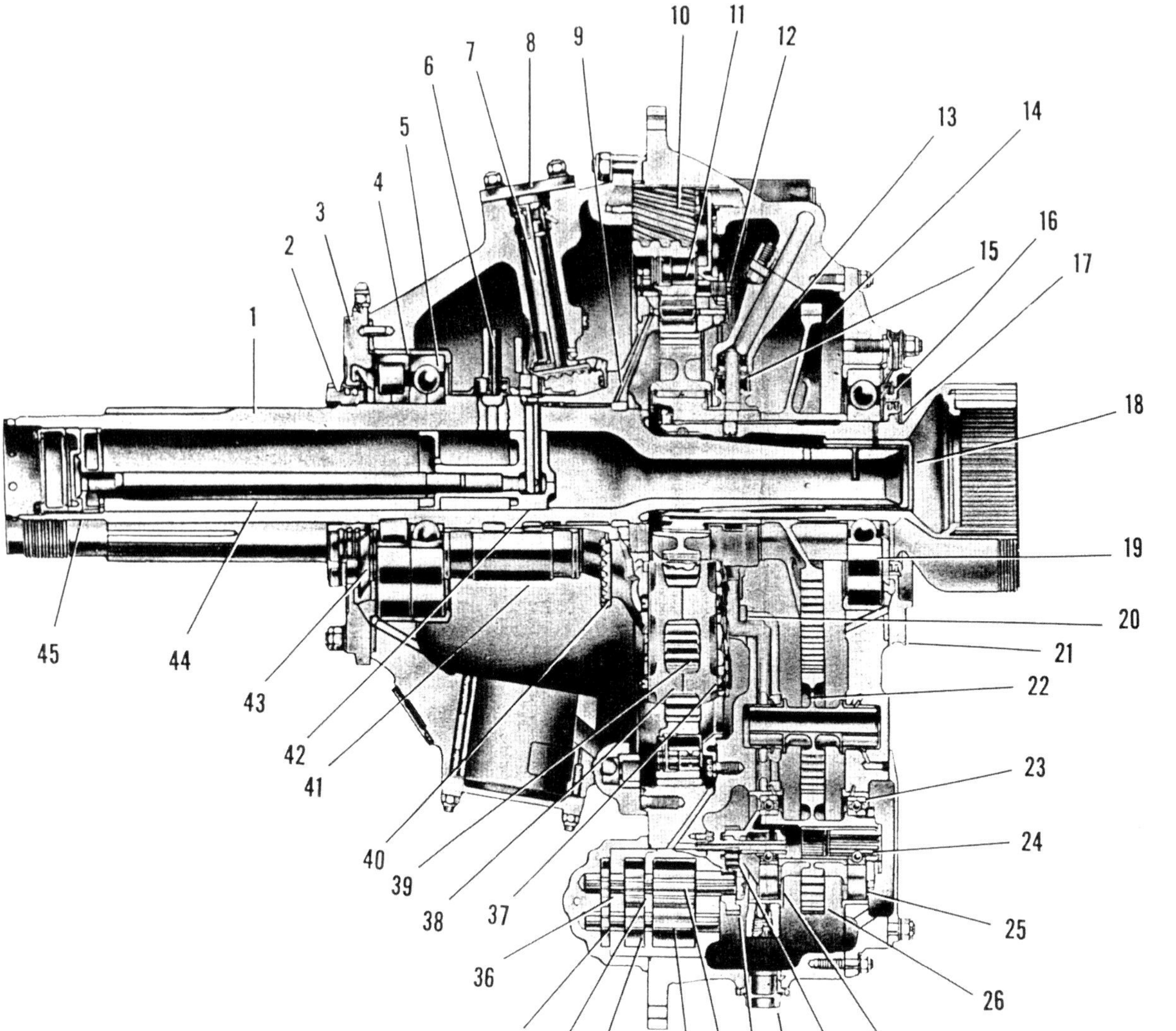

Reduction gearbox for the single rotation propeller drive. 1. Propeller Shaft; 2. Propeller Shaft Thrust Spacer; 3. Propeller Shaft Thrust Cover; 4. Propeller Shaft Roller Bearing; 5. Propeller Shaft Ball Bearing; 6. Governor Front Oil Transfer Tube; 7. Governor Drive Shaft Gear; 8. Governor Mounting Pad Cover; 9. Reduction Pinion Shaft Front Support; 10. Reduction Fixed Gear; 11. Torquemeter Piston; 12. Reduction Pinion Shaft Nut; 13. Reduction Drive Shaft Oil Transfer Bracket; 14. Power Take-off; Drive Gear; 15. Reduction Drive Shaft Oil Transfer Tube; 16. Reduction Drive Shaft Oil Slinger; 17. Reduction Drive Shaft; 18. Reduction Drive Shaft Plug; 19. Extension Shaft Ball Bearing; 20. Reduction Drive Shaft Oil Transfer Bearing Elbow; 21. Extension Shaft Ball Bearing Outer Cover; 22. Power Take-off Intermediate Drive Gear; 23. Power Take-off Drive Shaft Gear Rear Ball Bearing Liner; 24. Power Take-off Adapter; 25. Power Take-off Drive Shaft Gear Rear Ball Bearing; 26. Power Take-off Drive Shaft Gear; 27. Power Take-off Drive Shaft Gear Front Ball Bearing; 28. Front Oil Pump Intermediate Drive Gear; 29. Reduction Drive Gear Case Oil Drain Plug; 30. Front Oil Pump Drive Gear; 31. Front Oil Scavenge Pump Driving Impeller; 32. Front Oil Scavenge Pump Driven Impeller; 33. Front Oil Pressure Pump Driven Impeller; 34. Front Oil Pressure Pump Driving Impeller; 35. Torquemeter Pump Driven Impeller; 36. Torquemeter Pump Driving Impeller; 37. Reduction Pinion Shaft Rear Support; 38. Reduction Pinion Shaft Front Support; 39. Reduction Pinion; 40. Governor Drive Gear; 41. Propeller Shaft Oil Transfer Bearing; 42. Propeller Shaft Rear Plug; 43. Propeller Shaft Oil Slinger; 44. Propeller Oil Feed Tube; 45. Propeller Shaft Front Plug. *(AN 02A-10HC-3 Handbook Overhaul Instructions Models R-4360-45 and -47 Aircraft Engines)*

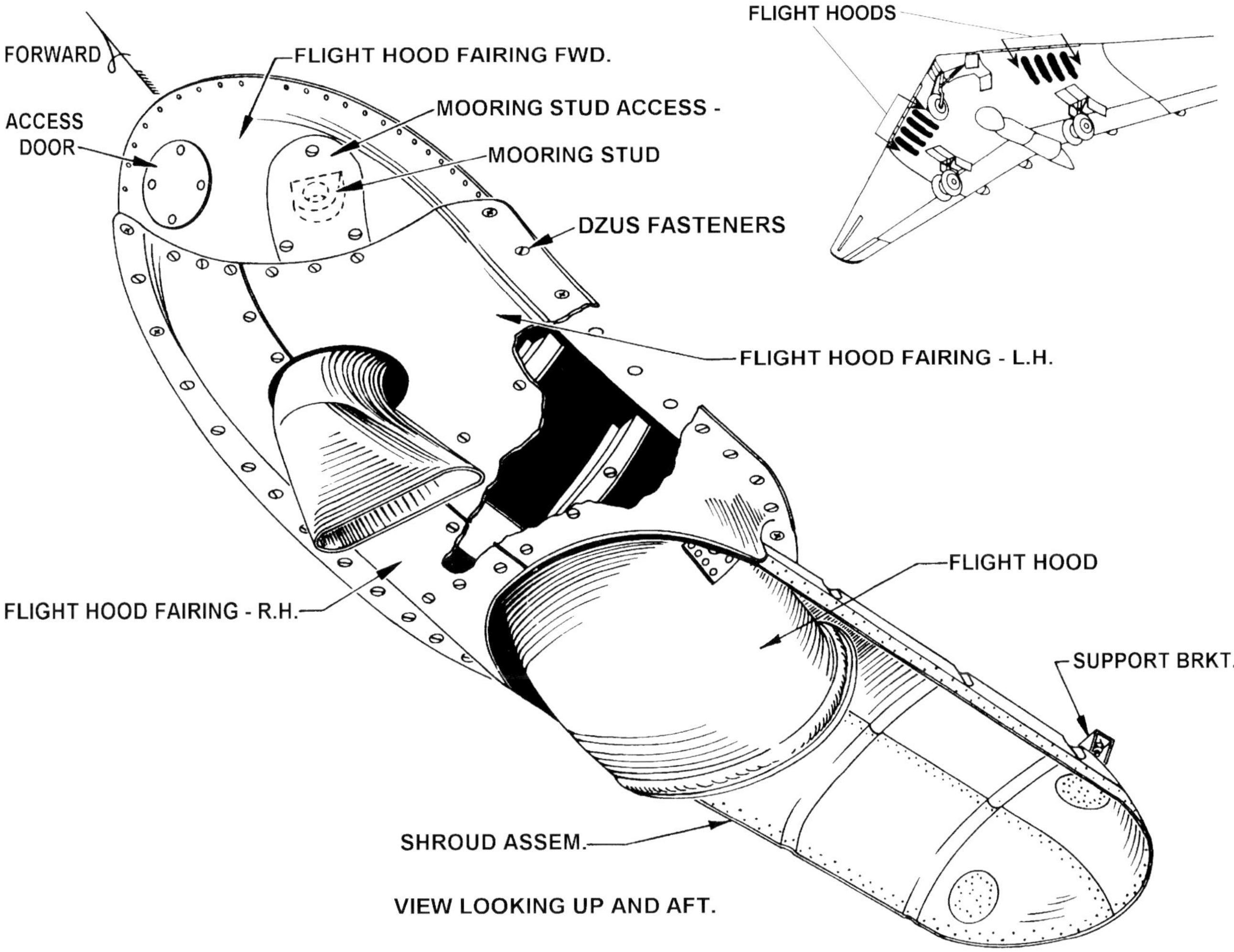

This line drawing shows one of the four flight hood assemblies–one for each engine. The smaller exhaust pipe ahead of the flight hood is an exhaust outlet for waste-gate gasses. *(Erection & Maintenance Manual for the XB-35. Courtesy of National Air & Space Museum)*

heat for outer wing anti-icing (outboard engines), and cabin heat (inboard engines). Exhaust gas from each engine was routed through a heat exchanger and diverted into a pair of turbosuperchargers, eventually escaping through the applicable exhaust outlet (flight hood) or waste pipe. As described, under certain conditions determined by the flight engineer, the exhaust gas may be directed through only one of the two turbosuperchargers. The heat exchanger's primary function was to decrease the temperature of the exhaust gas flowing into the turbosupercharger. It should be remembered that at the time of the B-35's development, high temperature alloys capable of handling the R-4360's stratospheric exhaust gas temperature had not been developed. Also, there were no good ways to manufacture internally cooled turbine buckets. Cool ram-air, after passing through and collecting heat from a heat exchanger, normally discharges into the slipstream through an overboard waste gate assembly. But it could be diverted for use in outer wing anti-icing or cabin heating.

The exhaust pipe coupling in the exhaust port of each cylinder is equipped with a steel liner and four studs for securing it to an appropriate exhaust pipe header. These headers were constructed of stainless steel and encased within

continued on page 381

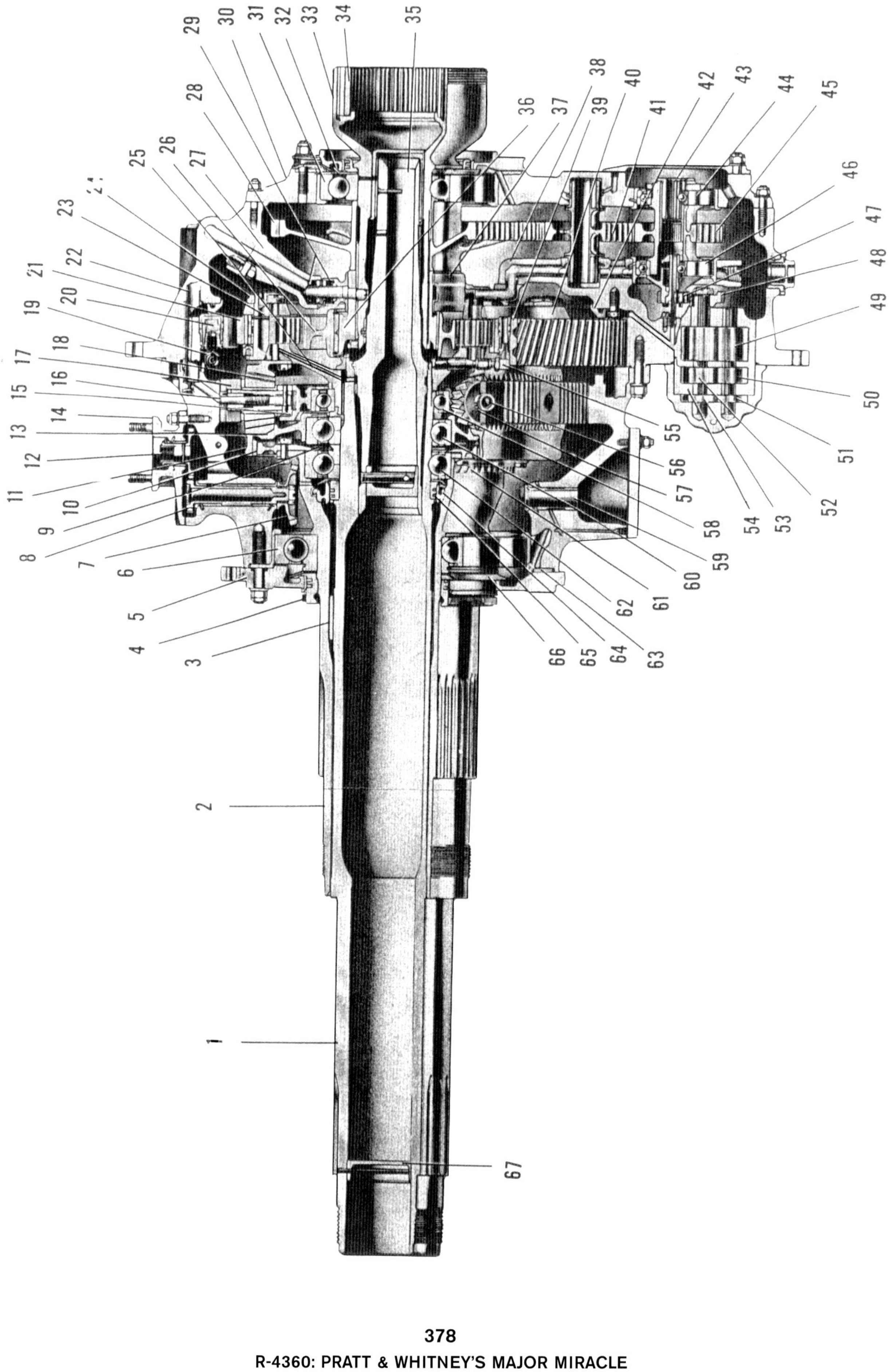
1
2
3
4
5
6
7
8
9
10
11
12
13
14
15
16
17
18
19
20
21
22
23
25
26
27
28
29
30
31
32
33
34
35
36
37
38
39
40
41
42
43
44
45
46
47
48
49
50
51
52
53
54
55
56
57
58
59
60
61
62
63
64
65
66
67

Opposite: Reduction gearbox for the dual rotation propeller drive. 1. Inner Propeller Shaft; 2. Outer Propeller Shaft; Outer Propeller Shaft Inner Bearing; 4. Thrust Nut; 5. Thrust Cover; 6. Outer Propeller Thrust Ball Bearing; 7. Governor Intermediate Drive Spur Gear; 8. Governor Intermediate Drive Gear; 9. Spacer; 10. Outer Propeller Shaft Coupling; 11. Outer Propeller Shaft Drive Gear; 12. Governor Drive; 13. Outer Propeller Shaft Drive Gear Support; 14. Governor Adapter; Mounting Pad; 15. Reverser Pinion Cage Support; 16. Reverser Pinion Cage Oil Transfer Gland; 17. Reverser Pinion Cage; 18. Reverser Pinion; 19. Torquemeter Oil Feed Hose; 20. Torquemeter Piston (Master); 21. Reduction Pinion; 22. Pressure Oil Transfer Ferrule; 23. Reduction Pinion Shaft Valve; 24. Reduction Pinion Shaft Nut; 25. Reduction Gear Support; 26. Reduction Drive Gear; 27. Pressure Oil Transfer Bracket; 28. Power Take-off Drive Gear; 29. Oil Transfer Bearing Oil Seals; 30. Reduction Drive Case Cover; 31. Reduction Drive Shaft Thrust Ball Bearing; 32. Reduction Drive Shaft Oil Slinger; 33. Reduction Drive Shaft; 34. (Axial-Spline Bushing); 35. Inner Propeller Tail Shaft; 36. Reduction Drive Shaft Coupling; 37. Reduction Drive Shaft Oil Transfer Bearing; 38. Reduction Drive Shaft Oil Transfer Elbow; 39. Torquemeter Piston; 40. Reduction Drive Fixed Gear; 41. Power Take-off Intermediate Drive Gear; 42. Reduction Drive Pinion Rear Support; 43. Power Take-off Drive Adapter; 44. Power Take-off Drive Pinion Rear Ball Bearing; 45. Power Take-off Drive Shaft Gear; 46. Power Take-off Drive-Pinion Front Ball Bearing; 47. Front Oil Pump Intermediate Drive Gear; 48. Front Oil Pump Drive Gear; 49. Front Oil Scavenge Pump Idler Gear; 50. Front Oil Pressure Pump Idler Gear; 51. Torquemeter Pump Idler Gear; 52. Front Oil Pump Gear; 53. Torquemeter Pump Gear; 54. Front Oil Pump Pressure Section; 55. Torquemeter Oil Pressure Take-off Elbow; 56. Reverser Pinion Shaft; 57. Reverser Pinion; 58. Reverser Case Ball Bearing; 59. Propeller Shaft Case Bottom Accessory Drive Bearing; (Unused); 60. Outer Propeller Shaft Drive Gear Support Bearing; 61. Governor Drive Gear; 62. Inner Propeller Shaft Thrust Ball Bearing; 63. Outer Propeller Shaft Ball Bearing Spacer; 64. Inner Propeller Shaft Oil Seal Rings Liner; 65. Inner Propeller Shaft Oil Seal Rings; 66. Outer Propeller Shaft Ball Bearing Oil Slinger; 67. Inner Propeller Shaft Front Plug. *(Handbook Overhaul Instructions Models R-4360-17 and -21 Aircraft Engines)*

This head-on view shows the nose wheel strut. Also of interest in this shot is the offset for the pilots' bubble canopy. The scoops located above the propeller support pylons feed cooling air to the propeller reduction gear oil cooler. *(Courtesy of National Archives & Records Administration)*

Opposite: Characteristic black streaks from exhaust gasses issuing from the flight hoods located under the wing can be seen in this underneath photo. (Courtesy of National Archives & Records Administration)

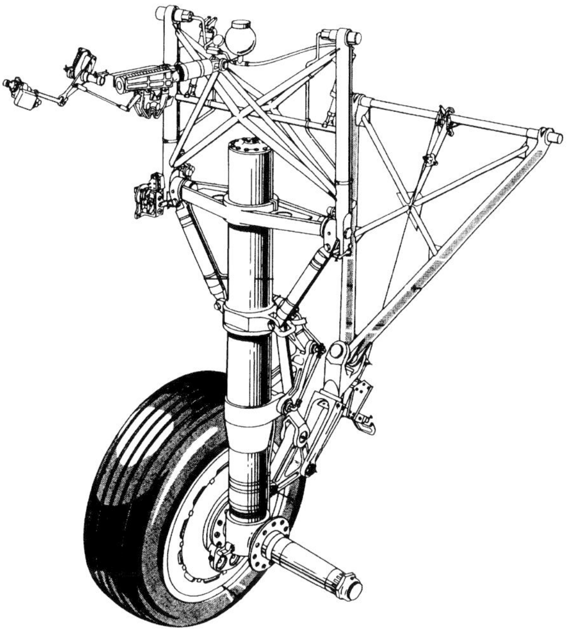

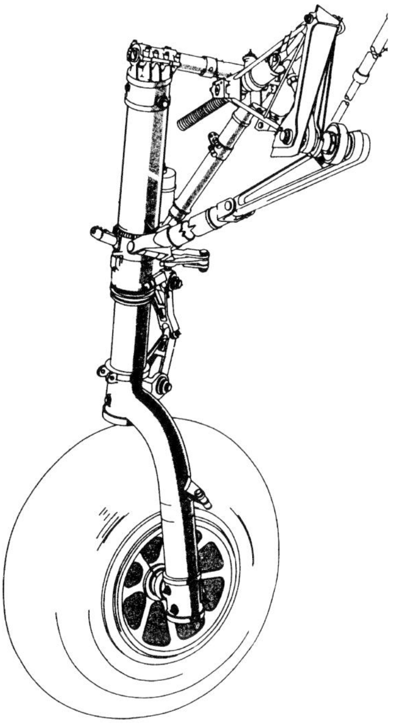

Above left: Conventional tri-gear with dual-wheel main wheels for the XB-35. *(Erection & Maintenance Manual for the XB-35. Courtesy of National Air & Space Museum)* *Above right:* This is the problematic nose gear. Severe wobble problems plagued this unit during its brief life. It retracted sideways. *(Erection & Maintenance Manual for the XB-35. Courtesy of National Air & Space Museum)*

continued from page 377
removable, individually formed, cooling shrouds. They were installed somewhat differently on the inboard and outboard engines and, because of the close fitting tolerance necessary, were not ordinarily interchangeable—even from one cylinder configuration to another. From the headers, the exhaust gas was routed into a collector ring similarly encased within cooling shrouds. On inboard engines, exhaust gasses then passed through a transition chamber directly into the heat exchanger. On outboard engines, exhaust gasses were first routed through a tailpipe. The exhaust pipes aft of the heat exchangers and leading to the turbosuperchargers were also encased in cooling shrouds. Exhaust gasses were dumped overboard under the wing near the leading edge. Photographs from the time of the B-35's test flights show prominent black streaks under the wing due to exhaust stains.

Landing Gear

Perhaps due to Northrop's exposure to Douglas products, the main gear bore all the hallmarks of a DC-4 main gear. This meant dual wheels were supported on a central oleo strut that retracted forwards. The nose gear was equally conventional, featuring a single strut and single wheel that retracted sideways. B-35s were plagued with nose wheel wobble problems, a common malady for nose wheel installations. This problem was never satisfactorily resolved—possibly because of more urgent issues that needed to be addressed.

Summary

In summary, the B-35 was a failure. After the expenditure of considerable funds and countless man-hours, the project never went anywhere. Treading new technological territory is always wrought with land mines and, in the case of the B-35, these land mines came to haunt the aircraft. It was a brave attempt to buck the prevailing trend

Northrop XB-35 Parameters *(Ref. 7-1):*	
Wingspan	170 feet, 27-degree sweepback
Wing area	4,000 square feet
Length	53 feet, 1 inch
Height	20 feet, 1 inch
Empty weight	89,560 pounds
Gross weight	180,000 pounds
Max. weight	209,000 pounds
Bomb load	16,000 pounds with 8,150 range at 183 mph or 51,070 pounds with 720 mile range at 240 mph
Max. speed	391 mph at 35,000 feet
Cruise speed	183 mph
Max. range	8,150 miles
Service ceiling	40,000 feet
Engines	R-4360-7 (outboard contra-rotating) R-4360-11 (inboard contra-rotating) R-4360-17 (outboard contra-rotating) R-4360-21 (inboard contra-rotating) R-4360-45 (outboard, single rotation) R-4360-47 (inboard, single rotation) All engines were single stage, with a variable speed supercharger augmented by dual CH-1 turbosuperchargers.
Number built	15 (only 5 actually flew)

and for that Northrop deserves credit. Further exacerbating the XB-35's problems was the political intrigue that surrounded it. W. Stuart Symington, the first secretary of the Air Force, put his faith in the B-36, which signed the death knell for the B-35. Over the years a considerable amount of controversy has surrounded Symington, particularly with regard to the underhanded way in which he dealt with Jack Northrop. It was Symington who ordered that all flying wings be scrapped. That must have been a painful task for Northrop employees who had poured their hearts and souls into this project. But by the end the day, the B-35 was simply ahead of its time, particularly regarding controllability issues. Little could be salvaged from the program; however, some R-4360-17s were converted to R-4360s for other uses.

Douglas TB2D-1, XTB2D-1 Skypirate

(Ref. 7-19 and 7-20)

Five companies bid on a Navy requirement for a "Dive Bomber Torpedo Attack," or simply BT for short. Of the five companies that bid on this contract and built prototypes, three were powered by the R-4360, one was powered by a Pratt & Whitney R-2800, and the successful bidder, an aircraft that was later designated AD Skyraider, was powered by a Wright R-3350. Douglas entered the XTB2D-1 for this Navy contract.

Weighing in at up to 27,000 pounds, the TB2D was a massive aircraft, particularly in light of the fact that it was single engined, albeit powered by the equally massive R-4360-8. Taking into consideration its size and weight, the performance of the TB2D was remarkable. The new Essex class carriers introduced after the start of World War II offered aircraft manufacturers an opportunity to employ larger and heavier designs of which the TB2D was one of many. Designed to replace stalwarts such as the Grumman TBF/TBM and Curtiss SB2C, the mission for the TB2D was essentially eliminated due to the stellar performance of the aforementioned aircraft. TBMs and SB2Cs were so successful at sinking Japanese aircraft carriers that the need for the TB2D was significantly reduced. Nevertheless, two prototypes were built. Both were powered by an R-4360-8 driving contra-rotating Hamilton Standard Super Hydromatic propellers. It should be noted that Hamilton Standard contra-rotating propellers were referred to as "Super." Dash eight R-4360s were a single stage, variable speed engine fitted with a torque meter and, of course, dual rotation.

The first of two XTB2D-1s built had its maiden flight in April 1945. A tricycle landing gear was employed which in conjunction with the contra-rotating prop that eliminated torque reaction made for an easy ground handling aircraft. Unusual for a single-engined aircraft was the fact that a fully feathering propeller was employed. However, use of a massive eight-blade contra-rotating propeller made this feature mandatory. Without it an engine failure would be disastrous due to the massive plate area drag caused by the propeller. The dual-wheel nose gear used the engine mount as a support and folded rearwards during the retraction cycle. The main gear folded outwards and was stowed just inboard of the break for the wing fold. The wings also housed the internal fuel tanks with a capacity of 774 total gallons. An additional 300 gallons was available

This head-on view of the Douglas XTB2D-1 Skypirate gives a good idea of the size of the Hamilton Standard "Super" Hydromatic propeller. Although fitted with the same "Super" Hydromatic propeller as the Northrop XB-35, the Skypirate never suffered from propeller vibration. This is further testimony to the flawed installation in the XB-35. *(Courtesy of Pratt & Whitney)*

with droppable tanks. And as befits an R-4360, a 60-gallon oil tank was provided situated behind the pilot. A pair of elliptical oil coolers, one in each wing root, provided the necessary temperature control.

An "open stack" siamesed exhaust system was fitted. Racers employing the R-4360 in Unlimited Class air racing also utilize siamesed exhaust systems for enhanced engine power. In other words, for each bank of four cylinders, two cylinders are paired off for the exhaust resulting in a total of 14 stacks. Siamesing has a number of advantages including better exhaust scavenging by taking advantage of exhaust pulses and a simpler and lighter system results. Induction air for the downdraft PR-100-B2-3 carburetor was supplied via a ram-air scoop located on top of the cowl just in front of the pilot's cockpit.

Replacing engines, especially piston engines, was and still is an ongoing chore. Anything that makes that chore easier is a tremendous boon to the maintenance crew. As engines became more complex, strict requirements were placed upon the airframe manufacturers for the time to perform an engine change. The Navy placed a one-hour time limit on changing the TB2D's R-4360-8. This condition was met; however, it has to be assumed that the requirement was for a complete QEC being exchanged for another complete QEC. The time-consuming tasks are items such as removing and replacing exhaust systems, and accessories such as starters, generators, etc. In the case of the TB2D, the exhaust is part of the QEC. Even so, Douglas engineers showed ingenuity when they designed in a simple work platform that attached to the nose bowl and the wing root. Similar type work platforms were designed for many flying boats; however, in that case it wasn't simply a nicety but an essential part of the aircraft.

Of the two prototypes built, one obvious distinguishing feature between the two was the size of their vertical stabilizer. The first prototype was fitted with a massive tail whose top stood 18 feet, 7 inches from the ground. A shorter 16-foot, 6-inch tail was fitted to the second prototype. Considering the vast amount of engineering effort that went into the TB2D, it's a shame that neither of them was saved for posterity. Worse yet, these noble and classic-looking aircraft almost disappeared into history never to be heard from again.

Douglas XTB2D Parameters *(Ref. 7-1):*

Wingspan	70 feet (36 feet, wings folded)
Wing area	605 square feet
Length	46 feet
Height	16 feet, 7 inches (short tail), 18 feet, 6 inches (tall tail)
Empty weight	17,839 pounds
Max. weight	27,366 pounds
Useful load	9,527 pounds
Internal fuel	774 gallons
Max. speed	377 mph at 14,200 feet
Range	1,200 miles
Ceiling	21,200 feet (500 fpm climb)
Rate of climb (S/L clean)	2,720 fpm
Engine	R-4360-8 (single stage, variable speed, automatic boost control)
Number built	2, BuNo's 36933 and 36934. Order for additional 23 aircraft cancelled.

Above: The three-quarter right-side view shows how large the XTB2D-1 Skypirate was. Note the dual-wheel nose gear and massive size of the vertical stabilizer. *(Courtesy of National Archives & Records Administration)*

Right: Left side view of the R-4360-8 that powered the Skypirate. Of note in this photograph are the #60 (inner) and #70 (outer) dual-propeller shafts. The rectangular shaped box attached to the side of the supercharger is the Eclipse automatic engine control. *(Courtesy of Pratt & Whitney)*

N933

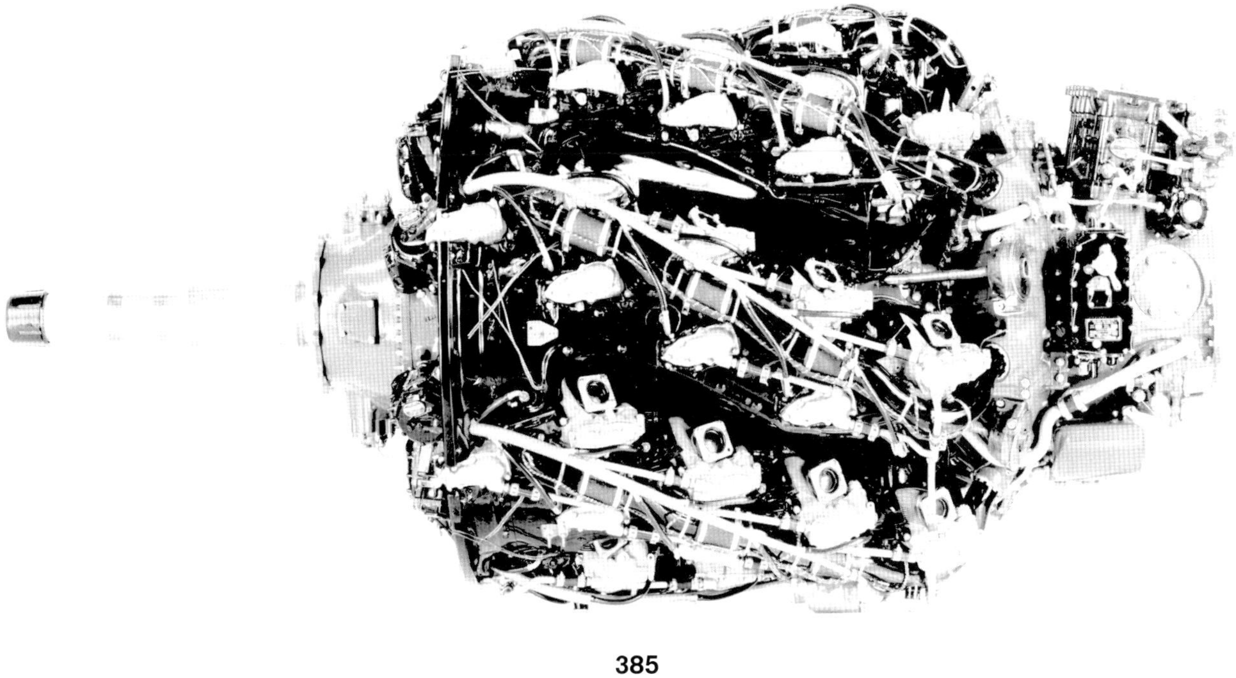

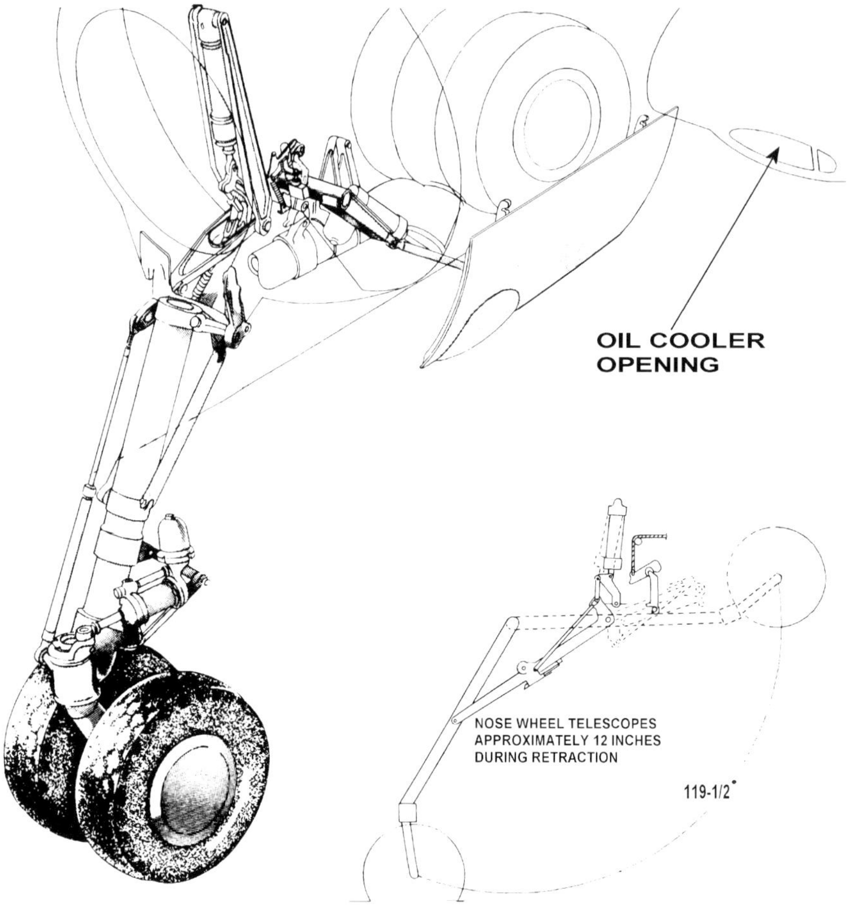

Dual-wheel nose gear for the Douglas TB2D Skypirate. Attached to the engine mount is the oleo strut that compressed as it retracted for more compact stowage. Also noticeable in this line drawing is the elliptical opening in the wing leading edge for cooling air to the oil cooler. *(Erection & Maintenance Manual for Douglas TB2D-1)*

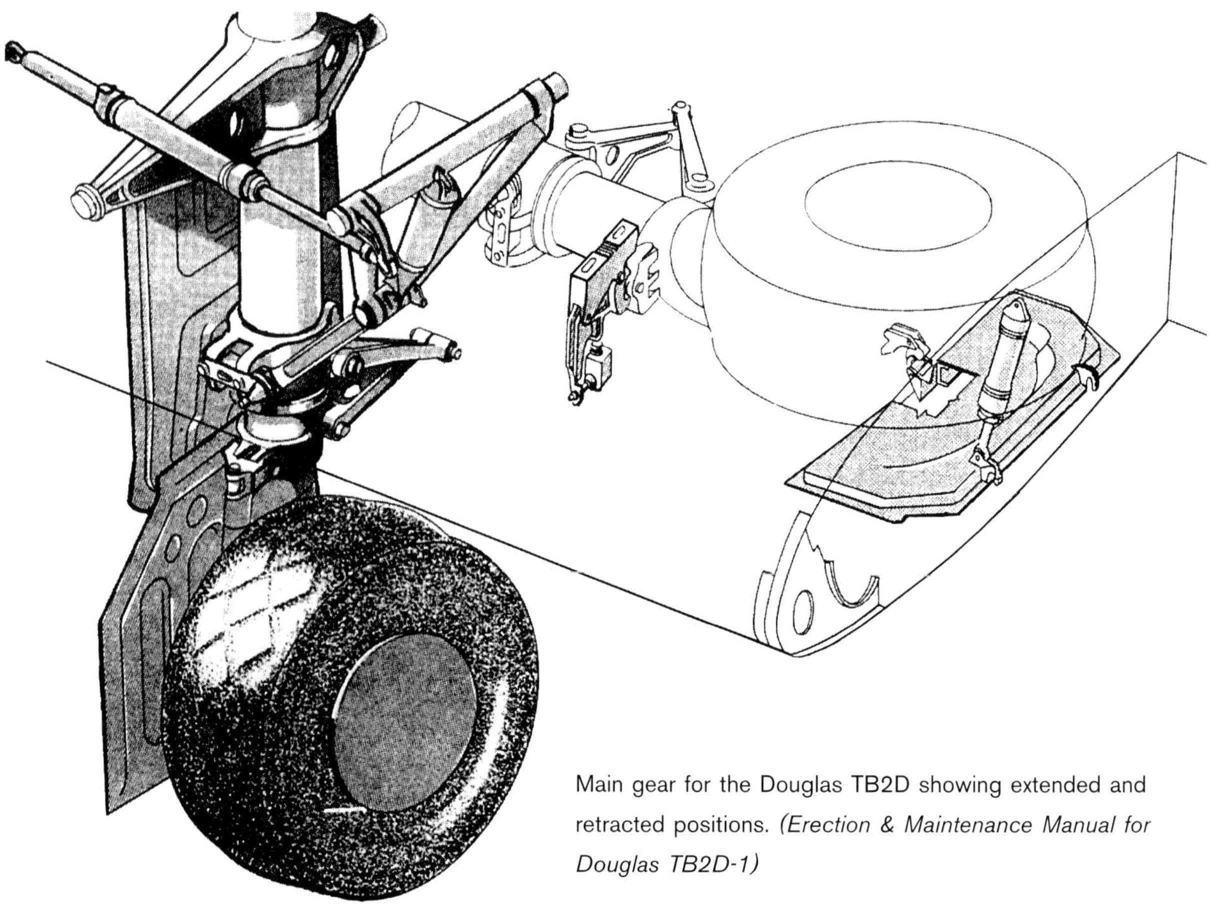

Main gear for the Douglas TB2D showing extended and retracted positions. *(Erection & Maintenance Manual for Douglas TB2D-1)*

Work stand installed on the Skypirate. With a maintenance headache like the R-4360, such innovative ideas offered some welcomed relief. It was also a contractual requirement to change out a QEC in a specified time. *(Courtesy of National Archives & Records Administration)*

Vultee XA-41, Convair A-41

(Ref. 7-1 and 7-21)

One of many prototype aircraft conceived during World War II, the Consolidated Vultee XA-41, was proposed as a close support aircraft for the Army Air Force. The Navy also took an interest in it. First flight occurred on February 11, 1944, at Lomita, California. It was briefly tested by the Army Air Force at Eglin Field, Florida, and by the Navy at Patuxent River, Maryland. Apparently not much of a performer, the single prototype was soon relegated to serving as a test aircraft for Pratt & Whitney.

Boeing XF8B-1

(Ref. 7-1, 7-21, and 7-22)

The Boeing XF8B-1 was conceived during World War II as a carrier-based fighter-bomber. Like all R-4360-powered aircraft developed during World War II, the XF8B-1 never entered service prior to V-J Day. As a consequence only three examples of the XF8B-1 were built. The R-4360-10 was essentially the same as the R-4360-13 that powered the XP-72. In other words, it was a two-stage, variable speed engine. The primary difference was the -10 mounted the auxiliary blower on the rear of the engine unlike the -13 for the

The Consolidated Vultee XA-41 was an uninspired design that was briefly tested by the Army Air Force and Navy before ending its days as a test platform for Pratt & Whitney. *(Courtesy of Pratt & Whitney)*

XP-72 that mounted its auxiliary blower remotely behind the pilot with drive being provided via a long extension shaft. Otherwise, all performance parameters and features are the same for these two engines.

Boeing's designation for the XF8B-1 was Model 400. Even though it was single-engined aircraft, its bomb load, carried internally, could match or exceed many multi-engine medium bombers of the same era. Over 3,000 pounds could be carried internally. As with many aircraft developments during World War II, progress was amazingly fast. Initial contracts were executed on May 4, 1943. A scant 18 months later, on November 27, 1944, the first prototype slipped the surely bonds on its maiden flight. The next two aircraft were built after the war. Like many other promising aircraft, the Navy dropped it like a hot potato.

Boeing XF8B-1 Parameters *(Ref. 7-1):*

Wingspan	54 feet
Wing area	489 square feet
Length	43 feet, 3 inches
Height	16 feet, 3 inches
Empty weight	13,519 pounds
Max. weight	20,508 pounds
Max. speed @ sea level	340 mph
Max. speed	432 mph at 26,900 feet
Normal range	1,305 miles
Max. range	2,708 miles
Ceiling	37,500 feet
Rate of climb	2,800 feet per minute
Engine	R-4360-10
Number built	3

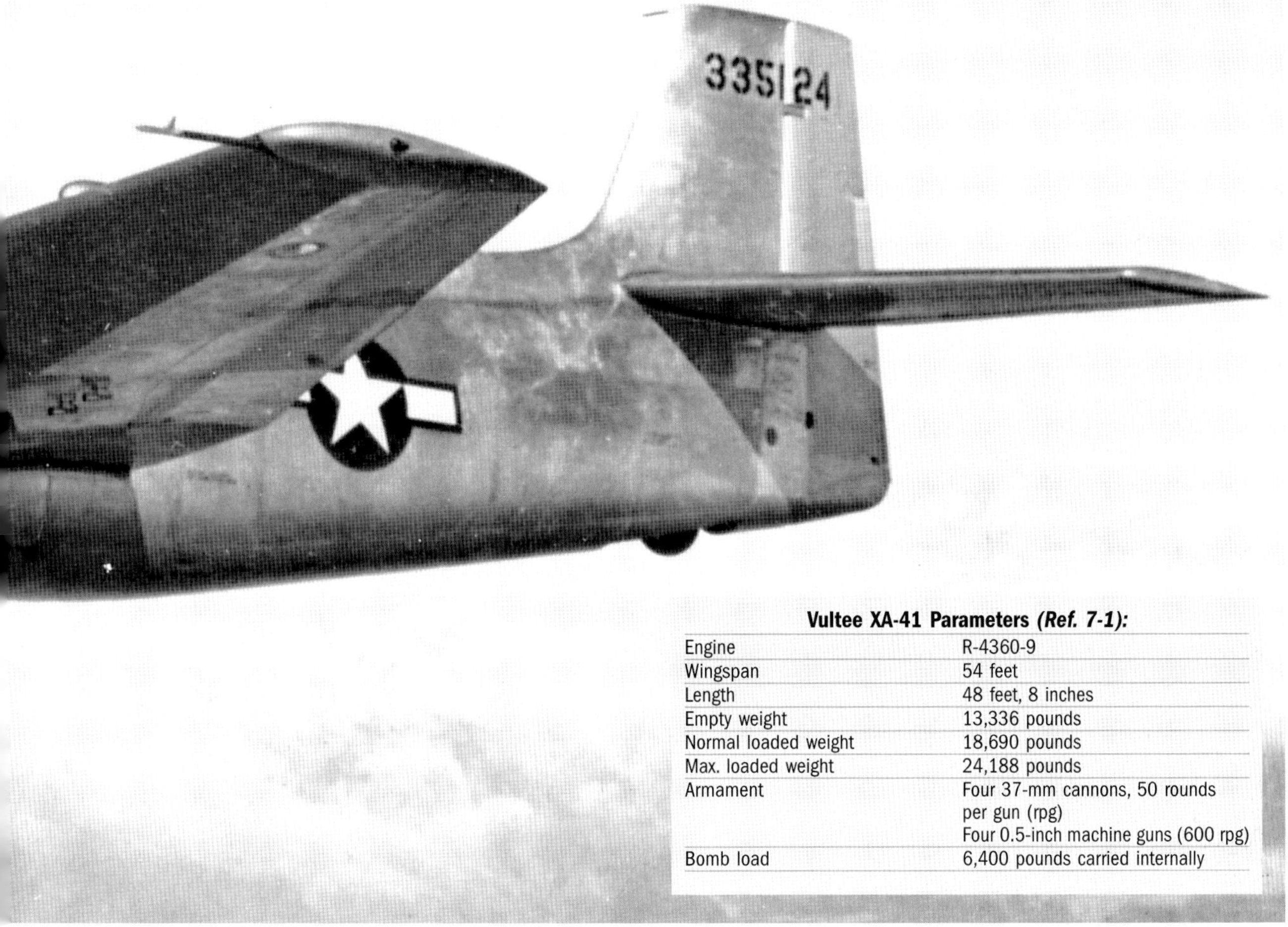

Vultee XA-41 Parameters *(Ref. 7-1)*:

Engine	R-4360-9
Wingspan	54 feet
Length	48 feet, 8 inches
Empty weight	13,336 pounds
Normal loaded weight	18,690 pounds
Max. loaded weight	24,188 pounds
Armament	Four 37-mm cannons, 50 rounds per gun (rpg) Four 0.5-inch machine guns (600 rpg)
Bomb load	6,400 pounds carried internally

Developed at the behest of the Navy, the Boeing XF8B-1 never went into production. Three prototypes were built. A complex dual-rotation engine with two-stage supercharging powered it. *(Courtesy of Warren Bodie)*

Another futile Curtiss development was the XBTC-2, powered by a dual-rotation engine driving a Curtiss propeller. *(Courtesy of Pratt & Whitney)*

Curtiss XBTC-2

(Ref. 7-1 and 7-23)

Even though the last large procurement contract Curtiss managed to win was for the venerable P-40, it wasn't through lack of trying for follow-on orders. It seems as though a plethora of designs and prototypes came from Curtiss, yet this once proud and respected company just could not do anything right. Included in this sad litany of failed designs was the XBTC-2. With the exception of its Curtiss contra-rotating propeller driven by an R-4360-8 or R-4360-14, there was nothing out of the ordinary about this aircraft. Although it's easy to criticize Curtiss for developing mediocre designs, they also suffered from the vagaries of its customers—the U.S. Navy and Army Air Force. In the case of the XBTC-2, it was originally designed as a single-seat torpedo bomber, but then a change in Navy battle tactics and the requirement to carry a second crewmember threw a wrench into the original design. Joining a long list of aircraft conceived during World War II and powered by the R-4360, a single prototype was built and then it disappeared into the sunset never to be seen or heard from again—until now.

Curtiss XBTC-2 Parameters *(Ref. 7-1):*

Wingspan	50 feet
Length	38 feet, 7 inches
Load	3963 pounds
Max. speed	386 mph
Max. range	1,245 miles
Ceiling	28,000 feet
Engine	R-4360-8 or R-4360-14
Number built	1

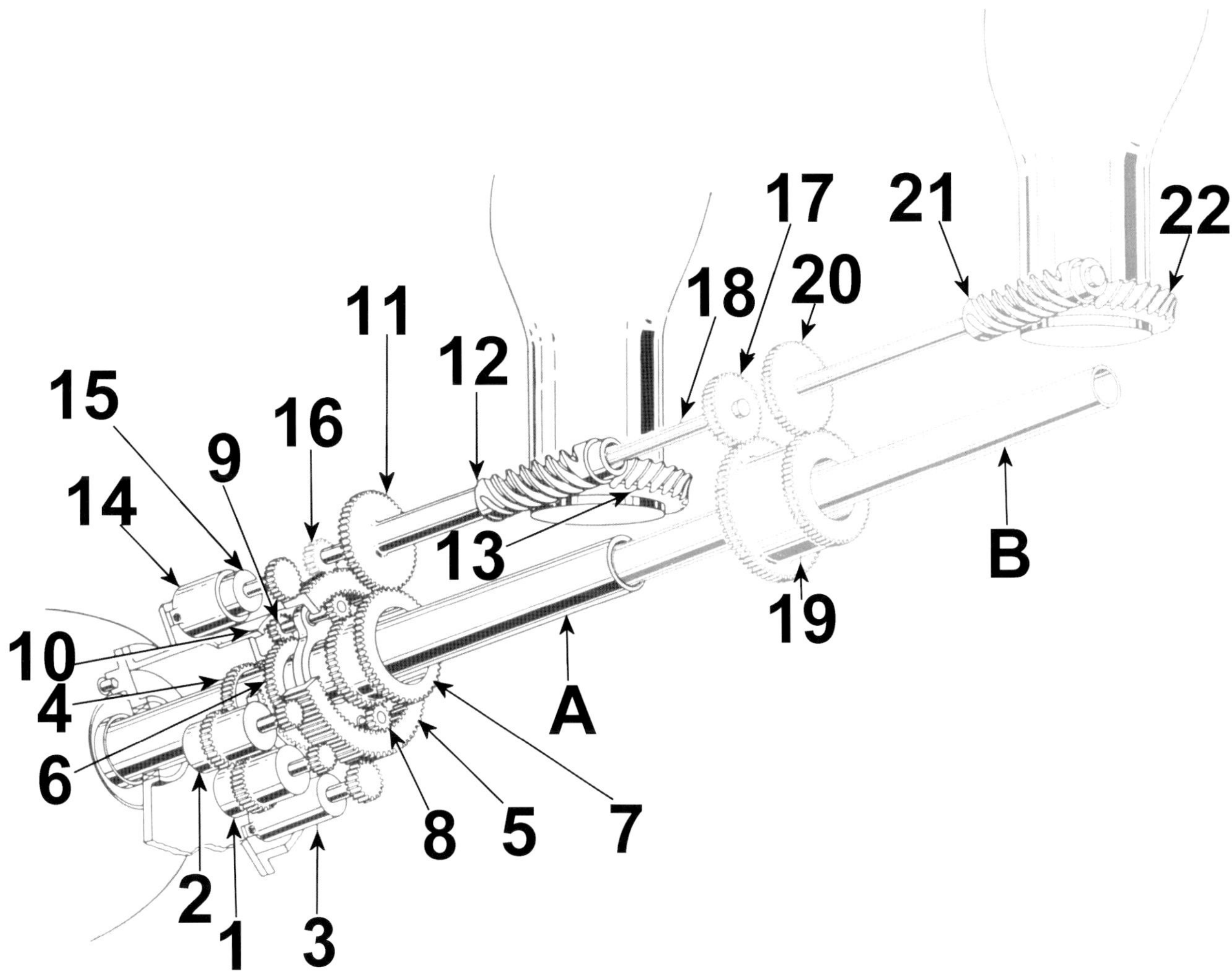

Even though the XBTC-2 didn't have much to commend it, its propeller is worth taking a close look at. The outboard planet gears (16 and 17) are mounted on shaft (18), operating through the pinion and worm assemblies (11 and 12), in the inboard hub. Rotation of the moveable ring gear (5) is transmitted through these planet gears (16 and 17) to the outboard pitch change sun gear (19), which is coupled to the outboard blade gear (22) through pinion gear (20) and worm gear (21). The sizes of planet gears (16 and 17) are selected to hold the outboard pitch change sun gear (19) stationary with respect to the outboard hub upon completion of the control signal from the propeller governor.

Blade angles and rates of pitch change of the inboard and outboard propellers can be selected to equalize the thrust loading between the propellers. A. Inboard Propeller Shaft; B. Outboard Propeller Shaft; 1. Increase Pitch Clutch; 3. Fixed Pitch Brake; 4. Propeller Shaft Gear; 5. Moveable Ring Gear; 6. Propeller Shaft Sun Gear; 7. Inboard Pitch Change Sun Gear; 8. Inboard Pitch Change Planet Gear; 9. Inboard Reaction Planet Gear; 10. Inboard Fixed Ring Gear; 11. Inboard Pitch Change Pinion Gear; 12. Inboard Blade Worm Gear; 13. Inboard Blade Gear; 14. Electric Motor (Feathering); 15. Electric Motor Clutch; 16. Outboard Planet Gear; 17. Outboard Planet Gear; 18. Outboard Planet Gear Shaft; 19. Outboard Pitch Gear; 20. Outboard Pitch Change Pinion Gear; 21. Outboard Blade Worm Gear; 22. Outboard Blade Gear. *(Courtesy of Pratt & Whitney)*

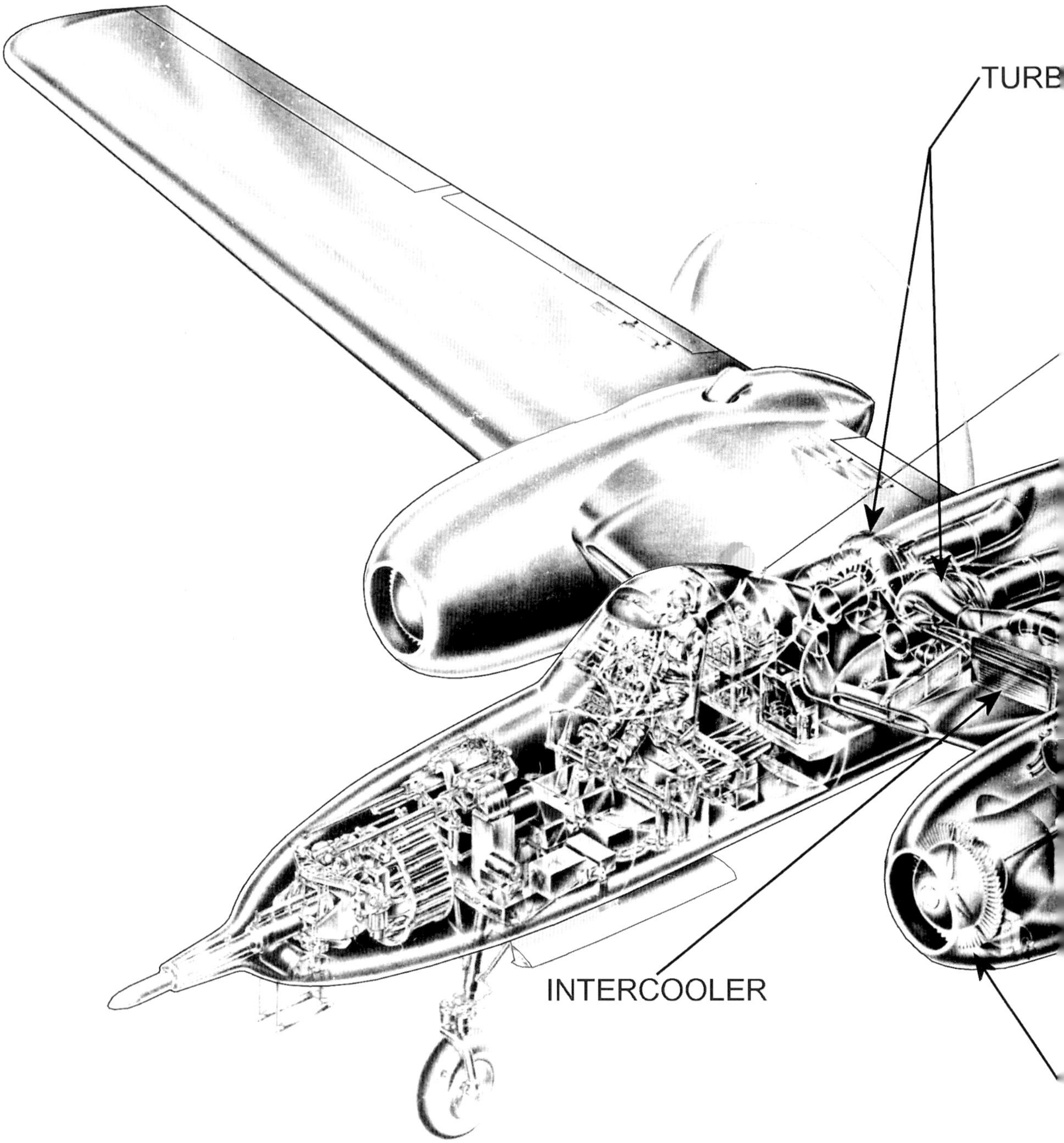

Curtiss XP-71

(Ref. 7-24, 7-25, and 7-26)

Yet another Curtiss design that turned out to be a paper airplane, the XP-71, didn't even achieve prototype status. This massive twin-engined fighter carrying a two-man crew in a pressurized compartment would have sported a fearsome array of firepower. Armament would have included two 37-mm cannons and one 75-mm cannon. The recoil from this formidable armor would have been awesome if it had ever been mounted in the aircraft and fired. Weighing as much as some four-engined bombers of the time, the Army Air Force wisely decided that this was simply too much. The proposed R-4360-23 was never built; it would have been a production version of the R-4360-3. With turbosupercharging and dual-rotation propellers in a pusher configuration, it bucked the trend for twin-engined fighters. Judging by the disparity of the thrust line and the wing trailing edge, one could envision

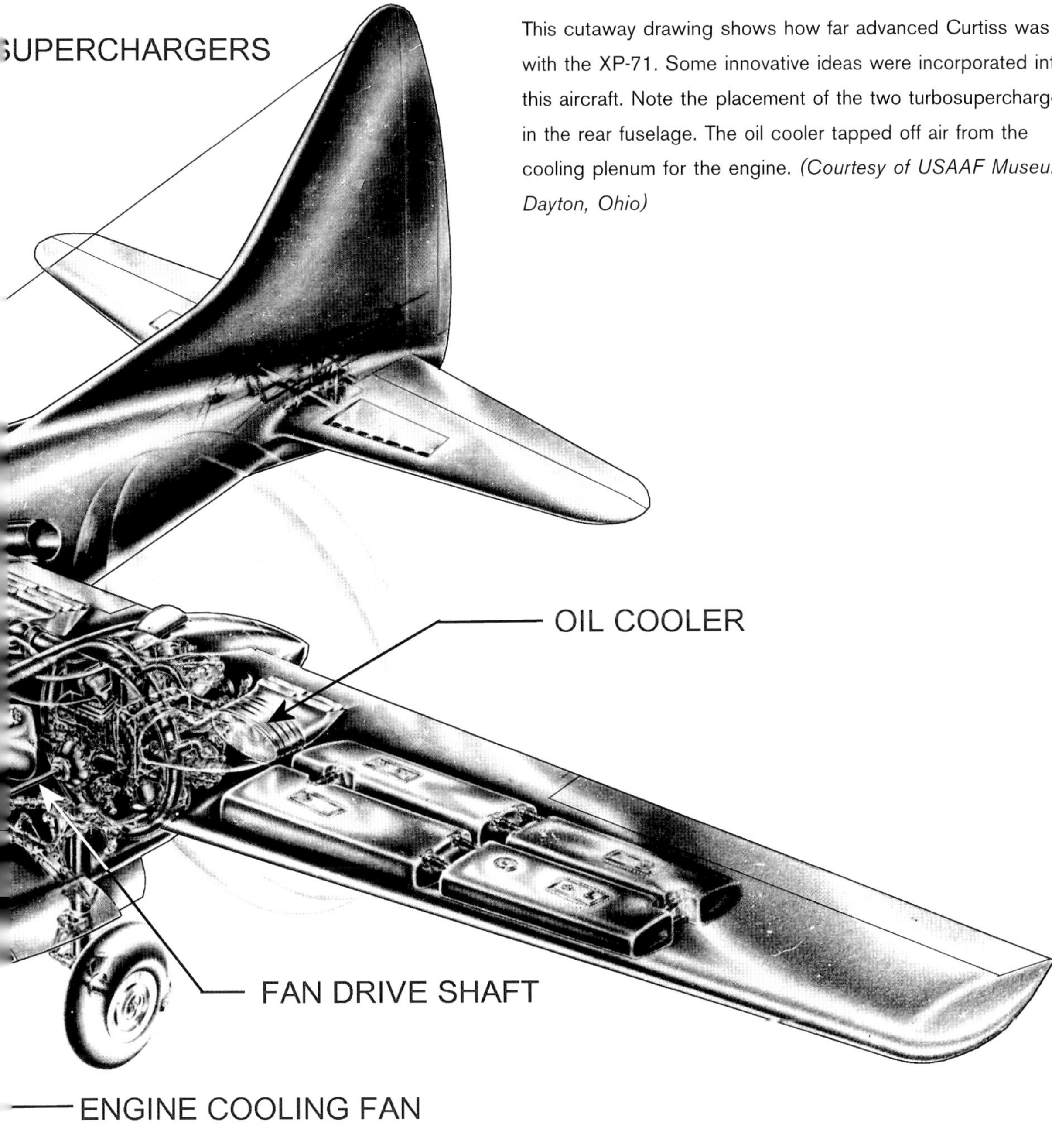

This cutaway drawing shows how far advanced Curtiss was with the XP-71. Some innovative ideas were incorporated into this aircraft. Note the placement of the two turbosuperchargers in the rear fuselage. The oil cooler tapped off air from the cooling plenum for the engine. *(Courtesy of USAAF Museum, Dayton, Ohio)*

problems of the kind that plagued the XB-35 and to a lesser extent the B-36. The project was cancelled on August 26, 1943.

Nevertheless, taking a close look at the XP-71 reveals some unusual and innovative ideas. The engines were buried within the wing structure. Being a pusher demanded some augmented cooling which was achieved via a variable speed fan mounted in the nose of the nacelle. Drive for the fan was through a long extension shaft emanating from the rear of the engine. A leading-edge scoop inboard of each nacelle provided induction air and cooling air to the intercooler. Each scoop was divided into two ducts. Surprisingly for a multi-engine aircraft, the two turbosuperchargers were mounted in the fuselage behind the pilot—a very unusual but innovative arrangement. A rectangular intercooler was mounted behind the main spar and exhausted its cooling air on top of the wing at the trailing edge. The engine-cooling fan discharged into a plenum. Tapped off the plenum was a duct that went outboard of the

Curtiss XP-71 Parameters *(Ref. 7-1)*:	
Wingspan	82 feet, 3 inches
Length	61 feet, 10 inches
Height	19 feet
Wing area	602 square feet
Empty Weight	31,060 pounds
Normal gross weight	39,950 pounds
Max. gross weight	46,950 pounds
Engine	(2) R-4360-3 and/or R-4360-23
Max. speed @ 25,000 feet	428 mph
Landing speed	97 mph
Rate of climb	12.5 minutes to 25,000 feet
Service ceiling	40,000 feet
Range	3,000 miles
Fuel capacity	1,940 gallons

nacelle and fed cooing air to the elliptical oil cooler. This cooling air exhausted in a similar fashion to the intercooler air, which was over the top of the wing at the trailing edge.

Over the years historians have attempted to explain how such a huge and wealthy industrial empire like Curtiss could fall from grace so precipitously. As usual in these situations, poor management has to take the brunt of the blame. This once fine company apparently could not do anything right. Even when handed a plum contract to manufacture P-47Ds, they managed to make a pig's ear out of the project. Curtiss-built P-47s were so bad due to quality control problems they were not sent to any battlefronts. Instead they were retained in the U.S. for training. The XP-71 was simply another doomed project in a long line of failures after the success of the P-40.

Lockheed XR6O-1 (Constitution), Lockheed R6V-1 Model 89

(Ref. 7-1, 7-27, and 7-28)

Built at the request of the U.S. Navy, just two Constitutions were manufactured. Designed as a large four-engined transport, the Constitution was perhaps underpowered even with its four massive R-4360s providing the necessary motive power. The pressurized fuselage was made up from a classic "double bubble," or two circular cross sections joined to create a figure eight. The intersection of the two circles created the floor of the two-deck design. It was ultimately hoped to sell the Constitution as a commercial airliner with Pan Am and American Overseas Airlines as the first customers. This never occurred, as both carriers cancelled out. As a commercial carrier it would have had a crew of 11 and 128 passengers.

Originally powered by four R-4360-18s augmented by a GE Model BH-3 turbosupercharger, the -18s were soon replaced by -22Ws capable of producing 3,500 hp with ADI. The additional 2,000 hp was surely needed with this underpowered aircraft. An interesting aspect to the engine installation was the use of a driveshaft to power a remote accessory gearbox buried in the wing. This idea was reminiscent of the way the British designed their QECs (powerplants). It had the advantage that if an engine needed to be changed it was not necessary to change out all the accessories such as generators, vacuum pumps, hydraulic pumps, etc. Additionally, if the need arose, it was easier to gain access to these accessories in flight. Chapter 4 describes the various turbosupercharging system envisioned for the Constitution. To recap, three companies submitted designs: Turbo Engineering Corporation, General Electric, and surprisingly, Lockheed Aircraft Corporation. Of the three, Lockheed's was the most advanced and innovative but also came with the most risk of problems during development. Wisely, Lockheed chose to go with GE. *(Ref. 7-29, 7-30, 7-31, 7-32, 7-33, and 7-34).*

Perhaps Lockheed came to the realization that expecting the four over-worked R-4360s to offer any kind of performance was simply asking too much. A follow-on design would have been powered by four turbo-props rated at 5,000 hp driving contra-rotating propellers. This idea never got beyond the conceptual stage. Sold to the civilian market in 1955, both Constitutions

Constitution Parameters *(Ref. 7-1)*:	
Wingspan	189 feet
Length	156 feet
Height	50 feet
Wing area	3,610 square feet
Weight loaded	189,000 pounds
Engine	R-4360-18 later replaced by -22Ws
Range	5,000 miles
Long range @ economy cruise	7,500 miles

Above: Built for the U.S. Navy, the Lockheed Constitution proved to be a disappointment primarily because it was underpowered. *(Courtesy of Pratt & Whitney)*

Below: Front view of the number four nacelle for the Lockheed Constitution. Of note is the large four-blade Aeroproducts propeller, ram induction scoop on top, and lower scoop for oil cooling and intercooler air. *(Courtesy of Pratt & Whitney)*

Underside view of a Constitution nacelle. Note the nicely designed exhaust outlet, which offered some degree of jet thrust. *(Courtesy of Pratt & Whitney)*

met sad ends after going through numerous owners. One ended up making an emergency landing at Opa Locka Airport near Miami. It sat there for many years going downhill due to vandalism, fire, and exposure to the harsh south Florida climate. Eventually someone had the bright idea of turning it into a nightclub. It was dismantled for its final trip. However the whole deal fell apart and the aircraft was scrapped. The other Constitution ended up in Las Vegas, Nevada, where it too ended up being scrapped.

Douglas XC-74 and C-74

(Ref. 7-1, 7-35, and 7-36)

World War II illuminated a number of logistical concerns. Primary among these concerns was the availability of a purpose built cargo aircraft capable of carrying heavy loads over long distances. Even though stalwarts such as the C-47 and C-46 performed sterling duty, they were never designed for the mission of cargo hauling. They were instead simply adaptations of existing airliners. Rather belatedly, a number of designs were rushed into production to alleviate this deficiency.

After the devastating attack on Pearl Harbor on December 7, 1941, a number of aircraft projects were put into high gear. Few of these projects, however, actually got to see combat in World War II. One of these projects was the C-74 transport. It was also, confusingly, referred to as the DC-7, not to be mistaken for the commercial DC-7, a totally different aircraft.

Layout was quite conventional, although it did incorporate the then new technology of laminar flow wing airfoil sections. Cargo loading and unloading was accomplished via the unusual

Left: Number four nacelle with cowl flaps and exit doors for intercoolers wide open. *(Courtesy of Pratt & Whitney)*

Below: Passenger view of 7,000 mighty horsepower. The two Constitutions soldiered on for the Navy before entering commercial service where they quickly went downhill before being scrapped; one in Las Vegas and the other at Opa Locka Airport. *(Courtesy of Pratt & Whitney)*

If the Douglas C-74 had been available during World War II, it would have been a valuable asset. Even though it was under development throughout World War II it didn't enter service until hostilities had ended. *(Courtesy of National Archives & Records Administration)*

method of using a pair of hoists built into the roof of the fuselage and lifting its payload through a rectangular hole formed in the base of the fuselage. Once inside, the cargo could be placed in the appropriate position by rolling the hoists along a monorail. Each hoist had a capacity of 8,000 pounds or 16,000 pounds when used together. Another hoist built into the forward side door could also load up 4,500 pounds of cargo. Typical loads could include: 10 R-3350s on cradles; 15 Allison V-1710s on cradles; two T-9E1 tanks; two 75-mm howitzers or 105-mm howitzers with tractors; various combinations of jeeps, ammunition trucks, and 1½ ton trucks; 90-mm AA guns; and three complete Bell P-39s or two complete Republic P-47s. Another unusual C-74 feature was its cockpit glazing. The pilot and co-pilot each had a bubble canopy, giving the C-74 a bug-eyed look. Later C-74s dispensed with this unusual cockpit arrangement and blended the two bug-eyes into one unit. Only 14 C-74s were manufactured and of these 12 saw service. The fate of the other two was as follows—one crashed and the other was used for destructive testing at Wright Field. Pan American Airways expressed an interest in a commercial version of the C-74. Orders for 26 aircraft were placed but then cancelled.

Douglas engineers used a very conservative approach to installing the four R-4360s *(Ref. 7-37)*. The fact that they were "open stack" engines (no turbosupercharger) made life easier for them. Being one of only a few multi-engined aircraft powered by R-4360s that was not pressurized would explain the lack of a turbo. The firewall, in particular, was such a huge diameter it raised the ire of Andrew Wilgoos at Pratt & Whitney. Upon seeing the first C-74 nacelle mock-ups he was clearly disappointed and expressed his feelings in a very cryptic and terse note in the margin of a letter sent from Douglas to Pratt & Whitney. His hand-written note read as follows *(Ref. 7-38):*

"*Perhaps this* (the letter from Douglas) *had better be filed as an example of why one should not bother to reduce frontal area of an engine. Do they* (Douglas) *use the engine cowl for a luggage compartment also? AW 8/2*"

A single oval oil cooler mounted under the nacelle and fed with air from a scoop in the cowl nose bowl took care of oil temperature issues. Oil for each engine was supplied by an 87.5-gallon

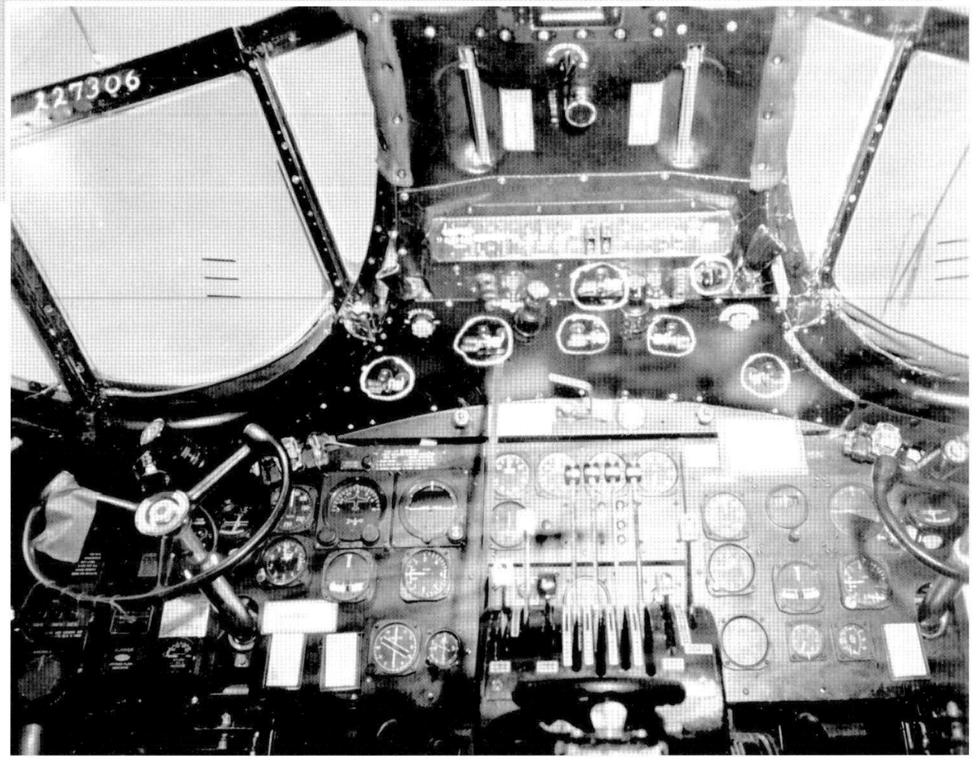

The Douglas C-74 was a fairly conventional design with one notable exception. The pilot and co-pilot had individual "bubble" canopies, giving the aircraft an odd-looking bug-eyed appearance. *(Courtesy of National Archives & Records Administration)*

A more conventional single canopy soon replaced the bug-eye canopies. This photograph shows the single canopy in the foreground and the bug-eye in the background. *(Courtesy of National Archives & Records Administration)*

C-74 QEC about to be mounted on the firewall. Notice the relatively small diameter of the engine compared to the diameter of the firewall—a design feature that got Andrew Wilgoos incensed. *(Courtesy of Pratt & Whitney)*

tank, which contained 82.5 gallons of oil. The exhaust was a relatively simple jet stack arrangement with the usual siamesing of "D" row with ""B row and "C" row with "A" row. Induction ram air came from a duct built into the inside leading edge of the cowl nose bowl. Variations for induction were cold ram, filtered, and warm. Each engine drove a four-blade Curtiss Electric propeller. All things considered, the C-74 and C-124 engine installation represented the most uninspired of all the R-4360 mounting schemes. This is somewhat surprising considering that the DC-6, a smaller sibling to the C-74 in many respects, had an excellent QEC design for its R-2800s. Many design features were clearly inspired by the DC-6. The landing gear for instance was of a similar design. In fact, all the Douglas tricycle-gear piston transports used similar design concepts.

C-74 Parameters *(Ref. 7-1)*:

Wingspan	173 feet, 2 inches
Length	124 feet, 1½ inches
Height	43 feet, 8 inches
Cruising speed	300 mph
Max. range	7,800 miles
Engines	R-4360-15, -27, -49, or -49A*

* The -27 was a production version of the -15.
The -49 used ADI.
The -49A was similar to the -49 except it used longer connecting rods.

Not one of the better R-4360 nacelle designs. As Andrew Wilgoos noted, there was enough space within its confines to create a luggage compartment. *(Courtesy of National Archives & Records Administration)*

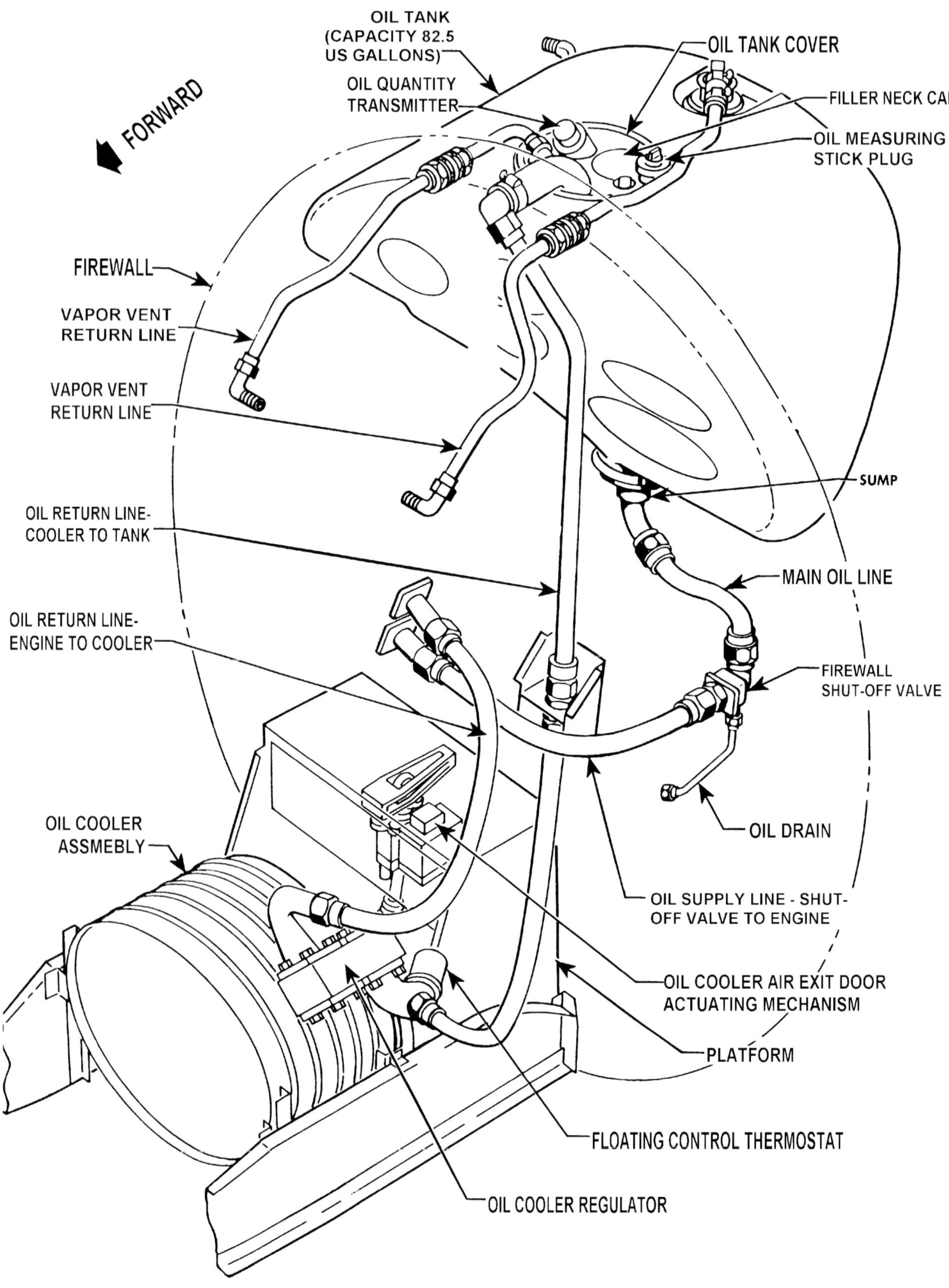

The C-74 oil cooler is mounted in the lower cowl. *(AN 01-40NVA-2 Erection & Maintenance Instructions for Army Model C-124 Airplanes)*

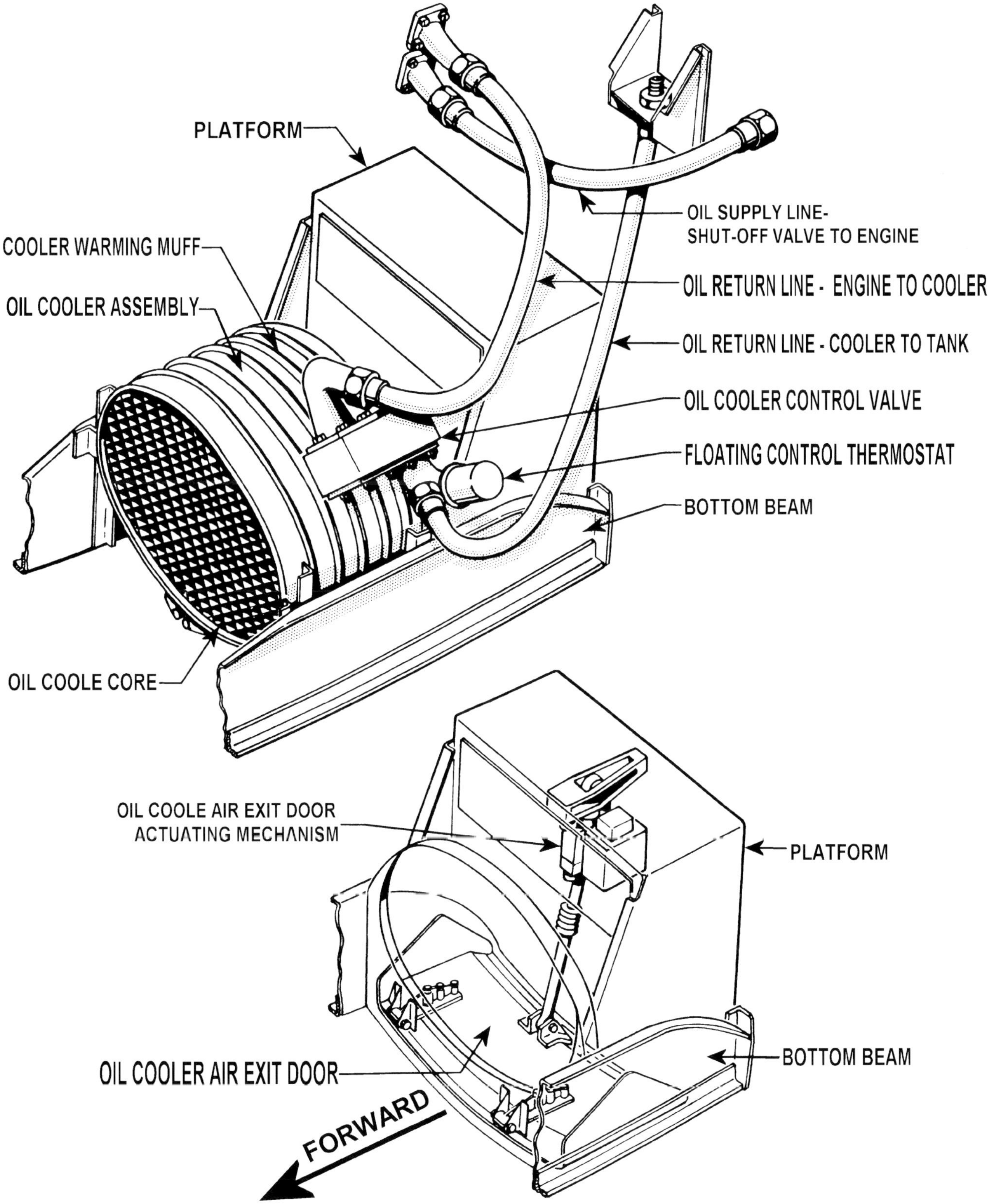

C-74 oil cooler details and cooling flap. *(AN 01-40NVA-2 Erection & Maintenance Instructions for Army Model C-124 Airplanes)*

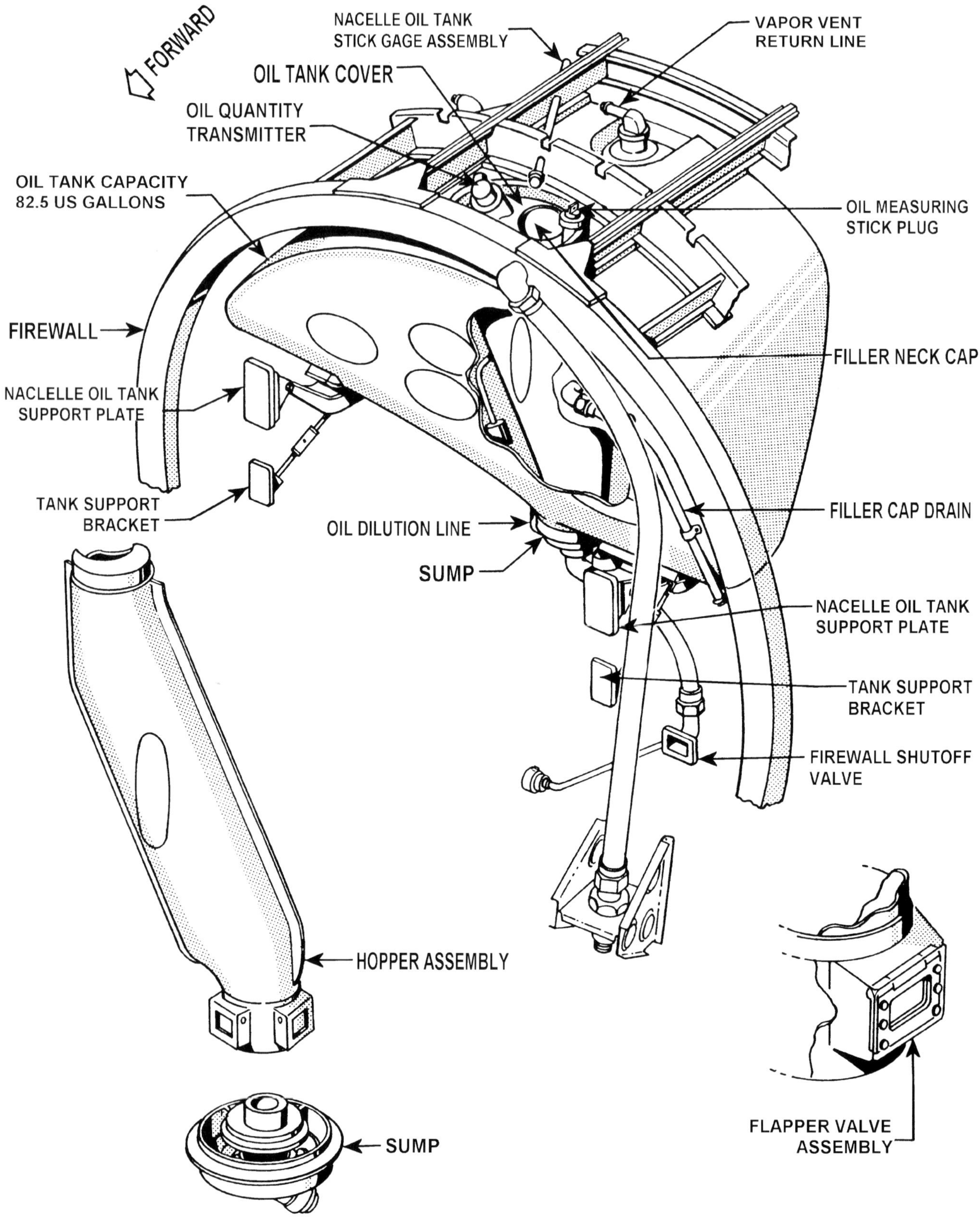

The C-74's oil tank and its associated plumbing are mounted behind the firewall. *(AN 01-40NVA-2 Erection & Maintenance Instructions for Army Model C-124 Airplanes)*

This photograph illustrates why the C-124 loading system was far more practical than the earlier C-74. There is no way that a C-74 could have loaded the tank. Not shown in this photograph taken on May 22, 1950, are the 2½-ton 6 x 6 truck and M-24 gun that were also loaded on this particular aircraft. *(Courtesy of National Archives & Records Administration)*

Douglas C-124

(Ref. 7-1, 7-39)

As a derivative of the earlier C-74, the C-124 shared many common components such as the wing, in modified form, the landing gear, and the QEC design. As with the C-74, the C-124 was not pressurized. Perhaps the most noticeable departure from its C-74 predecessor was the method of loading cargo. Instead of the hoist pulling cargo up into the belly of the aircraft, the C-124 used a far more convenient clamshell opening on the nose with hydraulically operated ramps. In this way vehicles could drive on board instead of having to be winched aboard. However, the C-124 still retained the C-74's hoist in the aft fuselage. Another noticeable departure from the C-74, with the exception of the YC-124, was the use of three-blade rather than four-blade Curtiss Electric propellers.

The C-124 turned out to be a highly successful aircraft and continued in service until the 1970s. As an interesting aside, the C-74/C-124 had one of the worst QEC designs of any

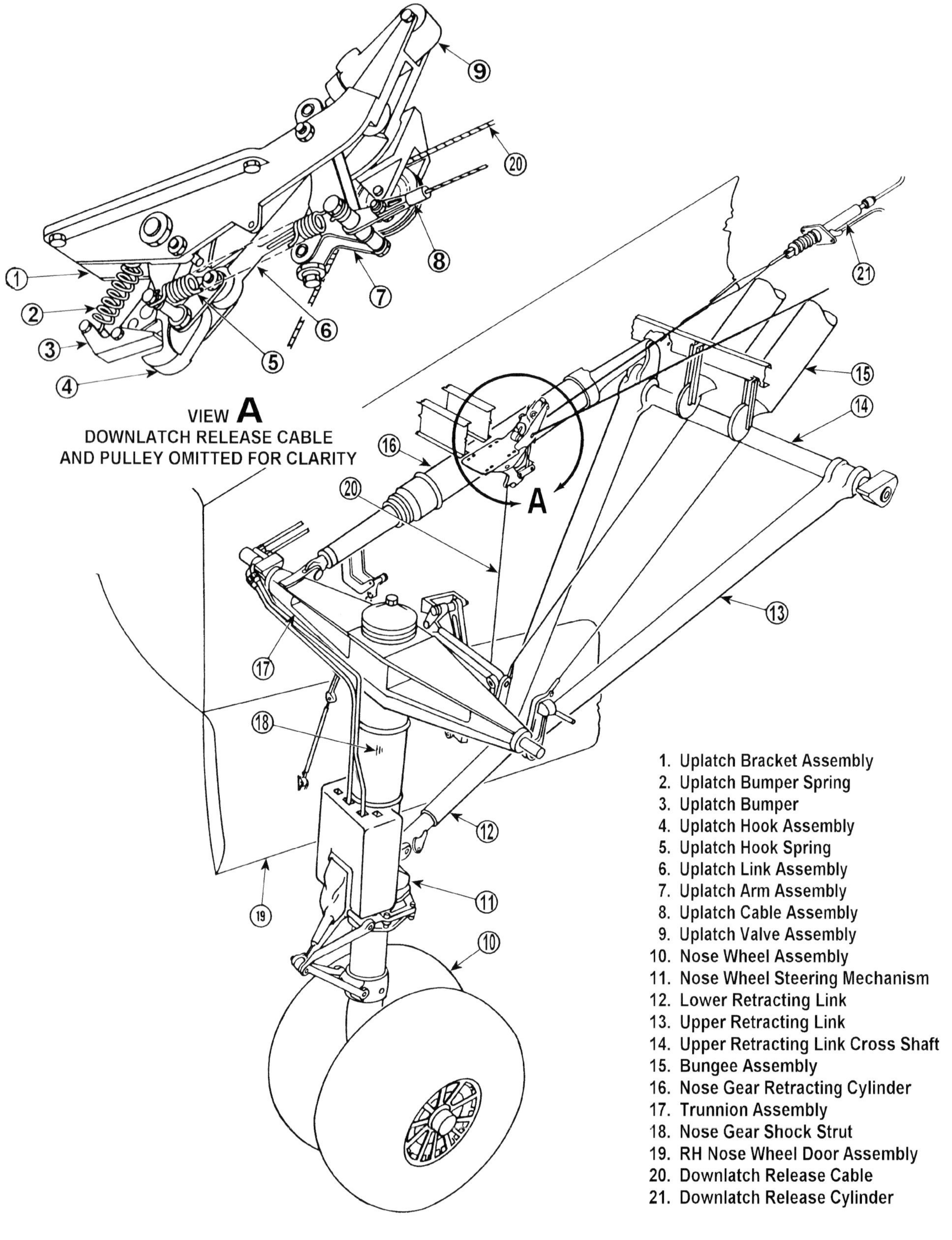

Landing gear for the Douglas C-124 was the same as that used on the similar C-74. This nose gear design is typical of Douglas with lots of springs and latches. *(AN 01-40NVA-2 Erection & Maintenance Instructions for Army Model C-124 Airplanes)*

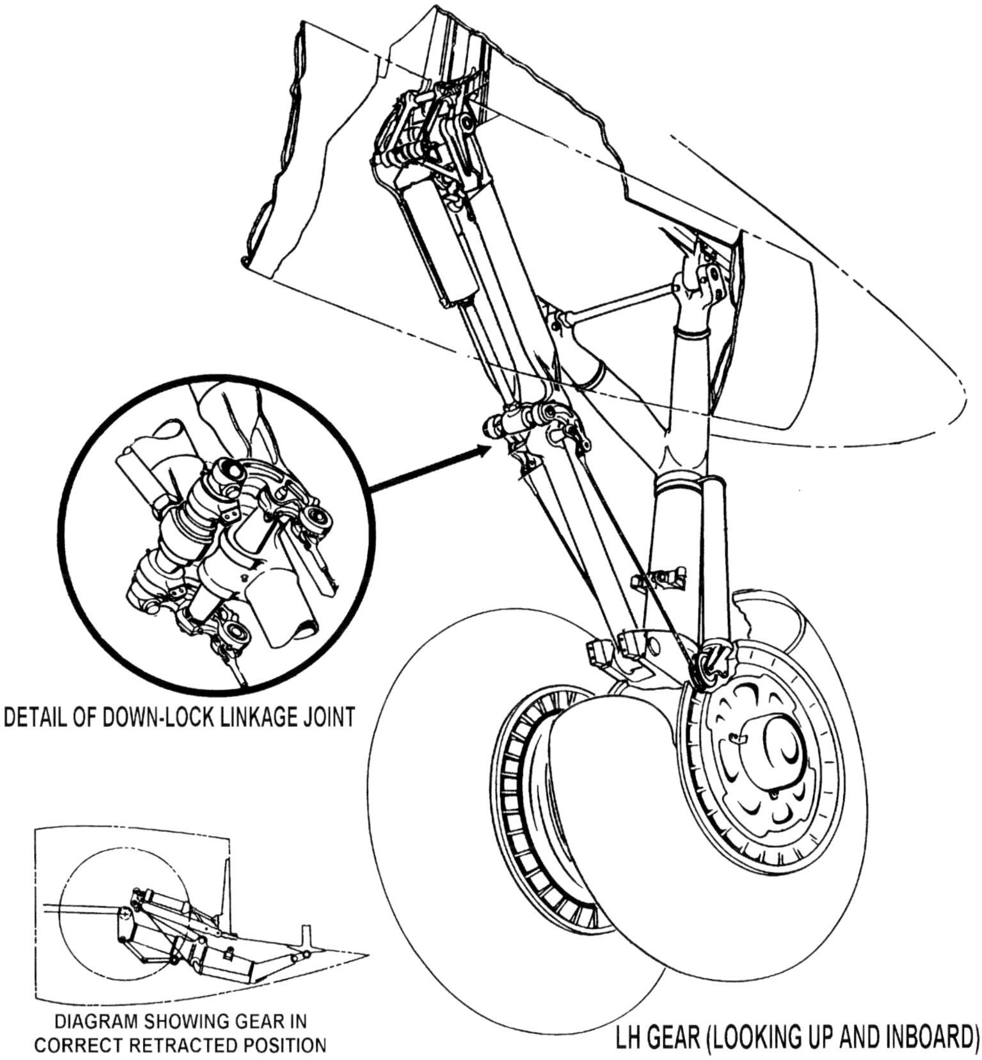

The C-74/C-124 main gear appears to be a scaled-up version of the DC-6/DC-7 unit. *(AN 01-40NVA-2 Erection & Maintenance Instructions for Army Model C-124 Airplanes)*

R-4360-powered aircraft and yet the R-4360-63As that powered the C-124Cs are now the preferred engine for racers *(Ref. 7-40 and 7-41).* Fitting an R-4360-63A into the C-124 was akin to installing a high-performance racecar engine into a Mack truck. Rated at 3,800 hp with ADI, the -63A was one of the better R-4360s. Notwithstanding its published maximum speed of 320 mph, the C-124 was always known as "Old Shakey" and a slow, albeit reliable, aircraft *(Ref. 7-42).*

C-124 Variations

YC-124

Prototype powered by four R-4360-35s driving four-blade propellers. First flight occurred on November 27, 1949.

C-124A

Production model powered by four R-4360-20Ws driving three-blade Curtiss Electric propellers (16 feet, 7 inches diameter). Entered service in May 1950.

YC-124B

One aircraft was converted to Pratt & Whitney YT-34-P1 turbine power. Each engine was rated at 5,000 hp. It also featured a pressurized flight deck, production C-124s were un-pressurized.

C-124C

Development of the C-124A. Powered by Ford-built R-4360-63 or -63A engines rated at 3,800 hp.

continued on page 412

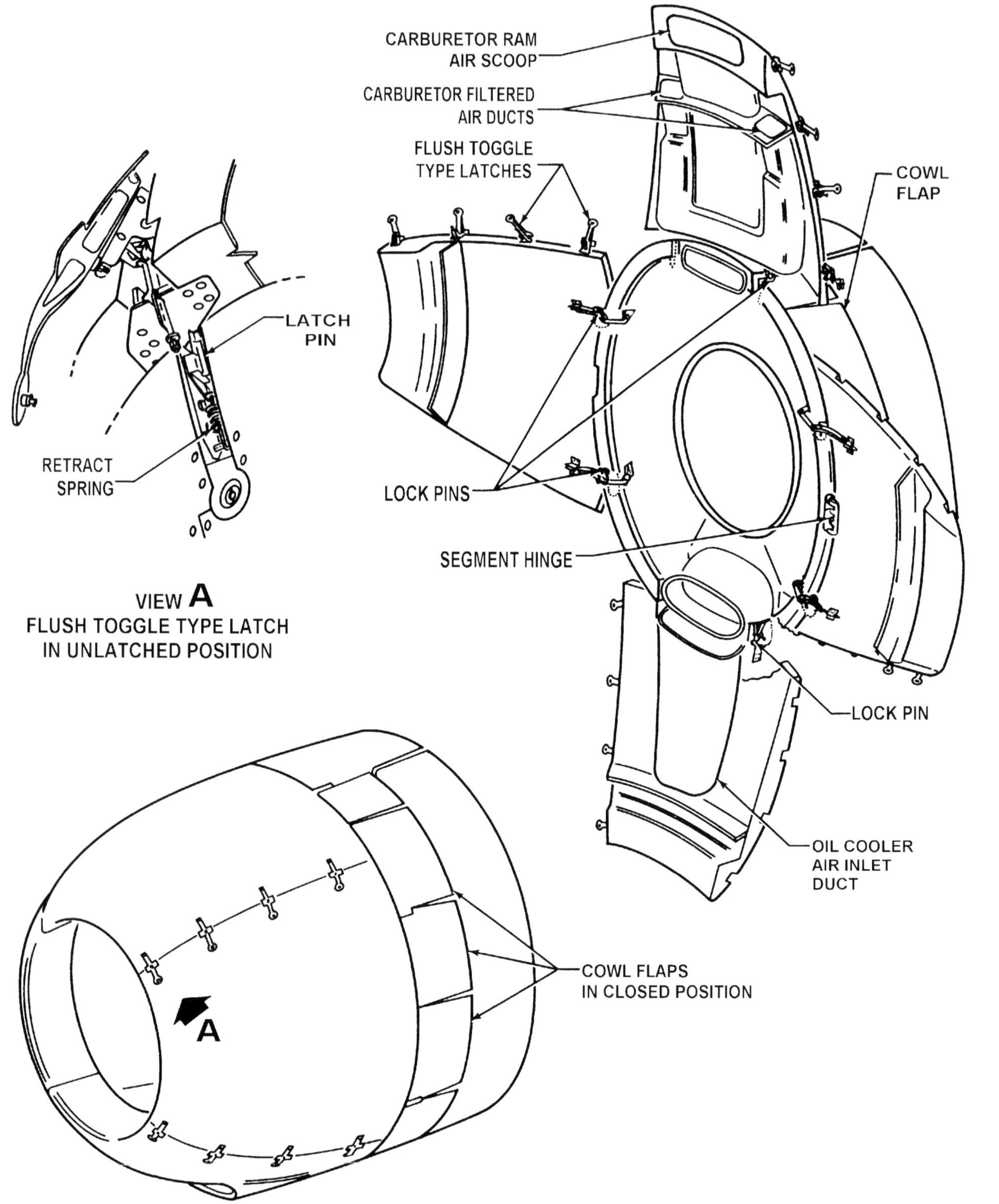

COWLING IN CLOSED POSITION

One hard lesson that came out of World War II was the importance of ease of maintenance. Removing cowling was an ongoing chore and if it was made difficult, maintenance was that much more demanding. The C-124 utilized the clamshell type cowl. Four sections could be easily unbuttoned and swung out of the way, leaving access to the engine and its accessories. *(AN 01-40NVA-2 Erection & Maintenance Instructions for Army Model C-124 Airplanes)*

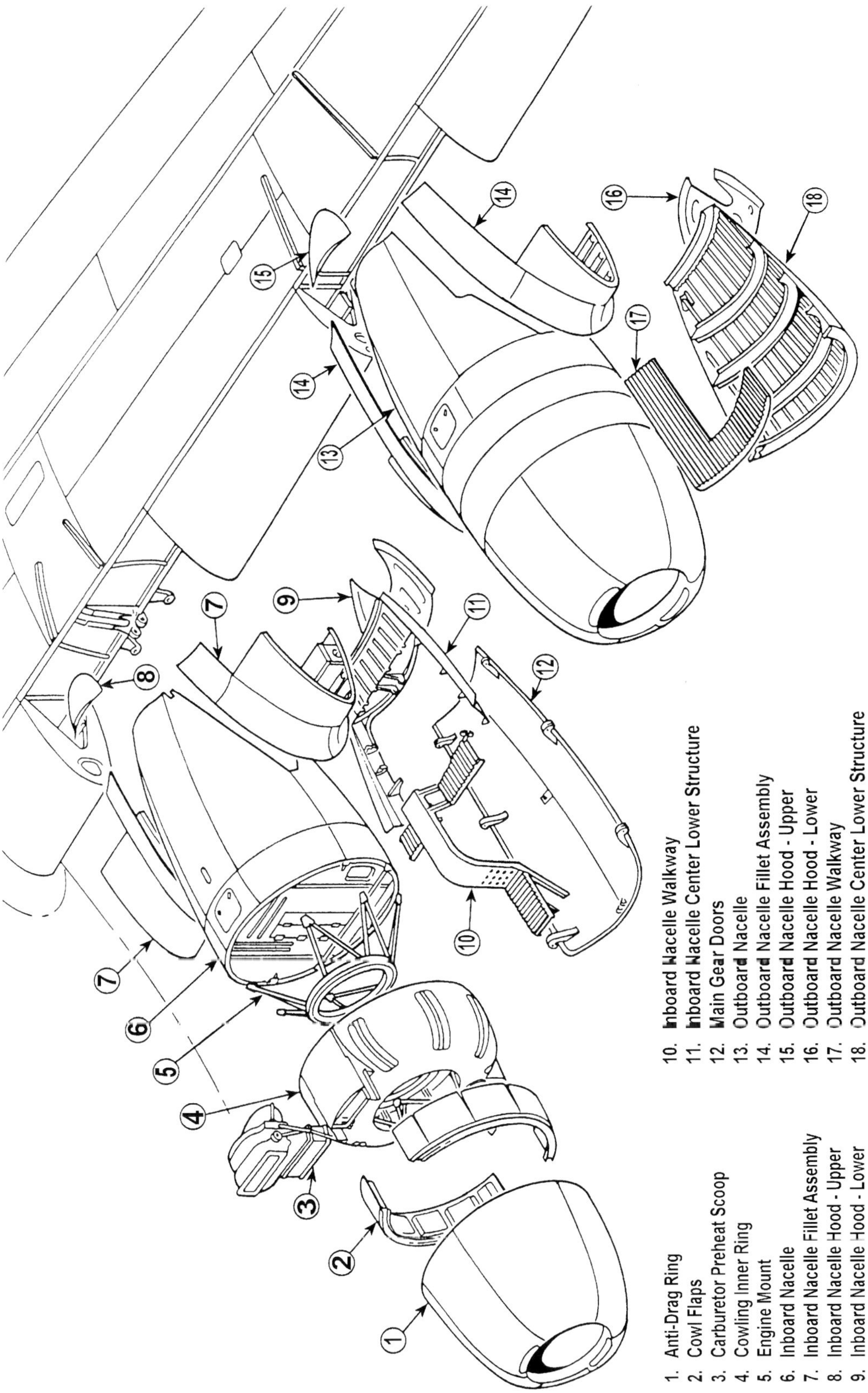

This line drawing gives a good perspective of the key sub-assemblies making up the C-124 nacelles. *(AN 01-40NVA-2 Erection & Maintenance Instructions for Army Model C-124 Airplanes)*

This C-124A was photographed at Alexandria, Virginia, on March 23, 1953. Note the relative size of the C-47 in the right side of the photograph. *(Courtesy of National Archives & Records Administration)*

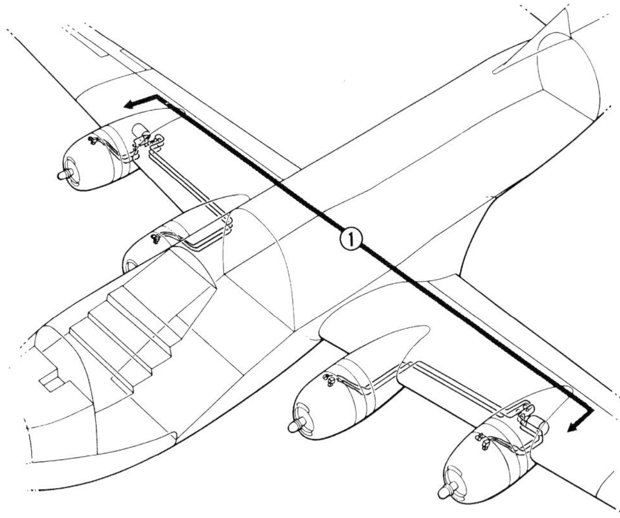

With the exception of the YC-124, all C-124 variations featured ADI. One tank per wing fed both engines on that particular wing. *(AN 01-40NVA-2 Erection & Maintenance Instructions for Army Model C-124 Airplanes)*

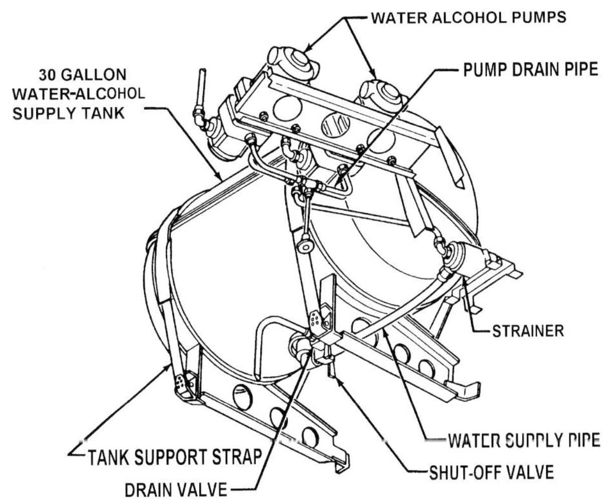

This line drawing shows how the 30-gallon ADI tank was mounted in the number one and number four nacelles. *(AN 01-40NVA-2 Erection & Maintenance Instructions for Army Model C-124 Airplanes)*

C-124 Parameters *(Ref. 7-1)*:	
Wingspan	173 feet, 3 inches
Length	127 feet, 2 inches
Height	48 feet, 3 inches
Max. speed	320 mph
Payload	50,000 pounds
Weight loaded	175,000 pounds
Engines	R-4360-20W (C-124A), R-4360-63* or -63A* (C-124C)

*-63s used a Chandler Evans carburetor and a -63A used a Bendix PR-100 carburetor. Other than that they were identical.

Fairchild R4Q-1, Fairchild C-119 "Packet," Fairchild C-120

(Ref. 7-1, 7-43, 7-44, and 7-45)

In a similar fashion to the C-74/C-124 described, the C-119 was a derivative of an earlier design—the C-82. Design of the C-82 was initiated in 1941, and the prototype first flew in 1944. Although some production models had been built prior to the end of hostilities, it was another project that didn't get into active service before the end of the war. Powered by a pair of Pratt & Whitney R-2800s, the C-82 was underpowered with these engines. However, it was apparent that Fairchild had a good design if only it had more power. Of course, replacing the R-2800s with R-4360s easily solved that problem. The first C-119 was simply a re-engined C-82 powered by R-4360s. This aircraft made its first flight in late 1947. Production C-119s differed from C-82s in a number of respects, the most noticeable being the relocation of the flight deck of the C-119 to the nose. This change facilitated parachute and cargo dropping resulting from the improved vision ahead. Another significant change implemented on the C-119 was to widen the fuselage by 14 inches. To handle its far greater load carrying capability, stronger wings were designed. An electrically operated monorail system that ran the length of the fuselage allowed discharge of "paracans" (PARAchute-dropped CANnisters) through a hatch in the bottom of the fuselage. Access to the fuselage for loading was via a pair of vertically hinged doors at the rear. The doors could be removed to facilitate parachute dropping.

An interesting variation on the C-119 theme was the one-off C-120 "Pack Plane," so named because of the disposable cargo compartment. A standard C-119B was pulled off the manufacturing line and modified into the Pack Plane. Capable of flying with or without the cargo pod, this versatile aircraft obviously did not impress the Air Force. The cargo pod had its own set of wheels for rolling around on the ground. The C-120 would taxi over the pod, or the pod could be towed under the C-120 fuselage. The pod would then be winched aboard the fuselage and locked into position ready for flight. With the normal location for the nose wheel now gone, Fairchild re-engineered the landing gear with "nose" gear assemblies protruding forward of each engine nacelle. The main gear remained unmodified. The C-120 made its first flight on August 11, 1950.

Above: The genealogy of the Fairchild C-119 went back to the R-2800-powered C-82. Even though the C-119 was a totally redesigned aircraft, Fairchild realized that the C-82 could use some additional horsepower. Therefore a standard C-82 was modified with R-4360 power as a proof of concept aircraft. *(Courtesy of National Archives & Records Administration) Right:* A monorail system incorporated within the C-119 allowed "paracans" to be dropped through the bottom of the fuselage. *(USAF T/O 1C-119B-2 Handbook Maintenance Instructions USAF Series C-119B, C-199C Aircraft. December 1, 1954)*

CQ-769
Fairchild Packet
C-119A

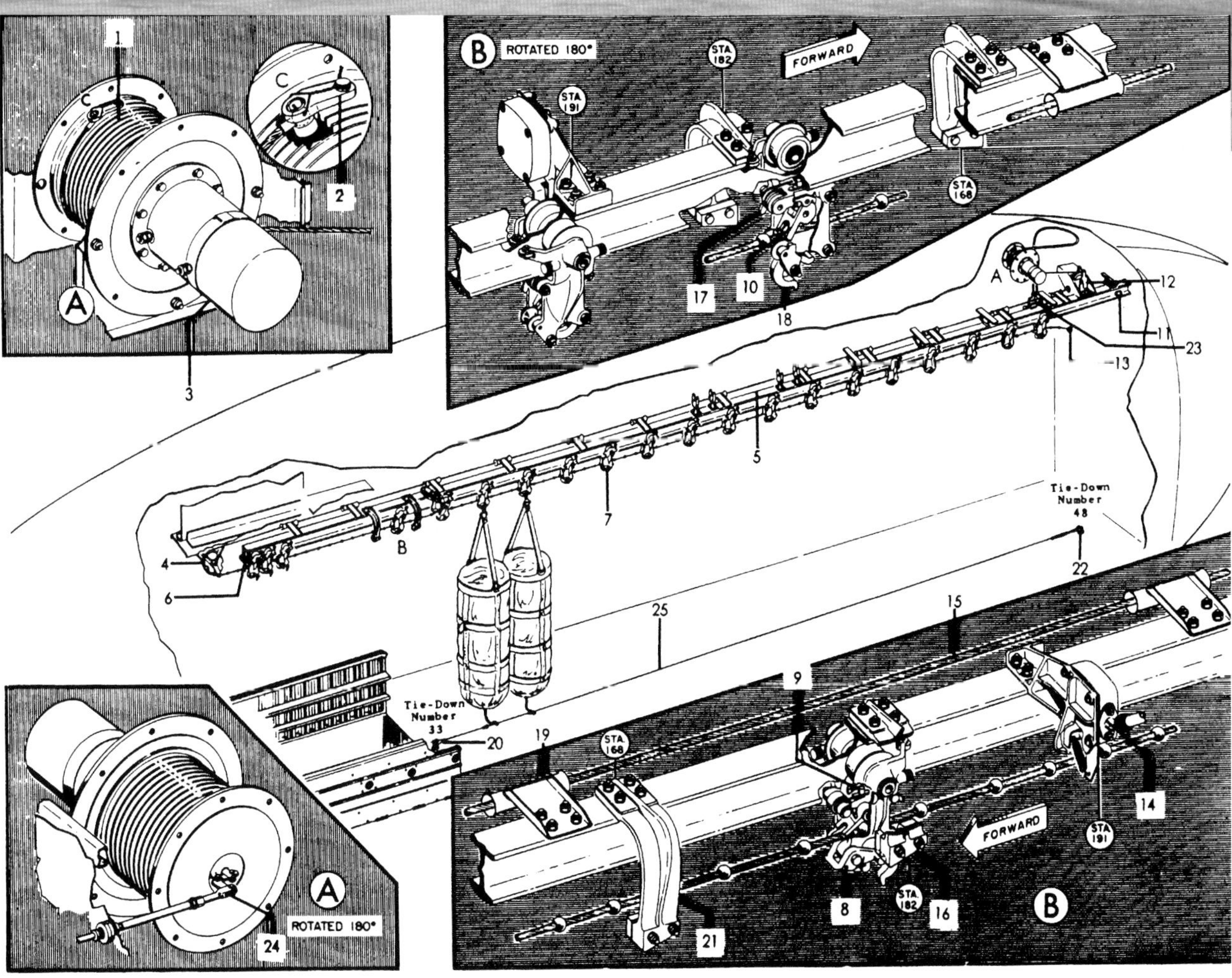
B ROTATED 180°
STA 191
STA 182
FORWARD
STA 168
A ROTATED 180°
Tie-Down Number 48
Tie-Down Number 33
FORWARD
STA 168
STA 182
STA 191

Two R-4360s in shipping cans are being loaded aboard a C-119. The vertically hinged doors could be removed if necessary for flight operations. *(Courtesy of National Archives & Records Administration)*

Another experiment tried on the C-119 was the use of rubber-tracked landing gear. Like the other aircraft this innovation was used on, it never went anywhere and was soon dropped. The C-119 also proved to be a good airframe for one of the early deployments of the so-called gun ships that bristled with every form of offensive armament for ground support.

Engine Installation

Like the larger C-124, the C-119 was not pressurized. With this in mind, it made little sense to turbosupercharge the engines. This allowed for a far lighter QEC, better reliability, and one less major sub-system to worry about. The flip side of this was less altitude capability and performance at altitude. But with no pressurization these were moot points. The seven-cylinder banks dumped into a collector exhaust system with two discharges, one either side of the nacelle. The fact that a jet stack design was not incorporated (it used a collector ring design) negated the possibility of using a tuned exhaust system. This being the case, all four cylinders making up a bank were simply manifolded together to discharge into the collector ring. A ram-air scoop located on the wing leading edge supplied cooling to the circular oil cooler located with the circular section situated longitudinally. Normally, oil temperature is controlled via two primary functions: (i) a bypass valve on the cooler directs the appropriate amount of oil through the cooler, that is, with a low oil temperature most of the oil is bypassed around the cooler and as oil temperatures increase more oil is diverted through the core of the cooler, and (ii) cowl flaps on the exit side of the

An interesting variation on the C-119 theme was the C-120–basically a standard C-119 with the lower fuselage chopped off. A removable cargo pod could then be attached. It was capable of flying with or without the cargo pod. *(Courtesy of National Archives & Records Administration)*

With the normal location for a nose wheel now gone, Fairchild engineers came up with an innovative solution by simply adding another wheel within the nacelle and extending it forward in order to perform the function of a nose wheel. *(Courtesy of National Archives & Records Administration)*

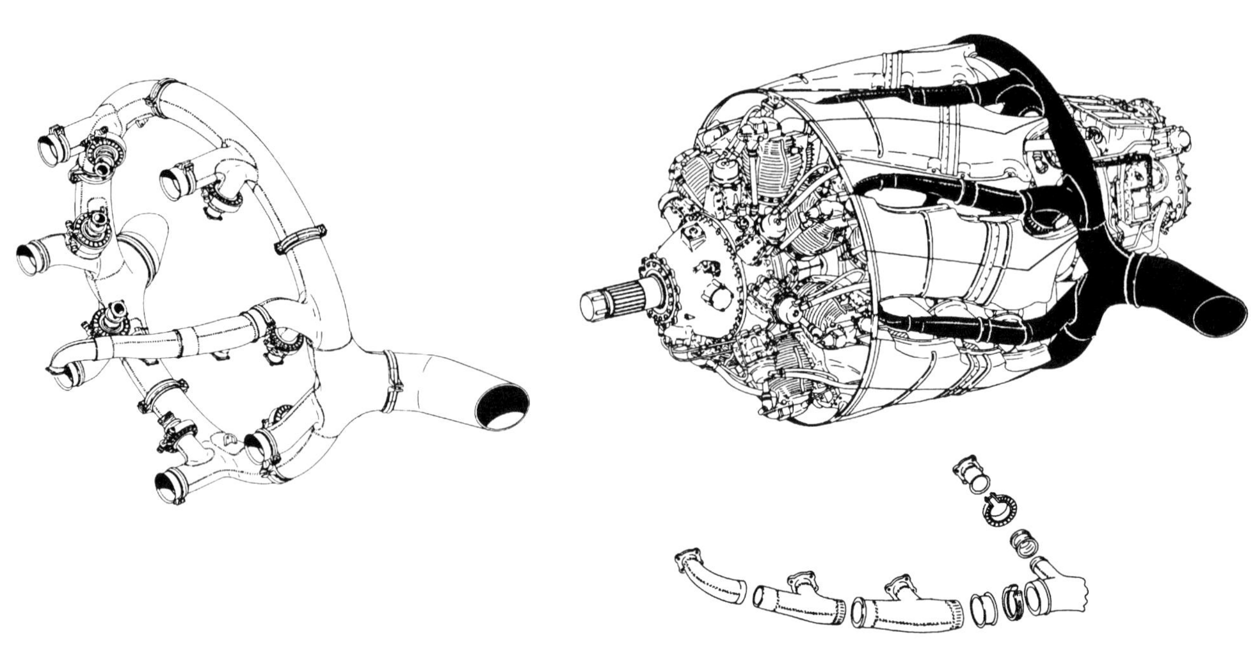

Being an "open stack" engine simplified the entire QEC package. A collector ring discharged exhaust through two tailpipes. However, a jet stack siamesed system would have offered greater performance. *(USAF T/O 1C-119B-2 Handbook Maintenance Instructions USAF Series C-119B, C-199C Aircraft. December 1, 1954)*

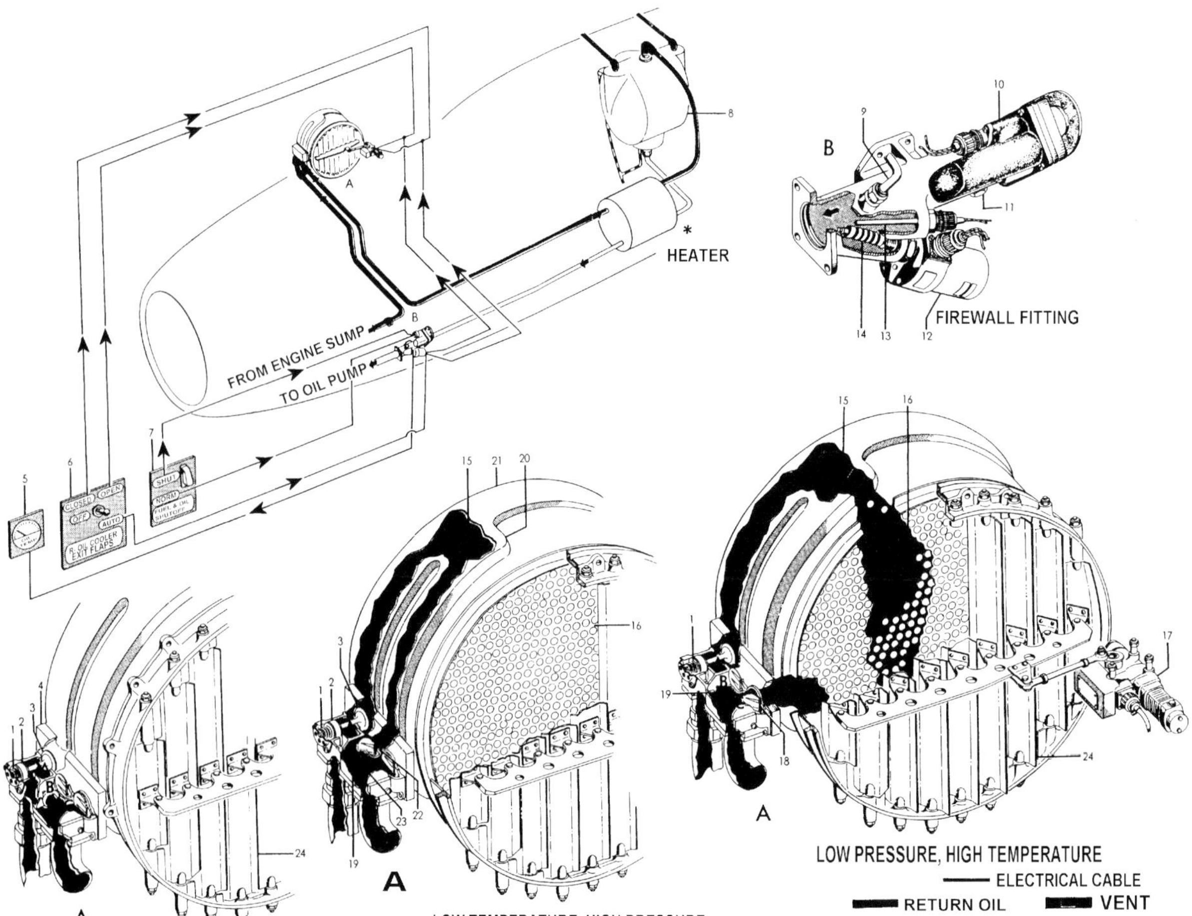

C-119 oil cooling system. *(USAF T/O 1C-119B-2 Handbook Maintenance Instructions USAF Series C-119B, C-199C Aircraft. December 1, 1954)*

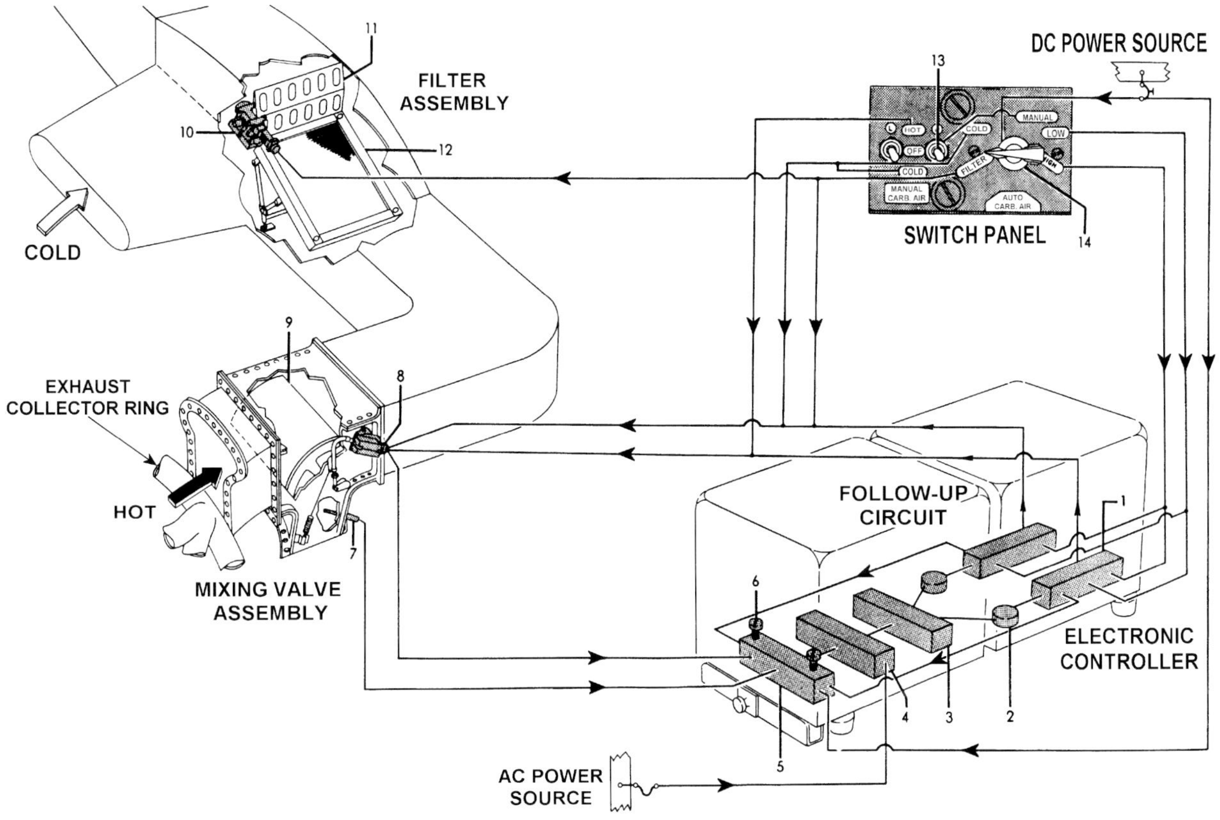

This is the induction air system for a C-119. The flight engineer had the choice of filtered, cold, or warm air. *(USAF T/O 1C-119B-2 Handbook Maintenance Instructions USAF Series C-119B, C-199C Aircraft. December 1, 1954)*

oil cooler ducting control the mass airflow through the cooler core. The C-119 used a slight variation on this traditional method. Rather than using cowl flaps for mass airflow control, the C-119 setup uses oil-cooler exit flaps integral with the cooler. An actuator opens and closes the exit flaps according to oil temperature.

This design is reminiscent of prewar luxury cars that used radiator shutters to control water temperature, except the C-119s shutters are on the discharge side. Induction air comes in from a similar ram-air scoop located on the opposite side of the nacelle to the oil cooler. A complex design of relays, servo motors, electronic controllers, and switch panels allow the engines to run with cold ram air or, cold and rammed filtered air. A mixing valve controls induction air temperature by allowing hot air from the exhaust system to mix with cold air. The landing gear offered no surprises with a dual main gear and a single nose wheel. The nose wheel design looked remarkably similar to the C-123's. This is not surprising considering that both aircraft came from the same company. One C-119 was experimentally fitted with a four-wheel main gear bogie. This idea was not pursued. One has to assume that it was tried in order to reduce unit-bearing stress, therefore opening more landing strips, and particularly semi-prepared rough landing fields. Wright R-3350s powered the last versions of the C-119, although most former flight crews concur that the R-4360-powered version was better.

C-119 Variations

C-119A

Prototype modified from a standard C-82 to test the R-4360 powerplants and new configuration.

C-119B

Production version of C-119A powered by two R-4360-20s.

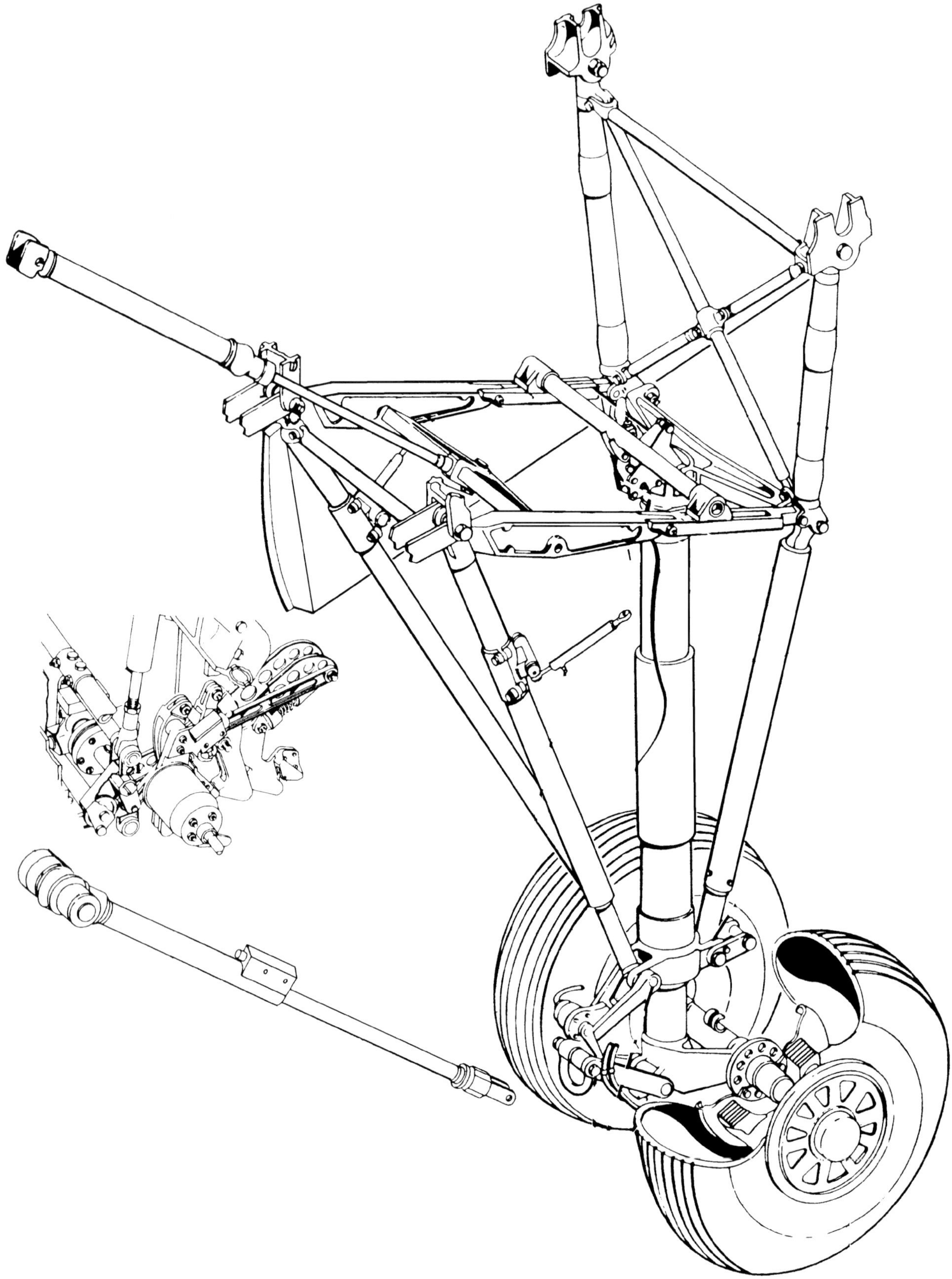

Dual wheel C-119 main gear which retracted forward. *(USAF T/O 1C-119B-2 Handbook Maintenance Instructions USAF Series C-119B, C-199C Aircraft. December 1, 1954)*

C-119 Parameters* *(Ref. 7-1):*	
Wingspan	109 feet, 3 inches
Length	86 feet, 6 inches
Height	26 feet, 3 inches
Empty weight	39,942 pounds
Payload	27,500 pounds
Max. T/O weight	73,150 pounds
Wing loading	50.5 pounds per square foot
Cruising speed	200 mph
Rate of climb	740 feet per minute
Range	1,950 miles
Engine	R-4360-20W

* These parameters are for the R-4360-20W-powered C-119C

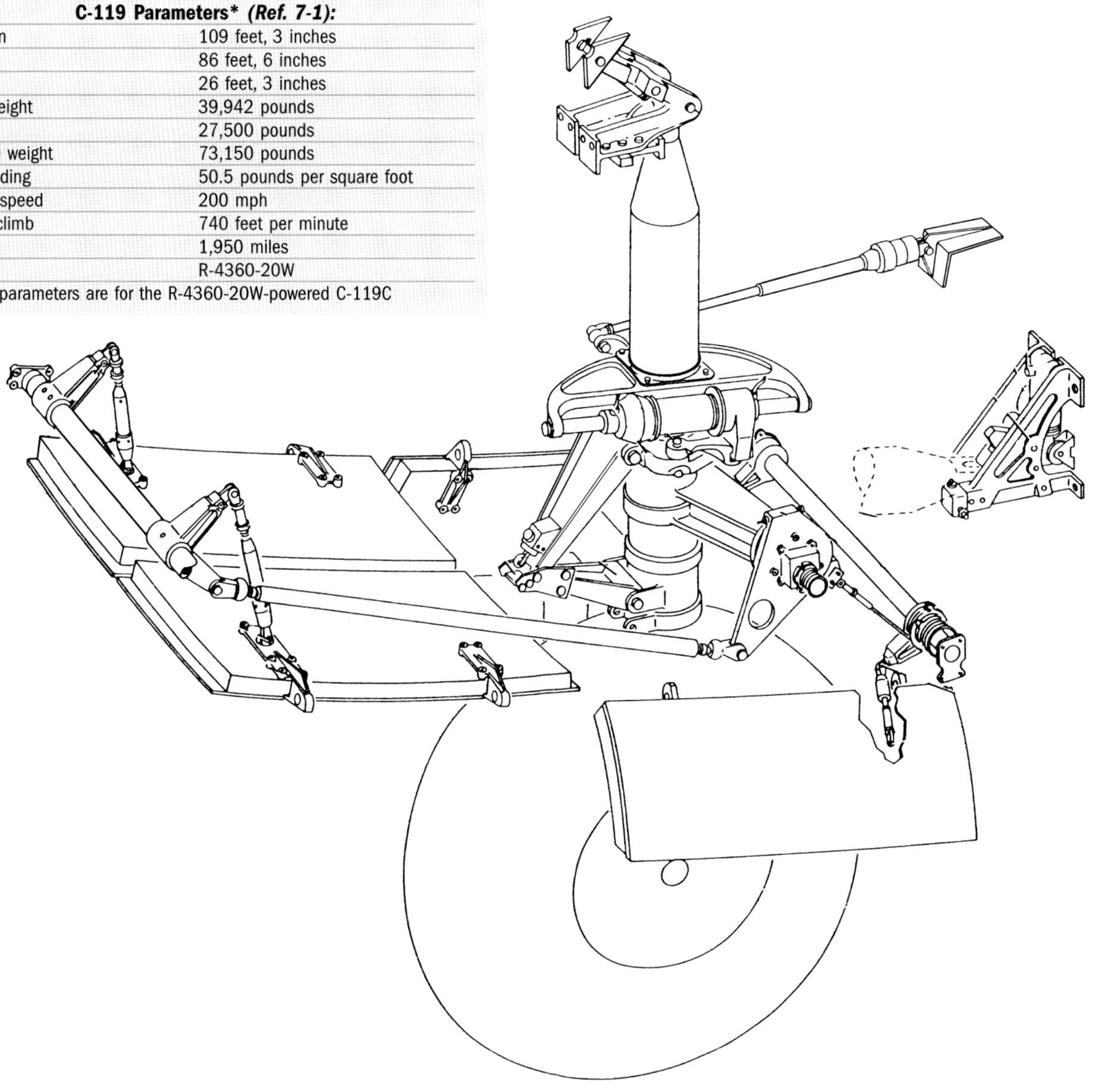

Typical Fairchild nose gear design for the C-119. Not surprisingly, it's a similar design to the Fairchild C-123 Provider. *(USAF T/O 1C-119B-2 Handbook Maintenance Instructions USAF Series C-119B, C-199C Aircraft. December 1, 1954)*

C-119C
Powered by a pair of R-4360-20Ws. Dorsal fins added to tail booms.

R4Q-1
Marine Corps version of C-119C. Both versions are identical.

C-119F
Similar to C-119C except powered by Wright R-3350s and Hamilton Standard propellers. Small ventral fins added.

R4Q-2
Marine Corps version of C-119F.

C-119G
Same as C-119F except fitted with Aeroproducts propellers.

C-119
Major redesign with larger span wing, new tail assembly, new landing gear and powered by R-3350s.

The landing gear of a C-119 was modified with a four-wheel bogie. It was not pursued any further. *(Courtesy of National Archives & Records Administration)*

Martin XP4M-1, P4M-1, P4M-1Q Mercator

(Ref. 7-1, 7-46, 7-47, and 7-48)

During the transition from piston engines to gas turbines, a number of military aircraft utilized both forms of power. Typically, the piston engine was the primary power and the high power-to-weight ratio of the gas turbine was utilized for high gross take-offs and emergency combat power. For the early generation gas turbines and late generation piston engines this was a good combination. Gas turbines, particularly early ones, suffered from very high specific fuel consumption rates, whereas the piston engine was relatively fuel efficient. Furthermore, a gas turbine operates on just about anything that burns, so it was no problem to run the jets on aviation gasoline. And so it was with the large and elegant Martin Mercator. Although it didn't appear so, it was a four-engined aircraft—two burning and two turning, or two piston engines and two gas turbines. Unlike some aircraft that used a combination of jet and piston power, the Mercator was

designed from the get-go for both forms of powerplant. Aircraft such as the KC-97, B-50, and B-36 had jets added almost as an afterthought. Now the question arises: where to mount the jets? In the case of the Mercator, an Allison J33 was fitted inside each engine nacelle. In the course of normal cruise when the jets were not in operation, a streamlined fairing covered the jet engine intake. For jet engine operation, a large clamshell door opened, allowing ram air to enter the single-stage centrifugal compressor of the gas turbine.

In most twin-engined aircraft, the landing gear folds into the nacelle. This was impossible

The gas turbine installation for the Mercator was neat and compact. Each Allison J33 was housed in the piston engine nacelle. When the jets were operating, large clamshell doors opened to allow air to the jet's single stage centrifugal compressor. This in-flight shot shows the doors open, indicating that the jets are running. Also of note are the partially faired main wheels buried within the wing structure. *(NAVAER 01-35EH-502 Martin P4M Mercator Maintenance Manual)*

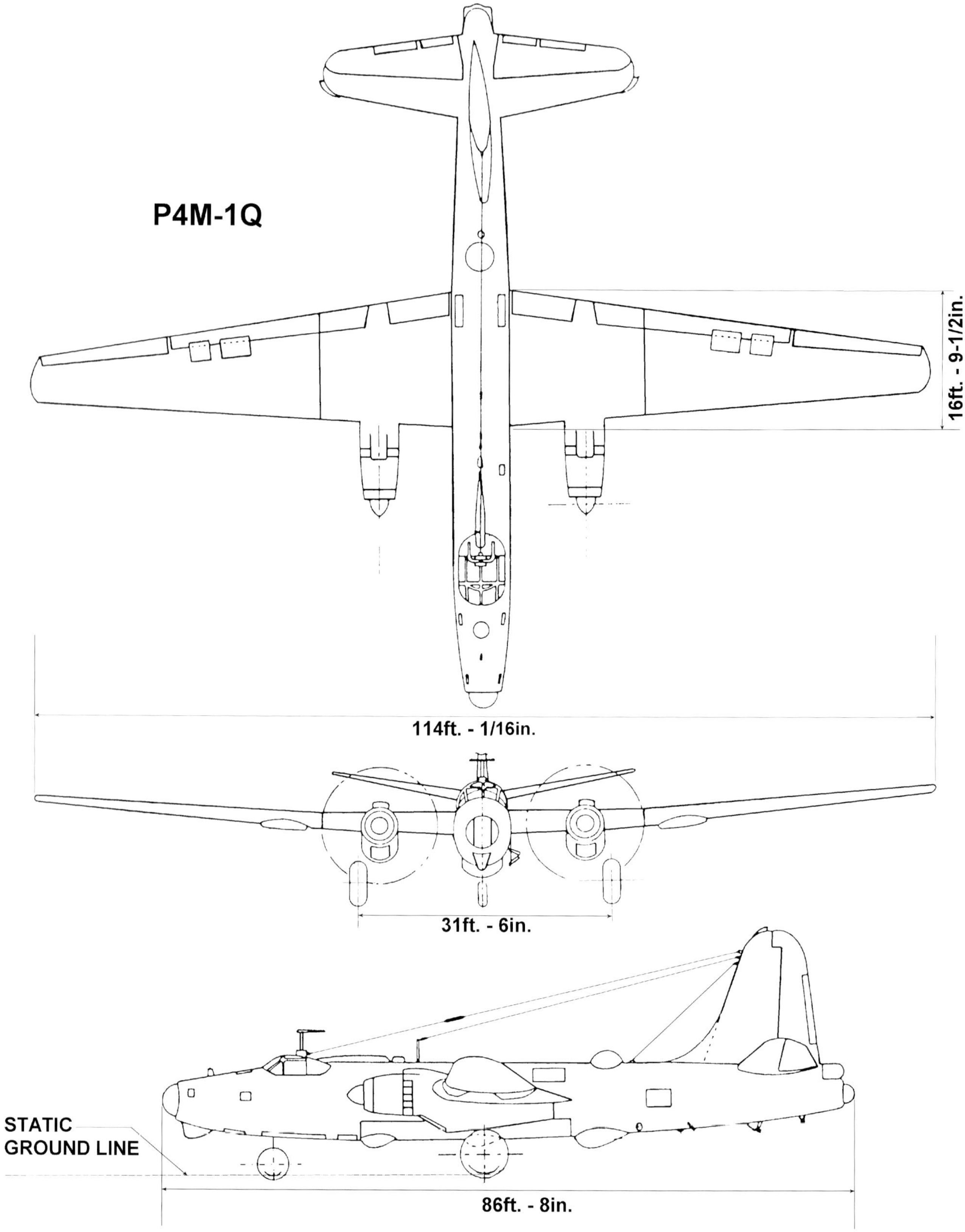

The Martin P4M-1Q Mercator is a little known aircraft, yet it performed essential surveillance missions for the United States during the height of the Cold War. This line drawing shows the auxiliary gas turbine air doors closed. *(NAVAER 01-35EH-502 Martin P4M Mercator Maintenance Manual)*

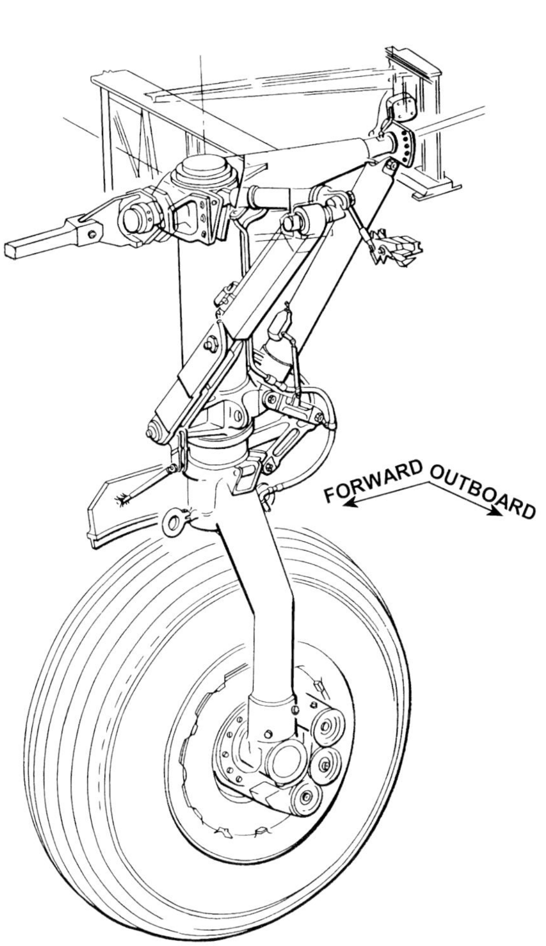

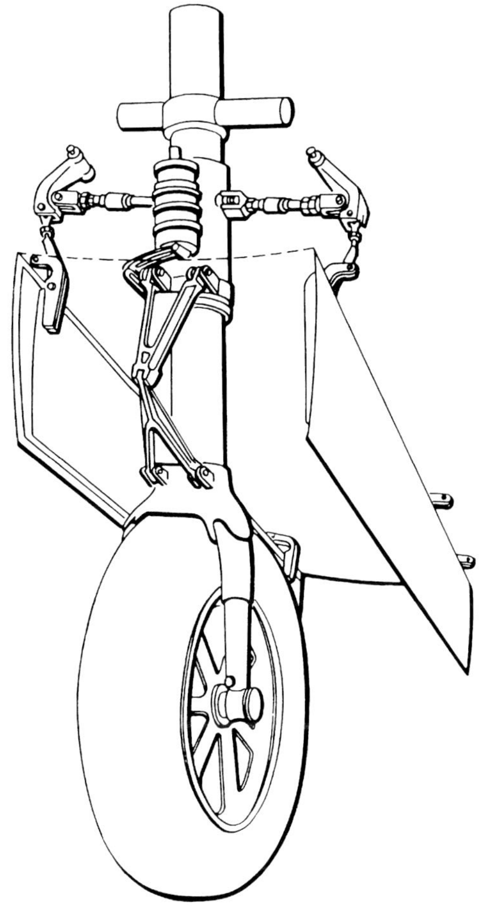

Above left: Martin didn't have much choice when it came to designing the main gear. It had to fold sideways, which demanded a single wheel. A dual wheel setup would have protruded into the airstream even more than the single wheel did. *(NAVAER 01-35EH-502 Martin P4M Mercator Maintenance Manual) Above right:* The Mercator nose gear was a conventional unit that retracted rearwards. *(NAVAER 01-35EH-502 Martin P4M Mercator Maintenance Manual)*

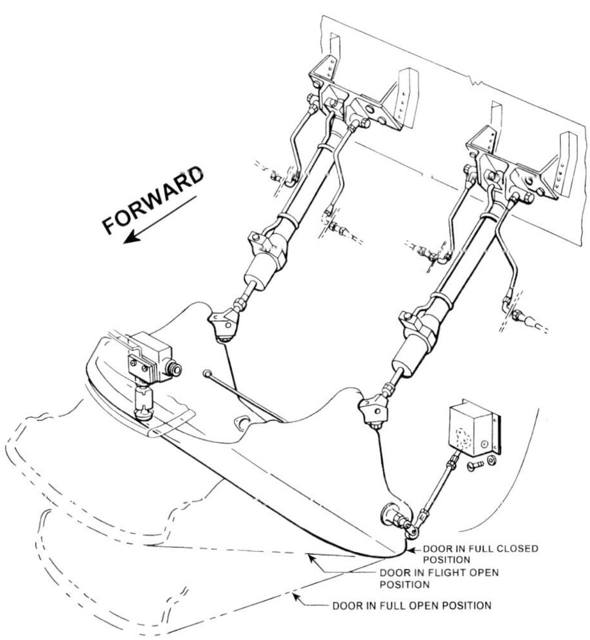

Two hydraulic rams actuate the gas turbine air doors. *(NAVAER 01-35EH-502 Martin P4M Mercator Maintenance Manual)*

in the case of the Mercator due to the jet. Martin's solution was a similar one to that employed by the Consolidated B-24 Liberator—the main wheels folded outboard of the nacelles. With the relatively thin wing employed by the Mercator, a compromise was necessary by having half the main wheel exposed. Fairings in front of and behind the main wheels were an attempt to reduce parasitic drag. Ironically, one of the aircraft replaced by the Mercator was the Consolidated PB4Y, which was a variation on the B-24 Liberator theme. The nose gear was a conventional

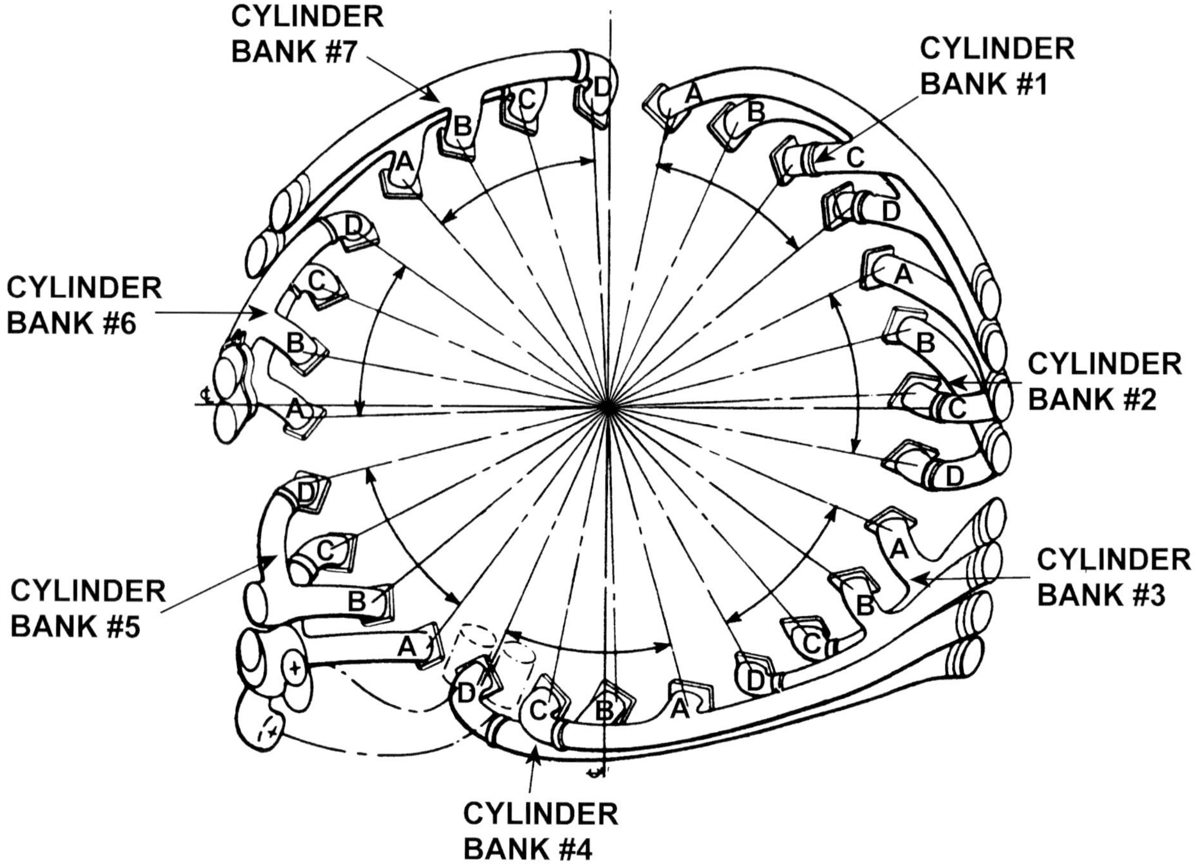

Being an open stack engine allowed Martin engineers to design a siamesed exhaust system. This line drawing shows the seven cylinder banks and cylinders A through D for each bank. *(NAVAER 01-35EH-502 Martin P4M Mercator Maintenance Manual)*

single wheel unit that retracted rearward. A retractable tailskid made up the fourth component of the landing gear, or alighting gear, to use Navy terminology.

The R-4360-20s that powered the Mercator were open-stack engines, meaning they were not turbosupercharged. This allowed the use of a siamesed exhaust system for improved exhaust scavenging. A conventional ram-air scoop located on top of the cowl supplied induction air. Vertical flow splitters in the entrance of the scoop ensured a smooth airflow at the carburetor deck during any time the aircraft was yawed. Two large circular oil coolers mounted in the leading edge and supplied with cooling air via ducts kept oil temperatures under control. An oil tank mounted behind the firewall supplied the necessary lubrication. Cowl flap design, again, did not venture into unknown territory. The arrangement was a conventional electrically driven process. An electric motor drove a flexible driveshaft through 90-degree gearboxes and jackshafts, which in turn operated the cowl flaps.

Although it was only built in small numbers, the significant combat history of the Mercator has been shrouded in secrecy. Nevertheless, it still achieved more than its small production run would indicate. In addition to two prototypes, 19 production aircraft were manufactured. The original mission profile for the Mercator was that of Navy patrol bomber. Lockheed had built a number of successful aircraft for this requirement: The PV-1, PV-2, and the P2V. The Mercator was designed for this requirement; however, in 1951 all surviving Mercators were converted to P4M-1Qs, or spy ships. With its huge fuselage, the

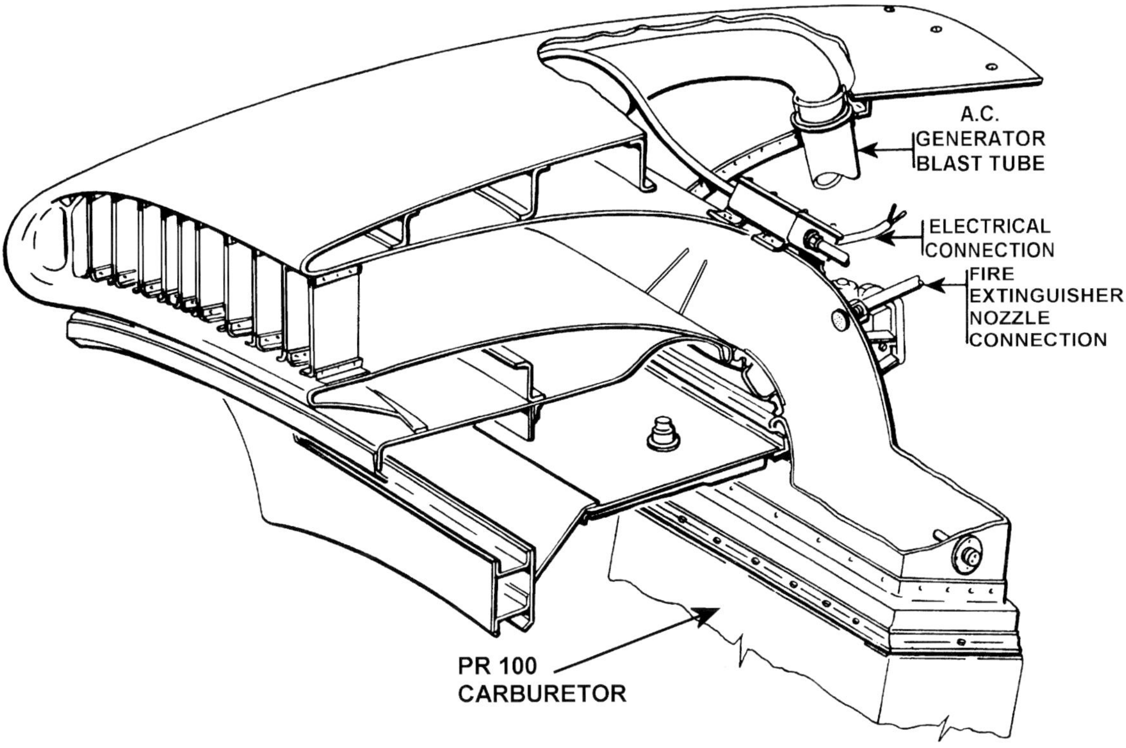

The Mercator's ram-air induction scoop used vertical flow splitters. This stops any stalling of the airflow if the aircraft is yawed. It also helps to straighten out the airflow from the propeller slipstream. *(NAVAER 01-35EH-502 Martin P4M Mercator Maintenance Manual)*

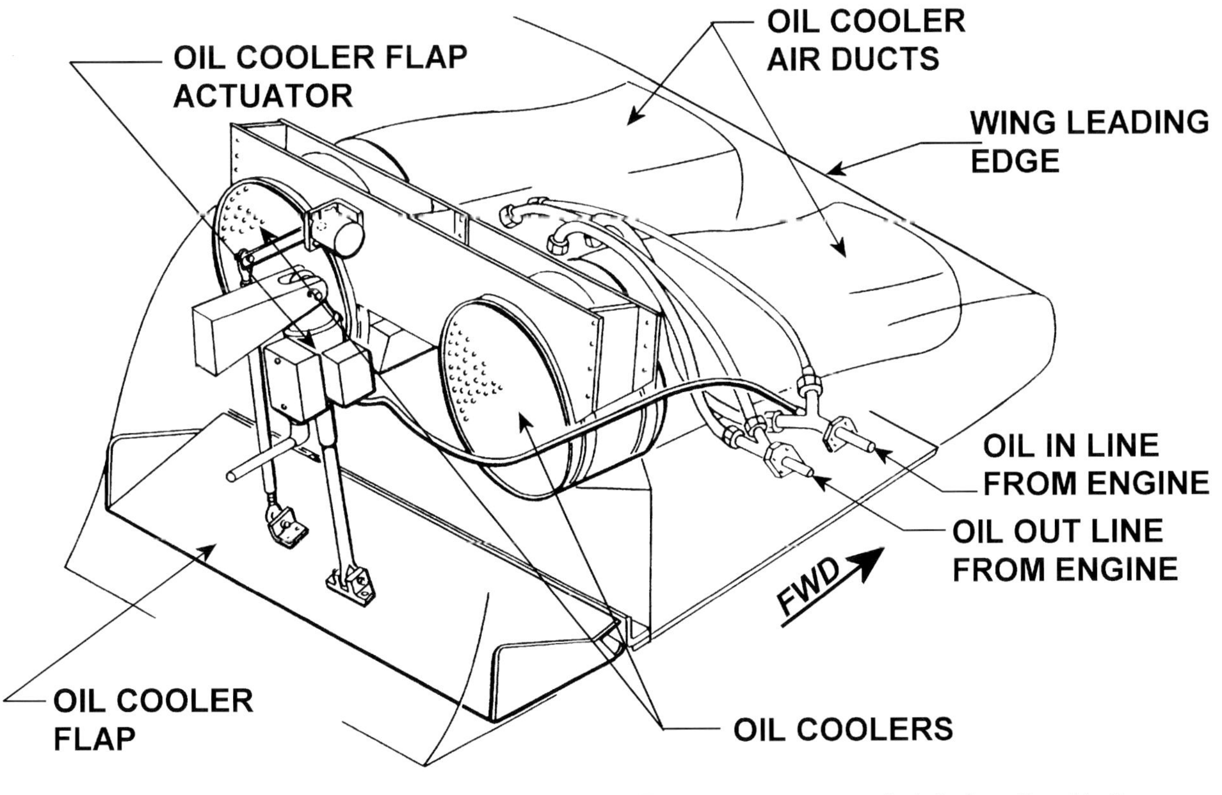

Dual circular oil coolers were mounted in the wing leading edge. Temperature was controlled via the adjustable flap. *(NAVAER 01-35EH-502 Martin P4M Mercator Maintenance Manual)*

This line drawing shows the Mercator oil tank and its attachment points. *(NAVAER 01-35EH-502 Martin P4M Mercator Maintenance Manual)*

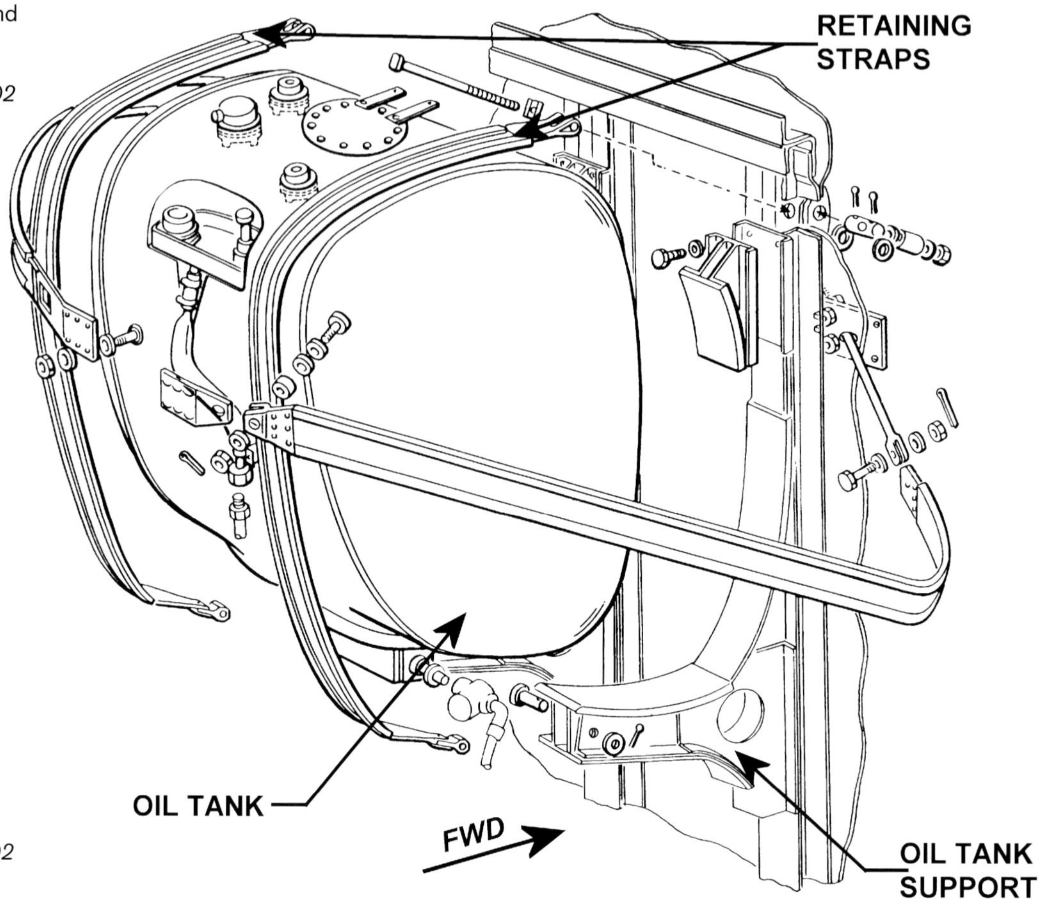

Opposite: Cowl flaps were actuated via an electric motor (power unit) that drove flexible shafting and "tee" gearboxes that in turn drove screw jacks. *(NAVAER 01-35EH-502 Martin P4M Mercator Maintenance Manual)*

P4M-1 was far more suitable for this role than its competitor the P2V, which soldiered on in the role it was designed for. As a spy ship, the P4M-1Q performed admirably from 1951 until its withdrawal in 1960. In yet another case of bureaucratic vandalism, no Mercators survived even though the last one was flown to Alameda and therefore would have made an ideal candidate for conservation. But this last survivor was soon scrapped.

The veil of secrecy is slowly being lifted on this beautiful looking aircraft, which offers an insight to its fascinating service career. It patrolled the former Soviet Union and China. During these dangerous operations, a number of P4M-1Qs were attacked and at least one was shot down by a MiG-15 with all on board being lost. This incident occurred on August 23, 1956, off the coast of China near Shanghai.

For such a large aircraft, its performance was impressive, particularly with two turning and two burning. With all fires lit, the P4M-1 was capable of a top speed of 419 mph.

Martin P4M-1 Mercator Parameters *(Ref. 7-1)*:

Wingspan	114 feet, 1/16 inches
Wing area	1,311 square feet
Length	86 feet, 8 inches
Height	29 feet, 2 inches
Max. weight	88,378 pounds
Internal fuel	4,200 gallons
Max. speed	419 mph at 14,200 feet (2 turning, 2 burning, i.e., all four engines operating)
Max. range	2,000 miles at max. gross weight, or 4,230 miles with max. fuel
Ceiling	25,262 feet
Engines	Two R-4360-20 (single stage, two speed) and two Allison J33 turbo jets rated at 3,800 pounds of thrust each
Number built	2 prototypes and 19 production aircraft

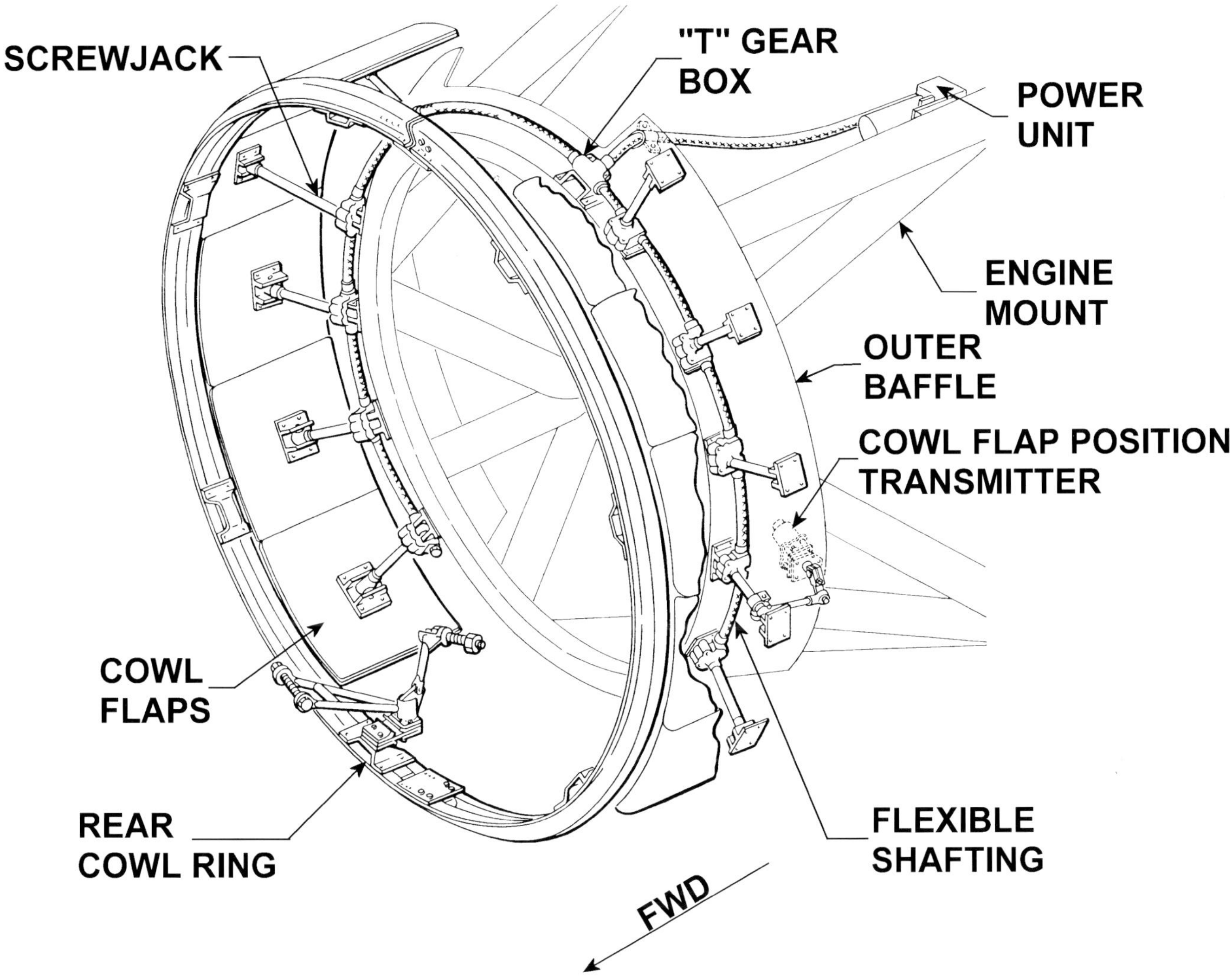

Convair B-36A, B, D, E, F, H, J, Convair XC-99, Convair Model 37

(Ref. 7-1, 7-49)

The aviation landscape is littered with the futile and failed attempts of building a large (over 200-foot wingspan) aircraft, particularly those powered by piston engines. Convair's mighty B-36 Peacemaker was the only one that succeeded where so many others had failed. Often derided because of its size and relatively slow speed, the B-36 was in fact an engineering marvel. Following a similar rationale to the Hughes HK-1, it was feared in the United States that Britain would be successfully invaded by Germany. If that had occurred, a bomber with sufficient range to carry a bomb load to England and return would have been necessary. It was this requirement that prompted the Army Air Force to let out a bid for a bomber with a 10,000-mile range with a 10,000-pound payload. As it turned out, neither of the two contenders, the B-35 and B-36, achieved this lofty goal. But both aircraft still managed some pretty impressive performance figures with regards to bomb load and range.

In 1948, a B-36B flew a mission of 7,500 miles with a 10,000-pound bomb load and 5,000 pounds of simulated ammunition that took a mind-numbing 36 hours at an average speed of 193.5 knots *(Ref. 7-50).* Although conceived in the early 1940s long before Convair even knew of the existence of the atomic bomb or the Manhattan Project, its primary mission ended up being that of a nuclear bomber. Early nuclear

Convair's XB-36 sitting on the ramp. Note the huge single-wheel main gear–soon replaced by a four-wheel bogie design. *(Courtesy of National Archives & Records Administration)*

devices were huge, weighing in the neighborhood of 21 tons and measuring over 24 feet long. It was the physical size of atomic bombs that contributed to the downfall of the B-36's competitor, the Northrop B-35. The B-35's bomb bay simply could not handle such a huge weapon even though it had sufficient payload capacity.

Development

B-36 development zigzagged from front burner to back burner status a number of times. Initially the B-36 program was a key project, but as Convair's more immediate products, such as the B-24 and B-32 swung into full production, emphasis had to be placed on these aircraft. As a result, engineers and development staff were siphoned off. Early design studies in 1942 showed the B-36 in a form that was close to the one placed into production. The most noticeable difference between the early design studies and production aircraft was the use of twin tails mounted on a horizontal stabilizer with a distinctive dihedral. Production aircraft used a horizontal stabilizer with no dihedral and a single tail.

After execution of the contract, Convair was obligated to deliver the first two XB-36 aircraft in mid 1944. As it turned out, that ambitious target was missed by over two years. Poor

workmanship among other schedule delaying issues plagued the XB-36 program. Inexperienced production and supervisory personnel contributed to this sad state of affairs. Incorrectly set rivets and dull drill bits causing excessive burring of rivet holes were just a couple of the maladies affecting quality. Worse yet was the fact that B-36s were designed with a laminar flow NACA 63 series airfoil. This airfoil section relied on a tolerance of plus or minus 1⁄32 of an inch for the B-36's huge wing—a miniscule tolerance taking into consideration its massive size. XB-36 wing skins were found to be far out of this tolerance range; consequently large panels required re-skinning.

First flight of the B-36 occurred on August 8, 1946, almost a year after V-J Day. However, even before the first flight, development problems surfaced. The fabric-covered magnesium flaps were early casualties. Due to the significant turbulence owing to the pusher configuration, fabric was ripped off and the magnesium support arrangement suffered structural damage. Stouter but much heavier ones made from aluminum soon replaced these early flaps. Weighing 200,000 pounds and carrying a crew of nine and 8,000 gallons of fuel, the first flight went off relatively trouble free. As would be expected with such a huge, complex, and state-of-the-art aircraft, early test flights revealed a number of problems. Similar problems that plagued the B-35 also afflicted the B-36. Again, the pusher configuration was the culprit. Severe propeller induced vibration caused structural damage. Unlike the B-35 where the

continued on page 436

Little wonder that only a handful of runways around the world could handle the enormous weight of the XB-36, particularly when the majority of that load was only distributed on two tires. *(Courtesy of National Archives & Records Administration)*

That's more like it! When the B-36's main gear was redesigned into a four-wheel bogie, it opened up far more runways capable of handling its weight. *(Courtesy of National Archives & Records Administration)*

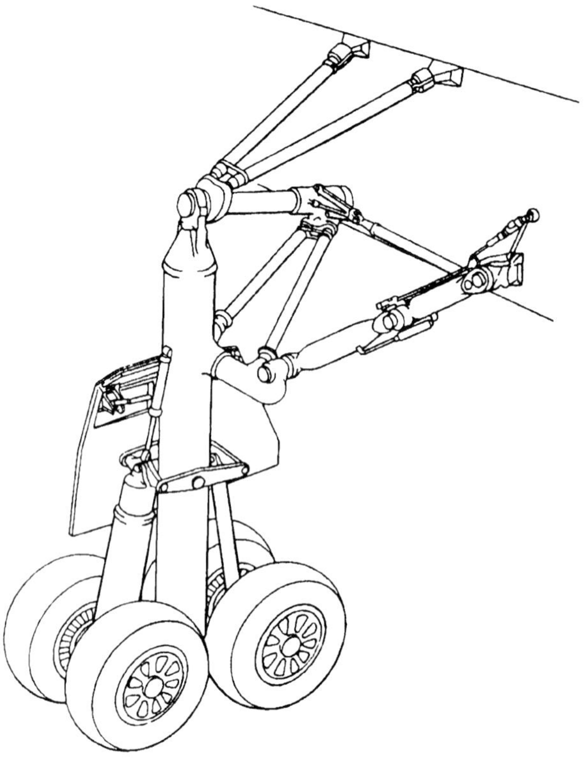

1. GEAR IS DOWN AND LOCKED

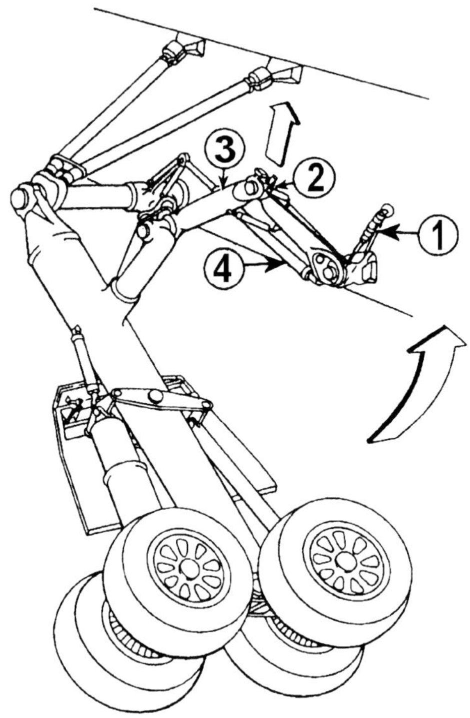

2. AUXILLIARY CYLINDER (1) RETRACTS TO UNLOCK SIDE BRACE LATCH (2) AND BREAKS SIDE BRACE (3) UPWARD. MAIN ACTUATING CYLINDER (4) BEGINES TO EXTEND, AND GEAR BEGINS TO RETRACT

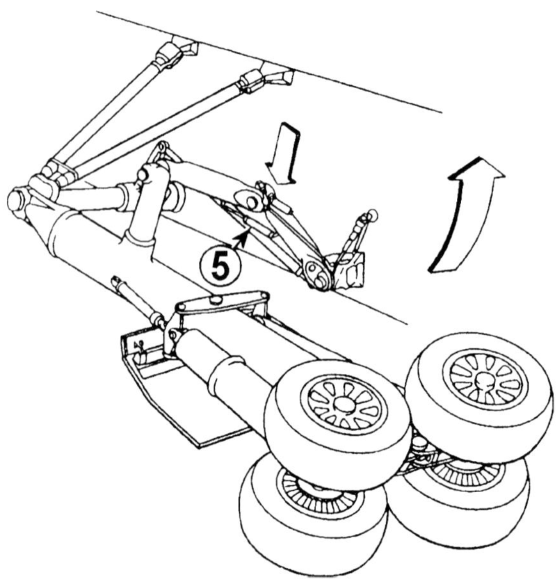

3. MAIN ACTUATING CYLINDER EXTENSION CONTINUES; SIDE BRACE BEGAINS TO STRAIGHTEN. SNUBBER (5) DAMPENS SIDE BRACE MOVEMENT.

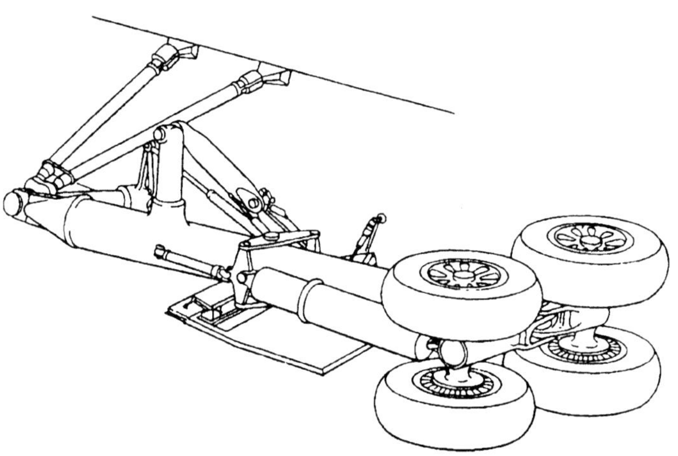

4. GEAR IS FULLY RETRACTED; SIDE BRACE IS STRAIGHT AND AUTOMATICALLY LOCKED BY LATCH.

This series of line drawings shows the retraction sequence of the B-36's main gear. Even though it was stouter than the prototype's single wheel, it could exhibit fragile tendencies during high gross weight landings. *(AN 01-5EUE-2 Handbook Erection and Maintenance Instructions USAF Series B-36F Aircraft. March 27, 1951)*

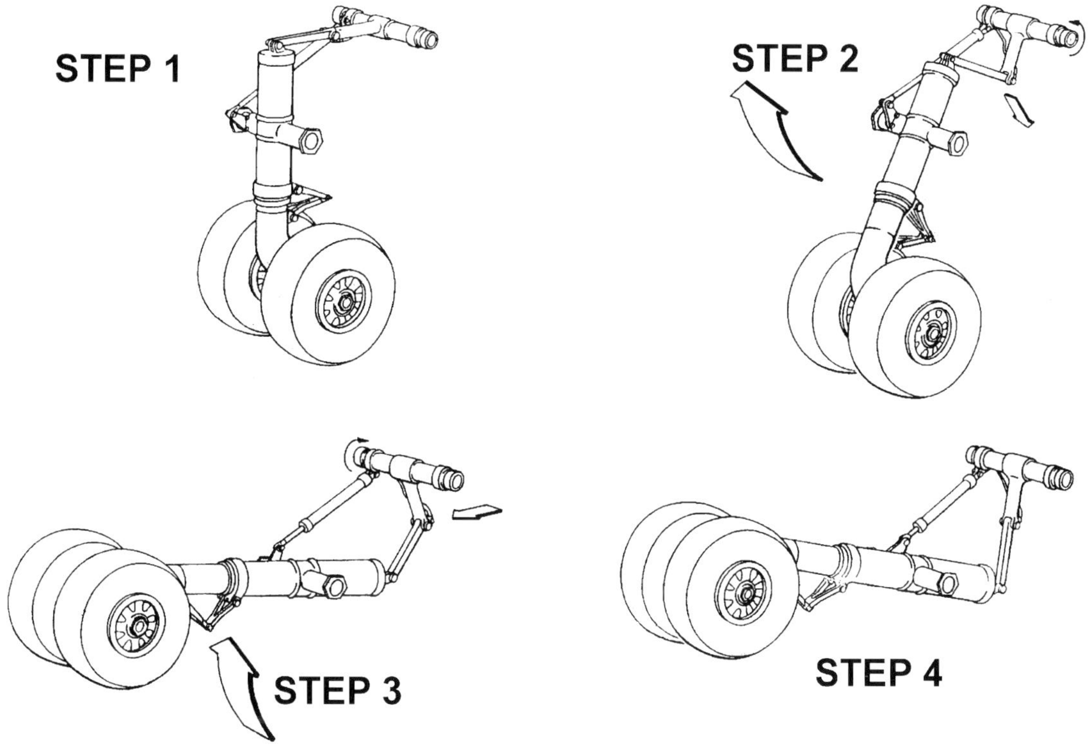

A dual-wheel nose design was used throughout the B-36's life. This retraction sequence is typical of the ilk. *(AN 01-5EUE-2 Handbook Erection and Maintenance Instructions USAF Series B-36F Aircraft. March 27, 1951)*

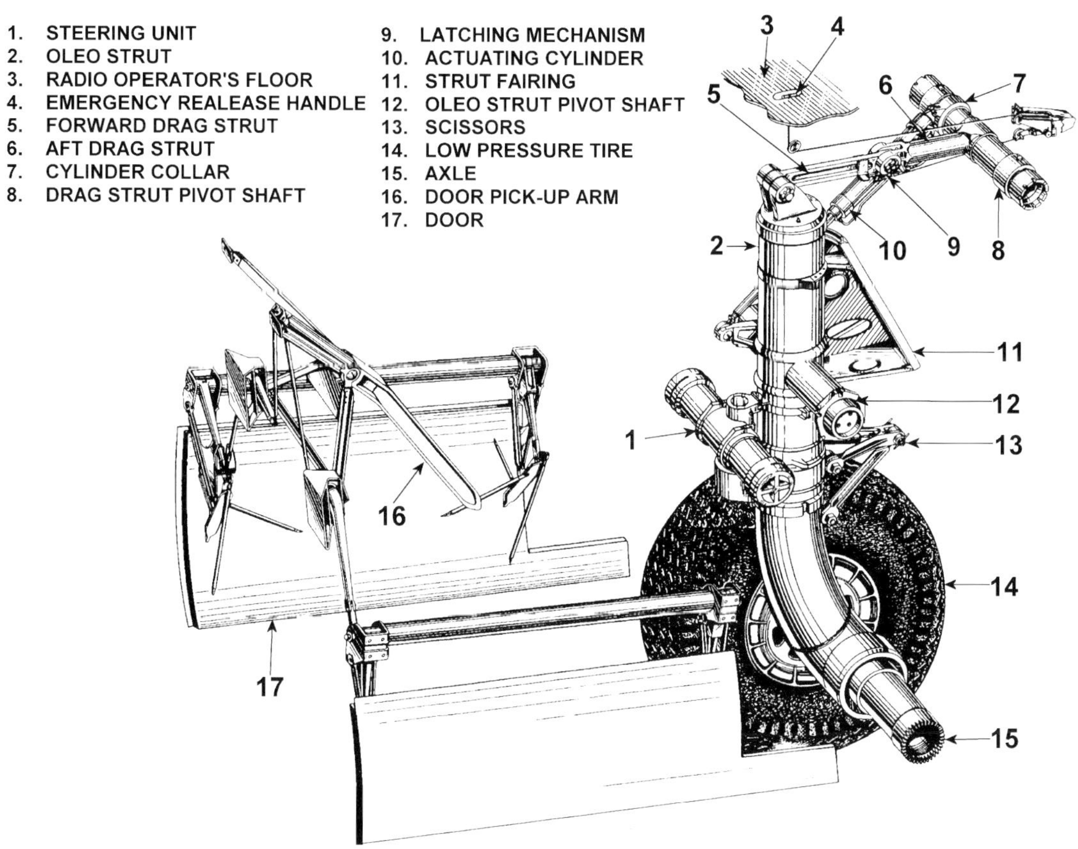

A different perspective on the major components that make up the B-36's nose gear. *(AN 01-5EUE-2 Handbook Erection and Maintenance Instructions USAF Series B-36F Aircraft. March 27, 1951)*

Experiments with main landing gear didn't end with the four-wheel bogie. In an effort to further reduce its bearing loads, a tracked gear was fitted to the XB-36. It flew once. (Courtesy of National Archives & Records Administration)

Above: Taking off on its one and only flight with tracked landing gear, the XB-36 looks very ungainly with this unusual modification. *(Courtesy of National Archives & Records Administration) Left:* During its one and only flight with the tracked landing gear, it's doubtful if it was even retracted. *(Courtesy of National Archives & Records Administration)*

In order to give the B-36 additional performance, four J35s were experimentally attached in pods under the outer wing panels. Production aircraft used the more powerful J47. *(Courtesy of National Archives & Records Administration)*

continued from page 429

vibration was all but insoluble, Convair fixed the XB-36's vibration maladies by considerable beefing up of the wing structure, particularly in the trailing edge. Engine over-heating was another early problem until a more efficient cooling fan could be developed.

A more serious problem was the main landing gear. Utilizing a huge single wheel measuring 110 inches in diameter, unit-bearing stresses on runways were so intense that only a handful of airports were capable of handling the XB-36. Of course, this situation would make the aircraft all but worthless as a weapon of war. As if this was not a significant enough problem, exacerbating the situation was the fact that this single-wheel landing gear wasn't up to the task and was prone to failure. Wisely, a four-wheel bogie replaced the impractical single wheel. Not only was this stronger and more reliable, it now allowed the aircraft to operate off most USAF bases. Nose gear design was a conventional two-wheel unit. However, landing gear development and experiments did not end with the single wheel and four-wheel bogies. The XB-36 was experimentally fitted with a tracked rubber endless belt gear. Its purpose was to further reduce unit-bearing load on runways. It flew once. That gives an indication of how successful it was.

A wing trailing edge sweepback of three degrees was part of the B-36's design. The propeller disks were parallel with the trailing edge—possibly to counteract adverse yaw in the event of an engine failure or shutdown. Additionally, the three propellers on a wing are in alignment. Having the propeller disk parallel with the trailing edge would also result in less propeller vibration induced by turbulence swirling off the wing.

From the lessons learned on the XB-36, significant changes were made to the YB-36, which first flew over a year after the XB-36, on December 4, 1947. The cockpit roof of the YB-36 was raised, crew stations were relocated, and a nose turret was installed. So significant were the changes made to the YB-36 that the XB-36 was deemed unsuitable as a service aircraft. Consequently, it had a relatively short life and was scrapped in the early 1950s.

Even when powered by six R-4360s, each developing over 3,500 hp, performance was lacking. In order to rectify this situation, the simple expedient of attaching a pair of General Electric J47s to each outer wing panel went a long way to rectify this shortcoming. Making life even

J47 jet pods are the same as those used on the contemporary B-47. A bracing strut was required with the J47 pods, while the J35 pods did not have this feature. *(Courtesy of National Archives & Records Administration)*

Above: Nice overhead shot of J35s being installed on the B-36. No sway bar was installed on this proof-of-concept application. *(Courtesy of National Archives & Records Administration) Opposite:* Front view of J35s hanging on the B-36 wing. Only one B-36 was fitted with J35s; production aircraft were powered by the similar but more powerful J47. Note that both engines in this shot have a protective covering for the vulnerable compressor blades. *(Courtesy of National Archives & Records Administration)*

simpler for everyone, inner jet pods from the recently introduced Boeing B-47 were used, each one being braced by a sway bar. With the additional four jets, each producing 5,200 pounds of thrust, high gross take-offs and combat performance were considerably improved. However, the jet assist idea used J35s, sans sway bars, on the first aircraft converted. Again, the pods came from the XB-47, which also flew initially powered by J35s.

A project that came tantalizingly close to fruition was the B-36C. This would have been a tractor configuration, but interestingly the engines would have been mounted in the same position as the pushers. Propeller drive would have required an extension shaft with the reduction gearing mounted in the extended nacelle. The R-4360-51 passed its 150-hour type test, so everything was looking good to proceed, until the rug was pulled out from under the B-36C in 1950. Aircraft on Convair's production line being built as B-36Cs were converted to B-36Bs. It's worth noting the vast improvement in altitude capability and fuel economy of the B-36C compared to a B-36B.

Table 7-3 shows the specific fuel consumption (pounds of fuel burned per HP per hour) of a standard B-36B boosted by a pair of General Electric BH-1 turbosuperchargers compared to the specific fuel consumption of the planned B-36C powered by R-4360-51 VDT engines boosted by a General Electric CHM-2 in series with a pair of CH-9 turbosuperchargers *(Ref. 7-51).*

This table illustrates succinctly and graphically the potential improvement in performance the VDT engine offered. What's even more surprising is the altitude capability of the VDT variant. The B-36 has often been criticized as being too slow. True as far it goes. But how many Cold War fighters were even capable of reaching 50,000 feet? No use having a hot rod if it cannot get to the altitude of the enemy.

(Table 7-3)

	POWER		RPM		ALTITUDE		SPECIFIC FUEL CONSUMPTION*	
	B-36B	B-36C (VDT POWER)	B-36B	B-36C (VDT POWER)	B-36B	B-36C (VDT POWER)	B-36B	B-36C (VDT POWER)
Take-off & Military Power	3,800	4,300	2,800	2,800	31,000	45,503	0.840	0.66
Normal Cruise Power	2,200	3,500	2,600	2,600	38,000	50,000	0.730	0.60
Max. Cruise Power	2,800	2,450	2,400	2,400	42,800	53,800	0.448	0.38
Min. Cruise Fuel Consumption	1,285	1,880	1,400	1,700	39,000	54,200	0.425	0.36

* Pounds of fuel burned per HP per hour.

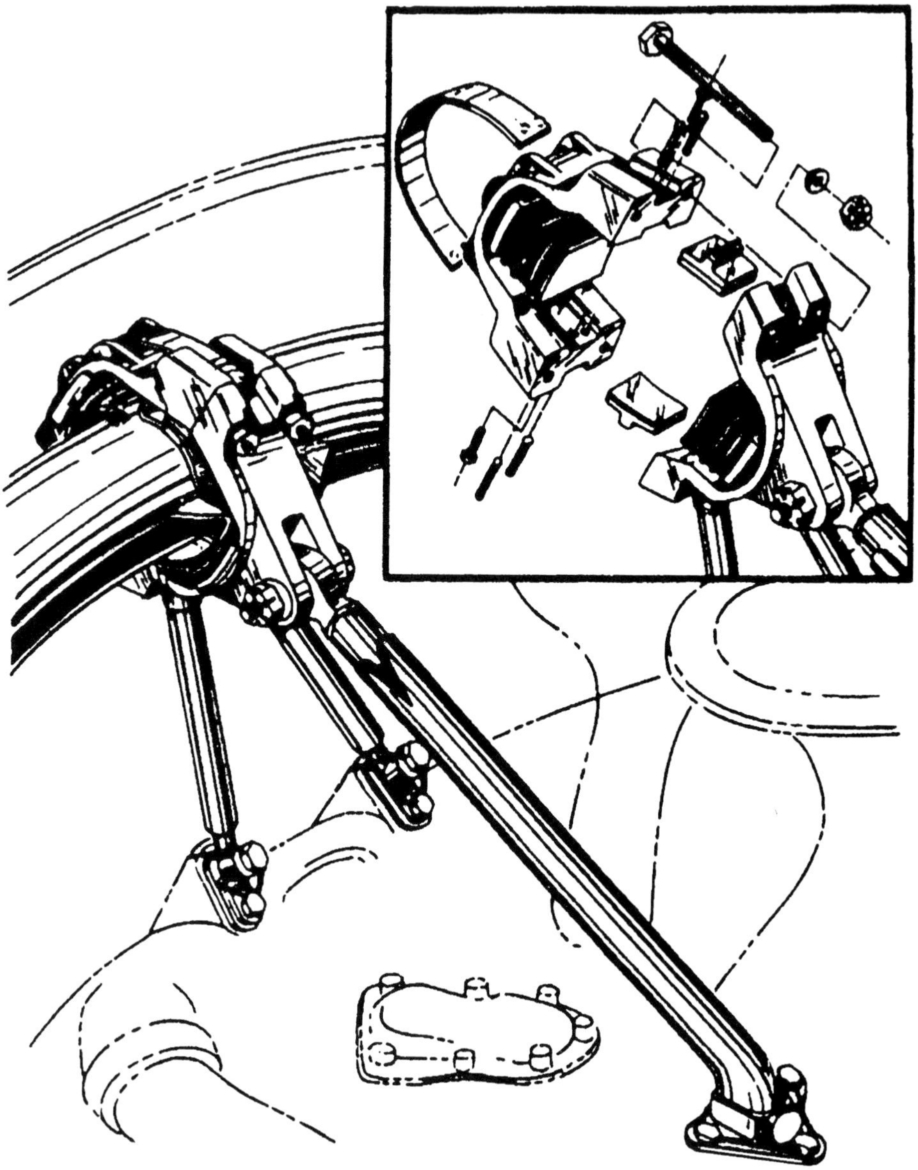

One of the seven engine mounts used to support each R-4360 in the B-36. *(AN 01-5EUE-2 Handbook Erection and Maintenance Instructions USAF Series B-36F Aircraft. March 27, 1951)*

Technical Overview

(Ref. 7-52, 7-53, 7-54, and 7-55)

Engine Mounting

Mounting six huge engines with their attendant and equally heavy accessories such as propellers, turbos, intercoolers, oil coolers, etc., on a wing made from thin-gauge aluminum is testament to an engineering achievement rarely equaled. In a similar fashion to a tractor-configured aircraft, the R-4360s were mounted on a chrome-moly tubular mount—with one big difference. The tractor-configured aircraft typically attaches the mount to a monocoque structure that ties into the wing spar. However, due to the B-36's buried pusher configuration, the chrome-moly tubular engine mount is cantilevered directly off the rear wing spar.

Air Requirements

A huge quantity of air is consumed by each of the B-36's six R-4360s. These requirements are: (i) induction air, (ii) cooling air for the cylinders, (iii) cooling air for the air-to-air intercoolers, and (iv) cooling air for the oil cooler. All of the above mentioned air requirements come from the wing leading edge via a pair of "eyebrow" intakes, which fed upper and lower ducts. Maximizing the

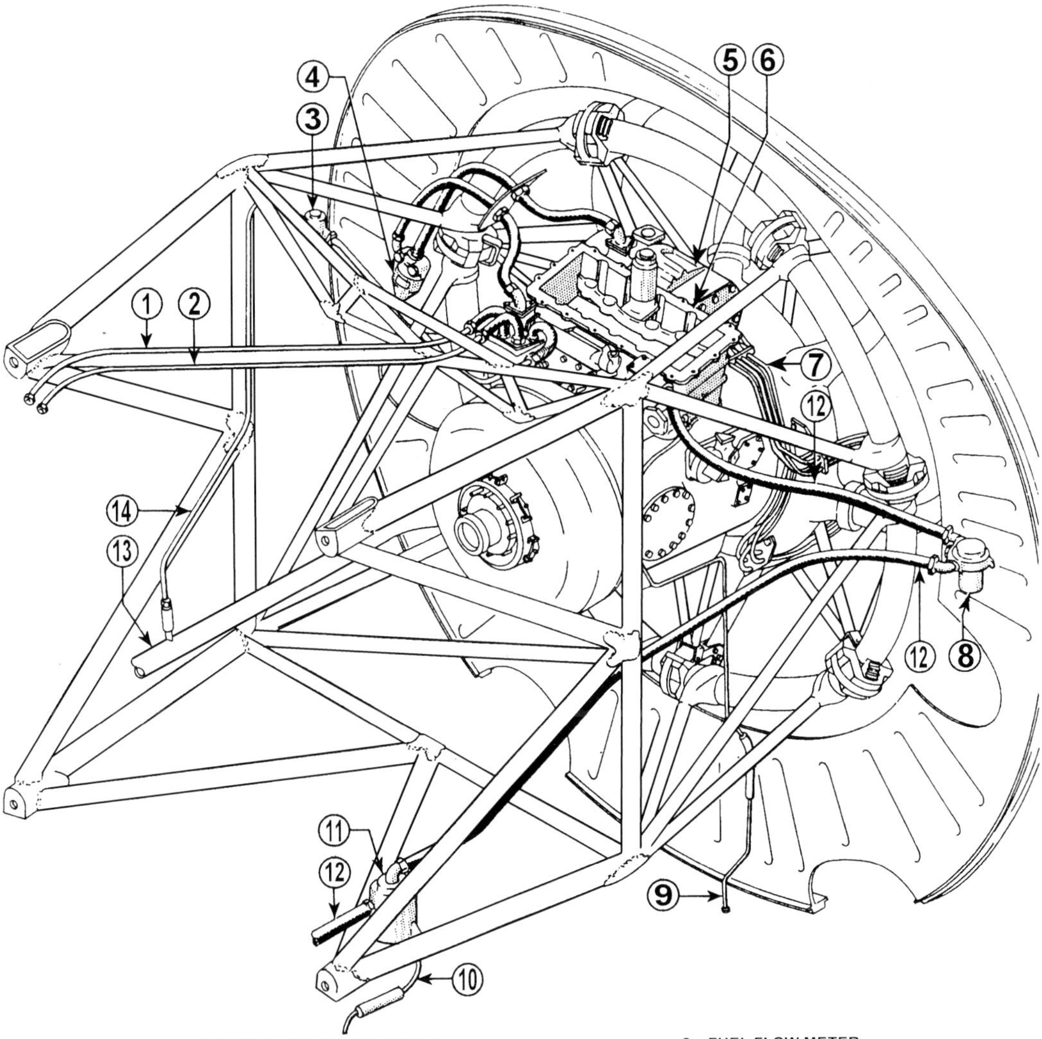

This chrome-moly tubular structure attached the QEC to the rear of the wing spar. *(AN 01-5EUE-2 Handbook Erection and Maintenance Instructions USAF Series B-36F Aircraft. March 27, 1951)*

efficiency of these various ducts was paramount in establishing the performance of the B-36.

Cooling

Even though temperatures at 30,000 feet and greater can dip down to minus 60 degrees or less, it's a little appreciated fact that it is under these conditions that overheating is more likely to occur. This seemingly intuitive concept is caused by the lack of air density compared to sea level conditions. The three primary factors affecting cooling airflow are ambient air temperature, air speed, and perhaps most importantly, air density.

The upper duct is divided into three compartments. The main center duct feeds into the engine-driven cooling fan and two side ducts feed cooling air to the dual rectangular air-to-air intercoolers. Cooling air from the intercoolers is discharged overboard on top of the wing via adjustable louvers to control mass airflow. In order to assist with mass airflow through the cowl, the engine-driven fan has six blades. Early B-36s used a

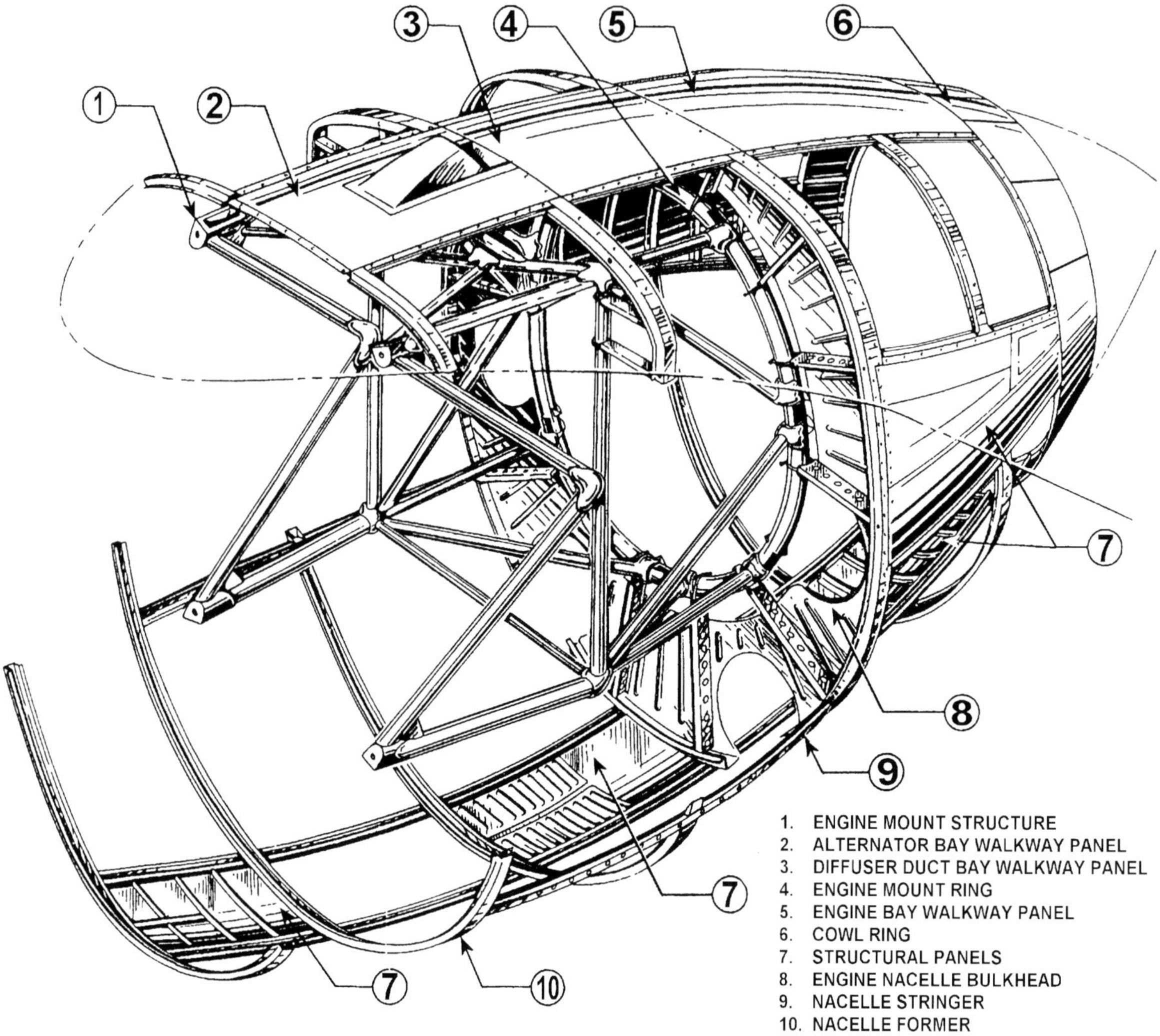

The nacelle design combines chrome-moly tubular space frame and monocoque construction techniques. *(AN 01-5EUE-2 Handbook Erection and Maintenance Instructions USAF Series B-36F Aircraft. March 27, 1951))*

Factors affecting B-36 engine cooling are illustrated well in this cartoon. They include: air density, air speed, ambient air temperature, position of air plug, engine fan speed, mixture, and oil temperature. Multiply this by six and this only takes engine cooling into account. It is easy to see why B-36 flight engineers were very busy. *(Aircraft Performance Engineer's Manual for B-36 Aircraft Engine Operation. January 1952)*

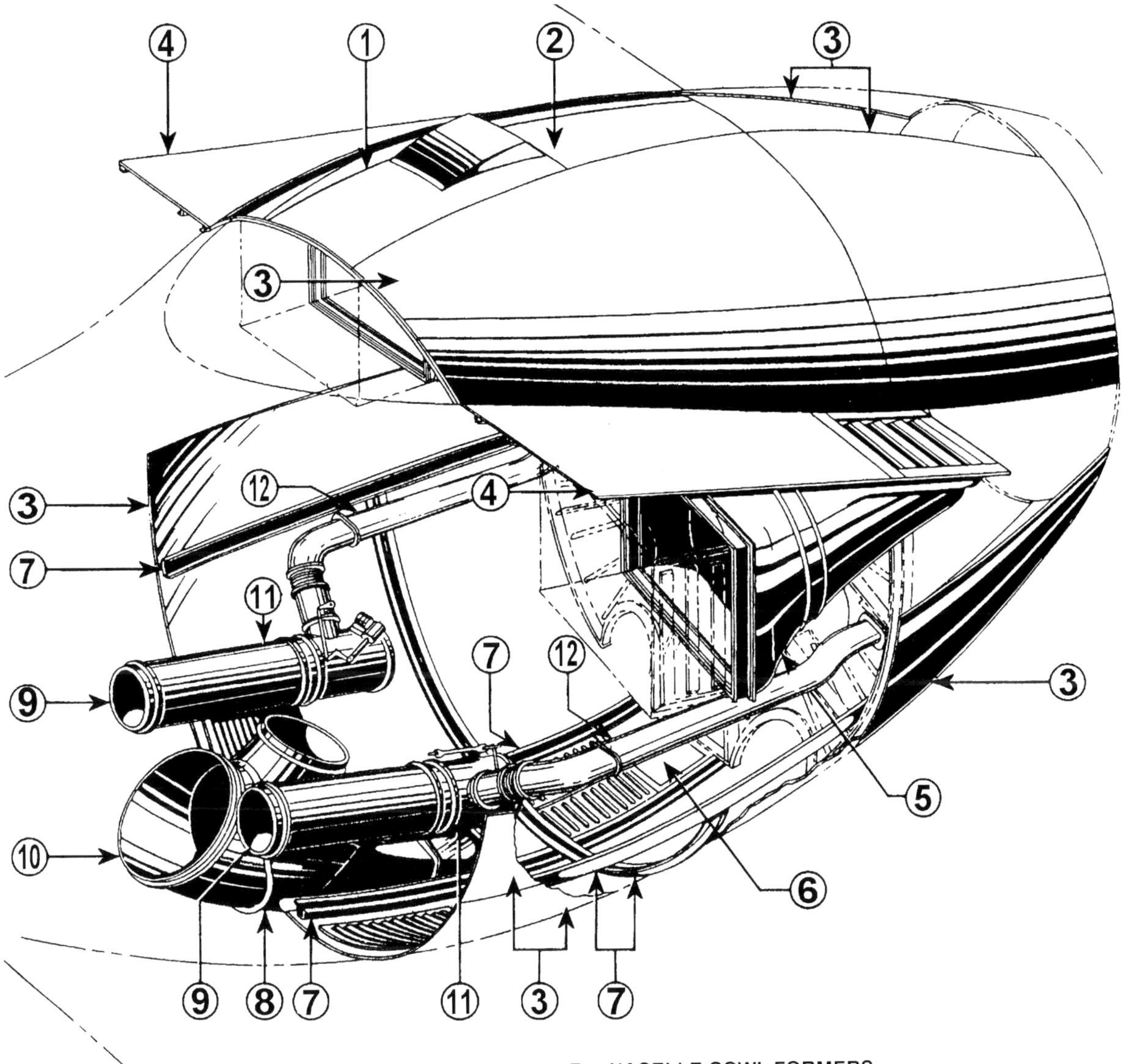

1. ALTERNATOR BAY WALKWAY PANEL
2. DIFFUSER DUCT BAY WALKWAY PANEL
3. ENGINE AND ACCESSORY COWL PANEL
4. INTERCOOLER ACCESS PANELS
5. INTERCOOLER OUTLET DUCTS
6. NACELLE STRUCTURE PANEL
7. NACELLE COWL FORMERS
8. OIL COOLER FLIGHT FLAP ASSEMBLY
9. INTERMEDIATE TURBO AIR INLET DUCTS
10. INTERMEDIATE OIL COOLER EXIT DUCTS
11. AFT TURBO AIR INLET DUCTS
12. CARBURETOR PREHEAT DUCTS

The B-36 nacelle was made up from these major sub-assemblies. *(AN 01-5EUE-2 Handbook Erection and Maintenance Instructions USAF Series B-36F Aircraft. March 27, 1951)*

multi-blade fan with a fixed contra-vane assembly. Driven through hydraulic couplings, two speeds were available—high and low. In the event an engine needed to be shut down in flight, a brake locked the fan position to prevent windmilling.

The engine cooling system performs five basic functions: (i) cooling for the engine, (ii) the exhaust system, (iii) propeller, (iv) turbosuperchargers, and (v) various electrical actuators. Mass airflow is controlled via a so-called "plug" that performs the same function as cowl flaps in a tractor configuration. It controls a movable annular member, or "plug" as it was commonly known as, that opens and closes the gap between the rear cowl and forward propeller spinner. The plug is electrically controlled and installed in each nacelle. Moving the plug forward (toward the wing leading edge) controlled mass airflow over the engine for an increase or aft for a decrease of flow. The units in the nacelle, which position the

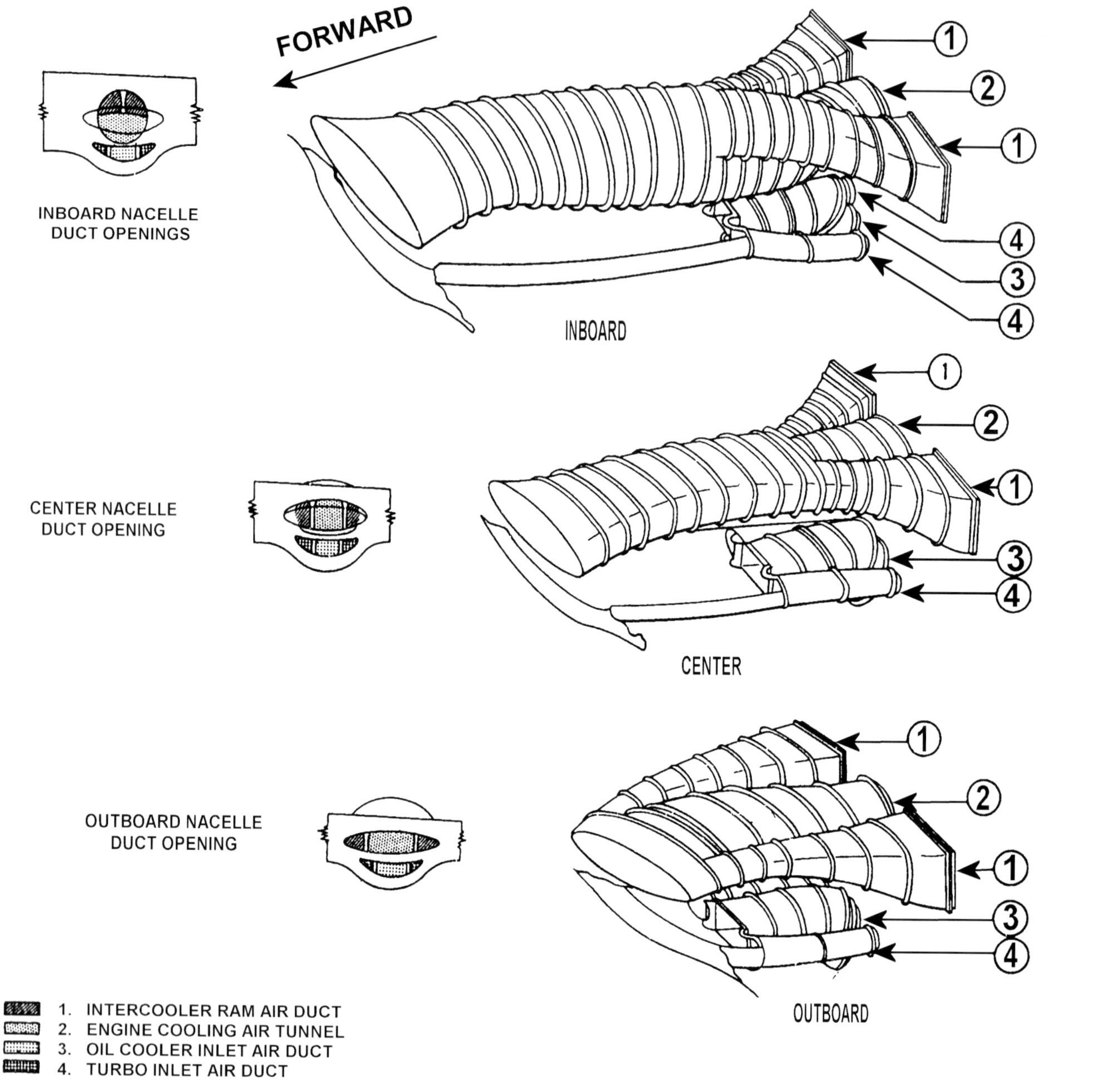

The fact that the R-4360 powering the B-36 was set so far back on the wing, in fact they were cantilevered behind the spar, meant that a complex ducting system was required. Furthermore, these ducts performed critical functions; therefore, it was imperative that they exhibited first-class aerodynamic qualities. *(AN 01-5EUE-2 Handbook Erection and Maintenance Instructions USAF Series B-36F Aircraft. March 27, 1951)*

air plug, are an actuator, five flex shafts, and four screw jacks. A three-way control switch for the operation of each air plug is located on the engineer's control panel. The air plug is a ring-shaped assembly with an airfoil cross section. It's mounted on four screw jacks located radially around the inside of the trailing cowl ring of each nacelle. The principal parts of each screw jack are: (i) gearbox, (ii) shaft, (iii) trunnion, and (iv) a case enclosing the shaft. Each flex shaft fitting in the gearbox and each end of the shaft is ball bearing mounted. The trunnion, attached to the air plug mounting brackets, is internally threaded to mate with the worm-geared shaft. Thus, when the shaft rotates, the trunnion travels fore or aft, moving the air plug in the same direction. Observers

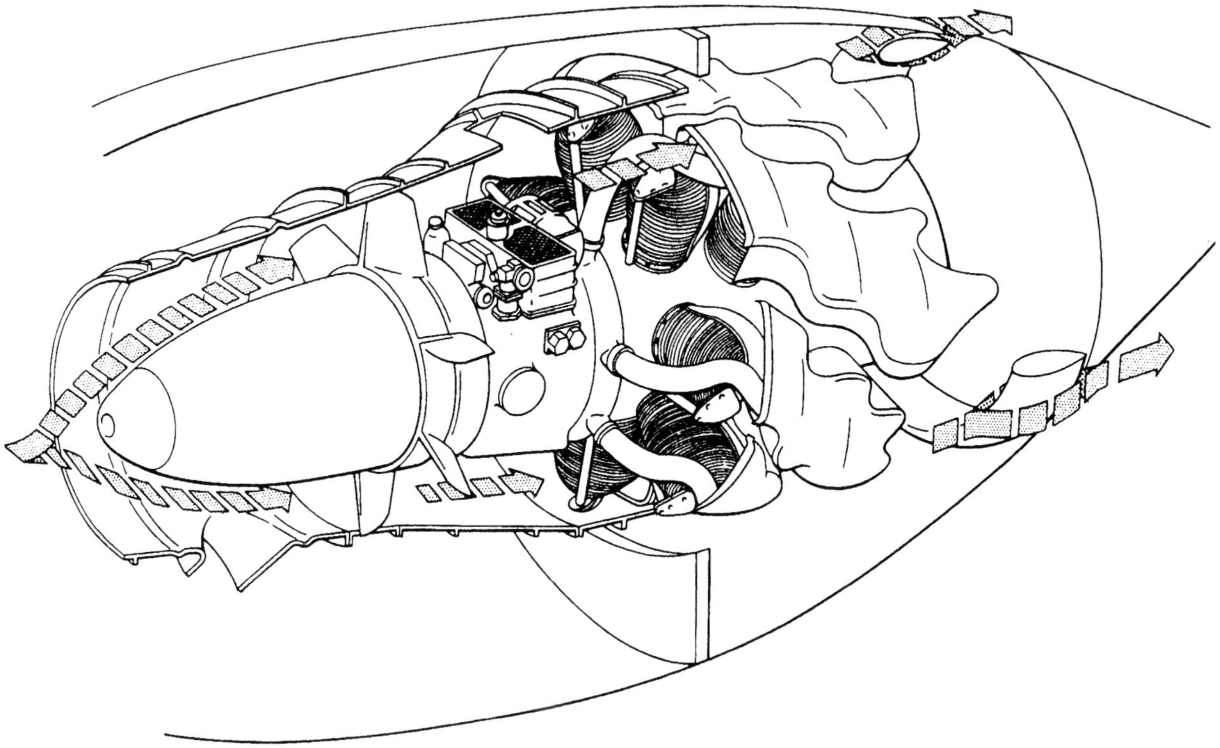

The leading-edge intakes were split into upper and lower ducts. These were then sub-divided for their various functions. The center portion of the upper duct served to supply cooling air to the engine driven fan. Although the B-36 cooled well, the massive PR-100 carburetor or 100-28 master control unit masked the cylinders. It would tend to make those cylinders run hotter than the others. The line drawing shows the later six-blade fan. *(AN 01-5EUE-2 Handbook Erection and Maintenance Instructions USAF Series B-36F Aircraft. March 27, 1951)*

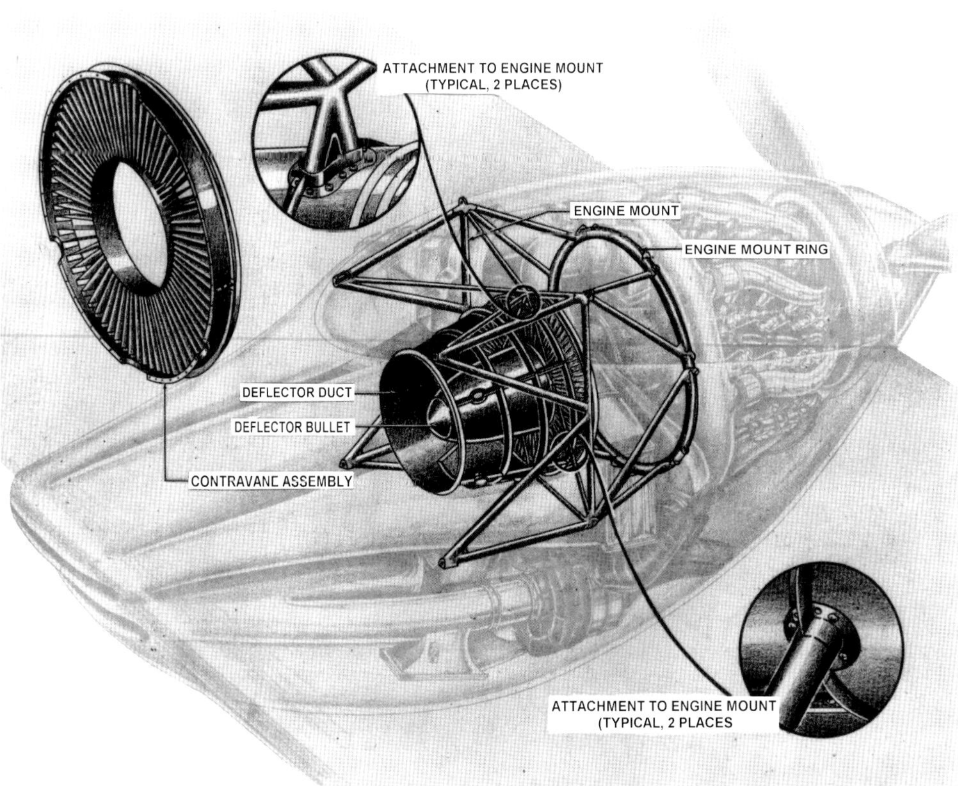

This phantom view shows the early multi-blade cooling fan used in the XB-36. This view also shows the relative positions of the turbos, ducting, and the tubular engine mount. *(Courtesy of Pratt & Whitney)*

Above: This photograph of a number four engine going through a typical smoky start also illustrates the air plug that controls mass airflow through the engine. In this shot, the plug is positioned at its wide-open setting. Also of interest in this photograph are the flap position markings shown as curved lines just ahead of the air plug. Zero-, 10-, 20-, and 30-degree flap settings are shown. *(Courtesy of Pratt & Whitney) Opposite top:* This drawing shows the diamond configuration to assist observers inside the aircraft to determine the position of the air plug. *(Aircraft Performance Engineer's Manual for B-36 Aircraft Engine Operation. January 1952)*

inside the aircraft can determine the position of the air plug via diamonds painted on the air plug.

A diffuser, basically a baffle that surrounds the engine, acts to optimize and direct cooling airflow. Tapped off the diffuser, a pipe directs cooling air to the Curtiss propeller.

The lower duct provides cooling air to the oil cooler and ram induction air to the pair of General Electric turbosuperchargers. Like the upper duct, it is divided into three compartments. The center duct feeds the oil cooler and the pair of side ducts feed the suction side of the turbosupercharger compressors.

Oil System

Each engine has its own oil supply—a whopping 200 gallons per engine. Included in the oil systems are an oil dilution system for cold weather operation, a sophisticated oil cooling system, and a propeller torque indicating system plus all the necessary ancillary equipment needed such as vent systems, hoses, tubing, and controls.

Oil Cooling

Keeping oil temperatures under control during flight is relatively straightforward. Cooling air via the dedicated duct from the leading edge is

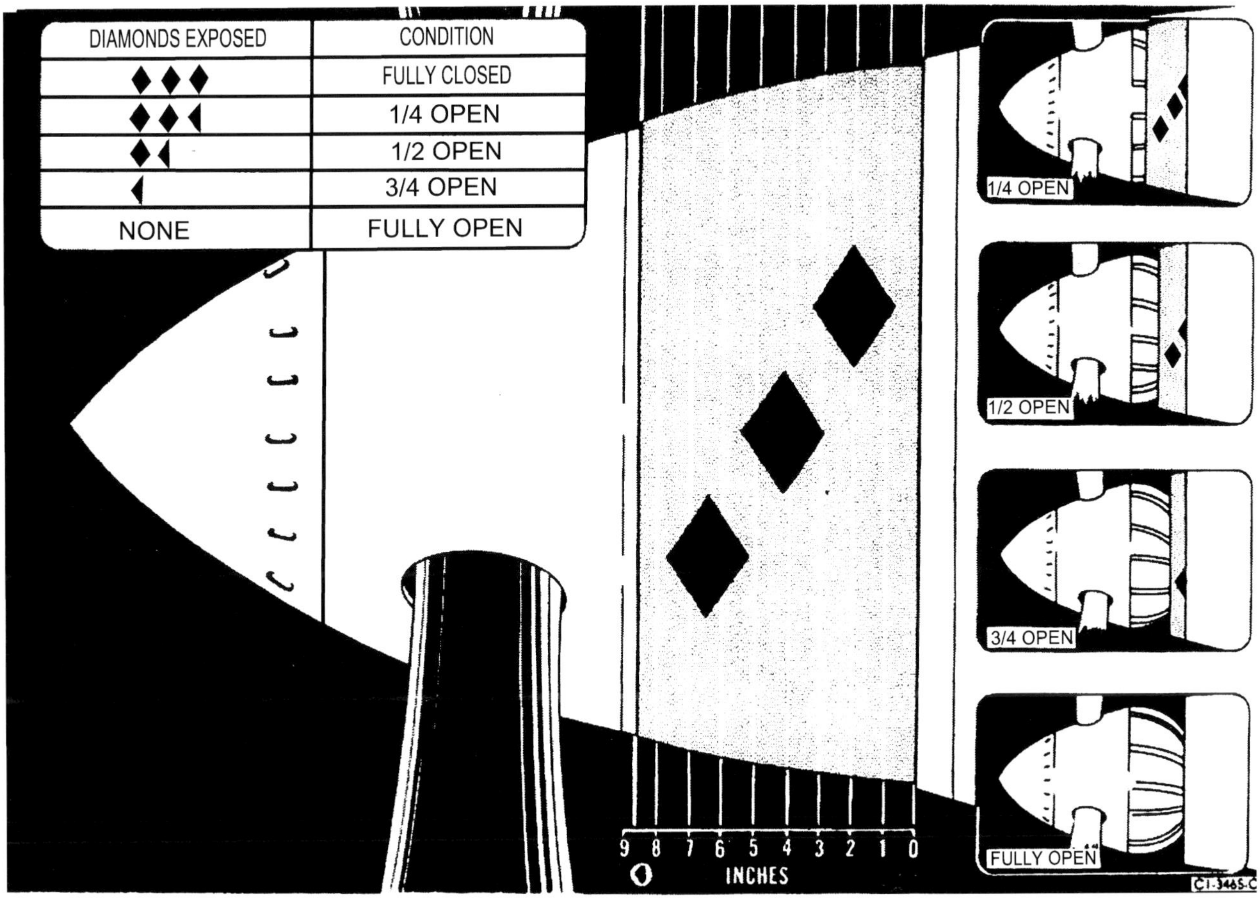

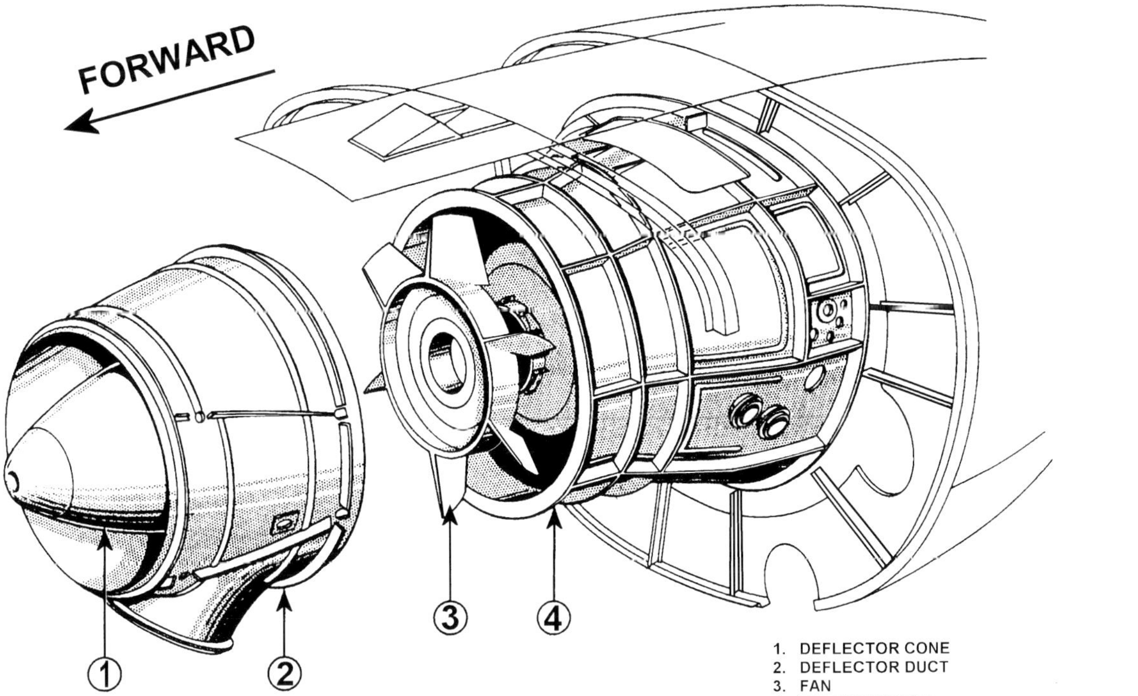

1. DEFLECTOR CONE
2. DEFLECTOR DUCT
3. FAN
4. DIFFUSER DUCT

A carefully designed diffuser duct surrounds the engine. This airtight duct ensures that air pumped by the cooling fan is not wasted. After cooling the engine, air is discharged through the air plug, making a slight contribution to the overall thrust generated by the engine. Therefore power absorbed by the fan is not totally wasted. *(AN 01-5EUE-2 Handbook Erection and Maintenance Instructions USAF Series B-36F Aircraft. March 27, 1951)*

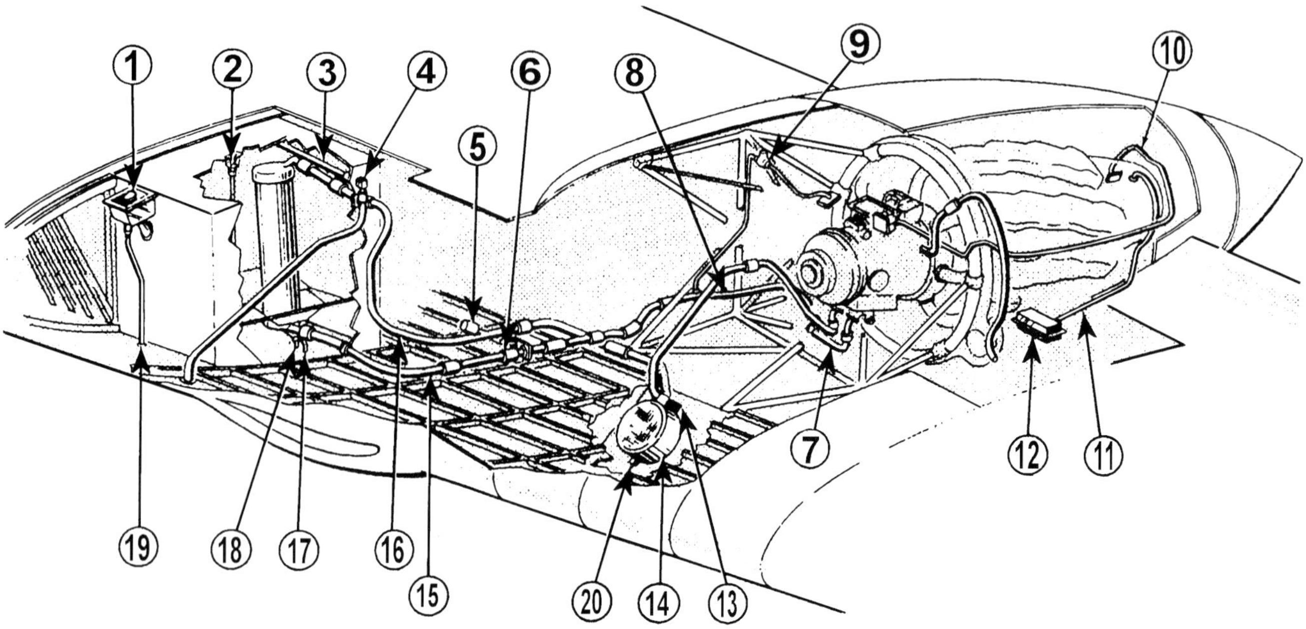

With its long-range missions and the range capability of the B-36, it was essential that an adequate oil supply and oil cooling system be incorporated. With its six 200-gallon tanks, a total of 1,200 gallons of oil was available to keep the R-4360s running. A larger, circular oil cooler was mounted in the lower duct. If the aircraft was deployed to a cold climate such as Alaska, an oil dilution system was an option to the flight engineer for an easy start in frigid conditions. *(AN 01-5EUE-2 Handbook Erection and Maintenance Instructions USAF Series B-36F Aircraft. March 27, 1951)*

directed through the massive circular oil cooler. The heated air is dumped overboard via an adjustable flap under the wing. However, during ground operations, this system would not work—insufficient air would flow over the cooler. This problem was resolved in a clever way by closing off the adjustable flap and opening a connecting duct to the main engine-cooling duct. In this way, cooling is sucked through the cooler due to the influence of the engine-cooling fan.

Turbosupercharger System

For its intended mission of long-range high-altitude bombing, turbosupercharging was the most practical way to accomplish these goals. Boost was provided to the engine-driven supercharger by a pair of General Electric CH-1 turbosuperchargers operating in parallel. Like the Hughes XF-11 and XB-35, both of which employed a similar turbosupercharging system, one turbo could be shut down under cruise conditions. This was accomplished through the simple expedient of a butterfly valve in the exhaust feeding one of the turbosupercharger's turbines. In this way the engine was operated on a single turbosupercharger.

Operating on one turbo not only saved wear and tear on a turbo, but more importantly it allowed the remaining turbo to operate more efficiently. At the critical altitude of single turbo operation, the other turbo would be brought online. Induction air was fed into the turbosupercharger's centrifugal compressor via the outer segments of the lower duct. Air discharged from the pair of turbo compressors, now heated due to compression, was fed through air-to-air inter-

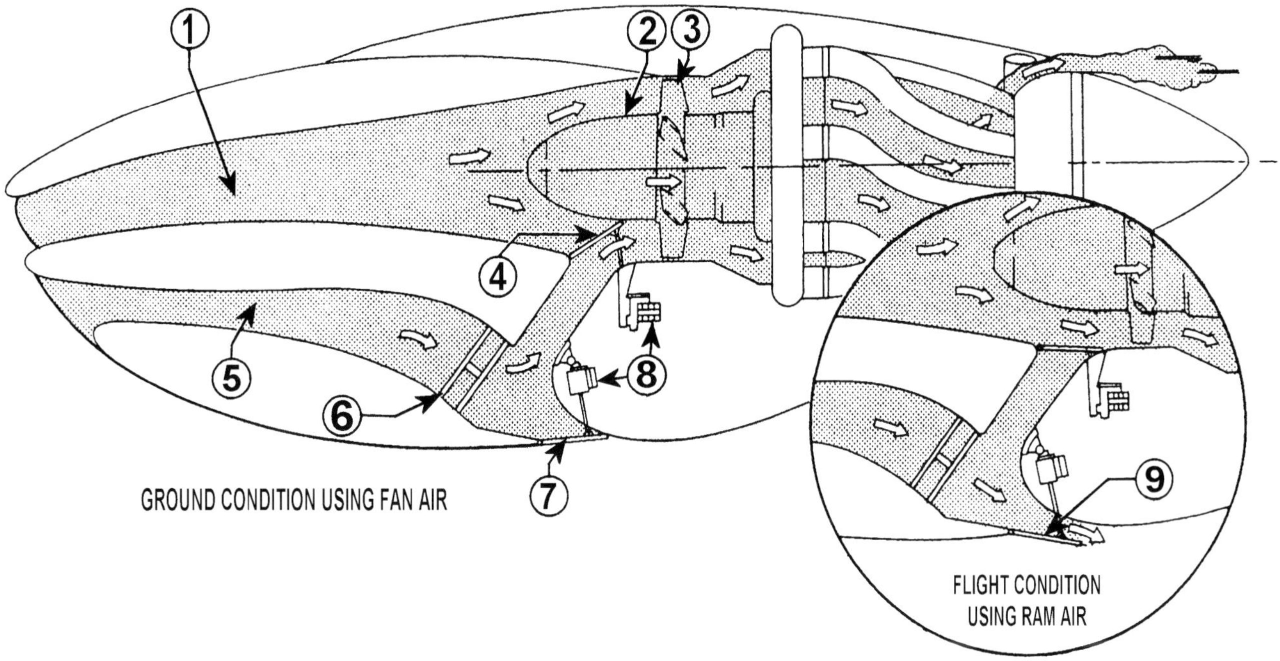

1. ENGINE COOLING AIR TUNNEL
2. DEFLECTOR CONE
3. ENGINE COOLING FAN
4. GROUND COOLING DOOR (OPEN)
5. OIL COOLER INLET DUCT
6. OIL COOLER
7. FLIGHT COOLING DOOR (CLOSED)
8. ACTUATORS
9. FLIGHT COOLING DOOR (OPEN)

In-flight cooling for the B-36's engines was more than adequate. However, ground operations could have posed a problem if it were not for some innovative solutions. To keep oil temperatures in the green during ground operations two doors; a ground-cooling door and a flight-cooling door, were actuated. The ground door was opened and the flight door closed. In this way a forced draft, generated by suction from the engine driven fan, flowed through the oil cooler. *(AN 01-5EUE-2 Handbook Erection and Maintenance Instructions USAF Series B-36F Aircraft. March 27, 1951)*

coolers and then into the PR-100 carburetor, or Bendix fuel injection master control unit if the aircraft was fuel injected—which later B-36s were—via a "Y" duct.

A shroud built around the exhaust collector ring provided heated air for propeller de-icing. Turbosuperchargers are capable of producing far more boost pressure than is healthy for the engine, particularly at low altitudes where the air density is greater. As with all other turbo installations (except the VDTs), a waste gate is provided to bleed off exhaust gasses before they enter the turbine, thus reducing the power output of the turbine and consequently reducing the discharge pressure of the compressor. Exhaust gas temperature is reduced via a heat exchanger, thus protecting the turbine buckets from potentially damaging temperatures.

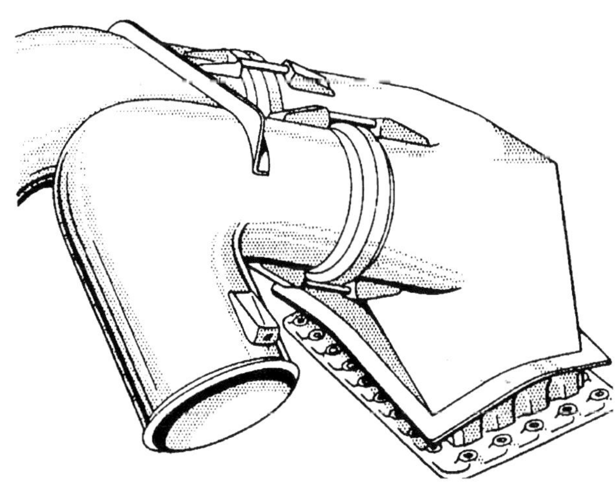

Discharge from the CH-1 compressors fed into a "Y" duct bolted to the PR-100 (carbureted) or 100-28 master control (fuel injected). *(AN 01-5EUE-2 Handbook Erection and Maintenance Instructions USAF Series B-36F Aircraft. March 27, 1951)*

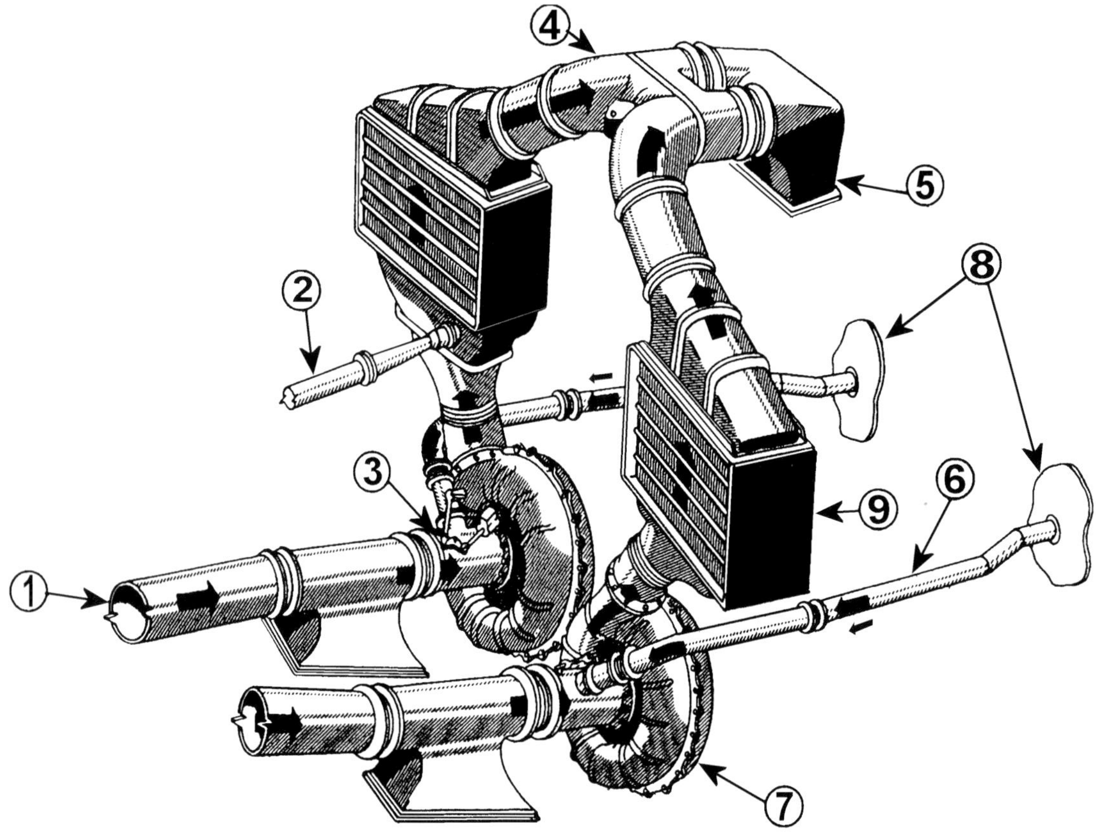

1. TURBO AIR INTAKE
2. CABIN PRESSURAIZATION TAKE-OFF
3. CARBURETOR PREHEAT SHUTOFF VALVE
4. "Y" DUCT
5. CARBURETOR SCOOP
6. CARBURETOR PREHEAT DUCT
7. TURBOSUPERCHARGER IMPELLER SECTION
8. ENGINE BULKHEAD

A pair of GE CH-1 turbosuperchargers running in parallel provided high-altitude capability for the B-36. The two outer sections of the upper leading-edge duct provided induction air. The CH-1s discharged into a pair of air-to-air intercoolers. *(AN 01-5EUE-2 Handbook Erection and Maintenance Instructions USAF Series B-36F Aircraft. March 27, 1951)*

Turbosupercharger Control System

A Minneapolis-Honeywell type B-4 regulator system controlled turbo operation. The components of the control system for each nacelle are an amplifier, a pressuretrol, a pressure limiter, a turbo governor, and a waste gate servomotor. Manual control is accomplished by the turbo boost selector installation at the pilot's and the engineer's stations. Electrical signals are received by the amplifier from the turbo boost selector, the pressuretrol, and the governor. These signals are then interpreted, amplified, and relayed by the amplifier to the waste gate servomotor, which positions the two waste gates in the nacelle accordingly. Positioning of the waste gates controls the speed of the turbos and consequently regulates carburetor deck pressure. If the electronic controls fail, an emergency control system prevents excessive carburetor deck pressure by mechanically opening the waste gates. The pressure limiter accomplishes emergency control, which is integral with the waste gate motor as a unit.

Turbo-boost Selector

The turbo-boost selector provides a means for selecting turbo boost simultaneously for all engines of the aircraft. This selector is the manual control unit of the B-4 regulator system and was used to select the carburetor deck pressure necessary to produce the desired manifold pressures for all flight conditions. All other control units of the regulator system, except those that

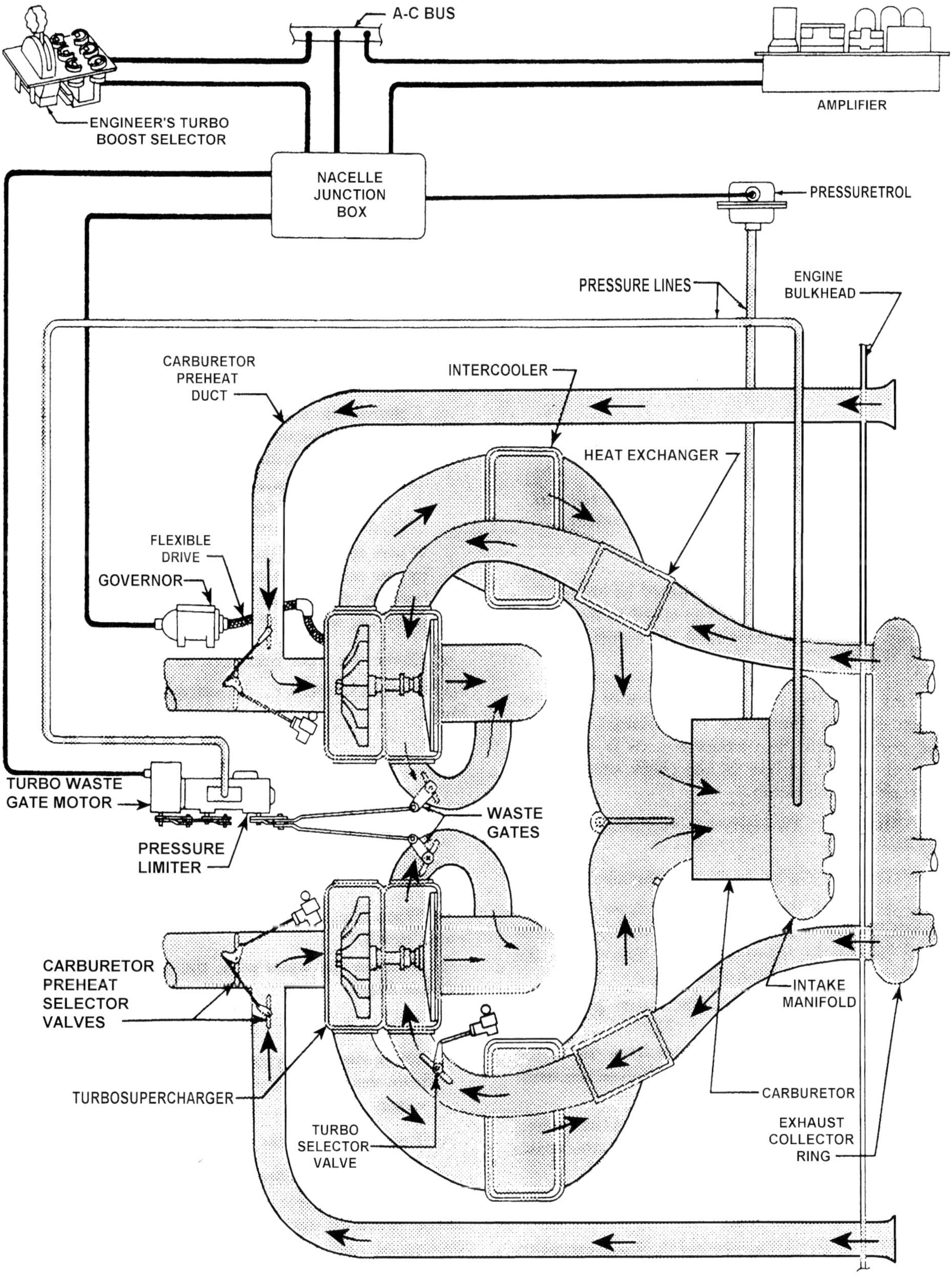

This diagram illustrates the circuitry for exhaust gasses and induction air. It also shows the flight engineer's turbo boost controls. The turbo selector valve also selects single-turbo operation by cutting off the exhaust to one turbo, the lower one in this diagram. *(AN 01-5EUE-2 Handbook Erection and Maintenance Instructions USAF Series B-36F Aircraft. March 27, 1951)*

act as protective controls, operated to hold pressure at the level selected by the turbo-boost selector. The engineer's turbo-boost selector was designed to operate six engines simultaneously through the use of a single lever. Six calibrated potentiometers were installed to the right of the turbo-boost selector on the flight engineer's table. Individual adjustments of signals were therefore possible to the individual waste gates, thus allowing for slight variations in turbo output. As a back up, a remote turbo-boost selector was installed on the pilot's pedestal.

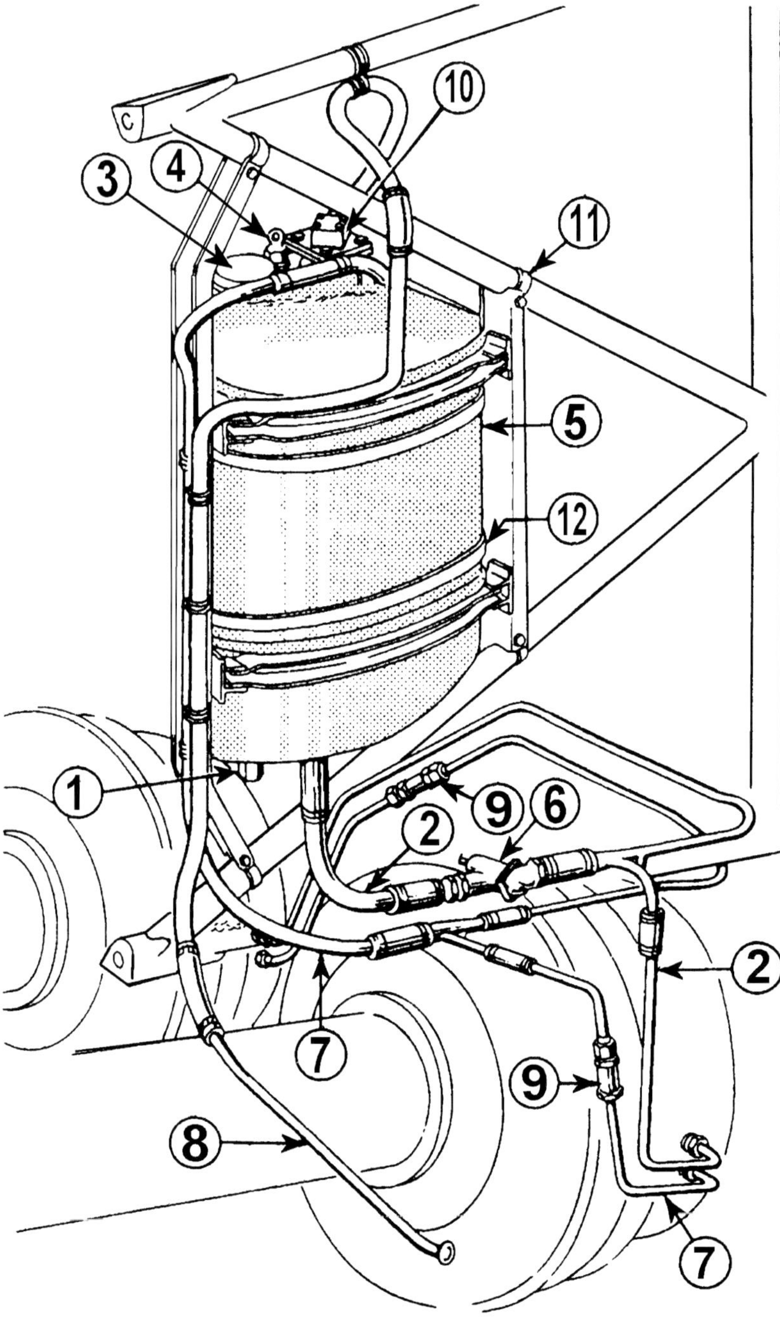

1. DRAIN VALVE
2. OIL SUPPLY LINE
3. FILLER CAP
4. SOUNDING ROD
5. RESERVOIR
6. ANTI-LEAK VALVE
7. OIL RETURN LINE
8. VENT LINE
9. CHECK VALVE
10. BARAMETRIC VALVE
11. BRACKET CLAMPS
12. RUB PATCH STRAPS

Wisely, a separate dry sump lubrication system was utilized for the B-36's turbosuperchargers. Some applications used engine oil to lubricate the turbo(s). Although this scheme simplified the installation, it also exposed the engine to contamination in the event of a turbo failure. And, of course, the reverse was true; if the engine failed, then the turbo was exposed to contamination. Having two separate systems eliminated this scenario. *(AN 01-5EUE-2 Handbook Erection and Maintenance Instructions USAF Series B-36F Aircraft. March 27, 1951)*

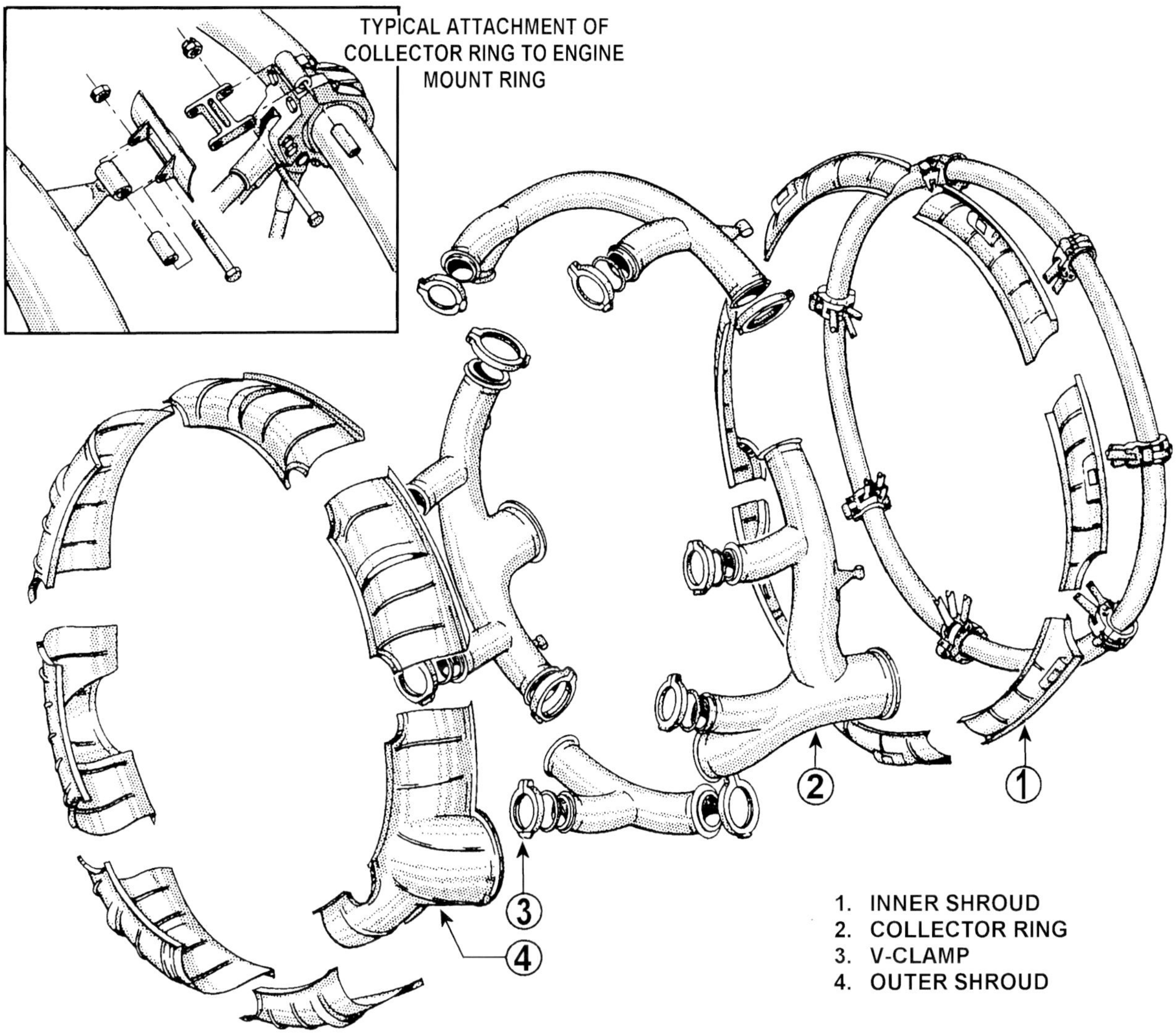

A conventional manifolded design with a collector ring made up the B-36 exhaust system. Two discharges fed the CH-1 turbosuperchargers. *(AN 01-5EUE-2 Handbook Erection and Maintenance Instructions USAF Series B-36F Aircraft. March 27, 1951)*

Turbo Lubrication

An 11-quart dry sump system, independent from engine oil, lubricated the dual turbosuperchargers. It was located between the dual turbosuperchargers. A pressure pump squirted oil on the rolling element bearings and a scavenge pump returned the oil to the tank. No cooler was employed.

Engine and its Exhaust

Powering the turbosupercharger turbine section is engine exhaust gases—its lifeblood. Each bank of four cylinders was manifolded into a single pipe. Seven of these manifolded pipes joined up with a collector ring. Two outlets in the collector ring, one for each turbo, fed the turbine section of the turbosupercharger. One of these outlets had a turbo selector valve for the purpose of shutting down a turbosupercharger during cruise conditions. Exhaust gasses were dumped overboard via two pipes, one for each turbo, under the nacelle.

ADI System

ADI (anti-detonation injection) has traditionally been an excellent method to improve engine

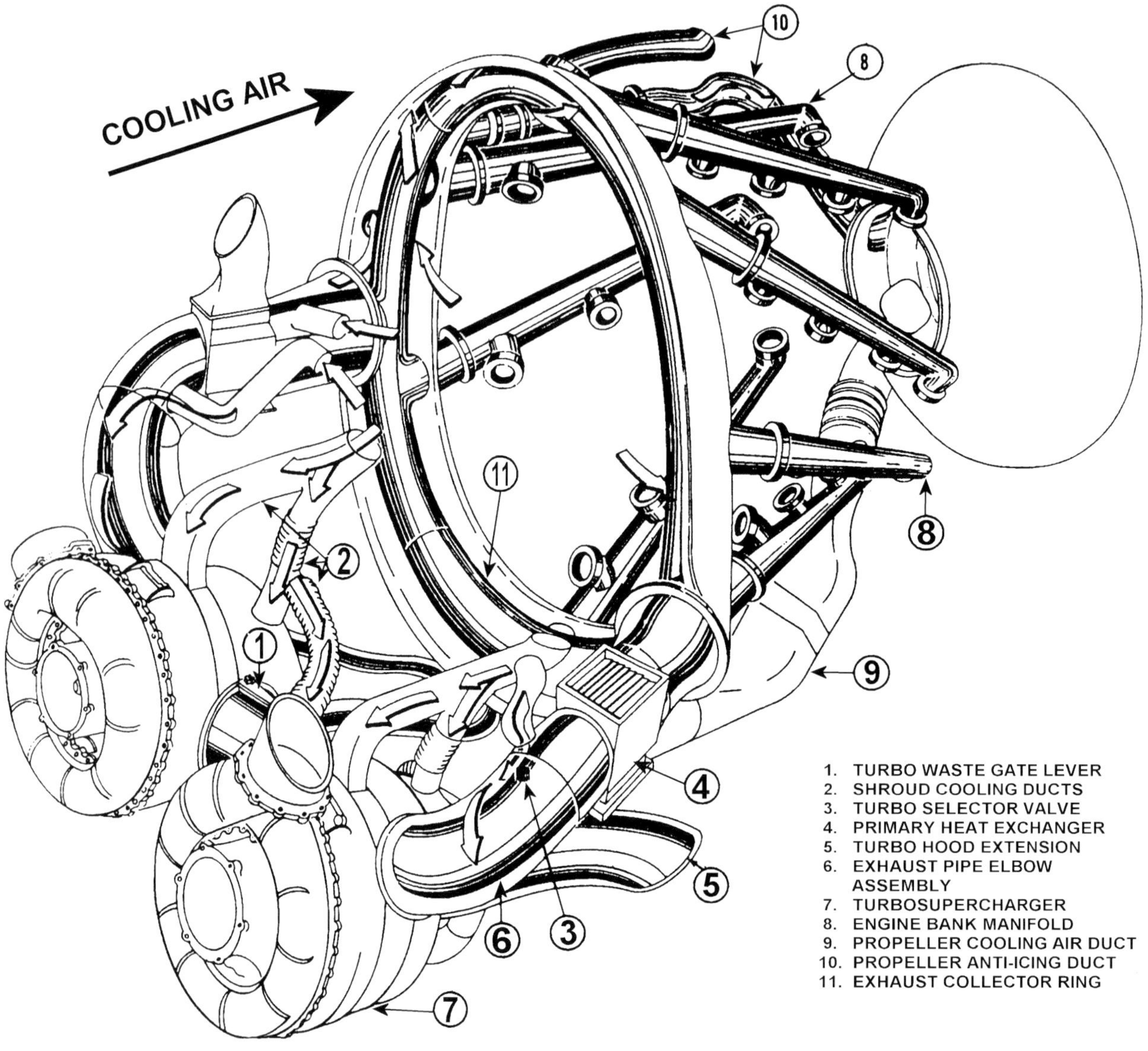

This is the entire "hot" section of the B-36 turbosupercharging system. The heat exchangers are of special interest. They perform several functions: They reduce the temperature of the exhaust gasses in order to protect the turbine buckets from excessive heat and they provide heat for de-icing. *(AN 01-5EUE-2 Handbook Erection and Maintenance Instructions USAF Series B-36F Aircraft. March 27, 1951)*

horsepower, particularly at lower altitudes. In other words, it's ideal for takeoff where high power at low altitude is required. For a typical B-36 powered by an R-4360-53, manifold pressure could be raised to 64.5 in. Hg. for takeoff power with the use of ADI. A non-hesitating ADI injection system incorporating constant ADI flow is employed on each engine to provide additional power during high performance takeoffs. Each of the six systems (one for each R-4360) consists of the following: an ADI tank equipped with submerged ADI pump, a pressure regulating valve, a pressure switch, an ADI regulator, a derichment valve attached to the carburetor, an ADI pressure gauge, an individual control switch that can be operated by a gang switch, and the interconnecting tubing and wiring. The system is operated at manifold pressures above 44 in. Hg. Placing the control switch on the flight engineer's control panel in the "on" position activates it. This energizes a circuit through the manifold pressure switch and through the pump relay and ADI is pumped through the ADI regulator into the engine. ADI pressure in the regulator actuates a pressure switch, which completes a circuit from the manifold pressure switch to energize the derichment solenoid valve on the carburetor. Thus ADI is injected into the supercharger simultaneously with the derichment solenoid valve operating which produces the desired fuel/air ratio to meet additional take-off power requirements. In this case it is somewhere in the neighborhood of 400 additional horsepower. Sufficient ADI fluid is stored in the nine-gallon tank for approximately five minutes operation.

At manifold pressures below approximately 50 in. Hg., the carburetor fuel-head power enrichment valve is closed, thereby preventing fuel flow through the derichment valve. Consequently, fuel flow cannot be deriched, and the ADI injection system has very little effect on engine operation. As power is increased above 50 in. Hg. with the ADI system on, proper operation is indicated by the following conditions:

I. The ADI pressure gauge reading increases to 26 to 28 psi.
II. The fuel flow decreases.
III. Torquemeter pressure increases.
IV. Cylinder head temperature does not increase as power is advanced to 64.5 in. Hg. of manifold pressure.

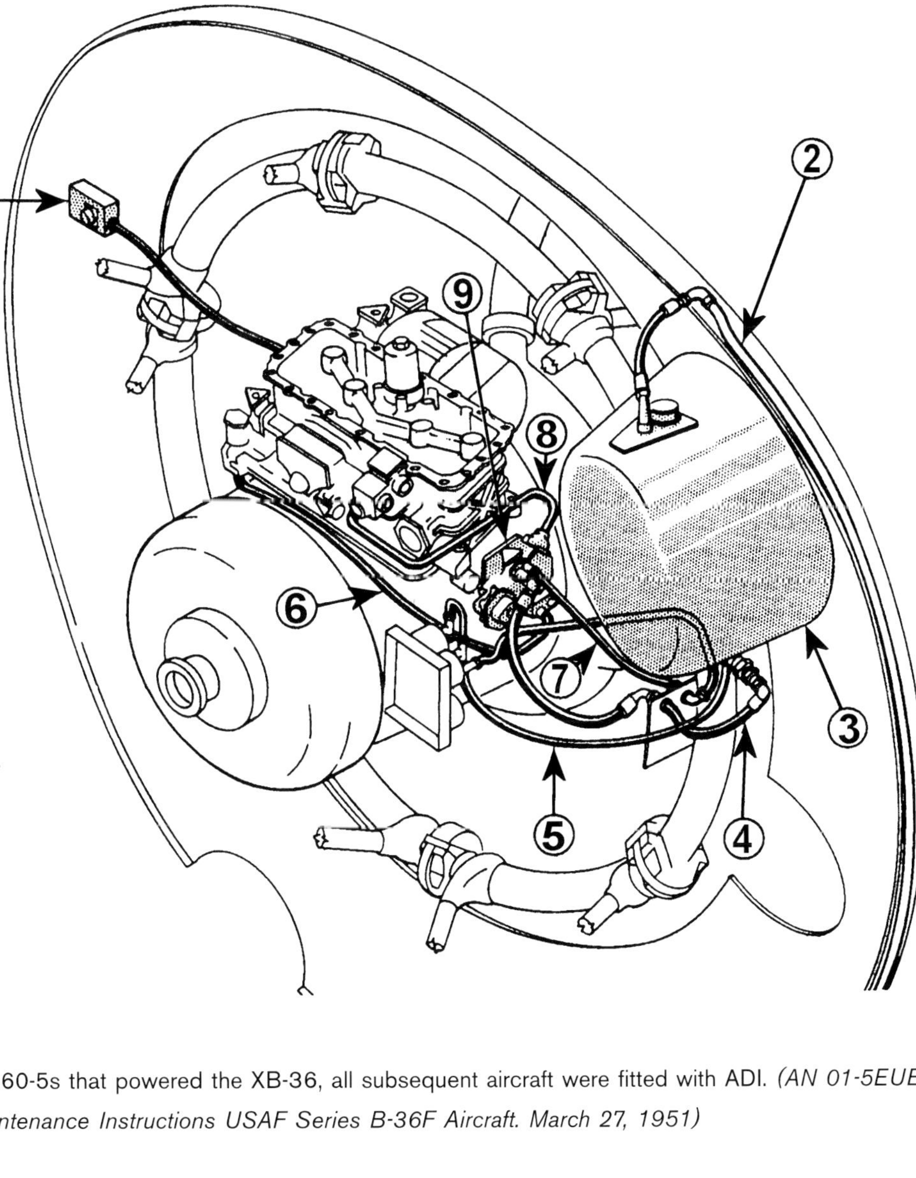

1. MANIFOLD PRESSURE SWITCH
2. WATER TANK VENT LINE
3. WATER TANK
4. WATER SUPPLY LINE
5. PUMP BALANCE LINE
6. ELECTRICAL LEAD TO DERICHMENT SOLENOID VALVE
7. WATER PRESSURE TRANSMITTER LINE
8. WATER TRANSFER LINE
9. WATER REGULATOR

With the exception of the R-4360-5s that powered the XB-36, all subsequent aircraft were fitted with ADI. *(AN 01-5EUE-2 Handbook Erection and Maintenance Instructions USAF Series B-36F Aircraft. March 27, 1951)*

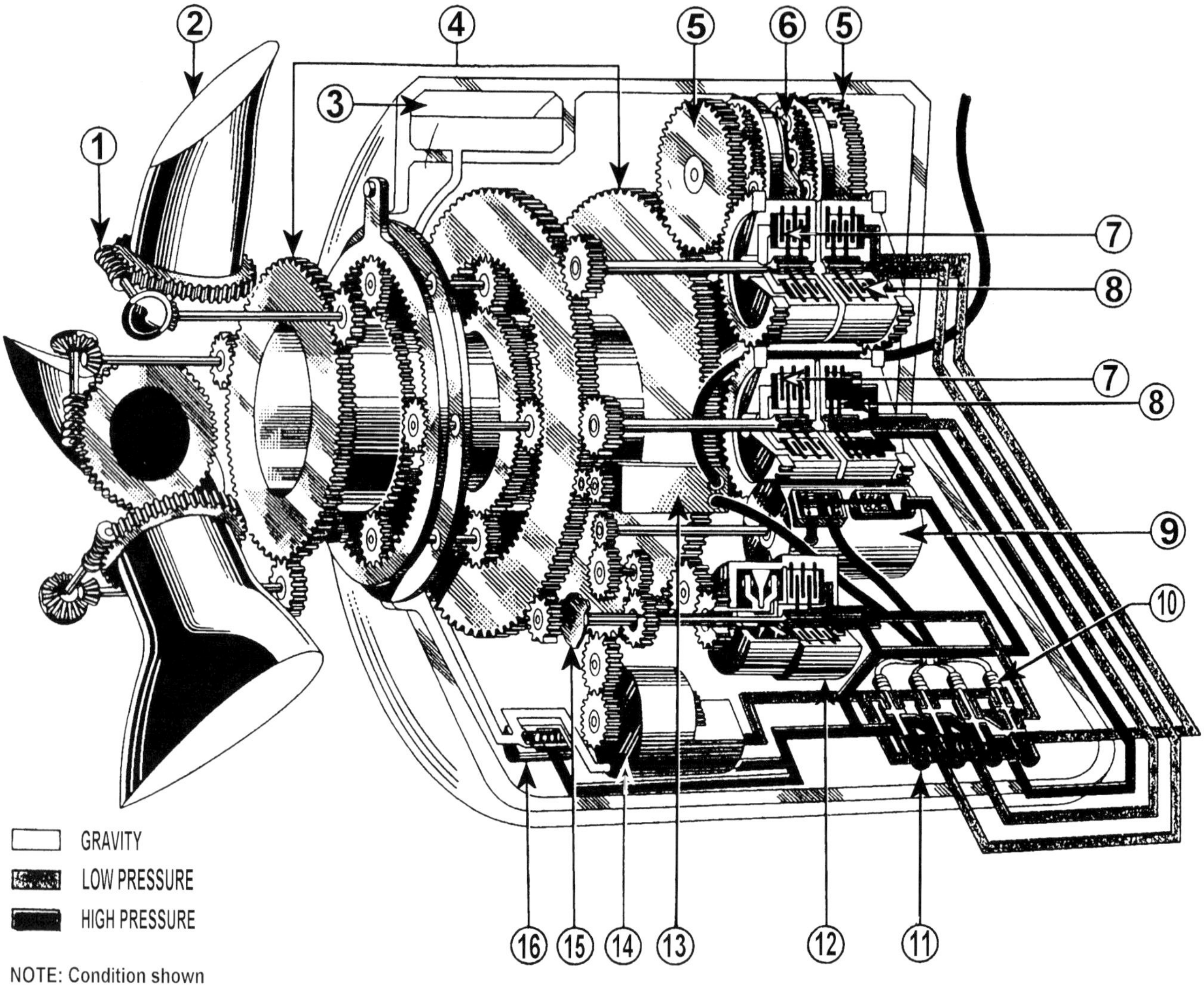

1. BLADE ROTATION GEARING
2. PROPELLER BLADE
3. RESERVOIR
4. INTERGEARING
5. CLUTCH DRIVE GEAR
6. REDUCTION GEARING
7. HIGH SPEED CLUTCHES
8. LOW SPEED CLUTCHES
9. FEATHER MOTOR
10. SOLENOIDS
11. SELECTOR VALVES
12. BRAKE
13. LIMIT SWITCH ASSEMBLY
14. PRESSURE AND SCAVENGING PUMP
15. PIN CLUTCH
16. RESERVOIR SHUT-OFF VALVE

The six huge Curtiss Electric propellers used the R-4360's power to change pitch via a series of clutches and a gear train. *(AN 01-5EUE-2 Handbook Erection and Maintenance Instructions USAF Series B-36F Aircraft. March 27, 1951)*

V. Water pressure does not drop below 24 psi as power is increased to 64.5 in. Hg. of manifold pressure.

VI. Engine operates smoothly without detonation or backfire.

Propeller

It's not often realized how much power is required to change the pitch of a propeller, particularly ones as large as those fitted to the B-36. A major contributing factor to this enormous power requirement is the phenomenon known as centrifugal turning force. In other words, a powerful turning moment tries to force the propeller into fine pitch. To overcome centrifugal turning moment (CTM), a powerful hydraulic ram, as used in most Hamilton Standard propellers, is used. Or a powerful electric motor with compounded reduction gearing, as used in Curtiss Electric propellers, is used. But even these methods would be

Heat collected from the engine's exhaust system furnishes hot air for propeller anti-icing. *(AN 01-5EUE-2 Handbook Erection and Maintenance Instructions USAF Series B-36F Aircraft. March 27, 1951)*

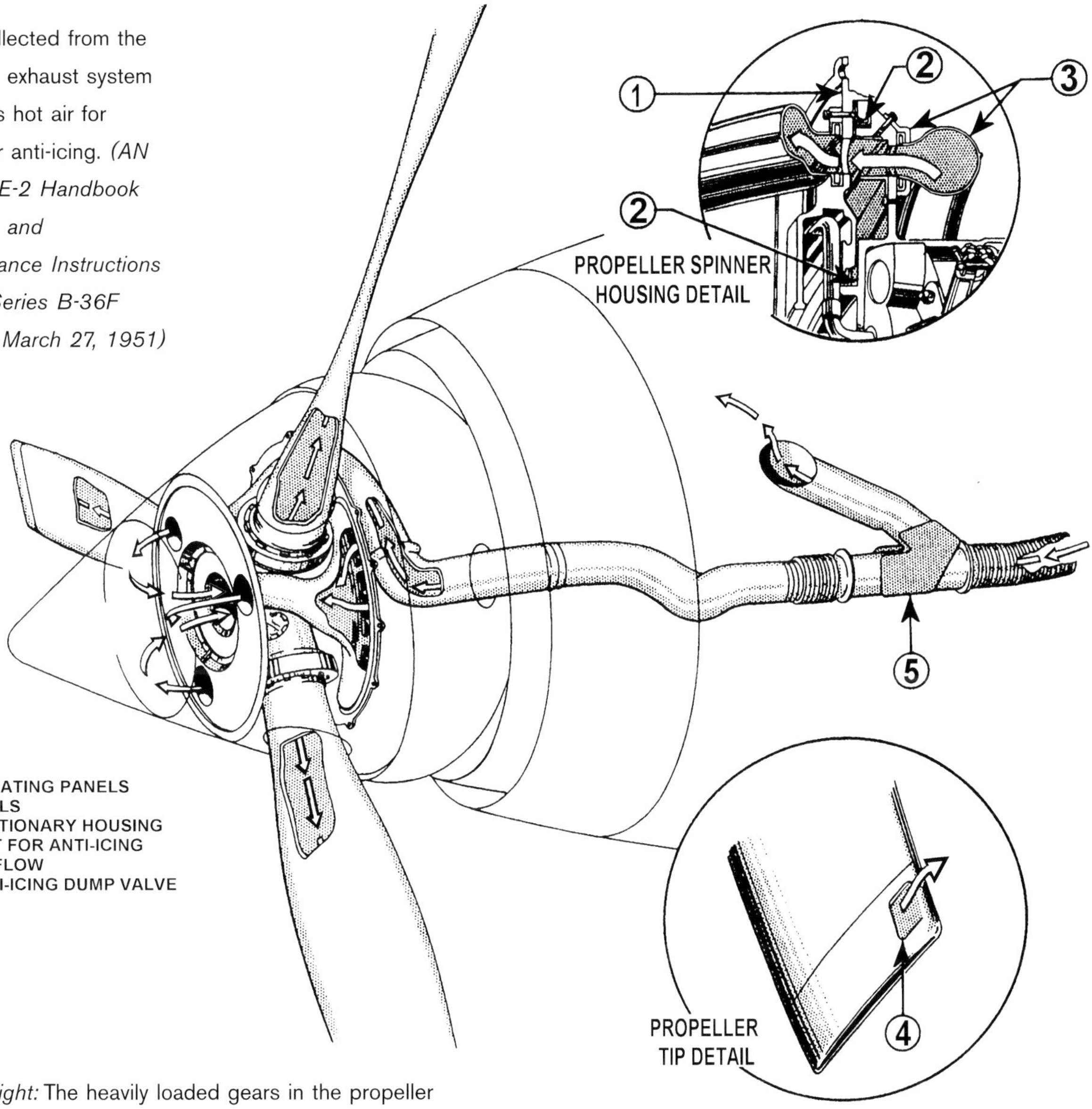

Below right: The heavily loaded gears in the propeller could generate heat. To keep things cool, a bleed-air pipe, connected to the engine diffuser duct, supplied cooling air to the propeller. *(AN 01-5EUE-2 Handbook Erection and Maintenance Instructions USAF Series B-36F Aircraft. March 27, 1951)*

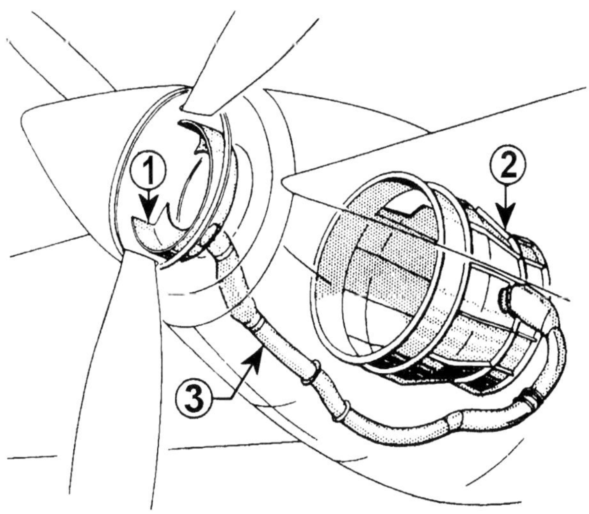

impractical for the B-36. One of its requirements was that of fast pitch change. CTM is a natural tendency for propellers to go into fine pitch. This force can be considerable, especially when dealing with the B-36's massive propeller driven by upwards of 3,800 hp. With so much horsepower on tap, it made sense to utilize engine power to control the substantial 19-foot diameter, three-blade propeller. And that's what the B-36's unique propeller system used. Manufactured by

Curtiss-Wright, they had constant speed, full feathering, and reversing capability.

The propellers are Curtiss-Wright Model C736SP-A1, constant-speed, full-feathering, reversible, three-blade. Pitch changes are accomplished through a series of hydraulic clutches and reduction gearing driven by the propeller shaft, such as engine power. Pitch change rate is approximately 2½ degrees per second during normal operation, or when feathering, reversing, or returning from reverse pitch a faster rate of 45 degrees per second is available. Both pitch change rates are based on a propeller speed of 900 rpm or an engine speed of 2,400 rpm. Automatic-constant-speed operation and manual operation are also possible in reverse pitch through the automatic reversal of normal pitch change control leads in the reverse pitch range. Cooling of the propeller pitch changing gearing is accomplished by air routed from the diffuser duct to the gear housing.

Propeller anti-icing is accomplished by passing hot air from the engine exhaust collector ring shrouding to the propeller hot air distributor ring. It is then directed into the propeller hub, through the propeller blades, and centrifuged through exit ports in the blade tips. The system is controlled by means of a hot air bypass valve operated by an actuator. When the propeller anti-icing system is operating, the valve directs hot air to the propeller hot air distributor ring. When the system is not in operation, the valve passes hot air overboard through an exit duct in the engine cowling.

Five major assemblies make up the propeller: (i) hub, (ii) attachment assembly, (iii) blades, (iv) pitch change mechanism, and (v) miscellaneous parts, including the thermal anti-icing components.

(i) The forward end of the hub is extended and is used to mount the pitch change mechanism.

(ii) The propeller attachment assembly consists of front and rear cones that mount on an SAE #60 (#70 for aircraft powered by the R-4360-53) propeller shaft. A bronze shaft-nut is equipped with internal splines in order to mate with the shaft-nut lock, which also serves as a heated air conductor and a shield for the thermal anti-icing system.

(iii) The type 1129 blades give the propeller a diameter of 19 feet. Blade assembly components consist of the blade nut, blade nut seals, and a split bearing stack to take the enormous centrifugal loads.

(iv) The pitch-changing mechanism is composed of mechanical, hydraulic, and electrical components. The mechanical components consist of (a) the drive gear assembly, which transmits rotation of the propeller shaft to the pitch change clutches; (b) the intergearing assembly, which conveys the rotation of the clutch output gear to the worm drive system; and (c) the hub worm drive system, which changes the angle of the three blades simultaneously. Hydraulic portions of the mechanism consist of a gear-type high-pressure pump, a scavenge pump, a low-pressure regulator valve, solenoid-operated selector valves, hydraulically actuated clutches, a hydraulically actuated brake, a sump, a reservoir, and a drain valve. Electrical components of the pitch-change mechanism consist of the (electrical) connector stack, the limit switch assembly, the selector valve solenoids, the feather motor, and the feather motor switch. The feather motor is used to complete the feathering cycle, to start the unfeathering cycle, and to supply the power for pitch change when the propeller is static.

(v) Miscellaneous equipment includes the hot air distributor of the power unit, ducts for the hub assembly, spinner mounting flange, shaft-nut locking sleeve, and spinner assembly.

To describe operation of the propeller during cruise it's best to describe it by assuming an operating condition. For an over-speed condition, the following steps explain the sequence of automatic corrective action.

I. A decrease RPM impulse is transmitted by the contactor.

II. The impulse travels to the propeller connector stack and through the high blade angle limit switch in the limit switch assembly to the decrease RPM selector valve solenoid. The individual cutout limit switches are connected in each operating electrical circuit to prevent blade angle change beyond their predetermined limit.

III. The energized solenoid diverts the high-pressure oil from the brake to the normal rate decrease RPM clutch (increase pitch).

When no pitch change is required, the high-pressure oil passes through the selector valves to the brake and prevents rotation of the movable ring gear. If the oil pressure should fail, the oil would no longer supply the braking force, nor would it continue to hold the flyball brake inoperative. The centrifugal force in the flyball brake would then force the brake plates together, replacing the force normally supplied by the high oil pressure. The flyball brake holds the propeller in the fixed-pitch position until propeller speed falls below

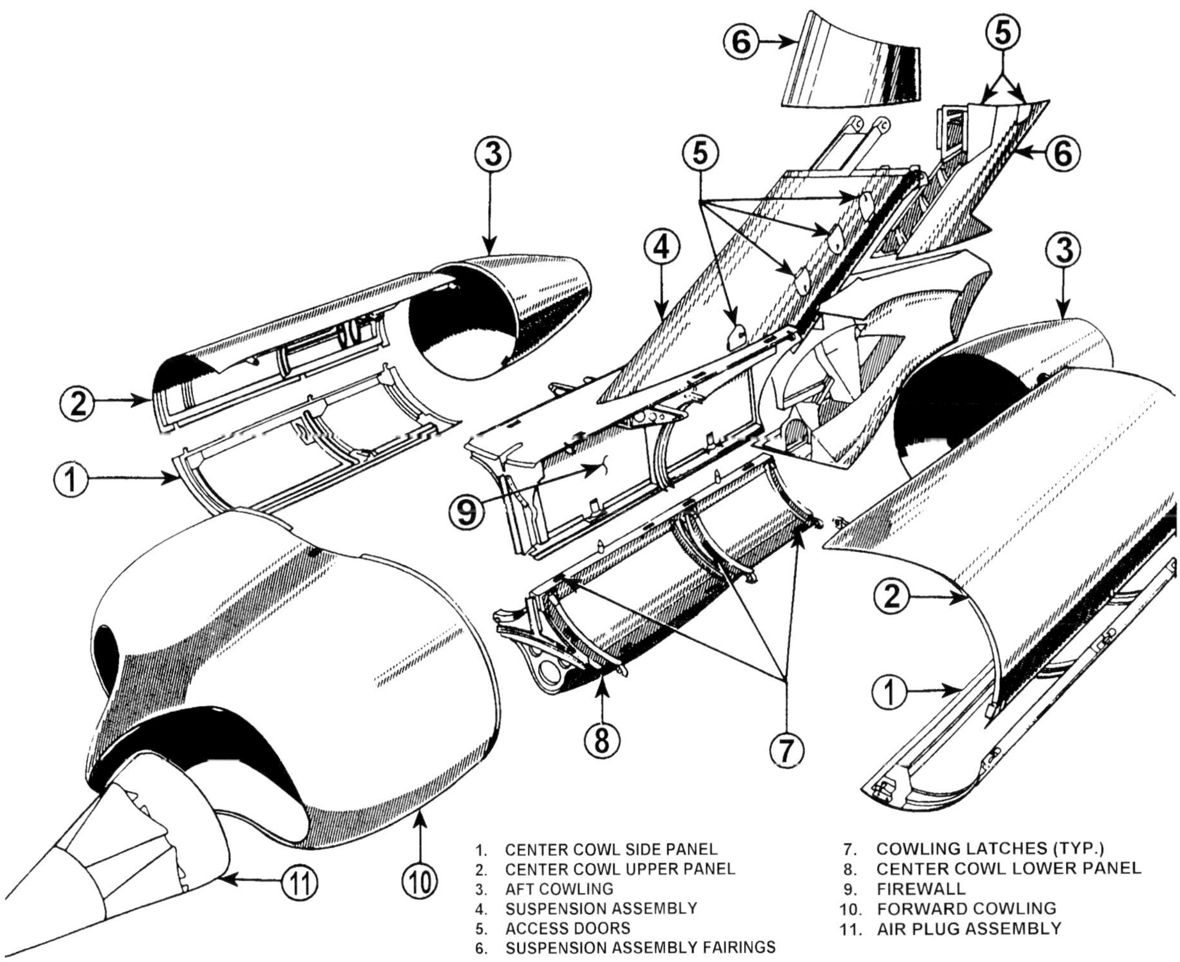

This line drawing shows the main cowl panels and suspension system for the pair of GE J47s attached to each B-36 wingtip. *(AN 01-5EUE-2 Handbook Erection and Maintenance Instructions USAF Series B-36F Aircraft. March 27, 1951)*

approximately 200 rpm and the feather motor can overcome the centrifugal force of the flyballs.

IV. Hydraulic pressure acting upon the clutch plates causes them to engage and transmit the rotation of the clutch housing to the drive shaft.

V. Rotation of the drive shaft is transmitted through the intergearing assembly and causes the blade angle change.

Taking advantage of electrical, hydraulic, and engine power, the B-36's propellers were unique in the world of piston engines and represented another fine example of the wonderful engineering that went into this fabulous aircraft.

Turbojet Installation

Even though the proposed VDT-powered B-36C offered some decent performance figures, it was apparent that the quick and easy way to additional performance was by way of jet assist. It was first test flown in October 1948 with a pair of Boeing XB-47 jet pods with J35s installed. After some relatively minor teething trouble, such as vibration that was fixed with a sway bar, it was a very successful modification. When service B-47s were upgraded with GE J47s, the B-36 followed suit still using the B-47 pod assembly. Only the prototype installation had J35s. All service B-36s had the four additional jets installed. Even the ones not originally designed for jets or manufactured without them had the J47s retrofitted. With the additional power of the

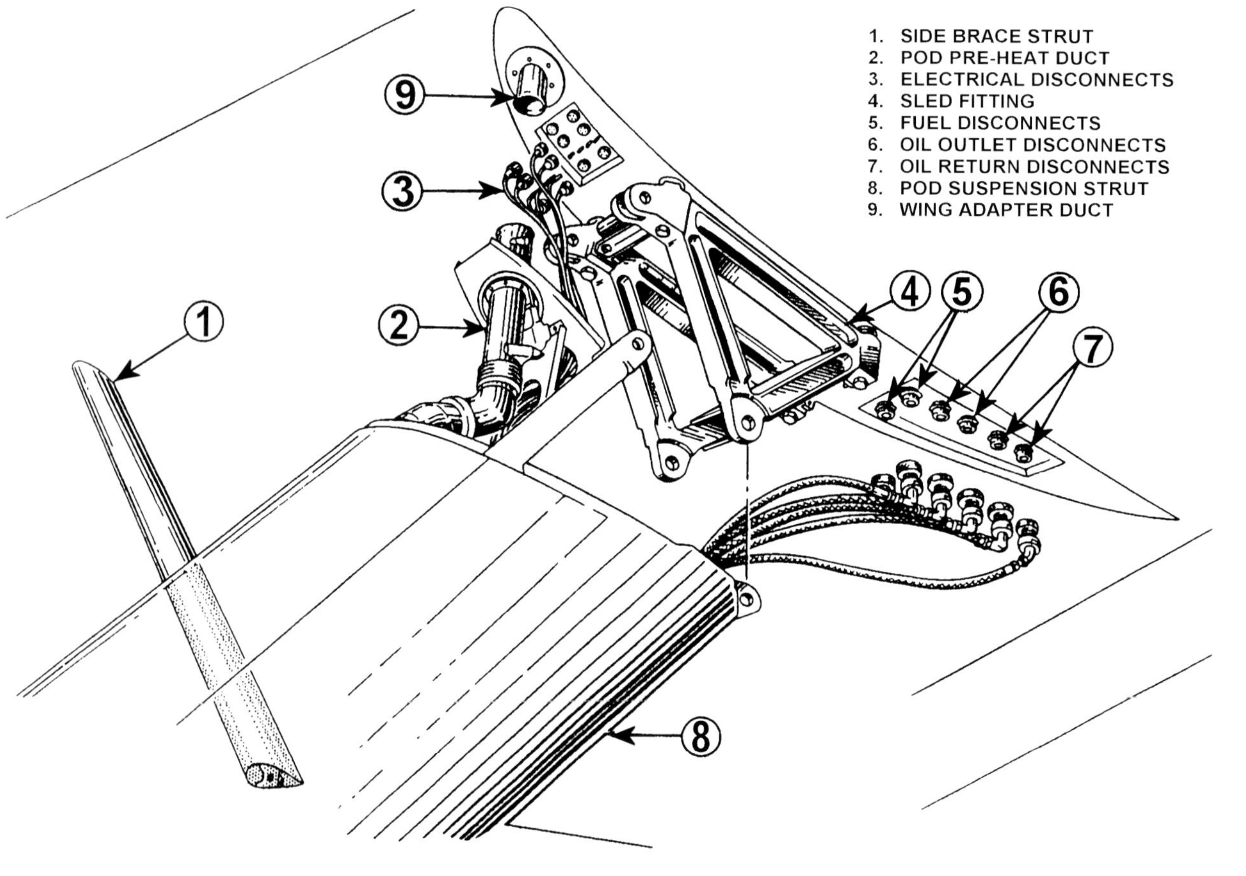

This illustration shows how the pylon was attached to the wing. *(AN 01-5EUE-2 Handbook Erection and Maintenance Instructions USAF Series B-36F Aircraft. March 27, 1951)*

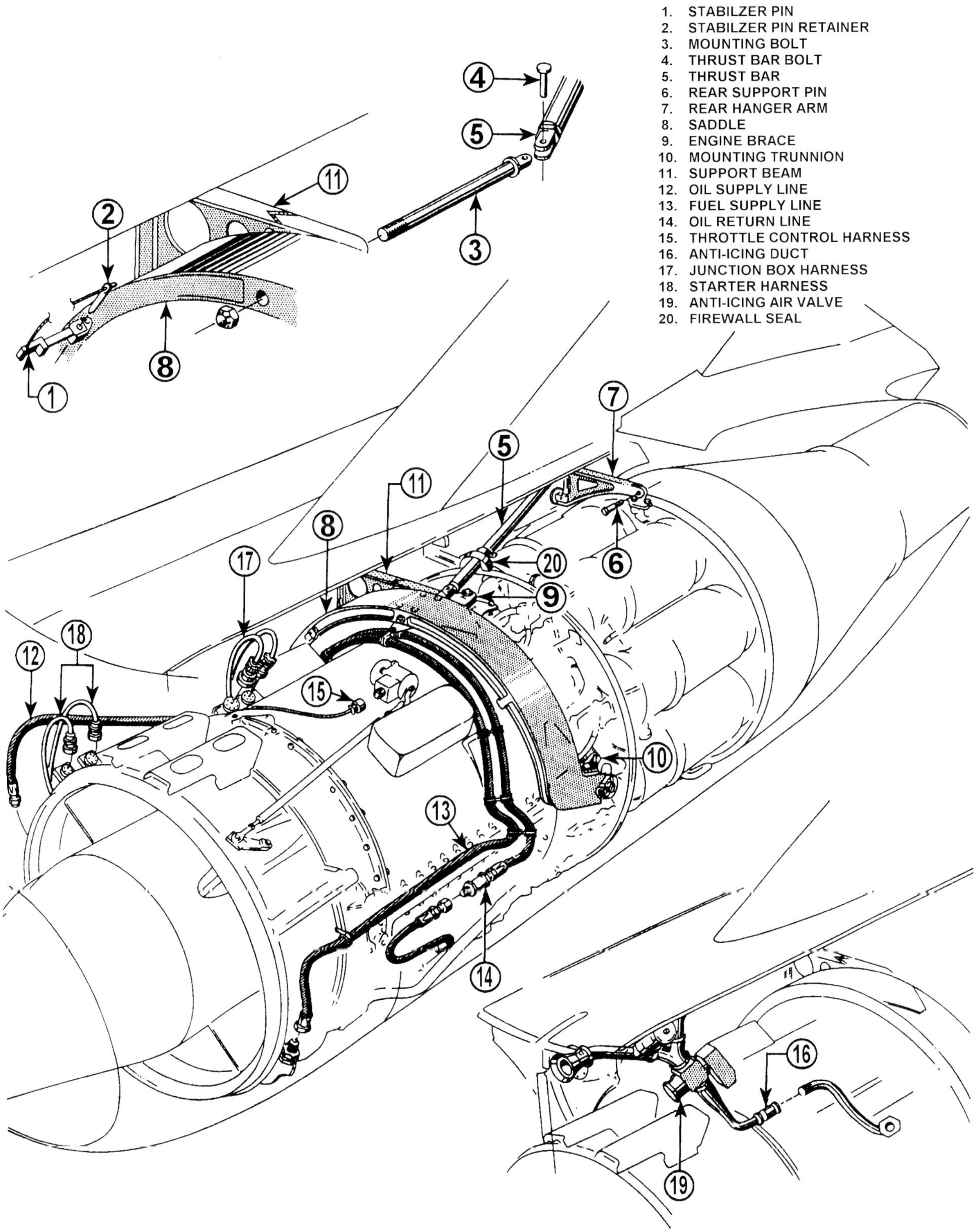

This illustration shows how the J47s were attached to the pylon assembly. *(AN 01-5EUE-2 Handbook Erection and Maintenance Instructions USAF Series B-36F Aircraft. March 27, 1951)*

four General Electric J47s, speed at altitude increased to over 435 mph with six turning and four burning—not too shabby for such a huge and heavy piston-powered behemoth. Making the retrofit even easier was the fact that the jet pods were essentially the same as those used on production Boeing B-47s. Relatively minor changes were made to the B-47 jet pod design to accommodate the B-36, but overall the modification proved to be remarkably easy and trouble free. Installation of the air plug, described below, represented the most significant alteration. Notwithstanding the foregoing—which perhaps paints a rosy picture—hanging a pair of J47s, each of which weighed in at 2,500 pounds plus cowling and attachments, was a testament to the structural integrity of the B-36. In addition, at full power, each jet pod was exerting a thrust load of over 10,000 pounds on the mount structure.

No major re-engineering of the aircraft was required for fitting the jet pods. In fact, they were interchangeable left to right and between aircraft. Primarily used for takeoff and emergency combat power, the jets spent most of their life simply hanging on the wing and doing nothing. Without some sort of fairing, they would create a significant amount of drag and windmill. With infrequent operation, they would also be more vulnerable to FOD (foreign object damage). A clever "air plug" was incorporated into the design to address these issues. When not in operation, the air plug closed off the J47's intake. Made up from eight doors that look like the petals of a flower, the air plug was an ingenious design. An electric actuator motor drove a screw jack that opened and closed the air plug.

Manufacturing

Any large aircraft is a challenge to manufacture simply due to the enormous sizes that need to be handled. In the case of the B-36 this problem was especially challenging. The height of the vertical stabilizer in particular gave Convair fits. Their solution was to tip the aircraft on its tail via an "A" frame structure attached to the nose gear in order to get the aircraft out of the assembly building and

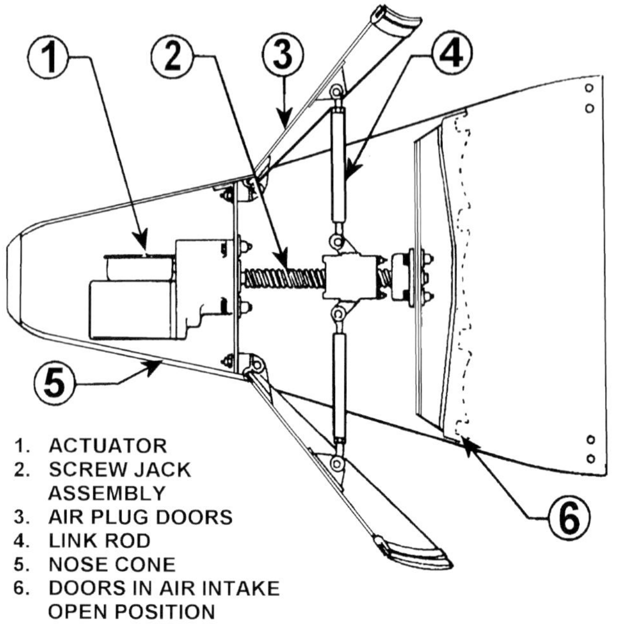

The J47s were only used for takeoff and combat conditions. The jets were shut down the majority of the time. In order to reduce drag and FOD, an electrically operated air plug was incorporated into the design. *(AN 01-5EUE-2 Handbook Erection and Maintenance Instructions USAF Series B-36F Aircraft. March 27, 1951)*

under the door. B-36s were manufactured in Convair's Fort Worth, Texas, facility. New engines were shipped from Pratt & Whitney in East Hartford or Ford Motor Company in Chicago and prepared for installation at Fort Worth.

Would Haves, Could Haves, and Should Haves

If it had not been for the introduction of far more capable jet powered aircraft such as the B-47 and B-52, the B-36 could have been developed into a very capable aircraft. The VDT program as described in Chapter 4 would have made the B-36 into a veritable hot rod. Another variation that has received little or no publicity was the plan to use a triple turbo set-up utilizing a pair of two-stage General Electric CHM-2 turbos in series with a single-stage CH-8. Engine exhaust gasses fed into the CH-8. Exhaust from the CH-8 was divided into two branches to drive the pair of CHM-2 turbos. Intake air from the leading edge

continued on page 467

This early production B-36B does not have the J47 jet pods. 120 B-36Bs were later converted to B-36D status, which included the jet pods. *(Courtesy of National Archives & Records Administration)*

B-36Ds with J47 jet pods rolling down the manufacturing line. *(Courtesy of National Archives & Records Administration)*

Hanging vertically, this engine will be transitioned to horizontal in preparation for installing in the aircraft. Note the engine cooling diffuser. *(Courtesy of National Archives & Records Administration)*

Top: Hanging at 45 degrees. *(Courtesy of National Archives & Records Administration) Above:* Judging by the high-tension ignition, this is probably a -25 or -41. *(Courtesy of National Archives & Records Administration)*

A beautiful shot of a B-36B banking to the left with its gear down. *(Courtesy of National Archives & Records Administration)*

continued from page 462

fed into the pair of CHM-2s, first then discharged into the CH-8. In other words, it was a crossover system. Calculations indicated that the B-36 could have cruised comfortably in excess of 50,000 feet, far out of the reach of contemporary fighters of the time. This design concept did not proceed beyond the proposal stage.

The following is a brief synopsis of the various B-36 models derived and paraphrased from the *1953 Jane's All The World's Aircraft.*

XB-36

The first prototype first flew August 8, 1946. Originally fitted with two 110-inch diameter single main gear wheels. Later equipped with four-wheel main gear bogies introduced on the B-36A. In 1950 the XB-36 was fitted with a tracked landing gear. It flew once with this gear. Powered by R-4360-5s.

YB-36

Production prototype. Roof of crew compartment raised above fuselage top line for improved vision, relocation of crew stations, and installation of nose turret. First flew on December 4, 1947. Powered by R-4360-5s.

B-36A

First production model. Single main wheels of prototypes replaced by four-wheel bogies to reduce structure weight and improve weight distribution on runways. Twenty-two built. Originally without armament and used for training and type familiarization. Modified into RB-36E. Powered by R-4360-5s.

B-36B

Second production model. Fully equipped for combat with full armament. First flew on July 8, 1948. Last of 130 B-36Bs modified by Convair to RB-36D standard with addition of four J47s and installation of latest radar and electronic equipment. Delivered in December 1951. Powered by R-4360-41 with ADI and J47s.

B-36C

Proposed VDT powered version, tractor configuration. Cancelled June 1948. Aircraft under construction were converted to B-36B status. Planned powerplants; R-4360-51 VDTs.

B-36D

Third production model. B-36B with additional power primarily to increase speed over target area. Fitted with four General Electric J47 turbojet engines in pairs in "pods" under outer wings to supplement the six 3,500-hp R-4360-41A or 3,800-hp R-4360-53 engines. Prototype, a converted B-36B with four Allison J35 engines, first flew on March 26, 1949. Over-target speed of B-36D with J47 turbojets increased to over 435 mph. Has new snap-action bomb bay doors instead of sliding type used in earlier models. Maximum gross weight 358,000 lbs.

RB-36D

Long-range strategic reconnaissance version of B-36D. Same defensive armament. Fourteen cameras in rear bomb bay. First flight on December 18, 1949. Powered by R-4360-41As or R-4360-53s plus four J47s.

RB-36E

B-36A modified for strategic reconnaissance. Re-engined with six 3,500-hp R-4360-41 or 3,800-hp R-4360-53 engines and fitted with additional jet power as for B-36D.

B-36F

Fourth production model. Six 3,800-hp Pratt & Whitney R-4360-53 engines, plus four J47 jet engines.

RB-36F

Long-range reconnaissance version of B-36F.

B-36H

Fifth production model, incorporating new two-place flight engineer's station. Six 3,800-hp Pratt & Whitney R-4360-53 engines, four J47 turbojets.

NB-36H

Modified from a tornado damaged B-36H, the NB-36H was designed to carry a nuclear reactor aloft, which it did.

The table opposite summarizes the main differences between later models of the B-36.

The following detailed information applies to the B-36H paraphrased from the *1953 Jane's All The World's Aircraft:*

Another beautiful shot of the B-36B. *(Courtesy of National Archives & Records Administration)*

(Table 7-4)

BOMBERS	RECIP. ENGINES (6)	ENGINEER'S STATION	WING FUEL TANKS	DESIGN GROSS WT. (LBS)	PRESSURIZED CREW COMPTS.	BOMB BAYS	BOMBING SYSTEM	CREW
B-36D	R-4360-41	SINGLE	8	357,500	2	4	K-1	15
B-36F	R-4360-53	SINGLE	8	357,500	2	4	K-1	15
B-36H	R-4360-53	DUAL	8	357,500	2	4	K-1	15
B-36J	R-4360-53	DUAL	10	410,000	2	4	K-1	15
RECON. BOMBERS								
RB-36D & E	R-4360-41	SINGLE	8	357,500	3	2	CONV.	22
RB-36F	R-4360-53	SINGLE	8	357,500	3	2	CONV.	22
RB-36H	R-4360-53	DUAL	8	357,500	3	2	CONV.	22

Type
Ten-engined Heavy Bomber.

Wings
Shoulder-wing cantilever monoplane. NACA laminar-flow wing section. Aspect ratio 11:1. Wing mounted slightly forward of mid point of fuselage. All metal structure with stressed skin. Leading-edge sweepback 15 degrees, 6.5 minutes, trailing edge sweepback 3 degrees. Gross wing area 4,772 square feet (443.3 square meters). Statically balanced ailerons with controllable trimtabs. Electrically operated trailing-edge flaps in three sections on each side of fuselage. Total flap area 519 square feet (48.2 square meters). Heated surface anti-icing.

Fuselage
Circular section all-metal structure.

Tail Unit
Cantilever monoplane type. All-metal structure. Horizontal stabilizer span 73 feet, 5 inches (22.38 meters). Total horizontal area 978 square feet (90.85 square meters). Total vertical area 542 square feet (50.34 square meters). Thermal anti-icing in leading edges of tailplane and fin.

Landing Gear
Retractable tricycle type. Main gear consists of two four-wheel bogies on single shock-absorber struts, each unit retracting inwards into wing. Twin nose wheel gear raised forward into fuselage. Hydraulic retraction. Wheel track 46 feet (14 meters), wheel base 59 feet (18.0 meters).

Powerplant
Six 3,800-hp Pratt & Whitney R-4360-53 28-cylinder radial air-cooled piston engines and four General Electric J47 turbojet engines (5,200 pounds static thrust each). Piston engines mounted as pushers aft of rear spars and drive Curtiss Electric three-blade constant speed, full feathering and reversing propellers with hollow steel blades and thermal anti-icing. Propeller diameter 19 feet. Each engine fitted with two turbosuperchargers. Inlets for induction and cooling air in and below leading edge of wings. Turbojet engines paired in pods under the outer wings. Wing fuel tanks with total capacity of over 21,000 U.S. gallons. Oil capacity over 1,200 U.S. gallons.

Accomodation
Crew of 16, including five-man relief crew. Pressurized crew compartments forward and aft of bomb bay with pressurized intercommunication tunnel 85 feet long, 25-inch diameter on left side of fuselage and below wings. Four-wheel truck for passage through tunnel. Thermal anti-icing and defrosting for pilot's and bombardier's compartments and for gun-sighting blisters in rear crew compartment. Total pressurized fuselage volume 3,924 cubic feet.

Armament and Equipment
Six retractable remotely controlled turrets, each mounting twin 20-mm cannons, plus two 20-mm cannons on flexible mounting in nose and two in radar controlled tail turret. Four section bomb bay with total volume of 12,300 cubic feet (348 cubic meters). Maximum bomb load 84,000 pounds (38,140 kilograms.)

Dimensions
Wingspan: 230 feet
Length: 163 feet
Height: 46 feet, 9 inches

Weights and Loadings
Max. gross weight: 357,500 pounds (410,000 pounds for B-36J)
Wing loading: 75 pounds per square foot

Performance
Max. speed: over 435 mph (six turning and four burning)
Stalling speed: 95 mph
Service ceiling: over 45,000 feet
Max. design range: 10,000 miles
Takeoff to clear 50 feet: 5,000 feet

The cargo-carrying version of the B-36 was the C-99. Even though only one was built, it earned its keep throughout the 1950s by hauling cargo for the USAF. Although this photograph shows it sporting the single wheel landing gear, like its sibling the B-36, the C-99 was soon converted to the four-wheel bogie design. *(Courtesy of National Archives & Records Administration)*

Convair C-99

(Ref. 7-56)

Following on the heels of the B-36, Convair developed a transport version of the B-36, which was designated C-99. Using the flying surfaces, engines, and landing gear of the B-36, the C-99 employed a totally redesigned fuselage. Only one was built, and it operated from 1949 to 1957. Although the concept of a transport version of the B-36 was envisioned in 1942, due to the more pressing needs of World War II it was placed on the back burner. The double-decked fuselage was capable of hauling 50 tons or 400 troops. First flight occurred on November 24, 1947, a little over one year after the XB-36. Like the XB-36, the C-99 started life with the single-wheel main gear. This was replaced as soon as the more practical four-wheel bogie became available. However, the C-99 never inherited its sibling's four J47 jets. As with any large aircraft, it had to be fully loaded to be economically viable. This situation did not always occur, resulting in higher than normal operating costs based on a per-ton-mile basis. The sole C-99 has survived—barely. As of this writing it's in very poor condition, having suffered the ravages of time, being stored outdoors, and the inevitable vandalism.

It appears that the C-99's flight deck could double as a living room! The seat between the pilot and co-pilot is the all-important flight engineer's position. *(Courtesy of National Archives & Records Administration)*

Republic XF-12, XR-12 Rainbow

(Ref. 7-1, 7-57, 7-58, 7-59, 7-60, and 7-61)

This remarkable aircraft was another victim of cancellation in the austere postwar years. One has to wonder if very capable aircraft cancelled during the late 1940s, such as the XF-11 and XF-12, were sacrificed for B-36 production. Conceived as a high-altitude, high-speed reconnaissance aircraft, it achieved these goals admirably. The XF-12's top speed of over 500 mph was faster than many fighters of the era. Contained within its slim and graceful fuselage were three camera compartments along with 18 high-intensity flash bulbs for night photography. This advanced aircraft was designed during World War II and made its maiden flight on February 4, 1946. Two prototypes were built and delivered to the Army Air Force. On September 18, 1947, the independent Air Force was created and at the same time many aircraft went through re-designations. In the case of the XF-12 (Experimental Foto No. 12), it changed to XR-12 (Experimental Reconnaissance No. 12). Notwithstanding the fact that the Rainbow was a highly capable aircraft, the program was cancelled in favor of the B-50 modified for the reconnaissance role.

On September 1, 1948, the Rainbow made a name for itself by photographing the entire width of the United States in one flight. Departing the Air Force Flight Test Center at Muroc, California, and heading west over the Pacific until 40,000

One of the nicest R-4360 installations was the one used for the Republic XF-12 Rainbow. Of interest in this shot is the good view of the engine cooling fan and leading-edge intakes for the oil cooler, intercooler, and induction. *(Courtesy of Warren Bodie)*

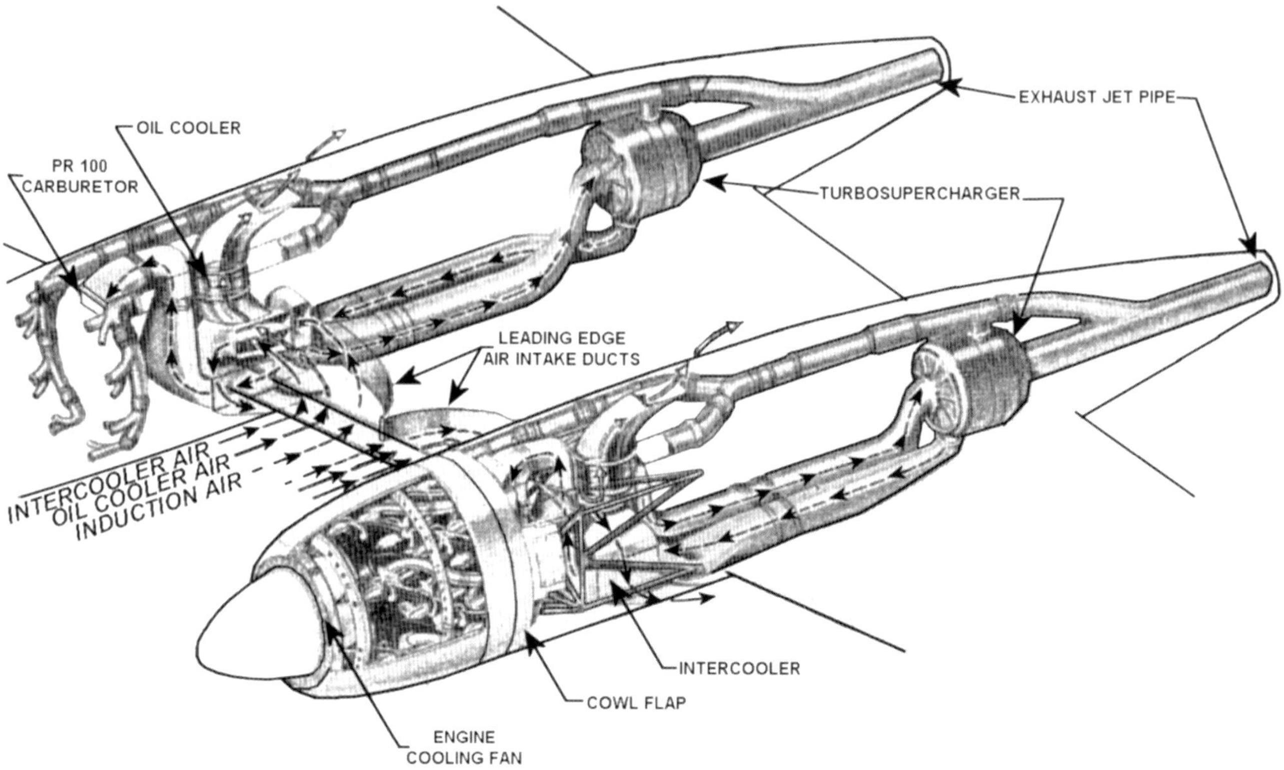

This artist's rendering shows the Rainbow's induction, cooling, and turbosupercharging systems. However, it's surprising to see that the artist only shows one turbo per engine instead of two. A wide, narrow rectangular slot between the nacelles provided all air requirements except for engine cooling. *(The Republic Rainbow Parts I & II. Aeroplane. December 6, 1946, and December 13, 1946)*

feet had been attained performed this amazing accomplishment. At this point the aircraft made a 180, heading east across the country taking pictures for the entire route. A continuous 325-foot-long roll of film was consumed with 390 exposures. At its 40,000-foot altitude, the field of vision was 490 miles wide. Despite this amazing feat proving its capability, the XR-12 program was already doomed for cancellation in favor of the B-50, the same fate that befell the XF-11. One XR-12 crashed while flying out of Eglin Air Force Base, and in yet another example of bureaucratic vandalism the remaining XR-12 was used as target practice at the Aberdeen Proving Grounds. One has to wonder how these pinhead bureaucrats get to such high executive levels where they can wield their enormous power to the detriment of all?

As with the Lockheed Constitution and Convair C-99, thought was given to producing a commercial version of the Rainbow. Pan Am and other airlines placed orders; however, after the cancellation of the military Rainbow, non-recurring costs such as tooling would need to be amortized over far fewer aircraft, making it financially unviable. For the same reason, Boeing was able to sell its Model 377 as a commercial airliner thanks to the military orders placed for the C-97, which amortized their non-recurring costs.

From a design perspective, the Rainbow was full of innovative features. Being in competition with Howard Hughes' equally capable XF-11, it's not surprising that R-4360-37s with dual General Electric turbosuperchargers powered both aircraft. However, in the case of the Rainbow, GE BM turbosuperchargers were employed, one being a BM 4 and the other being a BM 5. And as with the XF-11, single turbo operation was possible with the XF-12. Surprisingly, the turbos were lubricated with engine oil, a feature considered by

Another nice shot illustrating the graceful lines of the nacelles, which extend well beyond the wing trailing edge. Once again, we have the bureaucratic vandals to thank for not preserving this remarkable example of the engineer's art. It boggles the mind that high-ranking Air Force officers would have this aircraft scrapped. *(Courtesy of National Archives & Records Administration)*

other airframe designers but deemed too risky due to possible failure of the turbo system, which would then place the engine in jeopardy. Likewise, an engine failure would then contaminate the turbo. Production XF-12s were intended to have ADI. As it turned out, only the second aircraft had this feature. Originally, R-4360-31s with contra-rotating prop drive were slated to power the Rainbow. However, this engine didn't power either of the two aircraft built. The Hughes XF-11 first flew with -31s and almost killed Howard Hughes due to a propeller malfunction. It may have been this experience that changed the engine designation for the XF-12.

All air requirements including oil cooling, induction air, and intercooler air were taken in from the leading edge of the wing. A wide slot between the engine nacelles provided the necessary air intakes. Engine cooling was augmented with a two-speed engine-driven fan. Instead of the usual cowl flaps, a sliding ring was used to control mass airflow through the cowl. Oil capacity was rather diminutive (for an R-4360) with 54 gallons housed in a cylindrical tank mounted behind the firewall. Although the Rainbow's fuselage shape was close to aerodynamic perfection, the same could not be said for its visibility through the nose. A unique windshield arrangement for landing and other conditions where excellent visibility is essential was ingeniously incorporated.

Its GE turbosuperchargers, mounted in tandem, exhausted through the rear nacelle, which extended well beyond the wing trailing edge. In this way, exhaust energy was utilized for additional thrust. Exhaust for the pair of waste gates was located about halfway along the nacelle on the bottom. Perhaps as a forerunner to the Concorde, which featured a drooping nose for better pilot visibility during the takeoff and landing

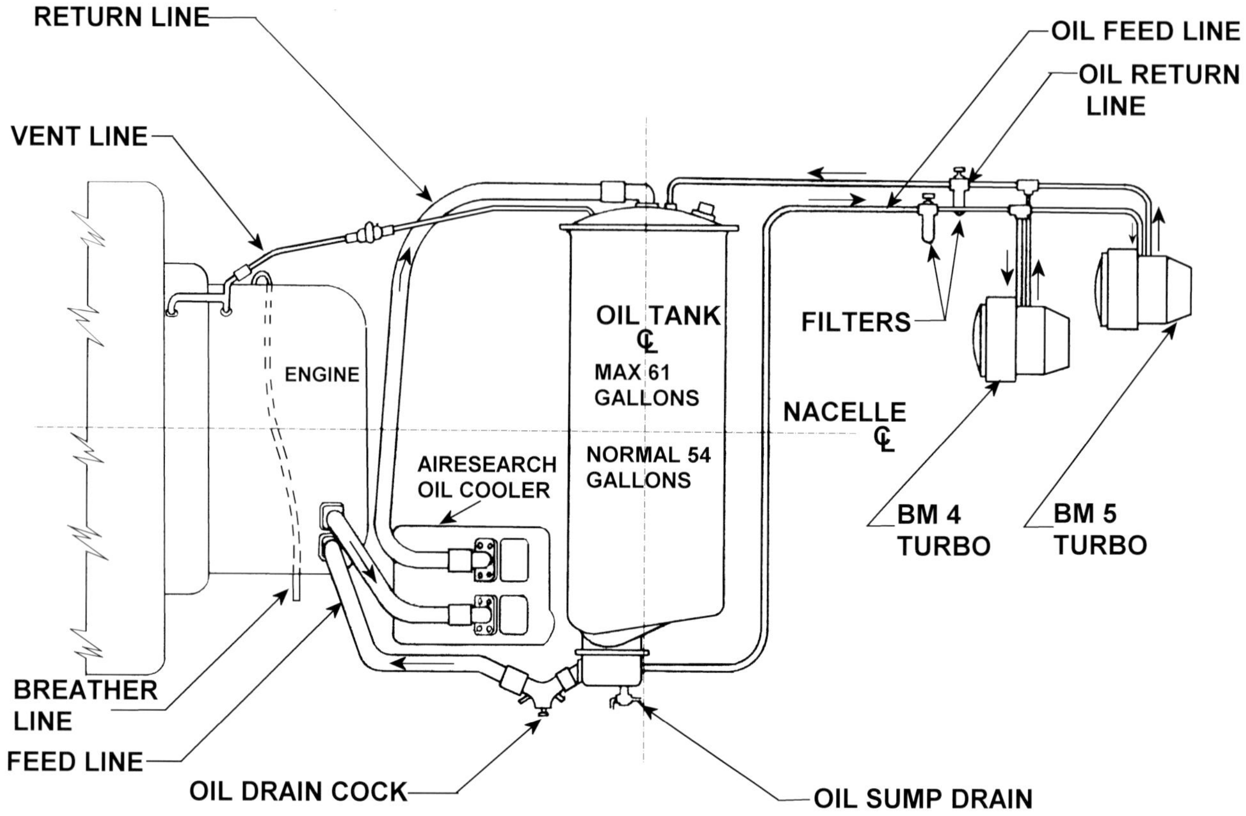

This line drawing shows the XF-12's lubrication system. Surprisingly, the dual turbosuperchargers shared the same oil as the engine. It's quite probable that if this aircraft had entered service this retrograde feature would have been corrected by having independent systems for the engine and turbos. *(XF-12 Flight Manual)*

phase, the upper Plexiglas panels in the nose rolled down and the pilot could now look through a flat glass panel. Without this feature, the pilot could be disoriented and the optics of the streamlined panels would not be optimal. If development of the Rainbow had continued, it would have been a candidate for VDT power. With a potential 4,300 hp on tap per engine, the Rainbow would have been a remarkable hot rod.

Another unusual feature of the Rainbow was its wing spar to fuselage juncture. It is normal practice to run the spar through the fuselage. In this way a continuous load path is preserved. Problem is the spar can also interfere with the internal space available within the fuselage, particularly with a compact design such as the Rainbow's. Republic engineers solved this problem by using a reinforced fuselage frame to deal with the carry-through loads. In this way the fuselage was not encumbered with a massive spar passing through it. Another consequence of the compact design of the Rainbow was where to house the main landing gear. For multi-engined piston-powered aircraft, designers typically house the gear in the nacelle. With its pencil-slim nacelles this would have been impractical for the Rainbow, so the single-wheel main gear folded inwards towards the fuselage. A reinforced box

XR-12 Parameters *(Ref. 7-1):*

Wingspan	129 feet, 2 inches
Wing area	1,640 square feet
Length	98 feet, 9 inches
Height	28 feet, 7 inches
Empty weight	73,420 pounds
Weight loaded	110,000 pounds
Max. gross	137,000 pounds
Fuel capacity	5,505 gallons
Max. speed	500 mph @ 40,000 feet
Cruising speed	375 to 450 mph
Combat radius	1,715 miles
Combat range	3,375 miles
Service ceiling	42,000 feet
Rate of climb	2,200 feet per minute at sea level, combat weight and max. power

section designed into the rear spar took care of the loads.

The first flight was not without its excitement. According to L. L. Brabham, Republic Aviation's famed test pilot, the rate of acceleration took him by surprise, and in order to avoid exceeding the gear retraction speed of 250 mph, he was forced to put the aircraft into a steep climb. Further adding to the first flight's excitement was a failure of the electric drive motor for the flaps. Due to this failure, the entire first flight was flown with half flap. On another test flight, Brabham applied METO (maximum except take off) power at 35,000 feet in level flight. Under these conditions the aircraft exceeded the do-not-exceed dive speed of a P-47 Thunderbolt!

In a typical Rainbow mission, the aircraft would take off at a gross weight of 110,000 pounds. After reaching its cruise altitude of 40,000 feet, the cruise speed would stabilize at 375 mph and as fuel burned off the speed would increase. At the end of the mission aircraft weight would have been reduced to 81,000 pounds and cruise speed had increased to a fighter-like 450 mph.

Boeing XB-44

(Ref. 7-1, 7-62, 7-63, and 7-64):

It is often assumed that the Boeing XB-44 was the prototype for the B-50. In fact, the B-44 was a very different aircraft from the B-50. Modified from a B-29 (B-29A-5-BN, S/N 42-93845) that was flown to East Hartford where the necessary modifications took place, the XB-44 employed a different philosophy from the B-50 for altitude performance. B-50s took advantage of turbosupercharging in conjunction with an engine-driven single stage supercharger as the chosen method to obtain altitude performance. The XB-44, on the other hand, used an engine-driven two-stage

Converted from a standard production B-29, the performance of the XB-44 powered by R-4360-33s was dramatically improved over its forebeer. *(Courtesy of National Archives & Records Administration)*

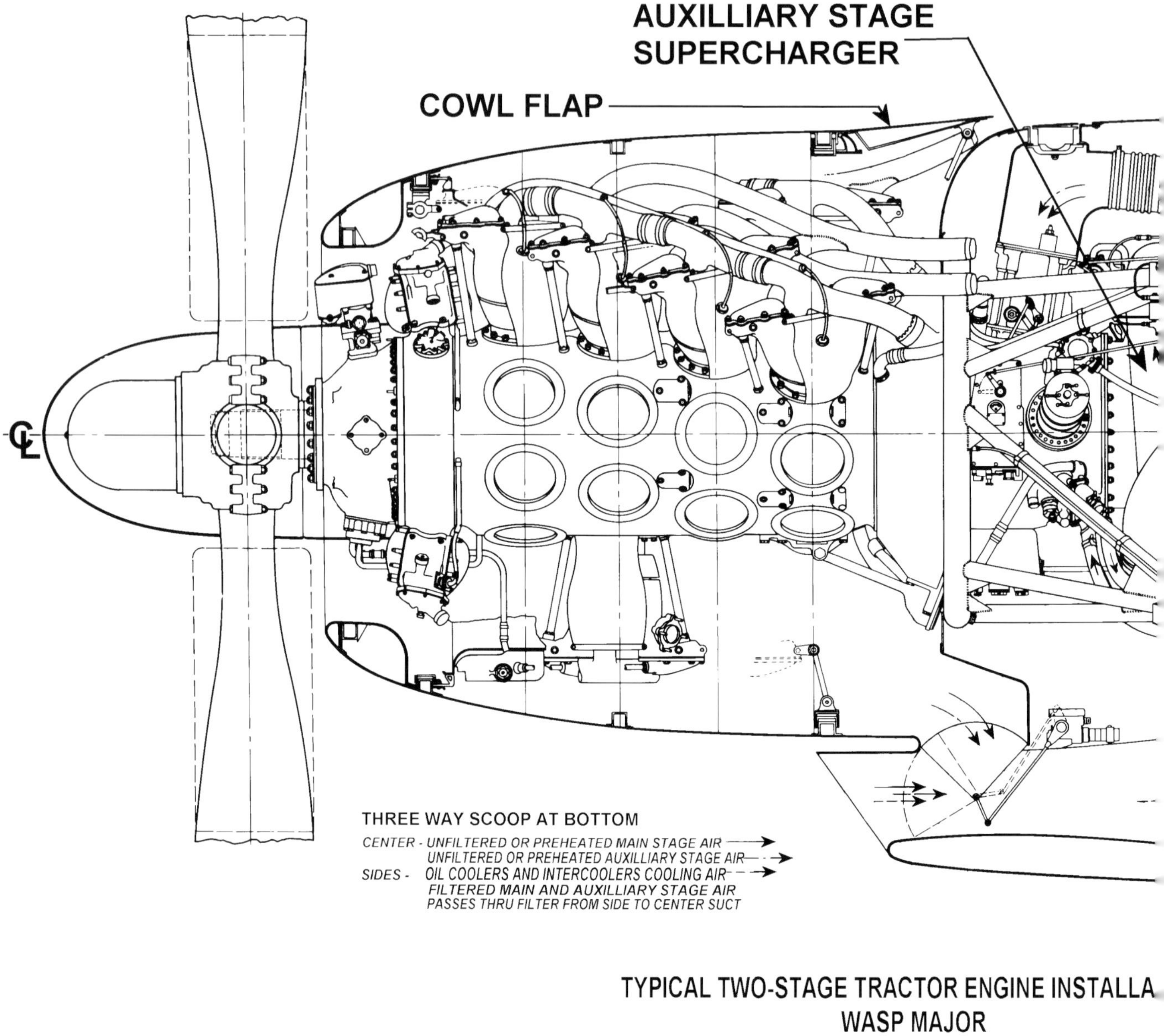

This drawing shows the XB-44's QEC design featuring a gear-driven two-stage R-4360-33 with air-to-air intercoolers. In retrospect it could be argued that this is how the follow-on B-50 should have been powered. *(Courtesy of Pratt & Whitney)*

supercharger. Therefore, the QEC design of these two aircraft was quite different. Notwithstanding the foregoing, the nacelles of the XB-44 and B-50 look very similar, both featuring a very tight cowl with a large scoop under the nacelle. Furthermore, the scoop was divided into three compartments on both aircraft, again giving credence to the theory that both aircraft were similarly powered. One distinguishing feature differentiated the XB-44 cowl design from that of the B-50. The fact that the XB-44 was an "open stack" engine, i.e., it did not employ an exhaust driven turbosupercharger, resulted in the XB-44 employing bumps in the cowl flaps to accommodate the exhaust jet stacks. An efficient siamesed system was employed. In contrast, B-50s had

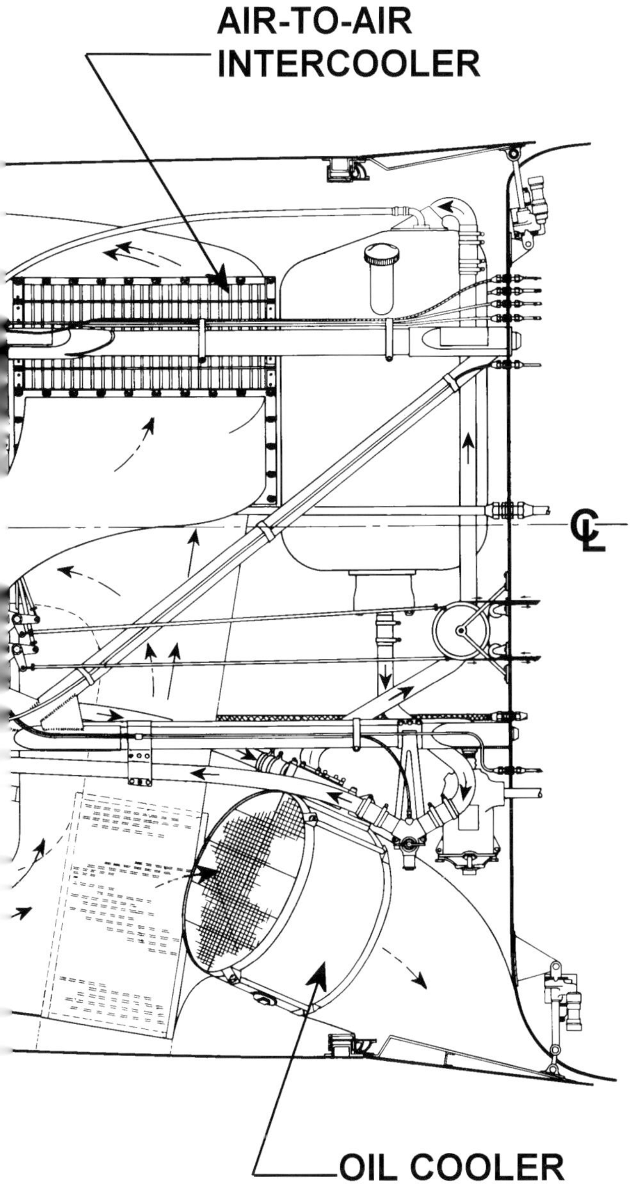

provided for, located at approximately the two o'clock and four o'clock positions. Each one discharges into an air-to-air intercooler and then finally into the PR-100 carburetor. It made for a beautifully compact installation. One wonders why Pratt & Whitney later gave up on gear driven multi-stage superchargers in favor of the more problem prone turbosupercharger installations.

It was no secret that the Boeing B-29 was suffering under handicap of an underdeveloped powerplant in the form of the Wright R-3350. Although the R-3350 would later go on to be the personification of efficiency; during its embryonic years it was nothing short of a disaster. With this in mind the XB-44 program was initiated in mid 1943 when Pratt & Whitney offered to modify a B-29 with R-4360 power. A contract was awarded in July 1944, and the first flight of the aircraft occurred in May 1945. The XB-44 was capable of cruising speeds about 50 to 60 mph faster than a production B-29. Furthermore, the reliability of the R-4360 was infinitely superior to the R-3350. Due to the fact that the program got off the ground in 1944, Pratt & Whitney probably knew that an R-4360-powered B-29 was unlikely to participate in World War II; even so the Army Air Force felt the program had enough merit to continue with it. Initially it was to be designated B-29D but then changed to XB-44. At the conclusion of World War II or shortly thereafter, it should be remembered that many superb projects were cancelled, including aircraft such as the very capable Hughes XF-11 and Republic Rainbow.

smooth cowl flaps. Induction air for the auxiliary supercharger came in from the center section of the lower nacelle scoop; the two outer sections provided cooling air to the pair of air-to-air intercoolers and the pair of oil coolers. Air from the center section makes a 90-degree bend and flows up into the suction side of the auxiliary blower. Two discharges for the auxiliary supercharger are

XB-44 Parameters *(Ref. 7-1)*:

Wingspan	141 feet, 3 inches
Length	99 feet
Height	29 feet, 7 inches
Weight	140,000 pounds (max. gross weight)
Armament	Two 0.50-caliber machine guns in the tail
Engines	R-4360-33
Crew	11 maximum, although it usually flew with a smaller test crew
Max. speed	392 mph @ 25,000 feet
Cruising speed	282 mph
Range	2,400 miles
Service Ceiling	39,000 feet

Above: The XB-44 used the classic A/C and B/D siamesed open stack exhaust system. The large scoop provided all air requirements–induction air, oil-cooling air, and cooling air for the dual intercoolers. Note the brush block holder for the Curtiss Electric propeller mounted over the propeller shaft. *(Courtesy of National Archives & Records Administration)*

Right: Starting from the horrible R-3350 installation it's hard to image that such a clean installation for the R-4360 could be achieved. *(Courtesy of National Archives & Records Administration)*

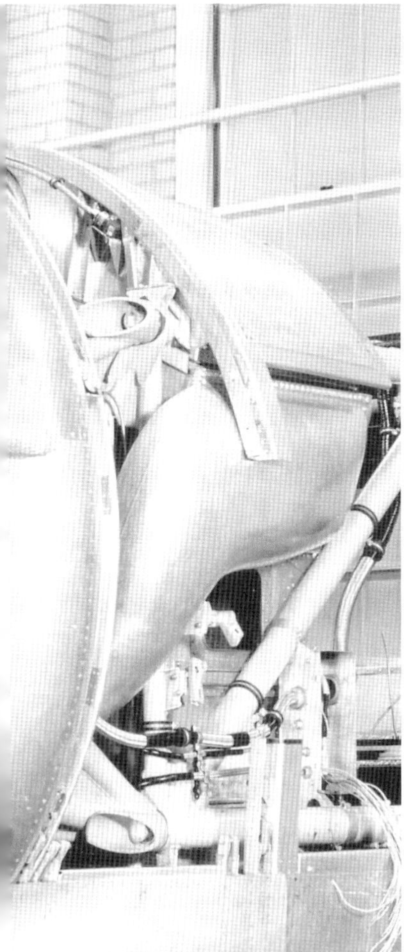

Above: Pratt & Whitney engineers were forced into using the basic nacelle dimensions of the B-29's Wright R-3350. Nevertheless, they pulled off the seemingly impossible by converting a pig's ear into a silk purse. *(Courtesy of National Archives & Records Administration)*

Above: This ¾ left rear view of the QEC shows one of the large rectangular intercoolers. The auxiliary supercharger driven off the rear of the engine had two discharges. Each one fed into an intercooler. *(Courtesy of National Archives & Records Administration)* *Below:* This gives another view of the compact design of the XB-44's nacelle. *(Courtesy of National Archives & Records Administration)*

Boeing B-50A, B, D, H, YB-50C, Boeing B-54A, RB-54A

(Ref. 7-1, 7-65, 7-66, and 7-67)

Continuing on from the pioneering work performed on the successful XB-44, the next iteration of an R-4360-powered B-29 was the B-50A, although it was originally intended to be designated B-29D. However, to simply write off the B-50 as just an R-4360-powered B-29 would be missing the point. Although its ancestry was clearly rooted with the B-29, notwithstanding their almost identical appearances, the B-50 incorporated significant changes and improvements. In fact, 75 percent of the B-50 was changed from the B-29. Apart from the different nacelles, the most obvious visual clue was the B-50's much larger vertical stabilizer, which stood 5 feet higher than the B-29's. However, this enlarged vertical stabilizer could have caused headaches for maintenance people because the increased height could not be accommodated in standard hangars of the day. Boeing fixed this issue by allowing the vertical stabilizer to hinge, thus allowing access to hanger space. Taking advantage of newer and stronger materials, a considerable amount of structural weight was eliminated. For instance, the wing was fabricated from 75-S aluminum, which was 16 percent stronger than the equivalent material used for the B-29. A weight savings of 650 pounds for the wing assembly alone resulted. Despite having powered nose wheel steering (the B-29's was castoring), the B-50s landing gear was also lighter. With 3,500 hp per engine on tap compared to the B-29's 2,200 hp, a power increase of 62 percent was realized. Not only did this result in a top speed increase of approximately 30 mph, but also an increased gross weight of about 35,000 pounds was possible over the B-29. After it was made obsolete by aircraft such as the B-47 and B-52, the B-50 soldiered on in the refueling role and was redesignated KB-50.

As with its predecessor the B-29, most of the B-50's systems were electrically operated rather than hydraulically operated. This included the landing gear. One B-50, designated EB-50B, was experimentally fitted with rubber-tracked landing gear. It seems that in the late 1940s it was almost a fad to fit rubber-tracked landing gear to aircraft. Part of the rationale was the significant increase in aircraft weight over a relatively short period of time. At the end of World War II, aircraft weighing 100,000 pounds or more were rare. Yet just a few short years later gross weights would double, consequently taxing the capabilities of runways to support such tremendous weights. As a result, in a similar fashion to the B-36, a rubber-tracked landing gear was experimentally installed on a B-50. It was apparently as successful as that fitted to the B-36—no production B-50s were fitted with this device.

Engine Installation

Although the nacelles of the B-50 looked the same as those fitted to the XB-44, looks can be deceiving. Unlike the XB-44's two-stage gear-driven supercharger, the B-50's R-4360s were boosted by a General Electric CH-7 single stage turbosupercharger in series with a gear-driven single-stage, single-speed supercharger integral with the engine. The CH-7 was mounted on the lower left side of the nacelle with a single exhaust outlet on the same side. Engine exhaust was manifolded into a large collector ring, which then powered the turbine section of the CH-7. As related in Chapter 4, horrendous problems were experienced with CH-7 turbosuperchargers resulting in numerous attempts at a fix. Discharged air from the CH-7 was routed through a large rectangular air-to-air intercooler and then through to the PR-100 carburetor. A single circular oil cooler was mounted on the lower right of the nacelle.

Like its predecessor the XB-44, the large scoop under the nacelle was divided into three compartments. The left fed induction air to the suction side of the turbosupercharger compressor, the center compartment provided cooling air to the intercooler, and the right side provided

Although it was a B-29 look-alike, the B-50 was significantly changed from its predecessor, thus justifying the new designation. The obvious visual clues are the taller vertical stabilizer and different engine nacelles. The photograph shows a B-50D. *(Courtesy of National Archives & Records Administration)*

cooling air to the oil cooler. The two inboard turbo installations (numbers 2 and 3) provided bleed air for cabin pressurization. All B-50 variants were powered by the R-4360-35. An interesting idiosyncrasy of the B-50 QEC in comparison to its sibling the C-97 and its commercial counterpart the B-377 was the fact that it was a mirror image of the latter. In other words, the

B-50's turbosupercharger was mounted on the left side of the nacelle as viewed from the rear. The C-97 mounted its turbosupercharger on the right as viewed from the rear. No plausible explanation has been forthcoming for this seeming anomaly, but there must have been a good reason for it. As a last valiant gasp at eking out more life from the venerable B-50, the B-54 was proposed.

This VDT-powered version would have exhibited remarkable performance characteristics. Alas, the farthest it progressed was the test flying of one VDT engine in the number three position of a B-50. This harrowing tale is related in more detail in Chapter 4.

Carburetor Preheat and Surge Bleed Valve

Although we normally think of cooling induction air through intercoolers and/or aftercoolers, it is possible to over-cool induction air. The ideal temperature is in the range of 70 to 80 degrees Centigrade. Surging of the turbo compressor is another common problem then and even now with gas turbine compressors. The B-50 power packager accomplished these requirements through a carburetor preheat and surge-bleed valve downstream from the turbo compressor. It is attached to the supercharger ram-air duct of each power package (QEC). A servomotor power unit connected to the valve actuator gearbox by a flexible driveshaft actuates the valve. A wire mesh screen over the valve discharge opening prevents foreign material from jamming the valve in the open position. The valve has two functions: (i) when carburetor preheat is required, the valve opens from 9 to 20 square inches, recirculating a portion of supercharged air for additional carburetor-air temperature, and (ii) for attenuation of compressor surge. At engine speeds below 2,300 rpm and 44 (+ 0.5/- 4.5) inches of manifold pressure, the valve opens to 7 square inches, recirculating a portion of supercharged air to stabilize the surge of compressor discharge pressures.

Carburetor Preheat

Control of carburetor preheat is through an "on/off" preheat switch and a preheat selector switch on the flight engineer's switch panel. With the preheat switch "on"and the preheat selector switch in "automatic," operation of the valve is derived from two resistance elements in the intercooler pressure duct, which sense the temperature of the air. These resistances are part of a resistant bridge that comprises the automatic

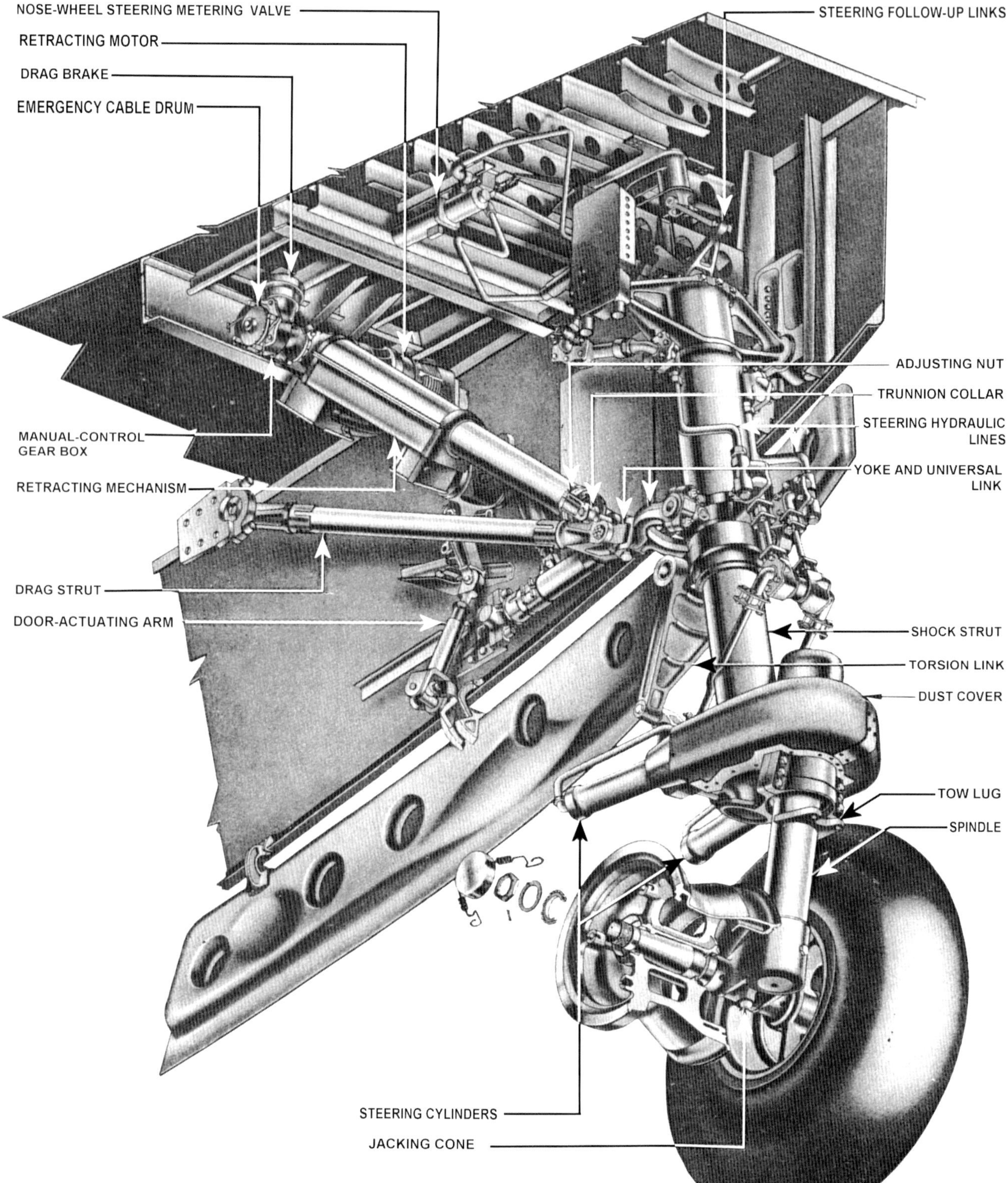

Above: A big improvement over the B-29's castering nose-gear was the power-assisted nose-wheel-steering of the B-50. As with all Boeing piston-powered aircraft, most systems were electrically operated instead of utilizing hydraulics. This included the electrically actuated gear retraction mechanism. *(AN 01-20ELA-2 Erection and Maintenance Instructions for Boeing B-50 Aircraft. Revised February 5, 1948)* *Opposite:* Main landing gear design was similar to that used on the B-29. Loads were shared between the front and rear spars. *(AN 01-20ELA-2 Erection and Maintenance Instructions for Boeing B-50 Aircraft. Revised February 5, 1948)*

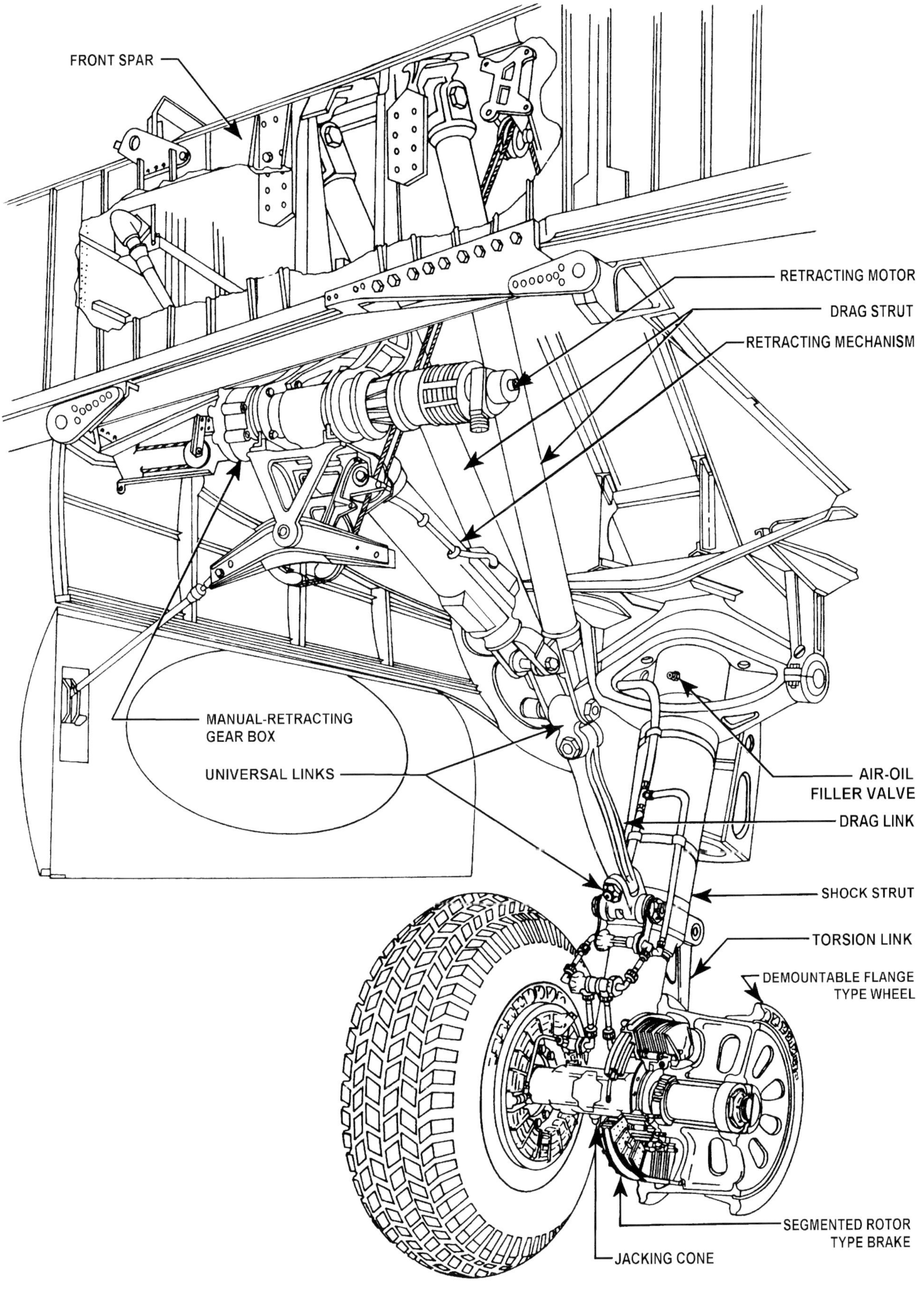
FRONT SPAR
RETRACTING MOTOR
DRAG STRUT
RETRACTING MECHANISM
MANUAL-RETRACTING
GEAR BOX
UNIVERSAL LINKS
AIR-OIL
FILLER VALVE
DRAG LINK
SHOCK STRUT
TORSION LINK
DEMOUNTABLE FLANGE
TYPE WHEEL
SEGMENTED ROTOR
TYPE BRAKE
JACKING CONE

Right: In a similar fashion for the B-36, a B-50 was experimentally fitted with a tracked landing gear for use on unimproved landing strips. The tri-pod bolted to the side of the fuselage holds a camera to record the results of this unique gear in action. *(Courtesy of National Archives & Records Administration)*

Below: Apart from a small bump in the gear doors, it's amazing that Boeing engineers managed to stuff such a huge landing gear into the wheel (track?) well. *(Courtesy of National Archives & Records Administration)*

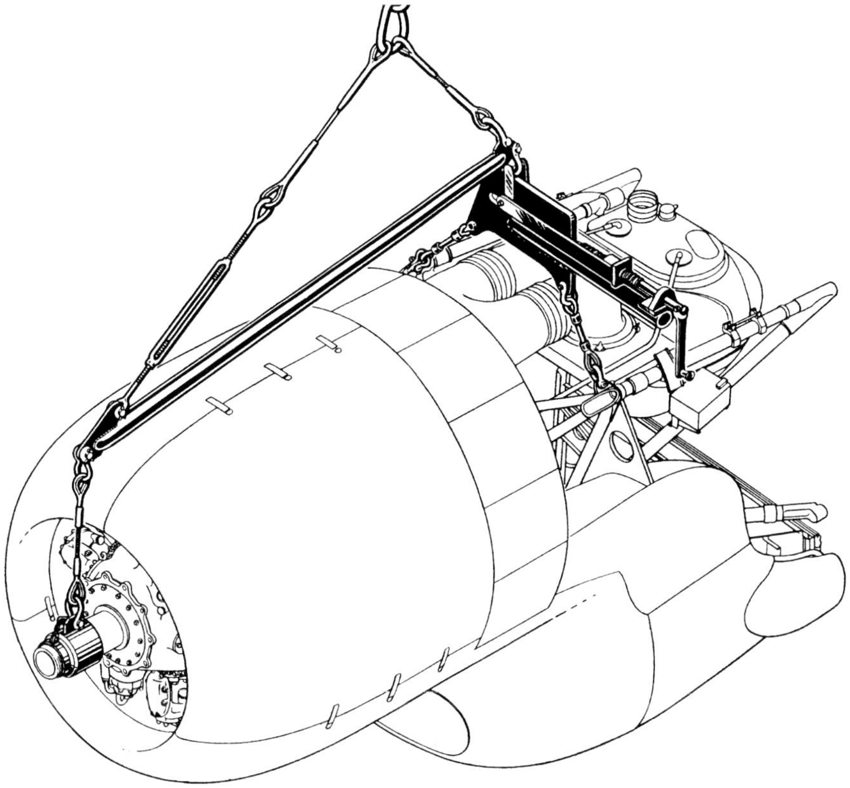

If a complete QEC was ready to go, it was a relatively easy job to change engines. However, if the QEC was not built up and just the engine was changed, a significant amount of work was required. *(AN 01-20ELA-2 Erection and Maintenance Instructions for Boeing B-50 Aircraft. Revised February 5, 1948)*

control of carburetor-air temperature. Variation of ampere output of the resistances with temperature changes unbalances the bridge. This causes a sensitive relay to operate, which operates the power servo unit to reposition the valve and balance the bridge.

B-50 Parameters *(Ref. 7-1):*

Max. speed	385 mph at 25,000 feet, 391 mph at 30,000 feet
Cruising speed	235 mph
Stalling speed	136 mph
Service ceiling	37,000 feet
Initial climb rate	2,225 feet per minute
Combat radius	2,193 miles with 10,000 pounds of bombs
Max. range	5,230 miles
Takeoff ground run	5,940 feet at sea level
Takeoff over 50-foot obstacle	7,425 feet at sea level
Wingspan	141 feet, 3 inches
Length	99 feet
Height	32 feet, 8 inches
Wing area	1,720 square feet
Empty weight	81,050 pounds
Combat weight	120,500 pounds
Max. take-off	168,708 pounds
Max. internal bomb load	20,000 pounds
Armament: Four 0.50-inch machine guns in forward dorsal turret, two 0.50-inch machine guns in rear dorsal turret, two 0.50-inch machine guns in forward ventral turret, two 0.50-inch machine guns in rear ventral turret, two 0.50-inch machine guns, and one 20-mm cannon in tail turret.	

Surge Bleed

With the preheat switch "off" and with the engine operating below 2,300 rpm and 44 (+ 0.5/- 4.5) inches of manifold pressure, the valve is automatically positioned at 7 square inches of opening. Above 2,300 rpm a surge-bleed synchronizing switch opens, which automatically closes the valve, or if manifold pressure is greater than 44 (+ 0.5/- 4.5) inches of mercury, a surge-bleed throttle switch, in series with the surge-bleed synchronizing switch, opens and automatically closes the valve.

B-50 Variations *(Ref. 7-49)*

B-50A

First production version. Four-blade Curtiss Electric fully feathering propellers.

TB-50A

Modified B-50A for bombing and navigation training. For details see under TB-50D.

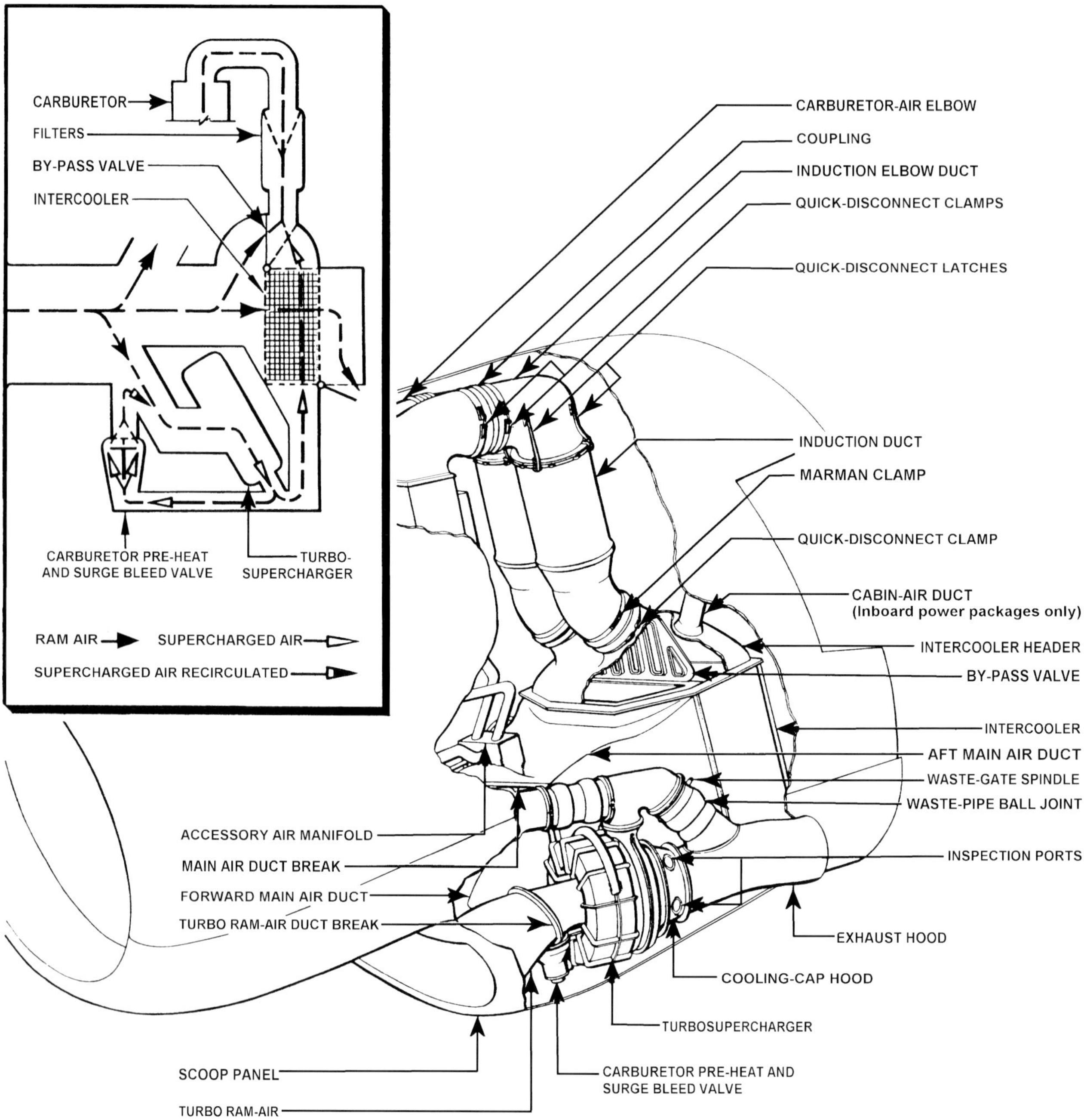

Schematic of the B-50's induction system. *(AN 01-20ELA-2 Erection and Maintenance Instructions for Boeing B-50 Aircraft. Revised February 5, 1948)*

B-50B

Structural changes allowing an increase in gross weight to 164,000 pounds to permit greater load and range. Subsequently modified to RB-50B status.

EB-50B

Experimental aircraft fitted with tracked-tread landing gear—soon abandoned.

RB-50B

Photoreconnaissance version of B-50B. Still retained bombing equipment but now fitted with nine cameras installed in four stations. Photo flash bombs for night photography carried in bomb bays. Reconnaissance equipment operated remotely from forward flight deck. Also has improved radar, flight refueling equipment, and weather recording instruments. Additional fuel can

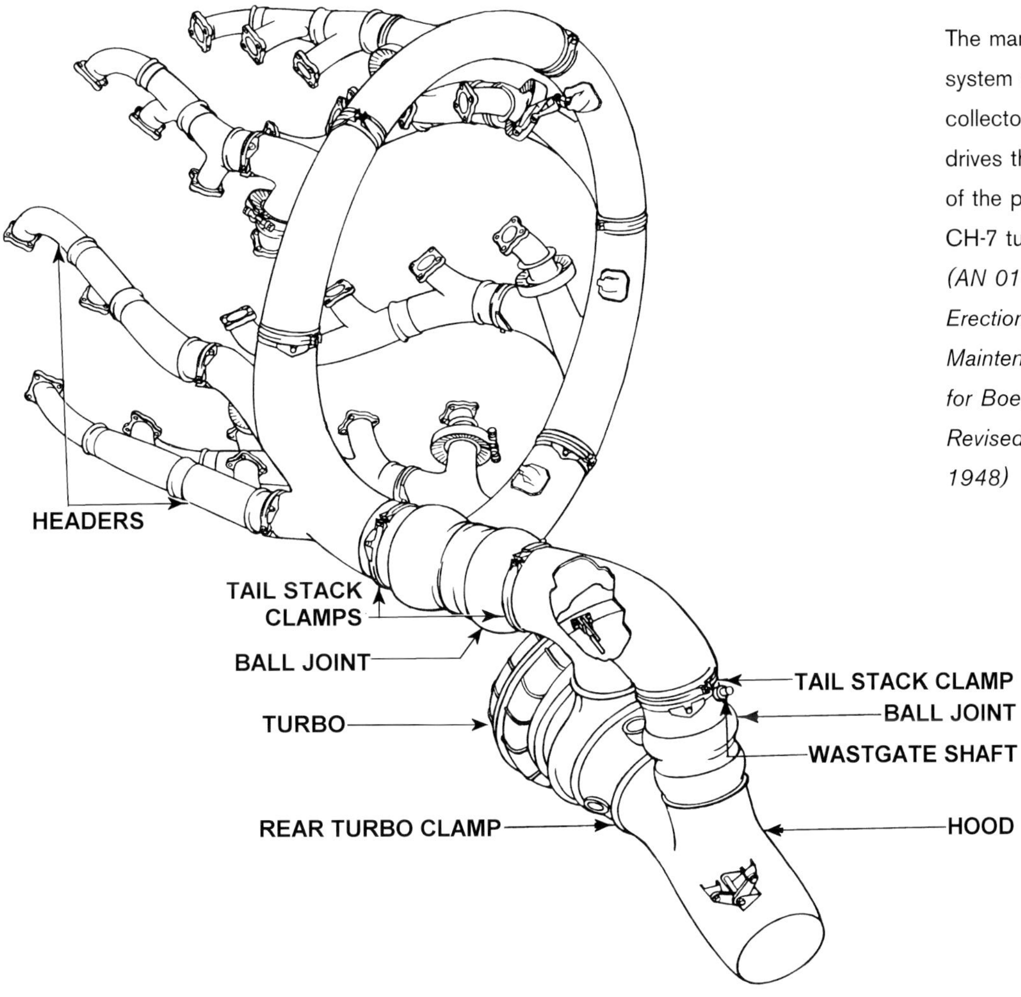

The manifolded exhaust system dumps into a collector ring, which then drives the turbine section of the problematic GE CH-7 turbosupercharger. *(AN 01-20ELA-2 Erection and Maintenance Instructions for Boeing B-50 Aircraft. Revised February 5, 1948)*

This production-line photo shows the QECs being built up. They are identifiable as B-50 units by the location ot the CH-7 turbosupercharger: C-97s and B-377s located the turbo on the opposite side of the nacelle. *(Courtesy of National Archives & Records Administration)*

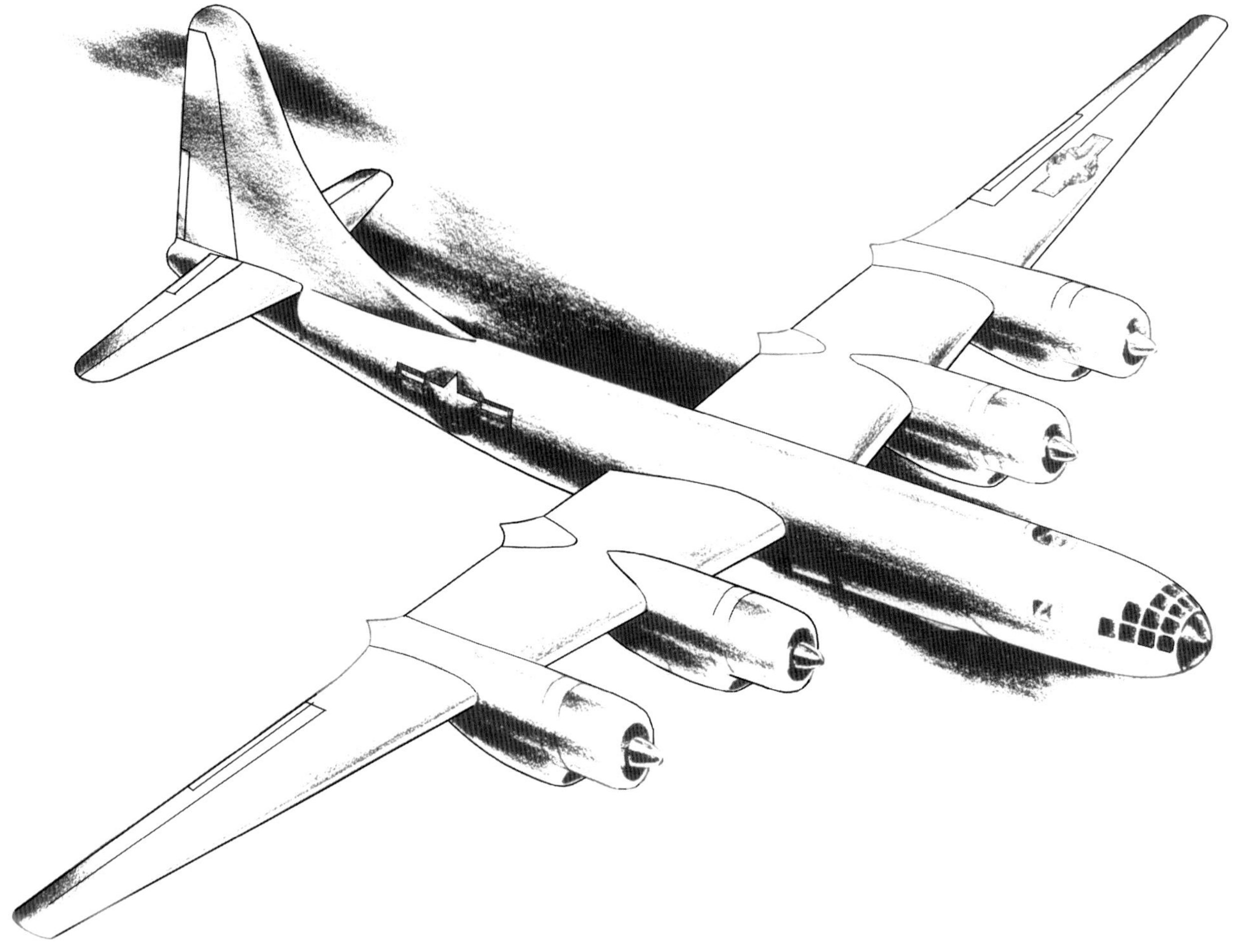

What could have been. An artist's concept of the VDT-powered B-54. *(Courtesy of National Archives & Records Administration)*

be carried through a pair of 700-gallon droppable wing tanks, as first introduced with the B-50D.

YB-50C

This was the stillborn version to be powered by VDT engines. One B-50 had a single nacelle modified for VDT power and was extensively test flown in 1949.

B-50D

Development of the B-50B with increased bomb and/or fuel load. Can carry two 700-gallon droppable wing tanks or two 4,000-pound bombs on support struts outboard of the outer (numbers 1 and 4) nacelles. Later fitted with an optically flat bomb aimer's panel in nose glazing. Additional changes included the radar installation.

TB-50D

Modified B-50D to train triple-duty crewmembers who go on to serve as combined bombardier/navigator/radar operators in Boeing B-47 jet bombers. All armament removed. Normal flight crew stations in forward fuselage, plus provisions for two student navigators and one instructor. Rear bomb bay sealed and used for installation of a large part of the electronic equipment used in training. Rear crew compartment contains installations for one instructor and two student radar operators.

RB-50E
Photographic reconnaissance version of B-50D.

RB-50F
Same as RB-50E except for added SHORAN radio. Late in World War II, the U.S. developed a navigation system based on radio transponders and similar to Gee-H, known as SHORAN (Short Range Air Navigation). It operated at a higher frequency, around 1 meter/300 MHz, giving it greater accuracy than Gee-H. It was too late to see service in the conflict.

RB-50G
Photographic reconnaissance version of B-50D.

KB-50
134 B-50As, RB-50Bs, and B-50Ds converted to airborne refueling role.

TB-50H
Production version of TB-50D. First TB-50H flew for the first time on April 29, 1952. First TB-50H delivered to USAF on September 6, 1952, last on February 26, 1953.

KB-50
Same as KB-50 but auxiliary wing tanks replaced by General Electric J47 jets.

B-54
VDT powered version of the B-50. None built.

Boeing C-97, KC-97E, YC-97A, YC-97B, KC-97F, KC-97G

(Ref. 7-1, 7-49, 7-68, and 7-69)

Using the wings, tail surfaces, landing gear, and QECs (with one notable exception being the turbos mounted on the opposite side of the nacelle) of the B-50, Boeing developed the long-lived C-97 and KC-97 series of aircraft. Interestingly, Boeing swapped sides when it came to mounting the turbo and oil cooler. The C-97/KC-97 mounted the turbosupercharger on the right side of the QEC as opposed to the left side. This was done in the case of the B-50. Likewise, the oil cooler was mounted on the left side rather than the right side in the case of the B-50.

Carrying on from its B-29 heritage, the first three XC-97 and 10 YC-97 prototypes were powered (underpowered?) by Wright R-3350s. Various models of the R-4360, all of which were turbosupercharged, powered all subsequent aircraft starting with the C-97A. It's often assumed that the commercial Boeing Model 377 Stratocruiser was a development of the C-97. In fact, it was the other way around. Boeing put plans in place as early as 1942 for a postwar commercial airliner based on the B-29. Ironically, relatively few 377s were manufactured but 888 C/KC-97s were built.

The primary difference between the B-50 and C-97 was the fuselage. C-97s basically used the bottom two-thirds of the B-50 fuselage with another larger fuselage grafted on top. Both fuselages formed the classic "double bubble." Boeing engineers placed the floor at the intersection of the two circular section fuselages. In this way the floor acted as a primary load-bearing member, not only to retain cargo loads, but also to contain the considerable loads due to pressurization. It was capable of carrying a variety of loads including three loaded 1½-ton trucks or two light tanks, which could be driven on board via a ramp in the rear fuselage. One hundred thirty-four fully equipped troops or 83 stretcher cases could also be loaded. However, it was as an aerial refueler that the newly designated KC-97 found its niche.

But it wasn't all smooth sailing. As new jets came into the Air Force inventory, the KC-97 was stretched to its limits trying to maintain a speed that was above the stall speed of the receiving aircraft. Various flying methods were tried including the so-called tobogganing technique. Once an aircraft was connected to the flying boom, the KC-97 pilot would put his aircraft into a shallow dive in order to maintain air speed. All the time METO power was applied to

A factory-fresh KC-97F is rolled out of Boeing's assembly building at Seattle, Washington. *(Courtesy of National Archives & Records Administration)*

the R-4360s. Of course, this kind of treatment did not go over well with the over-worked R-4360s. Relief, to a limited degree, came in the form of a General Electric J47 under each wing. The J47s replaced the underwing long-range tanks. Jet-assisted aircraft were modified in 1964 and redesignated KC-97L. Fuel was stored in the lower section of the double bubble. The upper section of the double bubble afforded the crew spacious accommodations. In fact it was almost as large as its contemporary, the C-99.

KC-97s continued in service until the 1970s and were the last of the R-4360-powered aircraft used by the USAF.

C-97 Variations

C-97A

First production version delivered to USAF on October 14, 1949. Powered by R-4360-27s, R-4360-35s, or R-4360-65s.

C-97 Parameters *(Ref. 7-1)*:

Wingspan	141 feet, 3 inches
Length	110 feet, 4 inches; 117 feet, 5 inches for KC-97
Height	38 feet
Empty weight	82,000 pounds
Weight loaded	142,500 pounds
Landing weight	121,000 pounds
Max. speed	350 mph; 400 mph with GE J47s
Cruising speed	300 mph
Operating range	3,750 miles
Service ceiling	30,000 feet

This underside view of the nacelle shows how the three scoop compartments are split up to provide air to the intercooler, oil cooler, and induction air. *(T/O 1C-97A-10 Erection and Maintenance Instructions for Boeing C-97 Aircraft)*

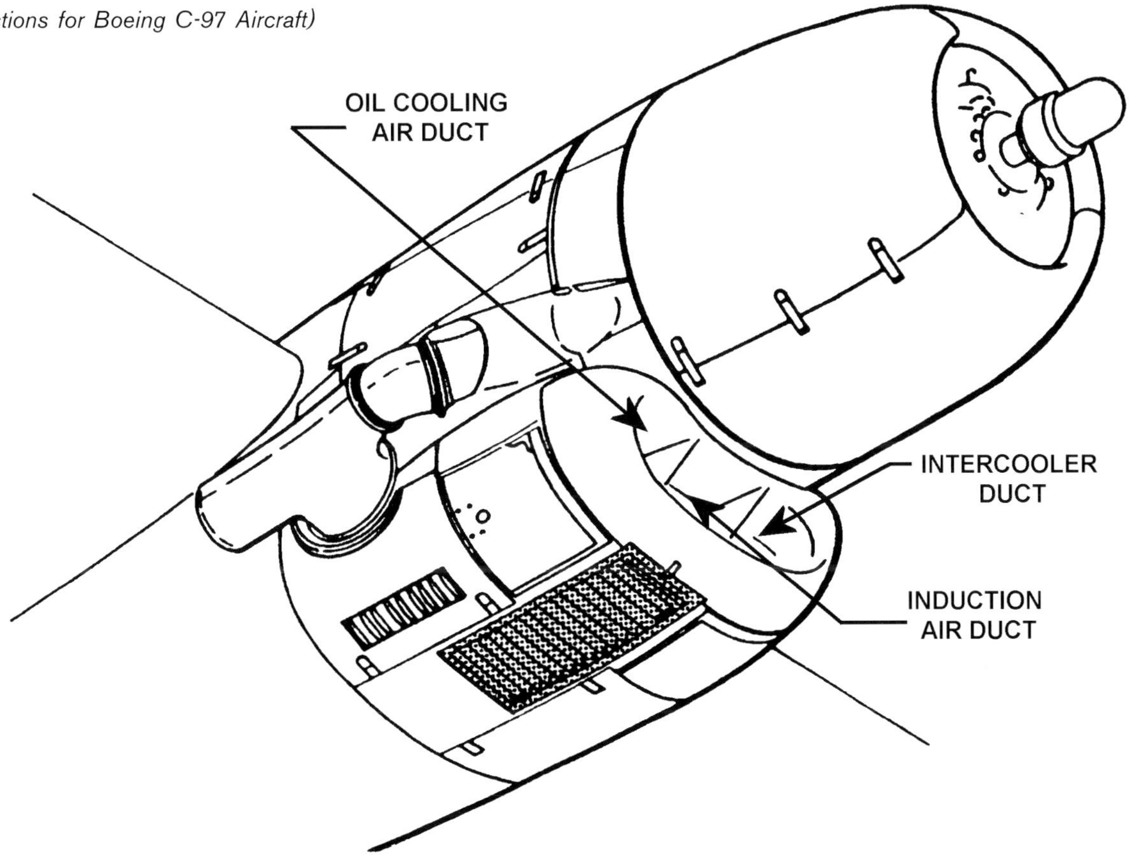

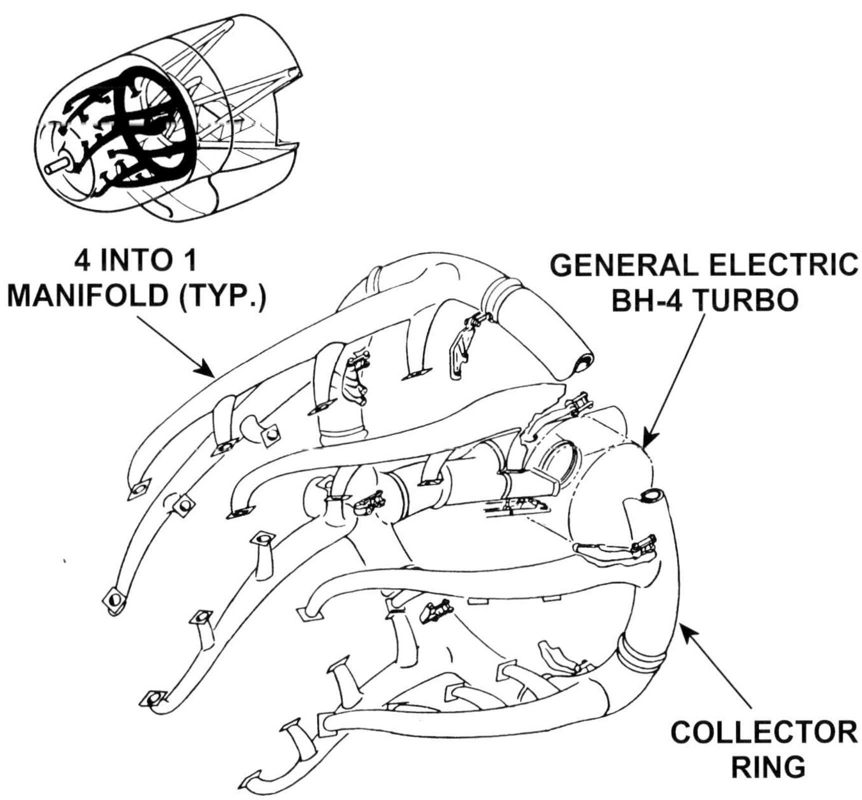

The C-97's exhaust system was similar to the B-50s with the exception of having the outlet that feeds the turbo on the opposite side. *(T/O 1C-97A-10 Erection and Maintenance Instructions for Boeing C-97 Aircraft)*

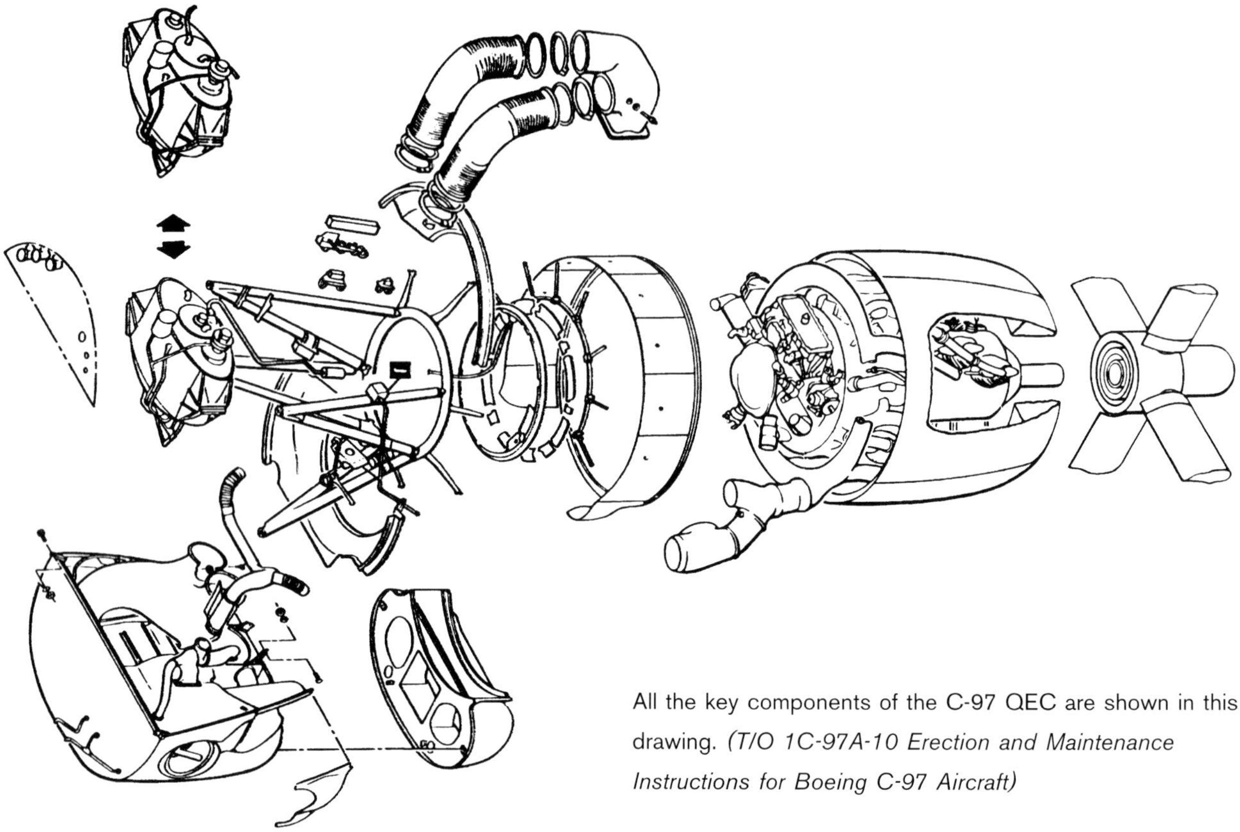

All the key components of the C-97 QEC are shown in this drawing. *(T/O 1C-97A-10 Erection and Maintenance Instructions for Boeing C-97 Aircraft)*

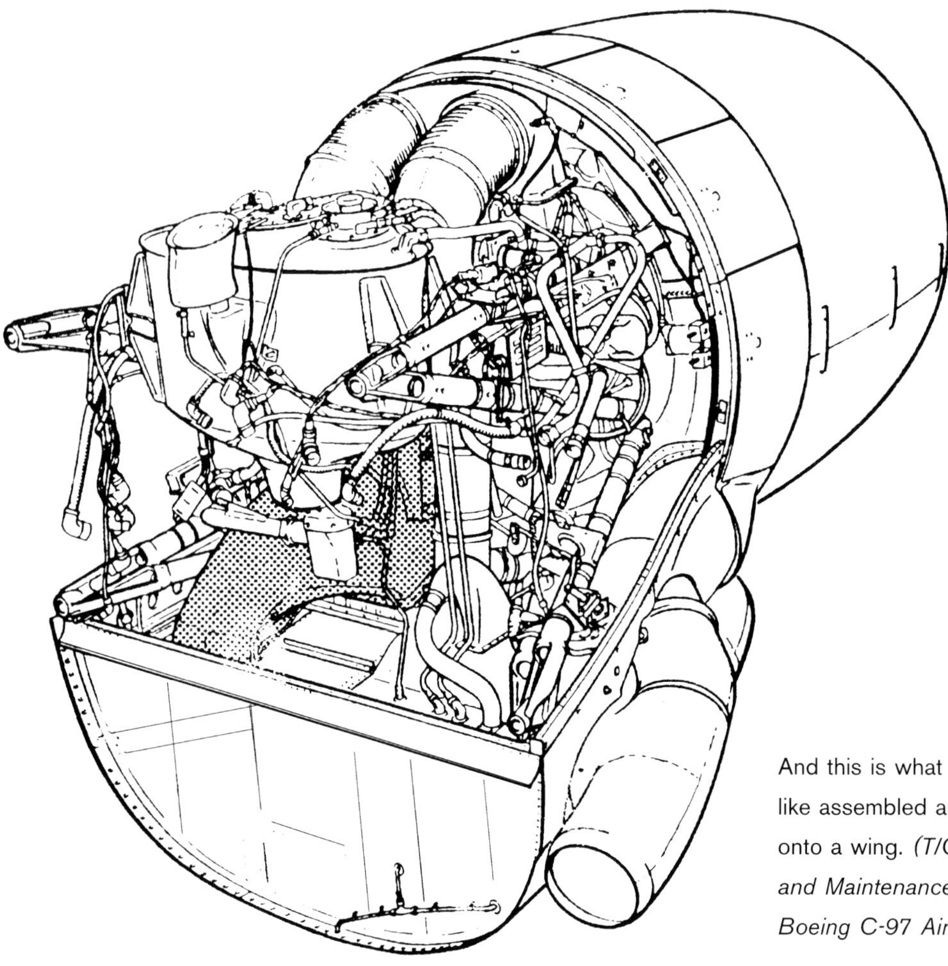

And this is what the C-97 QEC looks like assembled and ready to be bolted onto a wing. *(T/O 1C-97A-10 Erection and Maintenance Instructions for Boeing C-97 Aircraft)*

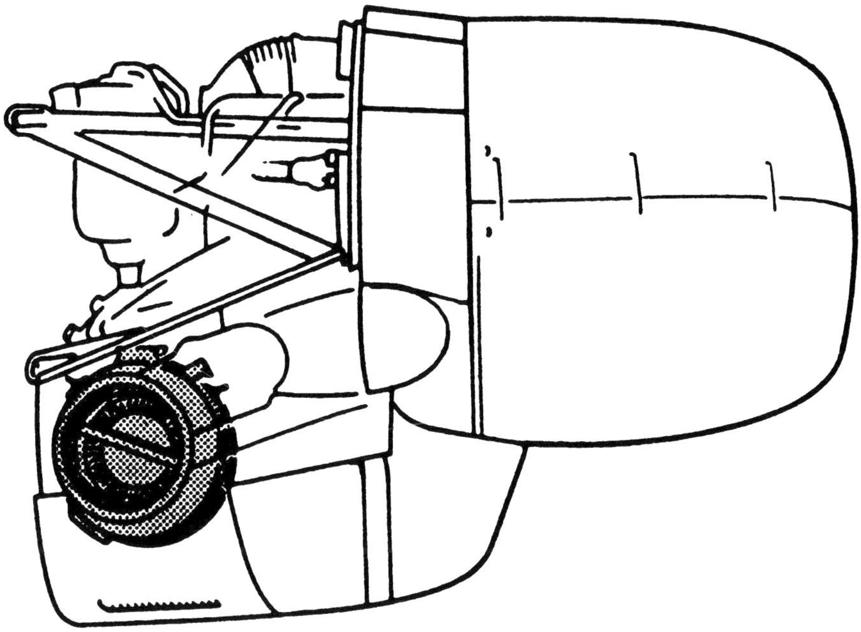

Upper mount attach points are shown in this side view. Two lower attach points, hidden by the cowl, make up the four-point mount. *(T/O 1C-97A-10 Erection and Maintenance Instructions for Boeing C-97 Aircraft)*

C-97 QEC during manufacture. The hooded baffles on the engine have been removed. Also of note in this photo is the intercooler radiator core visible on the lower right. *(Courtesy of National Archives & Records Administration)*

Rear view of the R-4360-59 mounted in the C-97 QEC. *(Courtesy of Pratt & Whitney)*

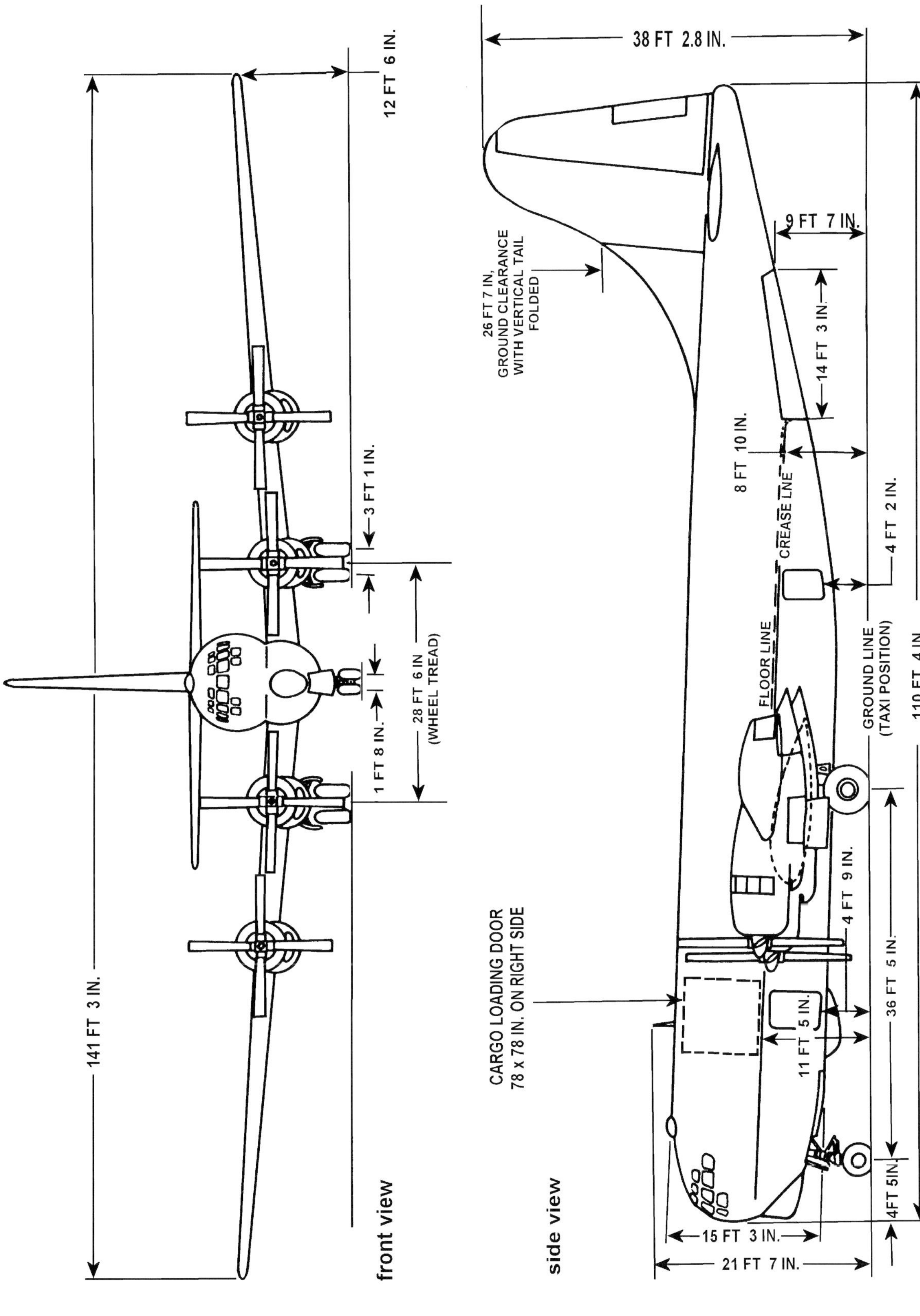

Key dimensions of the C-97. *(T/O 1C-97A-10 Erection and Maintenance Instructions for Boeing C-97 Aircraft)*

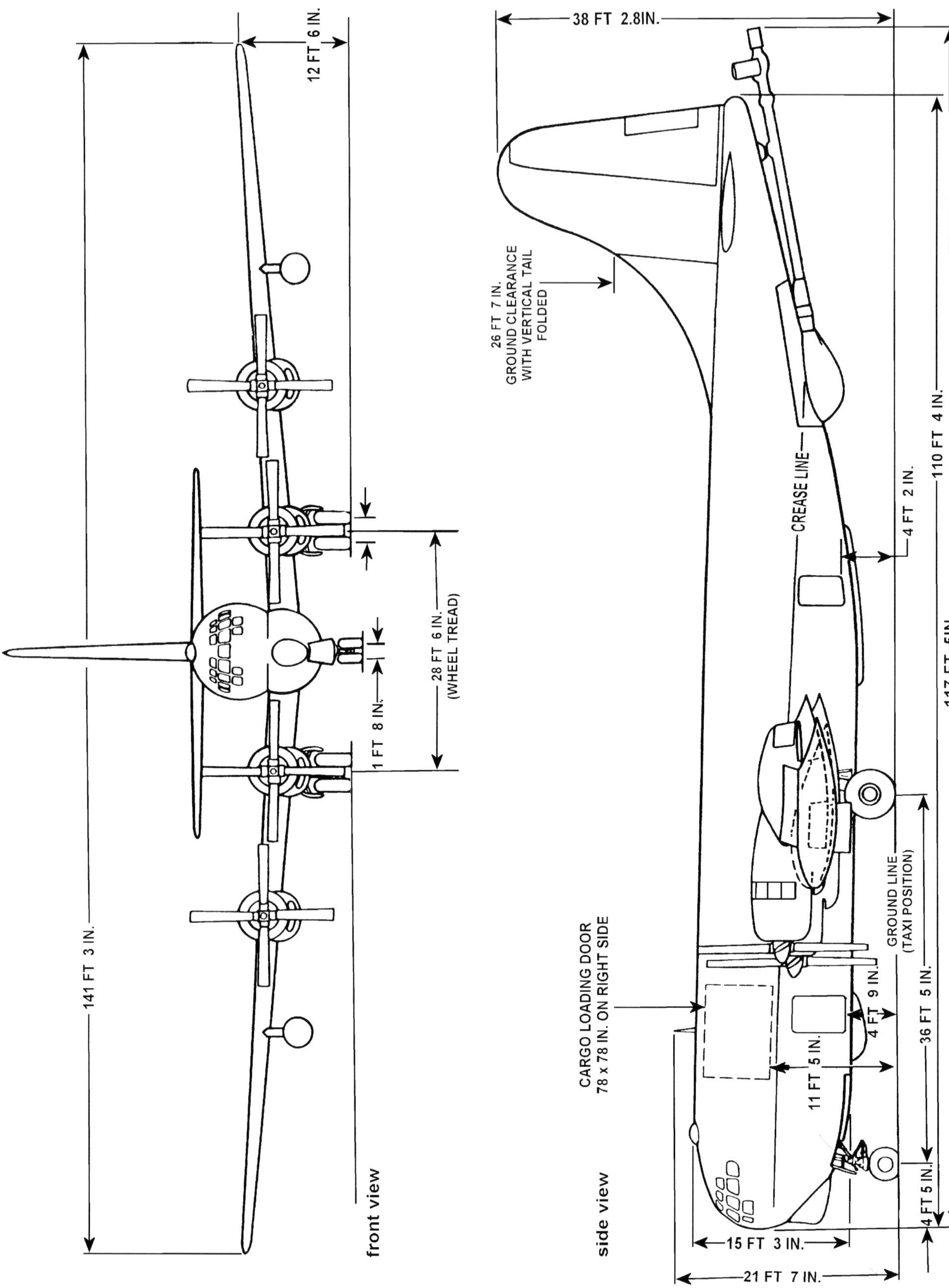

Key dimensions of the KC-97. The primary difference is the flying boom for refueling. *(T/O 1C-97A-10 Erection and Maintenance Instructions for Boeing C-97 Aircraft)*

Numerous KC-97 QECs and GE J47s in a depot ready for installation. *(Courtesy of National Archives & Records Administration)*

General Electric J47 jets were retrofitted to existing KC-97s to create the KC-97L. *(Courtesy of National Archives & Records Administration)*

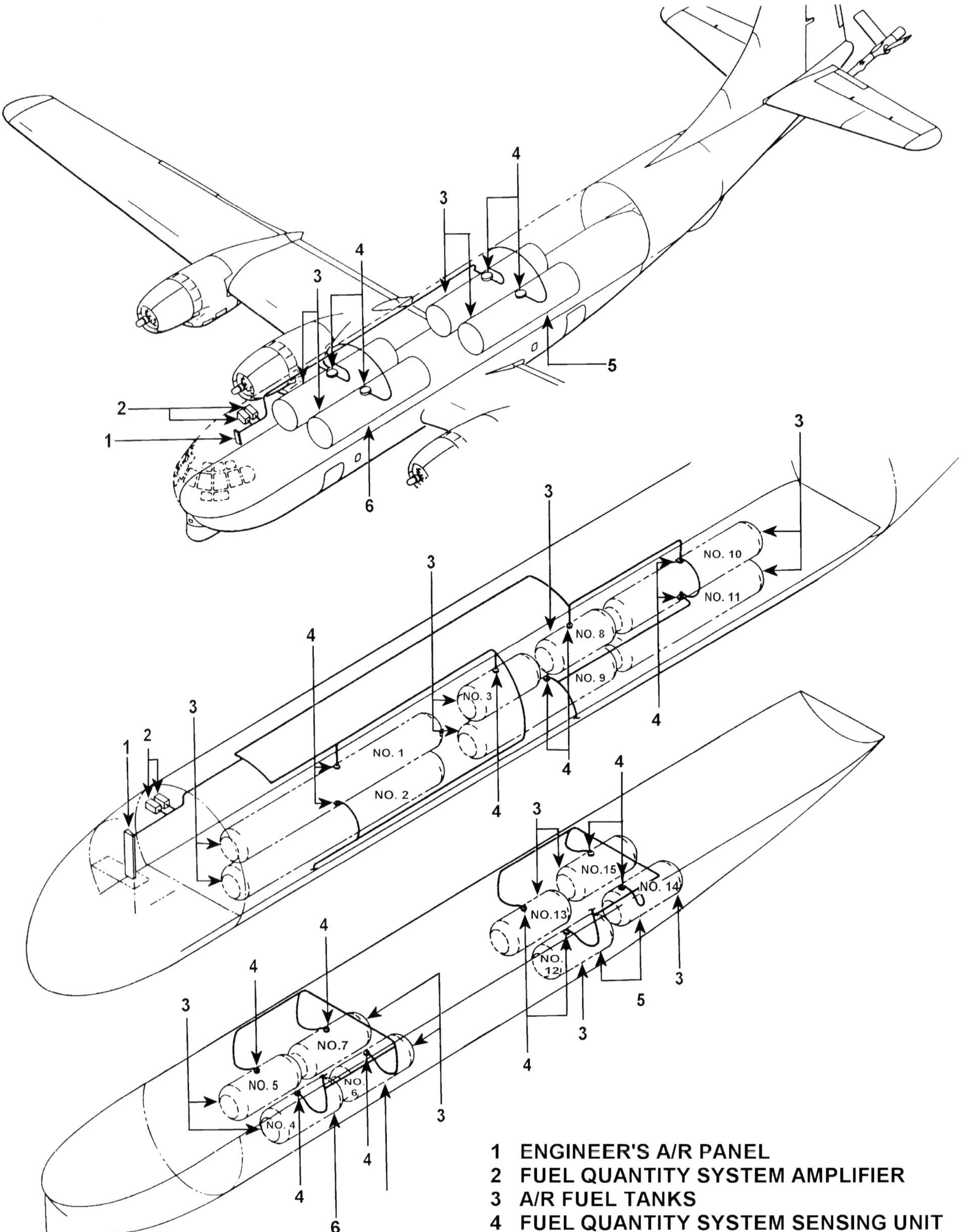

Fuel tank storage and fuel-valving system for the KC-97. *(T/O 1C-97A-10 Erection and Maintenance Instructions for Boeing C-97 Aircraft)*

YC-97B

Personnel transport, similar to commercial Model 377 Stratocruiser complete with spiral staircase, lower deck lounge, and rear upper deck galley. Capable of accommodating 80 passengers, plus 17,000 pounds of cargo. Only one aircraft built. Powered by R-4360-27s.

C-97C

Similar to C-97A. New radio antenna, heavier floor, higher payload. First delivery of C-97C made to USAF in February 1951. Powered by R-4360-27s, R-4360-35s, or R-4360-65s.

C-97D

Only three C-97Ds delivered to perform duty for the newly formed Strategic Air Command as mobile command posts. Used as staff transports and living quarters for Air Force brass in overseas training missions.

KC-97E

Multi-purpose transport and tanker. Has permanent equipment installed for tanker operations but can be rapidly converted to cargo or troop carrier, or ambulance. Pod carrying aerial refueling boom and controls is detachable. Additional fuel tanks and pumps normally used for aerial refueling and carried on the upper deck are removable. First production KC-97E delivered to USAF in July 1951. Powered by R-4360-35s, R-4360-59s, or R-4360-65s.

Well laid out flight deck of the C-97 showing the pilot, co-pilot, flight engineer, and radio operator. *(Courtesy of National Archives & Records Administration)*

Head-on view of the C-97, which illustrates its "double bubble" fuselage design. *(Courtesy of National Archives & Records Administration)*

KC-97F

Similar to KC-97E. Convertible airborne refueler or transport. Powered by R-4360-59s.

KC-97G

Development of KC-97F. Change in location of refueling tanks and related equipment, including the boom, so that it is not necessary to remove them when the aircraft is used as a transport. Provision for underwing long-range tanks.

KC-97H

Test bed for Pratt & Whitney T34 turboprop.

KC-97L

KC-97 modified with the addition of two GE J47 jets.

From the foregoing, it can clearly be seen that this wonderful engine powered many military aircraft—far more than most folks would realize. Coming in at the cusp of the gas turbine's entry

into general use for military applications, few wanted anything to do with the complex and temperamental R-4360 at the time. However, over time we tend to forget the less than desirable aspects of the R-4360 and focus instead on its wonderful charisma. Nevertheless, it's well worth the effort to take a closer look at its many attributes. And don't let us forget that R-4360-powered aircraft held the line between us and communist domination for several crucial years during the Cold War.

References

7-1 *Jane's All The World's Aircraft*. 1948.

7-2 Internal Goodyear Aircraft Report GER-2569.

7-3 *Jane's All The World's Aircraft*. 1949.

7-4 Kowalski, Bob. *Martin AM-1/1-Q Mauler.*

7-5 AN 01-35EF-2 Martin Mauler Erection & Maintenance Manual.

7-6 Bodie, Warren M., *Republic's P-47 Thun-*

derbolt: From Seversky to Victory. Widewing Publishing.

7-7 Republic XP-72/P-72 Erection & Maintenance Manual.

7-8 *Jane's All The World's Aircraft 1945/1946.*

7-9 http://www.vectorsite.net/avmars.html

7-10 White, Graham. *R-2800 – Pratt & Whitney's Dependable Masterpiece*. Society of Automotive Engineers, Warrendale, Pennsylvania. 2001.

7-11 Lewis, B. M., and C. T. McKinnie. *Report No. PWA WC. 47 R-4360-37 Engine Installation in XF-11 Airplane*. Pratt & Whitney Aircraft, East Hartford, Connecticut. April 3, 1947.

7-12 Sara Clark Files, Box 14, Airplane F-11. File DY5/D5 Hughes. Located in National Archives & Records Administration, College Park, Maryland.

7-13 Levins, Harry. *The History Of An Airline.*

7-14 http://www.aafo.com/goose/

7-15 *Hughes Flying Boat Manual*. Hughes Tool Company Aircraft Division.

7-16 Information obtained from original engineering drawing of the HK-1 via The Evergreen Aviation Museum. Detail drawings: 10-610 0009 Instal. Cowl Flap, 10-520 0951 Assembly, Nacelle, 10-520 0953 Instal. Powerplant & Nacelle.

7-17 Interviews and correspondence with Mike Wright of the Evergreen Museum.

7-18 Erection & Maintenance Manual for the XB-35.

7-19 Erection & Maintenance Manual for Douglas TB2D-1.

7-20 Kowalski, Bob, *Douglas XTB2D-1 Skypirate*.

7-21 http://www.wpafb.af.mil.museum/

7-22 http://warbirdsresourcegroup.org/NARG/xf8b-1.html

7-23 http://www.csd.uwo.ca/~pettypi/elevon/gustin_military/db/us/BTCCURTI.html

7-24 Angelluci, Enzo, with Peter Bowers, *The American Fighter*. Orion Books, New York.

7-25 Curtiss engineering cutaway drawing, Perspective Inboard Profile XP-71 drawn by G. L. Flanders. December 16, 1943.

7-26 http://www.wpafb.af.mil/museum/research/p71.htm

7-27 http://1000aircraftphotos.com/Transports/1642.htm

7-28 http://www.air-and-space.com/Lockheed%20R6V%20Constitution.htm

7-29 Report No. 4217 dated June 12, 1943. *Comparison of General Electric and Lockheed Turbosuperchargers*. Prepared by E. C. Posner, approved by C. L. Johnson. Lockheed Aircraft Corporation, Burbank, California.

7-30 Report No. 4342 dated September 14, 1943. *Preliminary Specifications for Exhaust Driven Turbo Supercharger.* Prepared by P. A. Beck, approved by C. L. Johnson and Hall Hibbard. Lockheed Aircraft Corporation, Burbank, California.

7-31 Report No. 4360 dated September 14, 1943. *Turbo Supercharger Design for a Transport Airplane.* Prepared by P. A. Colman, approved by W. W. Beman. Lockheed Aircraft Corporation, Burbank, California.

7-32 Report No. 4575 dated March 17, 1944. *Preliminary Specifications for Turbosupercharger.* Prepared by P. A. Beck and R. C. Andrews, approved by C. L. Johnson and H. L. Hibbard. Lockheed Aircraft Corporation, Burbank, California.

7-33 Report No. 4651 dated 3-4-44. *Comparison of Turbo Supercharger.* Prepared by W. W. Merrill, approved by P. A. Colman. Lockheed Aircraft Corporation, Burbank, California.

7-34 Report No. 4916 dated October 4, 1944. *Preliminary Specifications for Turbo Supercharger.* Prepared by R. C. Andrews and P. A. Beck, approved by C. L. Johnson and H. L. Hibbard. Lockheed Aircraft Corporation, Burbank, California.

7-35 http://www.theaviationzone.com/

factsheets/c74.asp
7-36 http://www.air-and-space.com/Douglas%20C-74.htm
7-37 *AN 01-40NT-2 Erection & Maintenance Instructions for Army Model C-74 Airplanes.* April 4, 1947.
7-38 Hand written note by Andrew Wilgoos on letter from Douglas Aircraft to Pratt & Whitney.
7-39 http://www.wpafb.af.mil/museum/modern_flight/mf9.htm
7-40 AN 01-40NVA-2 *Erection & Maintenance Instructions for Army Model C-124 Airplanes.*
7-41 AN 01-40NVA-4 *Group Assembly Parts List for Army Model C-124 Airplanes.*
7-42 http://www.theavaitionzone.com/
7-43 http://www.wpafb.af.mil/museum/fta/fta1197.htm
7-44 USAF T.O. 1C-119B-2 *Handbook Maintenance Instructions* USAF Series C-119B, C-199C Aircraft. December 1, 1954.
7-45 AN 01-115CCA-2 *Erection & Maintenance Instructions for USAF C-119B.*
7-46 http://tanks45.tripod.com/Jets45/Histories/Martin-P4M/Martin-P4M-Mercator.htm
7-47 NAVAER 01-35EH-502 *Martin P4M Mercator Maintenance Manual.*
7-48 Ginter, Steve, *Martin P4M-1/-1Q Mercator.*
7-49 *Jane's All The World's Aircraft*. 1953.
7-50 452.1—*Consolidated B-36—Long Range Mission 1948-49*. USAF report located at NARA.
7-51 Susag, M. P., and Hopper, P. S., Report TR-46. *Estimated Performance of a Wasp Major R-4360-C3 Variable Discharge Turbosupercharged Engine*. Pratt & Whitney Aircraft Division of United Aircraft Corporation, East Hartford, Connecticut. March 7, 1947.
7-52 *Aircraft Performance Engineer's Manual for B-36 Aircraft Engine Operation*. January 1952.
7-53 AN 01-5EUB-2 *Handbook Erection and Maintenance Instructions USAF Series B-36B Aircraft*. May 5, 1949.
7-54 AN 01-5EUE-2 *Handbook Erection and Maintenance Instructions USAF Series B-36F Aircraft*. March 27, 1951.
7-55 AN 01-5EUG-1 *Flight Handbook USAF Series B-36H Aircraft.*
7-56 *Handbook Flight Operating Instructions USAF XC-99 Airplane Serial Number 43-52436 MFG Model Designation 37*. March 31, 1949.
7-57 http://www.aerofiles.com/_repub.html
7-58 http://www.fas.org/irp/program/collect/xr-12.htm
7-59 Brabham, L. L., *Flying Combat Aircraft of USAAF.* 1975.
7-60 Marthason, A., Aeronautical Engineering. *The Republic Rainbow Parts I & II.* Aeroplane. December 6, 1946, and December 13, 1946.
7-61 *USAAF XF-12 Flight Manual.*
7-62 Engineering installation drawing for XB-44 located in Pratt & Whitney archives.
7-63 http://www.wpafb.af.mil/museum/research/bombers/b4/b4-6.htm
7-64 Nine photographs located in Pratt & Whitney archives showing details of XB-44 QEC design.
7-65 AN 01-20ELA-2 *Erection and Maintenance Instructions for Boeing B-50 Aircraft.* Revised February 5, 1948.
7-66 http://home.att.net/~jbaugher2/b50.html
7-67 AN 01-20ELB-2 *Erection and Maintenance Instructions for Boeing RB-50E and RB-50F Aircraft.* Revised October 1, 1951.
7-68 T.O. 1C-97A-10 *Erection and Maintenance Instructions for Boeing C-97 Aircraft.*
7-69 Document No. D6-3399. *Analysis for Engine Cooling Data With 34G60 Propellers and Revised Nacelle Configuration. Model No. C/KC-97*. Boeing Aircraft Company, Seattle, Washington. Approved April 27, 1959.5

CHAPTER EIGHT

Commercial Applications

If airline executives were aviation enthusiasts, then the R-4360, or Wasp Major in commercial speak, would have seen more applications. As it was, the R-4360 saw relatively few commercial applications. Operating such a large and complex powerplant was difficult to justify from a financial perspective. Wright Aeronautical's R-3350 turbo compound was as far as the airlines were willing to go with regards to complexity. It could also be argued that it was still an overly complex engine that required an inordinate amount of maintenance. With bills to pay and shareholders to keep happy, the R-4360 was not, to use modern vernacular, a good fit.

Although the military used the R-4360 very successfully into the late 1970s, they had a number of conditions in their favor. Even though it may rankle us as taxpayers and regardless of what is proclaimed to the contrary, the military essentially operates with a blank check. Also, the fact that so many R-4360s were operated by the military allowed an economy in numbers the airlines could never achieve. In this way, facilities such as SAAMA or San Antonio Air Materials Administration kept a production line operating on nothing but overhauling R-4360s. With this kind of economy of scale, the R-4360 made perfect sense for the military. It fulfilled a requirement for a high horsepower engine and being the last of the line as far as piston engines went, it represented the most sophisticated engine of this type ever used by the armed forces. The fact that so many R-4360s were in service made maintenance chores somewhat easier due to the huge spare parts inventory and expertise built up over the years of servicing these engines.

Boeing built the only U.S. commercial aircraft powered by the R-4360—the Model 377. However, a number of manufacturers proposed R-4360-powered airliners. None came to fruition, not even to the point of a prototype being built. Aero-Sud took the plunge after World War II and developed an R-4360-powered airliner, albeit of dubious usefulness. Wright Aeronautical's R-3350 turbo compound represented the R-4360's commercial competition. Both rated at approximately the same power, give or take a hundred or so at 3,500 hp, the R-3350 was far more successful in the commercial arena. Why was this? Given the R-3350's notorious reputation for unreliability, particularly in the early models, the answers are not easy to arrive at. Although the R-4360 was far more complex, in some respects routine maintenance was far easier. As an example, changing a cylinder on a 3350 could be the nightmare from Hell. The same job on the R-4360 was far easier; mainly because the mechanic had more room to work with because the 4360 only had seven cylinders per row as apposed to the 3350's tightly packed nine cylinders per row.

The R-3350 had one major advantage over the 4360—turbo compounding. With its three power recovery turbines (PRTs), the R-3350 could achieve remarkable fuel economy. If the R-4360 had also been equipped with PRTs, could it have offered serious competition to the R-3350? Hard to say, but it's probably a fair assumption to make that a PRT-equipped R-4360 would have offered better time between overhaul than its Wright counterpart. At best, the R-3350 was a "fragile" engine, in other words, it was a flight engineer's engine. If a turbo-compounded R-4360 had been installed in, say, the Lockheed Constellation or Douglas DC-7, would this engine have made these aircraft better? Probably so. As it was the only U.S.-built commercial aircraft powered by the R-4360 was the Boeing Model 377 fitted with turbosuperchargers. A turbo-compounded engine would have been lighter, more fuel efficient, and easier to maintain at a slight cost of altitude performance. It should be kept in mind that at 3,500 hp, the R-3350 was producing 1.04 hp per cubic inch. At this specific power an R-4360 would be rated at 4,555 horsepower. The additional 4,000 hp per aircraft, if PRT R-4360-powered in the likes of a DC-7, for instance, would have offered a tremendous performance boost. Or, the R-4360s could have been operated at considerably lower power for added longevity. Either way would have been a better proposition.

It should also be borne in mind that Pratt & Whitney was smart enough to realize that the end of the piston era was approaching, something that was overlooked at Wright. For this and many other reasons, Wright Aeronautical, at one time one of the largest industrial empires in the world, faded away into obscurity. Had the R-4360 received the kind of intense development heaped upon the R-3350, then 4,800-hp commercial Wasp Majors would have been powering airliners of the 1950s. This number is arrived at through figuring the specific power of the R-3350, at its zenith, reached 1.1 hp per cubic inch. Using this same specific power for the R-4360 gives the aforementioned 4,800.

In the context of this book, commercial applications are those engines not sold to the military, therefore designated "Wasp Major." However, several engines were developed that were clearly for the benefit of the Air Force or Navy. These engines are included in this chapter rather than the Military Applications chapter.

Applications are described in the order they appear in Table 8-1.

(Table 8-1)

(Ref. 5-3)

WASP MAJOR	APPLICATIONS
X Wasp	Vultee Model 85 Vought 326
TSB1-G	Vought F4U-1 WM
TSB2-G	None
TSB3-G	Boeing Model 377 Republic RC-2
B4	None
B5	Boeing Model 377 Model 10 & 19
B6	Boeing Model 377 Aero Spacelines B-377 Mini Guppy Aero Spacelines B-377 Super Guppy Aero Spacelines B-377 Pregnant Guppy
B6C	Boeing Model 377
B7	None
B7C	None
VSB11-G	Aero-Sud-Est SE-2010
VSSB21-G	None
B12	None
B13	Aero-Sud-Est SE-2010
B14	Aero-Sud-Est SE-2010
TSC1-G	None
C2	None
C3	None
C4	None (Military)
C5	None (Military)
C6	None (Military)
C7	None (Military)
C8	None (Military)
C9	None (Military)
C-01	None
C-02	None
C-03	None
C-04	None
C-05	None (Military)
C-06	None (Military)
VSSC21-G	None
VSC11-G	None
C12	None
C13	None
C14	None
C15	None
CB1	Boeing Model 377
CB2	Boeing Model 377 (C-97)
CB3	None
CB4	None
CB5	Boeing Model 377 for North West Airlines
CB11	None
CB12	None
CB13	None

Vultee Model 85

The Vultee Model 85 was the first of the test mules and the first aircraft powered by an R-4360. Although it had a short life, it paved the way for

One of the early test mules used by Pratt & Whitney to try out various concepts was this Vought VS-326. *(Courtesy of Pratt & Whitney)*

many follow-on aircraft. The history of this groundbreaking aircraft is described in Chapter 2.

Vought VS-326

(Ref. 8-1 and 8-2)

Only so much testing can be performed on an aircraft engine before it's time to get it in the air powering a real aircraft. To accommodate Pratt & Whitney's need for a flying test bed, Vought, which was located almost next door to Pratt & Whitney, developed a test vehicle with the uninspiring name of "VS-326" and a second one called "VS-326A." Despite its bland name, the VS-326 was an elegant-looking aircraft featuring, among other things, a pressurized cabin. Over its life the VS-326 was equipped with a number of variations on the R-4360 theme, such as turbosupercharging and remote gear-driven superchargers. First flown in June 1943, the first VS-326 had a relatively short life, being scrapped in 1946. The second VS-326A fared little better. It first flew in August 1944 and by the late 1940s ended its days as a ground-bound R-4360 test rig. Rationale for the brief lives of these test aircraft may be explained by the fact that more capable military aircraft came into production that could fulfill the needs of Pratt & Whit-

Below: This ¾ right front view shows the sheer elegance of the VS-326. Surprisingly, even though this was one of the first aircraft powered by the R-4360, it also employed what appears to be the tightest cowling arrangement for any aircraft powered by this engine. *(Courtesy of Pratt & Whitney)*

ney far more adroitly than the VS-326. Even so, a number of significant concepts were first tried in the 326. One noticeable feature of the VS-326 is the large scoop on top of the rear fuselage. This ram-air scoop fed induction air to a variety of turbosuperchargers including a sophisticated two-stage turbo developed by Rudolph Birman.

General Electric BH-1s, as used in the Convair B-36 and Northrop B-35, were flight tested in this resourceful aircraft. Fan cooling was another concept tested. Judging from the engineering drawings, various fan concepts were tested in this versatile aircraft—concepts that were carried over into the B-35 and B-36. Other concepts such as the Birman turbosupercharger never panned out. One has to assume that the

Left: Again, this left side view illustrates well the elegance of the VS-326. Note the small size of the windows of the pressurized cabin. *(Courtesy of Pratt & Whitney)*

Below: Of note in this ¾ left rear view of the VS-326 is the scoop located behind the cockpit. It was possibly used to supply induction air for the version with a remote gear-driven supercharger. This assumption is arrived at because this engine installation clearly has an open stack exhaust system, consequently negating any possibility of a turbosupercharger. *(Courtesy of Pratt & Whitney)*

This concept of the VS-326 shows a turbosupercharger installation in the rear fuselage. It appears that all air requirements are derived from the underbelly scoop. *(Courtesy of Pratt & Whitney)*

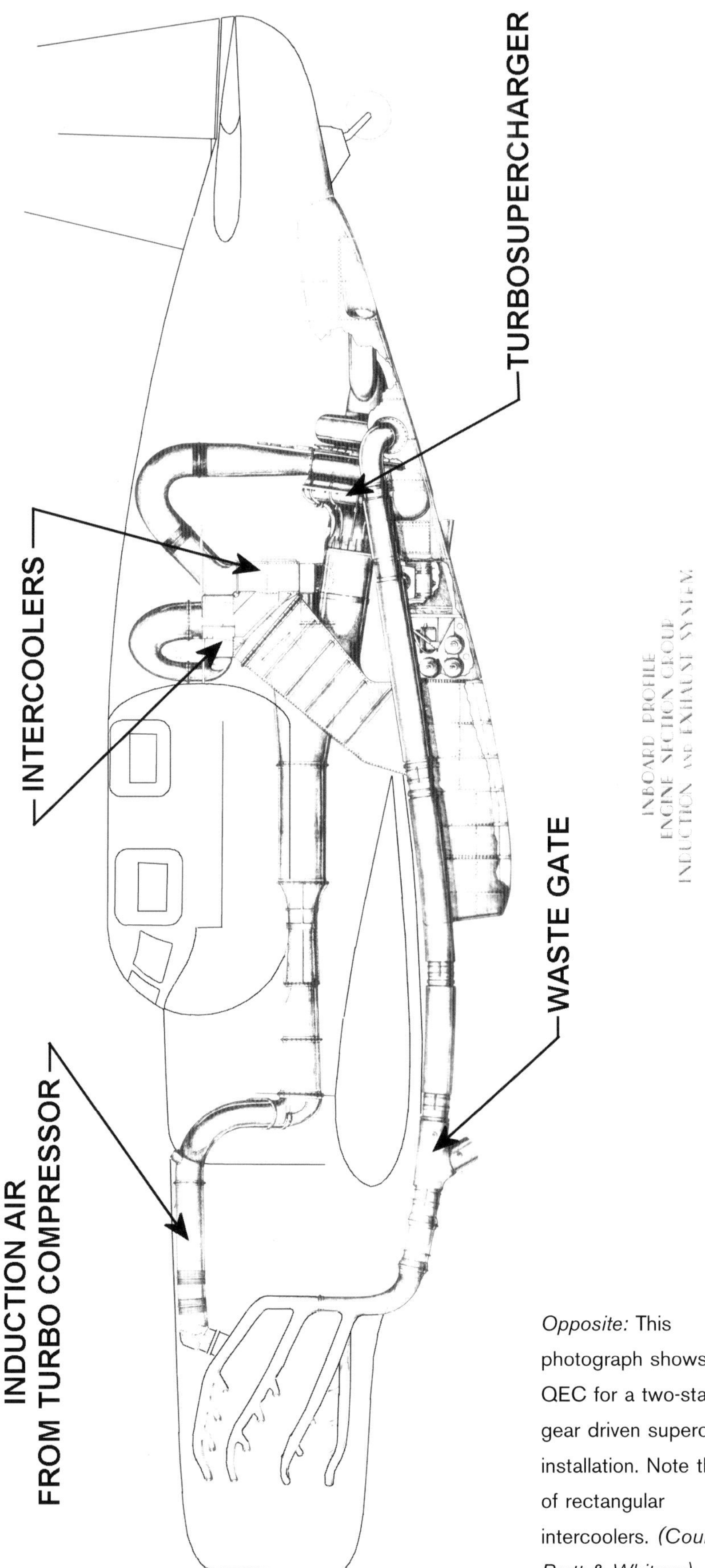

Opposite: This photograph shows a QEC for a two-stage gear driven supercharger installation. Note the pair of rectangular intercoolers. *(Courtesy of Pratt & Whitney)*

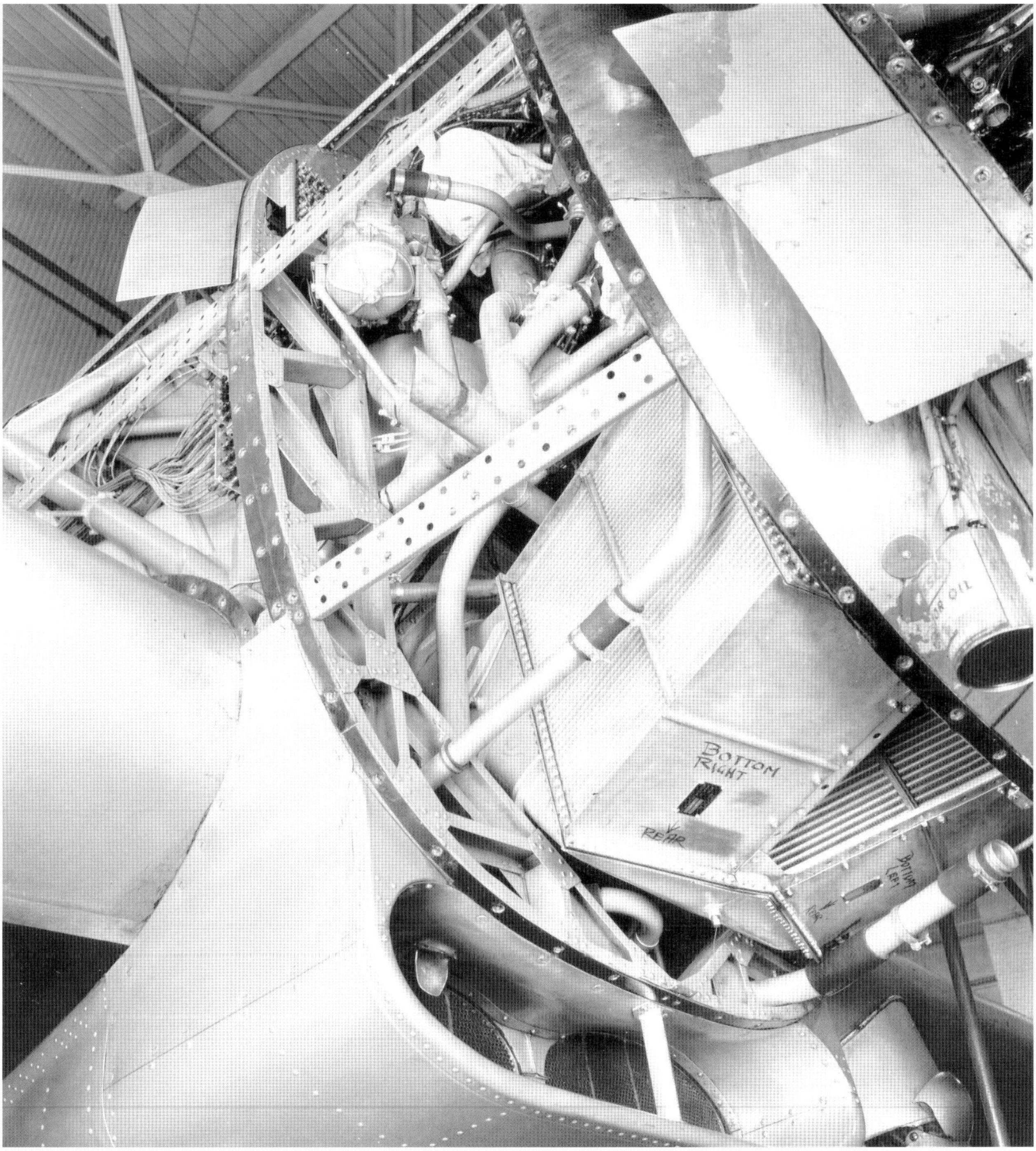

Birman experiments were undertaken at the instigation of the Navy. During World War II, the Navy set Birman up in business as the "Turbo Engineering Company" in order to develop an alternate turbosupercharger to the then universally used General Electric turbo. It's unclear as to why Birman's turbos never saw production, as they appeared to offer considerable advantages over their GE competitor. In fact, Pratt & Whitney adapted Birman's mixed flow compressor design for all of its later supercharger designs on the R-2800. Testing of two-stage superchargers was also performed in the VS-326; however, numerous problems were encountered. Low oil pressure in the nose case was one of these, which turned out to be cold oil congealing in the galleries, consequently blocking oil flow. Another was oil scavenging, or lack thereof. Report

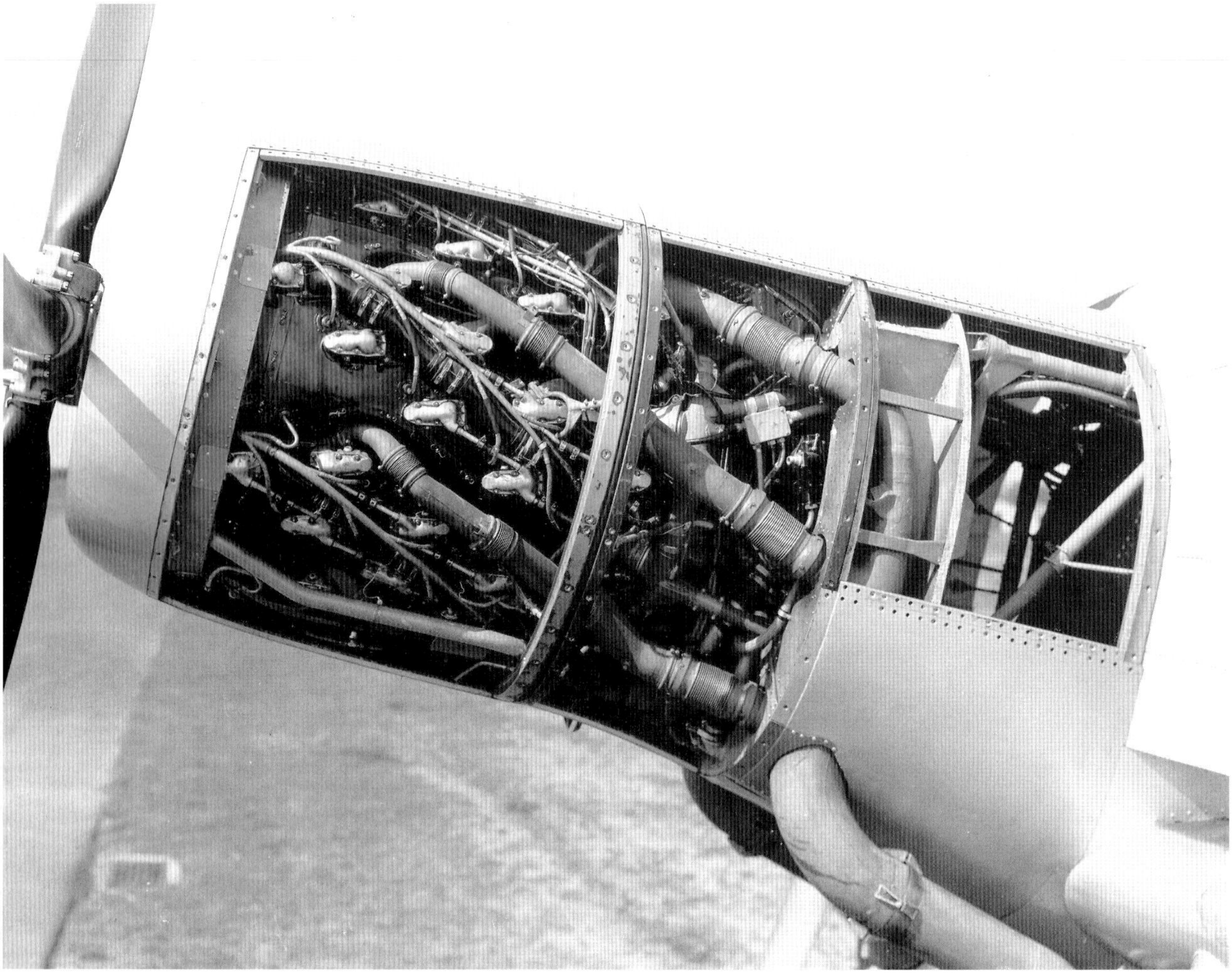

With the cowl panels removed, one can see just how tightly packed the R-4360 installation was in the VS-326. *(Courtesy of Pratt & Whitney)*

number T-129 indicated that the engine powering this aircraft at the time was X-109, an early two-stage engine. By installing scavenge pumps in each power section, the scavenge problems were alleviated. This would imply that the gerotor-type pumps used for this purpose were incorporated due to this series of testing.

Interestingly, of all the R-4360 installations, the VS-326 appears to have employed the tightest cowling arrangement. Hard to believe that over 3,000 hp lurks under the cowl of the VS-326. It's unfortunate that no VS-326s were preserved. Although it was a beautiful looking aircraft, aesthetics do not necessarily make for a good aircraft, notwithstanding the old saying, if it looks right... Judging by the narrow and spindly looking main gear, this aircraft must have been a handful on the ground, so this may have contributed to its relatively short life. Regardless of what the reasons are, the VS-326 made significant contributions to R-4360 development.

Boeing Model 377 Stratocruiser

(Ref. 7-49, 8-4, and 8-5)

At the conclusion of World War II, aircraft manufacturers went from feast to famine. After working their facilities to capacity they were faced with a tremendous glut of surplus aircraft on the market. Their primary salvation would be in commercial aircraft whose development had been placed on the back burner during the duration.

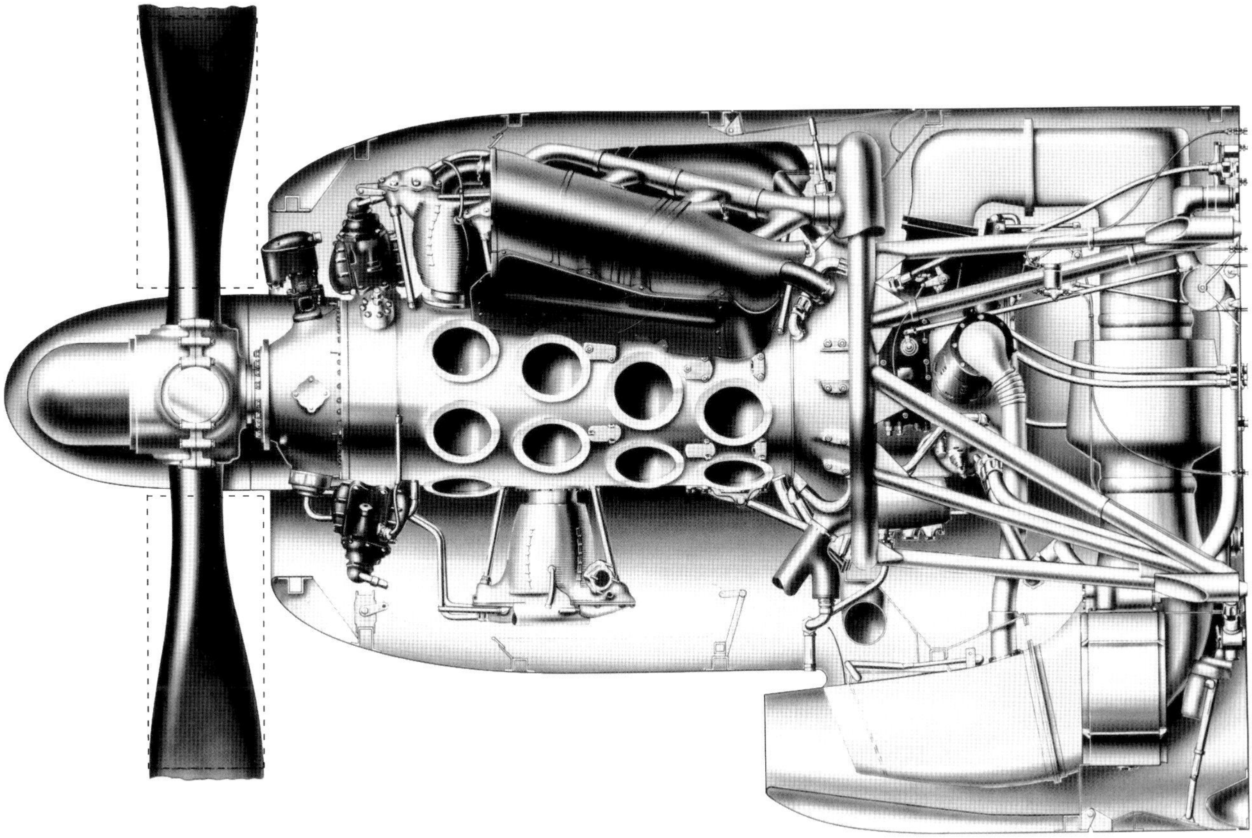

Above: This artist's impression gives an idea of the layout of the B-377 engine installation. Although it was a good clean installation, it is quite apparent from this illustration that the cowl diameter could have been shrunk down. Was this a throwback to the B-377's B-29 heritage in that Boeing was obliged to use the same firewall dimensions used for the R-3350? We will never know, but to optimize the size of the cowl/firewall for the R-4360 would have entailed a major redesign of the nacelle. *(Courtesy of Pratt & Whitney) Next page:* This evocative photo of a Boeing 377 was taken over Lower Manhattan in the 1950s, the golden age for piston-powered airliners. *(Courtesy of Pratt & Whitney)*

Model 377 Design

Design of the Model 377 Stratocruiser was initiated in 1944 and the first flight occurred on July 8, 1947. With its genealogy going back to the B-29, a considerable amount of design time was saved; nevertheless, it would be unfair to call the Stratocruiser a B-29 with different engines and a bigger fuselage. In fact, nothing could be further from the truth. However, it embodied many lessons learned from the problematic B-29 program. Additionally, like its siblings the B-50 and the B-377's military counterpart, the C-97, it could take advantage of newer and better materials. QEC design closely followed that of the C-97. It had the turbosupercharger on the right side of the nacelle compared to the B-50, which mounted the turbo on the left side of the nacelle. The design philosophy of the B-377's QEC was similar to the C-97. A large scoop under the nacelles performed three functions: (i) supplied air to the turbosupercharger compressor, (ii) ducted cooling air to the oil cooler, and (iii) fed cooling air to the large rectangular air-to-air intercooler. Commercial operators could not afford the luxury of intense maintenance of the kind showered upon military aircraft. Pratt & Whitney redesigned some of the problematic areas such as the exhaust system for improved durability. A massive system such as that fitted to the R-4360 went through huge dimensional excursions due to heating and cooling. To alleviate the inevitable cracking and other maladies that normally result

BOEIN

STRATOCRUISER
NX 90700

from these conditions, a spring-loaded ball joint joined the manifolds to the cylinder heads.

Customers

Flown initially by United Airlines (six aircraft), Pan American Airlines (28 aircraft), Northwest Airlines (10 aircraft), and BOAC (British Overseas Airways Corporation—10 aircraft), 377s soon migrated to secondary airlines. Two types of propellers were available, the Hamilton Standard and Curtiss Electric. It would appear that United was the only airline to opt for Curtiss Electric. In fact, when BOAC was forced into purchasing additional 377s due to the Comet disasters and the delayed introduction of the Bristol Britannia, they converted the United aircraft from Curtiss Electric to Hamilton Standard.

One advantage of the double bubble fuselage design was the ability to add another deck to the already capacious interior. Most airlines used the lower deck as a bar with a spiral staircase joining the two decks. Being the "jumbo" of its era it was very popular with passengers. However, the same enthusiasm could not be extended to the airlines that operated the 377. Records indicate that it was a bear to maintain. See Chapter 9 for additional insight as to why this was the case.

Safety Record

The safety record of the 377 is, by any measurement, abysmal. Of the 56 built, 12 were involved in crashes, an astounding 21 percent of the total manufactured. Today, none have survived.

(Table 8-2)

Hull-loss Accidents:	11 with a total of 139 fatalities
Other occurrences (hull-loss):	0 with a total of 0 fatalities
Unfiled occurrences (hull-loss):	1 with a total of 0 fatalities
Hijackings:	0 with a total of 0 fatalities
Selection of incidents:	0 with a total of 0 fatalities

It should be noted that two of the accidents in Table 8-2 involved Model 377s modified into Aero Spacelines Guppys and one of these was further modified with turbo-prop power.

Below are some brief summaries of B-377 accidents that involved mechanical problems. All

Judging by the suits and ties in this photo, one would have to assume that this may be a sales and marketing pitch. Regardless of why these folks are gathered around the engine, it offers a good idea of how the R-4360 was installed in the B-377. *(Courtesy of Pratt & Whitney)*

This is an early version of the B-377, recognizable by the high-tension ignition. Most, if not all, production B-377s were fitted with low-tension ignition. This particular aircraft is fitted with Curtiss Electric propellers. Customers had the choice of Curtiss Electric or Hamilton Standard. *(Courtesy of Pratt & Whitney)*

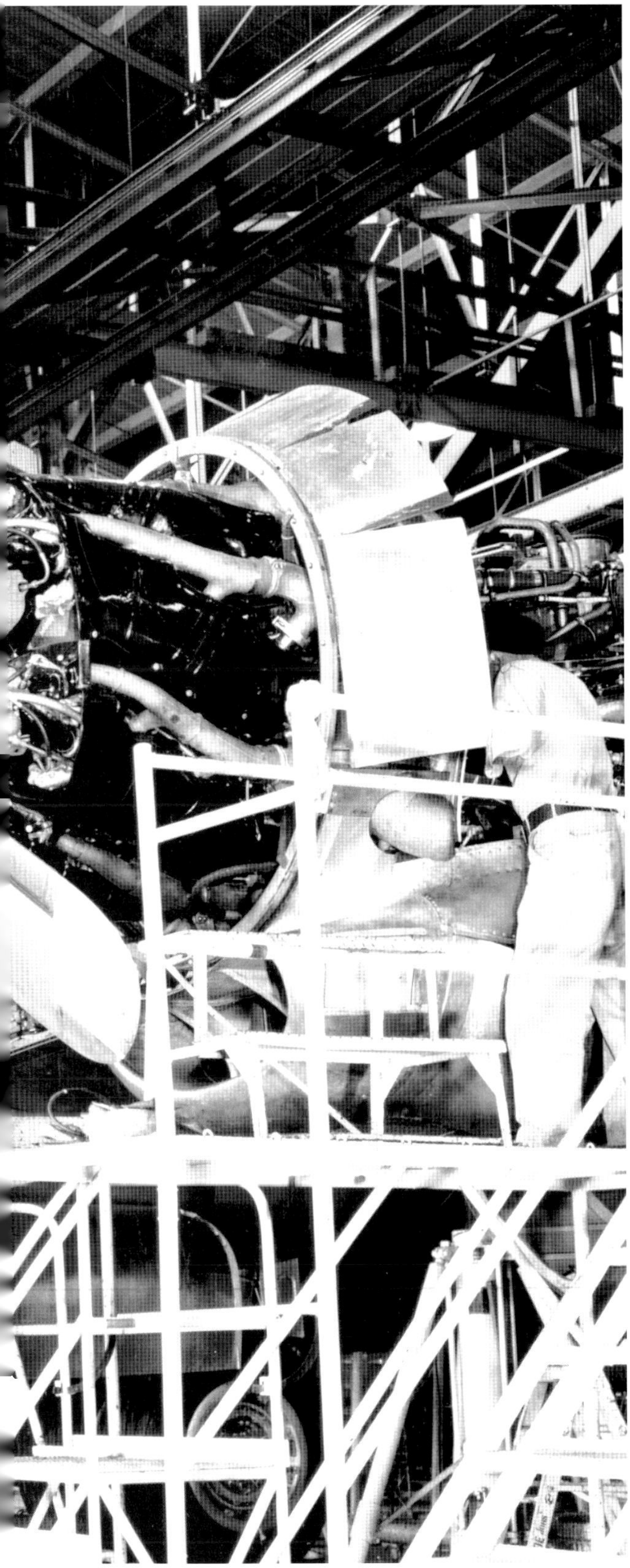

(Table 8-3)

Date:	29 April 1952
Type:	Boeing 377 Stratocruiser 10-26
Operator:	Pan American World Airways
Registration:	N1039V
Msn / C/n:	15939
Year built:	1949
Crew:	9 fatalities / 9 on board
Passengers:	41 fatalities / 41 on board
Total:	50 fatalities / 50 on board
Airplane damage:	Written off
Location:	Near Carolina (Brazil)
Phase:	Cruise
Nature:	International Scheduled Passenger
Departure airport:	Rio de Janeiro
Destination airport:	New York, N.Y.
Flight number:	202

other B-377 accidents appear to have been caused by human error.

Remarks

Flight 202, a Boeing Stratocruiser on a flight from Buenos Aires to New York, departed Rio de Janeiro at 02.43h for an off-airways direct flight to Port of Spain, Trinidad. Last radio contact with the flight was at 06.16 when the crew reported at FL145 abeam Barreiras. The aircraft was later found to have crashed in dense jungle.

Probable Cause: "The separation of the No. 2 engine and propeller from the aircraft due to highly unbalanced forces, followed by uncontrollability and disintegration of the aircraft for reasons undetermined."

(Table 8-4)

Date:	26 March 1955
Time:	11.12 PST
Type:	Boeing 377 Stratocruiser 10-26
Operator:	Pan American World Airways
Registration:	N1032V
Msn / C/n:	15932
Year built:	1949
Crew:	2 fatalities / 8 on board
Passengers:	2 fatalities / 15 on board
Total:	4 fatalities / 23 on board
Airplane damage:	Written off
Location:	35 miles W of Oregon (USA)
Phase:	Cruise
Nature:	Domestic Scheduled Passenger
Departure airport:	Portland International Airport, OR (PDX)
Destination airport:	Honolulu International Airport, HI (HNL)

Remarks

The No. 3 engine and propeller tore loose from the wing, causing severe control difficulties. The aircraft was eventually ditched 35 miles off the

The complex induction system of the B-377 is revealed in this line drawing. *(Courtesy of Pratt & Whitney)*

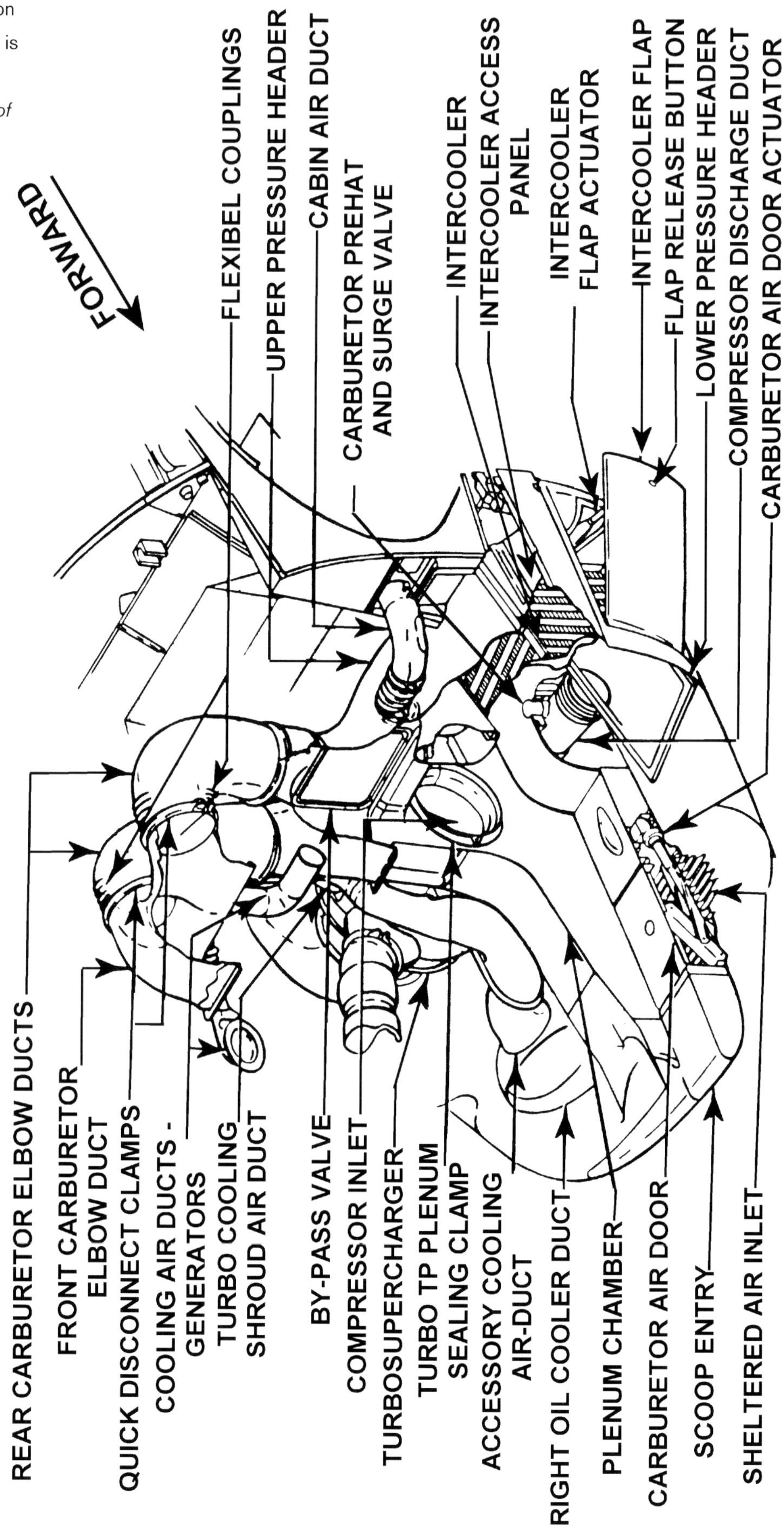

Oregon coast. The Boeing, named *Clipper United States*, sank after 20 minutes in water of about 1600 meters deep.

Probable Cause: "Loss of control and inability to maintain altitude following failure of the No. 3 propeller which resulted in wrenching free No. 3 power package."

Follow-up / safety actions

The Civil Aeronautics Administration issued an advisory that Stratocruiser hollow steel propeller blades be replaced with solid metal blades.

	(Table 8-5)
Date:	16 October 1956
Time:	06.15
Type:	Boeing 377 Stratocruiser 10-29
Operator:	Pan American World Airways
Registration:	N90943
Msn / C/n:	15959
Year built:	1949
Crew:	0 fatalities / 7 on board
Passengers:	0 fatalities / 24 on board
Total:	0 fatalities / 31 on board
Airplane damage:	Written off
Location:	Pacific Ocean (Pacific Ocean)
Phase:	Cruise
Nature:	Domestic Scheduled Passenger
Departure airport:	Honolulu International Airport, HI (HNL)
Destination airport:	San Francisco-International Airport, CA (SFO)
Flight number:	6

Remarks

PanAm Stratocruiser N90943 was on a round-the-world flight from Philadelphia to San Francisco with stops in Europe, Asia, and the Pacific. The aircraft took off from Honolulu at 20.26 HST for the last leg of the flight to San Francisco. The flight was cleared via Green Airway 9, track to 30 deg N, 140d eg W at FL130 and then at FL210 to San Francisco. About halfway, at 01.02h, the crew requested a VFR climb to FL210, which was approved. Immediately after reaching this altitude (at 01.19h) the No. 1 engine oversped. Reduction of airspeed didn't help and the prop could not be feathered, so the engine was cut. As the aircraft was losing altitude, a ditching seemed imminent. U.S. Coast Guard weather station November was contacted at 01.22 about the possible ditching. Climb power was then applied to the remaining engines. The No. 4 engine, however, was only developing partial power at full throttle. Despite these problems the crew managed to maintain altitude at 5,000 feet at an airspeed of 135 knots. Remaining fuel was insufficient, however, to reach San Francisco or fly back to Honolulu. The crew decided to orbit the cutter November and wait for daylight to carry out the ditching. Meanwhile, the cutter laid electric water lights to illuminate a track for the aircraft. At 02.45 the No. 4 engine backfired and failed. The prop was feathered. At 05.40 the captain contacted the cutter again about the intended ditching time and descended to 900 feet. The ditching was carried out at 06.15 with full flaps, gear up, and at a speed of 90 kts. The fuselage broke off aft of the main cabin door. The tail section swung to the left, trapping the life raft launched from the main cabin door. Some three minutes after all occupants had been rescued, at 06.32h the aircraft sank at position 30 deg 01.5'N, 140 deg 09'W.

Probable Cause: "An initial mechanical failure which precluded feathering the No. 1 propeller and a subsequent mechanical failure, which resulted in a complete loss of power from the No. 4 engine, the effects of which necessitated a ditching."

	(Table 8-6)
Date:	08 November 1957
Time:	ca 01.27
Type:	Boeing 377 Stratocruiser 10-29
Operator:	Pan American World Airways
Registration:	N90944
Msn / C/n:	15960
Year built:	1949
Crew:	8 fatalities / 8 on board
Passengers:	36 fatalities / 36 on board
Total:	44 fatalities / 44 on board
Airplane damage:	Written off
Location:	Pacific Ocean (Pacific Ocean)
Phase:	Cruise
Nature:	Domestic Scheduled Passenger
Departure airport:	San Francisco-International Airport, CA (SFO)
Destination airport:	Honolulu International Airport, HI (HNL)
Flight number:	7

Remarks

Boeing 377 Stratocruiser *Clipper Romance of the Skies* carried out a round-the-world flight originating in San Francisco, with its first stop being Honolulu. The aircraft departed San Francisco at

19.51h GMT for a 10-hour flight. Latest position report over the Pacific was at 01.04h. Some days later, on 14 November, the U.S. Navy carrier *Philippine Sea* located bodies and parts of wreckage some 940 miles east of Honolulu and 90 miles North of the intended track.

Probable Cause: "The Board has insufficient tangible evidence at this time to determine the cause of the accident. Further research and investigation is in process concerning the significance of evidence of carbon monoxide in body tissue of the aircraft occupants."

Variations

Five different versions of the B-377 were manufactured as follows.

Model 377-10-26

(Pan American) Seats for 61 day-passengers on the main (upper) deck with 25 seats available when 18 are made up to sleep 27. Maximum seating accommodation in coach class was 81 to 86.

Model 377-10-29

(Pan American) Seats for 61 day-passengers on the main deck with 25 seats available when 35 berths are made up to sleep 45. Maximum seating accommodations in coach class was 81 to 86. PanAm acquired these aircraft with the purchase of World Airways.

Model 377-10-30

(North West Airlines) Converted in 1953 to seat 83 in coach class.

Model 377-10-34

(United Airlines) One complete luxury compartment aft and seats for 56 on main deck with 20 seats available when 17 berths are made to sleep 26.

Model 377-10-28

(BOAC) Private stateroom aft, luxury compartment forward, and seats for 55 on main deck with 20 seats available when 17 berths are made up to sleep 26. This model was originally ordered by Scandinavian Airlines System (SAS) but sold before completion to BOAC. The original SAS internal furnishings were retained by BOAC.

This photo shows the optional Curtiss Electric propellers identifiable by their rounded tips. *(Courtesy of Pratt & Whitney)*

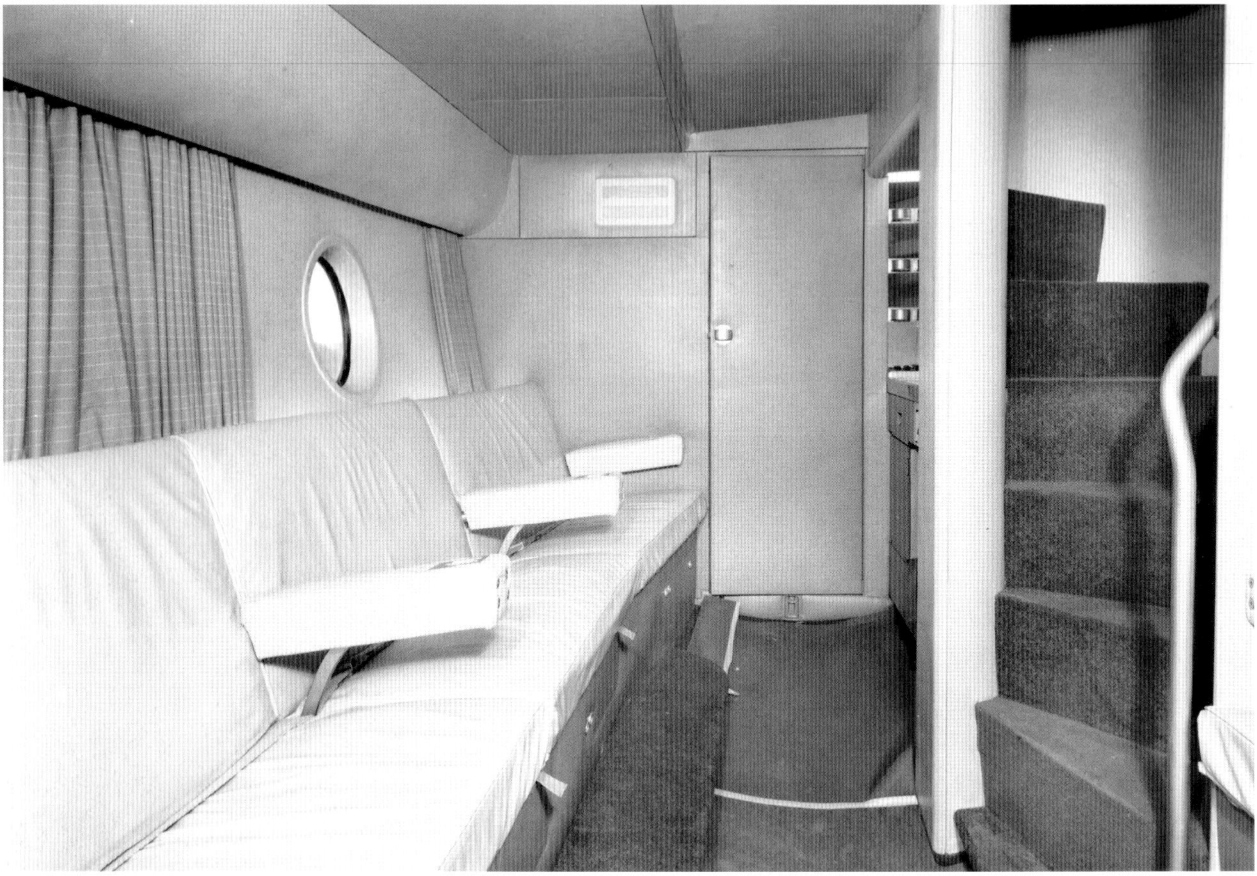

Boeing used a YB-97 to mock up the lower deck of a B-377. The idea of a spiral staircase in an aircraft was quite revolutionary. *(Courtesy of National Archives & Records Administration)*

Model 377-10-32
(BOAC) Seats for 12 in the lounge on the lower deck and seats for 60 on the main deck with seats for 28 when 16 berths are made up to sleep 24.

Boeing Model 377 Parameters:	
Wingspan	141 feet, 3 inches
Length	110 feet, 4 inches
Height	38 feet, 3 inches
Weight empty	83,500 pounds
Designed disposable load	59,000 pounds
Designed normal takeoff weight	145,800 pounds
Max. landing weight	121,700 pounds
Wing loading	80 pounds per square foot
Max. speed	375 mph at 25,000 feet
Max. cruising speed	340 mph (1,900 hp per engine) @ 25,000 feet
Landing speed	93 mph
Rate of climb at sea level	1,100 feet per minute
Rate of climb on three engines	500 feet per minute
Service ceiling	over 32,000 feet
Three-engine ceiling	21,000 feet
Take-off run to clear 50 feet	5,400 feet @ 135,000 pounds
Landing run from 50 feet	5,640 feet @ 105,000 pounds
Landing run from 50 feet with reverse thrust on two engines	4,830 feet

Aero Spacelines B-377 Mini Guppy, B-377 Super Guppy, B-377 Pregnant Guppy

(Ref. 8-7 and 8-8)

One interesting outgrowth of the Model 377 and C-97 series of aircraft were the "Guppys"—Mini, Super, and Pregnant. However, it was only the Mini Guppy and Pregnant Guppy that retained the R-4360 for power, the Super was powered by Allison 501-D22C gas turbines, and even the Pregnant was later converted to gas turbine power as well, therefore is out of the realm of this book. Initially developed in the early 1960s to transport out-sized rocket parts from California to Florida, the Mini soon developed into an even larger and more capable aircraft. Jack Conroy, who founded Aero Spacelines, could see the requirement of such an aircraft. He was astute

The innovative and bulbous Mini Guppy parked outside the Tillamook Museum in Oregon. *(Courtesy of Tillamook Aviation Museum)*

enough to realize that B-377s and KC-97s would fit the bill for modification, and furthermore, the price was right—these aircraft were then relatively cheap. By literally chopping off the upper fuselage, Aero Spacelines designed a bulbous new fuselage measuring 18 feet diameter on the inside. Hinging the nose behind the cockpit allowed cargo a straight shot into the fuselage. Super Guppys sported an even larger fuselage measuring 25 feet tall, 25 feet wide, and 111 feet long. Surprisingly, the Mini only weighed 1,500 pounds more than a B-377. It has to be assumed that by removing all the internal equipment not required for the cargo-hauling role (including pressurization equipment) saved considerable weight.

After the space program was seriously curtailed in the early 1970s, Guppys saw service in other areas. Airbus Industries in particular liked the concept due to its widely dispersed manufacturing facilities and need to transport large assemblies, such as wings, from England to France. If imitation is the ultimate form of flattery, then the Guppy has surely been flattered. Airbus Industries built its own version of a Guppy based on one of its own Airbus aircraft. Most, if not all, Guppys have now been taken out of service, although a few still survive in museums so that we can all marvel at these graceful giants of the sky.

Aero-Sud-Est SE-2010

(Ref. 8-9 and 8-10)

It's surprising that the French would embark upon a major aircraft project in 1942. At this time France was firmly under the jackboot of the German occupation force. Nevertheless, initial studies were undertaken during this difficult time for a large pas-

Aero-Sud-Est SE-2010 in flight. This grainy, poor quality photo illustrates the conventional lines of this unsuccessful commercial aircraft. Seems like a waste of four good R-4360s. *(Courtesy of Pratt & Whitney)*

senger aircraft. An updated design with far more capability soon overtook the initial design. Problem was the French Gnome radials of the day were totally incapable of powering this fairly large aircraft. The only powerplant remotely capable of flying this beast was, of course, the R-4360. The prototype's maiden flight occurred on April 2, 1949, by this time officially designated SE-2010 Armagnac. SE-2010s were outfitted with three levels for sleeping berths; however, this requirement was soon dropped to make more room for (presumably) paying passengers. When Air France refuses to buy a French aircraft, it's a bad omen. And so it was, Air France didn't want the thing. With its potentially largest customer not interested, production was cut back to nine aircraft. Even then the launch customer, Transports Aeriens, only flew SE-2010s for eight months before parking them due to the costs associated with operating these aircraft. A conglomerate of French companies named SAGETA used the parked 2010s for supplying material to the Indo-China conflict in support of French forces. It was in this role that the SE-2010 achieved a modicum of success.

Aero-Sud-Est SE-2010 Parameters:

Wingspan	160 feet, 9 inches
Length	130 feet
Wing area	2,540.2 square feet
Empty weight	99,035 pounds
Max. T/O weight	170,858 pounds
Max. speed	329 mph
Cruising speed	282 mph
Service ceiling	22,309 feet
Range with max. payload	1,522 miles
Max. fuel range (20,723 payload)	3,181 miles

Its huge fuselage measured over 15 feet in diameter resulting in a handicap in the form of

This is a ¾ left rear view of the VSB11-G that powered the Aero-Sud-Est SE-2010. This engine was the commercial equivalent of the R-4360-4. The rectangular-looking box attached to the supercharger housing is the Eclipse automatic engine control. VSB11-Gs were single stage, variable speed. *(Courtesy of Pratt & Whitney)*

Aero-Sud-Est SE-2010 QEC on a test stand. Note its similarity to the XB-44s. It would also appear that the SE-2010 was fitted with Curtiss Electric propellers judging by the brush block attached to the nose case. *(Courtesy of Pratt & Whitney)*

parasitic drag. Even though the closest and only comparison to the SE-2010 was the Boeing 377, itself not exactly a paragon of good streamlining, it could still easily outperform the French aircraft. In one respect, however, the SE-2010 was an improvement over the B-377—it did not feature a turbosupercharger. Instead, its VSB11-G engines were open stack. Supercharging was accomplished through a single-stage variable speed blower. It appears that the open stack engines were less of a maintenance headache than the turbosupercharged variants. Cowl design, however, closely followed that of the B-50/C-97/B-377. That meant it had a large chin scoop that contained all the ductwork for the various air and cooling requirements. Being open stack meant the scoop only provided cooling air for the oil cooler plus induction air. In addition, smaller ducts within the main duct fed cooling air to accessories such as generators, etc.

Paper Airplanes

At the conclusion of World War II, aircraft manufacturers scrambled to get on board the commercial bandwagon. Convair had labored long and hard to develop the B-36, so it was a relatively easy stretch to see how this huge aircraft could employ an enlarged fuselage and perform the role of a commercial airliner. Even though the C-99 was successfully deployed by the Air Force, Convair narrowly missed the boat with commercial operators. Pan Am came closest to purchasing the commercial version of the C-99. Alas, it never happened. Probably with good reason, the postwar airline travel boom never happened. And the aircraft that did enter service were, for the

An interesting feature of this photograph is the divided tailpipe for each exhaust stack. *(Courtesy of Pratt & Whitney)*

most part, rather pedestrian designs such as the DC-4 and DC-6. With its huge capacity it would have been tough for PanAm to fill the required number of seats to make money. The British also latched onto the notion of developing huge commercial aircraft. Two that immediately come to mind are the Bristol Brabazon and Saunders Roe Princess. Even after the passage of nearly 60 years, it still seems astounding that the British, who were all but bankrupt after World War II, thought they could fill such a huge aircraft with passengers. As if they hadn't learned their lesson with these two aircraft, the same mistake was repeated with the Concorde. Shame on me once...

Republic

Republic Aviation had a superb military aircraft on their hands with the XF-12 Rainbow. But without enlarging its fuselage, it would be difficult to see how this sleek aircraft could have been adapted to carry passengers across the Atlantic—at least without being stuffed into the fuselage like sardines in a can.

Northrop

Thank goodness the Northrop B-35 never got into the commercial arena. If it had, the Comet disasters would have paled in comparison to this worthless death trap. Even so, Northrop entertained the idea of building a commercial version of the B-35.

Lockheed

Lockheed's Constitution may have made an acceptable commercial airliner if it had more power. As the U.S. Navy found out, although the Constitution was basically a good aircraft it

Above: Left side view of the accessory section. *(Courtesy of Pratt & Whitney) Opposite:* Right side view of the accessory section. Compared to the QEC of, say, a Boeing 377, this one is significantly simpler. Makes one wonder if Boeing would have been better off to use a gear-driven two-stage supercharger. *(Courtesy of Pratt & Whitney)*

was simply too big for four R-4360s to haul aloft and exhibit anywhere near halfway decent performance.

By the early 1960s all remaining B-377s had been converted to cargo duties or scrapped. So ended the days of the R-4360 powering passenger-carrying aircraft.

Retrospective

At the end of the day, R-4360s did not enjoy commercial success. Why was this? Let's compare it to its closest competition, the Wright R-3350 turbo compound. Unlike Pratt & Whitney's R-4360, Wright's R-3350 was a commercial success due to several factors. The R-3350 was a classic case of winning the battle and losing the war. Tremendous resources were poured into perfecting the R-3350, whereas Wright's competitors dropped piston engine development in favor of gas turbine development. The result was the R-3350 turbo compound with its three power recovery turbines that significantly improved fuel economy. What if a turbo-compounded R-4360 had been produced? Turbo compounding engineering studies were made of the R-4360. No doubt its fuel economy would have been comparable or better than the R-3350's. Wisely on Pratt & Whitney's part, this development did not take

place. Instead, after about 1949, engineers were siphoned off to work on gas turbines. The long-term payoff not only paid rich dividends but also probably saved the company. Wright won the postwar commercial piston engine battle but lost the war by fading away into insignificance when it realized too late that it did not have a competitive gas turbine.

References

8-1 *Flight Test of R-4360 Two-Stage Engine No. X-109, Installation No. B3A in VS-326 Aircraft.* Report No. T-129. Pratt & Whitney Aircraft Division of United Aircraft Corporation, East Hartford, Connecticut. January 1, 1945 to May 3, 1945.

8-2 Moran, Gerard P., *Aeroplanes Vought.*

8-3 http://www.utdallas.edu/library/special/aviation/LTVhistory.html

8-4 *Installation of R-4360-TSB3-G Engines Boeing Model 377 Airplane.* Report No. WC.42. Western Field Engineering Completely Revised January 3, 1949. Pratt & Whitney Aircraft Division of United Aircraft Corporation, East Hartford, Connecticut.

8-5 http://www.aviation-history.com/boeing/377.html

8-6 http://aviation-safety.net/database/1970/700512-0.htm

8-7 *Jane's All The World's Aircraft*. 1967.

8-8 http://www.allaboutguppys.com/mgt/377mgtf.htm

8-9 Stroud, John. *European Transport Aircraft Since 1910*. Putnam, London, 1966.

8-10 http://www.aviafrance.com/609.htm

CHAPTER NINE

Racing Applications

If the old adage "there ain't no replacement for displacement" held true, then R-4360-powered aircraft would have walked away with every racing trophy. In reality, the R-4360's racing career saw its ups and downs with notable successes interspersed with dramatic failures. The post World War II era saw the R-4360's greatest accomplishments particularly in the capable hands of Cook Cleland. While it's always tempting to play what-ifs, the fact of the matter is that the R-4360 as a racing powerplant has been ironically eclipsed by the considerably smaller Rolls-Royce Merlin, which displaces a mere 1,650 cubic inches. Even though capable of producing well over 4,000 hp, the weight of the R-4360 is not in its favor. The greater weight of the engine demands a larger and heavier airframe, which in turn demands a greater fuel load. All this adds up to increased induced drag, which is further increased in a pylon course where loadings can exceed 4 Gs. Heavier aircraft bleed off considerable energy when negotiating a turn and consequently lose a considerable amount of speed. However, if the R-4360 had benefited from the intense racing development as the Rolls-Royce Merlin then its racing success would have, in all likelihood, been significantly better. As it was, during the 1949 to 1964 hiatus from Unlimited Class air racing, the Merlin underwent intense development at the hands of the unlimited hydroplane racers whereas the R-4360 lay dormant as a racing powerplant. This still holds true

today with present-day race-prepared Merlins being modified beyond all recognition from their original configuration.

Postwar

The Goodyear F2G Corsair represented the hottest fighter in the Navy inventory at the conclusion of World War II. Structurally similar to the F4U Corsair, or FG-1 in Goodyear parlance, the F2G was powered by a single-stage R-4360. At the termination of World War II, surplus aircraft flooded the market at prices of less than scrap value. When air racing resumed in 1946 it was natural for these military hot rods to fall into the hands of racers. Being a state-of-the-art military aircraft, the F2G was not available for the 1946 National Air Races held in Cleveland, Ohio. How-

Tragedy struck the 1947 Thompson Trophy Race when pilot Tony Janazzo was overcome by carbon monoxide and crashed in race number 84, an F2G-1. *(Photo courtesy of Warren M. Bodie)*

Right: Dick Becker came in second place with this XF2G-1, race number 94, in the 1947 Thompson Trophy Race. Cook Cleland used this same aircraft, in highly modified form, to win the 1949 Thompson. *(Photo courtesy of Warren M. Bodie)*
Below: Cook Cleland won the 1947 Thompson in race number 74, an XF2G-1. *(Photo courtesy of Warren M. Bodie)*

NX5577N

X5590N

ever, a highly decorated World War II Navy pilot by the name of Cook Cleland persuaded Admiral "Bull" Halsey to release two F2Gs for racing *(Ref. 9-1)*. Halsey was more than happy to comply with Cleland's request. Although this may sound remarkable in today's environment, it should be remembered that in the immediate postwar years a lot of acrimonious political wrangling was taking place between the services. The independent Air Force was in the future and the Army was vying for tax dollars to fund its own programs. The political climate got even worse when the independent Air Force was created on September 18, 1947. Racing was an excellent avenue for publicity, so it wasn't a difficult task for Cleland to persuade Halsey to part with the two F2Gs. And as it turned out, it was excellent value for money. By 1949 five F2Gs had been used as racers.

Post World War II Racing F2Gs

Five F2Gs were used for racing after World War II *(Ref. 9-2)*.

Race #74 NX5577N (F2G-2 / BuAer 88463) flown by Dick Becker
Race #94 NX5590N (XF2G-1 / BuAer 14693) flown by Dick Becker & Cook Cleland
Race #84 NX5588N (F2G-1 / BuAer 88457) flown by Tony Janazzo
Race #57 N5588N (F2G-1 / BuAer 88458 & 88457?) flown by Ben McKillan
Race #18 NX91092 (XF2G-1 / BuAer 14694) flown by Ron Puckett

Taking into account the fact that all these F2G variants were powered by an early development R-4360-2 or -4 engine, their accomplishments are all the more remarkable. It's difficult to determine what, if any, engine modifications were made, but it's a safe bet to say that the manifold pressure regulator was either modified, disabled, or removed. A key airframe modification was the extension of the ram-air induction scoop. Often described as adding 1,000 hp, this statement is obviously an exaggeration. However, it's probably fair to say that it made a significant contribution to the R-4360's power output. From contemporary photographs it appears that all five F2Gs had modified induction scoops ranging from rude, crude, and primitive to quite sophisticated.

1947 Thompson Trophy Races – Cleveland, Ohio

(Ref. 9-3 and 9-4)

Cook Cleland, Dick Becker, Tony Janazzo, and Ron Puckett entered and flew four F2Gs. Cleland easily won the 300-mile race (20 laps and 15 miles per lap) with a race average of 396.13 mph. Dick Becker came in second and Ron Puckett dropped out on the 19th lap after engine failure. Tragedy struck this race when Tony Janazzo inexplicably augured in at over 400 mph. Toxicology reports indicated large amounts of carbon monoxide in his system—he wasn't wearing an oxygen mask. With the open stack design of the F2G's exhaust system it's easy see how this unfortunate event occurred.

1948 Thompson Trophy Races – Cleveland, Ohio

(Ref. 9-3 and 9-4)

Nineteen-fourty-eight could be called the year of "triptane." In fact, this rather potent fuel proved to be the undoing of F2G racing efforts that year. It should be reiterated that a high-performance fuel does not increase the power of an engine per se. However, it does allow a higher manifold pressure to be used before the onset of detonation, resulting in more potential power output. Depending on how it was blended with gasoline and how much tetraethyl lead was contained in the fuel determined the performance number of triptane. By any measure, however, the stuff was pretty remarkable. Performance numbers of over 270 have been reported. Charles Kettering determined that triptane needed a close look, so in 1943 he instigated an intense study of this fuel.

Ron Puckett flying race number 18, an XF2G-1, suffered engine failure on the 19th lap of the 1947 Thompson Trophy Race. Note that the standard induction scoop is used. Puckett fared better in the 1949 race when he came in second behind Cleland. *(Photo courtesy of Warren M. Bodie)*

General Motors built a pilot plant capable of producing 150 gallons per day—sufficient for research purposes but totally inadequate to support an air force. Sponsored by the Army, NACA performed exhaustive testing with this wonder fuel. Its promise was so great that none of the laboratory engines had sufficient mechanical strength to stand up to the rigors of running triptane to its full potential *(Ref. 9-5)*.

Cook Cleland and Dick Becker managed to obtain sufficient triptane for their bid at the 1947 Thompson Trophy. Alas, it turned out to be their undoing. Three laps into the race, Becker suffered a massive backfire in the induction system. The ram-air induction scoop along with part of the cowling blasted off, forcing Becker to make an emergency landing. Then it was Cleland's turn to suffer the same fate except he lasted four full laps when his backfire occurred on the fifth lap. Over the ensuing years these costly backfires have been written off as running too lean, which may be true. However, the PR-100 carburetor is a sensitive and complex fuel-metering device. A drastic change in fuel such as the use of triptane, would demand significant changes be made to the setup of the PR-100. The only way to determine what these changes should be would be to run the carburetor on a flow bench. It's not known if these measures were taken.

1949 Thompson Trophy Races – Cleveland, Ohio

(Ref. 9-3 and 9-4)

After being burned by the use of triptane the year before, the F2G teams wisely decided to use an alternate fuel. Sohio promised 500 gallons of 130/170 PN fuel to teams that would display its logo on their aircraft. After the previous years' experiences, I'm sure that the F2G drivers thought about this for about one New York second.

Ben McKillan took the initial lead but was soon overtaken by Cook Cleland and Ron Puckett; they finished in this order. Since the races had been resurrected after World War II, the safety record had been going downhill. Numerous crashes, bailouts, and emergency landings had started people questioning the wisdom of racing such thoroughbreds around a closed circuit. The 1949 race turned out to be the final straw. A tragic accident occurred on lap two when Bill Odum, flying *Beguine* (a highly modified P-51), crashed

Summary of F2G Race Results 1947, 1948, and 1949

(Ref. 9-3)

1947 Thompson Trophy Race

Pilot	Race #	Qualification:	Race average	Place
Cook Cleland	#74	401.70 mph	396.13 mph	1st
Dick Becker	#94	400.94 mph	390.13 mph	2nd
Ron Puckett	#18	371.42 mph	out on lap 19 – engine failure	DNF
Tony Janazzo	#84	372.42	crashed - CO poisoning.	DNF

1948 Thompson Trophy Race*

Pilot	Race #	Qualification:	Race average	Place
Cook Cleland	#74	417.42 mph	out on lap 5, backfire blew ram scoop off	DNF
Dick Becker	#94	400.94 mph	out on lap 3, backfire blew ram scoop off	DNF

* The year of the triptane fiasco.

1949 Thompson Trophy Race

Pilot	Race #	Qualification:	Race average	Place
Cook Cleland	#74	407.21 mph	397.07 mph	1st
Ron Puckett	#18	373.52 mph	393.55 mph	2nd
Ben McKillan	#57	396.28 mph	387.59 mph	3rd

Ben McKillan's race number 57, an F2G-1, initially took the lead in the 1949 Thompson but was soon overhauled by Cleland and Puckett. This aircraft was recently restored to flying condition. *(Photo courtesy of Warren M. Bodie)*

After the triptane disaster of 1948, Cook Cleland came storming back to win the 1949 Thompson with the wingtips of his F2G-1 clipped. Note the nicely designed ram-air induction scoop that is extended to the leading edge of the nose bowl. Most F2G racers went through this modification. *(Photo courtesy of Warren M. Bodie)*

into a house killing, Odum plus a mother and child. After this episode Unlimited Class air racing ended—until 1964.

Post 1949

After the tragedy of the 1949 races, Unlimited racing went into a hiatus not to re-awaken until 1964. Things pretty much picked up where they left off in 1949. In other words, racers singled out World War II surplus aircraft to race, although it should be kept in mind that at this time the term "warbird" had not entered our vernacular. Furthermore, the cost of these former warriors was still reasonable if not downright cheap—definitely not the case now. In fact, more and more racers are now being re-converted back to military stock aircraft. And perhaps that is the way it should be. The problem is no new Unlimited aircraft are being developed to take their place—but that's another story for another day. One of the first attempts to go the "ain't no replacement for displacement" route in the "modern" era was Lyle Shelton when he managed to shoehorn a Wright R-3350 into an F8F Bearcat. This achievement was all the more remarkable when you consider how tight the original R-2800 was in the Bearcat. Keep in mind that the Bearcat was considered to be the smallest possible airframe built around the R-2800.

As racing became more competitive, new means were sought to develop a faster racer. This time the venue was Reno, Nevada, which was a perfect setting if ever there was one. Good weather and sparsely populated—at least in 1964. The 1964 race was held under less than ideal conditions. Operating out of Sky Ranch Airport, which only had dirt strips, this inauspicious resurrection would eventually metamorphize into a major sporting event. At least the 1964 Sky Ranch event got the ball rolling. For 1965 and all subsequent races, the event moved to Stead Airport, named after Bill Stead. This former military base would turn out to be the ideal location to race from. As icing on the cake, mountains in the background would reverberate from the noise of the racers as they blasted their way around the 10-mile course, to this day known as the valley of speed.

Super Corsair

(Ref. 9-6)

Steve Hinton, world-renowned race pilot and test pilot for warbirds, came up with the idea to relive the glory days of the F2G. Below is a note I

This in-flight photo personifies the beauty and grace of the R-4360-powered *Super Corsair*. Considering the fact it was built from junk parts in less than a year, it represented a remarkable achievement. *(Photo courtesy of Frank B. Mormillo via Pete Law)*

This ground shot gives an idea of the huge size of the *Super Corsair. (Photo courtesy of Pete Law)*

received from Steve that relates his experiences with the "Super Corsair."

Jim Maloney and I went to Champlin Fighter Museum for their grand opening in 1981. It was a big party. Mr. Champlin has an F2G Corsair. It was the first time we had ever seen the machine. We had read about them. It is very impressive. Jim and I joked about making one of our museum Corsairs an air racer, after all we had 2. Jack Sandberg, friend, noted warbird guy, and air racer, said if we wanted to do it he would donate a new overhauled 4360-63A. A Corsair is a great Navy fighter but the wing is FAT and it isn't in the same league as a P-51 racer so we thought. As the evening progressed we joked with Dave Zeuschel (famous rebuilder of Rolls-Royce Merlins who was sadly killed in an F-86 accident) *and Frank Sanders about the possibilities. They both thought it would be a waste of time. Over the next few weeks we continued to think about 4,000-hp Corsair, it had to be fast. Cook Cleland ran his F2G at sea level, 427 mph. If we took that speed to Reno it would true out at 440 mph. One of the guys we were good friends with was Bruce Boland, he was a knowledgeable Lockheed aero guy who also was responsible for many Reno air racers including Daryl Greenamyer's* Conquest 1 *as well as the famous Red Baron racer. We asked Bruce what he thought of the idea of building an F2G racer, he too chuckled and the conversation didn't go very far. A few weeks passed and Bruce called back to talk about the F2G. He said that if we were to build a 4360 Corsair racer we would be very impressed with the speed and if we decided* *to do it he would help us. He found some flight test info and did some estimating and came up with the assumption that a Super Corsair could run 450 at Reno. The F2G would need modifications to the airframe but it was within our capabilities. We talked the project up with a lot of our friends. Frank Sanders and Lloyd Hamilton were experienced air race guys who we respected and they laughed. Frank said his stock Sea Fury would run away from our Super Corsair. Lloyd said that it would never go 400 at Reno. He even bet us $500 that our Super Corsair wouldn't do it. Even with all the negative from some of our friends we had a huge following who believed it could go fast. From May through August 1982 we had a full time crew who put in 14-hour days. We rebuilt a F4U-1 that had not flown since 1945, modified the airframe, rebuilt all the systems, and made an engine mount, cowling, and many other mods to install the 4360-63A with a Skyraider prop. We flew it three weeks before the races in 1982 and were really impressed. We experienced a lot of problems that took time to sort out such as hot cylinder heads, touchy flight controls, fabric covered control surfaces blowing apart, snatching ailerons, and many more. On one test flight Frank Sanders, flying his Sea Fury, chased Jim Maloney. This flight was the first time we ran the 4360 to METO power, 2,600 hp. Jim ran up to 49" and 2,700 rpm and from the practice area, 12 miles from Chino airport, Jim had a three-mile lead on Frank's Sea Fury that he was running at full power, 2,550 hp. Frank landed and*

This is a close-up of one of the *Super Corsair's* two oil coolers originally installed in a Grumman S2F Tracker. Each one was installed in a wing root, just like the F2G. Further assisting the oil-cooling requirement were spray bars, not shown in this photo. *(Photo courtesy of Pete Law)*

was really impressed. He said if the 4360 could make a Corsair go that fast it could make a Sea Fury go even faster. We made a believer out of him that day. A million stories later we made it to Reno. I qualified the Super Corsair at 413 mph and Lloyd Hamilton had to pay up. He dropped a bucket of nickels for his $500 bet. The 4360 is a magnificent piece of machinery. It is beautiful and even if you don't have a trained eye, you will appreciate the quality of this beautiful engine. When we first hit the starter it was a thrill. It runs smooth, is easy to work on, and it is really tough. We ran it at 3,800 hp for hours. We ran 4150 hp at Reno once. The Corn Cob carried the Corsair at a race average of 447 mph on one of the heat races and its fastest one lap speed was 459 mph. The Super Corsair was the National Champion in 1985 when it won with an average speed faster then any P-51 had ever gone at Reno before. Dave Zeuschel came up to me after that race and said, "I would have bet my wife your Corsair would never win Reno, but I would have bet my life it could never have gone faster than a Mustang." We ran the engine on the cylinder head temp red-line of 265 degrees C. Oil temp was not a problem; it ran at 85 degrees C. We ran 2 gpm of water spray over the 2 ea. Grumman S2F coolers. We also ran 2 gpm water spray in the cowl to help cool the cylinder temps. The A row, rear, was the hottest. We ran 67" and 2,800 rpm @ 256 lbs torque most of the time. We ran 70" @ 2,950 rpm once. We ran one engine eight Reno races. The engine was not stock; it was modified for racing. We replaced the nose case with a 4360-59B one, this gave us a 60-spline prop shaft, we moved the starter drive to fit the new mount, machined the pistons for a better oil control ring and cam ground them, we modified the no. 1 main bearing with a flange to prevent it from working out of its press fit in the case. Frank Sanders had two of these bearings work themselves loose on Dreadnought *and cause a failure. We thought it might have been a case of* Dreadnought *having a more stable airframe than the Corsair. The Skyraider prop we used moved the airframe around a lot. The Sea Fury had a larger tail and it held the airframe better. We think this caused the No. 1 bearing that also supports the prop nose case to overload and move out of its press fit into the case. We ran the engine de-riched and non de-riched. Non de-rich was what we used most of the time. We tried opening up the bore in the cylinders and cemachroming them, the thought was to make the engine better to handle the high temps. The engine smoked on the ground a lot with this setup. I am not sure it was any better than the stock clearances. Actually stock clearances seemed to work best. We choked the exhaust exits* (diameter) *from 2½ inches to 2¼ inches. This was done to recover thrust from the stacks; we actually had choked down bolt on exhaust pieces that we ran one day. We ran 3,800 hp and got a speed. That same day we attached the reducers and tried again; eight mph increase is what we saw. One experience I had was when we moved the engine breathers to an area where I could see them vent if I had a problem. What this did is put them in a higher-pressure area and the engine didn't breathe very good. It pressurized the oil tank and blew it up. I had 35 gallons of hot oil in the belly of the cockpit. We had a huge mess. Another experience we had was with some TRICK oil additive. We used some graphite additive, this sounded good but what it did was put metallic, graphite substance on the spark plugs. It was a pain in the *** because it fouled out the plugs all the time. We flushed out the system three times before the problem went away. All in all I had a good experience with the 4360. I liked it. The Super Corsair was lost due to a Master Rod failure that caused a huge fire. The engine we installed was a 20-year-old overhauled, "in the can" piece. It looked beautiful but we should have had the bearings replated. It is all hindsight now. We are happy that Kevin Eldridge lived through it and is fine now. I have a lot of great memories of the friends who worked on the Super Corsair with us and great memories of what it is like to dial up 4,000 hp.....*

The indefatigable Pete Law holding the all-important water regulator for the *Super Corsair.* This dynamo of an individual has been a fixture at the Reno Air Races since the mid 1960s. *(Photo courtesy of Pete Law)*

Genesis

The basic airframe for the *Super Corsair* started life as an F4U-1. By the time the Super Corsair project was being bandied around, the donor F4U-1 airframe was in derelict condition sitting in the back of the Planes of Fame Museum in Chino, California. Steve Hinton's business, Fighter Rebuilders, is located on the same premises. Fabrication of the aircraft took a remarkably short time—from May to August 1982. But as Steve indicated in his note, it required the help of a lot of people working 12-hour days on a volunteer basis. Bruce Boland performed much of the engineering work. Bruce was the brilliant Lockheed engineer who was intimately involved in the Reno race scene until his untimely death in 1993. Critical items such as installation of the R-4360 were left to Boland. Rather than re-invent the wheel, parts from other aircraft were adapted to fit and work on the *Super Corsair*. As an example, engine cowling is a very difficult design challenge. The problem is so many poor examples of cowl designs and most R-4360 cowls were just too bad to consider (the C-124 is an excellent example of a poor R-4360 installation) or they were optimized for turbosupercharged applications.

As often happens in these situations, Steve and Bruce had the proverbial epiphany when they realized that the Douglas A-26 cowl may be just what the doctor ordered. Even though the A-26 was R-2800 powered, it could be adapted for the *Super Corsair.* Besides which, as they say, great minds think alike—that was the path taken by Pratt & Whitney when designing the original 4360-pow-

P&W R-4360 FUEL/HORESPOWER DATA

BENDIX PR 100 B4 CARBURETOR, P/L 391469-11

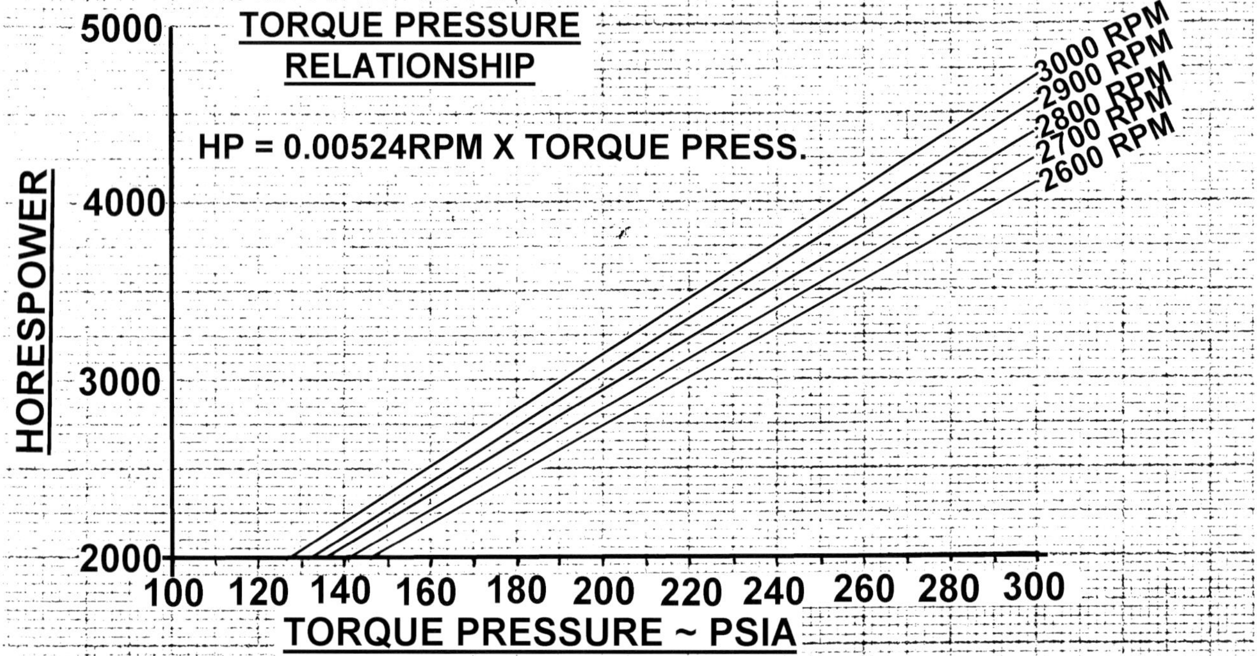

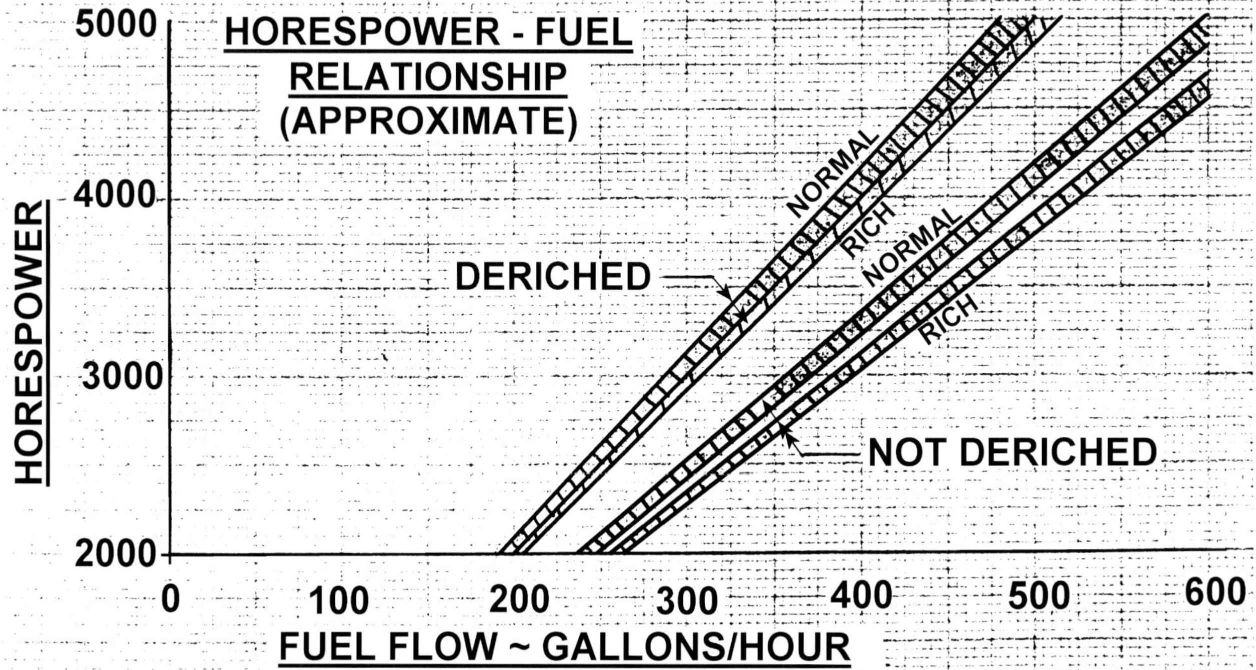

Pete Law is not only a first-rate mechanic, but he is a highly respected engineer. Retired from Lockheed's "Skunk Works" where he worked on top-secret projects, he has assisted the racers over the years by developing graphs such as this one that shows torque pressure versus horsepower. The lower half of the graph shows the relationship between fuel flow and horsepower. This graph also highlights the amount of derichment that takes place with the application of ADI. As an example, at 3,500 hp, without derichment, a fuel flow of approximately 475 gallons per hour is necessary. With the application of ADI and 3,500 hp, fuel flow drops to approximately 325 gallons per hour. *(Graph courtesy of Pete Law)*

ered Corsair: The Corsair WM. In the case of the Corsair WM, the stock F4U-1 cowl was elongating 12 inches to accommodate the 4360. The A-26 cowl was a better design for several reasons. It was a better fit and possibly more importantly, it had a nice induction ram scoop that picked up air at the leading edge, thus avoiding the modifications that Cook Cleland was forced into making for his F2G. It was now necessary to design an engine mount. Boland figured that a modified C-97 unit would work. Pete Law performed the thermal work and calculated that a pair of Grumman S2F Tracker coolers would work *(Ref. 9-7).*

Each oval-shaped cooler was mounted in the wing root in a similar fashion to the original F4U-1. Air intakes were drastically reduced in size in order to cut down drag. In fact, they looked remarkably similar to the Corsair WM oil cooler intakes. Assisting in the cooling chores were spray bars designed to spray water on the coolers. Spay nozzles were also used inside the cowl to assist in keeping head temperatures below meltdown figures. As it turned out, elevated head temperatures were a constant challenge. Water was pumped through the oil cooler spray bars at two gallons per minute. A similar flow rate was utilized for spraying onto the cylinders via spray nozzles attached to the inside diameter of the cowl nose bowl. ADI was essential to keep the engine alive at race powers, and again, Pete Law did the water regulator work. Pete also did the graph in Fig. 9-11 that gives the horsepower to fuel consumption ratio. In the same graph Pete graphed the torque pressures that relate to horsepower *(Ref. 9-8).*

As described in previous chapters, the -63A is fitted with a torque meter, therefore the pilot

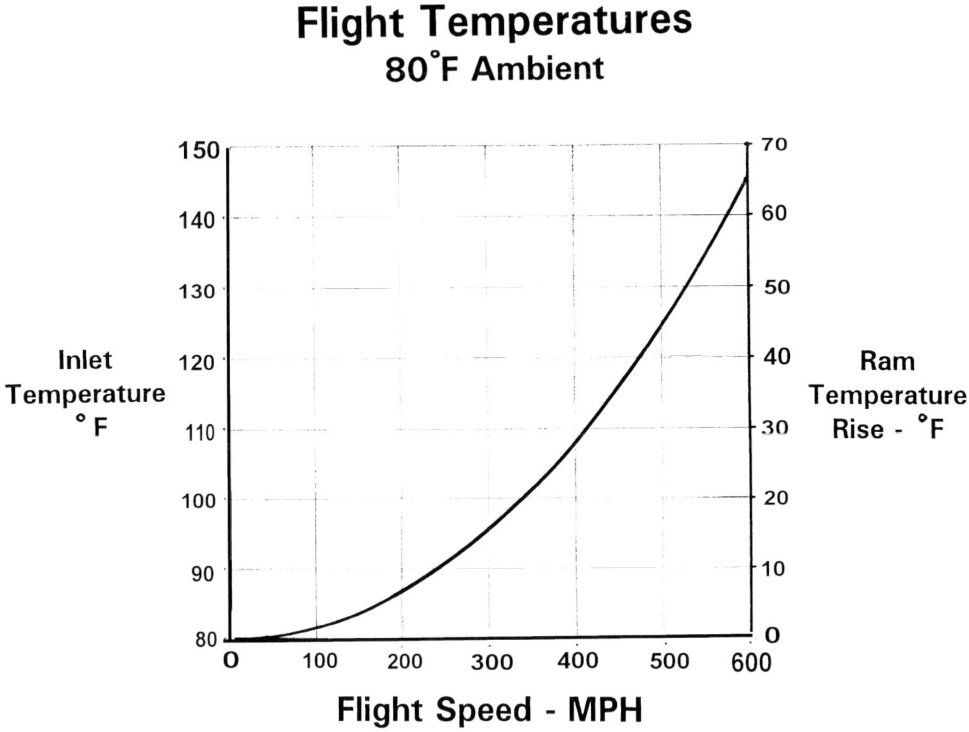

It is a little appreciated fact that at speed, induction temperature rises. Pete Law generated this graph illustrating the phenomenon. As an example, with an ambient temperature of 80 degrees, at 500 mph, the ram temperature rise is on the order of 45 degrees. *(Graph courtesy of Pete Law)*

knows exactly how much power the engine is generating. It's an often-overlooked fact that as air speed increases so does induction temperature. Up to 200 mph, the temperature rise is a modest 5 degrees; however, things take on a very different perspective beyond 200 mph. A top Unlimited racer is capable of flying well over 500 mph on the straightaways at Reno. As can be seen from the graph shown in Fig. 9-12, the ram temperature rise at these kinds of speeds is a significant 50 degrees or greater *(Ref. 9-9).*

Not surprisingly for such a radically modified aircraft, controllability problems became apparent with early flight-testing. It's a little appreciated fact that Corsairs employed fabric covered wooden ailerons. Although marginal for the stock aircraft, they turned out to be adequate. But that wasn't the case for the *Super Corsair.* Its 450-mph plus capability caused aileron snatch, therefore the wooden ones were replaced with ailerons made from metal and metal covered.

Race #1 Super Corsair Race Results

(Ref. 9-2 and 9-10)

1982 National Championship Air Races – Reno. Pilots: Steve Hinton & John Maloney

Race:	Speed	Position	Raced As
Qualification:	413.208 mph	4th Steve Hinton	Bud Light Special*
Gold Heat 1	338.558 mph	6th Steve Hinton	Bud Light Special
Gold Heat 2	382.880 mph	3rd John Maloney	Bud Light Special
Gold	362.496 mph	4th Steve Hinton	Bud Light Special

1983 National Championship Air Races – Reno. Pilot: Steve Hinton

Race:	Speed	Position	Raced As
Qualification:	408.331 mph	8th	Bud Light Special
Silver Heat 1	374.619 mph	3rd	Bud Light Special
Silver Heat 2	403.578 mph	1st	Bud Light Special
Silver	417.097 mph	1st	Bud Light Special

1984 National Championship Air Races – Reno. Pilot: Steve Hinton

Race:	Speed	Position	Raced As
Qualification:	424.015 mph	8th	Super Corsair
Heat 1A	417.327 mph	lst	Super Corsair
Heat 2A	408.631 mph	4th	Super Corsair
Heat 3A	422.652 mph	2nd	Super Corsair
Gold	413.686 mph	3rd	Super Corsair

1985 Bakersfield National Air Races – Shafter, California. Pilot: Steve Hinton

Race:	Speed	Position	Raced As
Qualification:	412.999 mph	4th	Super Corsair
Gold Heat	394.408 mph	3rd	Super Corsair
Gold	394.925 mph	3rd, cut pylon 5 on lap 3	Super Corsair

1985 National Championship Air Races – Reno. Pilot: Steve Hinton

Race:	Speed	Position	Raced As
Qualification:	431.944 mph	5th	Super Corsair
Heat 2A	421.007 mph	2nd	Super Corsair
Heat 3A	433.844 mph	2nd	Super Corsair
Gold	438.186 mph	1st new race record	Super Corsair

1986 National Championship Air Races – Reno. Pilot: John Maloney

Race:	Speed	Position	Raced As
Qualification:	422.006 mph	8th	Super Corsair
Heat 1A		Did Not Finish	Super Corsair
Heat 2B	314.637 mph	6th, 5 laps	Super Corsair

1987 National Championship Air Races – Reno. Pilot: John Maloney

Race:	Speed	Position	Raced As
Qualification:	431.607 mph	6th	Super Corsair
Heat 2A	399.565 mph	4th	Super Corsair
Heat 3A	419.800 mph	4th	Super Corsair
Gold	416.905 mph	4th	Super Corsair

1988 Wings of Victory Air Races, Hamilton Air Force Base, California. Pilot: John Maloney

Race:	Speed	Position	Raced As
Qualification	346.210 mph	10th	All Coast Super Corsair
Silver	352.630 mph	1st	All Coast Super Corsair

1988 National Championship Air Races – Reno. Pilot: John Maloney

Race:	Speed	Position	Raced As
Qualification:	442.319 mph	7th	All Coast Super Corsair
Heat 1A	445.072 mph	1st	All Coast Super Corsair
Heat 2A	420.177 mph	4th	All Coast Super Corsair
Heat 3A		Out mid-race	All Coast Super Corsair
Gold	368.126 mph	6th, 7 laps	All Coast Super Corsair

1989 National Championship Air Races – Reno. Pilot: John Maloney

Race:	Speed	Position	Raced As
Qualification:	437.583 mph	4th	All Coast Super Corsair
Heat 2A	402.090 mph	3rd	All Coast Super Corsair
Heat 3A	416.250 mph	5th	All Coast Super Corsair
Gold	406.265 mph	3rd	All Coast Super Corsair

1990 Texas Air Races, Sherman, Texas. Pilot: John Maloney

Race:	Speed	Position	Raced As
Qualification:	381.805 mph	4th	All Coast Super Corsair
Gold Heat	356.101 mph	4th	All Coast Super Corsair
Gold	376.479 mph	4th	All Coast Super Corsair

1990 Colorado National Air Races, Denver, Colorado. Pilot: John Maloney

Race:	Speed	Position	Raced As
Qualification:	418.671 mph	2nd	All Coast Super Corsair
Gold Heat		Out lap 5	All Coast Super Corsair
Gold	389.332 mph	3rd	All Coast Super Corsair

1990 National Championship Air Races – Reno. Pilot: John Maloney

Race:	Speed	Position	Raced As
Qualification:	426.653 mph	5th	All Coast Super Corsair
Heat 2A	406.945 mph	6th	All Coast Super Corsair
Heat 3A	415.224 mph	5th	All Coast Super Corsair
Gold	410.786 mph	5th	All Coast Super Corsair

1991 National Air Races – Reno. Pilot: John Maloney

Race:	Speed	Position	Raced As
Qualification:	434.207 mph	6th	All Coast Super Corsair
Heat 2A	377.832 mph	7th, 5 laps	All Coast Super Corsair
Heat 3A	410.179 mph	6th	All Coast Super Corsair
Gold	406.420 mph	7th, 7 laps	All Coast Super Corsair

1992 National Championship Air Races – Reno. Pilot: Kevin Eldridge

Race:	Speed	Position	Raced As
Qualification:	425.603 mph	7th	All Coast Super Corsair
Heat 1A	429.937 mph	2nd	All Coast Super Corsair
Heat 2B	423.164 mph	1st	All Coast Super Corsair
Heat 3A	411.291 mph	7th	All Coast Super Corsair
Gold	420.800 mph	6th	All Coast Super Corsair

1993 National Championship Air Races – Reno. Pilot: Kevin Eldridge

Race:	Speed	Position	Raced As
Qualification:	426.155 mph	6th	All Coast Super Corsair
Heat 2A		Out lap 4	All Coast Super Corsair
Heat 3A	377.522 mph	5th	All Coast Super Corsair
Gold	418.656 mph	5th	All Coast Super Corsair

1993 Kansas City Air Races. Pilot: Kevin Eldridge

Race:	Speed	Position	Raced As
Qualification:	377.73 mph	5th	All Coast Super Corsair
Heat 3	362.24 mph	1st, cut pylon 3 on pace lap	All Coast Super Corsair
Championship	375.84 mph	4th	All Coast Super Corsair

1994 Phoenix 500 Air Races. Pilot: Kevin Eldridge

Race:	Speed	Position	Raced As
Qualification:	428.104 mph	4th	All Coast Super Corsair
Heat 1A	355.440 mph	6th	All Coast Super Corsair
Heat 2B		Out lap 3, crashed after fire**	All Coast Super Corsair

*Initially sponsored by Budweiser Light. Sponsorship later shifted to All Coast Forest Products. Aircraft also known as Super Corsair.

** The R-4360-63A suffered a catastrophic failure when a master connecting rod let go. The ensuing fire forced pilot Kevin Eldridge to bail out, which he survived, albeit suffering significant injuries as he struck the horizontal stabilizer after exiting the aircraft.

Dreadnought

One of the most aerodynamically advanced piston-powered fighters appeared at the end of this golden era. This was the Hawker Fury and its naval version the Sea Fury. Powered by a Bristol Centaurus sleeve valve engine that was little understood in the U.S., it seemed a shame that such a potentially fast airframe should go to waste. It was this idea that formed the genesis of *Dreadnought*—an R-4360-powered Sea Fury. Developed and built by the late Frank Sanders, Frank's two sons, Dennis and Brian, have picked up where Frank left off.

Like the *Super Corsair* that preceded it, *Dreadnought* incorporated a modified Boeing C-97 mount. Using the C-97 ring mount, legs and firewall fittings were built around it. The stock Sea Fury cowl was an excellent design, so this was adapted by stretching it and making a homemade induction scoop. However, don't be fooled by the "homemade" description, everything about this aircraft was professional and first class. As we've seen in prior chapters, exhaust systems, particularly with an open stack engine, can make a significant difference. Using stock C-124 components for the B, C, and D rows, a scratch-built system for the A row was designed by Bob Smith and built by Frank Sanders. All this work was performed at the Sanders' facility. Bob Smith did the engineering calculations.

For maximum jet thrust, outlets were optimized at 2⅛ inches diameter—somewhat smaller than the *Super Corsair.* Airframe modifications were limited to extending the height of the vertical stabilizer by one foot, resulting in an additional two square feet of fin area. Empty weight of the aircraft is 10,400 pounds, up somewhat from a stock Sea Fury's empty weight of 8,997 pounds. Of course much of this weight is attributable to the additional weight of the engine; 3,811 pounds for the R-4360-63A compared to the weight of a Bristol Centaurus of 2,830 pounds. And then factor in the domino effect of a heavier mount and a much heavier propeller and it's easy to see why *Dreadnought* is 1,403 pounds heavier even after all military and other non-essential equipment was stripped out. To keep cylinder head temperatures in the green, a unique spray system was employed. Unlike *Furias*, the *Super Corsair,* and other radial powered racers that use nozzles, Dreadnought incorporates a modified anti-icing slinger ring built into the spinner. Normally used to sling anti-icing fluid via centrifugal force along the leading edge of the propeller, the Sanders used this principle to sling water into the cowl. Four nozzles are used. Dual spray bars are used for each oil cooler. Stock Bristol honeycomb coolers are used, one in each leading edge. Small tabs welded to the spray bars ensure that the spray bar water covers the entire cooler core.

Racing around pylons pulling upwards of 4 Gs or better can stress the airframe. No new news there. However, what is often overlooked is the fact that the propeller exerts not only the same G load but a significant gyroscopic load as well. This load is then transferred to the propeller shaft. As described previously, the propeller shaft is supported at its tail end inside the crankshaft at the front main bearing. Therefore, under the loads described, the front main can suffer undue stress. And this is what happened to *Dreadnought* in Heat 2A in 1984 at Reno. The engine seized due to a failed front main bearing induced by high G loads. Likewise in 1985 at Shafter, the engine had metal in the screens due to the same failure mode. At this point, Frank Sanders looked into the problem and figured out what was occurring. The fix required several iterations to get right. Initially, taper pins were driven at an angle into the steel backing of the bearing. Although

Owned and flown by Dennis and Brian Sanders, *Dreadnought*, the R-4360-63A-powered Hawker Sea Fury, has appeared at Reno and other racing venues for many years. *(Photo by author)*

Using exhaust energy to the fullest, *Dreadnought* uses the classic siamesed jet stack system. Seven outlets on each side of the aircraft are used. *(Photo by author)*

helpful, it clearly needed more. The definitive fix was to machine locking-tabs on the bearing and make a flange with mating tabs to lock onto the bearing. The flange was attached to the crankcase diaphragm via 10-32 through bolts. Frank derived inspiration for the modification from the front main bearing of a CB series R-2800, which featured a stouter design. However, mechanical failures were not over. In 1995 at Reno, the engine suffered a significant failure when a link rod failed and "centrifuged" the piston into the head. No easy fix for this one. The failure was caused by too many hours at race power settings and RPM. After posting 438.019 in qualifying at Reno in 2000 the aircraft was withdrawn due to engine problems. This time it turned out to be a cracked master rod. Fortunately, the impending failure was caught before the master rod let go. This put the aircraft out of commission until 2003. In the meantime, the engine was completely re-built; the power section was gone over by Precision Engines in Washington to zero time standards. Dennis and Brian rebuilt 28 new old stock (NOS) cylinders. Even though the cylinders were new, they still needed some additional tweaking such as replacing the valve guides with upgraded ones.

Finished shortly before the races, *Dreadnought* sounded crisp and healthy. It was notable that even on startup relatively little smoke came out of the stacks. This is a remarkable achievement for a 4360. Even though the engine sounded very crisp and healthy, it was producing a worrying amount of metal. Any newly overhauled engine produces a small amount of

This shot shows one of *Dreadnought's* two oil coolers. The pair of spray bars is clearly shown in this shot. The rectangular tabs are there to distribute the spray bar water over the entire face of the cooler core and therefore take full advantage of this essential feature. *(Photo by author)*

metal as it's part of the break-in process. However, this metal should diminish over time. In the case of *Dreadnought*, this was not occurring. Alas, after the Heat 2A race Dennis and Brian felt that something was coming apart inside the engine—it was now producing steel in the filters. Wisely, it was decided to call it a day and dismantle the aircraft for shipment back to Ione, California. Another modification made during its time off was the substitution of the slinger ring water spray system for 14 spray nozzles located inside the nose bowl.

Depending on competition, local conditions, etc., power settings can vary. However, according to Dennis, the hardest the engine has run is 80 in. Hg. at 3,200 rpm and a torque pressure of 238 pounds. It was possible for the engine to run even faster but even at 3,200 rpm, the propellers' tips, a modified Aeroproducts Skyraider unit, were exceeding the speed of sound and consequently the efficiency went down the tubes.

Temperatures Used For Racing

- cylinder head red-line; 265 degrees C
- induction; 80 to 90 degrees C
- oil; 80 degrees C

Fluid Flow Rates

- ADI – 135 gallons per hour
- spray bar – 1 gallon per min. per cooler = 2 gallons per minute total
- fuel flow – 450 gallons per hour at 3,100 rpm @ 75 in. Hg.

Race #8 Dreadnought Race Results

(Ref. 9-2 and 9-10)

1983 National Championship Air Races – Reno.
Pilot: Neil Anderson

Race:	**Speed**	**Position**
Race:	Speed	Position
Qualification:	446.392 mph	fastest
Heat 1C	418.225 mph	2nd
Heat 2C	435.584 mph	1st
Gold	425.242 mph	1st

1984 National Championship Air Races – Reno.
Pilot: Neil Anderson

Race:	**Speed**	**Position**
Qualification	442.747 mph	fastest
Heat 2A	429.857 mph	1st (engine seized after landing)

1985 Bakersfield National Air Races – Shafter, California.
Pilot: Frank Sanders

Race:	**Speed**	**Position**
Qualification	401.775 mph	(did not race due to metal in screens)

1985 National Championship Air Races – Reno.
Pilot: Neil Anderson

Race:	**Speed**	**Position**
Qualification:	443.129 mph	fastest
Heat 2A	424.796 mph	1st
Heat 3A	436.947 mph	1st
Gold	429.430 mph	2nd

1986 National Championship Air Races – Reno.
Pilot: Rick Brickert

Race:	**Speed**	**Position**
Qualification:	452.737 mph	fastest
Heat 2A	414.222 mph	1st
Heat 3A	416.787 mph	1st
Gold	434.488 mph	1st

1987 National Championship Air Races – Reno.
Pilot: Rick Brickert

Race:	**Speed**	**Position**
Qualification	448.698 mph	4th fastest
Heat 2A	441.291 mph	2nd
Heat 3A	444.920 mph	2nd
Gold	449.747 mph	2nd

1988 Wings of Victory Air Races, Hamilton Air Force Base, California.
Pilot: Rick Brickert

Race:	**Speed**	**Position**
Qualification:	399.110 mph	2nd fastest
Gold	383.106 mph	3rd

1988 National Championship Air Races – Reno.
Pilot: Rick Brickert

Race:	**Speed**	**Position**
Qualification:	458.920 mph	3rd fastest
Heat 2A	449.149 mph	1st
Heat 3A	457.014 mph	2nd
Gold	451.202 mph	2nd

1989 National Championship Air Races – Reno.
Pilot: Rick Brickert

Race:	**Speed**	**Position**
Qualification:	456.584 mph	3rd fastest
Heat 2A	432.954 mph	2nd
Heat 3A	441.287 mph	3rd
Gold	427.871 mph	2nd

1990 National Championship Air Races – Reno.
Pilot: Rick Brickert

Race:	**Speed**	**Position**
Qualification:	448.673 mph	4th fastest
Heat 2A	434.810 mph	3rd
Heat 3A	443.057 mph	3rd
Gold	441.715 mph	4th

1991 National Championship Air Races – Reno.
Pilot: Dennis Sanders

Race:	**Speed**	**Position**
Qualification:	441.203 mph	4th fastest
Heat 2A	412.375 mph	4th
Heat 3A	425.603 mph	4th
Gold	426.507 mph	5th

1992 Denver Air Races.
Pilot: Dennis Sanders

Race:	**Speed**	**Position**
Qualification:	392.770 mph	3rd fastest
Heat 1B	378.864 mph	5th
Gold	405.177 mph	2nd

1992 National Championship Air Races – Reno.
Pilot: Brian Sanders

Race:	**Speed**	**Position**
Qualification:	443.585 mph	3rd fastest
Heat 2A	444.005 mph	1st
Heat 3A	443.615 mph	3rd
Gold	442.495 mph	2nd

1993 National Championship Air Races – Reno.
Pilots: Brian & Dennis Sanders

Race:	**Speed**	**Position**
Qualification:	436.921 mph	2nd fastest
Heat 2A	439.845 mph	2nd
Heat 3A	440.346 mph	2nd
Gold	450.619 mph	2nd

1993 Kansas City Air Races.
Pilot: Dennis Sanders

Race:	**Speed**	**Position**
Qualification:	414.21 mph	2nd fastest
Heat 2	397.54 mph	1st
Gold	402.37 mph	2nd

1994 Phoenix 500 Air Races.
Pilot: Brian Sanders

Race:	**Speed**	**Position**
Qualification:	419.984 mph	5th fastest
Heat 1A	414.953 mph	1st
Gold	420.177 mph	2nd

1994 National Championship Air Races - Reno.
Pilot: Brian Sanders

Race:	Speed	Position
Qualification:	449.901 mph	4th fastest
Heat 2A		DNF
Heat 3A	416.592 mph	2nd
Gold	416.632 mph	2nd

1995 Phoenix 500 Air Races.
Pilot: Dennis Sanders

Race:	Speed	Position
Qualification:	413.103 mph	3rd fastest
Heat 1A	405.315 mph	2nd
Gold	426.056 mph	2nd

1995 National Championship Air Races - Reno.
Pilots: Brian & Dennis Sanders

Race:	Speed	Position
Qualification:	434.667 mph	5th fastest - Brian Sanders
Heat 1A	426.122 mph	1st - engine expired after finish of Heat 1A (Dennis Sanders pilot).

1997 National Championship Air Races - Reno.
Pilot: Brian Sanders

Race:	Speed	Position
Qualification:	438.426 mph	3rd fastest
Heat 2A	434.314 mph	3rd
Heat 3A	431.408 mph	3rd
Gold	441.467 mph	2nd

1998 National Air Races - Reno.
Pilot: Dennis Sanders

Race:	Speed	Position
Qualification:	433.283 mph	4th fastest
Heat 2A	426.061 mph	3rd
Heat 3A	439.551 mph	2nd
Gold	445.306 mph	2nd

1999 National Championship Air Races - Reno.
Pilot: Brian Sanders

Race:	Speed	Position
Qualification:	445.321 mph	4th fastest
Heat 2A	432.006 mph	3rd
Heat 3A	433.365 mph	3rd
Gold	441.221 mph	2nd

2000 National Championship Air Races - Reno.
Pilot: Brian Sanders

Race:	Speed	Position
Qualification:	438.019 mph	2nd fastest - Withdrew with engine problems and did not race.

2001 National Championship Air Races - Reno.
Terrorist attacks - no races.

2003 National Championship Air Races - Reno.
Pilot: Brian Sanders

Race:	Speed	Position
Qualification:	442.771 mph	4th
Heat 2A:	443.958*	5th
Heat 3A Gold		Did not start**

*Cut Pylon Outer 2 on Lap 6, +12 seconds penalty.
** Metal in screens.

Although they are tempting to use, *Dreadnought* has not been exposed to any exotic fuels such as triptane or nitrous oxide *(Ref. 9-11).*

Race #15 *Head Gorilla/Furias*

(Ref. 9-12, 9-13, and 9-14)

Introduced the same year as *Dreadnought*, *Head Gorilla,* as it was initially known, is another aircraft that started life as a Bristol Centaurus-powered Hawker Sea Fury FB Mk. II. A veteran of the Reno race scene, Lloyd Hamilton, undertook the work to modify the aircraft. Its first year of racing was in 1983, the same year as *Dreadnought*. Flown by its builder, Lloyd Hamilton, its introduction was inauspicious—it blew its engine during qualification at Reno in 1983. The following year Hamilton renamed the aircraft *Furias*, a name that it has retained to the present. In December 1997, Lloyd Hamilton passed away. *Furias* did not race again until 2000, when it was purchased by Bill Rogers and Dale Stolzer. Seven spray nozzles, one for each plenum, were incorporated into the cowl nose bowl for cylinder head temperature control. At race power settings, water is sprayed into the cowl. Power is transmitted to an Aeroproducts Douglas Skyraider propeller. Interestingly, as of 2003 the engine still retained the fluid coupling supercharger drive rather than the preferable "solid" gear drive. Several hundred horsepower are chomping at the bit to be unleashed via the gear drive. Another possible modification would be the incorporation of turning vanes inside the ram induction scoop where it makes a 90-degree transition from horizontal to vertical into the carburetor.

This shot of *Furias* shows the racer in its 2002 guise. Initially built by Lloyd Hamilton, since his passing it was purchased by Bill Rogers and Dale Stolzer. After sitting dormant for several years, the aircraft needed some maintenance and at the same time a new paint job (the one shown in this photo was applied). *Furias* follows all the design features of *Dreadnought,* including use of the same R-4360-63A and Aeroproducts Skyraider propeller. *(Photo by author)*

Furias differed from *Dreadnought* in a couple of minor areas. For instance, instead of using a de-icing slinger ring to spray water on the cylinders, *Furias* uses seven spray nozzles. This close-up of the cowl nose bowl shows three of the nozzles. *(Photo by author)*

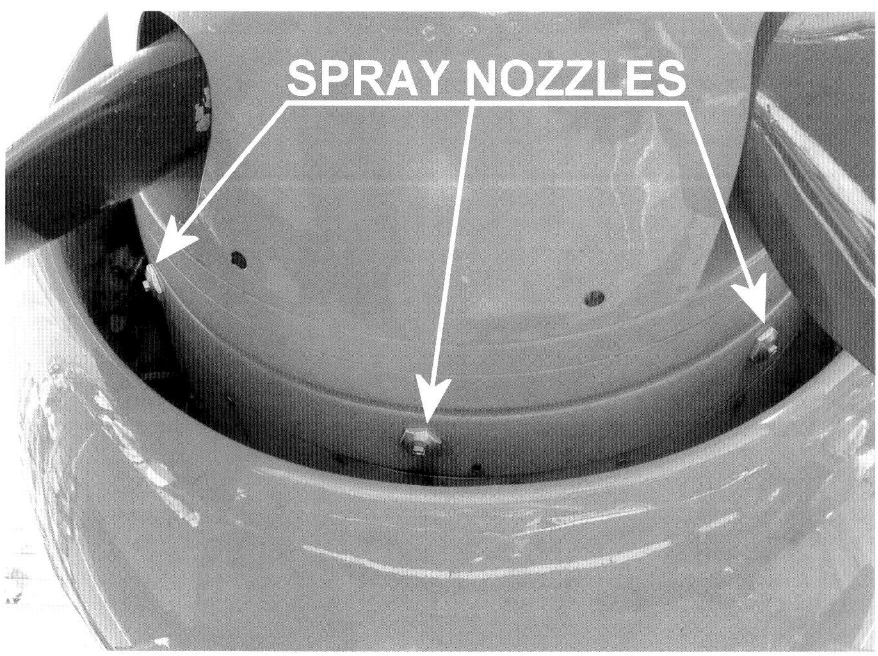

Race #15 Head Gorilla/Furias Race Results

(Ref. 9-2 and 9-10)

1983 National Championship Air Races – Reno.
Pilot: Lloyd Hamilton Raced as Head Gorilla

Race	Speed	Position
Qualification		Did not qualify – runaway prop blew engine.

1984 National Championship Air Races – Reno.
Pilot: Lloyd Hamilton Raced as Furias

Race:	Speed	Position
Qualification:	398.137 mph	11th fastest
Heat 1A	327.483 mph	7th
Heat 2B	344.423 mph	7th
Heat 3B		8th DNF – damaged aircraft due to engine backfire and blown cowling.

1985 National Championship Air Races – Reno.
Pilot: Lloyd Hamilton

Race:	Speed	Position
Qualification:	406.753 mph	11th fastest
Heat 1A	394.649 mph	2nd
Heat 2B	404.778 mph	2nd
Heat 3A	382.906 mph	5th
Gold	411.952 mph	4th

1986 National Championship Air Races – Reno.
Pilot: Lloyd Hamilton

Race:	Speed	Position
Qualification:	433.410 mph	5th fastest
Heat 2A	406.546 mph	2nd
Heat 3A	415.293 mph	2nd
Gold	429.374 mph	2nd

1987 National Air Races – Reno.
Pilot: Lloyd Hamilton

Race:	Speed	Position
Qualification:	418.548 mph	8th fastest
Heat 1A		Did not start due to a bearing failure after takeoff.

1988 Wings of Victory Air Races, Hamilton Air Force Base, California.
Pilot: Lloyd Hamilton

Race:	Speed	Position
Qualification:	377.739 mph	3rd fastest
Gold	366.528 mph	5th

1988 National Championship Air Races – Reno.
Pilot: Lloyd Hamilton

Race:	Speed	Position
Qualification:	424.081 mph	9th fastest
Heat 1A	410.444 mph	3rd
Heat 2B	411.963 mph	1st
Heat 3A	393.462 mph	3rd
Gold	403.632 mph	5th

1989 National Championship Air Races – Reno.
Pilot: Lloyd Hamilton

Race:	Speed	Position
Qualification:	417.022 mph	8th fastest
Heat 1A		DNF, blown engine – metal in screens.

1990 National Championship Air Races – Reno.
Pilot: Lloyd Hamilton

Race:	Speed	Position
Qualification:	347.256 mph	24th fastest
Heat 1C		DNF
Bronze	353.013 mph	2nd

1997 National Championship Air Races – Reno.
Pilot: Lloyd Hamilton

Race:	Speed	Position
Qualification:	386.425 mph	11th fastest
Heat 1B	387.601 mph	1st
Heat 2A	379.608 mph	7th
Heat 3B	395.592 mph	3rd
Gold	393.830 mph	6th

2000 National Air Races – Reno.
Pilot: Art Vance

Race:	Speed	Position
Qualification:	397.432 mph	17th fastest
Heat 1B	385.267 mph	5th
Heat 2C	385.359 mph	1st
Heat 3B	372.927 mph	6th
Silver		Did not start

2001 National Championship Air Races – Reno.
Terrorist attacks – no races.

2002 National Championship Air Races – Reno.
Pilot: Art Vance

Race:	Speed	Position
Qualification:	415.808 mph	7th fastest
Heat 1A	399.280 mph	3rd
Heat 2B	411.705 mph	2nd
Heat 3B	402.338 mph	2nd
Silver	382.288 mph	3rd

2003 National Championship Air Races – Reno.
Pilot: Gary Hubler

Race:	Speed	Position
Qualification:	445.420 mph	3rd fastest
Heat 2A	444.241 mph	4th
Heat 3A	437.758 mph	4th
Gold	439.944 mph	5th

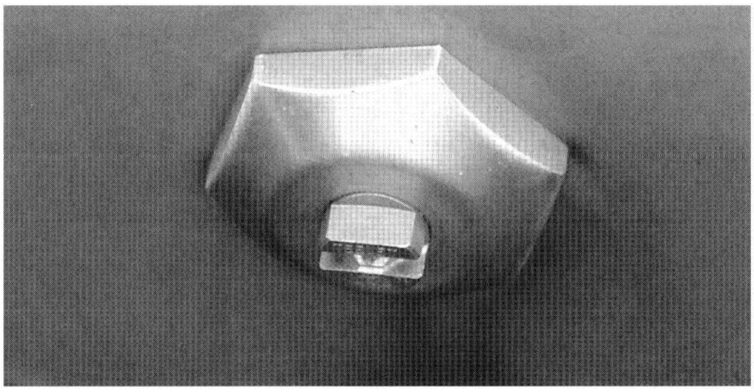

This is a close-up of one of the seven spray nozzles used by *Furias* to spray water on the cylinders. *(Photo by author)*

Below is a comparative summery of all the post-1949 R-4360-powered racers.

All Coast Super Corsair Race #1* (Also Bud Light Special) Summary

First year	1982
Fastest qualification	442.319 mph at Reno in 1988 flown by John Maloney
Fastest heat	445.072 mph at Reno in 1988 flown by John Maloney
Fastest Gold finish	438.186 mph at Reno in 1985 flown by Steve Hinton
Gold Wins	1 (1985 at Reno, 438.186 mph - new record at that time)
Heat wins	8 (includes Gold win from 1985)
Heats raced	49 (includes only 5 DNF)
Years raced	13 (1982 to 1994)

This is at least the second Unlimited Class racer to use Race #1. Daryl Greenamyer's F8F Bearcat, *Conquest*, also raced as #1.

Dreadnought Race #8 Summary

(Ref. 9-2 and 9-10)

First year	1983	
Fastest qualification	458.920 mph at Reno in 1988 flown by Rick Brickert	
Fastest heat	457.014 mph at Reno in 1988 flown by Rick Brickert	
Fastest Gold finish	434.488 mph at Reno in 1986 flown by Rick Brickert	
Gold wins	2	
	1983 at Reno; 425.24 mph flown by Neil Anderson,	
	1986 at Reno; 434.488 mph flown by Rick Brickert	
Gold Race 2nd places	14	
Heat wins	14 (includes 2 Gold wins)	
	Neil Anderson	5 wins
	Rick Brickert	4 wins
	Brian Sanders	3 wins
	Dennis Sanders	2 wins
Heats raced	54	
Years raced	13 (1983 to 1994, 2003)	

Dreadnought has finished a remarkable 1st or 2nd in over 50 percent of its heat runs.

Furias (Head Gorilla in 1983) Race #15 Summary

(Ref. 9-2 and 9-10)

First year	1983
Fastest qualification	433.410 mph at Reno in 1986 flown by Lloyd Hamilton
Fastest heat	429.374 mph at Reno in 1986 flown by Lloyd Hamilton
Highest Gold finish	2nd at Reno in 1986 (429.374) flown by Lloyd Hamilton
Heat wins	3 (two when flown by Lloyd Hamilton and one when flown by Art Vance)
Heats raced	30 (Lloyd Hamilton: 21, Art Vance: 7, Gary Hubler: 2)
Years raced	11 (1984 to 2000, 2002, 2003) (Missed 1991 to 1996, 1998 and 1999) (DNQ in 1983)

As of this writing (2003) it appears that Unlimited Class air racing is in its twilight years—at least as far as using modified World-War-II-era aircraft is concerned. Their value is now such that it simply does not make sense to modify an historic artifact for racing purposes. In the 1960s, it made perfect sense. Aircraft, engines, and other spares were cheap and plentiful. This is no longer the case. Currently, only two R-4360-powered racers are campaigned: *Furias* and *Dreadnought*. It's possible that a purpose-built racer will be built using R-4360 power. However, the monetary rewards, even if one wins, are so miniscule that it turns out to be a labor of love. And as we

This view shows the ram induction scoop used by *Furias*. It is interesting to note that the 90-degree bend that transitions induction into the carburetor does not have turning vanes. This alone would be worth several inches of manifold pressure at race speed. *(Photo by author)*

know, love goes flying through the window when the bills come due. Nevertheless, one advantage with an R-4360-powered racer is the amazingly low cost of procuring one of these complex engines. Like everything else, supply and demand dictates price. With very little demand, nice low time R-4360's can be purchased for a few thousand dollars.

References

9-1 Telephone interviews with Cook Cleland in 2002.

9-2 Veronico, Nicholas A., Grantham, A. Kevin. *Round-Engine Racers: Bearcats & Corsairs*. Specialty Press. 2002.

9-3 Huntington, Roger. *Thompson Trophy Racers: The Pilots and Planes of America's Air Racing Glory Days 1929-49.* Motorbooks International. 1989.

9-4 http://www.pylon1.com

9-5 Schlaifer, Robert and S. D. Heron. *Development of Aircraft Engines and Development of Aviation Fuels.* Harvard University, Boston. 1950.

9-6 Phone conversations and e-mails with Steve Hinton.

9-7 Phone conversations and e-mails with Pete Law.

9-8 Graph generated by Pete Law for the R-4360-powered racers.

9-9 Graph generated by Pete Law for calculating ram temperature rise.

9-10 Results were corroborated and corrected from previously published results by Pete Law.

9-11 All information on *Dreadnought* was obtained from Brian, Dennis, and Ruth Sanders.

9-12 Interviews with Art Vance, former pilot of *Furias*.

9-13 http://www.aafo.com/racing/news/00/04-02-00.htm

9-14 Conversations with Bill Rogers, current co-owner of *Furias*.

CHAPTER TEN

A Bridge Too Far?

Operating Difficulties

Being the last of the line for piston engines, it was only to be expected that the R-4360 pushed the envelope in many respects. Commercial operators in particular were sensitive to cost overruns due to maintenance troubles. In the late 1940s and early 1950s when BOAC was operating Boeing Model 377s, their huge losses were often blamed on the cost of keeping the R-4360s going. As is usually the case, this claim was only partly true. Mismanagement of this government-owned and controlled airline was responsible for the lion's share of its fiscal woes. Nevertheless, operating this huge engine at its limits in a commercial environment invariably provoked short times between overhauls and heavy maintenance in the interim. The only other engine in the R-4360's class was the R-3350 turbo compound.

Like the R-4360, this complex engine was the last of the line, which, in a similar fashion to the R-4360, pushed the technological envelope. Used more extensively for commercial use than the R-4360, the R-3350 was what flight engineers would describe as a "delicate" engine. By this it was meant that if the R-3350 turbo compound was treated carefully, with respect and all temperatures and pressures kept in the "green," it was a long-lived engine *(Ref. 10-1).* Abuse it and it wouldn't last one trip. Abuse, and this applies to any radial engine, would include rough handling of the throttles, incorrect mixture (such as too rich or too lean), over temping the oil and/or cylinders, and perhaps the worst operating sin for any "round" engine—reverse loading. Due to the dynamics of the master connecting rods and the loads imposed upon it, the design of the oil galleries and lubrication are optimized for the engine to be producing power. Reverse loading simply means the propeller is driving the engine instead of the engine driving the propeller. This condition would occur, for example, on short final when the runway is made and a careless pilot/flight engineer pulls the power back to idle and allows the propeller to windmill, which in turn drives the engine.

R-4360 overhaul times were generally in the 1,100- to 1,400-hour range and provided nothing failed in the intervening time. This overhaul time is comparable to an R-1830 as would be used to power, for instance, a DC-3. R-2800s enjoyed overhaul times of over 2,000 hours or about twice what the R-4360 could be expected to achieve. The 1,100 to 1,400 hours quoted would be for an engine with all the latest engineering changes incorporated and handled with care, conditions that did not always exist. As with all the big and complicated reciprocating engines, extreme care in handling, particularly by the flight engineer, was crucial to longevity. Put the R-4360 in the hands of a ham fisted or careless flight engineer and don't expect the engine to survive.

Ignition

With its genesis going back to the early 1940s, it should be no surprise that early R-4360s featured high-tension ignition, which is not a good idea for high-altitude flight. As has been explained on numerous occasions, as an aircraft ascends the atmosphere becomes less dense. With the protective insulation from the upper atmosphere diminishing, this reduced the dielectric strength and insulating value of the atmosphere also diminished. With upwards of 20,000 volts being generated in the magneto and then transmitting this high-tension to the appropriate spark plug, problems are bound to occur. Even under ideal conditions at sea level, where the insulating value of the atmosphere is at its greatest, high tension could jump from one ignition lead to another causing cross firing. Or, high tension could jump across contacts in the magneto. This issue was probably of greatest concern with the B-36 due to its mission requirement to fly at high altitudes.

Fried Valve Guides

(Ref. 10-2)

Excessive exhaust valve guide wear was a problem that plagued R-4360 operation throughout its life. When this occurred, high-temperature exhaust gasses leaked into the exhaust rocker box. Oil coming into contact with super heated valve components due to the exhaust leak turned to coke. One of the primary contributing factors to valve guide wear was the lead content of fuel used during the R-4360's heyday. Up to 5 cc of tetraethyl lead was used per gallon, which is great for improving fuel performance, but it created havoc with engine components. Lead would attach itself to the valve stem, which in turn would reduce the running clearance between the valve stem and its guide. Soon, rapid wear would occur due to lead buildup on the valve stem. Innumerable materials were tried in order to overcome this serious problem. As it turned out, it would require a composite guide made up with a tight fitting steel scraper ring to keep lead buildup to a minimum. Additional oil jets aimed in the rocker box plus additional cooling fins machined into exhaust rocker boxes made after and including the R-4360-35 alleviated the worst of these maladies.

Problem Log

(Ref. 10-3)

Pratt & Whitney kept a log of R-4360 problems for both the military and commercial operators. The following represents a (very) condensed summary of a few of these reports. As can be seen, it's difficult to pinpoint any one problem area, as they seem scattered throughout the engine. Obviously, if a design weakness had showed up through repeated failures it would have been immediately addressed. The root cause of the service difficulties varied greatly, including the following: poor manufacturing, poor flight engineer procedures, poor choice of materials, careless handling of parts, sub-standard overhauls, etc. Compared to other engines of its era, such as the R-2800, it seems that the R-4360 was more critical when it came to operating, maintenance, and overhaul. Fits, clearances, and tolerances were certainly tighter on the R-4360 than most, if not all, other radial aircraft engines. And why was this? With an engine as complex as the R-4360, it's difficult if not impossible to pinpoint any particular weak area. Possibly compounding the problem was the extensive use of turbosupercharging for altitude performance.

The following problem reports seem to call out a preponderance of turbosupercharged versions such as the -59 and -53. Certainly, it seems indisputable that turbo'd versions ran hotter and were subject to more abuse due to excessive manifold pressure. But even this does not tell the whole story. Perhaps at the end of the day it could be argued that the R-4360 simply needed more development time. It's fairly certain that if the R-4360 development engineers had had their druthers, some of the R-2180 design features would have been incorporated, especially the reverse positioning of the intake and exhaust ports. If the exhaust had exited straight through the top of the cylinder head as it did with the R-2180, the R-4360 would have possibly run cooler. Furthermore, a more efficient intake manifold design could have been incorporated, improving volumetric efficiency and consequently power. A built-up crankshaft with one-piece master rods would have contributed to the engine's durability. However, we'll never know the answer to these tantalizing questions. All we can do is study the records and make our judgments and assessments.

PROBLEM	Cracking of propeller shaft at reduction fixed gear support splines. Cracking is primarily confined to engines incorporating heavy support.
ENGINE MODELS	R-4360-20
CAUSE	Retaining nut is being found 5 degrees to 32 degrees loose.
CORRECTIVE ACTION	Provide sheared silver fit.

PROBLEM	Leakage at Propeller Shaft Thrust Plate - This is primarily a shop problem but is also being reported by UAL (United Air Lines), SAAMA (San Antonio Air Material)* and some Air Force. Leakage is between thrust rings and liner, around studs in thrust plate, and around O.D. of thrust plate. (FIGS 1 & 2)
ENGINE MODELS	R-4360-59-B6
CAUSE	
CORRECTIVE ACTION	Various ECs (engineering changes) provided new thrust plate seals. Provide silver flash on rings.

*SAAMA was major repair depot for military R-4360s.

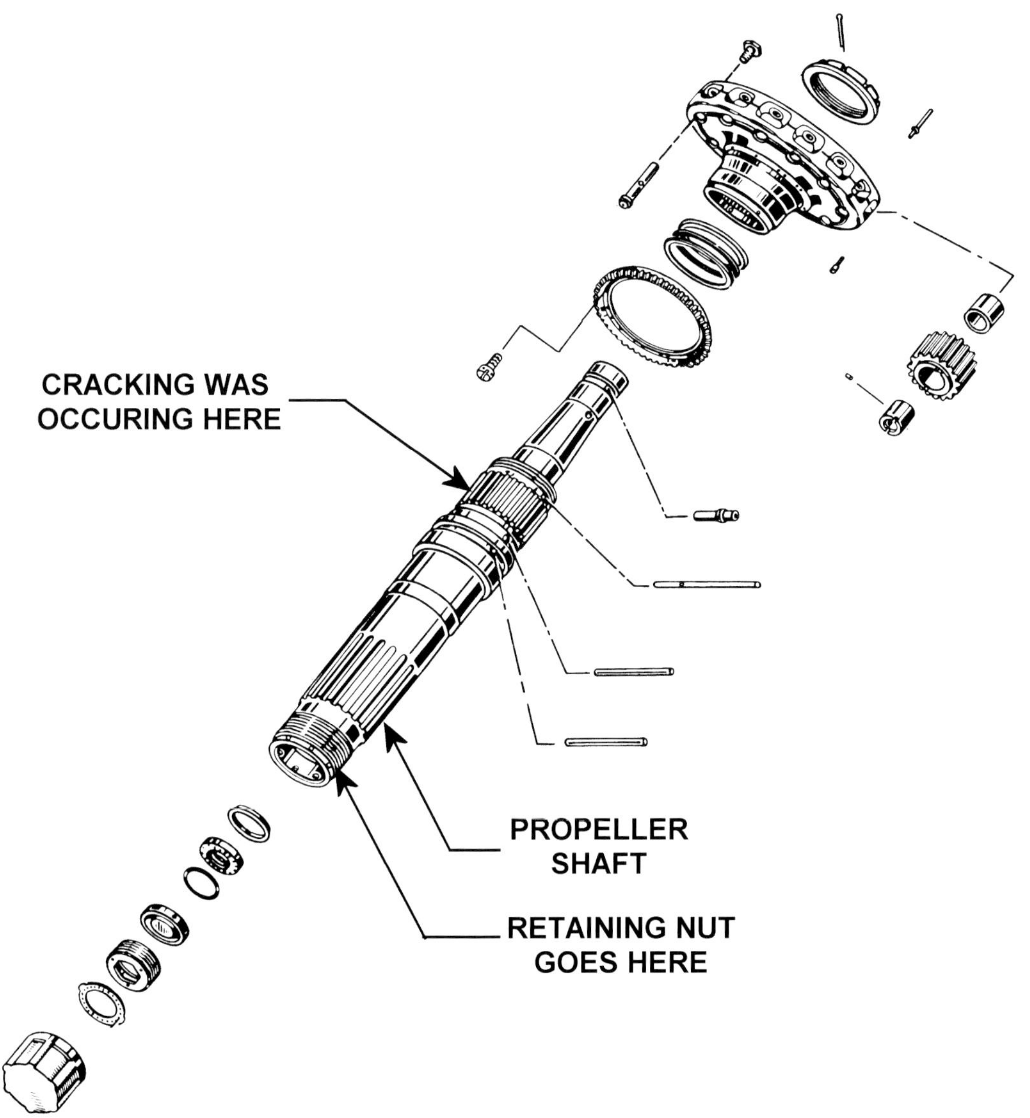

FIG 1 Considering the enormous loads that the propeller shaft is exposed to, it's hardly surprising that fatigue cracks should show up. *(Technical Manual Illustrated Parts Breakdown USAF Models R-4360-20WA and -35B Aircraft Engines dated March 1, 1966)*

FIG 2 Radial engines have always had the unenviable reputation for being oil leakers. Whenever the source of a leak could be determined, it was fixed. Beyond making an unsightly mess, it was a significant safety issue due to a potential fire hazard. *(Technical Manual Illustrated Parts Breakdown USAF Models R-4360-59B and -65 Aircraft Engines dated September 15, 1959)*

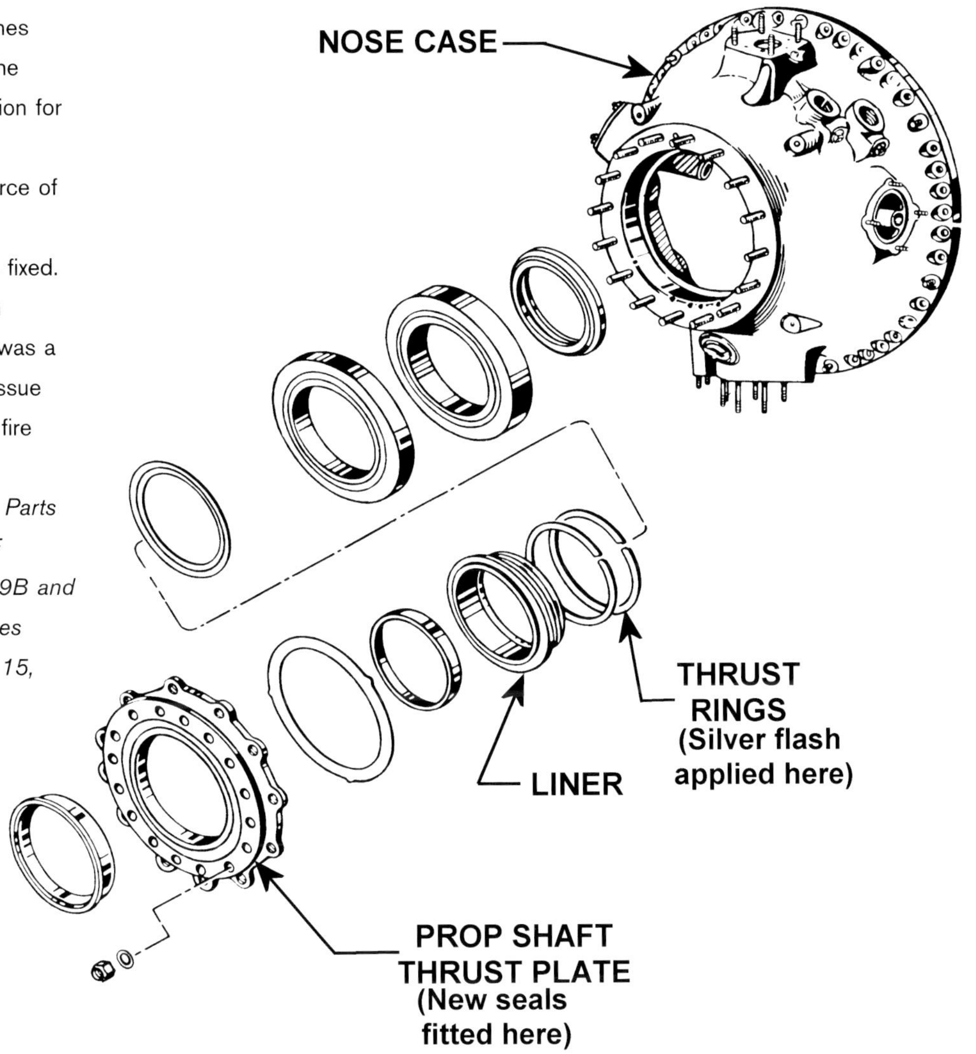

PROBLEM	Reduction drive gear chipping. Very high percentage of gears (part number quoted) is found at overhaul with chipping and pitting of the gear teeth. (FIG 3)
ENGINE MODELS	R-4360-20, B13 (B series with 0.425:1 reduction gear)
CAUSE	Chipping of teeth is due to excessive embrittlement of ends of teeth.
CORRECTIVE ACTION	Numerous ECs; reduce shot peening intensity. Provide grit blast and varnish graphite coating.

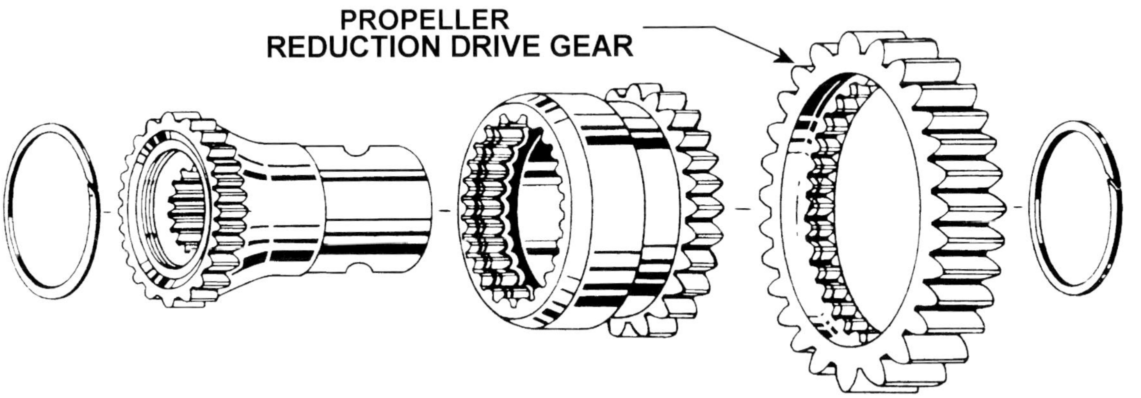

FIG 3 Arguably, one of the most heavily loaded gear trains in the R-4360 is the propeller reduction gear mounted on the crankshaft. Some -20s suffered chipping and galling of the reduction gear teeth. *(Technical Manual Illustrated Parts Breakdown USAF Models R-4360-20WA and -35B Aircraft Engines dated March 1, 1966)*

PROBLEM	Front Main Bearing Failure (FIG 4)
ENGINE MODELS	
CAUSE	Not determined. Failure usually occurs on test stand or within a few hours after being placed in service.
CORRECTIVE ACTION	Larger oil holes, new design bearing with "double taper."

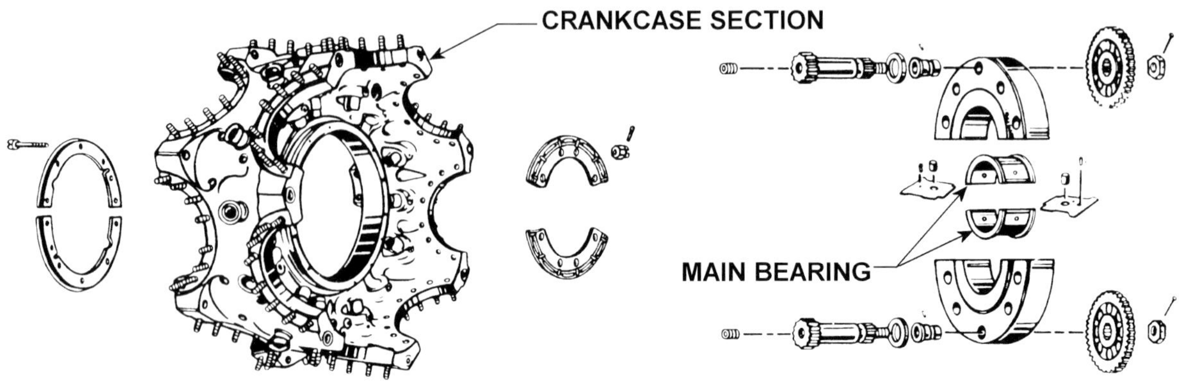

FIG 4 Crankshaft main bearings by their very nature operate under considerable stress. And the one that catches the heaviest load is the front main. Exacerbating the situation is the fact that it is lubricated last because oil is fed in from the supercharger end. In addition, the front of the crankshaft provides support for the rear of the propeller shaft. Gyroscopic and G loads generated by the propeller are consequently transferred to the front main. Little wonder that problems arose. *(Technical Manual Illustrated Parts Breakdown USAF Models R-4360-59B and -65 Aircraft Engines dated September 15, 1959)*

PROBLEM	Crankshaft Vibration Damper Wear (FIG 5) 1. Counterweight roller holes show wear during one run which makes it necessary to oversize holes or replace counter weights. 2. Counterweight floating bushing holes become badly galled during operation.
ENGINE MODELS	R-4360 "C" series and B6
CAUSE	1. Normal wear. 2. Fretting and galling of two steel surfaces.
CORRECTIVE ACTION	Numerous ECs. Provide replacement bushings. 1. Apply flash of silver to O.D. of counterweight bushings. 2. Bronze bushings for crank cheeks.

PROBLEM	Cracked Master Rod Caps. (FIG 6) Two types of failures reported: 1. Cracks originating at non-intersecting link pin holes. 2. Cracks originating at base of pilot shoulders where chrome plating has been used to regain pinch with the master rod.
ENGINE MODELS	R-4360-TSB3G - B6
CAUSE	1. Probably insufficient pinch fit of link pins. 2. Believed to be insufficient radius before chrome plate.
CORRECTIVE ACTION	Numerous ECs and service bulletin. Increase pinch fit of link pin to link pin holes. Copper plating to regain fit of cap and rod.

PROBLEM	Master rod link pin hole wear.
ENGINE MODELS	R-4360-B6, "C" series
CAUSE	1. Insufficiently tight fit of link pins to master rod. 2. Insufficient stiffness in non-intersected link pin hole area.
CORRECTIVE ACTION	Numerous ECs and service bulletin. Increase interference fit of all R-4360 link pins in the master rod. BOAC instigated a 0.003" interference fit.

PROBLEM	Master rod bearing failures.
ENGINE MODELS	All R-4360s
CAUSE	Operations, silver fatigue, chamfer at parting edges, insufficient oil supply, high silver stresses.
CORRECTIVE ACTION	ECs and service bulletin. 1. Better relief in the silver at the parting faces of the bearings. 2. Release of heavy walled master rod bearing with its lower silver stresses. 3. Service testing of high indium content bearing. 4. Service test of bearings having a lead-tin-antimony overlay.

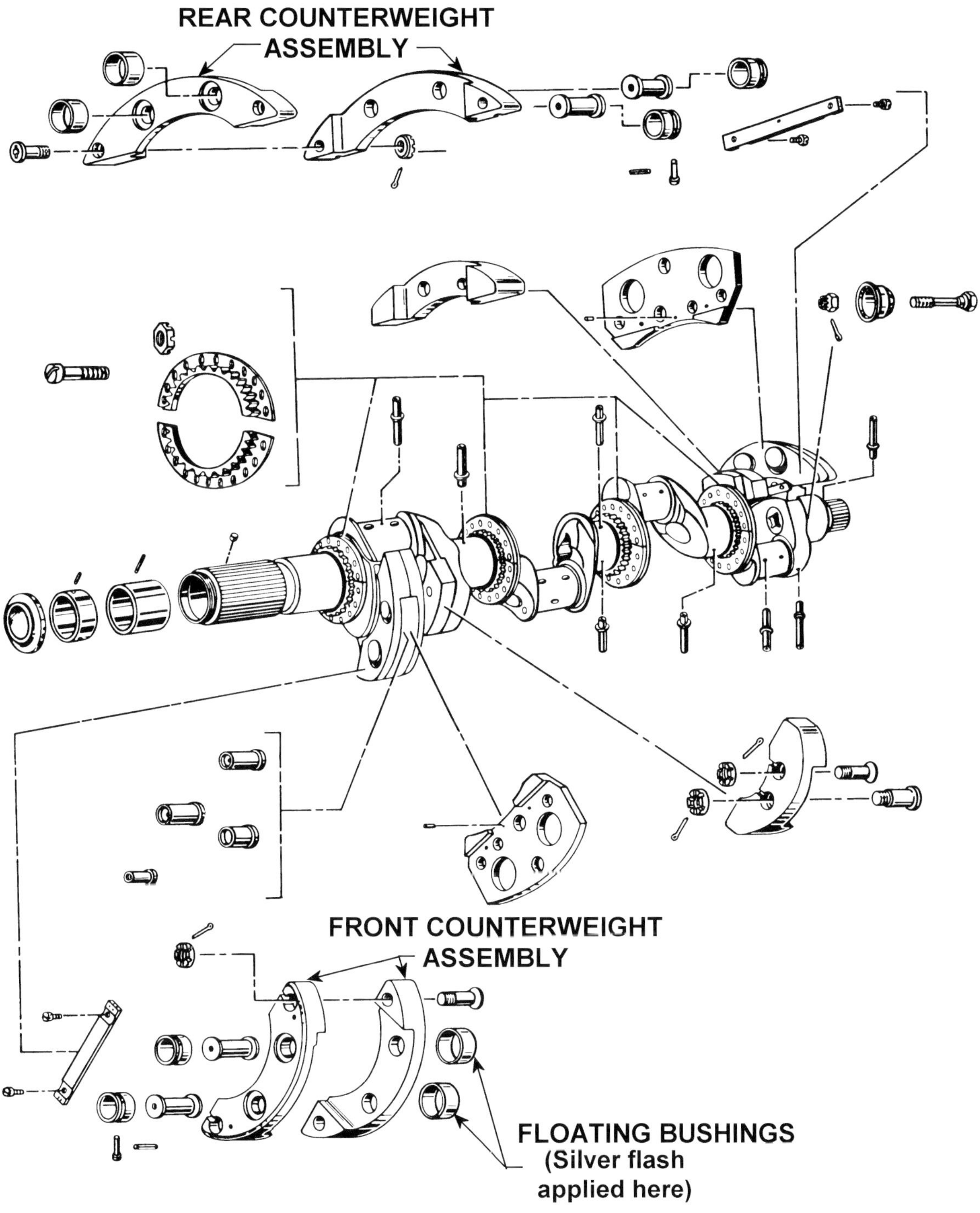

FIG 5 Dynamic counter weights are exposed to large centrifugal loads. It's easy to see how the counterweight roller holes get worn. The galling referred to in the report would, in all likelihood, be due to fretting. *(Technical Manual Illustrated Parts Breakdown USAF Models R-4360-59B and -65 Aircraft Engines dated September 15, 1959)*

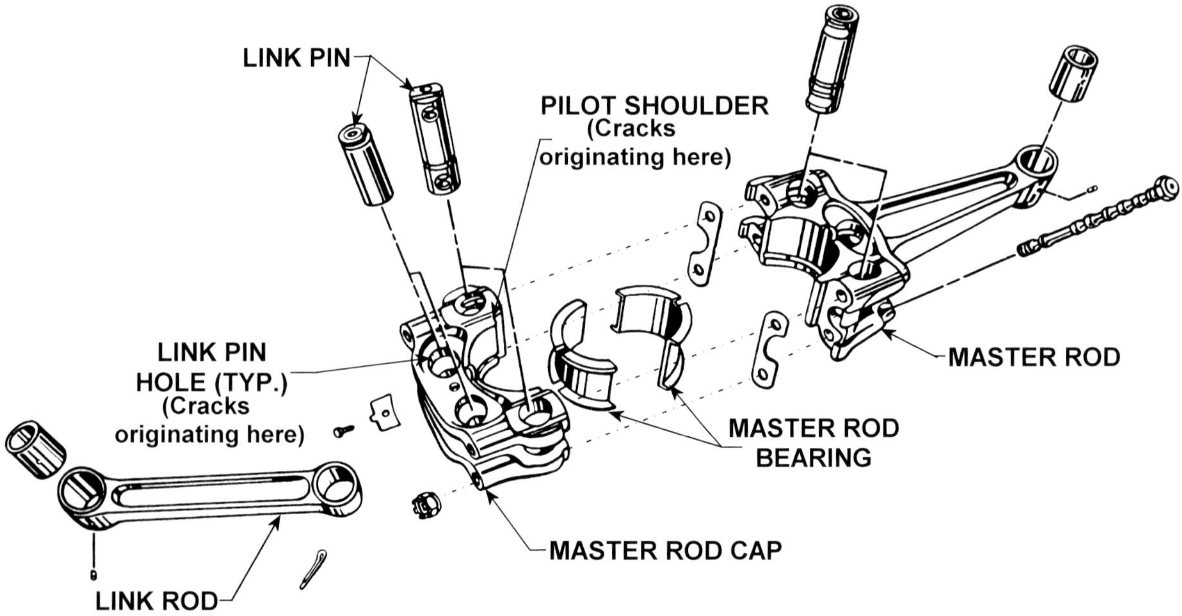

FIG 6 Whenever a sharp change in cross section occurs, a stress riser is generated. This no doubt contributed to the R-4360's cracked master rod caps. Another contributing factor, although not mentioned in the report, is the use of chrome plating, particularly to a critical part like a master rod. Hydrogen embrittlment is now a well-known problem with chrome plating; however, in the heyday of the R-4360 this phenomenon was less understood. *(Technical Manual Illustrated Parts Breakdown USAF Models R-4360-59B and -65 Aircraft Engines September 15, 1959)*

PROBLEM	Link pin failures.
ENGINE MODELS	R-4360-C, B6
CAUSE	Undetermined
CORRECTIVE ACTION	Most failures occurred through a radial oil hole which caused a fatigue progression having its focus at or near the bleed on the inside end. One failure originated at the outside blend. Redesign link pins with no oil holes.

PROBLEM	Exhaust cam distress. (FIG 7) Failures are associated with spalling of the pre-rise lobe.
ENGINE MODELS	R-4360-B6
CAUSE	Unknown
CORRECTIVE ACTION	Various including cam profile from R-4360-53 (B-36 engine), wider cam track, increased valve clearance.

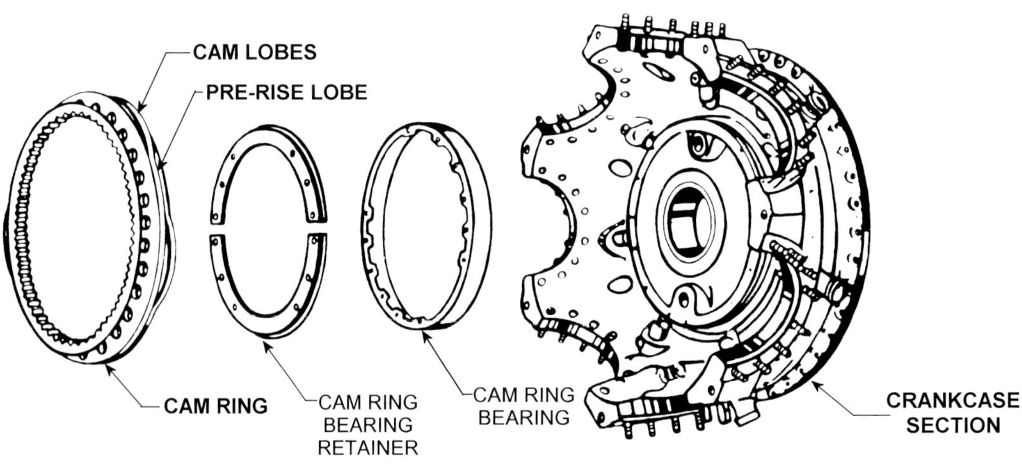

FIG 7 Exhaust cams traditionally take a heavier load than their intake counterparts. This is partly due to the substantial residual pressure that resides in the cylinder just prior to exhaust valve opening. The cam has to overcome this residual pressure consequently loading up the entire valve drivetrain. *(Technical Manual Illustrated Parts Breakdown USAF Models R-4360-59B and -65 Aircraft Engines dated September 15, 1959)*

PROBLEM	Crankcase cracking. (FIG 8) 1. At periphery of cylinder flange bolt heads. 2. From I.D. of cam retainer bolt holes. 3. From camshaft gear liner seat – "A" row.
ENGINE MODELS	R-4360-B6, "C" series
CAUSE	1. Indentation of bolt heads in case resulting in high stress concentrations. 2. Believed to be result of rough machining. 3. Galling of shaft gear liner to crankcase.
CORRECTIVE ACTION	ECs 1. Return three crankcases to operators for continued service with cracks in the cylinder bolt flange counter bores. 2. Correct "rough" hole condition in manufacturing. 3. This is the most widespread condition resulting in a high crankcase rejection rate. "Stretch" type cam drive shaft gear liners released to production. Test steel horseshoes under the cylinder flange bolts heads. Redesign center and intermediate cam bearing arrangement.

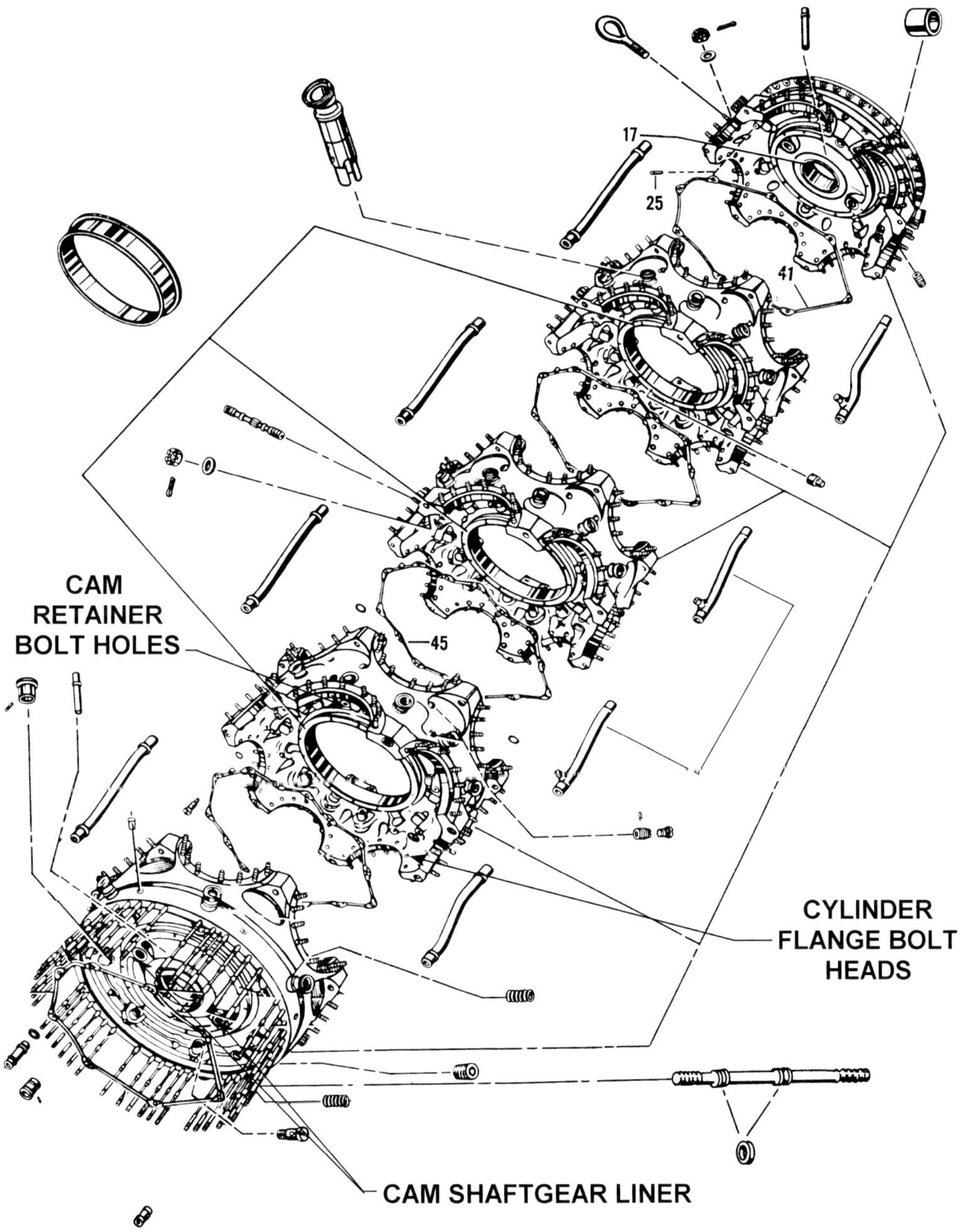

FIG 8 When having to compromise the conflicting requirements of an aircraft engine, those of lightweight and high horsepower, it is understandable that the more highly stressed components will suffer from stress cracking. So it was with the R-4360's crankcases sections. *(Technical Manual Illustrated Parts Breakdown USAF Models R-4360-59B and -65 Aircraft Engines dated September 15, 1959)*

PROBLEM	Cylinder head cracking around exhaust valve seat. (FIG 9)
ENGINE MODELS	R-4360 "C" series
CAUSE	Overheating and use of earlier material; PWA 621
CORRECTIVE ACTION	Improved operator handling of engine, i.e., do not overheat. Use of PWA 623 material for cylinder head.

PROBLEM	Exhaust rocker box cracking.
ENGINE MODELS	R-4360 "B" series
CAUSE	Believed to be high stress resulting from excessive clearance at rocker shaft spacer.
CORRECTIVE ACTION	Reduce clearance.

PROBLEM	Exhaust valve guide wear.
ENGINE MODELS	R-4360 all models
CAUSE	Lack of compatibility between cast iron valve guide and nickel valve. Any lack of lubrication, for example during start up, causes aggravated valve stem wear.
CORRECTIVE ACTION	Numerous valve guide materials tried including a so-called "bi-metal" guide. The upper half was made from bronze and the remainder cast iron. The bi-metal guide was unsuccessful. Later guides retained cast iron but incorporated a chamfer at the top edge to eliminate sharp edge.

PROBLEM	Exhaust valve face distress.
ENGINE MODELS	R-4360 all models
CAUSE	Valve seat distortion resulting in face cracking and burning.
CORRECTIVE ACTION	The so-called "flexible valve" was developed to conform to the valve seat. The flexible valve proved to be unreliable in service. However, the flexible valve proved to be successful in R-2800 operations.

PROBLEM	Intake valve seat and valve face wear.
ENGINE MODELS	R-460 all models
CAUSE	Valve spinning and high impact loading has been suggested as possible cause of spinning. Rotating valves is normally considered desirable but excessive spinning **cause**s galling and excessive wear.
CORRECTIVE ACTION	Numerous spring configurations tried. Pan Am had success by installing an O-ring in the intake valve guide.

PROBLEM	Exhaust manifold coupling spring seat cracking. (FIG 10)
ENGINE MODELS	R-4360-59, -53, CB
CAUSE	High temperature corrosion.
CORRECTIVE ACTION	1. Use Inconel material machined to a thicker dimension 2. "Aluminize" parts involved.

PROBLEM	Cylinder deflector wear and cracking. (FIG 11)
ENGINE MODELS	R-4360 "C" series
CAUSE	
CORRECTIVE ACTION	Redesign of deflectors (baffles). Add additional metal strip to the bottom of affected deflectors.

PROBLEM	Impeller loose on impeller shaft. (FIG 12)
ENGINE MODELS	R-4360-53 (-63, -20, and B13 use same assembly)
CAUSE	Unknown, suspected high impeller temperatures due to lack of fuel cooling plus adverse turbo supercharger and ambient temperatures.
CORRECTIVE ACTION	Numerous ECs and Service Bulletin. Increase pinch fit of the impeller and associated parts. Add 0.0005 in. silver plate to shaft. Modify tooling. It's interesting to note that even though the -3, -20 and B-13 use the same impeller assembly, they did not suffer the same failure. This was probably due to the fact that the -53, which powered the Convair B-36, was fuel injected and therefore did not have the advantage of enjoying the fuel cooling effect that the other engines had, which were carbureted.

PROBLEM	Blower drive gears – tooth spalling (FIG 13)
ENGINE MODELS	R-4360, TSB3G, B6, CB2, -35, -41, -49, -59, -61
CAUSE	Poor alignment of gear teeth (skew) – improper tooth form – inadequate back lash.
CORRECTIVE ACTION	Numerous ECs. Service Bulletins. Modified tooth form, incorporate 0.0007" to 0.0013" skew.

PROBLEM	Blower intermediate drive gear bearing wear & erosion.) (FIG 14)
ENGINE MODELS	R-4360
CAUSE	Lead washing – improper bond, erosion, poor lubrication, inadequate bearing size.
CORRECTIVE ACTION	Service Bulletins Change metallurgy of bearing from lead-indium to lead tin-indium.

PROBLEM	Hydraulic coupling failure – this is a major problem, which has accounted for a large percentage of R-4360-20 series engine failures. (FIG 15)
ENGINE MODELS	R-4360-20, -53, -63
CAUSE	Tablock washer failures and spanner nut loosening due to insufficient butt contact area – spanner nut thread interference. Inadequate bearing design or poor lubrication.
CORRECTIVE ACTION	Numerous ECs and numerous Service Bulletins. Improved tablock washer design, silver plate threads, eliminate fluid coupling altogether and use single speed blower.

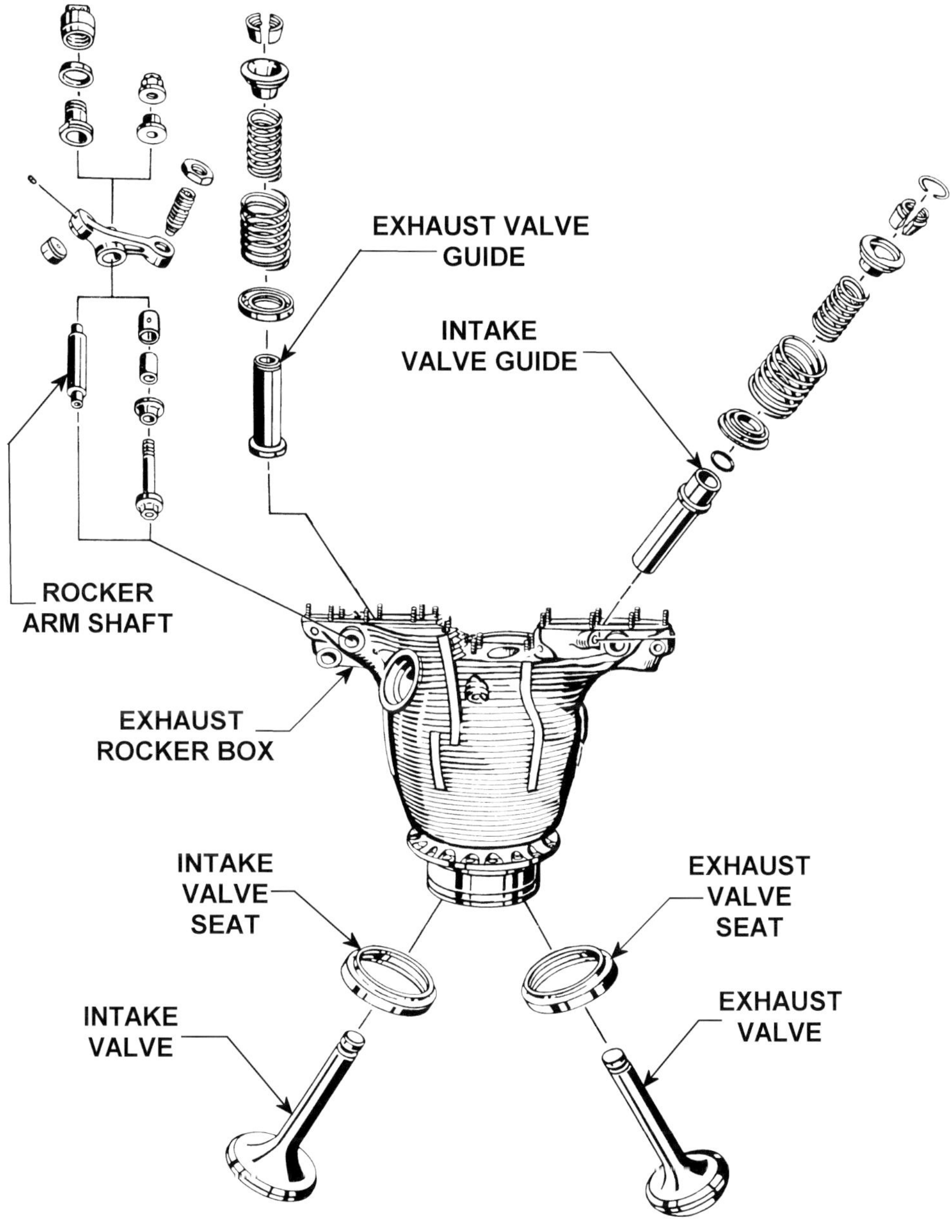

FIG 9 Cylinder problems were ongoing with the R-4360 despite its state-of-the-art design and manufacturing methods. If the engine was not handled carefully, especially with regards to overheating, cracking could and did occur, particularly around the exhaust valve seat. The exhaust rocker box was another highly stressed part of the cylinder design. The R-4360 was unique in having rocker boxes that were overhung, consequently adding a powerful bending moment to the cylinder assembly. Most radial engines experienced exhaust valve guide issues and the R-4360 was no different in this regard. R-4360s used a variety of guide designs including different materials and even the so-called bi-metal design of bronze and cast iron. This problem was never definitively fixed.

Another exhaust related issue was that of distorted exhaust valve seats. Once a leak occurred, it would not take long for the valve to be torched. The so-called "flex valve" was tried but without success. The theory was that the valve would flex and conform to the shape of the seat. It created even more serious problems when the valve head would separate from the stem. Many valvetrain designs incorporated a method to spin the valve during operation in order to even-out valve face wear. In the case of the R-4360, this could prove to be excessive and in some instances galling of the vale seat face occurred. Pan Am used an O-ring in the intake valve guide to reduce the spinning. *(Technical Manual Illustrated Parts Breakdown USAF Models R-4360-20WA and -35B Aircraft Engines dated March 1, 1966)*

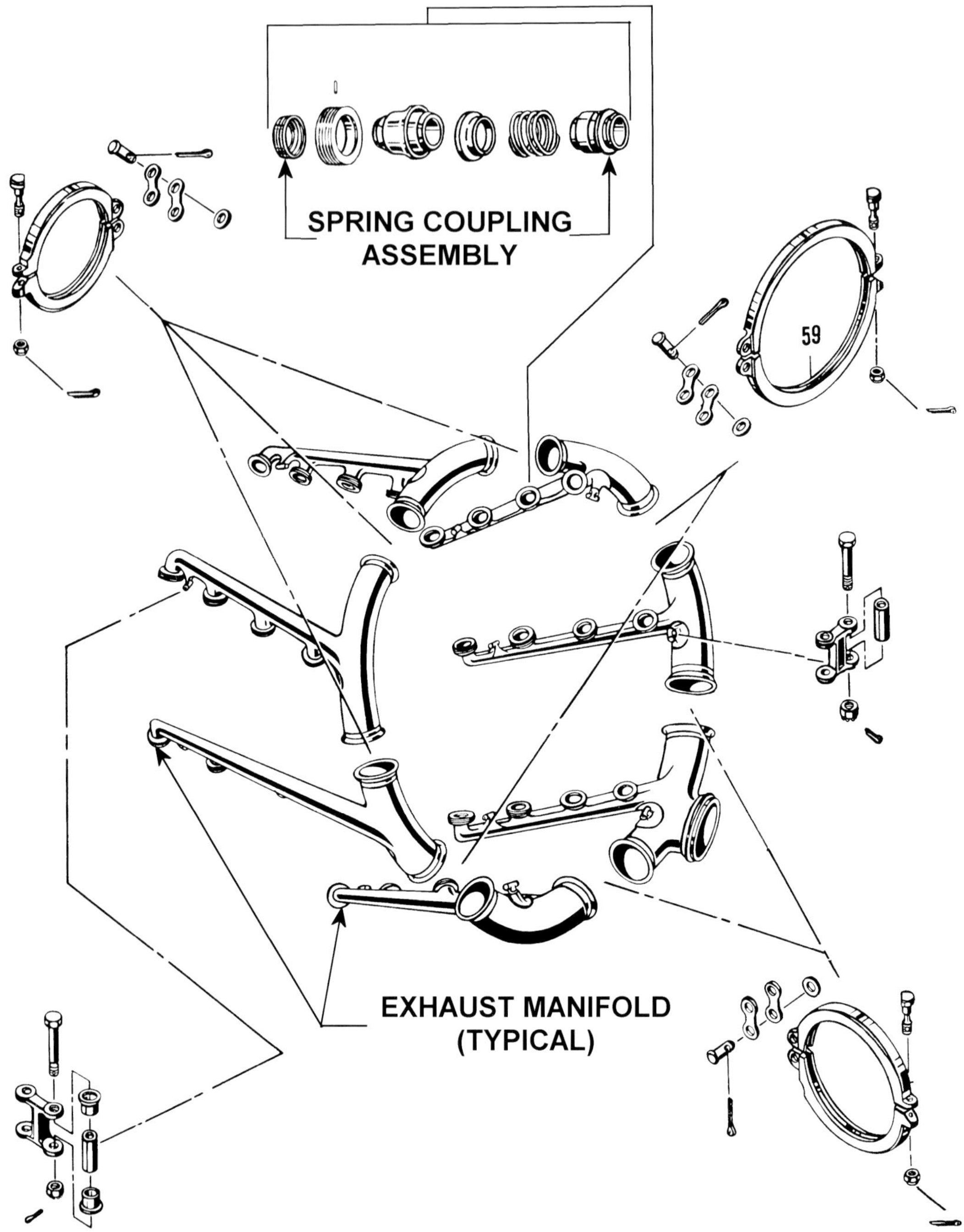

FIG 10 R-4360-59s and commercial CBs used a spring-loaded exhaust coupling to the cylinder head. Spring seats could crack, so beefing up the thickness of the seat material fixed the problem. *(Technical Manual Illustrated Parts Breakdown USAF Models R-4360-59B and -65 Aircraft Engines dated September 15, 1959)*

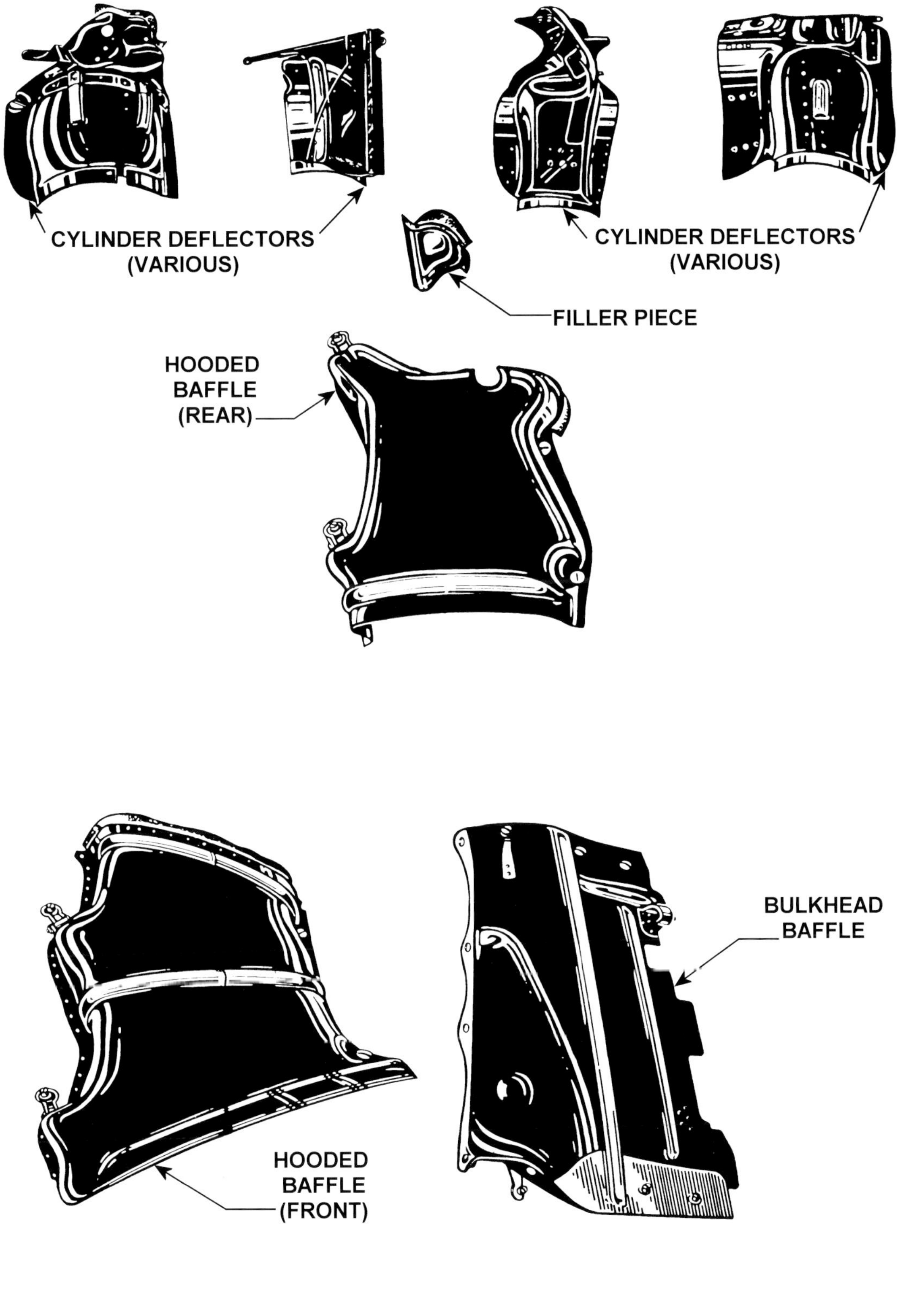

FIG 11 Cylinder baffles are subject to constant abuse from clumsy maintenance people and the constant vibration they undergo when the engine is running. Therefore, it's not surprising that stress cracks are an ongoing issue. Baffles perform a critical function and if they are allowed to deteriorate other more serious problems mushroom from this seemingly innocuous issue. *(Technical Manual Illustrated Parts Breakdown USAF Models R-4360-59B and -65 Aircraft Engines dated September 15, 1959)*

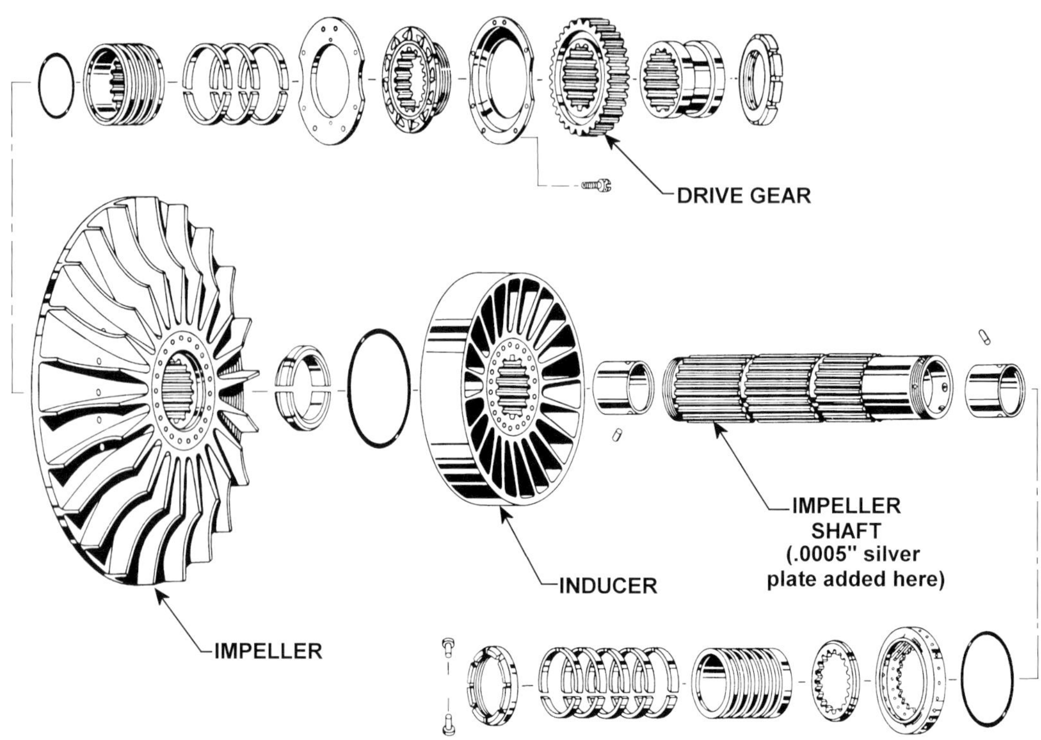

FIG 12 It's interesting to see that the impeller assembly for the R-4360-43 would come loose, yet the same assembly was used on the -20, -63, and the commercial B13 with no problem. As the report stated, the likely cause was overheating of the impeller assembly on the -43. This was a B-36 fuel-injected engine; therefore, no fuel was pumped through the impeller, as would be the case for the -20, -63, and B13. As a result the -20, -63, and B13 benefited from a cooling fuel flow. A number of tweaks were performed, all of which basically made the impeller a tighter fit on the shaft. *(Technical Manual Illustrated Parts Breakdown USAF Models R-4360-20WA and -35B Aircraft Engines dated March 1, 1966)*

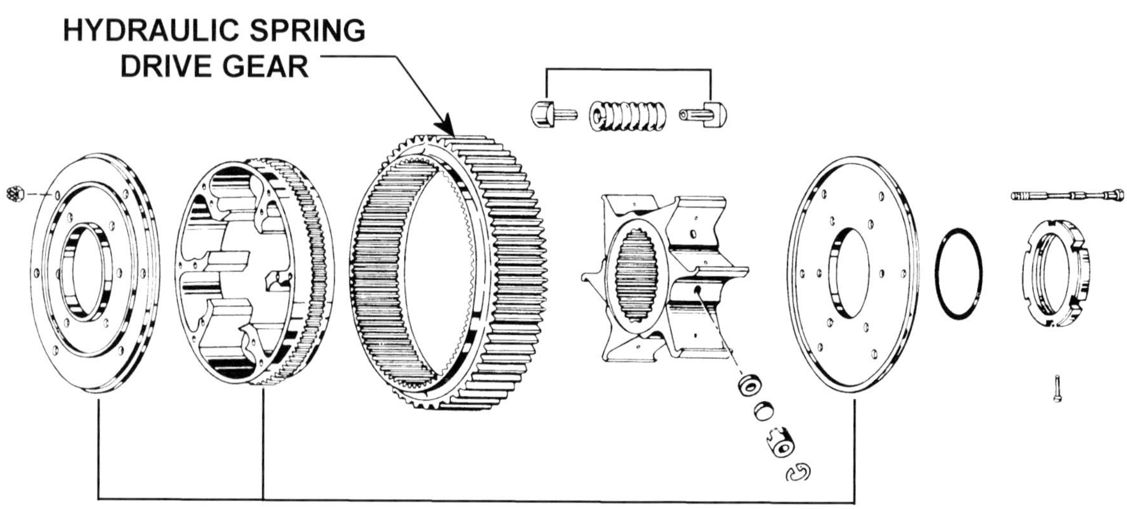

FIG 13 Supercharger drive requirements have always offered up a challenge to design engineers. In the case of the R-4360, as with most high-performance piston engines, gears would tend to gall. Tweaking the tooth form and introducing "skew" kept the problem under control. *(Technical Manual Illustrated Parts Breakdown USAF Models R-4360-59B and -65 Aircraft Engines September 15, 1959)*

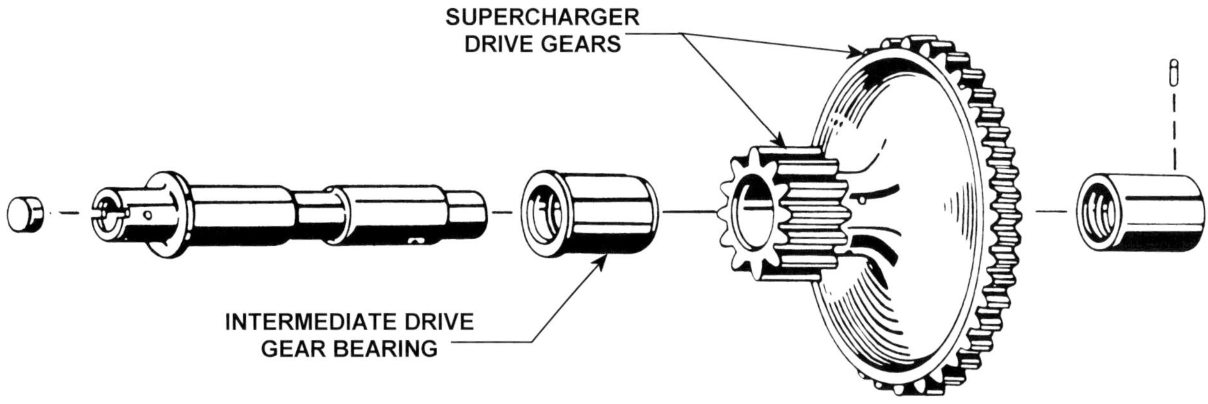

FIG 14 Depending on the blower ratio, the intermediate gear would rotate at approximately 9,000 rpm under heavy load. So it's not surprising that bearings would fail. Pratt & Whitney changed the bearing metallurgy from lead/indium to lead-tin-indium. *(Parts Catalog for Aircraft Engines Models R-4360-17, -21, and -25 Aircraft Engines dated August 22, 1947)*

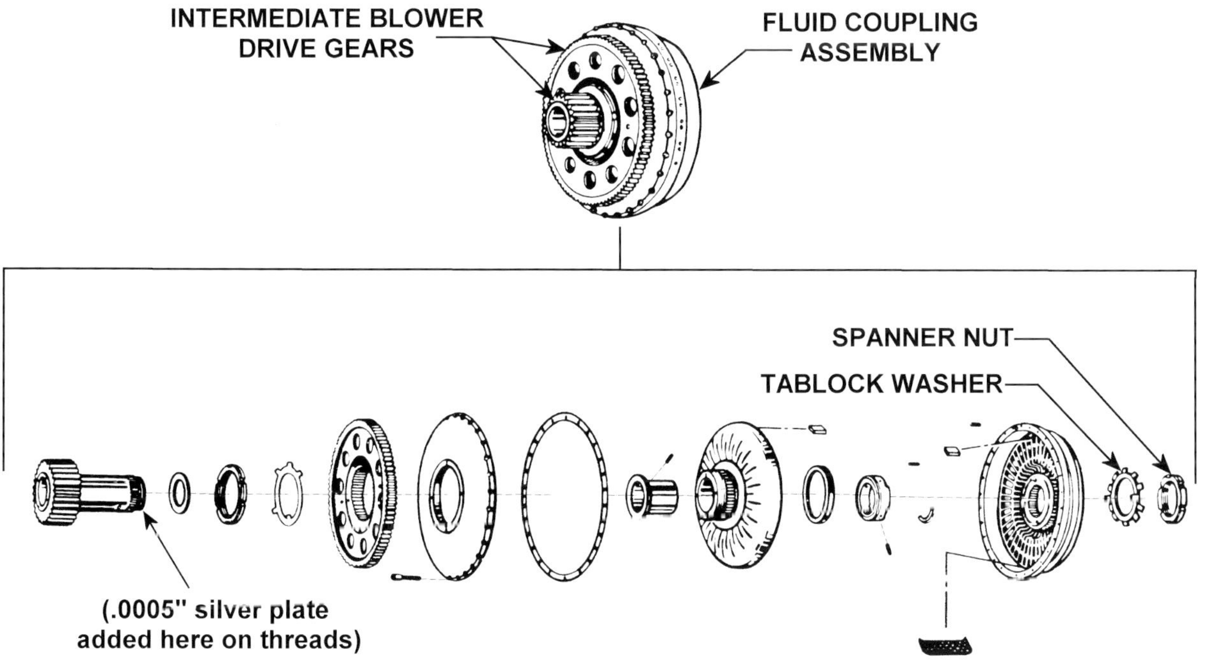

FIG 15 Looking at the tab lock design and the spanner nut retaining the fluid coupling, one can imagine how this assembly can back off. Part of the problem may have been attributable to careless handling of the engine. If the engine RPMs were suddenly backed off, fretting would be introduced at the spanner nut thread. Even today, this is a common problem with automobile and truck manual transmissions. Getrag and New Venture in particular has suffered this kind of problem with their 5-speed truck transmissions. *(Technical Manual Illustrated Parts Breakdown USAF Models R-4360-20WA and -35B Aircraft Engines dated March 1, 1966)*

PROBLEM	Collector case cracking at engine mount insert threads. (FIGS 16 & 17)
ENGINE MODELS	R-4360 TSB3G, B6, CB2, B13, -4 series, -20 series, -35 series, -41, -41A, -53, -59, -59B, -61, -63, -63A
CAUSE	Unsatisfactory engine support design. Excessive stress concentrations at bottom threads of insert.
CORRECTIVE ACTION	Use of longer thread insert with a tapered thread.

PROBLEM	Primary wire chaffing. (FIG 18)
ENGINE MODELS	R-4360 with low-tension ignition.
CAUSE	Teflon fiberglass insulated wire has poor abrasion resistance.
CORRECTIVE ACTION	1. Replace Teflon fiberglass insulated wire with Beldon Aircraft wire. 2. Teflon wrapped impregnated fiberglass insulation. 3. Packard fiberglass Teflon impregnated. 4. EC issued to provide improved scuff resistant Teflon impregnated wire with same O.D. and conductor capacity

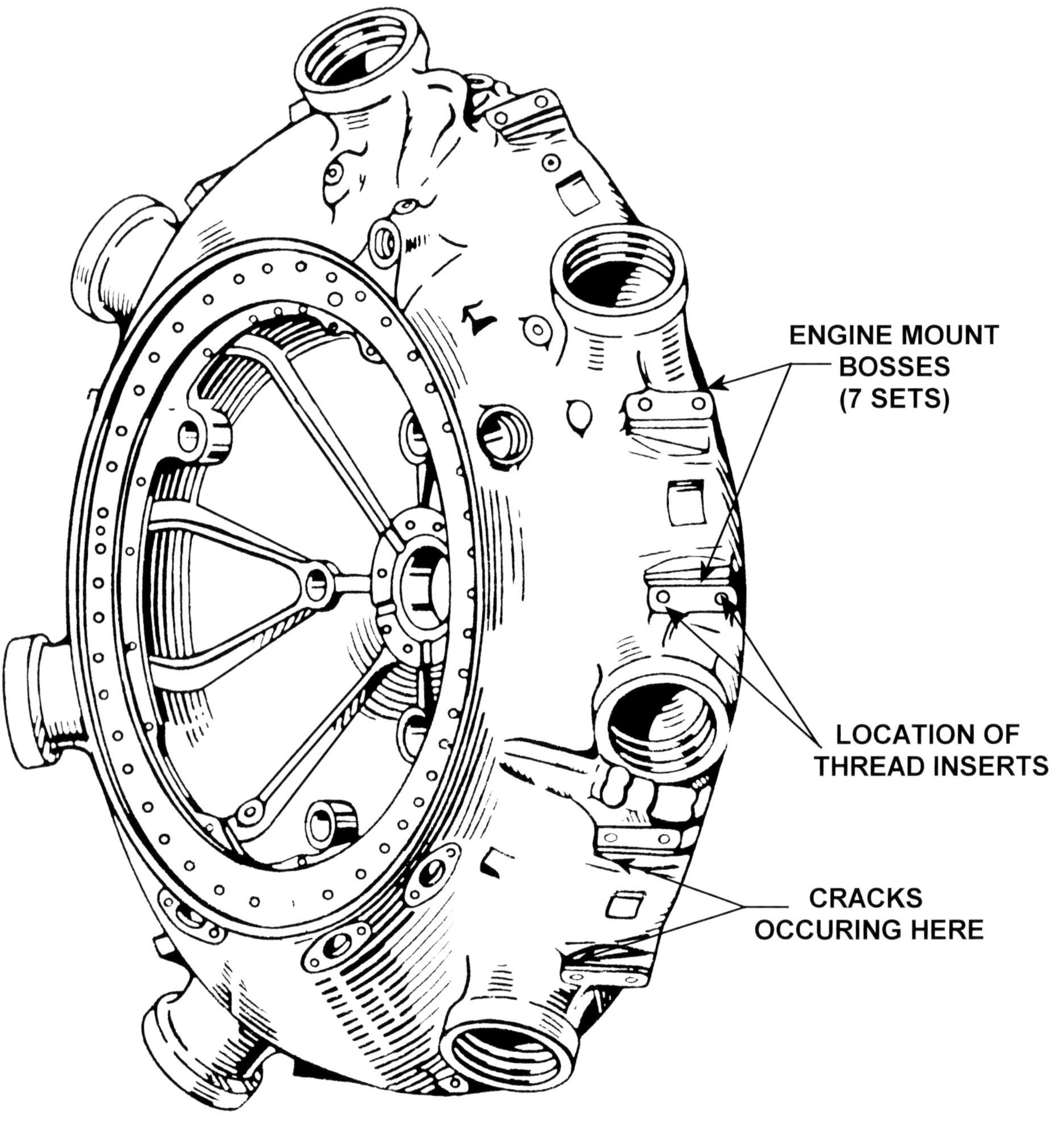

FIG 16 Considering the tremendous load carried by the R-4360's engine mounts, it was only to be expected that cracks should occur in this highly stressed location. *(Technical Manual Illustrated Parts Breakdown USAF Models R-4360-20WA and -35B Aircraft Engines dated March 1, 1966)*

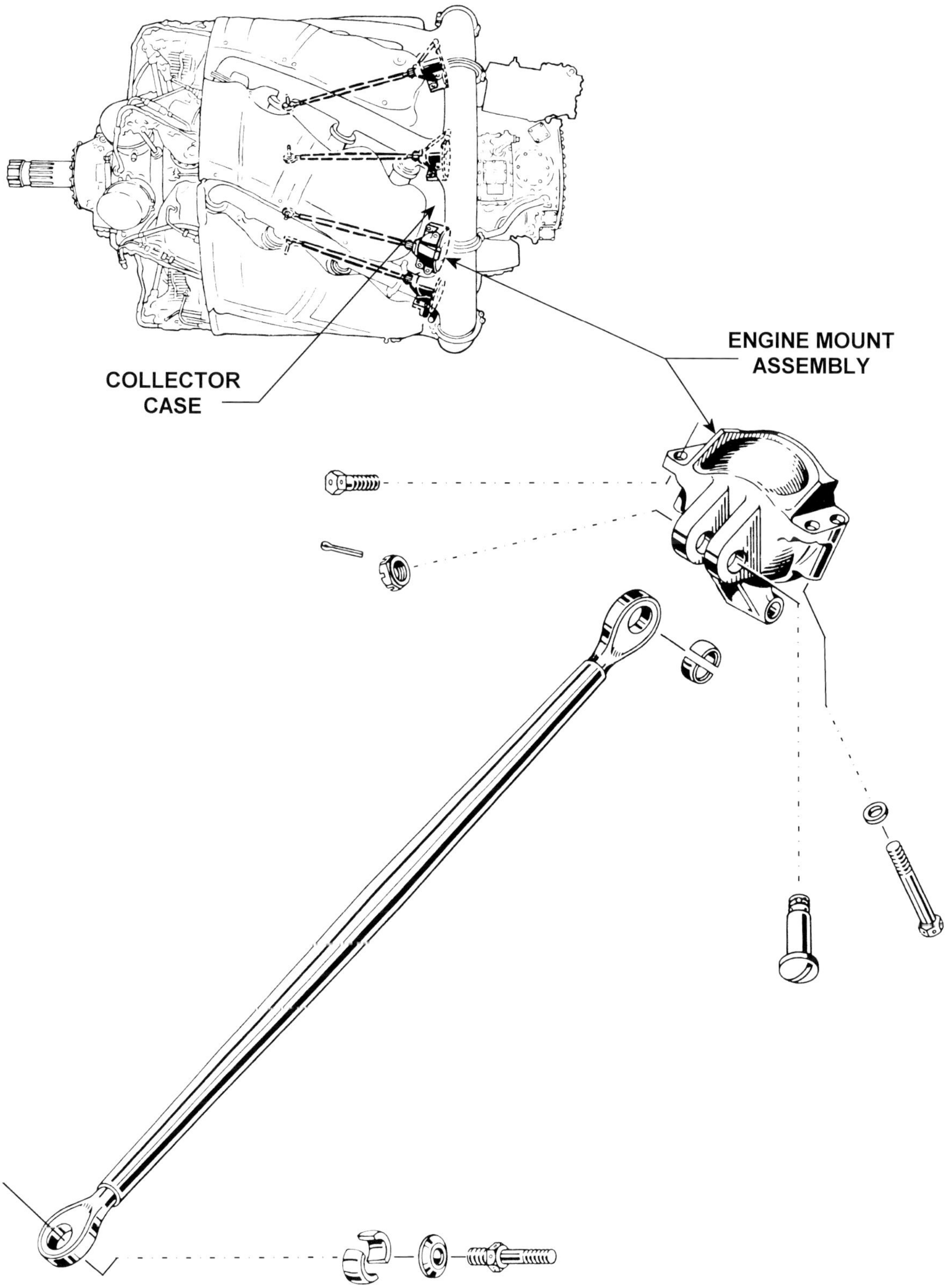

FIG 17 The drawing shows an R-4360-59's mounting system. Note how far forward the struts extend into the crankcase. *(Technical Manual Illustrated Parts Breakdown USAF Models R-4360-20WA and -35B Aircraft Engines dated March 1, 1966)*

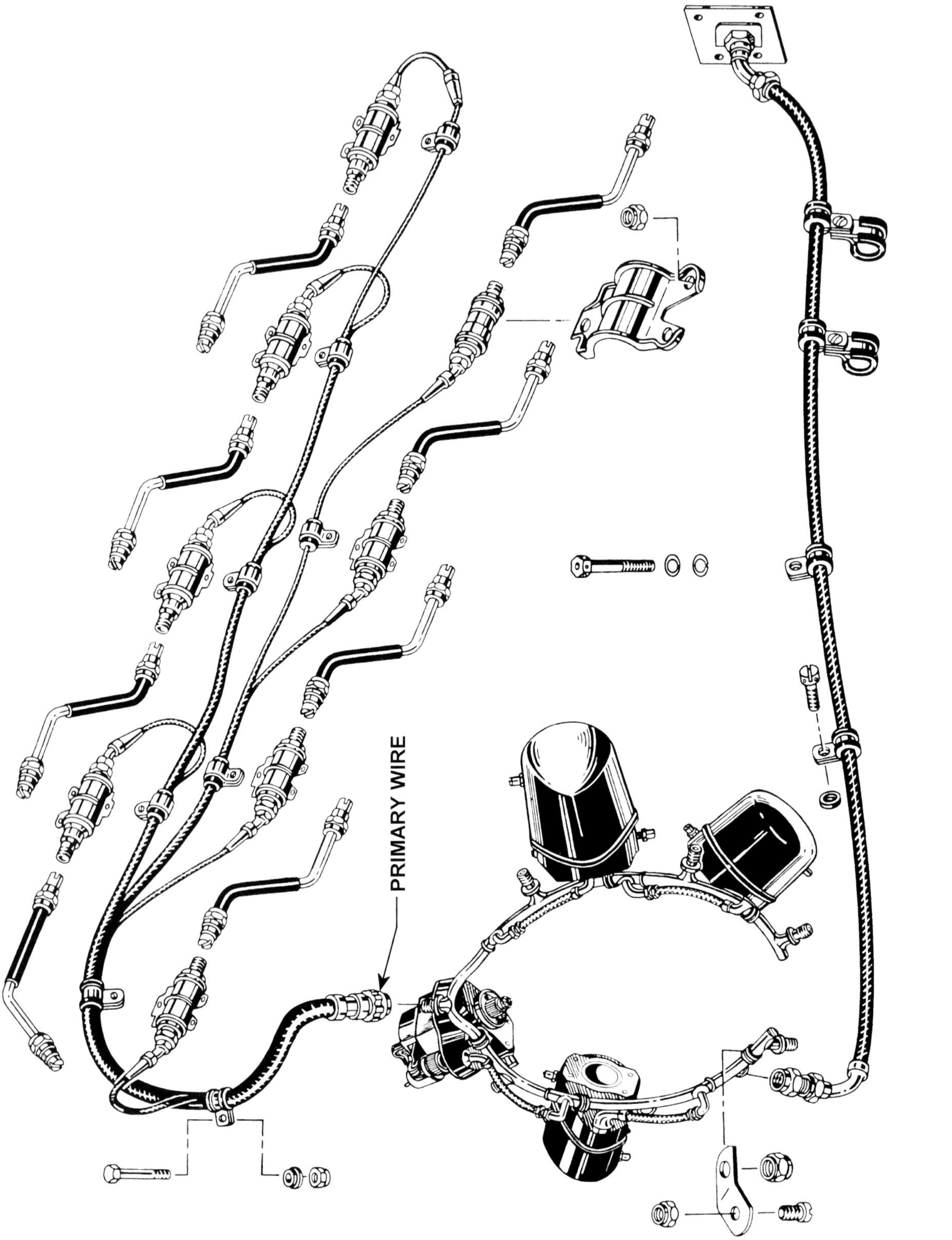

FIG 18 Scuffing was, and still is, a constant problem with aircraft wiring. That's why they are typically well protected and supported with Adel clamps and the like. *(Technical Manual Illustrated Parts Breakdown USAF Models R-4360-59B and -65 Aircraft Engines dated September 15, 1959)*

The foregoing offers a small insight into the problems facing Pratt & Whitney engineers plus maintenance personnel. As can be seen from this list, no recurring problem could be fingered for causing unreliability but rather the problems were spread throughout the engine. No doubt a satisfactory resolution would have been arrived at had Pratt & Whitney spent enormous resources working these issues. A significant contributor to the problems was the use of heavily leaded fuel during the waning days of the piston era. Upwards of 5 cc per gallon of TEL was mixed with the fuel. This massive amount of lead caused serious problems with exhaust valves, problems that were never really resolved satisfactorily.

Douglas C-124 War Stories

(Ref. 10-4)

Barry Meacham, a C-124 flight engineer, wrote the following wonderful anecdotes. His stories keep you on the edge of your seat as he relates the hair-raising stories of keeping Air Force C-124s airborne.

Barry J. Meacham TSgt USAF AFSC 43171R

Basic Training: Amarillo, TX, 1965

Tech School: Sheppard AFB, TX,1966

Crew chief on C124-C: Hickam AFB, HI, 8/66 – 8/69

Crew chief on C-131-D: Mather AFB, CA, 1969

Flight Engineer School: Richards-Gabor AFB, Kansas City, KS, 1970

Full time C-124-C: 945 M.A.W., 1970-72

Flight Engineer: Hill AFB, UT

SEA rated for Joint Mixed crews in Support of S.E.A.

Flight Engineer C-124-C: 185 ANG, Will Rogers Airport, Oklahoma City, OK, '72 – May '74

I have close to 2,000 hours in the C-124-C supporting worldwide missions with M.A.C. from 1970 – 1974.

The following stories are just a few of the many events that all the C-124-C crews experienced . . .

Over Temp Condition #1 Engine A/C Generator

We had just reached our assigned cruise altitude after taking off from Clark AFB in the Philippines. The minimum crew on a cargo trip is pilot, co-pilot, navigator, two flight engineers, and one loadmaster. We always had extra crewmembers. I was one of the flight engineers. We had just completed cruise checklist establishing the initial power settings to maintain 195 knots with our current gross weight. The MAC C-124-Cs were operated on long cross-country legs at a constant speed with horsepower reductions each hour. I was performing the scanner tasks while the other engineer was at the panel. We normally switched stations every two hours. An intercom call came down that #1 A/C generator was showing over temp condition. After checking all associated circuits it looked like we needed to shut down the engine.

Most of the crews I flew with did not have any direct on hands experience on C-124-Cs. I, on the other hand, had a lot of mechanical experience on this type. I advised the aircraft commander that I could go out to #1 engine through the wing crawl space and confirm the problem. If necessary I could remove the alternator, replace it with a cover cap, which would allow us to then restart the engine and continue the scheduled mission without aborting and returning to Clark. I went through my procedure several times and finally convinced him to let me go out into the wing.

I gathered up my lights, headset, gloves, tool bag, cover plate, and some rags and crawled out to #1 engine. With the engine still running, I opened the sliding door to the accessory section. The R-4360-63A QEC package used on "Old Shakey" is designed to interface with the firewall quick disconnects and engine mounts. Opening the 3-piece sliding door exposed all accessories on the back of the engine. Just ahead of the carburetor another stainless steel shield separates the actual power section from the accessory section. To the individual (me) working on an overheated alternator, the noise is very loud. What you have is a 28-cylinder, 3,000-hp engine driving a 17-foot, three-bladed Curtiss Electric prop exhausting all 28 cylinders over your head directly on top of the wing. The sheet metal is all that keeps the direct exhaust fire off your head. Let me assure you I will never forget that sound.

Back to the original issue. The alternator had in fact experienced a bearing failure getting real hot and proceeding to destroy itself. I advised the aircraft commander that the engine would need to be shut down so I could remove the alternator from its pad on the accessory section case and replace it with a pad cover plate. Then we could restart the engine while I checked the pad to ensure it was not going to leak engine oil. The 4360 package used on "Old Shakey" contains one 82.5-gallon oil tank per engine and cannot be refilled in flight. Running out of oil was a very real concern on long (12 hours plus) trips. Any significant oil leaks could use up your oil supply. I waited while the other flight engineer shut down the engine and feathered the prop. The sound of a 72-volt booster taking the normal 28-volt prop hub drive to a fully feathered position was interesting. The booster was right next to my head. The panel engineer advised me the engine was shut down. I returned his call with "I know." Things got a lot more civilized in the accessory section without that dude running.

- *The procedure is pretty straightforward.*
- *Disconnect the cooling inlet and outlet hoses.*
- *Disconnect and secure all wiring.*
- *With a 9/16-inch box end wrench break loose and loosen all the hold down nuts. It is necessary to loosen the nuts just enough to allow the alternator to be rotated around to the clearance holes. Once it has been rotated, you figure out how to pull the alternator off with its weight pinching your hand against the supports.*
- *Install the cover plate, clock it around, and tighten down the hold down nuts.*
- *Wrestle the heavy/hot alternator out of the accessory section and through the slide door.*
- *Restart the engine and leak check the cover plate.*

The plate held fine. I closed the door, which the panel engineer confirmed by his indicator light going out. I dragged myself through the wing catwalk and back into "P" compartment below the main cargo deck on "Old Shakey." The trip proceeded on schedule with a 1/4 reduction in A/C electrical power but all four engines running. The crew was impressed with my work and really appreciated not having to go back to Clark. We were on our way home.

If this even was not enough for one young flight engineer, I had another in-flight trip out through the wing crawl space on "Old Shakey." On this mission we had just taken off from Yokota AFB, Japan. As with all stops in Yokota, we had our scheduled load of cargo and fuel to island hop back to the U.S.A. M.A.C. crews will also remember that the personal baggage consisted of several thousand pounds of crew bags, electronic equipment, hibachi pots, and pinchino machines, etc. Yokota was the world's best shopping spot. This condition usually added several pounds to the aircraft gross weight. The departure was normal as we established climb power of 45 inches and 2,350 rpm. Thirty minutes later I established initial cruise power setting as calculated from the dash one manual. Things looked good for the first half of our flight. Just after E.D.P. (Equal Distant Point) the oil temperature started to climb on #3 engine. The oil temp at high limit now caused the engine oil pressure to start down. These are classic signs of an engine going away. I reviewed the engine analyzer for any signs of misfiring cylinders. Everything looked normal except hot oil. The engine was running fine. About this time my scanner, who had been watching the engine, announced that the oil cooler doors were fully closed. During normal cruise the doors should be in a trail position, which at 195 knots provides enough airflow to keep the engine oil temperature within normal limits. The cooler doors can be operated in auto or manual. I switched from auto and tried to open the door in manual. No go. Discussion with the aircraft commander and the other flight engineer lead us to conclude the electrical drive motor had run the doors shut and would not open them back up. Engine shut down was the only way to save the 4360. Then restart the engine for landing. I then told the group that I could go out to the engine, remove the jackscrew from the electric motor, install a manual jackscrew on the worm gear box, and manually open the doors to trail position. It was agreed and out I went into #3 accessory section. What was very different about my trip across the wing crawl space was that the right main gear was up in the well. This makes it a little more difficult to access the accessory section sliding door. I had no problem removing the motor drive shaft, installing the manual flex drive, and cranking the doors open while the other flight engineer advised on trail position. The engine oil temp came back into normal limits and the pressure came back up. All in a day's work.

I have one more story to relate. These events sound like testimonials on R-4360s and C-124-Cs. During my hours on "Old Shakey," I feathered (shut down) only three engines for mechanical failure. In all cases I knew the engine was real soft before it let go.

Case I: Involves a takeoff out of Mountain Home AFB, Idaho, during an A.F.T.P. (Additional Flight Training Period) with the 945 Air Force Reserve out of Hill AFB, Utah. During the initial run up the established field barometric power settings did not produce RPM and torque readings within specifications. Rather than spend several days at Mountain Home, we decided to make the takeoff. The aircraft (C-124-C) was very light and all we needed was to climb over one ridge then descend into the Salt Lake Valley on approach back to Hill. The engine (#1) did not like the takeoff roll, was backfiring at METO (maximum except take off) power, and really started to shake at the climb power reduction. The pilot called back on intercom to just try and keep it running for a couple more minutes. Several minutes later the aircraft made a hard turn left as the 4360 stopped rotating with the prop not feathered. The turn increased as the pilot yelled feather #1. With the prop feathered "Old Shakey" had enough rudder to stop the turn. Very exciting few moments. On the ground, the pilot came to me and said, "I guess we should not have pushed the engine that far without feathering." We landed without incident, without any more said.

Case II: My saddest flight ever in a C-124-C. Duty called for a minimum four-man crew to ferry the plane from Hill AFB to Davis Monthan AFB, Arizona. This would be the plane's last trip—to the boneyard. The aircraft had been stripped of anything we could use on our other airplanes. The engines had been changed out with ones that were high time off of other aircraft. Takeoff was not too bad. This was the lightest C-124-C ever. Just after establishing cruise power #3 started to show hot engine oil and oil pressure dropping. A few moments later the blower coupling let go and we feathered the engine. Thirty minutes out of DM, the #1 engine went junk, and I feathered it. Now on two engines, we descended into DM, AFB. It was as if the airplane was saying I am not going to the boneyard. Please, please don't take me to the boneyard. I walked away from the plane, turned, and looked back at those two feathered props. I was very sad.

C-124 – Heavy

The C-124 aircraft has a design gross weight of 185,000 lbs, war emergency of 205,000 lbs. Our normal M.A.C. missions associated with the worldwide flight status required max gross takeoffs. The aircraft basic weight, fuel load including reserves, crew/passenger weights were totaled on the TOLD card. Operations would then load freight to establish a gross weight of 185,000 lbs. Using the Dash-1 performance manual's field temperature/dew point/barometric/winds/primary conditions and others the takeoff distance and accelerate to V-1 speed were recorded on the TOLD card. The TOLD card also contains approach speed at various flap settings if the initial flight must be aborted just after takeoff. Full gross takeoffs are tricky on hot, humid days. Landings at gross weight are also very tricky.

The following true story contains the facts as best I can remember relating to a flight from Cam Ran Bay, Vietnam, to Yokota AFB, Japan. We had been directed to transport a large track-driven ditch digger from Vietnam to Japan. This machine had a very large round wheel with tooth-faced buckets welded to the wheel, which was lowered down to dig a ditch. I assisted the loadmaster by opening the clamshell doors and lowering the ramps. The machine operator started this rotary ditch digger up, put it in its lowest gear, and started up the ramps at full throttle. With diesel smoke pouring out, she slowly pulled herself up the ramp and into the cargo section of "Old Shakey." The loadmaster centered the machine up and chained it down.

This machine looked very heavy. During my walk around preflight I paid close attention to the gear struts. All three appeared to be lower than normal indicating the aircraft was really loaded. I then went to the loadmaster to double check the manifest weight on this machine. The numbers added up to the gross weight list on our orders. I went to the aircraft commander and told him the struts were down and the plane looked a little heavy. As we were discussing our options a VC rocket attack started out behind the ramp in and around a small set of hills. We were trying to decide what to do as the battle got hot and heavy. We were really not in harm's way, but loud explosions tend to get Air Force guys excited. After a few minutes the aircraft commander said, "Let's get the hell out of here."

Engine start, taxi, and run up were all normal. The max gross weight hot, humid weight TOLD card was presented to the co-pilot. On this card was the time to accelerate to V-1 and all other information the pilots needed. Our normal procedure was followed with the pilot asking for power. I set all four engines at field barometric and after verifying everything was in specification I announced go. The response back from the A/C was max power to which I responded max power. With my right hand full of all four throttles and my left hand close to the prop override levers, I set max power turning on the water injection at 45 inches. Max power with A.D.I. is 2,800 rpm and 62.5 inches mercury. "Old Shakey" started to slowly pick up speed. The engineer's panel also indicated altitude airspeed and the rate of climb. As I watched the engine instruments I noticed the speed was not matching the elapsed time since brake release. I thought well maybe I missed the actual brake release start time. Then the pilot said, "Meacham is the power okay?" My answer: "Yes, sir." We had now gone beyond the abort point and not broken ground. Finally after a significant ground roll, I saw the gear switches mounted on each strut light indicating the aircraft was off the ground. I looked at the vertical rate of climb indicator and saw it sitting at zero. The A/C called out gear up without the normal call of flaps 10. T/O flaps are set at 20 degrees. The three minutes at max power time had elapsed when I ask the pilot about METO power (2,600 rpm @ 45 inches). The response was, "No way, do not reduce power yet." A few more minutes went by and I announced the engines were getting pretty hot. The pilot called back they are going to be very wet if I reduce the power.

The forward facing engineer's panel on "Old Shakey" allows the engineer to look out a small side window. Side facing panels do not have windows. I much prefer the forward panels for the window access and that people were not bumping behind your seat to get in and out of the pilot's seats. As I watched all the temperature gauges go into their red zones and all found engines and the A.D.I. tanks go dry, I looked out my little side window to see what appeared to be large ocean waves just below the wing. "Old Shakey" was charging along as best she could W.O.T. (wide open throttle) in ground effect just over the Gulf of Tonkin. After what seemed like a very long time the pilot finally requested METO-power, which I very gently reduced the throttles to. The power reduction did stop the ever-increasing cylinder head and oil temperature, but they remained in the red zone. Climb power was finally established some time later, which allowed me to get the engines back to high end specification temperatures. Forty-five minutes later we were at only 5,000 feet, which was well below our planned cruise altitude. We did get the machine across the Gulf of Tonkin to Japan. The landing TOLD card I calculated was corrected off chart for what we were sure was a very heavy plane. After unloading the machine, the aircraft commander requested it be weighed. The new weight was 25,000 lbs more than what was shown on the manifest. He proceeded to base ops with this number. That afternoon a C-124C had apparently taken off at approximately 210,000 lbs. I wrote the engines up for B.P.O. inspection including pulling the oil screens. Everything looked okay. Just another testimony to a wonderful powerplant and a great airplane.

The following is an e-mail I received from Luke Roy who worked on B-36s during the 1950s. As you read between the lines from Luke's self effacing and modest style, it is very apparent that it required an inordinate amount of skill keeping R-4360-powered aircraft airborne *(Ref. 10-5).*

***War Stories From Luke Roy**, 4 July 2003*

G.

Per our previous "conversations," I will try to get something to disc. How much will depend upon how much fog and cobwebs I can clear from the memory bank.

Guess I wound up in the Heavy Bomber program, as that was the only prop driven aircraft school that was taking students when I went into the Service. I had been working small aircraft and doing some flying before the military. Worked for an air service in South Arkansas that bought and sold lots of light planes and built up a few duster and sprayer conversions. My job, along with turning wrenches etc was to receive and deliver most of the birds. Some hadn't flown in a while, so it got interesting on occasion.

Figured all this should get me into the Aviation Cadet program but didn't count on them proving I was in need of a guide dog because of one eye not being up to snuff.

It was B-36 or B-47. Jets were then a bad word to me so, me being a great dumb-ass, I took B-36. Career wise it was a bad move, but I have made lots of those.

Initial assignment at Travis was on the post flight crew. The airplanes were pulled into a dock area that covered the engine areas and provided a floor to allow work on the lower sides of the engines. The roof areas were supported by collapsible posts that had to be lowered to allow entry of the airplane and were then raised back into position to provide support against the Suisun Valley winds (considerable). The props were actually inside this containment and the props had to be positioned "long blade down" for entry. First heard this comment, I thought it a joke for the "new guy." Hell, I knew airplanes and knew all blades were the same length. Right?

The APG (airplane general) mechanics had responsibility for pumping the main gear struts up to assure the prop clearance on the provided channels when some floor plates were removed. The clearance on the inboard (number 3 and 4 engines) was very close. So it is pump struts to the max and check clearance. Things went along very well for a while but it finally bit me. We then had the -53 engines with direct cylinder injection, so to pressure check the lines, the engines had to be cranked thru several revolutions with the starter to pressurize the injection lines. I had checked two blades and had clearance but didn't reckon with the "infamous long blade" God, what a noise when that blade hit the dock. Not quite as much as the dock chief made chewing my ass though. As the prop shop guys were changing the prop, I had to ask about the blade lengths and was told that they could vary nearly ⅜ inch, from the long to the short. The blades were folded steel and welded along the trailing edge and across the tip. They were NOT all the same. They were hollow and hot air was piped thru them and out an exhaust port near the tip as an anti-ice measure. The source of this hot air has totally escaped my memory, as was the source of hot air that was pumped thru a double leading edge on the wings (see Chapter 7 for an explanation).

Another source of "amusement" usually occurred during this pressure check operation. Standing procedure was for the "p-leads" to be removed from the mags and a grounding wire inserted in its place to assure a dead mag. while in the dock. If my memory(?) is correct, there was a grounding device in the mag. that was supposed to do this when the p-leads were disconnected but didn't always, thus the requirement for the grounding wire. As with most GI operations, this didn't always get done. And also there was a mag. grounding check that was SUPPOSED to be done before engine shutdown, again to assure not having a hot mag. because of a bad p-lead. More than once an engine was fired up during this pressure check with guys laying all over it to check for leaks. Also when the engine started to fire, things (the airplane) would start to wiggle around and things could get real (exciting) in a hurry. Reread the long blade comment. The clearance got used up in a hurry and things, most of them heavy, could start to be flung around with great exuberance. Won't go so far as to say the engine guys would run, but they passed a few guys that were running. They would spend the next half hour or so in the coffee shack while the engine chief for that engine replaced me in the ass-chewing session. And usually the prop shop was on the way.

After I saw this happen the first time, I scared the hell out of myself, remembering how many mags I had timed on light planes without grounding them. Never had one fire on me but I guess it was just out house luck that I didn't get whacked pretty good. Guess also that a p-lead of a few feet length had a better chance of working than one that was about 100 feet long.

Was involved a bit in a "mystery" that had us all going for a while. On occasion the plane would come back with holes in the leading edge of the vertical and horizontal stabs and along the fuselage sides. Took us a while to find what was going thru the props as we could find no pieces inside the areas punctured. Finally talked to a gunner/scanner that told me that he had seen a "basketball size" ball of goop, which was his word for ice/oil mix that formed on the case breather vent on each engine. When it got big enough it would break off, go thru the prop and then thru anything in its path. Certainly was job security for the "tin benders" till a fix came out moving the vent to inside the cowling.

I think the -63 engines have individual exhaust stacks per cylinder with studs at the exhaust ports, which, in the early stages broke a lot. Called for a cylinder change if a stud broke. The -41 and -53, actually any 4360 with a collector ring exhaust, had a different system and port. There was some sort of a hi-temp. (Fiberglass?) seal at the port and a spring loaded coupling between the cylinder and the "header" that went to the collector ring. To put these couplings in, one had to compress the spring, wire it compressed, collapse the coupling, slip it in place with the seals and cut the wire. Again, as with a lot of GI operations, the wire sometimes would not get cut and a cylinder or two would get burned pretty bad. Finally a dictate came down that string would be used only so if forgotten it would burn loose.

Lot of problems were caused by the fact that it was a pusher airplane. The engines were numbered "airplane" wise. All airplane pieces were numbered this way. Engine pieces were numbered "engine" wise. The turbos were named "left" and "right" relative to the engine, as they were engine related. HOWEVER, the intercoolers that were between the turbo and the engine were labeled "left" and "right" but relative to the airplane. Of course, this led to many instances where the wrong part was worked on, including engines and props, and nearly any other part you can name. Had to laugh at an electrician once over this very thing. The fire warning light system was a real pain in the ass and always a source of a lot of work. This person admitted to working all day, on the wrong engine fire detector system, and in his words "Actually fixing things."

Probably the most frantic thing along this line that I recall was caused by the aforementioned prop shop troops. Airplane was due to change Nr 3 engine so prop shop had to pull the prop. Out they came and pulled the "afterbody." B-36 didn't have "spinners", they were called "afterbodies?" Then proceeded to break the prop retaining nut loose. About this time the crew chief asked the most embarrassing question as to why they were working on Nr 4 engine. Much shouting ensued. This work was duly noted in the AFTO Form 781(used to be the Form 1) under a red cross(grounding) condition and they retorqued the prop retainer, put it all back together, made the proper sign offs in the form and pulled Nr 3 prop. After the engine was changed, the prop was reinstalled, signed off, ground run etc and the airplane flew a normal mission. After the mission, the big screw-up was discovered. Shit hit the fan as it was required to fly a test flight if two props were changed. Since the first prop was actually loosened, the powers that be decided that constituted a "change" thus requiring a test flight. The aircraft commander that flew the mission was not test flight qualified, so the test requirement could not be signed for. Airplane had to be flown around the pattern by a qualified test flight crew to clear the paper trail. A few more folks to take my place in the ass chewing line.

We took the outfit to Guam in Jan 1955 for a normal 90-day TDY. Outfit that was due to relieve us (28th Bomb Wing at Rapid City, South Dakota) got snowed in and we stayed 4 months.

Remember a few things from there, other than how bad that San Miguel beer(?) could tear one's stomach apart. Had an airplane get some serious wingtip damage due to St. Elmo's Fire. Thing looked like a torch had been on it. The decision was made to remove a tip from an airplane in the Periodic Maintenance Dock to replace the damaged part. This wing tip was as big as a canoe and when they got it loose about four guys had hold of it for a while. Wind gust got it and they finally had to let go or go with it. I remember seeing this thing going down the ramp getting closer to a ball shape with each roll and three guys chasing it. Two airplanes without wing tips and a couple more troops in the ass-chewing line.

The B-36 did not have hydraulic pumps on the engines but instead had an electrical driven pump in the bomb bay area to run the gear. Again, if memory serves, the only thing hydraulic on the airplane was the gear and brakes. Maybe the bomb doors. A very large hydraulic control valve located in this area controlled the gear retract functions. One of these valves went bad and was replaced with another valve, yellow tagged, signed off kissed and blessed and all that. Only problem, it was wired backward at depot. This had to occur during some sort of big time alert, something had happened in Berlin or some other hot spot around the world and we had all the airplanes grossed out with iron bombs. When the power was applied after this valve change, the gear tried to retract. The main gear budged but as it had to go sideways the down locks held. The nose gear did retract, tearing the gear doors off and a lot of the skin was damaged in the area. With the airplane full of bombs and fuel, the air bags would not lift to far enough to extend the gear. The barracks were cleaned out and everyone that could be found was herded out to the beast and had to sit in the tail cone till enough weight was added to allow the air bags to work. At one time I knew how many folks were on board but that memory has since gone by the way. I was not on board. Reread the comment about the San Miguel beer.

This airplane was flown back to Travis with the nose gear down and sheet metal replacing the gear doors.

Fortunately, I left the B-36 program in Jan 1956 and went to the 303rd Air Refueling Sqdn. at Kindley AFB Bermuda. To KC-97 F & G airplanes. These airplanes had the -59B engines with PR-100 carburetors rather than the direct cylinder injection on the -53 engines. Single, rather than dual turbos.

We had some serious prop problems with the Ham-Standard steel blade props on this beast. Got to a point that a Magna-flux had to be done on them before each flight. We never lost any blades but I think MATS did on their C-97 airplanes. This always amazed me because we were supporting B-47's on what was then called "the reflex" program. They were rotated in and out of North Africa and England on a daily basis and we were in Bermuda to catch them going east. Another outfit caught the west bounds out of the Azores. During refueling ops the old 97 had to run at METO power and go downhill to keep the 47 above stall when it got close to full. One would have thought we would have had the real prop problems instead of MATS, whose power was at cruise most of the time.

It was amazing as to how heavy those Magna-Flux coils could be by the time you got to the fourth blade on the fourth prop. Then the blades had to be washed down with JP to remove the filings solution.

Finally were fitted with aluminum blade props about the time I returned to the B-36 program.

Couple memories from the 97 days. One was always aware when an engine was backfired on starting. Anytime you heard one, the guy who was running it had to be defined as the poor bastard would get stuck for beer at the club that night. This time Heinekens(sp). The engines were started on prime and then the mixture brought in and if you didn't get off the primer button at the right time, one could spend a lot of money at the club.

Another little thing that usually bit the new guy was how things had to be done on start up. The Start, Boost and Prime buttons were on the engineer's overhead panel. The throttle and mixture were forward and the engineer's instrument panel to the right. The essential inverter switch was at the far rear (right) of the panel. It had to be turned on after engine start to get instrument power. One had to use the left hand for start, boost and prime buttons in order to have the right hand to turn on the essential inverter. Then the mixture could be brought up with the right hand and then the other engines started. Have seen guys do the buttons with the right hand then not be able to reach the inverter switch with the left hand. That usually cost him some money at the club also.

I said it was fortunate that I left Travis when I did because shortly after one of the airplanes lost a rudder in flight and really ripped the fin up pretty bad. The airplane was landed and the crew got medals for it etc. The airplane was backed into a big hanger that had a cutout area in the overhead for fin/rudder clearance and a depot crew from Kelly worked on it about a month, rebuilding all the hinge areas etc and finally hung a new rudder. The airplane was declared complete and a crew was quickly assembled to get it out of the hanger and test flown. Two sets of yellow lines on the ramp. One set for nose in and one set for tail in. Tow tractor operator used wrong lines, hit hanger overhead, ripped rudder off, again. Everyone associated with the operation lost stripes, even guys just eating their lunch.

Lot of folks took my place at the ass chewing session over that one. Glad I was in Bermuda.

After I returned, I was a crew chief till the bird left. Just remember the last mission on the bird, number 3 engine lost a scavenge pump and pumped 160 gallons of oil overboard. Had to change the damned engine before they flew it to Davis-Monthan. Tried to get them to do it on five but no joy. What a waste.

The oil tanks held 200 gallons but normally were serviced to 150. The guy that broke me in serviced them to 160 gallons so I did too.

Bad thing about an engine change was having to remove and replace the hopper tank in the main tank. This was behind a stress panel about 24 inches in diameter and held in place with about 200 1/4-28 screws. This was ok unless the airplane was moved with this panel off cause then the screws didn't line up any more and sometimes the wing had to be jacked a bit to get it to fit.

Did win 20 bucks from some ex fighter mechanic assigned to me when the bird left. It was empty with a minimum fuel load and we went down to the runway with it. He wanted to see it break ground and wondered how far down the runway it will go. I told him 1,500 feet and he bet me 20 bucks it would go farther. They set power on the beast and when the brakes were released, the fuel sloshed once, the nose wheel came off the ground and old baby was flying in about 1,000 feet. Surely surprised him.

Fond memories NOT:

1. Crawling down the air tunnel to change a constant speed drive.

2. Crawling down the air tunnel to change a CSD shaft that sheared.

3. Anything to do with the CSD system because of the oil used in these damned things. It would eat all the paint off the gear under Nr 3 and Nr 4 if you had a leak. It would eat your fingernails off. And if you got it on your clothes, they were hors de combat.

4. Being on the ground cord during engine run up. One of those beasts at "atmospheric" would put so much vibration out that your diaphragm would vibrate in sympathy with them and you could not breathe. You had to move around till you found a spot that you could take a breath. Was rather scary the first time out.

5. Working on the relief can drain in the photo compartment after the line came off and flooded the area with pee. Yuck.

6. Installing a new blister in the rear compartment. The blisters only fit from the inside and were bigger than the entry hatch. Convair actually made a "blister stretcher" tool that would warp the blister out of shape enough to get it thru the hatch. I only know about this from the tech order. We never had one and no one in the outfit had ever seen one. We had to remove the band from the blisters, new and old, and replace the old frame on the new blister. Only about a hundred screws involved.

7. Most hated job. Putting the rudder control locks on. This consisted of a couple ropes that attached to the rudder and was then attached to the outer end of the horizontal stab. There was always several inches (it seemed) of oil all over the stab and you had to get up there and walk around to do this. In a good wind it took a couple of troops to push the rudder over to hook things up. Always scared me silly to do this as you were some distance above the ground and at night it was something else. How did I ever live this long?

I have seen some things written about these airplanes that make me wonder. Have a publication around here somewhere about engines. Think it was published at U. of Kansas. Author said the B-36 was the only successful shaft driven prop plane. The only shaft drive was a 6 inch extension on the nose case of the 4360. There was an adaptor quill on the rear to drive the fan and the alternators.

Also I have seen, I think in the Engine Historical Pub., a comment about a four hour startup procedure required on the 4360. This I know nothing about. We fired them up as required, when required.

Interesting thing about mag. work on the B-36 was that when you got both hands on the mag. you could not see it cause that portion of the cowling was structural and not removable and you had an arm around each side and it became a place to rest your nose, so it was hook up a light or an audio device to tell when the points broke. The -59 on the 97 was a lot easier cause they were out in the open. Guys with short arms could time them and could not time the ones on the B-36 cause they couldn't reach them.

Don't think I have done this much writing on any subject before. Fingers are getting tired so I best shut up. Hope you have a few laughs here and maybe something to think about.

Probably not what you wanted but send me some critical comments and maybe we can go another direction.

Hell, I have enjoyed this. Hope you do.

Luke.

Of course, I was delighted to receive these wonderful anecdotes written in such a jovial and entertaining manner. And yet, it points up to the fact that keeping these complex military machines alive and well took extraordinary skill and courage.

CONTIN ENTAL
DIVIS ION
0117
MILITARY AIR TRANSPO

In summary, the R-4360 could have, in all likelihood, enjoyed better reliability if more development had been focused on it. Gas turbine work took care of that lofty goal so the R-4360 was left to soldier on—warts and all. An unfortunate side of the R-4360 was its abysmal safety record powering Boeing B-377s. Several instances of runaway props on Boeing B-377s occurred. The typical scenario would be for a reduction gear failure to occur. The prop would go into fine pitch under the influence of centrifugal turning force. The overloaded propeller thrust bearing would fail under over heating conditions, and in time the propeller shaft would shear. At this point it was hoped that the prop missed any vital parts, like the fuselage, as it departed company from the airplane. By the early 1970s not a single example of a commercial R-4360-powered aircraft survived, except in the form of the Guppy. The rest had been scrapped.

References

10-1 Interview by author with numerous flight engineers and operators from Miami International Airports' infamous "Corrosion Corner."

10-2 Ryder, Earle A., *Recent developments in the R-4360 Engine*. Paper presented at the SAE Summer Meeting (French Lick, Ind.) Society of Automotive Engineers, Warrendale, Pennsylvania, June 1950.

10-3 All material described in the "Problem Log" was derived from a file in the Pratt & Whitney archives.

10-4 C-124 anecdotes as related by Barry Meacham.

10-5 B-36 and KC-97 anecdotes as related by Luke Roy.

C-124 sitting in the snow at Thule, Greenland.

CHAPTER ELEVEN

The Future

What Ifs

An often-asked question is: What would have occurred if gas turbines had not arrived on the aviation scene? The R-4360 offers an insight into the challenges facing the aviation industry for a 3,000-plus horsepower engine—maintenance personnel, airframe manufacturers, and flight crew. Each one of these disciplines was already stretched to the limit with the R-4360. The R-4360 would have paled in comparison to the even more complex engines envisioned as follow-ons. Examples such as the Lycoming XR-7755 and Wright Tornado come to mind. As it was, the R-4360 came close to bankrupting some airlines. These follow-on engines would surely have accomplished that.

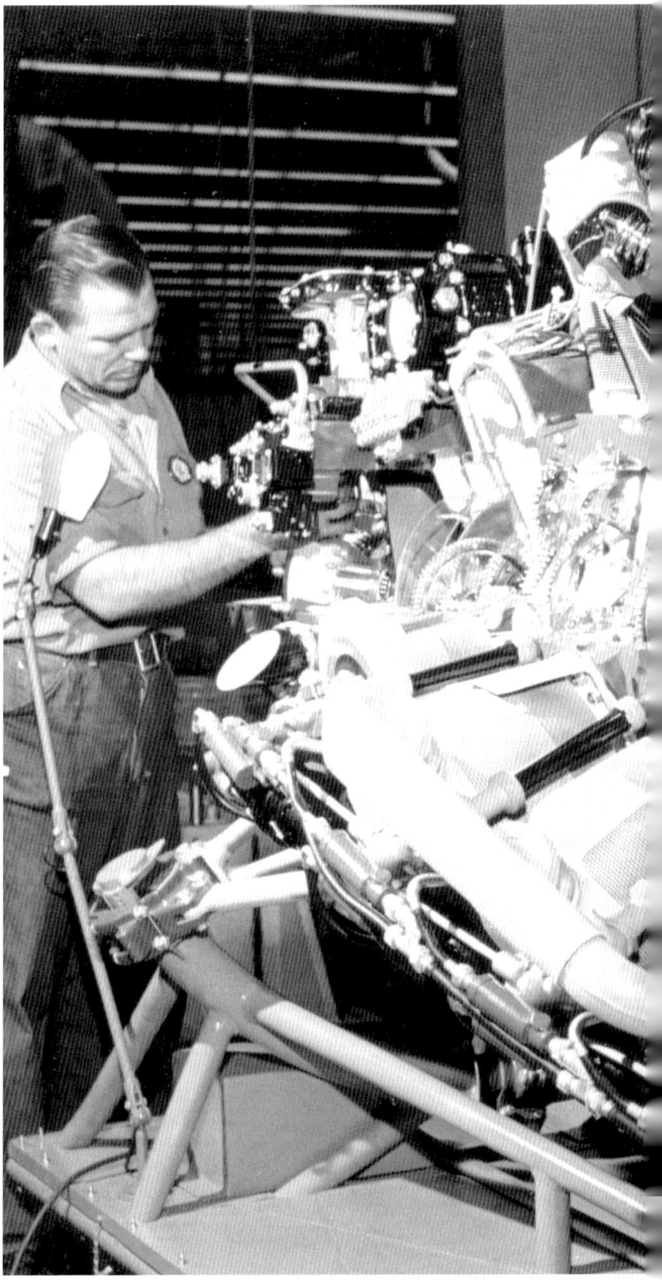

Endangered Species

Not surprisingly considering its charisma, nostalgia is a powerful force that keeps a handful of R-4360s alive in the form of displays in various museums around the world. Even fewer are actually fired up in anger, which puts it near the top of the endangered species list. In sharp contrast, engines such as the R-2800, R-1830, Rolls-Royce Merlin, etc., are assured of a secure future. In fact, a pretty significant cottage industry exists to keep these warriors from a bygone age in fine fettle. For various reasons, this is not the situation for the R-4360. Several rationales account for this state of affairs: (i) operating and overhaul costs are now through the roof, (ii) few R-4360-powered military aircraft have survived and those that have are typically large four-engined aircraft such as the Boeing B-50, an aircraft that would require cubic money to operate, and (iii) far fewer R-4360s were manufactured compared to other top-tier piston engines.

A handful of KC-97s were converted to cargo haulers or fire bombers. However, their operating costs and maintenance requirements made them less competitive than DC-4/6/7s, or even Lockheed Constellations—and that's saying something. The Connie was another bear to keep flying due to its complexity. But like everything else, if aircraft were available for the right price with a good supply of engines, then the available equipment would be used up like so much confetti at a wedding. Several KC-97s ended up in Miami International Airports' Corrosion Corner, due in no small part to the availability of large numbers of military surplus engines. Per normal

Now highly prized, Pratt & Whitney produced a number of cutaways. What better way to demonstrate the complexities lurking within the bowels of an R-4360. This photo appears to be a B-36 engine, identifiable by the nose case extension, possibly a -53. Producing these cutaways was an expensive and time-consuming process. As an example, look at how the cylinder fins have remained undamaged despite the additional machining operations required for the display purposes. *(Courtesy of Pratt & Whitney)*

Side view of an R-4360 cutaway. The high finish of all the machined parts is quite striking. It gives the effect of chrome plating. In fact these parts were polished and this is how production engines were made. *(Courtesy of Pratt & Whitney)*

operating practice for Corrosion Corner aircraft, they were typically operated over gross and in a hot climate, consequently putting additional strain on the already stressed-out turbosupercharged R-4360s. It's akin to powering a dump truck with an Indy 500 racecar engine and expecting it to survive day-in and day-out abuse. Unlike an R-2800, for instance, which can handle abuse, the R-4360 was a sensitive engine that did not possess the intestinal fortitude of its more erstwhile sibling. This resulted in overhaul times typically of less than 1,000 hours and that's if everything went well.

It was a rare occasion when an R-4360 reached 1,000 hours in cargo hauling operations. This all depended, of course, on the expertise of the all-important flight engineer. Unlike the R-2800 or R-1830, it simply did not make economic sense to overhaul the highly complex R-4360. Therefore, when the bottom of the R-4360 engine barrel was scraped clean, the KC-97s were unceremoniously scrapped. Thus ended a distinguished career for a wonderful aircraft whose roots went back to the desperate days of World War II and the problematic B-29.

Cutaways

A number of the major engine manufacturers were understandably proud of their products. To show off the technology and workmanship employed, cutaway engines were produced. Of course, it would also be a safe bet to say that they represented an excellent marketing opportunity. Regardless of their original intent, these cutaways are now a fascinating view into how these engines operated and the truly superb workmanship required to produce them. Pratt & Whitney made a number of R-4360 cutaways, which are now highly prized collector pieces.

Bob Odegaard rescued old race number 57 from an uncertain future. He not only saved this historically significant aircraft, he restored it to airworthy condition. *(Photo courtesy of Bob Odegaard)*

Many have found their way into museums where the public can push a button and marvel at all the various mechanisms in motion. All generations from young to old are usually awed by the advanced ideas and unique solutions to design problems inside this wonderful example of the engineer's art.

Old Soldiers Never Die–They Get Scrapped!

When R-4360-powered aircraft were finally retired, they suffered the indignity of mass scrapping with their wonderful R-4360s being sold off by the pound as so much junk. It wasn't until almost all surviving R-4360s had been scrapped that the realization hit home that a piece of engineering history was going down the tubes. Fortunately, a few R-4360- and R-4360-powered aircraft survived to be displayed in museums, but they are not representative of the thousands originally manufactured.

Charisma

As discussed in the racing chapter, several R-4360-powered Unlimited Class racers have been produced. Their open-stack exhaust systems exhibit that wonderful, unmuffled sound that only 28 huge, high-performance cylinders can produce. These racers offer the opportunity to see these engines operated at powers and speeds undreamt-of during their heyday. But running on

Next pages: Thanks to the efforts of the Berlin Airlift Historical Foundation, at least one KC-97 will be kept airworthy for the foreseeable future. Named *Angel of Deliverance*, this KC-97 will require major support to keep flying. *(Courtesy of Tim Chopp, Berlin Airlift Historical Foundation)*

UNITED STATES AIR FO

559595

the ragged edge of destruction puts these aircraft and their precious engines in harm's way.

Collecting

A few enterprising individuals, including myself, own their own personal R-4360s. It is hoped that several of these privately owned R-4360s can one day fire up in anger, wowing the crowds with their smoky starts, pyrotechnics, and unmistakable sound. Collecting, restoring, and running classic old aircraft engines is a great hobby, which hopefully burgeons into a popular pastime that keeps these warriors from a bygone age fired up rather than collecting dust in a museum.

Survivors Surviving R-4360-Powered Airworthy Aircraft

Currently, only five R-4360-powered aircraft are known to be flying. They are:

I. Ben McKillan's restored #57 racer owned by Bob Odegaard.

After the 1949 Thompson races it appeared as if the old racers were just so much junk. They were just left out in the elements to rot away. It has been reported that the last time number 57 flew was at an air show piloted by Cook Cleland in 1950. After that, it was left to the elements. Fortunately, number 57—the aircraft flown by Ben McKillan, which he won first place in the 1949 Tinnerman race—survived, but only by the skin of its teeth. This historically significant aircraft came very close to being scrapped on a number of occasions. At its nadir, the aircraft would, by any other standards, be totally beyond any hope of restoration. The aircraft sat at the Euclid, Ohio, airport from 1949 to 1954. It was during this time that its downward spiral started. From 1954 to 1962 the aircraft was stored at Joe Loecy's machine shop, but not under cover. In 1964 Cleland was offered sponsorship from the Martin Decker Company to mount an attempt on the word's air speed record for piston-powered aircraft. The aircraft was disassembled in preparation for the speed attempt and shipped to Pottstown, Pennsylvania. However, the sponsorship fell through.

A New Hampshire airline pilot, John Trainor, purchased the remains with the intention of making a static display out of it. However, Trainor lost his life in a P-51 accident and his heirs sold the remains of number 57 to Harry Doan in Florida. Doan started a restoration, but like his predecessor John Trainor, Harry Doan lost his

Looking resplendent after its painstaking restoration, race number 57 graces the Odegaard Aviation ramp. (Photo courtesy of Bob Odegaard)

life flying a Douglas Skyraider. Once again, old number 57 was on the move. Doan's wife sold the remains to Don Knapp in Miami, who then had it shipped to Texas for restoration. In an almost bizarre way, number 57 was jinxed. No sooner had Knapp purchased the aircraft and he too was killed in a P-51. The Lone Star Museum then picked up number 57. Lone Star started yet another restoration. However, the museum decided to sell it to Greg Morris of Hugoton, Kansas. This now brings the story up to 1996 when the current owner, Bob Odegaard, purchased what was left of the aircraft. In a testament to Bob Odegaard's restoration skills, he managed to pull off the seemingly impossible by restoring number 57 to flying condition.

Although the aircraft was originally powered by an R-4360-4, Bob wisely chose to upgrade the engine by incorporating various components. The power section is an R-4360-20 as is the nose case; however, the reduction gears are from an R-4360-35. And again a -35 blower is used. Although it would have been nice to use the -20 blower, it was 4½ inches longer than the -35 and as a result would not fit due to interference with the oil tank. Being a historically significant aircraft, it is flown at very conservative power settings: 47 in. Hg. at 2,700 rpm is used for takeoff. Head temperatures are a very cool 150 degrees C and this is with cowl flaps shut tight. Oil temperatures are similarly low with an exit temperature typically being 73 degrees C and oil inlet being 49 degrees C. Deservedly so, the aircraft has garnered a number of significant awards since its restoration. These include the prestigious Rolls-Royce Trophy for restorations at Reno in 1999 *(Ref. 11-1)*.

Hawkins & Powers KC-97 drops a load of fire retardant. This would appear to be a demonstration drop, as no fire is apparent. *(Courtesy of Hawkins & Powers)*

Six QECs were rescued from C-97G 53-3816 just prior to its scrapping. This photo shows one of the two tractor-trailer loads. *(Courtesy of Tim Chopp, Berlin Airlift Historical Foundation)*

II. Race #15 Furias described in Chapter 10.

III. Race #11 Dreadnought described in Chapter 10.

IV. Berlin Airlift Historical Foundation KC-97.

Berlin Airlift Historical Foundation is an organization dedicated to preserving the history of Cold War transport aircraft such as the DC-4 and KC-97. In 1996, BAHF acquired KC-97 52-2718 located at Moses Lake, Washington. The aircraft was ferried to Floyd Bennett Field, New York, in three legs. However, they were not three easy legs. Along the way, the nose wheels were converted to DC-8 wheels, and the aircraft was painted to represent YC-97A 45-59595, the 3rd prototype C-97, which was utilized in the Berlin Airlift.

In April 2003, Boeing C-97G 53-3816 was purchased for spares use. Six R-4360 QECs from that aircraft were among the essential parts donated by 3816. It is estimated that it will take at least another two years to complete the project in order to display the C-97 around the U.S. *(Ref. 11-2).*

V. KC-97 owned by Hawkins & Powers *(Ref. 11-3)*

Used as a fire bomber, this aircraft started life as

Immediately after World War II, Pratt & Whitney put on a major display for the public at its East Hartford facility. One of the display pieces was this one showing some of the variations on the R-4360 supercharger. From left to right they are: (i) single-stage, variable speed for -4, -8, -9, -9T, -14, -15, and -27 models, (ii) two-stage variable speed for -10 and -29 models, (iii) two-stage variable speed with remotely mounted auxiliary stage for -13 and -19 models. (Courtesy of Pratt & Whitney)

a KC-97 manufactured in 1952. After its military service it bounced around several civilian owners. Hawkins & Powers purchased the aircraft from a fish-hauling outfit in Alaska. The first order of business was to remove any remaining aerial refueling and military apparatus. Equipped with a 12-door slurry system, it's capable of carrying well over 3,500 gallons of fire retardant; however, it typically carries 3,000 gallons in the interest of preserving the engines.

Remarkably, Hawkins & Powers can reach a TBO of 1,400 hours with 30 percent of their engines. Cracked cylinder heads and failed main bearings are the most significant engine related problems. Failed master rod bearings are another source of failure. The plethora of bearing problems may be related to the fact that their engines have typically been in storage for a number of years. This prolonged storage causes the lead/indium to deteriorate, resulting in excessive bearing clearances. As of this writing, engines were not being overhauled. Instead, stocks of military overhauled engines in cans are being used up. When military overhauls are no longer available, the sad prospect of seeing this KC-97 grounded becomes a reality. Although it would be possible to overhaul the R-4360, as Precision Engines in Washington has proved, it would be cost prohibitive.

From time to time, racing projects have been announced with the intention of using an R-4360 for power. However, to date none of these projects have progressed beyond the planning stage.

Unfortunately, many variations and significant dash numbers of the R-4360 seem to have disappeared. For instance, it is doubtful if any of the two-stage variations have survived, which is a tragedy. This sophisticated supercharging system personified what could be done with piston engine technology. In a similar vein, it appears that none of the dual-rotation engines have survived along with their fascinating propellers. At least one dual-rotation gearbox for the XB-35 has survived in the Pratt & Whitney museum in East Hartford. In fact, it appears to have never been used.

In conclusion, it is my fervent hope that this book offers some insight into an engineering marvel from a bygone age. Somewhat akin to the steam age when triple-expansion engines the size of cathedrals ruled the waves, the R-4360 represented the largest and most sophisticated of its breed. And yet in a few short years, when gas turbines ruled supreme, the R-4360 was regarded as an anachronism to be derided and scoffed at. In other words, a similar fate that befell the wonderful triple-expansion engine when more efficient steam turbines arrived on the scene.

References

11-1 Interviews and conversations with Bob Odegaard, current owner of Race #57

11-2 Interviews and conversations with Timothy A. Chopp, president, Chairman Berlin Airlift Historical Foundation.

11-3 Interviews and conversations with various Hawkins & Powers personnel.

Glossary of Terms

ADI: Anti-detonation injection. A mixture of water and alcohol injected into the induction system that helps prevent detonation.

Automatic boost control: Same as manifold pressure regulator.

Blower: Used interchangeably with supercharger.

Collector ring: A pipe that joins the exhaust manifolds of all seven banks, (in the case of the R-4360) and collects them in a "collector ring." Typically used with turbosupercharged applications.

Contra rotating: Same as dual rotation. Two shafts that run in opposite directions on the same axis.

Counter rotating: Two shafts on different axis that rotate in opposite directions.

Dry: Means the engine is running or is rated without the application of ADI. See "Wet."

Dual rotation: Two propeller shafts on the same axis. See contra rotating.

Manifold pressure regulator: Performs same function as an automatic boost control. See automatic boost control.

Open stack: An engine that does not have an exhaust driven device such as a turbosupercharger. Instead, exhaust gasses are simply dumped overboard via a relatively simple exhaust system.

PRT: Power recovery turbine. See turbo compounding.

QEC: Quick engine change. Generally regarded as everything in front of the firewall. This would include, at a minimum, the engine and engine mount. Some QECs include the oil tank, oil cooler, turbosupercharger (if fitted), intercooler(s), and all associated lines and hoses.

Supercharger: Used interchangeably with blower.

TBO: Time between overhaul.

Turbo, Turbocharger, Turbosupercharger: These three terms are used interchangeably.

Turbo compounding: A power recovery system whereby exhaust gasses drive a turbine and the power from the turbine is fed directly back into the engine via drive shafts, reduction gearing, and/or fluid couplings.

USAAC: United States Army Air Corps.

USAF: United States Air Force—formed on September 18, 1947.

USAAF: United States Army Air Force, predecessor to the USAF.

VDT: Variable discharge turbine is a sophisticated turbosupercharging system developed by Pratt & Whitney for the R-4360.

Wet: Means the engine is running with, or rated with, ADI. See "Dry."

Index